A Handbook of Global Freshwater Invasive Species

Invasive non-native species are a major threat to global biodiversity. Often introduced accidentally through international travel or trade, they invade and colonize new habitats, often with devastating consequences for the local flora and fauna. Their environmental impacts can range from damage to resource production (e.g. agriculture and forestry) and infrastructure (e.g. buildings, road and water supply), to human health. They consequently can have major economic impacts. It is a priority to prevent their introduction and spread, as well as to control them. Freshwater ecosystems are particularly at risk from invasions and are landscape corridors that facilitate spread of invasives.

This book reviews the current state of knowledge of the most notable global invasive freshwater species or groups, based on their severity of economic impact, geographic distribution outside of their native range, extent of research, and recognition of the ecological severity of the impact of the species by the IUCN.

As well as some of the very well-known species, the book also covers some invasives that are emerging as serious threats. Examples covered include a range of aquatic and riparian plants, insects, molluscs, crustaceans, fish, amphibians, reptiles and mammals, as well as some major pathogens of aquatic organisms.

The book also includes overview chapters synthesizing the ecological impact of invasive species in fresh water and summarizing practical implications for the management of rivers and other freshwater habitats.

Robert A. Francis is Senior Lecturer in Ecology at King's College London, UK. He has broad research interests in aquatic, riparian and urban ecology and has been secretary of the British Ecological Society special interest group on invasive species since 2008.

A Handbook of Global Freshwater Invasive Species

Edited by
Robert A. Francis

Routledge
Taylor & Francis Group

LONDON AND NEW YORK

First published 2012 by Earthscan

2 Park Square, Milton Park, Abingdon, Oxforshire OX14 4RN
711 Third Avenue, New York, NY 10017

Routledge is an imprint of the Taylor & Francis Group, an informa business

First issued in paperback 2017

British Library Cataloguing in Publication Data
A catalogue record for this book is available from the British Library

Library of Congress Cataloging in Publication Data
A handbook of global freshwater invasive species / [edited by] Robert A. Francis.
    p. cm.
    Includes bibliographical references and index.
    1. Introduced freshwater organisms. I. Francis, Robert A.
    QH96.8.I57H36 2011
    333.95'23091692–dc23

                                    2011008847

ISBN 978-1-84971-228-6 (hbk)
ISBN 978-0-8153-7871-6 (pbk)

Typeset in Garamond
by Domex e-Data, India

Contents

List of illustrations   *ix*
Preface   *xv*
Contributors   *xvii*
List of abbreviations   *xxiii*

INTRODUCTION

Chapter 1   Invasive alien species in freshwater ecosystems: A brief overview   **3**
Robert A. Francis and Michael A. Chadwick

PART I — AQUATIC AND RIPARIAN PLANTS

Chapter 2   *Alternanthera philoxeroides* **(Martius) Grisebach (alligator weed)**   **25**
Shon S. Schooler

Chapter 3   *Crassula helmsii* **(T. Kirk) Cockayne (New Zealand pygmyweed)**   **37**
Anita Diaz

Chapter 4   *Eichhornia crassipes* **Mart. (Solms-Laubach) (water hyacinth)**   **47**
Xiaoyun Pan, Amy M. Villamagna and Bo Li

Chapter 5   *Heracleum mantegazzianum* **Sommier & Levier (giant hogweed)**   **57**
Jan Pergl, Irena Perglová and Petr Pyšek

Chapter 6   *Impatiens glandulifera* **Royle (Himalayan balsam)**   **67**
Christopher P. Cockel and Robert A. Tanner

Chapter 7   *Lagarosiphon major* **(Ridley) Moss ex Wager (curly water weed)**   **79**
Tobias O. Bickel

Chapter 8   *Lythrum salicaria L.* **(purple loosestrife)**   **91**
Keith R. Edwards

Chapter 9   *Myriophyllum aquaticum* **(Vell.) Verdcourt (parrot feather)**   **103**
Andreas Hussner and Paul D. Champion

Chapter 10 *Spartina anglica* C. E. Hubbard (English cord-grass)   113
           Philip D. Roberts

Chapter 11 *Tamarix* spp. (tamarisk, saltcedar)   125
           Anna Sher

PART II — AQUATIC INVERTEBRATES

Chapter 12 *Aedes albopictus* Skuse (Asian tiger mosquito)   137
           Paul Leisnham

Chapter 13 An overview of invasive freshwater cladocerans: *Bythotrephes longimanus*
           Leydig as a case study   149
           Angela L. Strecker

Chapter 14 Invasive freshwater copepods of North America   161
           Jeffery R. Cordell

Chapter 15 *Corbicula fluminea* Müller (Asian clam)   173
           Martina I. Ilarri and Ronaldo Sousa

Chapter 16 *Eriocheir sinensis* H. Milne-Edwards (Chinese mitten crab)   185
           Matt G. Bentley

Chapter 17 *Pacifastacus leniusculus* Dana (North American signal crayfish)   195
           Jenny C. Dunn

Chapter 18 Apple snails   207
           Robert H. Cowie and Kenneth A. Hayes

Chapter 19 *Potamopyrgus antipodarum* J. E. Gray (New Zealand mudsnail)   223
           Sarina Loo

PART III — FISH

Chapter 20 Bigheaded carps of the genus *Hypophthalmichthys*   235
           James E. Garvey

Chapter 21 *Cyprinus carpio* L. (common carp)   247
           Brendan Hicks, Nicholas Ling and Adam J. Daniel

Chapter 22   *Gambusia affinis* (Baird & Girard) and *Gambusia holbrooki* Girard (mosquitofish)   261
William E. Walton, Jennifer A. Henke and Adena M. Why

Chapter 23   *Pseudorasbora parva* Temminck & Schlegel (topmouth gudgeon)   275
Rudy E. Gozlan

Chapter 24   *Salmo trutta* L. (brown trout)   285
Angus McIntosh, Peter McHugh and Phaedra Budy

PART IV — AMPHIBIANS AND REPTILES

Chapter 25   *Rhinella marina* L. (cane toad)   299
Richard Shine

Chapter 26   *Eleutherodactylus coqui* Thomas (Caribbean tree frog)   311
Karen H. Beard and William C. Pitt

Chapter 27   *Rana [Lithobates] catesbeiana* Shaw (American bullfrog)   321
Antonia D'Amore

Chapter 28   *Trachemys scripta* (slider terrapin)   331
Gentile Francesco Ficetola, Dennis Rödder and Emilio Padoa-Schioppa

PART V — AQUATIC AND RIPARIAN MAMMALS

Chapter 29   *Castor canadensis* Kuhl (North American beaver)   343
Christopher B. Anderson, Nicolás Soto, José Luis Cabello, Guillermo Martínez Pastur, María Vanessa Lencinas, Petra K. Wallem, Daniel Antúnez and Ernesto Davis

Chapter 30   *Myocastor coypus* Molina (coypu)   357
Sandro Bertolino, M. Laura Guichón and Jacoby Carter

Chapter 31   *Neovison vison* Schreber (American mink)   369
Laura Bonesi and Michael Thom

PART VI — AQUATIC PATHOGENS

Chapter 32   *Bothriocephalus acheilognathi* Yamaguti (Asian tapeworm)   385
Anindo Choudhury and Rebecca Cole

Chapter 33   *Centrocestus formosanus* **Nishigori (Asian gill-trematode)**   **401**
Lori R. Tolley-Jordan and Michael A. Chadwick

Chapter 34   *Myxobolus cerebralis* **Höfer (whirling disease)**   **421**
Jerri L. Bartholomew

CONCLUSION

Chapter 35   **Management of freshwater invasive alien species**   **435**
Robert A. Francis and Petr Pyšek

Index   *447*

Illustrations

Plates

The following plates appear in a colour section between pp 216 and 217.

18.1   Shells of introduced ampullariids
18.2   Introduced *Pomacea canaliculata* crawling on the muddy bottom of a taro patch in Hawaii
18.3   Egg masses of introduced ampullariids
18.4   Abundant egg masses of *Pomacea canaliculata* indicating a major infestation in Hawaii

The following plates appear in a colour section between pp 288 and 289.

24.1   Colour variation associated with trout (*Salmo trutta*) from various freshwater situations in South Island, New Zealand, where they were caught
24.2   Impacts of non-native brown trout in invaded freshwater ecosystems

Figures

2.1   Illustration of alligator weed   26
2.2   Photograph of alligator weed showing papery white flowers and lanceolate leaves   27
2.3   Distribution of alligator weed   28
2.4   The combination of plant attributes and properties of the recipient environment that facilitate alligator weed invasions   30
3.1   *Crassula helmsii* growing (a) in terrestrial form and (b) in submerged form   38
3.2   Global distribution of *Crassula helmsii* based on countries where it has been recorded as growing in the wild   39
3.3   A hover-fly (Syrphidae) foraging for nectar on *Crassula helmsii* in Dorset, UK   40
3.4   Driver or passenger of change? *Crassula helmsii* carpeting the floor of (a) a temporary pond margin in the New Forest, UK and (b) reedbeds damaged by sika deer in Dorset, UK   44
4.1   Morphology and habitat of water hyacinth   48
4.2   Extensive mats of water hyacinth being sprayed with herbicide   48
4.3   Fibrous roots of water hyacinth that can form complex habitat for aquatic invertebrates and algae   48
5.1   Distribution of the three invasive *Heracleum* species in Europe   58
5.2   (a) The populations of *Heracleum mantegazzianum* in its native range in the forest belt of the Caucasus consist of scattered individual plants growing in species-rich tall herb communities; (b) in the invaded range in central Europe, it forms extensive stands with a high cover, suppressing species diversity of native plant communities   60
6.1   *Impatiens glandulifera* in its native habitat, the foothills of the Himalayas, Kaghan Valley, Pakistan   68
6.2   *Impatiens glandulifera* monospecific stand on the River Torridge, North Devon, UK   68
6.3   All records of *Impatiens glandulifera* occurrence across the British Isles to 2010   69
6.4   The global distribution of *Impatiens glandulifera*, highlighting the native and introduced range of the species   70
7.1   A single stem of *Lagarosiphon major*   80

7.2   Native and introduced distribution of *Lagarosiphon major*   80

7.3   A monospecific stand of *Lagarosiphon* in Lake Tarawera   82

8.1   Schematic of a typical *Lythrum salicaria* plant   92

8.2   Populations of *Lythrum salicaria* in: (a) its native Eurasian range and b) its invasive range   94

9.1   *Myriophyllum aquaticum*   104

9.2   Global distribution of *Myriophyllum aquaticum*   105

9.3   Regeneration capacity of *Myriophyllum aquaticum* from 25 plant fragments (n = 4)   105

9.4   Relative growth rates of *M. aquaticum* under different soil water regimes   106

9.5   Light and temperature response curves of net CO_2 gas exchange of emerged *M. aquaticum*   107

10.1   Line drawing detailing *Spartina anglica*   114

10.2   Worldwide distribution of *Spartina anglica*   115

10.3   The profile of intertidal mudflats: (a) with native vegetation prior to *Spartina* invasion and
        (b) after invasion by *Spartina*   118

11.1   Occurrence of species in the genus *Tamarix* within the US by state   126

11.2   Mature *Tamarix ramosissima* tree growing in Florence, Colorado, US   126

11.3   Near-monotypic *Tamarix* population growing along the Colorado River in Utah, US   127

12.1   Adult female *Ae. albopictus* after taking a blood meal   138

12.2   Larvae of *Ae. albopictus*   138

12.3   Artificial containers at an industrial site that collect rainwater and provide common habitat
        for *Ae. albopictus* egg and larval development   138

12.4   Traps used to monitor populations of *Ae. albopictus* and other *Aedes* mosquitoes   143

13.1   (a) *Bythotrephes longimanus*, with late stage embryos in its brood pouch, and (b) a fishing
        line from Lake Erie covered with *Bythotrephes*   150

13.2   Map of current distribution of *Bythotrephes* in its introduced range   151

13.3   Intercontinental invasions of cladoceran zooplankton species by biogeographic province   151

13.4   (a) Map of projected vulnerability of lakes to *Bythotrephes* invasion in the US and
        (b) percentage of lakes in dataset that are predicted to be suitable for *Bythotrephes* by
        hydrologic unit   154

14.1   *Pseudodiaptomus forbesi*, female with egg sacs   163

14.2   Columbia-Snake River drainage and stations sampled in 2009 for the introduced copepod
        Pseudodiaptomus forbesi on the Columbia and Snake Rivers, US   164

14.3   Percentage composition of the introduced copepod *Pseudodiaptomus forbesi* and indigenous
        taxa groups in reservoirs on the Columbia and Snake Rivers, US, September 2009   165

14.4   *Eurytemora affinis*, female with partial egg sac and attached spermatophore   166

14.5   Invasions by the *Eurytemora affinis* species complex from saline sources into freshwater lakes and
        reservoirs   167

14.6   *Limnoithona tetraspina*, female with egg sacs   167

15.1   *C. fluminea* showing: (a) the external and (b) internal view of shells   174

15.2   Total number of *C. fluminea* publications (a) per year and (b) per decade for each selected area   174

15.3   Percentage of *C. fluminea* publications per origin area of the study   175

15.4   Percentage of *C. fluminea* publications per selected theme   175

15.5   General pathways of ecological impacts mediated by *C. fluminea*   178

16.1   The Chinese mitten crab *Eriocheir sinensis*   186

16.2   A large number of Chinese mitten crab *Eriocheir sinensis* amassed at an obstacle to upstream
        migration on the River Elbe, Germany   188

17.1	A signal crayfish in a British river	197
17.2	Signal crayfish have a large chelae to body size ratio, giving them an advantage in competitive interactions with other species	198
17.3	Female signal crayfish have a high fecundity and females can carry between 100 and 400 abdominal eggs	198
19.1	*Potamopyrgus antipodarum* (New Zealand mudsnail)	224
19.2	New Zealand mudsnail populations can exhibit high densities	224
20.1	(a) Bighead carp (b) and silver carp from Illinois, US	236
20.2	Countries in which bighead and/or silver carp are probably established	236
20.3	Bighead carp foraging	237
20.4	Global aquaculture production in million metric tonnes of bighead and silver carp, 1950–2008	239
20.5	Intersection between the invaded Mississippi River System (including the Illinois River) and the Lake Michigan watershed	240
20.6	Ripe ovaries within an adult silver carp captured in the Illinois River	241
21.1	The global distribution of common carp (*Cyprinus carpio* L.)	248
21.2	Olive/bronze morph of wild common carp from the River Murray at Yarrawonga, Victoria, Australia (about 430mm fork length)	248
21.3	Orange morph of wild koi carp from New Zealand (565mm fork length, 4.18kg)	249
21.4	Koi carp-goldfish hybrid from New Zealand (355mm fork length, 1.02kg)	249
21.5	Common carp-goldfish hybrid from Australia (330mm fork length); note stout, highly serrated dorsal fin spine	249
21.6	Depressions from common carp feeding activity	253
21.7	Reduction of aquatic macrophyte cover, number of wildfowl and species richness with increasing biomass of common carp in two shallow lakes, Hennepin and Hopper, Illinois	254
22.1	A male and female *Gambusia holbrooki*; insert shows gonopodium of the male	262
22.2	A timeline for the introduction of *Gambusia affinis* and *Gambusia holbrooki* outside the US	262
22.3	The native and introduced geographic distributions of *Gambusia* spp. worldwide and in North America	265
23.1	Current distribution of topmouth gudgeon (*Pseudorasbora parva*) in its invasive range	277
23.2	(a) Female and (b) male topmouth gudgeon (*Pseudorasbora parva*)	278
24.1	Global distribution of brown trout within its native and introduced range	287
25.1	The cane toad, *Rhinella marina*	300
25.2	Areas in which cane toads occur naturally, and those to which the species has been deliberately or accidentally introduced	300
25.3	Current and predicted distribution of the cane toad in Australia	301
25.4	A native 'meat ant', *Iridomyrmex reburrus*, carrying away a metamorph cane toad beside a Northern Territory billabong	305
26.1	*Eleutherodactylus coqui* in potted nursery plant	312
26.2	Distribution of *Eleutherodactylus coqui* on the Big Island	312
26.3	*Eleutherodactylus coqui* on a *Heliconia* leaf	313
26.4	Areas treated for control on the Big Island	316
27.1	The American bullfrog (*Rana [Lithobates] catesbeiana*)	322
27.2	Global distribution of the American bullfrog	322
28.1	Feral individuals of (a) *Trachemys scripta elegans*; (b) *T. s. scripta*	332
28.2	Native range of *Trachemys scripta* and countries where feral slider terrapins have been recorded	332

28.3   Potential distribution of *Trachemys scripta* derived from physiological thresholds   334
29.1   North American beaver (*Castor canadensis*) swimming in a waterway on Tierra del Fuego Island   344
29.2   Map of the austral portions of Chile and Argentina, including the Tierra del Fuego and the Cape Horn Archipelagos, indicating the approximate dates of the range expansion for the North American beaver (*Castor canadensis*) since its initial introduction in 1946   344
29.3   Scientific literature citations regarding invasive beaver in Chile and Argentina per year in ISI, Chilean non-ISI and Argentine non-ISI peer-reviewed journals   348
30.1   *Myocastor coypus* (coypu)   358
30.2   Global distribution of coypu   358
30.3   Relationship of body weight to age of coypus in the native range (Argentina) and countries of introductions   361
30.4   (a) Coypu begging for food from one of the authors, and (b) sign warning the public not to feed animals (coypu are the target species), both in Germany   363
31.1   American mink   370
31.2   World distribution map of the American mink   370
31.3   Bank-side trap   373
31.4   Floating raft trap   374
32.1   World distribution of Asian fish tapeworm   386
32.2   Distribution of Asian fish tapeworm in North America, Hawaii and the Caribbean islands   386
32.3   Lifecycle of Asian tapeworm, *Bothriocephalus acheilognathi*   389
32.4   Morphology of the Asian tapeworm   392
33.1   Complex lifecycle of the gill trematode *Centrocestus formosanus* involving (A) a definitive bird (or mammal) host, (B) the first intermediate snail host and (C) an intermediate fish host   402
33.2   The global distribution, by country, of *Centrocestus formosanus* and its first intermediate snail host, *Melanoides tuberculata*   403
34.1   Lifecycle of *Myxobolus cerebralis* showing the two alternating spores: the myxospore that is released from the salmonid host and the actinospore released from the invertebrate host, *Tubifex tubifex*   422
34.2   Timeline for *Myxobolus cerebralis* detection both in Europe, from its native range, and into areas where it has been introduced   423
34.3   Distribution of *Myxobolus cerebralis*   424

Tables

1.1   A brief summary of some key concepts and hypotheses relating to alien and invasive species   5
1.2   A brief summary of some key characteristics relating to invasion success of alien species   6
8.1   Common methods used to control invasive *Lythrum salicaria* populations in North America, the population size for which each method is most useful and the advantages and disadvantages of each method   96
10.1   Synonyms and known common names for *Spartina anglica*   114
10.2   The effectiveness of control methods at reducing the densities of *S. angelica*   119
12.1   Countries by region in which *Ae. albopictus* was reported between 1979 and 2010   139
14.1   Planktonic copepods introduced to North American fresh waters   162

14.2   Densities of combined life history stages of *Pseudodiaptomus forbesi* in lower Columbia and
Snake River reservoirs sampled in September 2009   165

15.1   Fundamental abiotic variables determining the degree of invasion by *C. fluminea*   176

16.1   History of Chinese mitten crab invasion of Europe during the 20th century   186

16.2   History of Chinese mitten crab reports in North America   187

16.3   Recently recorded populations or new records of *Eriocheir sinensis*   188

17.1   European territories from which signal crayfish have been recorded in the wild,
along with the year of first record   196

18.1   Native and non-native ranges of introduced ampullariids in the wild   209

21.1   von Bertalanffy growth curve parameters for *Cyprinus carpio*, including koi carp from
New Zealand and common carp from Australia, France, Spain and Chile   251

24.1   Habitat suitability criteria for four primary life stages of brown trout   288

24.2   Examples of native fishes affected by non-native brown trout invasions   289

28.1   Countries or territories where feral sliders have been reported   332

29.1   Native and exotic terrestrial mammals of the Tierra del Fuego and Cape Horn archipelagos
(including both Argentina and Chile)   346

29.2   Historical timeline of benchmarks in research and application regarding the study and control
of North American beaver in the Magallanes region of Chile and their implications for the
sociopolitical process of linking academia with decision making   349

30.1   Vegetation consumed by coypus in various habitats in introduced and native ranges   359

30.2   Comparison between efforts and costs of the successful coypu eradication in England and
of the permanent control campaign in Italy   365

31.1   Examples of the range of densities of American mink   371

31.2   List of some of the projects for managing mink populations   375

33.1   The distribution of the first intermediate (*Melanoides tuberculata*), second intermediate (fishes)
and definitive hosts (birds) of the gill trematode *Centrocestus formosanus* (CF)   405

33.2   Families and species of fishes that serve as second intermediate hosts (gills infected) of
Centrocestus formosanus   409

Preface

This book was initiated following the 2009 conference 'Invasive species ecology and management: Linking research and practice' organized by the Invasive Species Special Interest Group of the British Ecological Society. Among the feedback from conference delegates was the observation that it was difficult to find up-to-date and comprehensive summaries of particularly notorious invasive species, especially for those who were not academics (and therefore with limited or no access to academic journals) but who were actively involved in controlling such species. It was noted that although very extensive and useful websites such as that of the Global Invasive Species Database of the Invasive Species Specialist Group of the International Union for Conservation of Nature (IUCN) exist, they sometimes have limitations in that the information provided in the various entries varies in quality, comprehensiveness and frequency of updates.

Consequently, the Handbook was proposed. The book is not intended to be a replacement for the invasive species databases that exist, but rather a comprehensive summary of selected species at time of publication, written by experts in the field. It should therefore provide a useful supplementary resource for researchers and managers of invasive species.

Why were freshwater species chosen in the first instance? As outlined in the introductory chapter (Chapter 1), many of the most invasive species are found in wetlands, and freshwater ecosystems are particularly vulnerable to invasion; this, alongside the importance of such systems for the direct and indirect ecosystem services and resources they provide to society, makes them a priority for the prevention and control of invasive species. In this book, freshwater ecosystems of all types and spatial scales are considered, from ponds to rivers and lakes, and riparian zones and estuaries are also included. Thirty three species were selected for inclusion in the volume based on their impact on biological diversity and/or human activities, and/or their illustration of important issues surrounding biological invasion. All are 'high profile' species that have been the focus of research and control efforts, and where possible species were selected from a broad range of taxa and geographical regions. There is much debate about what constitutes an 'invasive' species (e.g. as opposed to simply 'alien'), and two broad distinctions are made: those that are 'invasive' because of their detrimental impact on some aspect of their introduced location (whether ecological, economic or cultural), and those that spread (i.e. invade) rapidly when introduced. The two are sometimes related, but not always (see Ricciardi and Cohen, 2007, for discussion). Most of the species selections in this book are invasive in both senses, but some ally more with the former definition, and a few chapters (e.g. Chapter 13) with the latter, as in some cases impacts are not well understood. While a debate about the varying merits of 'invasive' terminology is interesting, it is not the focus of this volume. Instead, I hope that the contents of this book may inform the broader discipline of 'invasion ecology', whichever definition or interpretation is used.

Some people reading this book will be disappointed about the lack of inclusion of a species that they will consider essential to such a text. Much deliberation took place after the 2009 conference about what should be included in the volume, before the current line-up was decided upon. However, it is impossible to cover everything and some prominent species were naturally omitted. This does open the door to consideration of a second volume, though, and readers are encouraged to contact me (robert.francis@kcl.ac.uk) with suggestions of crucial species that were missed this time around.

For the case studies, each author was asked to discuss: (1) a summary of the history of the species and its introduction to host countries (including current distribution); (2) the ecological niche of the species, in both its native range and host range; (3) management efforts employed and how effective they have been in different situations; and (4) any challenges, controversies or differences of opinion that may exist regarding the presence or control of species in different countries. I did not wish to confine the chapters within a prescribed structure beyond this, as a structure appropriate for one species might not be for another, and I felt it was important to give chapter authors the creative freedom to construct their summaries as they saw fit. This was particularly the case for some of the chapters that present case study material (e.g. Chapter 29) or where a particular aspect of a species is not known (e.g. impacts; see Chapter 14).

In each chapter I have also not tried to impose the use of specific 'native/non-native' terminology on the authors, such that terms such as 'introduced', 'alien', 'exotic', 'non-native' and 'non-indigenous' are generally used interchangeably (see Pyšek et al, 2008). I consider that the essence of these terms is the same, i.e. any species moved outside an area where is it not naturally found, by human agency, qualifies for all of these terms. There are of course varying further distinctions that can be made as to whether the species is established in the wild or not (naturalized) and so on, but these distinctions are made at the discretion of the individual authors.

Finally, all chapters were peer reviewed and I must thank all those who kindly gave of their time and expertise to critically comment on the chapters: Alvaro Alonso, Przemek Bajer, Simon Baker, Mark Benedict, Tim Bonner, Seth Britch, Michael Chadwick, John Clayton, Francesco Ficetola, Mark Freeman, Andre Gassmann, Bill Granath, Lauren Harrington, Steve Johnson, Claude Lavoie, Carol Lee, Marta Lizzaralde, Andrew Mitchell, Evangelian Natale, Stefan Nehring, Nuria Polo-Cavia, Petr Pyšek, Jian-Wen Qiu, Frank Rahel, Greg Sass, Hana Skalova, Peter Sorensen, Jan Thiele, Amy Marie Villamagna, Norman Yan, Dan Yu and Eva Zahorskae. Special thanks must also go to Tim Hardwick at Earthscan for regular advice and consideration throughout production of the book.

Robert A. Francis
London, February 2011

References

Pyšek, P., Richardson, D. M., Pergl, J., Jarošík, V., Sixtová, Z. and Weber, E. (2008) 'Geographical and taxonomic biases in invasion ecology', *Trends in Ecology and Evolution*, vol 23, no 5, pp237–244

Ricciardi, A. and Cohen, J. (2007) 'The invasiveness of an introduced species does not predict its impact', *Biological Invasions*, vol 9, no 3, pp309–315

Contributors

Christopher B. Anderson, Sub-Antarctic Biocultural Conservation Program, Universidad de Magallanes, Punta Arenas, Chile AND University of North Texas, Denton, Texas, USA AND Omora Ethnobotanical Park (Universidad de Magallanes and Institute of Ecology and Biodiversity), Puerto Williams, Chile, christopher.anderson@umag.cl

Daniel Antúnez, Regional Office of Renewable Resources, Agriculture and Livestock Service, Ministry of Agriculture, Punta Arenas, Chile, wanderino_73@yahoo.es

Jerri L. Bartholomew, Department of Microbiology, Nash Hall 220, Oregon State University, Corvallis, OR 97331-3804,USA, bartholj@science.oregonstate.edu

Karen H. Beard, Department of Wildland Resources and Ecology Center, Utah State University, Logan, Utah, 84322-5230, USA, karen.beard@usu.edu

Matt G. Bentley, School of Marine Science and Technology, Newcastle University, Newcastle upon Tyne NE1 7RU, UK, m.g.bentley@ncl.ac.uk

Sando Bertolino, DIVAPRA (Department of Protection and Exploitation of Agricultural Resources) Entomology and Zoology, Via L. da Vinci 44, 10095 Grugliasco, Italy, sandro.bertolino@unito.it

Tobias O. Bickel, Biosecurity Queensland, Invasive Plant Science, Ecosciences Precinct, GPO Box 267, Brisbane QLD 4001, Australia, Tobias.Bickel@deedi.qld.gov.au

Laura Bonesi, University of Trieste, Piazzale Europa 1, 34127 Trieste, Italy, bonesi.laura@gmail.com

Phaedra Budy, US Geological Survey – Utah Cooperative Fish and Wildlife Research Unit, Watershed Sciences Department, Utah State University, Logan, Utah, USA, phaedra.budy@usu.edu

José Luis Cabello, Regional Office of Renewable Resources, Agriculture and Livestock Service, Ministry of Agriculture, Punta Arenas, Chile AND Master's of Science Program in Conservation and Management of Sub-Antarctic Environments and Resources, Universidad de Magallanes, Punta Arenas, Chile, jcabellocabalin@gmail.com

Jacoby Carter, US Geological Survey, National Wetlands Research Center, 700 Cajundome Blvd., Lafayette, Louisiana, USA, 70506, jacobycarter@usgs.gov

Michael A. Chadwick, Department of Geography, King's College London, Strand, London WC2R 2LS, UK, michael.chadwick@kcl.ac.uk

Paul D. Champion, National Institute of Water and Atmospheric Research (NIWA), PO Box 11–115, Hamilton 3251, New Zealand, p.champion@niwa.co.nz

Anindo Choudhury, Division of Natural Sciences, St Norbert College, 100 Grant Street, DePere, WI 54115, USA, anindo.choudhury@snc.edu

Christopher P. Cockel, Department of Geography, Queen Mary, University of London, Mile End Road, London E1 4NS, UK, c.p.cockel@qmul.ac.uk

Rebecca Cole, USGS National Wildlife Health Center, 6006 Schroeder Road, Madison, WI 53711, USA, rcole@usgs.gov

Jeffery R. Cordell, School of Aquatic and Fishery Sciences, 1122 Boat Street, University of Washington, Box 355020, Seattle, WA 98195-5020, USA, jcordell@u.washington.edu

Robert H. Cowie, Pacific Biosciences Research Center, University of Hawaii, 3050 Maile Way, Gilmore 408, Honolulu, Hawaii 96822, USA, cowie@hawaii.edu

Antonia D'Amore, Elkhorn Slough National Estuarine Research Reserve, 1700 Elkhorn Road, Watsonville, CA 95076, USA, nina@elkhornslough.org

Adam J. Daniel, Centre for Biodiversity and Ecology Research, Department of Biological Sciences, Faculty of Science and Engineering, The University of Waikato, New Zealand, adaniel@waikato.ac.nz

Ernesto Davis, Master's of Science Program in Conservation and Management of Sub-Antarctic Environments and Resources, Universidad de Magallanes, Punta Arenas, Chile

Anita Diaz, School of Conservation Sciences, Bournemouth University, Dorset House, Talbot Campus, Fern Barrow, Poole, Dorset BH12 5BB, UK, ADiaz@bournemouth.ac.uk

Jenny C. Dunn, RSPB, The Lodge, Potton Road, Sandy, Bedfordshire SG19 2DL, UK, Jenny.Dunn@rspb.org.uk

Keith R. Edwards, Department of Ecosystem Biology, Faculty of Science, University of South Bohemia, České Budějovice, Czech Republic, kredwards59@yahoo.com

Gentile Francesco Ficetola, Department of Environmental Sciences, University of Milano-Bicocca, Piazza della Scienza 1, 20126 Milano, Italy, francesco.ficetola@unimi.it

Robert A. Francis, Department of Geography, King's College London, Strand, London WC2R 2LS, UK, robert.francis@kcl.ac.uk

James E. Garvey, Department of Zoology, Fisheries and Illinois Aquaculture Center, Southern Illinois University, Carbondale, IL 62901, USA, jgarvey@siu.edu

M. Laura Guichón, Ecología de Mamíferos Introducidos, Departamento de Ciencias Básicas, Universidad Nacional de Luján, Rutas 5 y 7, 6700 Luján, Buenos Aires, Argentina, mlguichon@unlu.edu.ar

Rudy E. Gozlan, School of Conservation Sciences, Bournemouth University, Talbot Campus, Fern Barrow, Poole, Dorset, BH12 5BB, UK, rgozlan@bournemouth.ac.uk

Kenneth A. Hayes, Pacific Biosciences Research Center, University of Hawaii, 3050 Maile Way, Gilmore 408, Honolulu, Hawaii 96822, USA, khayes@hawaii.edu

Jennifer A. Henke, Department of Entomology, University of California, Riverside, CA 92521 USA, jennifer.henke@email.ucr.edu

Brendan J. Hicks, Centre for Biodiversity and Ecology Research, Department of Biological Sciences, Faculty of Science and Engineering, The University of Waikato, New Zealand, b.hicks@waikato.ac.nz

Andreas Hussner, Institute of Plant Biochemistry, University of Düsseldorf, Universitätsstraße 1, D-40225 Düsseldorf, Germany, Andreas.Hussner@uni-duesseldorf.de

Martina I. Ilarri, CIMAR-LA/CIIMAR – Centre of Marine and Environmental Research, Rua dos Bragas 289, 4050-123 Porto, Portugal AND ICBAS – Instituto de Ciências Biomédicas de Abel Salazar, Universidade do Porto, Departamento de Estudos de Populações, Laboratório de Ecotoxicologia, Lg. Prof. Abel. Salazar, 2, 4099-003 Porto, Portugal, milarri@ciimar.up.pt

Paul Leisnham, Department of Environmental Science and Technology, University of Maryland, College Park MD 20742, USA, Leisnham@umd.edu

María Vanessa Lencinas, Forest Resources Laboratory, Center for Austral Scientific Research (CADIC-CONICET), Ushuaia, Argentina, mvlencinas@gmail.com

Bo Li, Coastal Ecosystems Research Station of Yangtze River Estuary, Ministry of Education Key Laboratory for Biodiversity Science and Ecological Engineering, The Institute of Biodiversity Science, Fudan University, #220 Handan Road, Shanghai 200433, PR China, bool@fudan.edu.cn

Nicholas Ling, Centre for Biodiversity and Ecology Research, Department of Biological Sciences, Faculty of Science and Engineering, The University of Waikato, New Zealand, n.ling@waikato.ac.nz

Sarina Loo, Sustainable Water Environment Division, Department of Sustainability and Environment, PO Box 500, East Melbourne VIC 3002, Sarina.Loo@dse.vic.gov.au

Peter McHugh, Freshwater Ecology Research Group, School of Biological Sciences, University of Canterbury, Christchurch, New Zealand, pete.mchugh@canterbury.ac.nz

Angus McIntosh, Freshwater Ecology Research Group, School of Biological Sciences, University of Canterbury, Christchurch, New Zealand, angus.mcintosh@canterbury.ac.nz

Emilio Padoa-Schioppa, Department of Environmental Sciences, University of Milano-Bicocca, Piazza della Scienza 1, 20126 Milano, Italy, emilio.padoaschioppa@unimib.it

Xiaoyun Pan, Coastal Ecosystems Research Station of Yangtze River Estuary, Ministry of Education Key Laboratory for Biodiversity Science and Ecological Engineering, The Institute of Biodiversity Science, Fudan University, #220 Handan Road, Shanghai 200433, PR China, xypan@fudan.edu.cn

Guillermo Martínez Pastur, Forest Resources Laboratory, Center for Austral Scientific Research (CADIC-CONICET), Ushuaia, Argentina, cadicforestal@gmail.com

Jan Pergl, Institute of Botany, Academy of Sciences of the Czech Republic, Průhonice, Czech Republic, pergl@ibot.cas.cz

Irena Perglová, Institute of Botany, Academy of Sciences of the Czech Republic, Průhonice, Czech Republic, perglova@ibot.cas.cz

William C. Pitt, USDA/APHIS/WS/National Wildlife Research Center, Hilo Field Station, Hilo, Hawaii 96721, USA, Will.Pitt@aphis.usda.gov

Petr Pyšek, Institute of Botany, Academy of Sciences of the Czech Republic, Průhonice, Czech Republic AND Faculty of Science, Department of Ecology, Charles University Prague, Czech Republic, pysek@ibot.cas.cz

Philip D. Roberts, CABI, Nosworthy Way, Wallingford, Oxfordshire OX10 8DE, UK, p.roberts@cabi.org

Dennis Rödder, Biogeography Department, Trier University, Am Wissenschaftspark 25+27, 54286 Trier, Germany AND Zoologisches Forschungsmuseum Alexander Koenig, Adenauerallee 160, 53113 Bonn, Germany, roedder@uni-trier.de

Shon S. Schooler, CSIRO Ecosystem Sciences, Ecosciences Precinct, PO Box 2583, Brisbane, QLD, Australia, shon.schooler@csiro.au

Anna Sher, Department of Biology, Denver University, 2150 E. Evans Avenue #323, Denver, CO 80208, USA, anna.sher@du.edu

Richard Shine, Biological Sciences A08, University of Sydney, NSW 20906, Australia, rics@bio.usyd.edu.au

Nicolás Soto, Regional Office of Renewable Resources, Agriculture and Livestock Service, Ministry of Agriculture, Punta Arenas, Chile, nicolas.soto@sag.gob.cl

Ronaldo Sousa, CIMAR-LA/CIIMAR – Centre of Marine and Environmental Research, Rua dos Bragas 289, 4050-123 Porto, Portugal, AND CBMA – Centre of Molecular and Environmental Biology, Department of Biology, University of Minho, Campus de Gualtar, 4710-057 Braga, Portugal, ronaldo.sousa@ciimar.up.pt

Angela L. Strecker, School of Aquatic and Fishery Sciences, University of Washington, Seattle, WA 98105, USA, angelast@u.washington.edu

Robert A. Tanner, CABI E-UK, Bakeham Lane, Egham, Surrey TW20 9TY, UK, r.tanner@cabi.org

Michael Thom, Department of Biology, University of York, Heslington, York YO10 5DD, michael.thom@york.ac.uk

Lori R. Tolley-Jordan, Department of Biology, Jacksonville State University, 700 Pelham Rd. N Jacksonville AL 36265, USA, ljordan@jsu.edu

Amy M. Villamagna, Sustainable Development and Conservation Biology Program, University of Maryland – College Park, MD 20782, USA, amv@umd.edu

Petra K. Wallem, Center for Advanced Studies in Ecology and Biodiversity, Catholic University of Chile, Santiago, Chile, petra.wallem@gmail.com

William E. Walton, Department of Entomology, University of California, Riverside, CA 92521, USA, walton@ucr.edu

Adena M. Why, Department of Entomology, University of California, Riverside, CA 92521, USA, adena.why@email.ucr.edu

Abbreviations

APHIS	Animal and Plant Health Inspection Service
AUC	area under the curve
Bd	*Batrachochytrium dendrobatidis*
BSBI	Botanical Society of the British Isles
BTI	*Bacillus thuringiensis israelensis*
CAM	crassulacean acid metabolism
CBD	Convention on Biological Diversity
CHIRP	Coqui Hawaiian Integration and Reeducation Project
DDT	dichlorodiphenyltrichloroethane
DO	dissolved oxygen
EPA	Environmental Protection Agency
ET_0F	evapotranspiration (ET) expressed as a fraction of reference ET
FAO	Food and Agriculture Organization of the United Nations
FISK	Fish Invasiveness Scoring Kit
FONDEMA	Regional Fund for the Development of Magallanes
GARP	Genetic Algorithm for Rule-Set Production
GIS	geographical information system
GISD	Global Invasive Species Database
GISIN	Global Invasive Species Information Network
GISP	Global Invasive Species Programme
HIV	human immunodeficiency virus
IAS	invasive alien species
ILFA	Import of Live Fish Act
IPCC	Intergovernmental Panel on Climate Change
IPM	integrated pest management
IUCN	International Union for Conservation of Nature
LCR	Little Colorado River
LTSER	Long-Term Socio-Ecological Research
MHWN	mean high water neap
MHWS	mean high water spring
NDVI	normalized difference vegetation index
n. sp.	new species
NZBA	New Zealand Biosecurity Act
ppm	parts per million
ppt	parts per thousand
SAG	Agriculture and Livestock Service
SD	standard deviation
SEM	scanning electron microscopy
SMRT	sterile male release technique
SVL	snout–vent length
SWFL	south-western willow flycatcher
USDA	United States Department of Agriculture

USFWS	US Fish and Wildlife Service
USGS	United States Geological Survey
UV	ultraviolet
WFD	Water Framework Directive

Introduction

capacity to move with or against the predominantly unidirectional flows, overall vagility, reliance on aquatic media, habitat requirements etc.), but hydro-geomorphological characteristics of the river environment itself. For example, the spread of *Neogobius melanostomus* (round goby) from its host habitat of Lake Michigan to Wisconsin (US) tributaries was found to be best predicted by watershed area, slope and channel gradient. These factors reflect not just the limitations of the species to move upstream along high gradient channels, but also desirable habitat correlates such as bank-full width and coarseness of bed sediment (Kornis and Vander Zanden, 2010). This example helps to illustrate that longitudinal connectivity varies with location within the river network. As river networks display broad longitudinal gradients in channel size and slope, discharge, velocity, sediment calibre and transport, among other things, distributions may also be limited to certain network sections (e.g. headwater reaches vs. low gradient channels). These physical factors can be important for limiting system connectivity for a given invasive species (though some of this may be overcome during flood events) (e.g. Bodamer and Bossenbroek, 2008; DeGrandchamp et al, 2008). Nevertheless, longitudinal connectivity remains a key factor in the spread of alien species within lotic waterways and contributes to system invasibility (Leuven et al, 2009).

Both lotic and lentic systems experience lateral expansion and contractions; the expansion phase moderates connectivity between aquatic and terrestrial areas, across a riparian ecotone, though these dynamics can be more pronounced in lotic ecosystems (e.g. Tockner et al, 2000; Ward et al, 2002; Coops et al, 2003). Such connectivity is spatially and temporally variable, and can range from rare catastrophic events that completely rework both aquatic and riparian habitat while connecting many different landscape components (varying levels of flood pulse, Junk et al, 1989), to frequent minor cycles of expansion and contraction (flow pulse, sensu Tockner et al, 2000) that nevertheless allow for the exchange of materials between aquatic and riparian zones. This lateral connectivity is important for facilitating the movement of species to riparian areas and disassociated water bodies, such as disconnected channels and backwaters, or artificial water features such as drainage ditches. This process is well described for alien riparian plants (e.g. Predick and Turner, 2008), but other taxa may also be spread via flood events (Fowler et al, 2007). Lateral connectivity

and the flow processes driving it are closely related to disturbance of aquatic and riparian habitat, which also influences the invasibility of the ecosystem.

Disturbance

Disturbance lies at the heart of lotic freshwater ecosystems, mainly from water flow; though this is less of an ecosystem driver in lentic systems (e.g. Resh et al, 1988; Reice et al, 1990). The flow regime of a river system will determine the frequency, duration and intensity of many forms of disturbance, which are responsible for both the creation and destruction of physical habitat (e.g. shifting habitat mosaic, sensu Gregory et al, 1991; Stanford et al, 2005), the maintenance of ruderal species within riverine and riparian communities and suppression of competitors that may otherwise dominate (e.g. Tabacchi et al, 1996) and the entrainment, transportation and deposition of biotic and abiotic ecosystem components.

Disturbance can relate to both increases and decreases in flow, and can be both beneficial and detrimental to the spread and establishment of alien species (see Richardson et al, 2007). If a species introduced to a river system has a greater capacity to tolerate or recover from disturbance than some native species then it may be able to outcompete natives and become dominant. For example, alien crayfish *Orconectes neglectus* subsp. *chaenodactylus* (ringed crayfish) and *Pacifastacus leniusculus* (signal crayfish) demonstrate high tolerances to stream drying (such as may occur during a contraction phase, and particularly in ephemeral water bodies) compared to natives; in the first case a comparison of two weeks against two days for the native *Orconectes eupunctus* (Larson et al, 2009; see also Chapter 17). Likewise, alien species may demonstrate increased tolerance to stresses associated with fluvial disturbance such as high flow velocities, inundation, suspended sediments or burial (e.g. Tickner et al, 2001; Zardi et al, 2006). Several invasive riparian plants demonstrate the capacity to rapidly colonize newly created habitat, and characteristics such as abundance of propagules, early germination and rapid growth mean that newly deposited sediments may be colonized by such species, sometimes forming monotypic stands. Examples of this include *Impatiens glandulifera* (see Chapter 6), *Fallopia japonica* (Japanese knotweed) and *Heracleum mantegazzianum* (giant hogweed) (see Chapter 5).

In contrast, natural disturbance regimes may help to prevent the spread and establishment of alien species. For example, those that are competitively superior but not adapted to particular hydrogeomorphic stresses may not be able to compete with native species that are more sympathetic to the disturbance regime. Fausch et al (2001) found that the invasion success of *Oncorhynchus mykiss* (rainbow trout) was higher in river systems where the flow regime dictated a low probability of floods in months that coincided with fry emergence. Similarly, Predick and Turner (2008) found that a natural flow regime helped to limit alien plant frequency and abundance in riparian areas of the Wisconsin River, US, because those species were more sensitive to flood disturbance.

Landscape position

Freshwater ecosystems naturally form in landscape areas of low elevation, and they therefore represent sinks for a wide variety of materials, including water, sediment, nutrients, propagules and pollutants (Zedler and Kercher, 2004). Zedler and Kercher (2004) have argued that this 'sink' function increases wetland invasibility, though this mainly relates to those wetlands primarily fed by surface runoff, which is responsible for much of the transport of materials into the system. They also note that wetlands not fed primarily by surface runoff (such as high altitude fens and bogs) have lower numbers of alien species. Certainly the reception and storage of these materials in wetlands help to create conditions that can be favourable for invasions, for example high propagule pressure (a key factor in invasion success) (Elton, 1958; Lockwood et al, 2005), high nutrient flux (allowing competitors to maximize performance) and sediment delivery (to form new surfaces for taxa such as plants and invertebrates to colonize). Addition of such materials can be further exacerbated by anthropogenic activity, which can for example increase nutrient input into wetlands via the use of fertilizers within the catchment, or increase sediment entrainment due to soil erosion (Richardson et al, 2007).

Anthropogenic modification

Anthropogenic activity both within and around freshwater ecosystems can greatly affect invasion (Ervin et al, 2006; Richardson et al, 2007). In part this is due

to increased human-mediated introductions as noted above, but anthropogenic modification of freshwater ecosystems can also directly increase their invasibility. The most dramatic impacts on freshwater systems are associated with flow regulation, which usually involves hard engineering such as the construction of dams and reservoirs.

Such modifications influence both connectivity and disturbance. Flow regulation reduces the duration, intensity and extent of high water events, which can result in native communities losing their adaptive advantages and increasing the probability of successful alien establishment (e.g. Hobbs and Huenneke, 1992; Predick and Turner, 2008). This is reflected in changes to discharge and water level downstream of dams, where the corresponding changes in aquatic habitat, reduction of channel width, exposure of sediments and lower inundation stresses, can lead to changes in community composition, the colonization of aliens and increases in alien abundance (e.g. Richardson et al, 2007). For example, the invasive diatom *Didymosphenia geminate* has been found to bloom at higher frequency and densities in association with dam sites (Kirkwood et al, 2009). In general, low flow variability and lower than natural levels of fluvial disturbance have been linked to greater invasibility (Mortenson and Weisberg, 2010).

The installation of dams and creation of associated impoundments can also create artificial, relatively homogeneous ecosystems that can support alien species. For example, impoundments have been found to have far greater occurrences of alien species than natural lakes due to higher levels of connectivity, disturbance and environmental heterogeneity (Havel et al, 2005; Johnson et al, 2008), and may act as 'stepping stones' for the further spread of aliens. In contrast, the interruption of connectivity resulting from dams can also help to prevent the spread of aliens in river networks, acting as barriers for both upstream and downstream movement ('favourable fragmentation'; see Jackson and Pringle, 2010; Rood et al, 2010), and in some cases the potential spread of IAS has been raised as a serious possibility following dam removal (e.g. Kornis and Vander Zanden, 2010).

In essence, any anthropogenic modification that alters natural (and naturally dynamic) connectivity and disturbance regimes may increase the already notable invasibility of freshwater ecosystems; this observation comes too late to prevent such modifications to the large majority of global freshwater systems, but these factors should be considered for future rehabilitation or restoration (e.g. Richardson et al, 2007; Francis, 2009; Kornis and Vander Zanden, 2010; Mortenson and Weisberg, 2010). Clearly this subject is very complex, and more detailed discussions of patterns of invasion and the functioning of freshwater landscapes can be found in Tabacchi and Planty-Tabacchi (2005) and Richardson et al (2007).

Impacts of Freshwater IAS

IAS can have a range of impacts on freshwater ecosystems, though it is important to note that many alien species do not have a detrimental impact and indeed may be a beneficial addition (e.g. Colautti and MacIsaac, 2004). Determining this can be difficult as some impacts may not be obvious or easily quantified (e.g. Vilà et al, 2010). Further, many may only emerge following a time lag from the point of alien introduction (e.g. Leuven et al, 2009; see Table 1.1). As such, impacts are often underestimated (Gherardi, 2007; Ricciardi and Kipp, 2008). In some cases the impact of a species can be relatively simple and clear to observe, such as increased predation of natives or the spread of a host-specific disease. In other situations the impacts are less direct and/or may be connected and multifaceted (e.g. Didham et al, 2005; Vilà et al, 2010). The primary forms of impact are briefly covered here, but this discussion is not intended to be an exhaustive description of impacts (which will be apparent after reading the case studies that form this book), but rather an overview of some of the more prevalent impacts of IAS.

Changes in biodiversity and community composition

One of the most common impacts of IAS is the disruption of native species populations leading to population declines, changes in community composition, and in some cases species extirpation or extinctions. Predation or direct mortality (e.g. via spreading of an infectious disease or parasites) are common impacts of animal taxa leading to population decline (e.g. Mandrak and Cudmore, 2010; see Chapter 33), though competitive exclusion is also frequently documented (Kaufman, 1992; Mandrak and Cudmore, 2010; Michelan et al, 2010). Confirmed extinctions associated with impacts from IAS are rare, though some

of the most well-documented cases are the fish invasions of the Great Lakes in North America (Mandrak and Cudmore, 2010) and the contribution of *Pacifastacus leniusculus* (North American signal crayfish) to the extinction of *Pacifastacus nigrescens* (sooty crayfish) in its native range around San Francisco Bay (Gherardi, 2007; see Chapter 17). Not surprisingly, multiple invasions can interact to cause extirpation. For example, Johnson et al (2009) found that the combined effects of the invasive taxa *Orconectes rusticus* (rusty crayfish) and *Bellamya* (=*Cipangopaludina*) *chinensis* (Chinese mystery snails) in mesocosm experiments resulted in the extirpation of the native snail *Lymnaea stagnalis* (great pond snail).

A 'classic' example of a freshwater alien invasion leading to substantial change in community composition is the introduction of *Lates nilotica* (Nile perch) to Lake Victoria in the 1950s. Massive rises in the population of this species in the lake in the late 1970s and early 1980s resulted in a complete change in fish community to one dominated by aliens (predominantly *L. nilotica*). This change was accompanied by the loss of around 200 species endemic to the lake, mainly as a result of competition and predation (Kaufman, 1992; Gherardi, 2007). A similar effect is occurring in the Great Lakes in North America as a result of 35 alien introductions over the last 200 years (Mandrak and Cudmore, 2010).

Hybridization and genetic decline can also result from the introduction of an alien that is taxonomically related to native species, influencing biodiversity at the genotype level and the capacity of species to adapt to changing conditions. Examples include interbreeding between the invasive *O. rusticus* and the native *Orconectes propinquus* (and between the hybrid and *O. rusticus*) in Wisconsin, US, leading to loss of genetic diversity in the native species (Gherardi, 2007); and the hybridization of native fish species with invasive *Oncorhynchus mykiss* (rainbow trout) in North America (Simon and Townsend, 2003; Gherardi, 2007). This may be particularly significant when founder populations of alien species are based on very few individuals (e.g. Kalinowski et al, 2010; cf. Roman and Darling, 2007).

Changes in habitat

IAS can also change the quality of habitat for native species (e.g. sediment loads or quality, nutrient levels, macrophyte cover), as well as the ways in which species utilize that habitat or acquire resources (e.g. behavioural change). In some cases IAS can indirectly cause population decline by affecting feeding behaviours and food availability, limiting natives to less favourable habitat, changing diets to less suitable food sources, or by interrupting mating and reproduction. A good example of the latter can be found in the attraction of *P. leniusculus* males to chemical signatures of *Austropotamobius pallipes* (white-clawed crayfish) females, leading to mating disruption; see Chapter 17). As a perhaps more dramatic example, Leslie and Spotila (2001) found that the alien plant *Chromolaena odorata* (common floss flower) shaded the nesting sites of *Crocodylus niloticus* (Nile crocodile) in a wetland reserve to such a degree that only females were born (sex determination being temperature dependent), potentially skewing the sex ratio of the crocodile population to such an extent that eventual extirpation was possible.

Physical habitat disruption can result from changes to sediment dynamics, for example dense stands of alien riparian vegetation leading to greater sediment deposition and accretion that can create further habitat for alien species while compromising natives, and potentially even change river morphology (e.g. Hoffmann and Moran, 1988; Gordon, 1998; Richardson et al, 2007). Some species (e.g. *Eriocheir sinensis, P. leniusculus*; see Chapters 16 and 17) reduce available bank habitat by burrowing, weakening bank stability and sometimes leading to collapse.

More subtle forms of habitat modification can result from, for example, changes to water temperature (e.g. via shading by riparian or aquatic plants; Urban et al, 2009), water quality (Strayer, 1999), sediment chemistry (Urban et al, 2009), oxygen depletion (Kaufman, 1992), turbidity (Taylor et al, 1984) and nutrient levels (Arnott and Vanni, 1996), all of which may lead to increased stress and consequently reduced populations or ranges of native species. Often these changes are exacerbated by other impacting factors, such as flow regulation or pollution.

Changes in ecosystem function and resilience

The effects that IAS can have on the functioning of a freshwater system are varied and often linked to the biodiversity and habitat quality impacts noted above. These effects include changes in food webs and trophic interactions (e.g. Britton et al, 2010), organic matter processing and nutrient dynamics (Alonso et al, 2010), sediment dynamics (see Chapter 10), hydrology (Richardson et al, 2007), seral dynamics (Richardson et al, 2007) and seed bank decline (de Winton and Clayton, 1996). In some cases IAS impacts can drastically alter system functioning, such as the effect of *Pomacea canaliculata* (golden apple snail) on wetlands in Thailand, which exhibited a change from clear water and abundant macrophytes to a turbid, plankton-dominated state following introduction of the species (Carlsson et al, 2004; see Chapter 18).

IAS impacts can also affect the resilience of an ecosystem. Resilience can be defined in two ways that are often conflated in the literature, which Holling (1996) refers to as 'engineering resilience' and 'ecological resilience'. The former indicates the capacity of an ecosystem to resist disturbance (resistance) and to return to its original state (and how fast) following disturbance, and reflects traditional conceptualizations of ecosystems as existing naturally in steady state equilibria. This is now often thought of as ecosystem 'stability'. Ecological resilience refers instead to the level of disturbance that a system can absorb before changing to another state or 'stability domain' (i.e. structure) by changing the characteristics and processes that control system behaviour (such as the dramatic changes that can occur with lake eutrophication when a critical threshold of nutrient addition and algal population growth is reached; see Pahl-Wostl, 1995). IAS can affect both types of resilience (Richardson et al, 2007).

Freshwater communities are generally resilient as they are often driven by disturbance, though individual species respond differently to disturbances (e.g. floods) occurring at varying intensities, frequencies and durations. Smaller and more frequent disturbances are likely to be recovered from quickly (up to decades, for example, for riparian communities reconfigured by large floods) while larger disturbances, whether of high intensity, frequency or duration, are more likely to result in a change in state (such as the lake eutrophication example given above or a catastrophic flood that results in a change in channel morphology from single to multiple channel, or vice versa). The addition of alien species that change community composition can inherently reduce engineering resilience (as it is less likely that the original state will be returned to) (Richardson et al, 2007). Disturbances that alter functioning such as primary productivity or nutrient dynamics can also prevent the 'original' or pre-invasion state from re-establishing. Further, these events may also bring about shifts in state. Dent et al (2002) use *Tamarix ramosissima* (tamarisk) as an example of an IAS that colonizes exposed river sediments downstream of impoundments and subsequently alters the flood regime and increases sediment accumulation. In this work, the researchers were able to show a shift in state from a more dynamic 'native' system to the less dynamic one dominated by *T. ramosissima*. Whether the new stable state is less resilient than the original one is difficult to quantify in most cases. However, if the changes have resulted in a loss of some level of biodiversity or ecosystem function then this is certainly possible. In both cases, dramatic effects on ecosystem structure and function can be determined from species invasions, and the capacity of ecosystems to cope with further disturbance may be compromised.

Interruption of ecosystem services

All of the impacts on biodiversity, habitat and ecosystem function can also compromise freshwater ecosystem services. Pejchar and Mooney (2009) present a detailed review of the impacts of IAS on ecosystem services, which can be broadly summarized as direct (provisioning or resource acquisition) and indirect (regulating, supporting and cultural) services. Those impacts that relate most specifically to freshwater systems include losses to aquaculture (e.g. Lovell et al, 2006), obstruction of waterways, for example by extensive weed monotypes such as *Eichhornia crassipes* (water hyacinth) (Kasulo, 2000; see Chapter 4), interruptions to water acquisition, for example the clogging of water supply pipes by *Dreissena polymorpha* (zebra mussels) (Pejchar and Mooney, 2009), impacts on water quality and aesthetic appearance of water bodies (Pejchar and Mooney, 2009), increasing flood risk (Zavaleta, 2000) and limiting recreation and tourism (Pejchar and Mooney, 2009). This remains a nascent but very important aspect of IAS for further investigation, particularly given the economic losses that can result from both the impacts and control of IAS (Pimental et al, 2005).

Human health

Some freshwater IAS can have direct and indirect impacts on human health. These include species that act as vectors for human pathogens as well as the pathogens themselves (e.g. HIV (human immunodeficiency virus), West Nile virus, cholera). Perhaps the most obvious example of the former is the spread of invertebrates that act as vectors for diseases to which humans are susceptible, for example mosquitoes. Elton's (1958) discussion of the spread of *Anopheles arabiensis* in Brazil highlights this, and documents an extensive but successful control programme as a result; a useful lesson being perhaps that those IAS that do impact on human health are more likely to be subject to control and eradication measures. Other examples of species that act as vectors for human pathogens are the mosquitoes *Aedes albopictus*, *A. japonicus* and *A. aegypti* (Sanders et al, 2010; see Chapter 12), while other species may have more minor health impacts, such as skin irritation (e.g. *Heracleum mantegazzianum*; see Chapter 5). See McMichael and Bouma (2000) for a review of several aspects of invasive species and human health, and Pimentel et al (2005) for some estimates of costs for treating invasive human pathogens.

Changing Perspectives of Alien Species

Douglas (1966), in her anthropological studies of pollution and its role in different cultures, defined 'dirt' as 'matter out of place' (p36). In the last few decades, *invasive* alien species have been treated as 'species out of place': organisms that do not belong in their current environment (see Van Driesche and Van Driesche, 2000). With the growing awareness of the potential threats posed by alien species, this perception has begun to cross to alien species in general (see Simberloff, 2003; Brown and Sax, 2004; Stromberg et al, 2009). More recently, there has been explicit criticism of this treatment of IAS and aliens in general with suggestions that such attitudes represent a form of socially acceptable xenophobia (see Hettinger, 2001; Simberloff, 2003) and a somewhat subjective emotional attachment to 'native' species assemblages (Trudgill, 2008). This has included criticism of the negative (essentially antagonistic or warlike) terminologies frequently associated with alien species, such as 'invasive' itself, alongside 'aggressive', 'noxious' and so on, all of which may cloud the scientific objectivity that we need to understand such species (Stromberg et al, 2009; Lavoie, 2010).

Certainly for the majority of alien species that arrive in novel ecosystems, it would seem that innocuous assimilation results and once established these species often become functionally relevant members of the ecosystem. Moreover, some IAS generate unexpected benefits (e.g. Sax and Gaines, 2003; Goodenough, 2010), which logically reinforces the concept of 'assisted colonization' of threatened species (Sutherland et al, 2010). There is also certainly a debate to be had over not only our perceptions and responses to alien species but also their treatment by the popular media and how this may contrast with the scientific research (Stromberg et al, 2009; Lavoie, 2010). These discussions are endlessly blurred by conflicting opinions on what may constitute a 'native' species, how valid such a concept is, what values we attach to native species compared to alien species, what timescale is most appropriate for evaluation of nativeness and invasions and so on, right up to the meaning, values and nature of Nature itself (e.g. Hettinger, 2001; Simberloff, 2003; Castree, 2005). This includes such distinctions as when an alien species technically becomes invasive, and how the term 'invasive' itself is to be defined (see Colautti and MacIsaac, 2004).

Nevertheless, a practical caution must remain paramount: species that are invasive (i.e. leading to some form of ecosystem degradation) are of concern because they impact on human society via the disruption of ecosystem functioning and services, and freshwater ecosystems in particular are extremely precious and relatively vulnerable for the reasons discussed above. We cannot always predict the future impacts or effects of alien species, and although it is not always feasible or necessarily desirable to prevent the introduction of such species, or facilitate their removal, the importance of freshwater ecosystems to human society and the biosphere means that a conservative approach is needed.

Conclusion

The case is already well made for freshwater ecosystems to be a priority for global conservation on biodiversity grounds alone (Sala et al, 2000), and understanding and managing freshwater IAS are crucial components of this. This is a particularly applied science: Vörösmarty et al (2010) note that almost 80 per cent of the global human population have high levels of threat to water security (the vast majority of water coming from freshwater ecosystems), much of which is mitigated in more developed countries by technology and infrastructure. Water availability will be a key driver of future relations between nations and regions, and may engender both conflict and cooperation (e.g. Chapagain and Hoekstra, 2008). Within this context the facilitation of integral functioning of freshwater ecosystems remains a major challenge, and one that society simply must, in its own interest, devote sincere attention to.

This chapter has provided a brief overview of alien introductions and impacts in freshwater ecosystems, and the issue of management and control is considered in the concluding chapter of the book. The 33 case studies of notable freshwater IAS that form the core of the book help to illustrate some of the points covered here and are intended not only to provide readers with the latest understanding of these species, but also (for the majority) a critical analysis of control effectiveness. Mitigating the impacts of freshwater IAS is a crucial human endeavour for the 21st century, and it is hoped that this book will contribute, however modestly, to that endeavour.

References

Alexandrov B., Boltachev, A., Kharchenko, T., Lyashenko, A., Son, M., Tsarenko, P. and Zhukinsky, V. (2007) 'Trends of aquatic alien species invasions in Ukraine', *Aquatic Invasions*, vol 2, no 3, pp215–242

Alonso, A., Gonzalez-Muñoz, N. and Castro-Diez, P. (2010) 'Comparison of leaf decomposition and macroinvertebrate colonization between exotic and native trees in a freshwater ecosystem', *Ecological Research*, vol 25, no 3, pp647–653

Arnott, D. L. and Vanni, M. J. (1996) 'Nitrogen and phosphorus recycling by the zebra mussel (*Dreissena polymorpha*) in the western basin of Lake Erie', *Canadian Journal of Fisheries and Aquatic Sciences*, vol 53, no 3, pp646–659

Baker, H. G. (1965) 'Characteristics and modes of origin of weeds', in H. G. Baker and G. L. Stebbins (eds) *The Genetics of Colonizing Species*, Academic Press, New York, pp147–168

Balon, E. K. (1995) 'Origin and domestication of the wild carp, *Cyprinus carpio*: From Roman gourmets to the swimming flowers', *Aquaculture*, vol 129, pp3–48

Balon, E. K. (2004) 'About the oldest domesticates among fishes', *Journal of Fish Biology*, vol 65, pp1–27

Bodamer, B. L. and Bossenbroek, J. M. (2008) 'Wetlands as barriers: Effects of vegetated waterways on downstream dispersal of zebra mussels', *Freshwater Biology*, vol 53, no 10, pp2051–2060

Booth, B. D., Murphy, S. D. and Swanton, C. J. (2003) *Weed Ecology in Natural and Agricultural Systems*, CABI Publishing, Wallingford

Bossdorf, O., Auge, H., Lafuma, L., Rogers, W. E., Siemann, E. and Prati, D. (2005) 'Phenotypic and genetic differentiation between native and introduced plant populations', *Oecologia*, vol 144, pp1–11

Bostock, J., McAndrew, B., Richards, R., Jauncey, K., Telfer, T., Lorenzen, K., Little, D., Ross, L., Handisyde, N., Gatward, I. and Corner, R. (2010) 'Aquaculture: Global status and trends', *Philosophical Transactions of the Royal Society B: Biological Sciences*, vol 365, no 1554, pp2897–2912

Britton, K. O., Buford, M., Burnett, K., Dix, M. E., Frankel, S. J., Keena, M., Kim, M.-S., Klopfenstein, N. B., Ostry, M. E. and Sieg, C. H. (2010) 'Invasive species overarching priorities to 2029', in M. E. Dix and K. Britton (eds) *A Dynamic Invasive Species Research Vision: Opportunities and Priorities 2009–29*, USDA General Technical Report WO-79/83, USDA, Washington, DC, pp3–11

Brown, J. H. and Sax, D. F. (2004) 'An essay on some topics concerning invasive species', *Austral Ecology*, vol 29, no 5, pp530–536

Bunn, S. E. and Arthington, A. H. (2002) 'Basic principles and ecological consequences of altered flow regimes for aquatic biodiversity', *Environmental Management*, vol 30, no 4, pp492–507

Cambray, J. A. (2003) 'Impact on indigenous species biodiversity caused by the globalisation of alien recreational freshwater fisheries', *Hydrobiologia*, vol 500, pp217–230

Carlsson, N. O. L., Bronmark, C. and Hansson, L. A. (2004) 'Invading herbivory: The golden apple snail alters ecosystem functioning in Asian wetlands', *Ecology*, vol 85, no 6, pp1575–1580

Carlton, J. T. (1999) 'Molluscan invasions in marine and estuarine communities', *Malacologia*, vol 41, no 2, pp439–454

Castree, N. (2005) *Nature*, Routledge, Abingdon

Chapagain, A. K. and Hoekstra, A. Y. (2008) 'The global component of freshwater demand and supply: An assessment of virtual water flows between nations as a result of trade in agricultural and industrial products', *Water International*, vol 33, no 1, pp19–32

Colautti, R. I. and MacIsaac, H. J. (2004) 'A neutral terminology to define "invasive" species', *Diversity and Distributions*, vol 10, no 2, pp135–141

Colautti, R. I., Ricciardi, A., Grigorovich, I. A. and MacIsaac, H. J. (2004) 'Is invasion success explained by the enemy release hypothesis?', *Ecology Letters*, vol 7, no 8, pp721–733

Coops, H., Beklioglu, M. and Crisman, T. L. (2003) 'The role of water-level fluctuations in shallow lake ecosystems – workshop conclusions', *Hydrobiologia*, vol 506, no 1–3, pp23–27

Copp, G. H., Bianco, P. G., Bogutskaya, N. G., Eros, T., Falka, I., Ferreira, M. T., Fox, M. G., Freyhof, J., Gozlan, R. E., Grabowska, J., Kovac, V., Moreno-Amich, R., Naseka, A. M., Penaz, M., Povz, M., Przybylski, M., Robillard, M., Russell, I. C., Stakenas, S., Sumer, S., Vila-Gispert, A. and Wiesner, C. (2005) 'To be, or not to be, a non-native freshwater fish?', *Journal of Applied Ichthyology*, vol 21, no 4, pp242–262

Costanza, R., dArge, R., deGroot, R., Farber, S., Grasso, M., Hannon, B., Limburg, K., Naeem, S., ONeill, R. V., Paruelo, J., Raskin, R. G., Sutton, P. and vandenBelt, M. (1997) 'The value of the world's ecosystem services and natural capital', *Nature*, vol 387, no 6630, pp253–260

Costello, C. J. and Solow, A. R. (2003) 'On the pattern of discovery of introduced species', *Proceedings of the National Academy of Sciences of the United States of America*, vol 100, no 6, pp3321–3323

Courtenay, W. R. (2007) 'Introduced species: What species do you have and how do you know?', *Transactions of the American Fisheries Society*, vol 136, no 4, pp1160–1164

Cowie, R. H. (1998) 'Patterns of introduction of non-indigenous non-marine snails and slugs in the Hawaiian Islands', *Biodiversity and Conservation*, vol 7, no 3, pp349–368

Cowie, R. H. and Robinson, D. G. (2003) 'Pathways of introduction of nonindigenous land and freshwater snails and slugs', in G. M. Ruiz and J. T. Carlton (eds) *Invasive Species: Vectors and Management Strategies*, Island Press, Washington, DC, pp93–122

Crooks, J. A. (2005) 'Lag times and exotic species: The ecology and management of biological invasions in slow-motion', *Ecoscience*, vol 12, no 3, pp316–329

Crutzen, P. J. (2006) 'The "Anthropocene"', in E. Ehlers and T. Krafft (eds) *Earth System Science in the Anthropocene*, Springer, The Netherlands, pp13–18

Daehler, C. C. (1998) 'The taxonomic distribution of invasive angiosperm plants: Ecological insights and comparison to agricultural weeds', *Biological Conservation*, vol 84, pp167–180

Daehler, C. C. (2001) 'Darwin's naturalization hypothesis revisited', *American Naturalist*, vol 158, pp324–330

Daehler, C. C. (2003) 'Performance's comparisons of co-occurring native and alien invasive plants: Implications for conservation and restoration', *Annual Review of Ecology and Systematics*, vol 34, pp183–211

DeGrandchamp, K. L., Garvey, J. E. and Colombo, R. E. (2008) 'Movement and habitat selection by invasive Asian carps in a large river', *Transactions of the American Fisheries Society*, vol 137, no 1, pp45–56

Dent, C. L., Cumming, G. S. and Carpenter, S. R. (2002) 'Multiple states in river and lake ecosystems', *Philosophical Transactions of the Royal Society B: Biological Sciences*, vol 357, no 1421, pp635–645

de Winton, M. D. and Clayton, J. S. (1996) 'The impact of invasive submerged weed species on seed banks in lake sediments', *Aquatic Botany*, vol 53, no 1–2, pp31–45

di Castri, F. (1989) 'History of biological invasions with special emphasis on the Old World', in J. A. Drake, H. A. Mooney, F. di Castri, R. H. Groves, F. J. Kruger, M. Rejmánek and M. Williamson (eds.) *Biological Invasions: A Global Perspective*, John Wiley and Sons, Chichester, pp1–30

Didham, R. K., Tylianakis, J. M., Hutchison, M. A., Ewers, R. M. and Gemmell, N. J. (2005) 'Are invasive species the drivers of ecological change?', *Trends in Ecology and Evolution*, vol 20, no 9, pp470–474

Douglas, M. (1966) *Purity and Danger: An Analysis of Concepts of Pollution and Taboo*, Routledge, London

Drake, J. A., Mooney, H. A., di Castri, F., Groves, R. H., Kruger, F. J., Rejmanek, M. and Williamson, M. (1989) *Biological Invasions: A Global Perspective*, John Wiley and Sons, Chichester

Dudgeon, D., Arthington, A. H., Gessner, M. O., Kawabata, Z. I., Knowler, D. J., Leveque, C., Naiman, R. J., Prieur-Richard, A. H., Soto, D., Stiassny, M. L. J. and Sullivan, C. A. (2006) 'Freshwater biodiversity: Importance, threats, status and conservation challenges', *Biological Reviews*, vol 81, no 2, pp163–182

Duncan, R. P. and Williams, P. A. (2002) 'Darwin's naturalization hypothesis challenged', *Nature*, vol 417, p608

Egerton, F. N. (1983) 'The history of ecology: Achievements and opportunities, part one', *Journal of the History of Biology*, vol 16, no 2, pp259–310

Elton, C. S. (1958) *The Ecology of Invasions by Animals and Plants*, The University of Chicago Press, Chicago

Ervin, G., Smothers, M., Holly, C., Anderson, C. and Linville, J. (2006) 'Relative importance of wetland type versus anthropogenic activities in determining site invasibility', *Biological Invasions*, vol 8, no 6, pp1425–1432

Essl, F., Dullinger, S., Rabitsch, W., Hulme, P. E., Hülber, K., Jarošík, V., Kleinbauer, I., Krausmann, F., Kühn, I., Nentwig, W., Vilà, M., Genovesi, P., Gherardi, F., Desprez-Lousteau, M.-L., Roques, A. and Pyšek, P. (2011) 'Socioeconomic legacy yields an invasion debt', *Proceedings of the National Academy of Sciences of the United States of America*, vol 108, pp203–207

Fausch, K. D., Taniguchi, Y., Nakano, S., Grossman, G. D. and Townsend, C. R. (2001) 'Flood disturbance regimes influence rainbow trout invasion success among five holarctic regions', *Ecological Applications*, vol 11, no 5, pp1438–1455

Fowler, A. J., Lodge, D. M. and Hsia, J. F. (2007) 'Failure of the Lacey Act to protect US ecosystems against animal invasions', *Frontiers in Ecology and the Environment*, vol 5, pp353–359

Francis, R. A. (2009) 'Perspectives on the potential for reconciliation ecology in urban riverscapes', *CAB Reviews: Perspectives in Agriculture, Veterinary Science, Nutrition and Natural Resources*, vol 4, art 73

García-Berthou, E., Alcaraz, C., Pou-Rovira, Q., Zamora, L., Coenders, G. and Feo, C. (2005) 'Introduction pathways and establishment rates of invasive aquatic species in Europe', *Canadian Journal of Fisheries and Aquatic Sciences*, vol 62, pp453–463

Gherardi, F. (2007) 'Biological invasions in inland waters: An overview,' in F. Gherardi (ed) *Biological Invaders in Inland Waters: Profiles, Distribution, and Threats*, Springer, Dordrecht, pp3–25

Gherardi, F. (2010) 'Invasive crayfish and freshwater fishes of the world', *Revue Scientifique et Technique – Office International des Epizooties*, vol 29, no 2, pp241–254

Gippoliti, S. and Amori, G. (2006) 'Ancient introductions of mammals in the Mediterranean Basin and their implications for conservation', *Mammal Review*, vol 36, no 1, pp37–48

Gollasch, S. (2002) 'The importance of ship hull fouling as a vector of species introductions into the North Sea', *Biofouling*, vol 18, no 2, pp105–121

Golley, F. B. (1993) *A History of the Ecosystem Concept in Ecology*, Yale University Press, London

Goodenough, A. E. (2010) 'Are the ecological impacts of alien species misrepresented? A review of the "native good, alien bad" philosophy', *Community Ecology*, vol 11, no 1, pp13–21

Gordon, D. R. (1998) 'Effects of invasive, non-indigenous plant species on ecosystem processes: Lessons from Florida', *Ecological Applications*, vol 8, no 4, pp975–989

Gregory, S. V., Swanson, F. J., McKee, W. A. and Cummins, K. W. (1991) 'An ecosystem perspective of riparian zones', *BioScience*, vol 41, no 8, pp540–551

Havel, J. E., Lee, C. E. and Vander Zanden, M. J. (2005) 'Do reservoirs facilitate invasions into landscapes?', *BioScience*, vol 55, no 6, pp518–525

Hettinger, N. (2001) 'Exotic species, naturalisation, and biological nativism', *Environmental Values*, vol 10, no 2, pp193–224

Higgins, S. I. and Richardson, D. M. (1999) 'Predicting plant migration rates in a changing world: The role of long-distance dispersal', *American Naturalist*, vol 153, pp464–475

Hobbs, R. J. and Huenneke, L. F. (1992) 'Disturbance, diversity and invasion: Implications for conservation', *Conservation Biology*, vol 6, no 3, pp324–337

Hoffmann, J. H. and Moran, V. C. (1988) 'The invasive weed *Sesbania punicea* in South Africa and prospects for its biological control', *South African Journal of Science*, vol 84, no 9, pp740–742

Holling, C. S. (1996) 'Engineering resilience versus ecological resilience', in P. C. Schulze (ed) *Engineering Within Ecological Constraints*, National Academy of Engineering, Washington, DC, pp31–44

Holzapfel, E. P. and Harrell, J. P. (1968) 'Transoceanic dispersal studies of insects', *Pacific Insects*, vol 10, no 1, pp115–153

Hooper, D. U., Chapin, F. S., Ewel, J. J., Hector, A., Inchausti, P., Lavorel, S., Lawton, J. H., Lodge, D. M., Loreau, M., Naeem, S., Schmid, B., Setala, H., Symstad, A. J., Vandermeer, J. and Wardle, D. A. (2005) 'Effects of biodiversity on ecosystem functioning: A consensus of current knowledge', *Ecological Monographs*, vol 75, no 1, pp3–35

Hudina, S., Faller, M., Lucic, A., Klobucar, G. and Maguire, I. (2009) 'Distribution and dispersal of two invasive crayfish species in the Drava River basin, Croatia', *Knowledge and Management of Aquatic Ecosystems*, vol 394–395, art 09

Hughes, J. D. (2003) 'Europe as consumer of exotic biodiversity: Greek and Roman times', *Landscape Research*, vol 28, no 1, pp21–31

Hulme, P. E., Bacher, S., Kenis, M., Klotz, S., Kuhn, I., Minchin, D., Nentwig, W., Olenin, S., Panov, V., Pergl, J., Pyšek, P., Roques, A., Sol, D., Solarz, W. and Vila, M. (2008) 'Grasping at the routes of biological invasions: A framework for integrating pathways into policy', *Journal of Applied Ecology*, vol 45, no 2, pp403–414

Hulme, P., Pyšek, P., Nentwig, W. and Vilà, M. (2009) 'Will threat of biological invasions unite the European Union?', *Science*, vol 324, pp40–41

Isaäcson, M. (1989) 'Airport malaria: A review', *Bulletin of the World Health Organization*, vol 67, no 6, pp737–743

Jackson, C. R. and Pringle, C. M. (2010) 'Ecological benefits of reduced hydrologic connectivity in intensively developed landscapes', *BioScience*, vol 60, no 1, pp37–46

Johnson, P. T. J., Olden, J. D. and Vander Zanden, M. J. (2008) 'Dam invaders: Impoundments facilitate biological invasions into freshwaters', *Frontiers in Ecology and the Environment*, vol 6, no 7, pp359–365

Johnson, P. T. J., Olden, J. D., Solomon, C. T. and Vander Zanden, M. J. (2009) 'Interactions among invaders: Community and ecosystem effects of multiple invasive species in an experimental aquatic system', *Oecologia*, vol 159, no 1, pp161–170

Junk, W. J., Bayley, P. B. and Sparks, R. E. (1989) 'The flood pulse concept in river-floodplain systems', *Canadian Special Publication of Fisheries and Aquatic Sciences*, vol 106, pp110–127

Kalinowski, S. T., Muhlfeld, C. C., Guy, C. S. and Cox, B. (2010) 'Founding population size of an aquatic invasive species', *Conservation Genetics*, vol 11, no 5, pp2049–2053

Kasulo, V. (2000) 'The impact of invasive species in African lakes', in C. Perrings (ed) *The Economics of Biological Invasions*, Edward Elgar Publishing, Cheltenham, pp262–297

Kaufman, L. (1992) 'Catastrophic change in species-rich freshwater ecosystems', *BioScience*, vol 42, no 11, pp846–858

Keane, R. M. and Crawley, M. J. (2002) 'Exotic plant invasions and the enemy release hypothesis', *Trends in Ecology and Evolution*, vol 17, no 4, pp164–170

Kennedy, C. R. and Fitch, D. J. (1990) 'Colonization, larval survival and epidemiology of the nematode *Anguillicola crassus*, parasitic in the eel, *Anguilla anguilla*, in Britain', *Journal of Fish Biology*, vol 36, no 2, pp117–131

Kirkwood, A. E., Jackson, L. J. and McCauley, E. (2009) 'Are dams hotspots for *Didymosphenia geminata* blooms?', *Freshwater Biology*, vol 54, no 9, pp1856–1863

Kornis, M. S. and Vander Zanden, M. J. (2010) 'Forecasting the distribution of the invasive round goby (*Neogobius melanostomus*) in Wisconsin tributaries to Lake Michigan', *Canadian Journal of Fisheries and Aquatic Sciences*, vol 67, no 3, pp553–562

Kraft, C. E., Sullivan, P. J., Karatayev, A. Y., Burlakova, Y. E., Nekola, J. C., Johnson, L. E. and Padilla, D. K. (2002) 'Landscape patterns of an aquatic invader: Assessing dispersal extent from spatial distributions', *Ecological Applications*, vol 12, no 3, pp749–759

Larson, B. M. H. (2007) 'An alien approach to invasive species: Objectivity and society in invasion biology', *Biological Invasions*, vol 9, no 8, pp947–956

Larson, E. R., Magoulick, D. D., Turner, C. and Laycock, K. H. (2009) 'Disturbance and species displacement: Different tolerances to stream drying and desiccation in a native and an invasive crayfish', *Freshwater Biology*, vol 54, no 9, pp1899–1908

Lavoie, C. (2010) 'Should we care about purple loosestrife? The history of an invasive plant in North America', *Biological Invasions*, vol 12, no 7, pp1967–1999

Leppäkoski, E., Gollasch, S. and Olenin, S. (2002) 'Introduction', in E. Lepäkoski, S. Gollasch and S. Olenin (eds) *Invasive Aquatic Species of Europe: Distribution, Impacts and Management*, Kluwer Academic Publishers, Amsterdam, pp1–6

Leslie, A. J. and Spotila, J. R. (2001) 'Alien plant threatens Nile crocodile (*Crocodylus niloticus*) breeding in Lake St Lucia, South Africa', *Biological Conservation*, vol 98, pp347–355

Leuven, R. S. E. W., van der Velde, G., Baijens, I., Snijders, J., van der Zwart, C., Lenders, H. J. R. and de Vaate, A. B. (2009) 'The river Rhine: A global highway for dispersal of aquatic invasive species', *Biological Invasions*, vol 11, no 9, pp1989–2008

Liu, H. and Stiling, P. (2006) 'Testing the enemy release hypothesis: A review and meta-analysis', *Biological Invasions*, vol 8, no 7, pp1535–1545

Lockwood, J. L., Cassey, P. and Blackburn, T. (2005) 'The role of propagule pressure in explaining species invasions', *Trends in Ecology and Evolution*, vol 20, no 5, pp223–228

Lonsdale, W. M. (1993) 'Rates of spread of an invading species – *Mimosa pigra* in Northern Australia', *Journal of Ecology*, vol 81, no 3, pp513–521

Lovell, S. J., Stone, S. F. and Fernandez, L. (2006) 'The economic impacts of aquatic invasive species: A review of the literature', *Agricultural and Resource Economics Review*, vol 35, pp195–208

Lowe, S., Browne, M., Boudjelas, S. and De Poorter, M. (2004) *100 of the World's Worst Invasive Alien Species: A Selection from the Global Invasive Species Database*, IUCN, New Zealand

Mack, R. N. (1999) 'The motivation for importing potentially invasive plant species: A primal urge?', in D. Eldridge and D. Freudenberger (eds) *People and Rangelands: Building the Future Vols 1 and 2*, VI International Rangeland Congress Inc, Aitkenvale, pp557–562

Mack, R. N. (2003) 'Global plant dispersal, naturalization, and invasion: Pathways, modes and circumstances', in G. M. Ruiz and J. T. Carlton (eds) *Invasive Species: Vectors and Management Strategies*, Island Press, Washington, DC, pp3–30

Mack, R. N. and Lonsdale, W. M. (2001) 'Humans as global plant dispersers: Getting more than we bargained for', *BioScience*, vol 51, no 2, pp95–102

Mandrak, N. E. and Cudmore, B. (2010) 'The fall of native fishes and the rise of non-native fishes in the Great Lakes Basin', *Aquatic Ecosystem Health and Management*, vol 13, no 3, pp255–268

McMichael, A. J. and Bouma, M. J. (2000) 'Global changes, invasive species and human health', in H. A. Mooney and R. J. Hobbs (eds) *Invasive Species in a Changing World*, Island Press, Washington, DC, pp191–210

Meyerson, L. A. and Mooney, H. A. (2007) 'Invasive alien species in an era of globalization', *Frontiers in Ecology and the Environment*, vol 5, no 4, pp199–208

Michelan, T. S., Thomaz, S. M., Mormul, R. P. and Carvalho, P. (2010) 'Effects of an exotic invasive macrophyte (tropical signalgrass) on native plant community composition, species richness and functional diversity', *Freshwater Biology*, vol 55, no 6, pp1315–1326

Minchin, D. (2006) 'The transport and the spread of living aquatic species', in J. Davenport and J. L. Davenport (eds) *The Ecology of Transportation: Managing Mobility for the Environment*, Springer, Amsterdam, pp77–97

Mortenson, S. G. and Weisberg, P. J. (2010) 'Does river regulation increase the dominance of invasive woody species in riparian landscapes?', *Global Ecology and Biogeography*, vol 19, no 4, pp562–574

Naiman, R. J. and Décamps, H. (1997) 'The ecology of interfaces: Riparian zones', *Annual Review of Ecology and Systematics*, vol 28, pp621–658

Naylor, R. L., Williams, S. L. and Strong, D. R. (2001) 'Aquaculture: A gateway for exotic species', *Science*, vol 294, no 5547, pp1655–1656

Padilla, D. K. and Williams, S. L. (2004) 'Beyond ballast water: Aquarium and ornamental trades as sources of invasive species in aquatic ecosystems', *Frontiers in Ecology and the Environment*, vol 2, no 3, pp131–138

Pahl-Wostl, C. (1995) *The Dynamic Nature of Ecosystems: Chaos and Order Entwined*, John Wiley and Sons, Chichester

Parmakelis, A., Russello, M. A., Caccone, A., Marcondes, C. B., Costa, J., Forattini, O. P., Sallum, M. A. M., Wilkerson, R. C. and Powell, J. R. (2008) 'Historical analysis of a near disaster: *Anopheles gambiae* in Brazil', *American Journal of Tropical Hygiene and Medicine*, vol 78, no 1, pp176–178

Pejchar, L. and Mooney, H. A. (2009) 'Invasive species, ecosystem services and human well-being', *Trends in Ecology and Evolution*, vol 24, no 9, pp497–504

Pimentel, D., Lach, L., Zuniga, R. and Morrison, D. (2000) 'Environmental and economic costs of nonindigenous species in the United States', *BioScience*, vol 50, no 1, pp53–65

Pimentel, D., Zuniga, R. and Morrison, D. (2005) 'Update on the environmental and economic costs associated with alien-invasive species in the United States', *Ecological Economics*, vol 52, no 3, pp273–288

Predick, K. I. and Turner, M. G. (2008) 'Landscape configuration and flood frequency influence invasive shrubs in floodplain forests of the Wisconsin River (USA)', *Journal of Ecology*, vol 96, no 1, pp91–102

Pusey, B. J. and Arthington, A. H. (2003) 'Importance of the riparian zone to the conservation and management of freshwater fish: A review', *Marine and Freshwater Research*, vol 54, no 1, pp1–16

Pyšek, P. (1998) 'Is there a taxonomic pattern to plant invasions?', *Oikos*, vol 82, pp282–294

Pyšek, P. and Prach, K. (1993) 'Plant invasions and the role of riparian habitats: A comparison of 4 species alien to central Europe', *Journal of Biogeography*, vol 20, no 4, pp413–420

Pyšek, P. and Richardson, D. M. (2007) 'Traits associated with invasiveness in alien plants: Where do we stand?', in W. Nentwig (ed) *Biological Invasions, Ecological Studies 193*, Springer-Verlag, Berlin and Heidelberg, pp97–125

Pyšek, P., Bacher, S., Chytry, M., Jarosik, V., Wild, J., Celesti-Grapow, L., Gasso, N., Kenis, M., Lambdon, P. W., Nentwig, W., Pergl, J., Roques, A., Sadlo, J., Solarz, W., Vila, M. and Hulme, P. E. (2010a) 'Contrasting patterns in the invasions of European terrestrial and freshwater habitats by alien plants, insects and vertebrates', *Global Ecology and Biogeography*, vol 19, no 3, pp317–331

Pyšek, P., Jarošík, V., Hulme, P. E., Kühn, I., Wild, J., Arianoutsou, M., Bacher, S., Chiron, F., Didžiulis, V., Essl, F., Genovesi, P., Gherardi, F., Hejda, M., Kark, S., Lambdon, P. W., Desprez-Loustau, A.-M., Nentwig, W., Pergl, J., Poboljšaj, K., Rabitsch, W., Roques, A., Roy, D. B., Shirley, S., Solarz, W., Vilà, M. and Winter, M. (2010b) 'Disentangling the role of environmental and human pressures on biological invasions across Europe', *Proceedings of the National Academy of Sciences of the United States of America*, vol 107, pp12157–12162

Rahel, F. J. (2007) 'Biogeographic barriers, connectivity and homogenization of freshwater faunas: It's a small world after all', *Freshwater Biology*, vol 52, no 4, pp696–710

Reice, S. R., Wissmar, R. C. and Naiman, R. J. (1990) 'Disturbance regimes, resilience, and recovery of animal communities and habitats in lotic ecosystems', *Environmental Management*, vol 14, no 5, pp647–659

Rejmánek, M. (1996) '1996: A theory of seed plant invasiveness: The first sketch', *Biological Conservation*, vol 78, pp171–81

Resh, V. H., Brown, A. V., Covich A. P., Gurtz M. E., Li, H. W., Minshall G. W., Reice, S. R., Sheldon, A. L., Wallace, J. B. and Wissmar, R. C. (1988) 'The role of disturbance in stream ecology', *Journal of the North American Benthological Society*, vol 7, no 4, pp433–455

Ricciardi, A. and Cohen, J. (2007) 'The invasiveness of an introduced species does not predict its impact', *Biological Invasions*, vol 9, no 3, pp309–315

Ricciardi, A. and Kipp, R. (2008) 'Predicting the number of ecologically harmful exotic species in an aquatic system', *Diversity and Distributions*, vol 14, no 2, pp374–380

Ricciardi, A. and MacIsaac, H. J. (2000) 'Recent mass invasion of the North American Great Lakes by Ponto-Caspian species', *Trends in Ecology and Evolution*, vol 15, no 2, pp62–65

Richardson, D. M. and Pyšek, P. (2006) 'Plant invasions: Merging the concepts of species invasiveness and community invasibility', *Progress in Physical Geography*, vol 30, no 3, pp409–431

Richardson, D. M., Holmes, P. M., Esler, K. J., Galatowitsch, S. M., Stromberg, J. C., Kirkman, S. P., Pyšek, P. and Hobbs, R. J. (2007) 'Riparian vegetation: Degradation, alien plant invasions, and restoration prospects', *Diversity and Distributions*, vol 13, no 1, pp126–139

Roman, J. and Darling, J. A. (2007) 'Paradox lost: Genetic diversity and the success of aquatic invasions', *Trends in Ecology and Evolution*, vol 22, no 9, pp454–464

Rood, S. B., Braatne, J. H. and Goater, L. A. (2010) 'Favorable fragmentation: River reservoirs can impede downstream expansion of riparian weeds', *Ecological Applications*, vol 20, no 6, pp1664–1677

Sala, O. E., Chapin, F. S., Armesto, J. J., Berlow, E., Bloomfield, J., Dirzo, R., Huber-Sanwald, E., Huenneke, L. F., Jackson, R. B., Kinzig, A., Leemans, R., Lodge, D. M., Mooney, H. A., Oesterheld, M., Poff, N. L., Sykes, M. T., Walker, B. H., Walker, M. and Wall, D. H. (2000) 'Biodiversity – Global biodiversity scenarios for the year 2100', *Science*, vol 287, no 5459, pp1770–1774

Sanders, C. J., Mellor, P. S. and Wilson, A. J. (2010) 'Invasive arthropods', *Revue Scientifique et Technique – Office International des Epizooties*, vol 29, no 2, pp273–286

Sax, D. F. and Gaines, S. D. (2003) 'Species diversity: From global decreases to local increases', *Trends in Ecology and Evolution*, vol 18, no 11, pp561–566

Simberloff, D. (2003) 'Confronting introduced species: A form of xenophobia?', *Biological Invasions*, vol 5, no 3, pp179–192

Simberloff, D. (2005) 'Non-native species do threaten the natural environment!', *Journal of Agricultural and Environmental Ethics*, vol 18, no 6, pp595–607

Simon, K. S. and Townsend, C. R. (2003) 'Impacts of freshwater invaders at different levels of ecological organisation, with emphasis on salmonids and ecosystem consequences', *Freshwater Biology*, vol 48, no 6, pp982–994

Stanford, J. A., Lorang, M. S. and Hauer, F. R. (2005) 'The shifting habitat mosaic of river ecosystems', in J. Jones (ed) *International Association of Theoretical and Applied Limnology, Vol 29, Pt 1, Proceedings*, International Association of Theoretical and Applied Limnology, Stuttgart, pp123–136

Strayer, D. L. (1999) 'Effects of alien species on freshwater mollusks in North America', *Journal of the North American Benthological Society*, vol 18, no 1, pp74–98

Stromberg, J. C., Chew, M. K., Nagler, P. L. and Glenn, E. P. (2009) 'Changing perceptions of change: The role of scientists in *Tamarix* and river management', *Restoration Ecology*, vol 17, no 2, pp177–186

Sutherland, W. J., Clout, M., Cote, I. M., Daszak, P., Depledge, M. H., Fellman, L., Fleishman, E., Garthwaite, R., Gibbons, D. W., De Lurio, J., Impey, A. J., Lickorish, F., Lindenmayer, D., Madgwick, J., Margerison, C., Maynard, T., Peck, L. S., Pretty, J., Prior, S., Redford, K. H., Scharlemann, J. P. W., Spalding, M. and Watkinson, A. R. (2010) 'A horizon scan of global conservation issues for 2010', *Trends in Ecology and Evolution*, vol 25, no 1, pp1–7

Tabacchi, E. and Planty-Tabacchi, A. M. (2005) 'Exotic and native plant community distributions within complex riparian landscapes: A positive correlation', *Ecoscience*, vol 12, no 3, pp412–423

Tabacchi, E., Planty-Tabacchi, A. M., Salinas, M. J. and Décamps, H. (1996) 'Landscape structure and diversity in riparian plant communities: A longitudinal comparative study', *Regulated Rivers: Research and Management*, vol 12, no 4–5, pp367–390

Taylor, J. N., Courtenay, W. R. Jr and McCann, J. A. (1984) 'Known impacts of exotic fishes in the continental United States', in W. R. Courtenay and J. R. Stauffer (eds) *Distribution, Biology, and Management of Exotic Fishes*, The Johns Hopkins University Press, Baltimore, pp322–355

Thébaud, C. and Debussche, M. (1991) 'Rapid invasion of *Fraxinus ornus* L. along the Herault River system in Southern France – the importance of seed dispersal by water', *Journal of Biogeography*, vol 18, no 1, pp7–12

Tickner, D. P., Angold, P. G., Gurnell, A. M. and Mountford, J. O. (2001) 'Riparian plant invasions: Hydrogeomorphological control and ecological impacts', *Progress in Physical Geography*, vol 25, no 1, pp22–52

Tockner, K., Malard, F. and Ward, J. V. (2000) 'An extension of the flood pulse concept', *Hydrological Processes*, vol 14, no 16–17, pp2861–2883

Toy, S. J. and Newfield, M. J. (2010) 'The accidental introduction of invasive animals as hitchhikers through inanimate pathways: A New Zealand perspective', *Revue Scientifique et Technique – Office International des Epizooties*, vol 29, no 1, pp123–133

Trudgill, S. (2008) 'A requiem for the British flora? Emotional biogeographies and environmental change', *Area*, vol 40, no 1, pp99–107

Urban, R. A., Titus, J. E. and Zhu, W. X. (2009) 'Shading by an invasive macrophyte has cascading effects on sediment chemistry', *Biological Invasions*, vol 11, no 2, pp265–273

Van Driesche, J. and Van Driesche, R. (2000) *Nature Out of Place: Biological Invasions in the Global Age*, Island Press, Washington, DC

Vilà, M., Basnou, C., Pyšek, P., Josefsson, M., Genovesi, P., Gollasch, S., Nentwig, W., Olenin, S., Roques, A., Roy, D., Hulme, P. E. and DAISIE partners (2010) 'How well do we understand the impacts of alien species on ecosystem services? A pan-European, cross-taxa assessment', *Frontiers in Ecology and the Environment*, vol 8, pp135–144

Vitousek, P. M., DAntonio, C. M., Loope, L. L. and Westbrooks, R. (1996) 'Biological invasions as global environmental change', *American Scientist*, vol 84, no 5, pp468–478

Von Holle, B. and Simberloff, D. (2005) 'Ecological resistance to biological invasion overwhelmed by propagule pressure', *Ecology*, vol 86, no 12, pp3212–3218

Vörösmarty, C. J., McIntyre, P. B., Gessner, M. O., Dudgeon, D., Prusevich, A., Green, P., Glidden, S., Bunn, S. E., Sullivan, C. A., Liermann, C. R. and Davies, P. M. (2010) 'Global threats to human water security and river biodiversity', *Nature*, vol 467, no 7315, pp555–561

Ward, J. V., Malard, F. and Tockner, K. (2002) 'Landscape ecology: A framework for integrating pattern and process in river corridors', *Landscape Ecology*, vol 17, pp35–45

Welcomme, R. L. (1988) *International Introductions of Inland Aquatic Species*, FAO Fisheries Technical Paper 294, FAO, Rome

Wiens, J. A. (2002) 'Riverine landscapes: Taking landscape ecology into the water', *Freshwater Biology*, vol 47, no 4, pp501–515

Williamson, M. H. and Brown, K. C. (1986) 'The analysis and modelling of British invasions', *Philosophical Transactions of the Royal Society of London B*, vol 314, pp505–522

Williamson, M. and Fitter, A. (1996) 'The varying success of invaders', *Ecology*, vol 77, no 6, pp1661–1666

Wilson, M. A. and Carpenter, S. R. (1999) 'Economic valuation of freshwater ecosystem services in the United States: 1971–1997', *Ecological Applications*, vol 9, no 3, pp772–783

Zambrano, L., Martinez-Meyer, E., Menezes, N. and Peterson, A. T. (2006) 'Invasive potential of common carp (*Cyprinus carpio*) and Nile tilapia (*Oreochromis niloticus*) in American freshwater systems', *Canadian Journal of Fisheries and Aquatic Sciences*, vol 63, no 9, pp1903–1910

Zardi, G. I., Nicastro, K. R., Porri, F. and McQuaid, C. D. (2006) 'Sand stress as a non-determinant of habitat segregation of indigenous (*Perna perna*) and invasive (*Mytilus galloprovincialis*) mussels in South Africa', *Marine Biology*, vol 148, no 5, pp1031–1038

Zavaleta, E. (2000) 'The economic value of controlling an invasive shrub', *Ambio*, vol 29, no 8, pp462–467

Zedler, J. B. and Kercher, S. (2004) 'Causes and consequences of invasive plants in wetlands: Opportunities, opportunists, and outcomes', *Critical Reviews in Plant Sciences*, vol 23, no 5, pp431–452

Part I

Aquatic and Riparian Plants

2

Alternanthera philoxeroides (Martius) Grisebach (alligator weed)

Shon S. Schooler

Introduction

Alternanthera philoxeroides (Amaranthaceae), commonly known as alligator weed, is a perennial stoloniferous herbaceous plant that is primarily associated with aquatic habitats, but can spread into moist terrestrial environments (Julien and Bourne, 1988). The stems are hollow when mature with a pair of opposing lanceolate leaves at each node (Figure 2.1). Each node can also produce roots, usually when in contact with soil or water. The plant grows prostrate along the ground, rooting at the nodes, or across the water's surface, anchored to the shore. Roots are initially fibrous and thicken when covered with soil. Over time, alligator weed develops an extensive underground root system. Reproduction is primarily by vegetative means and both stem nodes and root fragments can produce new plants. Alligator weed rarely produces viable seeds in its native range and viable seeds have not yet been found throughout its introduced range. This is probably due to the hybridization of fertile diploid ancestors that has produced sterile polyploid populations, which were subsequently introduced into new habitats around the world (Sosa et al, 2008). Flowers are small white balls with a papery texture and are attached to the nodes by a stalk (Figure 2.2). The plant's morphology is highly plastic, which, along with its clonal growth habit,

allows it to colonize a wide range of habitats (Geng et al, 2007; Wang et al, 2009).

Growth and morphology are dependent on habitat and environmental conditions. Alligator weed is commonly described as an amphibious plant due to its ability to root at the water's edge and grow across the surface, which is aided by thick hollow stems for floatation and thin fibrous roots that extract nutrients from the water column (Julien et al, 1992). This process creates thick mats that can grow up to 70m from the shoreline (Julien et al, 1992), and can become detached and float to new locations (Zeigler, 1967). It is moderately tolerant of salinity, which it achieves through increasing leaf thickness (Longstreth et al, 1984), and which allows it to grow in estuarine habitats (Julien et al, 1992). Alligator weed also grows in moist terrestrial habitats as a prostrate plant with thin stems and thick roots that contain extensive stores of carbohydrate (Wilson et al, 2007). Root material builds up over time to create large reserves in the soil; over 7.3kg m^{-2} dry biomass has been collected at terrestrial sites in Australia with a greater than 20 year invasion history (Schooler et al, 2008). Observations of above-ground biomass are not indicative of below-ground biomass, with established sites having ten times more biomass below ground than above (Schooler et al, 2008).

Figure 2.1 *Illustration of alligator weed showing: (a) aquatic form with stoloniferous growth habit and thin fibrous roots, (b) terrestrial form with thickened tap root, (c) inflorescence and (d) individual flower*

Distribution and Introduction Pathways

Alligator weed originates from the Paraná River region of South America, which extends from Brazil through northern Argentina (Maddox, 1968) (Gopurenko, D., unpublished data). Although it can grow in terrestrial habitats, alligator weed prefers moist floodplain areas in warm temperate and subtropical climates that experience regular inundation. It is considered a serious weed in

Source: Photograph by S. Schooler, copyright CSIRO

Figure 2.2 *Photograph of alligator weed showing papery white flowers and lanceolate leaves*

the US, China, Australia, New Zealand, Indonesia, India and Thailand (Julien and Bourne, 1988; Julien et al, 1995) (Figure 2.3). It was first detected in the US in 1897 (Zeigler, 1967), New Zealand in 1906 (Roberts and Sutherland, 1989), China in the 1930s (Wang et al, 2005) and Australia in 1946 (Hockley, 1974). It has also been found in France and Italy, although it is not yet considered a pest in these countries. Climate modelling indicates that many other countries would provide suitable habitat for alligator weed should it be introduced, such as most of southeast Asia, southern Africa and southern Europe (Julien et al, 1995).

The primary method of introduction into new countries has historically been through the dumping of ballast material (Zeigler, 1967; Roberts and Sutherland, 1989). However, introduction into Australia is deemed more likely to be from ship cargo due to the late date of introduction (Julien and Bourne, 1988), and an aquatic plant collector has been implicated in at least one introduction into an Australian inland site. Genetic

analyses indicate that genetic diversity is very low throughout both its native and introduced range, which is a product of vegetative propagation and few introduction events (Gopurenko, D., unpublished data). For example, there appears to be only one genotype in China, suggesting a single introduction event into the country (Xu et al, 2003; Wang et al, 2005), whereas genetic analyses indicate several introductions into the US (Wain et al, 1984) and Australia (Gopurenko, D., unpublished data).

Once naturalized, alligator weed is usually spread within catchments by flooding events, in mud on animals' hooves, and on machinery. Long distance dispersal is probably entirely by human means, usually accidentally on machinery such as earthmoving equipment and mowers. However, alligator weed was also purposefully cultivated in many backyard gardens in Australia by members of the Sri Lankan community who had mistaken alligator weed for a traditional culinary herb, mukunuwenna or sessile joy weed (*A. sessilis*) (Gunasekera and Bonila, 2001).

Impacts

Alligator weed invades agricultural areas and blocks drainage and irrigation channels, causing problems on agricultural land (Spencer and Coulson, 1976; van Oosterhout, 2007). The plant is a significant weed of rice crops in China resulting in an annual estimated loss of $75 million (Shen et al, 2005). In Australia, infestations in turf farms have prevented sales due to contamination of material (van Oosterhout, 2007). In addition, the plant can cause photosensitization and liver damage in cattle, although this appears to be infrequent and may be caused by a combination of high alligator weed ingestion rates and an unknown secondary factor (Bourke and Rayward, 2003).

Alligator weed also has environmental impacts, such as altering evapotranspiration rates, plant and insect communities, decomposition and nutrient cycling, and increasing the abundance of disease vectors. Evapotranspiration by alligator weed increases water loss compared with evaporation over open water and evapotranspiration over native floating species, thus reducing water retention in infested areas (Boyd, 1987; but see Allen et al, 1997). A removal experiment found that reduction in alligator weed cover resulted in increased biomass of native plant species (Allen et al, 2007). In addition, a study comparing alligator weed

Distribution of Alligator weed

 Native range
Introduced: noxious weed status
Introduced: not considered a pest

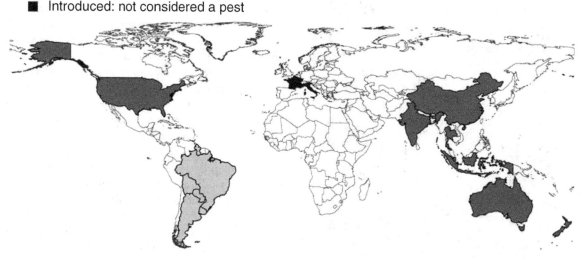

Note: Its distribution within countries is generally restricted to floodplains in warm temperate and subtropical habitats, such as south-eastern US, southern China and eastern Australia.

Figure 2.3 *Distribution of alligator weed showing countries that comprise its: (a) native range, (b) introduced range where it is considered a serious weed and (c) introduced range where it is not currently considered a pest*

with native sedges around a New Zealand lake found that alligator weed altered insect community composition and abundance (Bassett et al, 2011). Alligator weed also alters decomposition processes. It was found to decompose more rapidly than two New Zealand native species, thus altering patterns of nutrient cycling (Bassett et al, 2010). Other concerns over alligator weed include water pollution from plant decomposition and an increase in disease vector (mosquito and snail) breeding areas (Sculthorpe, 1967; Spencer and Coulson, 1976).

Ecology and Invasion Mechanisms

In its native range, alligator weed grows in ditches, wetlands and along steams and rivers where it exhibits boom and bust dynamics (Sosa, A. and Julien, M., pers. comm.). It rarely covers large areas in monospecific stands for extended periods as it does throughout its introduced range. In terrestrial areas, drought conditions cause it to die back to its roots as it is

extensively fed on by numerous insect species and outcompeted by grasses (Vogt, 1973; Julien et al, 2011). When flooding occurs, the grasses die back and alligator weed grows rapidly, replenishing carbohydrate supplies in the roots. In aquatic situations, insect defoliation and competition from native plants, many of which are themselves serious aquatic weeds, such as salvinia (*Salvinia molesta*) and water hyacinth (*Eichornia crassipes*), appear to prevent the formation of large floating mats.

A number of studies have examined the invasion mechanisms by which alligator weed is able to proliferate throughout its introduced range. Mechanisms investigated include the release from natural enemies, resource acquisition and plant growth, phenotypic plasticity, clonal integration and tolerance of disturbance. A study comparing the effects of natural enemies in the native and introduced ranges found that alligator weed experiences a release from damage by natural enemies in Australia as compared with Argentina (Clech-Goods, 2009), thereby gaining an advantage in its introduced range. A second study examined the growth of alligator weed compared with that of a native congener species, lesser joy weed (*A. denticulata*), along a nitrogen gradient. It was predicted that alligator weed would grow more rapidly than the native species at high nitrogen concentrations, but more slowly under low nitrogen concentrations (Schooler et al, 2006), which is thought to be a common trade-off that differentiates invasive and non-invasive plants. However, the study found that alligator weed grew more rapidly at all nitrogen concentrations (Clech-Goods, 2009), suggesting it is able to more rapidly capitalize on available resources than native species, even when they are at low abundance.

One of the most obvious traits of alligator weed is its great diversity of growth forms. This was first recognized in 1967 during the development of biological control in the US (Zeigler, 1967). Subsequently it was found that alligator weed in the US exhibited two categories of morphological variation, one based on phenotypic plasticity and one based on genetic differences (Wain et al, 1984), whereas in China there were no genetic differences and all morphological differences were due to phenotypic plasticity (Geng et al, 2007). Phenotypic plasticity allows a plant to quickly respond to changes in environmental conditions. A number of studies have examined phenotypic plasticity and invasiveness in alligator weed. Alligator weed responds to above-ground damage by growing closer to the ground and allocating more resources to roots, which makes grazing and mowing ineffective for control (Jia et al, 2009). A laboratory study found that alligator weed exhibited greater plasticity across a moisture gradient than its congener species, sessile joy weed (*A. sessilis*), thereby allowing it to access resources more quickly (Geng et al, 2006). A field study found that alligator weed exhibited greater plasticity in response to habitat variation than native species across a range of habitats: swamp, moist field, marsh dunes and gravel dunes (Pan et al, 2006). Alligator weed also responds to soil phosphorus concentrations. At high soil phosphorus concentrations it accumulates more phosphorus in plant tissues and produces larger individuals, but with fewer vegetative propagules (stem nodes) (Guan et al, 2010). Therefore, the ability of alligator weed to rapidly change its morphology, produce hollow stems for buoyancy and respiration in aquatic habitats, produce denser stems for vertical growth under terrestrial conditions, thicken leaves to tolerate increased salinity, produce thick carbohydrate-rich roots in soil to tolerate extended durations of unsuitable conditions and respond rapidly after disturbances, and produce thin fibrous roots in water to acquire nutrients from the water column, all facilitate rapid responses to changing environmental conditions.

Clonal integration is another mechanism that allows alligator weed to rapidly take advantage of changing environmental conditions and new habitats. A laboratory study found that clonal integration did not directly increase its competitive ability against a competitor species, but allowed alligator weed to explore open space and quickly find suitable environments (Wang et al, 2008). A second study found that clonal integration facilitated the rapid expansion of alligator weed from terrestrial into aquatic environments (Wang et al, 2009). Clonal integration also allows connected alligator weed stems to grow more rapidly under resource poor (shaded) environments than stems not connected to a non-shaded plant (Xu et al, 2010). In addition, subsidies from the non-shaded plants were based on ramet need so that the plant subsidized shaded ramets more than unshaded ramets. These studies show that clonal integration allows alligator weed to rapidly take advantage of disturbances through colonization of new habitats and tolerance of unsuitable local conditions.

In addition to phenotypic plasticity and clonal growth, alligator weed has physiological mechanisms that allow it to tolerate and respond rapidly to flooding disturbance. Dark anaerobic conditions trigger a dormancy mechanism, termed 'hypoxic quiescence', that allows alligator weed stem propagules (nodes) to withstand long durations under water (Quimby and Kay, 1977). Dark to light transfers showed that the propagules began to grow as soon as light became available, physiologically triggered by an increase in oxygen concentration (Quimby et al, 1978). Thus, despite the absence of viable seed, alligator weed creates a 'propagule bank' in flooded environments, which begins to grow as soon as flooding recedes enough for light to penetrate to the substrate.

Much of alligator weed's invasive capability appears to be based on tolerance of unfavourable conditions and rapid response after disturbance to take advantage of open space and available resources; it can grow rapidly, modify its morphology to suit conditions and tolerate inundation, and transport resources through tissue to subsidize rapid growth into new habitats. This is an example of disturbance-mediated competition, which has been shown to allow alligator weed to dominate a

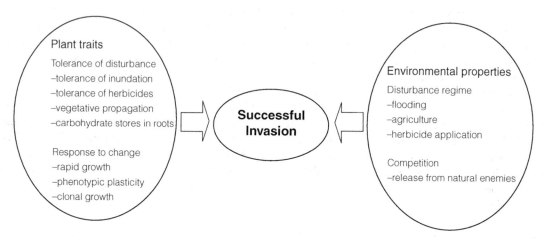

Figure 2.4 *The combination of plant attributes and properties of the recipient environment that facilitate alligator weed invasions*

superior competitor, kikuyu (*Pennisetum clandestinum*). In areas where flooding exceeded 28 days duration, alligator weed was able to displace kikuyu, but where flooding duration was less, kikuyu dominated the plant community (Schooler et al, 2010).

Therefore, it is not an innately superior competitive ability that drives alligator weed invasions, but rather its tolerance and flexibility in regard to disturbance events coupled with a release from damage by natural enemies (Figure 2.4). This makes alligator weed a particularly difficult weed to manage as humans are often the instigators of large environmental disturbances and most management strategies are based on non-selective disturbance events, such as physical removal and broad spectrum herbicide application.

Management Methods

Mechanical removal and chemical control

Management of alligator weed is difficult because the two most common control methods, mechanical removal and chemical control, are often not effective at eliminating infestations and can even exacerbate the problem (Sainty et al, 1997). Mowing has little suppressive effect and is likely to increase spread (Schooler et al, 2007; Jia et al, 2009). Mechanical

removal must be done carefully in order to contain and remove all propagules. Alligator weed reproduces and spreads primarily by adventitious rooting from stem nodes and root fragments (Julien et al, 1992; Ensbey and van Oosterhout, 2009) and, to be effective, excavating must remove all these fragments. If propagules are left in terrestrial areas, the disturbance usually results in bare patches for alligator weed to recolonize. In aquatic environments the remaining propagules increase plant spread downstream.

Although herbicides greatly reduce alligator weed biomass (Schooler et al, 2008), a portion of both stem nodes (Dugdale et al, 2010) and roots (Bowmer and Eberbach, 1993; Schooler et al, 2008, 2010) survive even after repeated treatments due to limited translocation through vascular tissue (Tucker et al, 1994). This makes alligator weed very difficult to eradicate once established. Glasshouse experiments indicate that a single application of glyphosate has no long term impact (>6 weeks) on root biomass compared with untreated controls (Schooler et al, 2007). In addition, alligator weed stores large amounts of carbohydrates in below-ground material, which is used to rapidly replace shoots and leaves after above-ground disturbances (Wilson et al, 2007). Follow-up treatments are generally less effective because plant surface area is reduced after the first application. Non-selective (or broad spectrum) herbicides may even facilitate growth and spread of alligator weed by removing competitors that are intolerant of herbicides, while remaining viable alligator weed propagules quickly produce new shoots and spread through the open habitat. Due to plant fragmentation, herbicides used in aquatic habitats can cause viable propagules to spread via water movement unless containment measures are in place (Dugdale et al, 2010).

Selective herbicides are useful in manipulating competition between alligator weed and monocot species. A two year field study found that a dicot selective herbicide (metsulfuron) increased competition from grasses and resulted in a significantly greater reduction to both above- and below-ground alligator weed biomass than a broad spectrum herbicide (glyphosate), while increasing both above- and below-ground biomass of desirable forage grass species (Schooler et al, 2008). In addition, selective herbicide application in conjunction with 40 days inundation was the only treatment that increased the abundance of a native Australian grass species, couch (*Cynodon dactylon*) (Schooler et al, 2010). Researchers have also found that a dicot selective herbicide (imazapyr) was more effective than a broad spectrum herbicide (triclopyr amine) in increasing biomass of native species in marshes in south-eastern US (Allen et al, 2007). Therefore, disturbances that selectively target alligator weed are more effective than broad spectrum disturbances.

Biological control

One management method that is both self-sustaining and selective is biological control. Classical biological control uses natural enemies from the native range to control an invasive species in its introduced range. It is usually the last resort in management strategies for invasive organisms due to its high initial cost and the difficulty of finding suitable agents that are both effective and host-specific (Culliney, 2005). Careful testing of host specificity reduces the risk that the introduced biological control agents feed on native and economically important plants (Pemberton, 2000). Biological control has led to the sustained management of many invasive plants, including alleviating environmental impacts (Van Driesche et al, 2010).

The first surveys for biological control agents for alligator weed were conducted by US researchers in South America in 1960–1961 (Vogt, 1973; Julien et al, 2011). Three insects were selected and subsequently released in the US. The leaf-eating flea beetle *Agasicles hygrophila* (Coleoptera: Chrysomelidae) was released

in 1964, the tip-distorting thrips *Amynothrips andersoni* (Thysanoptera: Phlaeothripidae) in 1967, and the stem-boring moth *Arcola malloi* (Lepidoptera: Pyralidae) in 1971. Releases of the flea beetle *A. hygrophila* and the moth *A. malloi* were later made in Australia in 1977, New Zealand in 1980, and of the flea beetle in Thailand (1981) and China (1986) (Julien and Griffiths, 1998). These agents controlled aquatic mats of alligator weed in the US, Australia, and China, but only in warmer regions where population development was not limited by the cold winters. In Australia, although the moth *A. malloi* established widely, the key agent controlling the weed is *A. hygrophila*. The flea beetle is particularly effective against aquatic mats because larvae bore into the hollow stems in order to pupate. These holes reduce buoyancy and introduce generalist pathogens, which cause the mat to rot and sink.

However, the current suite of biological control agents does not provide adequate control of alligator weed growing in terrestrial habitats or cooler climates (Julien and Bourne, 1988; Julien et al, 1995). Therefore, subsequent research has focused on the identification and testing of better adapted agents. The attempt to introduce a cold-hardy biotype of *A. hygrophila* to the US in 1979 was unsuccessful and a terrestrial insect, the flea beetle *Disonycha argentinensis* (Coleoptera: Chrysomelidae), was released in Australia and New Zealand but failed to establish in either country (Julien et al, 2011). Current efforts have identified and tested five additional insects, but none have been sufficiently host-specific to release (Schooler, unpublished data). A further five species, including a fungus, have been identified as potential biological control agents and testing is currently under way.

Management strategies

The most recent comprehensive strategy for control of alligator weed has been compiled in Australia (van Oosterhout, 2007). The strategy classifies infestations as either core or non-core (outlying). Core areas are those where eradication is not considered feasible and the main goal is control and containment. The strategy emphasizes the use of selective herbicides in terrestrial areas to reduce propagules and encourage monocots. Aquatic infestations in warmer climates (warm temperate and subtropical) are generally effectively controlled by the flea beetle.

The goal for outlying infestations is eradication. If the site is small, manual excavation is recommended. For larger areas the strategy is to mark the boundaries of the site and use a non-selective herbicide to both reduce the amount of alligator weed and create a bare patch around the infestation to make regrowth more evident for subsequent inspections. Recommended inspection times are every three to four months and should be treated as new growth appears. Once the infestation has been sufficiently reduced, then the remaining plants should be dug by hand. Since alligator weed can remain dormant for long periods, often until conditions are optimal for growth, it is recommended to continue inspections for five years after regrowth is no longer evident. Inspections should be conducted particularly after precipitation following an extended dry period.

Eradication has rarely been successful, even for small backyard infestations where alligator weed was grown for culinary purposes. Over 900 sites were identified across Australia, mostly in Victoria, and an intensive campaign to eradicate these infestations was instigated in 1997 (Gunasekera and Bonila, 2001; van Oosterhout, 2007). Concurrently a native species, lesser joy weed (*A. denticulata*), was promoted as a replacement for alligator weed. More than a decade later, after repeated treatments and inspections, fewer than 5 per cent of sites have been declared eradicated (no regrowth observed for five consecutive years), although 60 per cent of Victorian sites are currently under monitoring with no regrowth observed for one to four consecutive years (Dugdale, T., pers. comm.). It is not clear whether this is due to ineffectiveness of eradication methods, restrictions on what methods can be used due to potential off-target effects, ability to access sites or the rigour of the monitoring programme. Infestations in natural areas, particularly in wetlands and along stream banks, have also proved problematic and only a few small infestations have been successfully eradicated in Australia (Petroeschevsky, A. and Dugdale, T. pers. comm.). Due to this difficulty in eradicating new infestations, alligator weed is continuing to spread in Australia despite extensive coordinated efforts at control and containment (Burgin and Norris, 2008).

Challenges and Controversies

The three primary challenges in the management of alligator weed are: (1) developing methods to eradicate new infestations, (2) managing disturbance regimes to reduce habitat suitability, and (3) introducing selective control mechanisms to balance the reduction in herbivore loads. The ability to eradicate new infestations is crucial for controlling the spread of an invasive organism. The tenacity of alligator weed plants and propagules, particularly their tolerance of herbicides, makes eradication almost impossible. New tools, such as improved herbicides, would greatly assist containment efforts. Alligator weed is also extremely successful in tolerating disturbance and adapting to changing conditions. These traits are particularly problematic for control because human enterprise often creates broad-scale disturbances, which further facilitate the expansion of alligator weed infestations. This, in addition to the competitive advantage gained from reduced herbivore pressure, allows alligator weed to proliferate at the expense of native plant species. Therefore, from an ecological perspective, the most effective management strategies will simultaneously reduce the intensity and frequency of general disturbances, particularly inundation duration, and selectively increase damage to alligator weed, such as from biological control agents, selective herbicides or hand removal.

There are no controversies regarding the severity of alligator weed invasion. Several countries list alligator weed as one of their most troublesome pests. The primary dispute is usually how much effort to put into eradication and control when no method appears to be able to slow the progress of the invasion. In the face of such an intractable foe, many managers eventually decide it is not worth the effort and give up. Meanwhile, government representatives charged with controlling the invasion must enforce regulations of eradication and containment without the necessary methods available to enable land managers to do so. Therefore, the development of effective eradication methods, the ability and willingness to manage disturbance, and the search for new biological control agents are the key to improved control of alligator weed.

Acknowledgements

Thanks to Mic Julien, Rieks van Klinken, Rob Francis and an anonymous reviewer for providing constructive comments on earlier versions of this manuscript.

References

Allen, L. H., Sinclair, T. R. and Bennett, J. M. (1997) 'Evapotranspiration of vegetation of Florida: Perpetuated misconceptions versus mechanistic processes', *Soil and Crop Science Society of Florida Proceedings*, vol 56, pp1–10

Allen, S. L., Hepp, G. R. and Miller, J. H. (2007) 'Use of herbicides to control alligatorweed and restore native plants in managed marshes', *Wetlands*, vol 27, pp739–748

Bassett, I. E., Beggs, J. R. and Paynter, Q. (2010) 'Decomposition dynamics of invasive alligator weed compared with native sedges in a Northland lake', *New Zealand Journal of Ecology*, vol 34, pp324–331

Bassett, I. E., Paynter, Q. and Beggs, J. R. (2011) 'Differences in invertebrate community composition between an invasive herb (*Alternanthera philoxeroides*) and two native sedges in a New Zealand lake', *Insect Conservation and Diversity*, in press

Bourke, C. A. and Rayward, D. (2003) 'Photosensitisation in dairy cattle grazing alligator weed (*Alternanthera philoxeroides*) infested pastures', *Australian Veterinary Journal*, vol 81, pp361–362

Bowmer, K. H. and Eberbach, P. L. (1993) 'Uptake and translocation of 14C-glyphosate in *Alternanthera philoxeroides* (Mart.) Griseb. (alligator weed). II. Effect of plant size and photoperiod', *Weed Research*, vol 33, pp59–67

Boyd, C. E. (1987) 'Evapotranspiration/evaporation (E/Eo) ratios for aquatic plants', *Journal of Aquatic Plant Management*, vol 25, pp1–3

Burgin, S. and Norris, A. (2008) 'Alligator weed (*Alternanthera philoxeroides*) in New South Wales, Australia: A status report', *Weed Biology and Management*, vol 8, pp284–290

Clech-Goods, C. (2009) 'Relating resource availability and herbivore escape opportunity to plant community invasibility; creating integrated strategies to manage *Alternanthera philoxeroides* and *Phyla canescens*', PhD Thesis, University of Queensland, Brisbane, Australia

Culliney, T. W. (2005) 'Benefits of classical biological control for managing invasive plants', *Critical Reviews in Plant Sciences*, vol 24, pp131–150

Davis, M. A., Grime, J. P. and Thompson, K. (2000) 'Fluctuating resources in plant communities: A general theory of invasibility', *Journal of Ecology*, vol 88, pp528–534

Dugdale, T. M., Clements, D., Hunt, T. D. and Butler, K. L. (2010) 'Alligatorweed produces viable stem fragments in response to herbicide treatment', *Journal of Aquatic Plant Management*, vol 48, pp121–125

Ensbey, R. and van Oosterhout, E. (2009) *Alligator Weed*, NSW DPI Primefact 671, New South Wales Department of Industry and Investment, Orange, New South Wales, Australia, www.dpi.nsw.gov.au/__data/assets/pdf_file/0006/309246/alligator-weed.pdf

Geng, Y. P., Pan, X. Y., Xu, C. Y., Zhang, W. J., Li, B. and Chen, J. K. (2006) 'Phenotypic plasticity of invasive *Alternanthera philoxeroides* in relation to different water availability, compared to its native congener', *Acta Oecologica*, vol 30, pp380–385

Geng, Y. P., Pan, X. Y., Xu, C. Y., Zhang, W. J., Li, B., Chen, J. K., Lu, B. R. and Song, Z. P. (2007) 'Phenotypic plasticity rather than locally adapted ecotypes allows the invasive alligator weed to colonize a wide range of habitats', *Biological Invasions*, vol 9, pp245–256

Guan, B. H., An, S. Q., Schooler, S. and Cai, Y. (2010) 'Influence of substrate phosphorus concentration and plant density on growth and phosphorus accumulation of *Alternanthera philoxeroides*', *Journal of Freshwater Ecology*, vol 25, pp219–225

Gunasekera, L. and Bonila, J. (2001) 'Alligator weed: Tasty vegetable in Australian backyards?', *Journal of Aquatic Plant Management*, vol 39, pp17–20

Hockley, J. (1974) '... and alligator weed spreads in Australia', *Nature*, vol 250, pp704–705

Jia, X., Pan, X. Y., Li, B., Chen, J. K. and Yang, X. Z. (2009) 'Allometric growth, disturbance regime, and dilemmas of controlling invasive plants: A model analysis', *Biological Invasions*, vol 11, pp743–752

Julien, M. H. and Bourne, A. S. (1988) 'Alligator weed is spreading in Australia', *Plant Protection Quarterly*, vol 3, pp91–96

Julien, M. H. and Griffiths, M. W. (eds) (1998) *Biological Control of Weeds: A World Catalogue of Agents and their Target Weeds*, 4th edn, CABI Publishing, CAB International, Wallingford, UK

Julien, M. H., Bourne, A. S. and Low, V. H. K. (1992) 'Growth of the weed *Alternanthera philoxeroides* (Martius) Grisebach, (alligator weed) in aquatic and terrestrial habitats in Australia', *Plant Protection Quarterly*, vol 7, pp102–108

Julien, M. H., Skarratt, B. and Maywald, G. F. (1995) 'Potential geographical distribution of alligator weed and its biological control by *Agasicles hygrophila*', *Journal of Aquatic Plant Management*, vol 33, pp55–60

Julien, M., Sosa, A., Chan, R., Schooler, S. and Traversa, G. (2011) '*Alternanthera philoxeroides* (Martius) Grisebach – Alligator weed', in M. Julien, R. McFadyen and J. Cullen (eds) *Biological Control of Weeds in Australia 1960 to 2010*, CSIRO Publishing, Collingwood, Australia, pp18–35

Longstreth, D. J., Bolanos, J. A. and Smith, J. E. (1984) 'Salinity effects on photosynthesis and growth in *Alternanthera philoxeroides* (Mart.) Griseb.', *Plant Physiology*, vol 75, pp1044–1047

Maddox, D. M. (1968) 'Bionomics of an alligator weed flea beetle, *Agasicles* sp. in Argentina', *Annals of the Entomological Society of America*, vol 61, pp1300–1305

Pan, X. Y., Geng, Y. P., Zhang, W. J., Li, B. and Chen, J. K. (2006) 'The influence of abiotic stress and phenotypic plasticity on the distribution of invasive *Alternanthera philoxeroides* along a riparian zone', *Acta Oecologica*, vol 30, pp333–341

Pemberton, R. W. (2000) 'Predictable risk to native plants in weed biological control', *Oecologia*, vol 125, pp489–494

Quimby, P. C. and Kay, S. H. (1977) 'Hypoxic quiescence in alligatorweed', *Physiologia Plantarum*, vol 40, pp163–168

Quimby, P. C., Potter, J. R. and Duke, S. O. (1978) 'Photosystem II and hypoxic quiescence in alligatorweed', *Physiologia Plantarum*, vol 44, pp246–250

Roberts, L. I. N. and Sutherland, O. R. W. (1989) '*Alternanthera philoxeroides* (C. Martius) Grisebach, alligator weed (Amaranthaceae)', in P. J. Cameron, R. L. Hill, J. Bain and W. P. Thomas (eds) *A Review of Biological Control of Invertebrate Pests and Weeds in New Zealand 1874 to 1987*, CAB International Institute of Biological Control (CIBC), Wallingford, UK, pp325–330

Sainty, G., McCorkelle, G. and Julien, M. (1997) 'Control and spread of alligator weed *Alternanthera philoxeroides* (Mart.) Griseb., in Australia: Lessons for other regions', *Wetlands Ecology and Management*, vol 5, pp195–201

Schooler, S., Clech-Goods, C. and Julien, M. (2006) 'Ecological studies to assess the efficacy of biological control on populations of alligator weed and lippia', *Australian Journal of Entomology*, vol 45, pp272–275

Schooler, S. S., Yeates, A. G., Wilson, J. R. U. and Julien, M. H. (2007) 'Herbivory, mowing, and herbicides differently affect production and nutrient allocation of *Alternanthera philoxeroides*', *Aquatic Botany*, vol 86, pp62–68

Schooler, S., Cook, T., Bourne, A., Prichard, G. and Julien, M. (2008) 'Selective herbicides reduce alligator weed (*Alternanthera philoxeroides*) biomass by enhancing competition', *Weed Science*, vol 56, pp259–264

Schooler, S. S., Cook, T., Prichard, P. and Yeates, A. G. (2010) 'Disturbance-mediated competition: The interacting roles of inundation regime and physical and herbicidal control in determining native and invasive plant abundance', *Biological Invasions*, vol 12, pp3289–3298

Sculthorpe, C. D. (1967) *The Biology of Aquatic Vascular Plants*, St Martin's Press, New York

Shen, J. Y., Shen, M. Q., Wang, X. H. and Lu, Y. T. (2005) 'Effect of environmental factors on shoot emergence and vegetative growth of alligatorweed (*Alternanthera philoxeroides*)', *Weed Science*, vol 53, pp471–478

Sosa, A. J., Greizerstein, E., Cardo, M. V., Telesnicki, M. C. and Julien, M. H. (2008) 'The evolutionary history of an invasive species: Alligator weed, *Alternanthera philoxeroides*', in M. H. Julien, R. Sforza, M. C. Bon, H. C. Evans, P. E. Hatcher, H. L. Hinz and B. G. Rector (eds) *Proceedings of the XII International Symposium on Biological Control of Weeds*, CAB International, Wallingford, UK, pp435–442

Spencer, N. R. and Coulson, J. R. (1976) 'The biological control of alligatorweed, *Alternanthera philoxeroides*, in the United States of America', *Aquatic Botany*, vol 2, pp177–190

Tucker, T. A., Langeland, K. A. and Corbin, F. T. (1994) 'Absorption and translocation of C-14 imazapyr and C-14 glyphosate in alligatorweed *Alternanthera philoxeroides*', *Weed Technology*, vol 8, pp32–36

Van Driesche, R. G., Carruthers, R. I., Center, T., Hoddle, M. S., Hough-Goldstein, J., Morin, L., Smith, L., Wagner, D. L., Blossey, B., Brancatini, V., Casagrande, R., Causton, C. E., Coetzee, J. A., Cuda, J., Ding, J., Fowler, S. V., Frank, J. H., Fuester, R., Goolsby, J., Grodowitz, M., Heard, T. A., Hill, M. P., Hoffmann, J. H., Huber, J., Julien, M., Kairo, M. T. K., Kenis, M., Mason, P., Medal, J., Messing, R., Miller, R., Moore, A., Neuenschwander, P., Newman, R., Norambuena, H., Palmer, W. A., Pemberton, R., Panduro, A. P., Pratt, P. D., Rayamajhi, M., Salom, S., Sands, D., Schooler, S., Schwarzlander, M., Sheppard, A., Shaw, R., Tipping, P. W. and van Klinken, R. D. (2010) 'Classical biological control for the protection of natural ecosystems', *Biological Control*, vol 54, S2–S33

van Oosterhout, E. (2007) *Alligator Weed Control Manual*, NSW Department of Primary Industries, Orange, Australia

Vogt, G. B. (1973) *Exploration for Natural Enemies of Alligator Weed and Related Plants in South America*, US Army Corps of Engineers, Waterways Experiment Station, Aquatic Plant Control Program, Technical Report 3

Wain, R. P., Haller, W. T. and Martin, D. F. (1984) 'Genetic relationship among two forms of alligatorweed', *Journal of Aquatic Plant Management*, vol 22, pp104–105

Wang, B., Li, W. and Wang, J. (2005) 'Genetic diversity of *Alternanthera philoxeroides* in China', *Aquatic Botany*, vol 81, pp277–283

Wang, N., Yu, F. H., Li, P. X., He, W. M., Liu, F. H., Liu, J. M. and Dong, M. (2008) 'Clonal integration affects growth, photosynthetic efficiency and biomass allocation, but not the competitive ability, of the alien invasive *Alternanthera philoxeroides* under severe stress', *Annals of Botany*, vol 101, pp671–678

Wang, N., Yu, F. H., Li, P. X., He, W. M., Liu, J., Yu, G. L., Song, Y. B. and Dong, M. (2009) 'Clonal integration supports the expansion from terrestrial to aquatic environments of the amphibious stoloniferous herb *Alternanthera philoxeroides*', *Plant Biology*, vol 11, pp483–489

Wilson, J. R. U., Yeates, A., Schooler, S. and Julien, M. H. (2007) 'Rapid response to shoot removal by the invasive wetland plant, alligator weed (*Alternanthera philoxeroides*)', *Environmental and Experimental Botany*, vol 60, pp20–25

Xu, C., Zhang, W., Fu, C. and Lu, B. (2003) 'Genetic diversity of alligator weed in China by RAPD analysis', *Biodiversity and Conservation*, vol 12, pp637–645

Xu, C. Y., Schooler, S. S. and Van Klinken, R. D. (2010) 'Effects of clonal integration and light availability on the growth and physiology of two invasive herbs', *Journal of Ecology*, vol 98, pp833–844

Zeigler, C. F. (1967) 'Biological control of alligatorweed with *Agasicles* n. sp. in Florida', *Water Hyacinth Control Journal*, vol 6, pp31–34

3

Crassula helmsii (T. Kirk) Cockayne (New Zealand pygmyweed)

Anita Diaz

Origins, History of Introduction and Current Distribution

Crassula helmsii is an amphibious or aquatic perennial herb that grows in a range of habitats from slow flowing river margins, shallow lakes, ponds and temporary pools to the damp margins around water bodies (Kelly and Maguire, 2009). A full description can be found in Dawson and Warman (1987) but a brief summary is provided here. The plant consists of a mat of rounded stems that are a few millimetres in diameter and up to 1.5m long. They are branched to varying amounts (parts of longer stems deep in water are less branched) but in all cases the sessile leaves are in pairs, inserted at nodes along the stem and 4–24mm long and 0.5–2mm wide. Leaves are longer and thinner where light levels are lowest. Roots are produced at all but the most apical stem nodes. Small, white to pale pink, sweet-scented flowers are produced in the leaf axils of emergent stems. Figure 3.1 illustrates the differences in overall plant morphology between plants growing in submerged and terrestrial forms.

A native of New Zealand and Australia, *C. helmsii* has become widely naturalized across Britain and many countries in western and central mainland Europe (Margot, 1983; Dawson and Warman, 1987; EPPO, 2007). It has also been recorded in Russia and is considered as a likely problem invasive species in Florida and North Carolina (EPPO, 2007). Figure 3.2

indicates the countries where *C. helmsii* has been recorded as growing in the wild. Common names include New Zealand pygmyweed, Australian swamp stonecrop, swamp stonecrop and watercrassula. Latin synonyms include *Tillaea helmii* (T. Kirk), *Tillaea recuva* (Hook. f.) and *Crassula recurva* (Hook. f.) (Dawson and Warman, 1987). Within the Australian part of its native range *C. helmsii* occurs most frequently in still, shallow freshwater bodies, swamps and water margins across south-eastern Australia and Tasmania (Smith and Marchant, 1961; Toelken, 1981). By contrast, within New Zealand it does not seem to grow in open water but grows on a wide range of damp margins including those of coastal lagoons and in brackish conditions on salt marshes (Dawson and Warman, 1987). Flowering occurs during the summer months but over a shorter period in Australia (November and December) compared to New Zealand (November to March) (Kirby, 1964). It seems probable that different ecotypes exist within the native range but this has not been investigated.

The spread of *C. helmsii* beyond Australasia began in the 20th century apparently as a result of it being traded by garden centres and nurseries as an attractive perennial 'oxygenating plant' (Dawson and Warman, 1987). Trade in this species remains ongoing, despite recommendations for its regulation (Brunel, 2009). No comprehensive and reliable record exists of where the original source locations were for genotypes of

Note: During the summer growing and flowering season plants can change between these forms within a month of being subjected to a change in level of inundation. The stems in (b) were emergent three weeks before this photograph was taken and still show the signs of flowers and aborted flower buds produced when growing terrestrially.

Source: A. Diaz

Figure 3.1 Crassula helmsii *growing (a) in terrestrial form and (b) in submerged form*

C. helmsii sold by garden centres/nurseries and it is unknown how many different genotypes have become naturalized across the world. Most of our current understanding of naturalized populations of C. helmsii derives from research carried out in Britain over the last half century by a range of scientists, particularly Dr F. H. Dawson, and by conservation practitioners. Dawson has carried out the only reported genetic analysis of naturalized populations of C. helmsii. He surveyed populations from across Britain (using

allozymes) and found that all plants may have come from a single source, most probably the River Murray in southern Australia (Dawson, 1994). It would be valuable to develop this line of research further using modern, higher resolution DNA-based molecular genetic techniques, to investigate the origins of naturalized populations of C. helmsii at a global scale.

The first record for C. helmsii growing in the wild outside of its native range was of it growing in a pond in Greenstead in Essex, UK in 1956 (Laundon, 1961). It is, however, possible that it had been introduced into the wild in Britain earlier than this as C. helmsii was brought into England from Tasmania in 1911 and was on sale by the 1920s (CEH, 2004). There was a rapid spread of C. helmsii across Britain through the 1980s and 1990s resulting in it being officially registered by the Botanical Society of the British Isles (BSBI) in over 100 separate locations by the start of the 21st century and the number of locations has continued to grow steadily since then (Watson, 2001; BSBI, 2010; Lockton, 2010).

Biological Attributes

As is typical for species in the family Crassulaceae and ubiquitous in the genus *Crassula*, C. helmsii is succulent and possesses crassulacean acid metabolism (CAM) (Klavsen and Maberly, 2009). Most members of the family live in dry habitats and these traits are used as xerophytic adaptations, but in aquatic plants CAM confers an enhanced ability to take up carbon dioxide (CO_2) from the water or air around the stems (Newman and Raven, 1995; Klavsen and Maberly, 2009). The use of CAM by aquatic plants is a very phenotypically plastic trait; generally CAM is used less when the rate of photosynthesis is not limited by the availability of CO_2 because it is being limited by other factors (such as low light levels, low nutrients and low temperatures) (Boston and Adams, 1985). Recent study of C. helmsii in the field by Klavsen and Maberly (2009, 2010) has shown that use of CAM is an advantage to C. helmsii as it contributes around 18–42 per cent of the total carbon budget derived from CO_2 fixation. The authors conclude that this is likely to confer a competitive advantage to C. helmsii over other species without CAM and postulate that this may be one of the reasons for the broad ecological niches and invasiveness of C. helmsii.

The growth habit and growth form of C. helmsii are both also extremely phenotypically plastic. In terms of

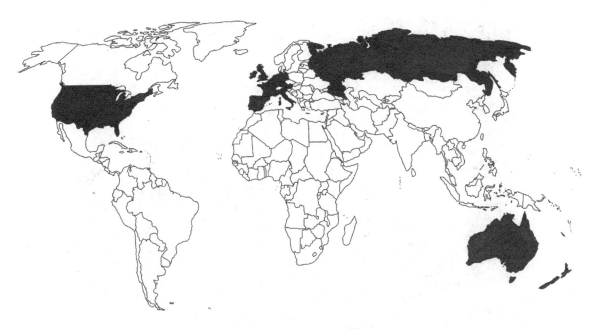

Note: Although entire countries are highlighted this does not mean that Crassula helmsii is present throughout all regions within the country.

Figure 3.2 *Global distribution of* Crassula helmsii *based on countries where it has been recorded as growing in the wild*

growth habit, *C. helmsii* can grow as either attached, rooted plants in terrestrial or shallow water environments or as free-floating mats in deeper water (Klavsen and Maberly, 2009). Growth form can vary continuously depending on habitat conditions but Dawson and Warman (1987) describe three basic forms, a deep water, shallow water and marginal growth form. Stems are longest and least branched when the plant is growing in deep water (>1m). The lower parts of such plants have anchoring roots, particularly long internodes and thin, sparsely distributed leaves. In shallow waters (<0.5m) the growth form is a mat of prostrate stems with vertical side shoots that can become emergent. Where *C. helmsii* grows in marginal habitats, its stems are often very short and form an emergent dense mat that flowers profusely. Tank experiments indicate that stems can change growth form within a month of a change of water levels, at least during the summer in southern England (Diaz, A., unpublished data). The possession of these abilities for phenotypic plasticity as a response to water levels is likely to be a key attribute enabling *C. helmsii* to both continue to occupy a fluctuating environment and to colonize a wide range of new environments.

Small, white to pale pink, sweet-scented flowers are produced in the leaf axils of emergent stems. No flowers are produced below the water and if water levels rise flower buds are aborted (see Figure 3.1). Flowering occurs in Britain from mid-summer (July) to autumn (October or sometimes even November if not flooded). During the summer and early autumn, flowers are visited by a number of species of hover-fly (Syrphidae) (Diaz, A., pers. obs.) (Figure 3.3). The flies typically forage for nectar on the mats of *C. helmsii* for several minutes at a time, moving every few seconds from one flower to the next. Usually the flies walk rather than fly between flowers and so tend to move between immediately adjacent flowers. Consequently, although they certainly transfer pollen between flowers, most of this pollen flow will be geitonogamous flow within the same plant, even if different genotypes exist in the population.

Seeds of *C. helmsii* are small, smooth surfaced and float, and so can easily be dispersed by water or airflows. However, it has been suggested that seed produced by *C. helmsii* naturalized in Britain may not be viable (Preston and Croft, 1997). Even without producing seed, *C. helmsii* can disperse and re-establish very

Figure 3.3 *A hover-fly (Syrphidae) foraging for nectar on* Crassula helmsii *in Dorset, UK*

effectively from vegetative fragments consisting of just a single node (Dawson and Warman, 1987). Fragments are produced either as a result of mechanical damage of the plants or naturally in the autumn when growth slows and the plant produces leafy laterals called 'turions' that are easily separated from the plant and spread by water and wind currents or in mud transported from one site to another (Dawson and Warman, 1987). A wide range of possible dispersal vectors have been proposed including machinery used in conservation management, personal footwear, water or equipment used to stock ornamental fish ponds, fishing tackle and free-ranging ponies, herons, geese and ducks (Dawson and Warman, 1987).

Ecological Niche

Habitat requirements

A striking feature of *C. helmsii* is its ability to grow in a wide range of environmental conditions although its growth is generally fastest in neutral, nutrient-rich conditions (Dawson and Warman, 1987; Leach and Dawson, 1999; Hussner, 2009). It can occupy aquatic and marginal habitats in alkaline neutral and acidic systems that can be either eutrophic or oligotrophic (Dawson and Warman, 1987; Leach and Dawson, 1999; Klavsen and Maberly, 2009). Furthermore, it is able to grow in brackish (but not fully saline) conditions (CEH, 2002), is tolerant to copper pollution (Küpper et al, 2009), is able to withstand desiccation (Kelly and Maguire, 2009) and is frost resistant as, although it appears to over-winter best when fully submerged (Dawson and Warman, 1987), it has been reported as surviving growing at several degrees below zero in terrestrial conditions in pots in a greenhouse (Kirby, 1965).

Within its native range, *C. helmsii* occurs in both still and slow flowing water bodies such as river margins, although experiments by Dawson and Warman (1987) demonstrated that growth rates were greater in flowing water if plants were rooted. Wave action is known to damage the stems and it is suggested that this may be why *C. helmsii* is most common in small water bodies (Lockton, 2010). Dawson and Warman (1987) suggest that the relative absence of *C. helmsii* in flowing water bodies in Britain may be because these tend to not have the soft, open sediments that are ideal for anchorage of *C. helmsii*. However, further investigation is needed of factors affecting the risk of invasion of marginal riverine habitats by *C. helmsii*.

Growth rates of *C. helmsii* are greatest under very light shade as it is damaged by high light levels (Hussner, 2009; Lockton, 2010). It can also grow under the shade of trees and other vegetation (Laundon, 1961; Dawson and Warman, 1987) but exhibits marked enhanced growth if shading is reduced. For example, although it grows within dense beds of *Phragmitis australis* at Studland Bay, Dorset, UK, its growth rate is greatly enhanced along paths through the reeds created by Sika deer (*Cervus nippon*) (Diaz, A., pers. obs.). These (non-native) deer make extensive use of the reedbeds as harbourage and create a network of paths through the reeds that are likely to be important corridors for the spread of *C. helmsii*. Their feet may also spread vegetative propagule fragments of *C. helmsii* to other habitats. Overall it is likely to be a combination of factors, particularly the availability of light and water, that determine growth rate and accumulation of biomass of *C. helmsii* during the growing season. In general this is likely to be greatest in very shallow water conditions that combine high humidity and good light availability. However, such conditions are often ephemeral due to changes in water levels and so annual accumulation of biomass of *C. helmsii* in any one location can be difficult to predict.

Ecological interactions and impacts

The rapid growth rates of *C. helmsii* and its tendency to form dense mats when growing conditions are optimum results in the species being capable of producing major fluctuations in the availability of dissolved oxygen, CO_2 and nutrients in the surrounding water. Major decreases in the availability of oxygen can directly kill fish (CEH, 2004) and potentially damage invertebrates and other members of the biotic community. Loss of these species may have knock-on effects through food webs. Also competition may be particularly high for light, nutrients and inorganic carbon as well as possibly for oxygen and root anchor points. The ability of *C. helmsii* to use CAM means that it is able to acquire CO_2 for photosynthesis almost continuously and so

may, at times, deplete the CO_2 available to other neighbouring aquatic plants. The perennial nature of *C. helmsii* also means that it can persist as a dense mat through the winter and so suppress the spring regrowth of other plant species.

Open water species reported as having reduced abundance as a result of the presence of *C. helmsii* include *Nymphaea* spp. (Swale and Belcher, 1982) and *Elodea* spp. (Cockerill, 1979), while species reported as almost entirely excluded from marginal open mud habitats by *C. helmsii* include *Damasonium alisma* (Watson, 2001), *Ludwigia palustris* and *Galium debile* (Dawson and Warman, 1987; Leach and Dawson, 1999). There is also some evidence that *C. helmsii* can change dominance hierarchies within water bodies (Leach and Dawson, 1999) or dominate to the virtual exclusion of other species (Dawson and Warman, 1987). However, most of these results are anecdotal rather then being based on more formal studies and the strength of evidence that *C. helmsii* outcompetes rare native plants in Britain has been questioned (Lockton, 2010). One study that quantitatively analysed the difference in species richness between 19 ponds in northwest England that had been invaded by *C. helmsii* with 9 ponds that had not, found no evidence of significant effect of invasion on either species richness of the adult plants or of the seed bank (Langdon et al, 2004). The authors acknowledge that this result may be affected by low statistical power and high variance in number of species found in ponds within each treatment group. However, Langdon et al (2004) also report an examination of the effect of *C. helmsii* on germination success of native plants and found that even very short, artificially created mats of *C. helmsii* negatively affected the germination of 6 out of 11 species tested that are used as egg-laying sites by newts (these were *Epilobium hirsutus*, *Lythrum salicaria*, *Mentha aquatica*, *Ranunculus sceleratus*, *Veronica beccabunga* and *Myosotis scorpioides*).

There have been very few robust, quantitative studies of the effects of *C. helmsii* on animals but a notable exception is Deegan and Ganf (2008), who used stable isotopes to examine the importance of species of aquatic and riparian plants to groups of invertebrate primary consumers. Results indicated that, within its native range at least, *C. helmsii* can provide up to 72 per cent of the organic carbon for Amphipod primary consumers and up to 50 per cent for Trichoptera primary consumers. More needs to be known of the relationship between *C. helmsii* and these invertebrate communities in invaded systems. It may be particularly interesting to investigate this in copper-polluted systems as Küpper et al (2009) have shown that not only is *C. helmsii* tolerant to copper, it hyperaccumulates it and can potentially provide a route for copper to become biomagnified up trophic levels. No direct evidence exists of *C. helmsii* having deleterious impacts on vertebrates but Langdon et al (2004) found that in experimental tanks smooth newt (*Lissotriton vulgaris*) eggs hatched at a later developmental stage when laid on *C. helmsii* than on their more usual choice of plant, *Nasturtium officinale* (common watercress). Further work is required to establish if this also occurs in the wild and has any fitness consequences.

Past and Current Approaches to Management

It is clear from what is known of the biology of *C. helmsii* and from previous failed attempts at management that mechanical cutting is not only ineffective but actively counterproductive as it risks future spread of vegetative propagules (CEH, 2004). Dredging (or hand-pulling of small populations) is recommended by the Environment Agency in the UK as an alternative physical method as *C. hemsii* is shallow rooted (Environment Agency, 2003) but this recommendation comes with the caveat that is vital to contain any vegetative fragments produced during this process.

Attempts to control aquatic macrophtyes by covering plants with sheets of black plastic or other dark material were first proposed by Dawson and Hallows (1983), applied specifically to *C. helmsii* by Dawson and Warman (1987) and have been used a number of times by conservation organizations (e.g. Bridge, 2005; Wilton-Jones, 2005). Broadly the technique appears to work if the plastic is left in place for at least several months (Kelly and Maguire, 2009) (longer is generally considered to be better but see Dawson and Warman, 1987). However, limitations include that the treatment often needs repeating as *C. helmsii* can easily recolonize from edges beyond the end of the plastic and that the method is time consuming, affects other species and is aesthetically unappealing and so difficult to use on very public parts of nature reserves. Other techniques that cause rapid

changes in physical conditions such as winter draining of ponds (Dawson and Warman, 1987) have been found to suppress growth of *C. helmsii* but only flooding with salt water (appropriate only in very specific circumstances) has been reported to eliminate it (Charlton et al, 2010). Freezing with liquid nitrogen has been found to be useful for treating small patches but is logistically difficult for large areas and attempts to apply high heat through using flame throwers have been found to be ineffective as insufficient heat is produced (EPPO, 2007). Similarly spraying with 'Waipuna' hot foam only succeeded in killing the top parts of the plant and so did not eradicate it (Bridge, 2005). The risk with all such wholesale disturbance treatments is that they may damage non-target plants and other wildlife more than the *C. helmsii* and/or that *C. helmsii* will be able to recolonize bare substrates and water bodies more quickly than non-target plants and so be a net beneficiary of the disturbance event.

No effective biological control agents are known for *C. helmsii* and very little is known in general about the ecological interactions of this species in its natural range (Gassmann et al, 2006). Consequently, there is currently little baseline information from which to start to consider the merit of developing new biological control methods. The use of herbivorous fish such as the grass carp *Ctenopharyngodon idella* as a means of controlling *C. helmsii* was explored in some detail by Dawson and Warman (1987), who found some indication of an overall increase in diversity of submerged macrophytes but limitations due to the fish eating non-target macrophytes and only feeding on *C. helmsii* in water temperatures above 16°C. Also, dense mats of *C. helmsii* can produce depletions in dissolved oxygen that kill the fish (CEH, 2004). Egyptian geese have been reported as eating *C. helmsii* (Warren, 2008) and anecdotal reports also suggest that swans may take small amounts but there is no quantitative evidence for birds providing significant control of growth of *C. helmsii*.

Currently the most widely used control method is to apply herbicides, although concerns over safety to other aquatic plants and animals produce significant constraints on what chemicals are permitted. Since Dawson and Warman (1987) reported that diquat was successful in controlling growth of *C. helmsii*, it has been used quite extensively (usually in the form of Reglone, e.g. Gomes, 2005 (work carried out in 2001)) for treating submerged forms of *C. helmsii*. However, diquat was withdrawn for use in aquatic systems by the European Union in 2001 (EU, 2001). Appeals are ongoing for an exception permitting the use of diquat for control of *C. helmsii* but in the meantime current advice is that submerged material can be chemically treated using dichlobenil in February and March while all parts of the plant are submerged (Dawson, 1996; CEH, 2004; Kelly and Maguire, 2009). The only remaining effective herbicides for emergent growth are those based on glyphosate. As this has impacts on other non-target aquatic species and is ineffective in winter temperatures in Britain, the Environment Agency recommendation is to apply it only between April and November and only to either fully terrestrial *C. helmsii* or to only the dry parts of emergent plants (Environment Agency, 2010).

Emerging Debates in Approaches to Management

Conservation practitioners faced with managing sites invaded by *C. helmsii* are increasingly having to decide where to target limited resources by weighing up the substantial financial burden and anticipated ecological costs to non-target native organisms of applying any management control of *C. helmsii* against the predicted ecological conservation benefits of taking action. A key challenge in managing *C. helmsii* is that it often regrows quicker after the application of control techniques than can non-target species. Past approaches to solving this have focused on searching for more aggressive control techniques but this has led to expressions of concern that conservation work can over-focus on achieving control of the invasive species without first establishing the ecological benefits gained through its control and the effect of management on non-target native species (Didham et al, 2005; Lockton, 2010).

Research to date has proved useful in giving a specific understanding of some key biological and ecological traits of *C. helmsii* but progress towards generating general predictive models of how much *C. helmsii* will impact on its invaded community has been far more limited. An important constraint has been the absence of specific theoretical frameworks within which to develop predictions. However, recent thinking on species roles in invaded systems might have major implications for approach to management of *C. helmsii*.

Two such lines of thought are: (1) the importance of whether the invasive species is a passenger, driver or somewhere in between in its relationship to the ecological changes affecting the system it has invaded (Didham et al, 2005; MacDougall and Turkington, 2005; Bulleri et al, 2010); and (2) the importance of an invasive species' ecological niche difference and fitness advantage compared to other species in the invaded system (MacDougall et al, 2009).

As an example of the first of the above lines of thought, predicting the ecological benefits gained through control of *C. helmsii* may be facilitated by considering whether the role played by *C. helmsii* in the invaded system is that of a driver of change, for example providing competitive exclusion of other species or if its role is purely that of a passenger that is abundant simply because it is able to grow well in the specific conditions provided but that has no effect on other species. If *C. helmsii* is purely a passenger then conservation efforts should be directed at changing the environmental conditions that favour its growth so that they are more suitable for native species, not at controlling *C. helmsii* directly, as only the former approach will actually result in any increase in the abundance of native species. For example Figure 3.4a shows a temporary pool in the New Forest, Hampshire, UK, the margin of which is densely invaded by *C. helmsii*, and where anecdotal evidence suggests *C. helmsii* is driving the loss of native species through competitive exclusion. In this case, application of herbicide might be an effective management option. By contrast, in the case of reedbeds in Dorset, UK, invaded by *C. helmsii* (described above and illustrated also in Figure 3.4b), the species may be acting purely as passenger, i.e. benefiting from being less grazed and damaged by deer than is the native reed but not having any competitive effects on the reed. If this is the case, then any efforts by the landowner to control its growth through applying herbicides are likely to be futile and possibly counterproductive in terms of overall conservation benefits.

The concept of invasive species as passengers and drivers of change describes two extreme ends of a spectrum and can perhaps most usefully be employed as a mental framework for considering species roles rather than a definitive classification, as roles can vary between systems and differ over time. For example, Bulleri et al (2010) studied the effect of the invasive

(a)

(b)

Source: A. Diaz

Figure 3.4 *Driver or passenger of change?* Crassula helmsii *carpeting the floor of (a) a temporary pond margin in the New Forest, UK and (b) reedbeds damaged by sika deer in Dorset, UK*

seaweed *Caulerpa racemosa* on the communities of Mediterranean reefs and found that *C. racemosa* is initially a passenger of ecological change that becomes dominant in disturbed communities because it is ecologically best adapted to exploit the disturbed habitat. However, once established it increases the amount of sediment accumulation, which favours algal turf forms over erect forms and results in it driving major ecological change. Consequently the advice to conservation managers would be that management intervention in the early stages of invasion should focus on addressing the disturbance effect directly but once *C. racemosa* is established as a driver then any conservation management regime will need to remove both the disturbance effect and the *C. racemosa* and to reverse (if possible) its ecological effects. In the same way, it is possible that some current conservation

management of *C. helmsii* is misfocused on controlling it in systems where it is a passenger and where it would be better to deal directly with the factor(s) enabling its invasion, while in other cases *C. helmsii* is, or has become once established, truly a driver of ecological change within its invaded system that needs to be directly targeted by control methods. In these cases it is also vital to understand what conditions enabled the establishment of *C. helmsii* in the first instance and whether it needs certain conditions to enter first as a passenger or whether it entered from the start as a driver. In systems where *C. helmsii* is a driver from the start, resistance may be futile; in every other case long term successful control of *C. helmsii* may be greatly enhanced by first understanding its ecological role and then applying the most appropriate management approach.

References

Boston, H. L. and Adams, M. S. (1985) 'Seasonal diurnal acid rhythms in two aquatic crassulacean acid metabolism plants', *Oecologia*, vol 65, pp573–579

Bridge, T. (2005) 'Controlling New Zealand pygmyweed *Crassula helmsii* using hot foam, herbicide, and by burying at Old Moor RSPB Reserve, South Yorkshire, England', *Conservation Evidence*, vol 2, pp33–34

Brunel, S. (2009) 'Pathway analysis: Aquatic plants imported in 10 EPPO countries', *EPPO Bulletin*, vol 39, no 2 pp201–213

BSBI (Botanical Society of the British Isles) (2010) 'Hectad distribution map of *Crassula helmsii* in Britain and Ireland', www.bsbimaps.org.uk/atlas/main.php, accessed 14 October 2010

Bulleri, F., Balata, D., Bertocci, I., Tamburello, L. and Benedetti-Cecchi, L. (2010) 'The seaweed *Caulerpa racemosa* on Mediterranean rocky reefs: From passenger to driver of ecological change', *Ecology*, vol 91, no 8, pp2205–2212

CEH (Centre for Ecology and Hydrology) (2002) '*Crassula helmsii* focus on control – an update', Natural Environment Research Council, UK

CEH (2004) 'Information sheet 11: Australian swamp stonecrop', Centre for Aquatic Plant Management, CEH, Natural Environment Research Council, UK, www.ceh.ac.uk/sci_programmes/documents/AustralianSwampStonecrop.pdf, accessed 2 August 2011

Charlton, P. E., Gurney, M. and Lyons, G. (2010) 'Large-scale eradication of New Zealand pygmy weed *Crassula helmsii* from grazing marsh by inundation with seawater, Old Hall Marshes RSPB reserve, Essex, England', *Conservation Evidence*, vol 7, pp130–133

Cockerill, D. (1979) '*Crassula helmsii*', *Botanical Society of the British Isles News*, vol 21, p18

Dawson, F. H. (1994) 'Spread of *Crassula helmsii* in Britain', in L. X. C. De Waal, L. E. Child, P. M. Wade and J. H. Brock (eds) *Ecology and Management of Invasive Riverside Plants*, John Wiley and Sons Ltd, Chichester, UK, pp1–13

Dawson, F. H. (1996) '*Crassula helmsii*: Attempts at elimination using herbicides', *Hydrobiologia*, vol 340, no 1–3, pp241–245

Dawson, F. H. and Hallows, H. B. (1983) 'Practical applications of a shading material for macrophyte control of water courses', *Aquatic Botany*, vol 17, pp113–118

Dawson, F. H. and Warman, E. A. (1987) '*Crassula helmsii* (T.Kirk) Cockayne: Is it an aggressive alien aquatic plant in Britain?', *Biological Conservation*, vol 42, pp247–272

Deegan, B. M. and Ganf, G. G. (2008) 'The loss of aquatic and riparian plant communities: Implications for their consumers in a riverine food web', *Austral Ecology*, vol 33, pp672–683

Didham, R. K., Tylianakis, J. M., Hutchison, M. A., Ewers, R. M. and Gemmell, N. J. (2005) 'Are invasive species the drivers of ecological change?', *Trends in Ecology and Evolution*, vol 20, no 9, pp470–474

Environment Agency (2003) 'Guidance for the control of invasive weeds in or near fresh water' www.environment-agency.gov.uk/commondata/105385/booklet_895604.pdf, accessed 2 October 2010

Environment Agency (2010) 'Managing invasive non-native plants', http://publications.environment-agency.gov.uk/pdf/GEHO0410BSBR-e-e.pdf, accessed 2 October 2010

EPPO (European and Mediterranean Plant Protection Organization) (2007) '*Crassula helmsii*', *Bulletin*, vol 37, no 2, pp225–229

EU (European Union) (2001) 'Review report for the active substance diquat', European Commission Directorate-General Health and Consumer Protection, http://ec.europa.ec/food/plant/protection/evaluation/existactive/list1_diquat_en.pdf, accessed 2 October 2001

Gassmann, A., Cock, M. J. W., Shaw, R. and Evans, H. C. (2006) 'The potential for biological control of invasive alien aquatic weeds in Europe: A review', *Hydrobiologia*, vol 570, pp217–222

Gomes, B. (2005) 'Controlling New Zealand pygmyweed *Crassula helmsii* in field ditches and a gravel pit by herbicide spraying at Dungeness RSPB Reserve, Kent, England', *Conservation Evidence*, vol 2, p62

Hussner, A. (2009) 'Growth and photosynthesis of four invasive aquatic plant species in Europe', *Weed Research*, vol 49, pp506–515

Kelly, J. and Maguire, C. M. (2009) 'New Zealand Pigmyweed *(Crassula helmsii). Invasive Species Action Plan*', Prepared for NIEA and NPWS as part of Invasive Species Ireland, www.invasivespeciesireland.com, accessed 29 September 2010

Kirby, J. R. (1964) '*Crassula helmsii* in Britain', *The Cactus and Succulent Journal of Great Britain*, vol 26, pp15–16

Kirby, J. R. (1965) 'Notes on *Crassula helmsii*', *The Cactus and Succulent Journal of Great Britain*, vol 27, pp9–10

Klavsen, S. K. and Maberly, S. C. (2009) 'Crassulacean acid metabolism contributes significantly to the *in situ* carbon budget in a population of the invasive aquatic macrophyte *Crassula helmsii*', *Freshwater Biology*, vol 54, pp105–118

Klavsen, S. K. and Maberly, S. C. (2010) 'Effect of light and CO_2 on inorganic carbon uptake in the invasive aquatic CAM-plant *Crassula helmsii*', *Functional Plant Ecology*, vol 37, no 8, pp737–747

Küpper, H., Götz, B., Mijovilovich, A., Küpper, F. C. and Meyer-Klaucke, W. (2009) 'Complexation and toxicity of copper in higher plants. I. Characterization of copper accumulation, speciation and toxicity in *Crassula helmsii* as a new copper accumulator', *Plant Physiology*, vol 151, no 2, pp702–714

Langdon, S. J., Marrs, R. H., Hosie, C. A., Hugh, A., McAllister, K. M., Norris, J. and Potter, A. (2004) '*Crassula helmsii* in UK ponds: Effects on plant biodiversity and implication for newt conservation', *Weed Technology*, vol 18, pp1349–1352

Laundon, J. R. (1961) 'An Australasian species of *Crassula* introduced into Britain', *Watsonia*, vol 5, pp59–63

Leach, J. and Dawson, H. (1999) '*Crassula helmsii* in the British Isles – an unwelcome invader', *British Wildlife*, vol 10, no 4, pp234–239

Lockton, A. J. (2010). 'Species account: *Crassula helmsii*', Botanical Society of the British Isles, www.bsbi.org.uk, assessed 2 October 2010

MacDougall, A. S. and Turkington, R. (2005) 'Are invasive species the drivers or passengers of change in degraded ecosystems?', *Ecology*, vol 86, no 1, pp42–55

MacDougall, A. S., Gilbert, B. and Levine, J. M. (2009) 'Plant invasions and the niche', *Journal of Ecology*, vol 97, pp609–615

Margot, J. (1983) 'La végétation aquatiques des "springputten" en forest de Meerdael: Evolution et presences floristiques', *Naturalistes Belges*, vol 64, pp119–221

Newman, J. R. and Raven, J. A. (1995) 'Photosynthetic carbon assimilation by *Crassula helmsii*', *Oecologia*, vol 101, pp494–499

Preston, C. D. and Croft, J. M. (1997) *Aquatic Plants in Britain and Ireland*, Harley Books, Colchester, UK

Smith, G. G. and Marchant, N. G. (1961) 'A census of aquatic plants of Western Australia', *Western Australian Naturalist*, vol 8, p15

Swale, E. and Belcher, H. (1982) '*Crassula helmsii*, the swamp stonecrop, near Cambridge', *Nature Cambridgeshire*, vol 25, pp59–62

Toelken, H. R. (1981) 'The species of *Crassula* L. in Australia', *Journal of Adelaide Botanical Garden*, vol 3, pp57–90

Warren, J. E. (2008) 'Egyptian geese eating New Zealand pigmyweed', *British Birds*, vol 101, no 4, p200

Watson, W. R. C. (2001) 'An unwelcome aquatic invader!', *Worcestershire Record*, issue 10, www.wbrc.org.uk/Worcrecd/Issue10/invader.htm, accessed 20 September 2010

Wilton-Jones, G. (2005) 'Control of New Zealand pygmyweed *Crassula helmsii* by covering with black polythene at The Lodge RSPB Reserve, Bedfordshire, England', *Conservation Evidence*, vol 2, pp63

4

Eichhornia crassipes Mart. (Solms-Laubach) (water hyacinth)

Xiaoyun Pan, Amy M. Villamagna and Bo Li

Biology of Water Hyacinth

Eichhornia crasspies (water hyacinth) (Figure 4.1) is an erect, free-floating, perennial clonal herb within the family Pontederiaceae, and is widely recognized as one of the world's most invasive aquatic weeds (Holm et al, 1991). Leaf morphology varies, with individuals in low density populations producing short, stocky petioles that may support vertical growth, while those in high density populations have longer petioles (Center et al, 2002). Leaves form whorls of six to ten, and an individual plant essentially consists of a series of attached rosettes (Center and Spencer, 1981). The plant has lavender flowers with a central yellow patch, which are presented in clusters on a single spike (Center et al, 2002). Flowering occurs over 14 days, concluding with the spike dipping into the water and releasing seeds; each rosette may release over 3000 seeds in a single year (Barrett, 1980; Center et al, 2002; Lu et al, 2007). Seeds have a long viability (15–20 years) (Gopal, 1987), and germinate best in moist conditions or in shallow water (Center et al, 2002; Lu et al, 2007). Once germinated, flowering may occur within 10–15 weeks (Barrett, 1980; Center et al, 2002).

Population increase is mainly via vegetative propagation (clonal growth) in the introduced range, due to meristem differentiation (Center and Spencer, 1981; Center et al, 2002). Due to its great reproductive capacity, a single plant can potentially produce up to 140 million daughter ramets every year if space and other abiotic conditions are suitable (Ogutu-Ohwayo et al, 1997).

The dense and intricate roots of water hyacinth cause individual plants to intertwine and create extensive, floating mats (Figure 4.2). Under these floating mats, the roots provide a complex structure near the surface of the water that promotes the growth of epiphytic invertebrates and algae (Figure 4.3) (Brendonck et al, 2003; Toft et al, 2003; Rocha-Ramirez et al, 2007) and provides refuge to other invertebrates and fish (Arora and Mehra, 2003).

Ecological Niche

Habitat

Water hyacinth is found in fresh water, ranging from temporary pools to substantial water bodies such as rivers, lakes and reservoirs. Water levels in the Amazon basin fluctuate considerably because of heavy, seasonal rainfall, leading to frequent connections between water bodies such as pools and lakes. This in turn facilitates the dispersal, establishment and growth of water hyacinth, both vegetatively and via seed (Barrett, 1989).

The species is a generalist and is broadly tolerant of a wide range of water levels, hydraulic stresses, pH, temperatures, nutrients and pollution (Gopal, 1987), partly explaining its wide distribution and invasion

Figure 4.2 *Extensive mats of water hyacinth being sprayed with herbicide*

Figure 4.3 *Fibrous roots of water hyacinth that can form complex habitat for aquatic invertebrates and algae*

Figure 4.1 *Morphology and habitat of water hyacinth: (a) the whole plant; (b) flower; (c) a population invading a river in Shanghai; and (d) dead plants polluting water and influencing environmental aesthetics*

success. In particular, the species may form dense free-floating mats if water velocity is low; these are then dispersed to other locations during rises in water level

or increases in flow velocity in response to (particularly seasonal) rainfall events (Li et al, 2004). In this way, flow variation helps to facilitate spread of the species through connected water bodies.

Nutrient requirements

While water hyacinth is tolerant of a wide range of nutrient concentrations, from clean, nutrient-poor rivers and reservoirs to highly polluted, sewage lagoons (Li et al, 2004), plants tend to grow faster and stronger in nutrient-rich freshwater habitats (e.g. Wilson et al, 2005); eutrophication may be considered the key factor in maintaining water hyacinth populations and facilitating spread (Musil and Breen, 1977). Calcium concentration also affects water hyacinth growth as new ramet production is limited in calcium-poor mediums and extremely low calcium conditions tend to increase sprout mortality (Talatala, 1974; Desougi, 1984).

Temperature

Water hyacinth is generally not tolerant of cold conditions (see Wilson et al, 2005), with its minimum, optimum and maximum temperatures being 12°C, 25–30°C and 33–35°C, respectively (Ramey, 2001). The plant can withstand frosts as long as rhizomes are not affected, and buds can tolerate temperatures of –5°C. A lengthy period of cold weather may lead to mortality, but regeneration from the seed bank is a frequent occurrence, provided that water conditions are suitable for germination and establishment (Gao and Li, 2004).

Salinity and pH

Water hyacinth is generally tolerant of both acid and alkaline conditions (pH 4.0–10.0) (Haller and Sutton, 1973), though optimum growth is observed in neutral waters (Gopal, 1987; Center et al, 2002). Water hyacinth cannot tolerate salinity above 1.6 per cent (Parsons and Cuthbertson, 2001), which limits its range in estuaries and coastal areas.

History and Geographic Distribution

Water hyacinth is native to the Amazon basin, but has become naturalized in many tropical and subtropical regions (Holm et al, 1991). It has mainly been introduced to other countries because of its value as an ornamental plant, initially into Europe (around 1879), then the US (1884) and into Australia and southeast Asia in the early part of the 20th century (see Lu et al, 2007 and references therein). Lu et al (2007) report that the species can be found in 62 countries, between the longitudes of 40°N

and 40°S. Further details are given below for some of the most heavily infested regions.

Africa

In tropical Africa, water hyacinth first appeared in the Nile River, Egypt, during the 1890s. In 1903 it was found in Natal Province, South Africa, and by the 1930s it was present in several lakes in Zimbabwe. Records indicate that the plant colonized the Congo and White Nile rivers in Sudan during the 1950s. Since then, water hyacinth has been found in the Pangani River and Lake Tanganyika in Tanzania, Lake Chivero in Zimbabwe (Rommens et al, 2003) and many freshwater lagoons within Benin, Côte d'Ivoire, Ghana, Mali, Malawi, Niger, Nigeria, Kenya and Uganda (Batanounya and El-Fikya, 1975). In 1989, water hyacinth was found in Lake Victoria, bordered by Uganda, Kenya and Tanzania (Kateregga and Sterner, 2009). Throughout the early 1990s, water hyacinth has spread throughout the lake and covered close to 80 per cent of the lake's shoreline (2200ha) and an additional 1800ha was covered by mobile floating mats, mostly in northern bays (Twongo et al, 1995).

China

Water hyacinth was probably translocated from Japan to Taiwan of China in 1901, and thence to mainland China in the 1930s (Li and Xie, 2002; Lu et al, 2007). It then expanded north via the Yangtze River and its tributaries, partly due to cultivation for animal feed until the 1980s (Lu et al, 2007). After this point, it spread from the ponds and small rivers where it was intentionally cultivated into larger rivers and lakes, where it has become a noxious weed (Gao and Li, 2004). Lu et al (2007) note that by 2004, its distribution covered 19 of China's provinces, from Hainan Province to Shandong Province.

USA

Water hyacinth is thought to have been introduced into the US in 1884, probably in Florida (Center et al, 2002). It has subsequently spread to many of the southeastern US states, as well as California, Hawaii and the Virgin Islands. Populations were also found in Arizona, Arkansas and Washington, though Ramey (2001) considers these to have been eradicated.

Ecological and Socioeconomic Impacts

Water hyacinth can be responsible for a range of ecological and socioeconomic impacts (e.g. Sinkala et al, 2002; Villamagna and Murphy, 2010). When water hyacinth clusters into mats it creates a canopy at the water's surface, which effectively blocks light from penetrating through the water column. A decrease in light infiltration limits the photosynthetic potential of phytoplankton and other submerged plants. Areas of high water hyacinth density often experience a decrease in dissolved oxygen because water hyacinth prevents oxygen exchange at the water's surface (Hunt and Christiansen, 2000), and hence plants release oxygen into the atmosphere rather than the water (Meerhoff et al, 2006), and photosynthetic activity is suppressed beneath water hyacinth canopies (Rommens et al, 2003). While this can lead to a decrease in algal production and therefore an increase in water clarity, water hyacinth is also known to trap detritus within its roots and cause a localized decrease in transparency (Gopal, 1987).

Water hyacinth acts as a sink for nutrients and water-soluble contaminants (Aoi and Hayashi, 1996; Rommens et al, 2003); therefore it affects biogeochemical cycles within water bodies, as well as the fate of other contaminants (Villamagna and Murphy, 2010). Water hyacinth is a fast growing plant with a high evapotranspiration rate (Gopal, 1987; Delgado et al, 1993), which can lead to negative ecological and socioeconomic effects in water-scarce regions, and can even modify the microclimate of invaded systems (Ding et al, 1995).

In addition to water quality, water hyacinth is known to alter food webs and energy flows within invaded ecosystems (Villamagna and Murphy, 2010). Changes in the primary production base from phytoplankton to floating plant, common in the case of water hyacinth invasions, can cause far reaching changes in phytoplankton, plant, invertebrate, fish and bird communities throughout the ecosystem. A loss of free-floating phytoplankton due to a decline in light infiltration can decrease the food available to many zooplankton species; however, epiphytic populations may increase as water hyacinth roots provide suitable and abundant substrate for colonization. Villamagna and Murphy (2010) suggest that zooplankton response to water hyacinth is probably dependent on existing algal concentrations, physiochemical conditions, the presence of predators and the effect that water hyacinth may have on them, and the spatial distribution of water hyacinth. Aquatic macroinvertebrates tend to benefit from water hyacinth mats as they provide substrate for colonization and sometimes ample epiphytic food resources (Masifwa et al, 2001). Water hyacinth also provides macroinvertebrates and some fish refuge from predators and a safe place to spawn (Johnson and Stein, 1979).

Although, water hyacinth is not known to provide a major food source for fish, the complexity of the plants roots does increase invertebrate and small or juvenile fish populations and thus indirectly provides larger fish with prey. Water birds are also affected directly via changes in the availability and location of plant structure and indirectly via changes in prey populations (i.e. fish and invertebrates). The presence of water hyacinth mats can affect the distribution and behaviour of water birds (Villamagna and Murphy, 2010), which can indirectly affect the rest of the ecosystem through changes in top-down population pressures (i.e. predation and competition).

Key socioeconomic impacts result from both the direct effects of the water hyacinth invasion itself and from the costs and benefits of control programmes (discussed below). Direct impacts may include blocked waterways, clogged water pipes, increased evapotranspiration and consequent reduction of water resources (Gopal, 1987), loss of native plant species, a potential increase in livestock feed, and decreased catchability of some fish species (Kateregga and Sterner, 2009). Extensive mats throughout shallow areas may also impede fish access to breeding and nursery grounds and drive decreases in population growth (Twongo and Howard, 1998; Villamagna and Murphy, 2010). Other potential changes include an increase in insects (e.g. mosquitoes) as a result of slow moving water, and a loss of aesthetic appeal (see Figure 4.1). The severity of the socioeconomic impact is related to the particular societal uses of the water body, with those that support greater human use experiencing the greatest impact. In China alone, the cost of water hyacinth invading water bodies throughout the country was estimated to be $1.5 billion per year (Li et al, 2004).

In most cases, it is one or more of these impacts that initiate a control programme (discussed in more detail below). The impacts associated with control programmes are the obvious costs of labour, equipment and materials, but also the often overlooked impacts of the control efforts themselves. For example, widespread herbicide spraying or water hyacinth mowing can cause drastic, although short term, decreases in dissolved oxygen. These decreases in dissolved oxygen can have localized, lethal or sub-lethal effects on fish and invertebrates that could negatively affect fishermen, the fishery and individuals that depend on fish as a source of protein. As a result of the negative ecological impacts of some control programmes, chemical-based programmes can be subject to water use restrictions and require water quality monitoring that may negatively impact domestic, agricultural and industrial water users and increase the economic cost of control. In addition, biological control programmes often require multiple introduction events in order to augment the agent population to the point where it can suppress and reduce the existing water hyacinth population (Center et al, 1999). Repeated introductions, including the captive breeding or translocation of biological agents, can greatly add to the long term cost of control (Villamagna and Murphy, 2010). Even physical control programmes can become expensive as a result of repeated treatments, equipment maintenance, labour costs and, in the case of removal programmes, the cost of land where water hyacinth can be brought to decompose.

Management

Reducing the input of nutrients into water bodies is probably the most sustainable and long term solution to managing existing water hyacinth populations and decreasing the spread to new areas (Musil and Breen, 1977). Lu et al (2007) propose an integrated management framework that incorporates principles of landscape and ecosystem ecology into integrated pest management, including the use of different control techniques and long term monitoring at varying spatial and temporal scales. Unfortunately, watershed-scale nutrient management is challenging and in many cases not pursued (Li et al, 2004; Gao and Li, 2004). Instead, water hyacinth, like other aquatic plants, is commonly controlled by physical, chemical and biological means. Each method has its advantages and disadvantages (Seagrave, 1988), and no single method seems to be appropriate in all cases. The density, extent and configuration of water hyacinth mats should be considered as well as weather patterns and the desired use of the water body in question (Thayer and Ramey, 1986; Gibbons et al, 1994). Once established, permanent eradication is extremely difficult due to the persistence of

the plant and the longevity of the seed bank; therefore repeated control efforts are usually needed.

Physical control

Small, local water hyacinth populations can be controlled by manually removing plants from the water or by using mechanized cutting (e.g. mowers) or dredging equipment that may or may not remove plant materials from the water. Physical removal and in-situ cutting immediately open space, but plant materials left to decompose in the water can decrease dissolved oxygen and ultimately alter trophic structure as a result of changes in nutrient and carbon balances (Villamagna and Murphy, 2010). While removal of water hyacinth is optimal, plants contain 90 per cent water (Gopal, 1987) and therefore heavy, making transport and disposal difficult and expensive. For example, water hyacinth removed from a water body in Florida, US, was found to weigh 200 tons per acre (Harley et al, 1996). In addition to the sheer weight of the plants, water hyacinth from contaminated waters deserves special attention to minimize the possible negative effects on human health and the potential spread of contamination into terrestrial ecosystems. This can further increase the cost of removal and complicate control efforts (Thayer and Ramey, 1986). Physical control programmes range from single-man removal or cutting to large-scale dredging operations and therefore the cost of physical control also varies drastically. In contrast to chemical control programmes, discussed below, physical control programmes are selective and are not generally accompanied by water use restrictions.

Chemical control

Chemical control programmes apply herbicides including glyphosate (Roundup), diquat, and 2,4-D amine to water hyacinth plants using a range of spraying apparatuses (Seagrave, 1988; Gutierrez et al, 1994; Lugo et al, 1998). Herbicides are considered less expensive and less labour intensive when compared to physically controlling the same amount of water hyacinth (Guitierrez-Lopez, 1993), but the actual cost of a chemical control plan depends on the equipment used to apply herbicides (e.g. backpack sprayer, boat, helicopter) and the frequency of treatments necessary (Villamagna and Murphy, 2010). The efficacy of herbicides varies according to how quickly it may be distributed through the plant to the roots; this may in part relate to the age of the plant, as younger plants translocate herbicidal chemicals through their system faster, though are generally less susceptible to the effects of the herbicide (Sculthorpe, 1985).

Aerially applied 2,4-D amine is considered to be the most effective chemical control and the efficacy is highest during periods of hot weather when translocation is most rapid (Gopal, 1987). 2,4-D amine is non-selective and works on broad-leaved plants and some monocots like water hyacinth, though it does not affect grasses. Toxicity is largely dependent on the formulation of 2,4-D amine. Ester formulations are more toxic to fish and aquatic invertebrates than salt formulations, therefore salt formulations are most appropriate for controlling water hyacinth (Tu et al, 2001). Glyphosate is also commonly used, though this is a non-selective herbicide that can kill plants within eight weeks when applied at a 2kg/ha dose, and it has significant non-target impacts (Gopal, 1987).

Although generally non-toxic to invertebrates and fish when sprayed at appropriate concentrations, non-selective herbicides kill algae and non-target plants (Seagrave, 1988) and can have far reaching effects on the ecosystem and increase the ecological costs of control (Villamagna and Murphy, 2010). Unlike harvesting efforts, plants treated with herbicides are left to sink and decompose, whereupon they release nutrients and contaminants previously absorbed by the plants. The decomposition process, especially in areas of extensive spraying, can lead to decreased dissolved oxygen concentration throughout the water column and especially in the benthic zone (Greenfield et al, 2007). Many negative effects associated with the mass decomposition of sinking water hyacinth plants can be avoided if small patches are sprayed with sufficient time between spray events to allow the ecosystem to rebound; however, this adds to the cost of control and is often overlooked to speed up the process.

Biological control

Biological control relies on the introduction of an organism that applies top-down pressure on the water hyacinth population. It is less toxic than chemical control, less labour intensive than physical controls, and generally considered to be a more long term solution to water hyacinth infestation than other options. While over 100 insect and fungi species have been screened as potential control agents, *Neochetina*

eichhorniae and *N. bruchi*, weevils native to South America, are the two most common biological controls for water hyacinth (Center et al, 1999; Sosa et al, 2007). *Neochetina eichhorniae* has been introduced to many countries and has become established throughout much of water hyacinth's range in southeast US. *Neochetina bruchi* is well established in Florida (Center et al, 1999). Although less common in Louisiana, Texas and California, it has been credited with suppressing water hyacinth populations in California (Center et al, 2002). Both weevils decrease water hyacinth populations by feeding on leaves, and by reducing buoyancy, which may cause plants to sink (Wilson et al, 2007; Villamagna and Murphy, 2010). Efficacy varies, however, and Center and Dray (2010) note that plant quality in particular influences the populations of these control agents and consequently their effectiveness; they therefore recommend partial herbicidal treatment to remove weaker water hyacinth plants and improve the quality of those remaining, thereby improving chances of successful biological control.

Niphographa albiguttalis, also known as *Sameodes albiguttalis* or water hyacinth moth is considered established in Florida, Louisiana and Mississippi. Reports suggest that the moth affects young and developing water hyacinth mats, but otherwise control is considered inconsistent and moderate (Julien and Griffiths, 1988). Other insects and plant pathogens have also been used with varying success throughout the introduced range (Coetzee et al, 2007a).

Although fish, such as grass carp and some tilapia species, are common biological control agents for non-native plants, there are no known fish that feed on water hyacinth enough to suppress populations (Gopal, 1987; Seagrave 1988). Considerable effort has been put towards developing fungal biological controls or mycoherbicides, including *Uredo eichhorniae*, a highly specialized, water hyacinth obligate fungal pathogen found throughout Brazil, and *Alternaria eichhorniae*, a globally distributed fungal species that was considered to have potential for future water hyacinth control (Charudattan et al, 1996; Barreto et al, 2000). However, Shearer (2008) notes that *U. eichhorniae* may find alternative hosts, and as such permission for release in the US (the main driver of research into this species) has not been forthcoming, though applications to study the species in quarantine have been made. Ray et al (2008) have demonstrated that combinations of fungal pathogens are most effective in causing disease in water hyacinth populations, but that *A. eichhorniae* is negatively affected by the presence of other pathogens, potentially limiting its use.

As with any biological control programme, host specificity is an important consideration. Agents that focus on the target plants but are able to utilize other resources when the host plant is in low density will be most successful and least likely to cause further ecological problems. The population size of the introduced control agent is an important consideration. For example, in some cases where *Neochetina* spp. were introduced, effective control of water hyacinth was not achieved for many years (Harley, 1990; Hill and Olckers, 2001; Coetzee et al, 2007a, 2007b; Wilson et al, 2007). In many situations, an integrated approach or combination of methods is used. Physical and chemical controls may provide quick relief from infestation, but they also pose potential ecological concerns when applied at large scales within a short period of time. In some cases, biological control is preceded by mechanical or chemical control methods to make initial conditions more suitable for biological control agents (Adekoya et al, 1993; Villamagna and Murphy, 2010). However, herbicides can negatively influence biological agent populations by reducing the availability of the target and focal species beyond sustainable levels.

Overall, the success of the control programme will depend on local conditions, the desired uses of the water body, and the size of water hyacinth populations. The effects of control programmes on local ecological and socioeconomic conditions are also an important consideration. While chemical, biological and some physical control efforts leave plants to decompose in the water, the ecological impacts are largely determined by the timing and spatial extent of decomposition. For example, large-scale cutting and herbicide applications are more likely to cause water quality (e.g. dissolved oxygen) or ecological (e.g. decrease in algae and invertebrate populations) problems than biological control or manual removal (Villamagna and Murphy, 2010).

Future Challenges

Potential response to global climate change

In general, there are strong interactions between climate change and plant invasions, including water

hyacinth. Future global climate is predicted to be warmer in many regions due to anthropogenic impacts, and water hyacinth, among other warm climate species, is likely to be affected by broad-scale climatic changes (Hellmann et al, 2008). For example, sexual recruitment in water hyacinth is largely limited by an absence of suitable conditions for seeding establishment in introduced habitats (Barrett, 1980). Although seeds are produced in introduced ranges, the absence of suitable germination sites means that vegetative reproduction dominates; a situation that may change as the hydrological regimes of freshwater systems adjust to changing climates (Barrett, 2000). In turn, genetic diversity may be enhanced in introduced water hyacinth populations, potentially influencing control efforts, as more control-tolerant genotypes may emerge (Barrett, 2000). Furthermore, water hyacinth populations may encroach into higher latitudes as a result of warmer winter temperatures and more pronounced seasonality. Climate change may also impact existing populations via changes in precipitation patterns and nutrient fluxes. In response to changing water hyacinth distributions and demographics, current management practices may need to be reconsidered and new strategies need to be developed.

Water hyacinth utilization and further invasions

Utilization as a form of control is a potential management strategy, and water hyacinth has been used for its ornamental value, as fodder, as biomass energy, as fertilizer, and as a material for building and other crafts (e.g. Mishima et al, 2008; Jafari, 2010). In particular, it has gained attention as a potential treatment for wastewater because it absorbs nutrients and contaminants effectively, due to fast reproductive cycles (Jafari, 2010). This is often difficult to implement for plant species and remains controversial (e.g. Coetzee and Hill, 2009), as utilization may facilitate species spread. However, where populations are well established and control is problematic, management efforts may seek to harness the plant's positive effects (e.g. nutrient and contaminant absorption, formation of complex habitat) and minimize the costs and damages associated with widespread, high density mats.

References

Adekoya, B. B., Ugwuzor, G. N., Olurin, K. B., Sodeinde, O. A. and Ekpo, O. A. (1993) 'A comparative assessment of the methods of control of water hyacinth infestation with regards to fish production', in A. A. Eyo and A. M. Balogun (eds) *Annual Conference of the Fisheries Society of Nigeria (FISON), Abeokuta (Nigeria)*, Proceedings of the Annual Conference of the Fisheries Society of Nigeria (FISON), 16–20 November 1992, pp181–184

Aoi, T. and Hayashi, T. (1996) 'Nutrient removal by water lettuce (*Pisitia stratiotes*)', *Water Science and Technology*, vol 34, pp407–412

Arora, J. and Mehra, N. K. (2003) 'Species diversity of planktonic and epiphytic rotifers in the backwaters of the Delhi segment of the Yamuna River, with remarks on new records from India', *Zoological Studies*, vol 42, pp239–247

Barreto, R., Charudattan, R., Pomella, A. and Hanada, R. (2000) 'Biological control of 346 neotropical aquatic weeds with fungi', *Crop Protection*, vol 19, pp697–703

Barrett, S. C. H. (1980) 'Sexual reproduction in *Eichhornia crassipes* (water hyacinth). II. Seed production in natural populations', *Journal of Applied Ecology*, vol 17, pp113–124

Barrett, S. C. H. (1989) 'Waterweed invasions', *Scientific American*, vol 260, pp90–97

Barrett, S. C. H. (2000) 'Microevolutionary influences of global changes on plant invasions', in H. A. Mooney and R. J. Hobbs (eds) *Invasive Species in a Changing World*, Island Press, Washington, DC, pp115–139

Batanounya, K. H. and El-Fikya, A. M. (1975) 'The water hyacinth (*Eichhornia crassipes* Solms) in the Nile system', *Egypt Aquatic Botany*, vol 1, pp243–252

Brendonck, L., Maes, J., Rommens, W., Dekeza, N., Nhiwatiwa, T., Barson, M., Callebaut, V., Phiri, C., Moreau, K., Gratwicke, B., Stevens, M., Alyn, N., Holsters, E., Ollevier, F. and Marshall B. (2003) 'The impact of water hyacinth (*Eichhornia crassipes*) in a eutrophic subtropical impoundment (Lake Chivero, Zimbabwe). II. Species diversity', *Archiv für Hydrobiologie*, vol 158, pp389–405

Center, T. D. and Dray, F. A. (2010) 'Bottom-up control of water hyacinth weevil populations: Do the plants regulate the insects?', *Journal of Applied Ecology*, vol 47, pp329–337

Center, T. D. and Spencer, N. R. (1981) 'The phenology and growth of water hyacinth (*Eichhornia crassipes* (Mart.) Solms) in a eutrophic north-central Florida lake', *Aquatic Botany*, vol 10, pp1–32

Center, T. D., Dray, F. A., Jubinsky, G. P., Grodowltz, M., de Anda, J., Shear, H., Maniak, U. and Riedel, G. (1999) 'Biological control of water hyacinth under conditions of maintenance management: Can herbicides and insects be integrated?', *Environmental Management*, vol 23, pp241–256

Center, T. D., Hill, M. P., Cordo, H. and Julien, M. H. (2002) 'Water hyacinth', in R. V. Driesche, S. Lyon, B. Blossey, M. Hoddle and R. Reardon (eds) *Biological Control of Invasive Plants in the Eastern United States*, USDA Forest Service, Washington, DC, pp41–64

Charudattan, R. E., Labrada, R., Center, T. D., Kelly-Begazo, C. (eds) (1996) *Strategies for Water Hyacinth Control*, FAO, Rome

Coetzee, J. A. and Hill, M. P. (2009) 'Management of invasive aquatic plants', in M. N. Clout and P. A. Williams (eds) *Invasive Species Management: A Handbook of Principles and Techniques*, Oxford University Press, Oxford, pp141–152

Coetzee, J. A., Byrne, M. J. and Hill, M. P. (2007a) 'Predicting the distribution of *Eccritotarsus catarinensis*, a natural enemy released on water hyacinth in South Africa', *Entomologia Experimentalis et Applicata*, vol 125, pp237–247

Coetzee, J. A., Byrne, M. J. and Hill, M. P. (2007b) 'Impact of nutrients and herbivory by *Eccritotarsus catarinensis* on the biological control of water hyacinth, *Eichhornia crassipes*', *Aquatic Botany*, vol 86, pp179–186

Delgado, M., Bigeriego, M. and Guardiola, E. (1993) 'Uptake of Zn, Cr and Cd by water hyacinths', *Water Research*, vol 27, no 2, pp269–272

Desougi, L. A. (1984) 'Mineral nutrient demands of the water hyacinths (*Eichhornia crassipes*) (Mart.) Solms in the White Nile', *Hydrobiologia*, vol 110, pp99–108

Ding, J. Q., Wang, R. and Fan, Z. N. (1995) 'Distribution and infestation of water hyacinth and the control strategy in China', *Journal of Weed Science*, vol 9, pp49–51

Gao, L. and Li, B. (2004) 'The study of a specious invasive plant, water hyacinth (*Eichhornia crassipes*): Achievements and challenges', (in Chinese), *Acta Phytoecologica Sinica*, vol 28, pp735–752

Gibbons, M. V., Gibbons, H. Jr and Sytsma, M. D. (1994) *A Citizen's Manual for Developing Integrated Aquatic Vegetation Management Plans*, Washington State Department of Ecology, Olympia, WA

Gopal, B. (1987) *Water Hyacinth*, Elsevier, Amsterdam

Greenfield, B. K., Siemering, G. S., Andrews, J. C., Rajan, M., Andrews, S. P. and Spencer, D. F. (2007) 'Mechanical shredding of water hyacinth (*Eichhornia crassipes*): Effects on water quality in the Sacramento-San Joaquin River Delta, California', *Estuaries and Coasts*, vol 30, pp627–640

Guitierrez-Lopez, E. (1993) 'Effect of glyphosate on different densities of water hyacinth', *Journal of Aquatic Plant Management*, vol 31, pp255–257

Gutierrez, E., Arreguin, F., Huerto, R. and Saldana, P. (1994) 'Aquatic weed control', *International Journal of Water Resources Development*, vol 10, pp291–312

Haller, W. T. and Sutton, D. L. (1973) 'Effects of pH and high phosphorus concentrations on growth of water hyacinth', *Hyacinth Control Journal*, vol 11, pp59–61

Harley, K. L. S. (1990) 'Management of water hyacinth (*Eichhornia crassipes*)', *Zimbabwe Science News*, vol 24, no 4–6, pp40–41

Harley, K. L., Julien, M. H. and Wright, A. D. (1996) 'Water hyacinth: A tropical worldwide problem and methods for its control', in H. Brown, G. W. Cussans, M. D. Devine, et al (eds) *Proceeding of the 2nd International Weed Control Congress*, Copenhagen, Denmark, pp639–644

Hellmann, J. J., Byers, J. E., Bierwagen, B. G. and Dukes, J. S. (2008) 'Five potential consequences of climate change for invasive species', *Conservation Biology*, vol 22, pp534–543

Hill, M. P. and Olckers, T. (2001) 'Biological control initiatives against water hyacinth in South Africa: Constraining factors, success and new courses of action', in M. H. Julien, M. P. Hill, T. D. Center and J. Q. Ding (eds) *Biological and Integrated Control of Water Hyacinth*, Eichhornia crassipes, ACIAR Proceedings 102, Australian Centre for International Agricultural Research, Canberra, Australia, pp33–38

Holm, L. G., Plucknett, D. L., Pancho, J. V. and Herberger, J. P. (1991) *The World's Worst Weeds: Distribution and Biology*, Kreiger Publishing Co., Malabar, Florida

Hunt, R. J. and Christiansen, I. H. (2000) 'Understanding dissolved oxygen in streams: Information kit', *CRC Sugar Technical Publication (CRC Sustainable Sugar Production)*, p27

Jafari, N. (2010) 'Ecological and socio-economic utilization of water hyacinth (*Eichhornia crassipes* Mart Solms)', *Journal of Applied Sciences and Environmental Management*, vol 14, no 2, pp43–49

Johnson, D. L. and Stein, R. A. (eds) (1979) *Response of Fish to Habitat Structure in Standing Water*, American Fisheries Society

Julien, M. H. and Griffiths, M. W. (1998) *Biological Control of Weeds: A World Catalogue of Agents and their Target Weeds*, 4th edition, CABI Publishing, New York

Kateregga, E. and Sterner, T. (2009) 'Lake Victoria fish stocks and the effects of water hyacinth', *The Journal of Environment and Development*, vol 18, pp62–78

Li, B., Liao, C. Z., Gao, L., Luo, Y. Q. and Ma, Z. J. (2004) 'Strategic management of water hyacinth (*Eichhornia crassipes*), an invasive alien plant', *Journal of Fudan University*, vol 43, pp267–274

Li, Z. and Xie, Y. (2002) *Invasive Alien Species in China*, Chinese Forestry Press, Beijing

Lu, J. B., Wu, J. G., Fu, Z. H. and Zhu, L. (2007) 'Water hyacinth in China: A sustainability science-based management framework', *Environmental Management*, vol 40, pp823–830

Lugo, A., Bravo-Inclan, L. A., Alcocer, J., Gaytan, M. L., Oliva, M. G., Sanchez , M. d. R., Chavez, M. and Vilaclara, G. (1998) 'Effect on the planktonic community of the chemical program used to control water hyacinth (*Eichhornia crassipes*) in Guadalupe Dam, Mexico', *Aquatic Ecosystem Health and Management*, vol 1, pp333–343

Masifwa, W. F., Twongo, T. and Denny, P. (2001) 'The impact of water hyacinth, *Eichhornia crassipes* (Mart.) Solms on the abundance and diversity of aquatic macroinvertebrates along the shores of northern Lake Victoria, Uganda', *Hydrobiologia*, vol 452, pp79–88

Meerhoff, M., Fosalba, C., Bruzzone, C., Mazzeo, N., Noordoven, W. and Jeppesen, E. (2006) 'An experimental study of habitat choice by Daphnia: Plants signal danger more than refuge in subtropical lakes', *Freshwater Biology*, vol 51, pp1320–1330

Mishima, D., Kuniki, M., Sei, K., Soda, S., Ike, M. and Fujita, M. (2008) 'Ethanol production from candidate energy crops: Water hyacinth (*Eichhornia crassipes*) and water lettuce (*Pistia stratiotes* L.)', *Bioresource Technology*, vol 99, pp2495–2500

Musil, C. F. and Breen, C. M. (1977) 'Application of growth kinetics to control of *Eichhornia-Crassipes* (Mart) Solms – through nutrient removal by mechanical harvesting', *Hydrobiologia*, vol 53, pp165–171

Ogutu-Ohwayo, R., Hecky, R. E., Cohen, A. S. and Kaufman, L. (1997) 'Human impacts on the African Great Lakes', *Environmental Biology of Fishes*, vol 50, pp117–131

Parsons, W. T. and Cuthbertson, E. G. (2001) *Noxious Weeds of Australia*, CSIRO Publishing, Collingwood, Australia

Ramey, V. (2001) '*Eichhornia crassipes*', Centre for Aquatic and Invasive Plants, University of Florida, www.plants.ifas.ufl.edu/node/141, accessed 24 January 2011

Ray, P., Kumar, S. and Pandey, A. K. (2008) 'Efficacy of pathogens of water hyacinth (*Eichhornia crassipes*) singly and in combination for its biological control', *Journal of Biological Control*, vol 22, no 1, pp173–177

Rocha-Ramirez, A., Ramirez-Rojas, A., Chavez-Lopez, R. and Alcocer, J. (2007) 'Invertebrate assemblages associated with root masses of *Eichhornia crassipes* (Mart.) Solms-Laubach 1883 in the Alvarado Lagoonal System, Veracruz, Mexico', *Aquatic Ecology*, vol 41, pp319–333

Rommens, W., Maes, J., Dekeza, N., Inghelbrecht, P., Nhiwatiwa, T., Holsters, E., Ollevier, F., Marshall, B. and Brendonck, L. (2003) 'The impact of water hyacinth (*Eichhornia crassipes*) in a eutrophic subtropical impoundment (Lake Chivero, Zimbabwe). I. Water quality', *Archiv für Hydrobiologie*, vol 158, pp373–388

Sculthorpe, C. D. (1985) *The Biology of Aquatic Vascular Plants*, Koeltz Scientific Books, Konigstein, West Germany

Seagrave, C. (1988) *Aquatic Weed Control*, Fishing New Books, Surrey, England

Shearer, J. F. (2008) 'Is classical biocontrol using fungi a viable option for submersed aquatic plant management?', *Journal of Aquatic Plant Management*, vol 46, pp202–205

Sinkala, T., Mwase, E. T. and Mwala, M. (2002) 'Control of aquatic weeds through pollutant reduction and weed utilization: A weed management approach in the lower Kafue River of Zambia', *Physics and Chemistry of the Earth*, vol 27, pp983–991

Sosa, A. J., Cordo, H. A. and Sacco, J. (2007) 'Preliminary evaluation of *Megamelus seutellaris* Berg (Hemiptera: Delphacidae), a candidate for biological control of water hyacinth', *Biological Control*, vol 42, pp129–138

Talatala, R. L. (1974) 'Some aspects of the growth and reproduction of water hyacinth (*Eichhornia crassipes* (Mart.) Solms)', *Southeast Asian Workshop on Aquatic Weeds*, 25–29 June, Malang

Thayer, D. and Ramey, V. (1986) 'Mechanical harvesting of aquatic weeds', A technical report from the Florida Department of Natural Resources (now the Department of Environmental Protection) Bureau of Aquatic Plant Management, Florida

Toft, J. D., Simenstad, C. A., Cordell, J. R. and Grimaldo, L. F. (2003) 'The effects of introduced water hyacinth on habitat structure, invertebrate assemblages, and fish diets', *Estuaries*, vol 26, pp746–758

Tu, M., Hurd, C. and Randall, J. M. (2001) *Weed Control Methods Handbook, The Nature Conservancy*, http://tncweeds.ucdavis.edu, version April 2001

Twongo, T. and Howard, G. (1998) 'Ways with weeds', *New Scientist*, vol 159, p57

Twongo, T., Bugenyi, F. W. B. and Wanda, F. (1995) 'The potential for further proliferation of water hyacinth in Lakes Victoria, Kyoga and Kwania and some aspects for research', *African Journal of Tropical Hydrobiology*, vol 6, pp1–10

Villamagna, A. M. and Murphy, B. R. (2010) 'Ecological and socio-economic impacts of invasive water hyacinth (*Eichhornia crassipes*): A review', *Freshwater Biology*, vol 55, pp282–298

Wilson, J. R., Holst, N. and Rees, M. (2005) 'Determinants and patterns of population growth in water hyacinth', *Aquatic Botany*, vol 81, pp51–67

Wilson, J. R. U., Ajuonu, O., Center, T. D., Hill, M. P., Julien, M. H., Katagira, F. F., Neuenschwander, P., Njoka, S. W., Ogwang, J., Reeder, R. H. and Van, T. (2007) 'The decline of water hyacinth on Lake Victoria was due to biological control by *Neochetina spp*', *Aquatic Botany*, vol 87, pp90–93

5

Heracleum mantegazzianum Sommier & Levier (giant hogweed)

Jan Pergl, Irena Perglová and Petr Pyšek

Introduction

There are not many invasive plant species in Europe as conspicuous as *Heracleum mantegazzianum* (giant hogweed). As a tall European herbaceous species with the ability to cause injuries to human skin, the plant reminds many people of John Wyndham's famous creatures from his book *The Day of the Triffids*. Like the triffids, *H. mantegazzianum* was first brought to gardens and parks all over Europe and remained inconspicuous for years. However, soon its harmful effects on human health and ability to escape from cultivation to colonize natural or semi-natural communities were discovered and *H. mantegazzianum* became a serious problem in Europe (Nielsen et al, 2005). The size of individual plants and their exotic appearance made this species popular among botanists and ecologists since the beginning of its spread, which resulted in a good knowledge of the history of its invasion as well as its ecology and biology (e.g. Ochsmann, 1996; Tiley et al, 1996; Page et al, 2006; Pyšek et al, 2007a, 2008). Giant hogweed receives attention from media in many countries and is often presented as a highly illustrative example of a serious plant invader.

The genus *Heracleum* includes about 65 species, the highest diversity of which is found in the Caucasus and China (Jahodová et al, 2007a). Three of them occur as invasive in Europe: *H. mantegazzianum*, *H. sosnowskyi* and *H. persicum* (Jahodová et al, 2007b) (Figure 5.1).

Despite intensive recent research, the taxonomic position of individual species has not been fully resolved, which makes the identity of some historical records unclear due to historically poor accessibility of regions within the native range, isolated research in Russia, and possible hybridization among hogweed species. *Heracleum mantegazzianum* is native to the western part of Greater Caucasus (Georgia, Russia), while the other two closely related invasive tall hogweeds came to Europe from Iran, Iraq and Turkey (*Heracleum persicum*, invasive in northern Europe in Scandinavia), and from the central part of Greater Caucasus (*Heracleum sosnowskyi*, now occurring in some of the former Soviet republics, for example Ukraine, and also reported from Hungary). All three species were introduced to Europe as garden ornamentals in the 19th century. Two of them (*H. mantegazzianum* and *H. sosnowskyi*) were also grown in some regions of Central and Eastern Europe as pasture crops; this practice resulted in large areas of abandoned fields being infested with hogweeds in the former USSR (for example Ukraine, Latvia) (Buttenschon and Nielsen, 2007; Jahodová et al, 2007a). These attempts, however, failed because the plants pose a high risk to human health due to their phototoxic sap, which is also dangerous to animals, and in the case of *H. sosnowskyi* due to the fact that its anise smell tainted dairy products.

The first record for *H. mantegazzianum* for Europe comes from a seed list of the Royal Botanical Gardens at Kew in the UK in 1817 (but under the name

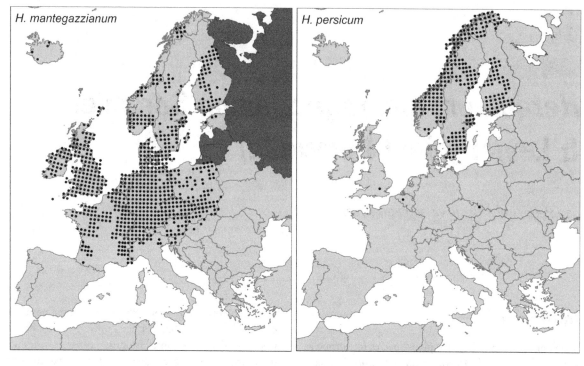

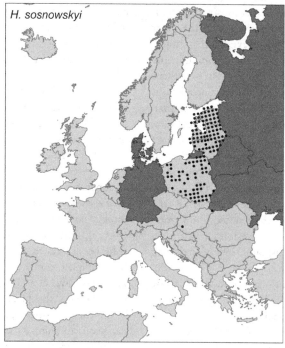

Note: Occurrence in 50 × 50km grid cells is shown. Dark-grey shading indicates countries from which the species is reported but detailed distribution in grid cells is not known.

Figure 5.1 *Distribution of the three invasive* Heracleum *species in Europe*

H. giganteum) (Jahodová et al, 2007a). Later it was introduced as an ornamental species to many European countries: The Netherlands, Switzerland, Germany, Ireland, Czech Republic and several others (Pyšek et al, 2008). Not surprisingly the species soon became fairly popular because of its magnificent stature, conspicuous flowering stems with showy large umbels of flowers, large and decorative leaves and overall exotic appearance, and was widely planted in many botanical as well as private gardens. At present it is reported as naturalized (forming self-reproducing populations in the wild) in 19 countries in Europe. In the second half of the 20th century it became naturalized also in North America, in the US and Canada; the first herbarium specimen documenting its occurrence in Canada was collected in 1949 (Page et al, 2006). In the North American continent it is distributed on the western coast in British Columbia, from the area of Vancouver and to Oregon. On the eastern coast, the most northerly documented occurrence of *H. mantegazzianum* is from central Ontario and the Island of Newfoundland; nevertheless the core of its distribution is from the area near the Canadian and US border (Toronto, Montreal) (Page et al, 2006). In the south, the distribution reaches to Pennsylvania along the coast, and to Illinois inland (USDA, 2010).

The enormous capacity of this species to escape from cultivation can be illustrated from two European countries with detailed data available on its invasion history and distribution. In the UK, the first record of escape from cultivation dates only 11 years after the species was introduced to Kew Gardens (1828, Cambridgeshire). Almost the same lag between the first record of planting and subsequent fast escape occurred in the Czech Republic, where it is reported as an ornamental planted in the garden of the Chateau Kynžvart in 1862 and only 15 years later as growing in the wild not far from the site of introduction. This speed of escape is symptomatic of the rapid spread that followed in many European countries (Pyšek et al, 2007b). The invasion of *H. mantegazzianum* is an exemplary case of the spread of an alien plant species, realized through a combination of deliberate planting that created foci in more distant areas, from which the species spread to the surrounding landscapes. Such foci have been crucial for invasion at large geographical scales, from regional to continental, where the invasion was driven by human-mediated dispersal, while at the

small, local scale, invasion dynamics are driven by species traits and availability of suitable habitats that the species can colonize (Pyšek et al, 2008).

Ecological Niche of the Species

In its native range *H. mantegazzianum* occurs in a wide range of habitats ranging from natural meadows on mountain slopes in the forest altitudinal belt to human-disturbed habitats at various altitudes, including lowland (Otte et al, 2007). In the relatively undisturbed sites, *H. mantegazzianum* forms a part of species-rich subalpine tall herb communities on wet and nutrient-rich soils and occurs in low densities, while in less natural habitats around settlements, along roads or on pastures, even in high mountains, it grows in dense populations resembling those in invaded regions. In Europe, *H. mantegazzianum* invades in particular semi-natural grassland communities, nutrient-rich sites, forest edges and human-made habitats (Pyšek and Pyšek, 1995; Thiele et al, 2007). The wide range of invaded habitats reflects the presence of historical sites in which it was planted such as old gardens or parks, their neighbourhood, and transport corridors along which seeds are dispersed, such as rivers, railways or road verges. Such places then serve as sources for subsequent invasion into adjacent semi-natural habitats (meadows, forest edges) and into margins of pastures and agricultural fields. If not constrained by local conditions, giant hogweed forms extensive populations harbouring thousands of individuals (Figure 5.2); however, more often it is found in smaller populations of a few individuals along linear landscape structures such as roadsides, watercourses or in abandoned gardens and parks and unmanaged meadows. Distribution data from the Czech Republic show that the actual distribution depends on mean January temperature and density of human population in the region; it is less frequent in less populated areas with a warm winter (Pyšek et al, 1998).

Heracleum mantegazzianum and its two congeners are the largest herbaceous plants in Europe, which makes them strong competitors in the plant communities invaded. Flowering stems of *H. mantegazzianum* reach up to 5m in height (commonly 2–4m), leaves are up to 2.5m long, terminal inflorescences up to 80cm in diameter, and the flowering stem up to 15cm in

Figure 5.2 *(a) The populations of* Heracleum mantegazzianum *in its native range in the forest belt of the Caucasus consist of scattered individual plants growing in species-rich tall herb communities; (b) in the invaded range in central Europe, it forms extensive stands with a high cover, suppressing species diversity of native plant communities*

diameter. The species is usually monocarpic: it persists in a form of a vegetative rosette for several years, then flowers and dies in the same year (Pergl et al, 2007). Although 12-year-old individuals have been found, *H. mantegazzianum* usually lives for 3 to 5 years in the invaded range. However, the lifespan in the native and invaded distribution range differs: plants from unmanaged sites in central Europe were shown to flower at an earlier age than plants from the native range of the Caucasus, which may increase their competitive ability during invasion by speeding up the

lifecycle (Pergl et al, 2006). *Heracleum mantegazzianum* reproduces by seed and its enormous fecundity is one of the key factors contributing to its invasion potential. The values of 10,000–20,000 seeds per plant seem to be the most common average in Europe, with maxima occasionally reaching around 50,000 seeds; the previously reported maximum estimate of 100,000 seeds produced by a single plant seems to be improbable (Perglová et al, 2006, 2007). Flowering plants of *H. mantegazzianum* possess a distinct architecture and under favourable conditions strong plants can develop several shoots that arise from the base of the main flowering stem. If severed, the plant is able to regenerate rapidly and develop several branches that stand for the main shoot (Pyšek et al, 2007c). The flowering plant usually bears many inflorescences (compound umbels). Each compound umbel consists of a number of umbellets, simple umbels that bear a large number of small flowers. The flowers are either hermaphrodite or male. The proportion of male flowers increases in higher order umbels, while the terminal umbel usually contains only hermaphrodite flowers (Perglová et al, 2006). The individual flowers are unspecialized, pollinated by a large number of pollinators, with bees, wasps, flies and beetles being the most frequent visitors (Grace and Nelson, 1981).

The species is considered to be self-compatible and protandrous, which promotes out-crossing. However, male and female flowering phases within individual umbels overlap to some extent; selfing is therefore possible and at the level of whole plant it is quite common. As a result, the ability for selfing together with full self-compatibility can play a significant role in invasion as a new population can be formed following the dispersal of a single plant. This is emphasized by the fact that there is no difference in the germination ability of selfed and out-crossed seed, or in the relative growth rate of seedlings (Perglová et al, 2007).

The seeds of *H. mantegazzianum* need to experience cold and wet stratification before germination to overcome dormancy. The length of stratification period required depends on outside conditions but at a constant temperature of 5°C it takes about two months. Over 90 per cent of seed germinated at optimum laboratory conditions (Moravcová et al, 2006, 2007). Temporal patterns of germination and persistence of seeds in the soil show that only a small proportion of seed persists in the soil until the following year, forming thus a short term persistent seed bank. The amount of living seed in the soil decreases sharply from almost 10 per cent after one year, to 3 per cent after two years and 0.5 per cent after three years (Moravcová et al, 2007). Seeds start to germinate early in the spring, first cotyledon leaves emerge soon after the snow melts and in the following two weeks the density of seedlings reaches maximum. Only a small proportion of seedlings survive to the stage of vegetative rosettes and only a few to the flowering stage. The density decreases from about 700–1700 seedling per square metre to 5–7 vegetative rosettes and only 0.5–1.0 flowering individuals per square metre (Pergl et al, 2007).

Management Efforts

Heracleum mantegazzianum has direct negative effects on human health and can form monospecific stands that reduce local biodiversity. The health risk caused by the phototoxic sap of *H. mantegazzianum* significantly lowers the recreational value of invaded areas and makes eradication at invaded sites even more difficult. The phototoxic sap with photosensitizing furanocoumarins is contained in all the tissues of *H. mantegazzianum*, including the seed. Ultraviolet (UV) radiation activates the phototoxic reaction with the sensitivity peak between half an hour and two hours after contact with the sap. The inflammatory reaction follows after about three days. After that, hyperpigmentation may occur on the affected skin which may last for several weeks and the skin may remain sensitive to UV for a long time (years). Nevertheless, the intensity of the reaction depends on an individual's sensitivity and some people and animals may be resistant. In case of contact with the sap, the skin should be washed as soon as possible and kept out of sunlight for at least 48 hours (Nielsen et al, 2005).

As far as the impact of *H. mantegazzianum* on biodiversity is concerned, established invading populations with a high cover reduce the number of species by up to 90 per cent compared to uninvaded vegetation in the vicinity of invading stands. The negative effects of *H. mantegazzianum* species on native plant species (both species diversity and vegetation composition) can be compared to that of rhizomatous species forming dense polycormons such as knotweeds (*Fallopia* spp.) (Hejda et al, 2009). The strong impact of *H. mantegazzianum* results from its ability to form homogeneous stands with a high cover. The invader is much taller than species of invaded resident communities, which results in effective suppression of native vegetation and short species in particular (Thiele et al, 2010).

The economic costs associated with the invasion of *H. mantegazzianum* are rarely documented. However, the estimates of direct costs on health systems and costs of realized control measures are available from Germany (Reinhardt et al, 2003). The annual costs extrapolated to the whole of Germany range between €6 million and €21 milllion, with a mean of €12 million. This total sum consists of €1–2 million for public health, €1 million in conservation areas and €2.5 million for eradications along roadways. Costs for eradication in rural areas are estimated to start at €5.5 million. In contrast there is a very limited profit (besides its decorative value) resulting from the persistence of *H. mantegazzianum* in the invaded regions: (1) usage of *H. mantegazzianum* by a limited number of beekeepers as a food supply for bees; and (2) usage as a fodder crop. In the case of fodder crop the estimates of dry mass vary between 5.7 to 15 tons per hectare and the nutritional value of leaf biomass is suitable for livestock having high organic digestibility (Buttenschon and Nielsen, 2007).

Any management action against *H. mantegazzianum* needs to be planned with the fact in mind that this species shows extremely high potential for regeneration (Pyšek et al, 2007c). The flowering plants, if damaged, may resprout and set seed within one month. Nevertheless, *H. mantegazzianum* is a strictly monocarpic species, although it is possible that adventive buds may be activated under some extreme circumstances and flowering plants can survive to the following year(s) (Perglová et al, 2007). But the ability of repeated flowering in consequent years induced by damage to plants and removal of flowering umbels, repeatedly reported in literature, has not been observed during thorough recent research nor confirmed experimentally (Pyšek et al, 2007c). Since both monocarpic (e.g. *H. sosnowskyi*) and polycarpic (e.g. *H. persicum* and the native *H. sphondylium*) species occur within the genus, occasional shifts from monocarpic to polycarpic behaviour cannot be excluded.

Heracleum mantegazzianum is a prominent alien species in Europe and as such it was a target species of an integrative EU project aimed at developing a sustainable control strategy with sufficient knowledge on its ecology and biology (the Giant Alien project, 2002–2005; see Pyšek et al, 2007a). The project was not only aimed at assessing the possibilities of the common control methods (chemical, mechanical) but focused on finding potential agents of biological control that might help to manage *H. mantegazzianum* populations in its invaded range. Biological control is based on introduction of natural enemies from the native to invaded range and is closely linked with the 'enemy release hypothesis'. This hypothesis explains the success of some invasive species as a result of escape from their specialized enemies that regulate plant species populations in the native range; if the species leaves its specialized enemies behind, it only has to cope with generalist herbivores and pathogens. This allows it to invest resources that are no longer needed for defence into growth and reproduction and gain a competitive advantage over native species in the region of introduction (Keane and Crawley, 2002). The use of biological control in Europe is still somewhat limited in the field; however, in greenhouses it is widely used.

To evaluate the possibility of biological control, the screening of existing enemies (i.e. herbivores and pathogens) in both distribution ranges was carried out, followed by tests of their host specificity and level of damage caused to hogweed plants. The surveys of mycobiota associated with *H. mantegazzianum* in both ranges revealed that there is a large number of species from a wide range of genera. Most of the identified pathogens have insignificant impacts and initially promising primary pathogens found in the Caucasus for *H. mantegazzianum* and not found in the invaded range (*Phleospora heraclei*, *Septoria heracleicola* and *Rampulariopsis* sp.) were also found on native *H. sphondylium*. The mycobiota of *H. mantegazzianum* are in general impoverished in its invaded range compared to the Caucasus and native *Heracleum* species in the Caucasus seem to share common mycobiota (Seier and Evans, 2007). More than 350 insect species were found on different *Heracleum* species, but the majority of them are only visitors and do not interact with the plant (Hansen et al, 2006, 2007). The species composition of herbivores significantly differs between the two ranges. Several groups of herbivores attacking *H. mantegazzianum* were found, including sap suckers, root or stem borers, leaf chewers or disease transmitters. Nevertheless no insect that feeds exclusively on *H. mantegazzianum* was found in the Caucasus (Hansen et al, 2007). To minimize the attack of enemies, *H. mantegazzianum* has evolved two types of defence system: (1) secondary

plant compounds, mainly including the furanocoumarins, and (2) trichomes. Although the toxicity of extracts from plants in the invaded range seems to be higher and plants have fewer and shorter trichomes than those in the native range, no differences were found in terms of investment into defence systems (Hattendorf et al, 2007).

Consequently, no suitable, efficient and specific pathogen or insect has been identified up to now (Cock and Seier, 2007), and the control and management of *H. mantegazzianum* needs to rely on mechanical or chemical methods. *Heracleum mantegazzianum* is sensitive to a wide range of total or selective herbicides containing glyphosate or triclopyr. Current knowledge of the species' biology and ecology can be used to minimize the costs needed for eradication and maximize the efficiency of the measures applied. Based on the life history of the species, several issues need to be kept in mind for control and eradication. It is generally known that prevention and early eradication is the least costly approach to alien species management. In the case of *H. mantegazzianum*, introduced as an ornamental species and sometimes still planted and offered by garden centres, raising public awareness is important to prevent further dispersal. Furthermore, the character of invaded habitats near human settlements and the extreme size of the plant makes it easily detectable and thus individual plants or small populations can be immediately recognized and rapidly destroyed. As *H. mantegazzianum* only reproduces by seed, it is crucial to target prevention and control measures to reduce seed set and dispersal. At invaded sites with small populations or individual plants, chemical treatment is the most efficient approach. Cutting does not kill the plants, but extends their lifespan by postponing the time of flowering into the following years. Because of the high regeneration potential, the only mechanical treatment that immediately kills the plant is cutting the tap root 15 cm below ground (Tiley and Philp, 1997; Pyšek et al, 2007c). In the case of flowering plants, removal of umbels is effective if done at the peak of flowering or the beginning of fruit formation. Umbels must be destroyed (burnt). The timing of umbel removal is crucial and subsequent cutting of regenerated flowering umbels as they emerge prevents the plant fruiting. Umbel removal later in the season reduces the possibility of regeneration, but highly increases the risk of release of ripe fruits when manipulating the umbels. Cutting whole flowering stems and leaving them at a site is not recommended as even flowers cut as early as at the phase of the end of stigma receptivity can produce viable seed providing they are connected to a stem (Pyšek et al, 2007c).

For large infestations, mechanical methods such as grazing and cutting may help to reduce the size of the populations and the amount of seeds produced, however, the timing and frequency is again crucial (Buttenschon and Nielsen, 2007). Moreover, mowing early in the season may help to increase the accessibility for later application of herbicide and reduce the leaf area of adult plants so that juvenile individuals may also be targeted. When a long term eradication programme is possible, only flowering plants need to be targeted until the population is depleted. This can be done by removal of umbels or by application of herbicide on reproducing plants early in the season (mid to late June in central Europe), when the plants show reduced regeneration capacity and do not produce viable seed. To conclude, small infestations can be completely eradicated; in the case of large infestations, the aim is mainly to prevent the species from spreading further by reducing the extent of invaded sites to a reasonable scale. Following eradication, the site should be monitored for at least seven to ten years for the occurrence of plants emerging from the seed bank to avoid recovery of the population. Such monitoring should also include a wider neighbourhood than the target area so that isolated plants be eliminated that could serve as a source of subsequent reinvasion. This is of crucial importance in areas neighbouring rivers and transport corridors (roads or railways).

Challenges and Controversies

Heracleum mantergazzianum is one of the invasive species that raises little controversy because there is an unambiguous consensus about its negative ecological and economic impacts. Because of the publicity this species has been receiving, public awareness of negative impacts has improved in recent decades, which has led to increased eradication and control efforts. Unfortunately, these efforts are often local and are not always applied for a period of time long enough to achieve complete eradication. Given the enormous reproductive capacity and efficient fruit dispersal that make the danger of rapid reinvasion of cleaned sites quite realistic, future management measures need to be designed for sufficiently large geographical scales. The apocalyptic future portrayed in the rock band Genesis's song *The Return of Giant Hogweed* from 1971 has not quite happened, but continued vigilance is the key to future success.

References

Buttenschon, R. M. and Nielsen, C. (2007) 'Control of *Heracleum mantegazzianum* by grazing', in P. Pyšek, M. J. W. Cock, W. Nentwig and H. P. Ravn (eds) *Ecology and Management of Giant Hogweed* (Heracleum mantegazzianum), CAB International, Wallingford, UK

Cock, M. J. W. and Seier, M. K. (2007) 'The scope for biological control of giant hogweed, *Heracleum mantegazzianum*', in P. Pyšek, M. J. W. Cock, W. Nentwig and H. P. Ravn (eds) *Ecology and Management of Giant Hogweed* (Heracleum mantegazzianum), CAB International, Wallingford, UK

Grace, J. and Nelson, M. (1981) 'Insects and their pollen loads at a hybrid *Heracleum* site', *New Phytologist*, vol 87, pp413–423

Hansen, S. O., Hattendorf, J., Wittenberg, R., Reznik, S. Y., Nielsen, C., Ravn, H. P. and Nentwig, W. (2006) 'Phytophagous insects of giant hogweed *Heracleum mantegazzianum* (Apiaceae) in invaded areas of Europe and in its native area of the Caucasus', *European Journal of Entomology*, vol 103, pp387–395

Hansen, S. O., Hattendorf, J., Nielsen, C., Wittenberg, R. and Nentwig, W. (2007) 'Herbivorous arthropods on *Heracleum mantegazzianum* in its native and invaded distribution range', in P. Pyšek, M. J. W. Cock, W. Nentwig and H. P. Ravn (eds) *Ecology and Management of Giant Hogweed* (Heracleum mantegazzianum), CAB International, Wallingford, UK

Hattendorf, J., Hansen, S. O. and Nentwig, W. (2007) 'Defence systems of *Heracleum mantegazzianum*', in P. Pyšek, M. J. W. Cock, W. Nentwig and H. P. Ravn (eds) *Ecology and Management of Giant Hogweed* (Heracleum mantegazzianum), CAB International, Wallingford, UK

Hejda, M., Pyšek, P. and Jarošík, V. (2009) 'Impact of invasive plants on the species richness, diversity and composition of invaded communities', *Journal of Ecology*, vol 97, pp393–403

Jahodová, Š., Fröberg, L., Pyšek, P., Geltman, D., Trybush, S. and Karp, A. (2007a) 'Taxonomy, identification, genetic relationships and distribution of large *Heracleum* species in Europe', in P. Pyšek, M. J. W. Cock, W. Nentwig and H. P. Ravn (eds) *Ecology and Management of Giant Hogweed* (Heracleum mantegazzianum), CAB International, Wallingford, UK

Jahodová, Š., Trybush, S., Pyšek, P., Wade, M. and Karp, A. (2007b) 'Invasive species of *Heracleum* in Europe: An insight into genetic relationships and invasion history', *Diversity and Distributions*, vol 13, pp99–114

Keane, R. and Crawley, M. (2002) 'Exotic plant invasions and the enemy release hypothesis', *Trends in Ecology and Evolution*, vol 17, pp164–170

Moravcová, L., Pyšek, P., Pergl, J., Perglová, I. and Jarošík, V. (2006) 'Seasonal pattern of germination and seed longevity in the invasive species *Heracleum mantegazzianum*', *Preslia*, vol 78, pp287–301

Moravcová, L., Pyšek, P., Krinke, L., Pergl, J., Perglová, I. and Thompson, K. (2007) 'Seed germination, dispersal and seed bank in *Heracleum mantegazzianum*', in P. Pyšek, M. J. W. Cock, W. Nentwig and H. P. Ravn (eds) *Ecology and Management of Giant Hogweed* (Heracleum mantegazzianum), CAB International, Wallingford, UK

Nielsen, C., Ravn, H. P., Cock, M. J. W. and Nentwig, W. (eds) (2005) *The Giant Hogweed Best Practice Manual: Guidelines for the Management and Control of an Invasive Alien Weed in Europe*, Forest and Landscape Denmark, Hoersholm, Denmark

Ochsmann, J. (1996) '*Heracleum mantegazzianum* Sommier & Levier (Apiaceae) in Deutschland Untersuchen zur Biologie, Verbreitung, Morphologie und Taxonomie', *Feddes Repertorium*, vol 107, pp557–595

Otte, A., Eckstein, R. L. and Thiele, J. (2007) '*Heracleum mantegazzianum* in its primary distribution range of the Western Greater Caucasus', in P. Pyšek, M. J. W. Cock, W. Nentwig and H. P. Ravn (eds) *Ecology and Management of Giant Hogweed* (Heracleum mantegazzianum), CAB International, Wallingford, UK

Page, N. A., Wall, R. E., Darbyshire, S. J. and Mulligan, G. A. (2006) 'The biology of invasive alien plants in Canada. 4. *Heracleum mantegazzianum* Sommier & Levier', *Canadian Journal of Plant Science*, vol 86, pp569–589

Pergl, J., Perglová, I., Pyšek, P. and Dietz, H. (2006) 'Population age structure and reproductive behaviour of the monocarpic perennial *Heracleum mantegazzianum* (Apiaceae) in its native and invaded distribution ranges', *American Journal of Botany*, vol 93, pp1018–1028

Pergl, J., Eckstein, L., Hüls, J., Perglová, I., Pyšek, P. and Otte, A. (2007) 'Population dynamics of *Heracleum mantegazzianum*', in P. Pyšek, M. J. W. Cock, W. Nentwig and H. P. Ravn (eds) *Ecology and Management of Giant Hogweed* (Heracleum mantegazzianum), CAB International, Wallingford, UK

Perglová, I., Pergl, J. and Pyšek, P. (2006) 'Flowering phenology and reproductive effort of the invasive alien plant *Heracleum mantegazzianum*', *Preslia*, vol 78, pp265–285

Perglová, I., Pergl, J. and Pyšek, P. (2007) 'Reproductive ecology of *Heracleum mantegazzianum*', in P. Pyšek, M. J. W. Cock, W. Nentwig and H. P. Ravn (eds) *Ecology and Management of Giant Hogweed* (Heracleum mantegazzianum), CAB International, Wallingford, UK

Pyšek, P. and Pyšek, A. (1995) 'Invasion by *Heracleum mantegazzianum* in different habitats in the Czech Republic', *Journal of Vegetation Science*, vol 6, pp711–718

Pyšek, P., Kopecký, M., Jarošík, V. and Kotková, P. (1998) 'The role of human density and climate in the spread of *Heracleum mantegazzianum* in the Central European landscape', *Diversity and Distributions*, vol 4, pp9–16

Pyšek, P., Cock, M. J. W., Nentwig, W. and Ravn, H. P. (eds) (2007a) *Ecology and Management of Giant Hogweed* (Heracleum mantegazzianum), CAB International, Wallingford, UK

Pyšek, P., Müllerová, J. and Jarošík, V. (2007b) 'Historical dynamics of *Heracleum mantegazzianum* invasion at regional and local scales', in P. Pyšek, M. J. W. Cock, W. Nentwig and H. P. Ravn (eds) *Ecology and Management of Giant Hogweed* (Heracleum mantegazzianum), CAB International, Wallingford, UK

Pyšek, P., Perglová, I., Krinke, L., Jarošík, V., Pergl, J. and Moravcová, L. (2007c) 'Regeneration ability of *Heracleum mantegazzianum* and implications for control', in P. Pyšek, M. J. W. Cock, W. Nentwig and H. P. Ravn (eds) *Ecology and Management of Giant Hogweed* (Heracleum mantegazzianum), CAB International, Wallingford, UK

Pyšek, P., Jarošík, V., Müllerová, J., Pergl, J. and Wild, J. (2008) 'Comparing the rate of invasion by *Heracleum mantegazzianum* at continental, regional, and local scales', *Diversity and Distributions*, vol 14, pp355–363

Reinhardt, F., Herle, M., Bastiansen, F. and Streit, B. (2003) *Economic Impact of the Spread of Alien Species in Germany*, Umweltforschungsplan, Des Bundesministeriums für Umwelt, Naturschutz und Reaktorsicherheit, Germany

Seier, M. K. and Evans, H. C. (2007) 'Fungal pathogens associated with *Heracleum mantegazzianum* in its native and invaded distribution range', in P. Pyšek, M. J. W. Cock, W. Nentwig and H. P. Ravn (eds) *Ecology and Management of Giant Hogweed* (Heracleum mantegazzianum), CAB International, Wallingford, UK

Thiele, J., Otte, A. and Eckstein, R. L. (2007) 'Ecological needs, habitat preferences and plant communities invaded by *Heracleum mantegazzianum*', in P. Pyšek, M. J. W. Cock, W. Nentwig and H. P. Ravn (eds) *Ecology and Management of Giant Hogweed (*Heracleum mantegazzianum*)*, CAB International, Wallingford, UK

Thiele, J., Isermann, M., Otte, A. and Kollmann, J. (2010) 'Competitive displacement or biotic resistance? Disentangling relationships between community diversity and invasion success of tall herbs and shrubs', *Journal of Vegetation Science*, vol 21, pp231–220

Tiley, G. E. D. and Philp, B. (1997) 'Observations on flowering and seed production in Heracleum mantegazzianum in relation to control', in J. H. Brock, P. M. Wade, P. Pyšek and D. Green (eds) *Plant Invasions: Studies from North America and Europe*, Backhuys, Leiden, The Netherlands

Tiley, G. E. D., Dodd, F. S. and Wade, P. M. (1996) '*Heracleum mantegazzianum* Sommier & Levier', *Journal of Ecology*, vol 84, pp297–319

USDA (United States Department of Agriculture) (2010) '*Heracleum mantegazzianum*, plant profile', http://plants.usda.gov/java/profile?symbol=HEMA17, accessed 30 November 2010

6

Impatiens glandulifera Royle (Himalayan balsam)

Christopher P. Cockel and Robert A. Tanner

Introduction

Impatiens glandulifera (Balsaminaceae) (Figure 6.1), commonly known as Himalayan Balsam, Indian Balsam, Policeman's Helmet, Jumping Jacks, Nuns, Bee-Bums, Poor Man's Orchid, Pink Peril, and perhaps erroneously Ornamental Jewelweed, is a prime example of adaptation and success in the plant world. Unfortunately, the consequence of this success has resulted in the species becoming an unwelcome alien invader in many regions of the world to which it has been introduced, particularly in Europe. Predominately a weed of riparian habitats, *I. glandulifera* has invaded riverbanks and lakesides where it forms dense monospecific stands (Shaw and Tanner, 2008) (Figure 6.2) enabling the plant to outcompete native species (Hulme and Bremner, 2006; Maskell et al, 2006). *Impatiens glandulifera* is included along with 17 other terrestrial plant species in Europe's Top 100 most invasive species (DAISIE, 2010).

Origin of Invasion

Impatiens glandulifera was first recorded in Europe from the UK in 1839, when seeds were sent to the Royal Botanical Gardens, Kew, from Kashmir by John Forbes Royle, the then curator of the botanical gardens in Saharanpur, Northern India (Beerling and Perrins, 1993). Initially prized by plant collectors for its

attractive pink and white zygomorphic flowers, and more recently by beekeepers for the profuse quantities of nectar the flowers produce, it was not long before the species became established throughout the UK countryside. *Impatiens glandulifera* now occurs throughout mainland Britain, in much of Ireland as well as more isolated localities of the UK, such as the Isles of Scilly, Shetland and Orkney (Beerling and Perrins, 1993) (Figure 6.3). To date, *I. glandulifera* has been introduced into 27 European countries (DAISIE, 2010) where it is widespread in 18 and invasive in 12 (CABI, 2004). The plant is also regarded as invasive in the US (USDA, 2010), Canada (Clements et al, 2008), New Zealand (Sykes, 1982), the Russian Far East (Markov et al, 1997), and Japan (Drescher and Prots, 2003) (see Figure 6.4). Pyšek and Prach (1995) describe the spread of *I. glandulifera* throughout Europe since the 1960s as massive, regardless of the date of introduction to a particular country.

Impatiens glandulifera was first thought to be a highly desirable addition to English gardens. In general, the early recognition of an undesirable/potentially invasive species is rare and often takes decades (Williamson, 1996) and this was most certainly the case with *I. glandulifera*. This may be due to a lag phase associated with the invasive nature of some plants, where initially the population expansion is slow while the species adapts to the new environment; subsequently, this is followed by the exponential phase that sees a

Figure 6.1 Impatiens glandulifera *in its native habitat, the foothills of the Himalayas, Kaghan Valley, Pakistan*

Figure 6.2 Impatiens glandulifera *monospecific stand on the River Torridge, North Devon, UK*

rapid explosion in the spread and occurrence of the species. The rate at which a species spreads is likely to be as much related to human activity (Perrings et al, 2002) and landscape factors, as it is to the biology and ecology of the species (Williamson et al, 2005). *I. glandulifera* has found its ideal conditions predominantly along the rivers of Europe and elsewhere, aided by the explosive release of its seeds coupled with hydrochorous dispersal. There is also evidence that

I. glandulifera is capable of invading disturbed deciduous woodland and ungrazed tall herb/ruderal/grassland habitats (e.g. Maskell et al, 2006; Andrews et al, 2009) as a result of human-assisted transportation of seed material in soil, by beekeepers and accidental release in garden waste and on vehicles (Kowarik, 2003).

In its native range, the foothills of the western Himalayas, *I. glandulifera* is commonly found in high altitude meadows (Sharma and Jamwal, 1988), at the fringe of deciduous woodlands (Blatter, 1927), on hillsides (Nasir, 1980) and near streams (Nair, 1977) at characteristic altitudes of between 2000m and 2500m above sea level (Beerling and Perrins, 1993; Kurtto, 1996) and as high as 3700m (Tanner, R. A., pers. obs., 2006–2010). In contrast to the colonization of the species throughout riparian systems in the invasive range, in the native range *I. glandulifera* is not normally confined to riparian habitats, but is a species of high altitude meadows. In the introduced range, seeds can become incorporated into the river system and are subsequently conveyed downstream to form new populations. However, in the Himalayas, during the monsoon season, seeds are washed down gullies and rivers but are rarely able to colonize riverbanks due to the extensive rock formations found along the side of rivers and the fast flowing movement of the water. Only where human settlements are established along the riverbanks and have disturbed and altered the composition of the bank structure can *I. glandulifera* gain a foothold.

Why Is *Impatiens glandulifera* Such a Successful Invader?

There appear to be many underlying reasons why *I. glandulifera* has been so successful in colonizing areas susceptible to invasion, such as riparian corridors, damp woodlands and wastelands. In common with other non-native invasive plants, it possesses many competitive advantages over native species. Williamson (1996) discusses three predictors of invasion success that have a statistical basis, namely, propagule pressure, suitability of habitat and previous invasion success. He also notes other potentially influential factors such as intrinsic rate of increase, modes of reproduction and genetic structure, abundance and range in the native habitat and climatic matching, which refers to the match between a plant's native habitat and that of its introduced

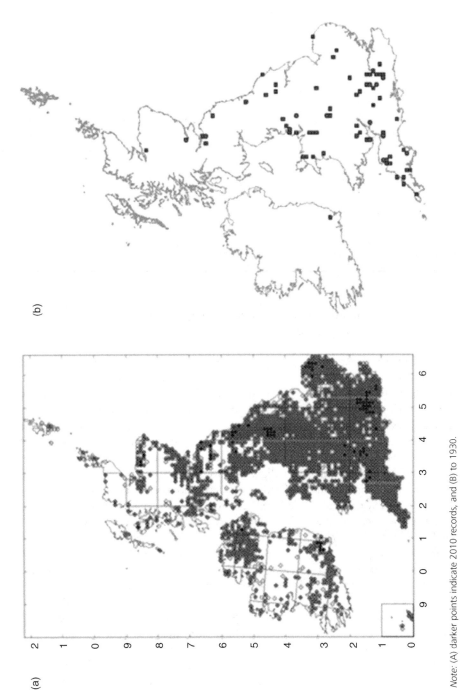

(a)

(b)

Figure 6.3 *All records of* Impatiens glandulifera *occurrence across the British Isles to 2010*

Note: (A) darker points indicate 2010 records, and (B) to 1930.

Source: BSBI

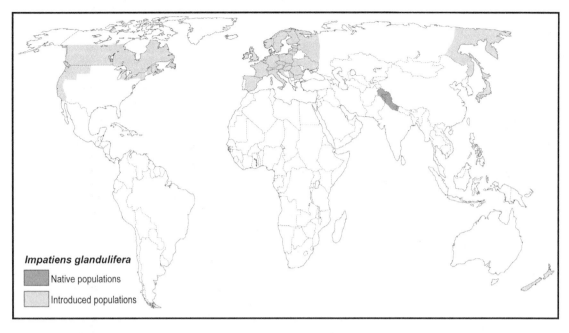

Source: CABI, 2004; DAISIE, 2010

Figure 6.4 *The global distribution of* Impatiens glandulifera, *highlighting the native and introduced range of the species*

range. Invasive plants often display early sexual maturity (Brock, 1999), which is certainly true of *I. glandulifera*.

As Europe's tallest annual herb (Chittka and Schürkens, 2001), *I. glandulifera* quickly grows taller (over 2m) than native species such as *Urtica dioica*. Extensive branching from the main stem ensures the population gains a monopoly of the aerial environment (Chittka and Schürkens, 2001). Plant biomass has been shown to explain 63 per cent of the variation in competitive ability, with plant height, canopy diameter, canopy area and leaf shape explaining most of the residual variation (Gaudet and Keddy, 1988). When the plant grows in dense monocultures, the population can produce a seed rain of up to 30,000 seeds per square metre (Cronk and Fuller, 2001), that are dispersed widely by autochory, up to 7m from the parent plant. The propagule pressure applied by the production and explosive dispersal of such a large number of seeds increases the probability that seed will find suitable habitats and environmental conditions for successful germination (Williamson, 1996). When populations are near water bodies, seeds are incorporated into the river system and conveyed downstream to form new populations. Stamp and Lucas (1983) describe the

explosive action of the seed capsules as ballistic, and Bond (1998) describes how the seed capsules of the genus explode in a fusillade at a mere touch lending the genus *Impatiens* (meaning impatient) its name. Stamp and Lucas (1983) conclude that seeds with ballistic properties are smooth and hence aerodynamic, as well as necessarily heavy to generate sufficient momentum in flight to cover some distance. Despite the highly effective dispersal mechanism that the species has evolved, the role of humans in unintentionally and intentionally extending the range of the species in soil material should not be underestimated.

In the introduced range, *Impatiens glandulifera*, along with *H. mantegazzianum*, has a preference for lower powered streams at lower altitudes and with finer sediment particle size (Dawson and Holland, 1999). Burton (1983) observed how in the London area, along rivers, *I. glandulifera* flourishes in soft riverbank soil where the existing vegetation is poorest and space is available for colonization, but also that the species can be found in other areas as a result of seedlings being flung from a garden plant. Only annual species of *Impatiens* have been successful in the UK (Grey-Wilson, 1980) most notably *I. glandulifera*, in part because the seeds are able to survive low temperatures in the winter, and indeed, rely on cold stratification to break dormancy (Mumford, 1988).

The lack of specialist natural enemies in the introduced range affords *I. glandulifera* a competitive advantage over native species enabling the plant to invest more into growth and fecundity and less in the secondary chemicals used to deter natural enemy attack. Herbarium specimens held at Kew and collected in Kashmir confirm that plants from the native range are smaller than those found growing along rivers in the UK, which in part can be attributed to natural enemy pressure, though other environmental factors including altitude, latitude, climate and competition with other species may all play a part.

Another crucial factor in invasion success is pollination by indigenous insects (Grey-Wilson, 1980). Generalized pollination strategies have been shown to be important in invasive species success, as with invasive grasses in the western US that have been successful because of their reliance on wind pollination (Brock, 1999). *Impatiens glandulifera* does not have this advantage, though its ability to lure pollinators away from native species with its high sugar nectar concentration has been demonstrated (Chittka and Schürkens, 2001). Studying bumblebee-pollinated plants along riverbanks in central Germany, Chittka and Schürkens (2001) observed that pollinators alter their visitation behaviour and favour *I. glandulifera* over European natives such as *Stachys palustris*, *Lythrum salicaria* and *Epilobium hirsutum*. Other studies, conducted in central Germany, have found no evidence that *I. glandulifera* outcompetes native plants for pollinators (Bartomeus et al, 2010). While there is a lack of evidence of self-pollination among the genus *Impatiens* (Grey-Wilson, 1980), as demonstrated with cultivated plants that require hand pollination in order to produce seeds, the self-pollinating abilities of *I. glandulifera* specifically have received little attention. However, Valentine (1978) observes that *I. glandulifera*, although self-compatible (via geitonogamy), is not able to automatically self-pollinate due to protandry.

Tolerance and Adaptation

Impatiens glandulifera is tolerant of a wide variety of soil textures and structures, and can be found on fine and course alluvium (Beerling and Perrins, 1993). It is also tolerant of a range of climates (Chittka and Schürkens, 2001) and soil pH values from relatively acidic to neutral (pH 5.0 to 8.0) (Grime et al, 1988) as well as a high to low nutrient level soil (Beerling and Perrins, 1993).

Populations of *I. glandulifera* in Europe exhibit more frost intolerance than populations in the Himalayas and this may be a limiting factor to the spread of the species in higher altitude areas in Europe (Beerling and Perrins, 1993). Plants of all ages are sensitive to frosts, though often populations are protected from the severity of late spring frosts and early autumn frosts due to growing in sheltered riparian habitats. Beerling and Perrins (1993) note that *I. glandulifera* can be susceptible to drought, though most populations in the UK persist during prolonged drought periods due to their proximity to water. Grime et al (1988) observe that *I. glandulifera* plants growing in dry habitats in the UK have shorter internodes and smaller leaves than those growing in damp conditions. However, this does not seem to impede the plant's ability to flower and produce viable seed (Cockel, C., pers. obs., 2008). Partial shade tolerance has also been observed (Beerling and Perrins, 1993).

Kollmann and Bañuelos (2004), looking at the adaptive response to environmental gradients, recorded variations in growth and phenology of the species from nine European regions, with northern populations (England, Denmark and Sweden) observed to have reduced above-ground biomass, but flower 10–15 days earlier than southern populations (Czech Republic, France and Switzerland). This may suggest the occurrence of genotypic differentiation according to latitude (Brock, 1999).

As previously mentioned, in the introduced range *I. glandulifera* is the host to an impoverished invertebrate fauna. In the UK, two species of aphid are known to feed on *I. glandulifera*, namely *Aphis fabae* (black bean aphid) and the rare *Impatientinum balsamines* (Hopkins et al, 2002). *Impatiens glandulifera* is also a food plant for the larvae of *Deilephila elpenor* (elephant hawk-moth) (Beerling and Perrins, 1993). Slugs, particularly *Arion* species, are known to feed on *I. glandulifera* including the germinating seeds and cotyledons (Prowse, 1998). However, their ability to negatively impact the plant's growth and fecundity appears to be minimal (Cockel, C., pers. obs., 2009). There are instances when *I. glandulifera* may be susceptible to viral infections resulting in reduced plant biomass, though flowering and seed production appear to be unaffected (Kollmann et al, 2007). There is little evidence of seed predation on mature plants. In the native range of the species the situation is very different with almost all populations being attacked by an array of natural enemies that help to keep the species in balance with the surrounding vegetation (Tanner et al, 2008). Throughout the Himalayas *I. glandulifera* is a host to both specialist and generalist arthropod species, including the generalist flea beetle, *Altica hemensis*, which is capable of causing complete skeletonization of the leaves when in high abundance, and the more specialist stem-boring beetles, *Metialma suturella* and *Languriophasma cyanea*. *Impatiens glandulifera* is also the host to a number of plant pathogens including both *Phoma* and *Septoria* leaf spots and an autoecious rust pathogen *Puccinia cf. komarovii*, which infects the stem and the leaves of the plant.

The northern limits of *I. glandulifera* in Europe appear to be regulated by the length of the growing season (Beerling and Perrins, 1993), though changing atmospheric conditions may see the range of *I. glandulifera* shift northwards (Brock, 1999). At the same time, the potential of climate change together with a decline in UK biodiversity may increase the susceptibility of ecosystems to invasion by non-native plant species (Manchester and Bullock, 2000). Dukes and Mooney (1999) emphasize the positive response that many invasive plants have displayed as a consequence of elevated CO_2 levels and increased nitrogen deposition, along with rising average temperatures, higher levels of precipitation, increased habitat fragmentation and altered disturbance regimes.

Though Cronk and Fuller (2001) and Grime et al (1988) note that *I. glandulifera* produces no persistent seed bank, year on year, seeds can remain viable for up to 18 months (Beerling and Perrins, 1993). If imbibed at 20°C, seeds can remain viable for up to three years (Mumford, 1988). Nozzolillo and Thie (1983) in their germination studies of North American *Impatiens* species *I. capensis* and *I. pallida*, noted that these species are vernal, that is their seeds germinate in the spring after a period of cold stratification, which is a characteristic shared with *I. glandulifera* seeds, which germinate synchronously in the spring. The over-wintering ability of *I. glandulifera* seeds is critical to its

success along rivers (Grime et al, 1988). *Impatiens glandulifera* seeds appear not to be buoyant (Beerling and Perrins, 1993), at least not once they become immersed in water (Pyšek and Prach, 1993), although they are sufficiently light in weight to be easily carried along in fast flowing water.

Problems and Impacts Associated with *I. glandulifera*

Although there has been much research into the impacts of *I. glandulifera* on native flora, and more recently its effect on pollinators, more research is needed at both the species and ecosystem levels. A considerable amount of research has investigated the impact of *I. glandulifera* on native plants, though to date there are no published studies on the impact of *I. glandulifera* on invertebrate populations. As *I. glandulifera* can displace native plant species it is feasible to suggest that where the species forms dense stands this can in turn deplete the local diversity of associated invertebrates.

The most significant negative impact *I. glandulifera* has on native plant species in riparian habitats is the ability to shade out and displace species that assist in riverbank stability (Dawson and Holland, 1999). This in turn could lead to increased bank erosion and sediment entrainment, which may negatively impact fish spawning grounds and invertebrate niches. In addition, dead plant material can become incorporated into the water body where it can block drains thereby increasing the risk of flooding. Riverbanks densely colonized by *I. glandulifera* have been shown to have reduced plant diversity by up to 25 per cent (Hulme and Bremner, 2006). Maule et al (2000) studied the impact of *I. glandulifera* in wooded habitats and demonstrate that *I. glandulifera* can successfully compete with native plants, including tree seedlings, with the potential to inhibit the regeneration cycle of woodlands. However, experiments conducted by Hejda and Pyšek (2006) in the Czech Republic demonstrate that *I. glandulifera* merely displaces tall native dominant nitrophilous species, such as *Urtica dioica*, leaving shorter native species unaffected.

As previously mentioned, as well as competing with native species for nutrients, water, light and physical space, *I. glandulifera* is successful in competing for pollinators by offering sweeter nectar (Chittka and Schürkens, 2001). Habitat loss and a reduction in native plant species are threatening pollinator communities (Bartomeus et al, 2010). Beekeepers have expressed an interest in *I. glandulifera* as a valuable food source for declining pollinator populations (Showler, 1989). However, caution must be expressed. A shift of pollinators from natives to non-natives could potentially lead to a reduction in the seed set of native plants resulting in a negative effect on their fitness. It could be argued that habitat conservation and the promotion of native species restoration would be more valuable than reliance on a non-native species, which could lead to further native species decline.

One potential impact, which has received little attention in the UK, is the impact of *Impatiens glandulifera* on the soil microbial community. Tanner (2008) highlighted that *I. glandulifera* has virtually no associated mycorrhizae, which are essential for the establishment of native plant species. This low dependency on soil microbes leads to a depletion of mycorrhizae under the invasive stands, as in the absence of a suitable host the mycorrhizae are unable to proliferate. As a result, native plant species are unable to recolonize invaded areas due to the changes in the soil mycobiota caused by the non-native. Thus, in the context of habitat restoration, any impacts must be evaluated and rectified to aid native species colonization and establishment.

The Scottish Executive in their November 2006 consultation on a proposal to amend the 1981 Wildlife and Countryside Act to include *I. glandulifera*, described *I. glandulifera* as not only shading out native plants but also detrimental to human pursuits by 'impeding access to riverbanks' for such activities as sport fishing (Scottish Executive, 2006, p42). There are also economic impacts resulting from the invasive behaviour of *I. glandulifera* (Dawson and Holland, 1999) brought on by higher riverbank maintenance costs as well as a reduction in habitat and landscape value, particularly in areas valued for certain species or habitat types. The UK Environment Agency (2003) estimated it would cost £150–300 million ($235–470 million) to eradicate the plant from the UK. Indeed, eradication is now practically impossible, but when attempting to control *I. glandulifera* costs can be as high as £10/m² ($16) using traditional methods and incorporating post-control habitat restoration (Tanner et al, 2008).

Legislation

Despite a 'plethora' of UK legislation aimed at reducing the impact of non-native species, this legislation only goes 'part of the way' toward achieving its goal and more effective enforcement is required (see Manchester and Bullock, 2000).

A range of policy instruments also exists at a European level designed to tackle the threat posed by invasive species as a result of transport in the process of international trade and commerce. However, existing customs and quarantine measures are principally aimed at preventing the spread of agricultural pests and diseases and have proved largely inadequate safeguards against species that threaten biodiversity (Hulme, 2007). As such, a more coordinated approach is encouraged among European nations to increase the effectiveness of policy and enforcement (Hulme, 2007).

In the UK, after public consultation during 2007 and 2008 on the review of Schedule 9 of the Wildlife and Countryside Act (1981), Defra and the Welsh Assembly extended the Act to include *Impatiens glandulifera* and an additional 37 plant species. The amendment of the Act, which came into force in April 2010, made it an offence to plant *I. glandulifera*, or otherwise cause the species to grow, in the wild.

Currently, the only plants prohibited from trade or cultivation in the UK are *H. mantegazzianum*, *F. japonica*, *Macrocystis pyrifera* and *Sargassum muticum*. *Impatiens glandulifera* seed is still commercially available in the UK and in the US, as are the cultivars *I. glandulifera* 'Mien Ruys', 'Wine Red' and the white-flowered *I. glandulifera* 'Candida'. However, there is evidence of national and local governments starting to take action to halt this trade (Washington Administrative Code (WAC) 16-752-610 and the November 2006 proposal by the Scottish Executive to include *I. glandulifera* in Schedule 9 of the Wildlife and Countryside Act (1981)). In addition, Defra and the Welsh Assembly have proposed that *I. glandulifera* and an additional 15 plant species should be banned from sale in the UK under Section 14ZA of the Wildlife and Countryside Act (1981).

On an international level, signatories of the 1992 Convention on Biological Diversity (CBD), under the auspices of the United Nations, are urged to prevent the introduction of and to control alien species that 'threaten ecosystems, habitats or species' (Article 8.h). The CBD also calls on signatories to 'rehabilitate and restore degraded ecosystems' (Article 8.f). The EC Water Framework Directive (WFD) (EC, 2000), which requires EU member states to collect and maintain information of significant anthropogenic pressures placed upon surface water bodies in relation to water quality, does not specifically mention invasive non-native species, but as invasive plants have the potential to reduce the ecological and conservation potential of river systems their impacts must be addressed in order to achieve the 'good ecological status' required by the directive.

Management Techniques

The key to controlling *I. glandulifera* is to prevent the plants from flowering and fruiting (Dawson and Holland, 1999). Repeated clearance every two weeks, early in the season is necessary to ensure plants do not set seed and that late germinating plants are not allowed to mature (Beerling and Perrins, 1993). Chemical and manual control methods can be effective when *I. glandulifera* is growing in discrete areas though both are labour intensive and costly. Chemical control of *I. glandulifera* can be effective, though care must be taken when applying chemicals around water and advice should first be sought from local agencies on which chemicals can be used. In Europe, chemicals that were once deemed safe for use around water bodies are now being banned, indeed in some countries any chemical application around water is illegal. The application of herbicides when the plants are in flower is reported to be ineffective at preventing the production of viable seed (Hejda, 2009).

Although removal experiments have shown a rapid response in terms of site species richness (Hulme and Bremner, 2006), other experiments (Cockel, C., unpublished data) suggest that removal over at least two seasons is necessary before a heavily invaded site can begin to recover due to the ability of *I. glandulifera* seeds to remain viable for up to 18 months (Beerling and Perrins, 1993). It should also be noted that the removal of *I. glandulifera* may simply present opportunities for common native ruderal species, such as the native, *Urtica dioica,* and other non-native invasive species, such as *H. mantegazzianum and F. japonica,* to flourish (Hulme and Bremner, 2006;

Cockel, C., unpublished data). The effectiveness of removal has also been questioned due to the effective transport of seeds along river corridors leading to rapid reinvasion (Hejda, 2009).

It is clear from the widespread occurrence of *I. glandulifera* and its continued spread that existing methods are failing to control this plant. For any control method to be successful, management must take place on a catchment scale, ideally in an upstream to downstream direction to limit reinvasion, though effective clearance is often complicated or rendered nearly impossible due to the division of land ownership, costs and the sheer scale of the problem. Localized control may be achieved, but if there are populations upstream left untreated, reinvasion is almost inevitable and any control will only be temporary. Since 2006, CABI UK, funded by a consortium of funders, including the UK Environment Agency, Defra, the Scottish Government and Network Rail, have undertaken research into the biological control of *I. glandulifera* using co-evolved host-specific natural enemies from the native range of the species. Classical biological control, defined as the utilization of natural enemies in the regulation of host populations (DeBach, 1964), is a sustainable, ecological management tool that can be applied on a catchment scale and integrated with current management methods. During 2010 research progressed substantially, due mainly to securing export of prioritized natural enemies from the Indian region of the Himalayas and successful infection of UK biotypes of *I. glandulifera* with the most promising agent – the *Puccinia* rust pathogen.

Sheep and cattle are known to feed on the leaves, stems and flowers of *I. glandulifera* (Beerling and Perrins, 1993; Navchoo and Kachroo, 1995), and although there is no evidence of widespread grazing in the UK, horses by the River Thames at Richmond upon Thames, Surrey, have been observed to feed on *I. glandulifera* plants (Cockel, C., pers. obs., 2005). As a management regime though, grazing at the water's edge may inevitably result in further disturbance and if permitted at the time of seed dispersal may lead to seed being transported to other uninvaded sites.

Discussion

Overall, the literature relating to *I. glandulifera* is comprehensive, though the tendency of the species to colonize urban riparian zones remains relatively unexplored (Pyšek, 1998) and this is a key research priority. The absence of research focusing on *I. glandulifera* until relatively recently, and the rapid expansion in the spread of the species throughout Europe over the last few decades in parallel with milder winters and the debate over climate change, opens another research possibility that has not yet been adequately addressed. As such, opportunities for research exist in examining the link between extended growing seasons and the growth and development of invasive annuals and consequent impacts upon native species (Beerling and Woodward, 1994).

In terms of appropriate management strategies, a first step for conservation practitioners is to decide on priorities. If the presence of an invasive species in a habitat is deemed detrimental to biological diversity, either on its own merits or by local people, control of that invasive species would appear to be an appropriate course of action. However, the physical effort involved in terms of time and resources might be better diverted to more pressing aims, such as habitat creation. Even after control measures have been decided upon, the manner in which these are carried out needs to be 'carefully considered in order that these actions by themselves are constructive and do not lead to further invasions. The time of year at which control measures are implemented is critical, ensuring that any efforts to control the spread of *I. glandulifera* are carried out sufficiently early in the season to ensure that manual removal of the plants does not create additional undue disturbance. The method of control also needs to be taken into account. While manual and mechanical control may overly disturb the riparian zone, opening up opportunities for further invasion, any chemicals applied, though effective, are likely to enter a watercourse.

Previous research has concluded that, in general, some form of intervention is justified as a means of controlling the spread and vigour of *I. glandulifera*, by reducing the seed bank and providing native species with opportunities to recolonize. However, there appears to be a lack of quantitative data regarding the various management options. One strategy that is often adopted by nature conservation groups is the labour-intensive practice of balsam bashing, though this may just add to the disturbance and in essence furrow the ground for seed-setting upstream populations. In addition, there is no known research on the impact of uprooting and disturbing an already diminished microbial community. Merely cutting the plants appears to be a largely ineffective method of control, as plants will quickly recover and still produce flowers. Far greater benefits are likely to result from the complete uprooting of a plant in April or May (followed by local composting) to enable other species to recolonize an area (Cockel, C., unpublished data). Whatever control strategy is chosen, continued monitoring and additional management is essential to ensure efforts are not wasted.

In the context of river rehabilitation and restoration methods such as allowing better access to a waterway, re-establishing hydrological diversity, attracting wildlife and providing an aesthetically pleasing habitat, the ecological function of the catchment may remain compromised by the presence of non-native invasive plants. While research into the validity of different management regimes regarding *I. glandulifera* and other invasive species can be problematic, opportunities exist for creativity in terms of management. Bearing in mind invasive species adaptations, further research could investigate native climate, sediment and nutrient characteristics, pests and other inhibitors such as the role of allelopathy in the native range (Sheppard et al, 2006). Further research opportunities also exist in the field of population genetics (Provan et al, 2007) and comparative genetics (plants and seeds) between *I. glandulifera*, the invasive species known in Europe, and the benign individual in its native range.

Acknowledgements

Thanks go to the National Environment Research Council (NERC), the UK Environment Agency, Defra, the Scottish Government and Network Rail for funding aspects of this research, and to Ed Oliver at Queen Mary for producing the species distribution map.

References

Andrews, M., Maule, H. G., Hodge, S., Cherrill, A. and Raven, J. A. (2009) 'Seed dormancy, nitrogen nutrition and shade acclimation of *Impatiens glandulifera*: Implications for successful invasion of deciduous woodland', *Plant Ecology and Diversity*, vol 2, pp145–53

Bartomeus, I., Vila, M. and Steffan-Dewenter, I. (2010) 'Combined effects of *Impatiens glandulifera* invasion and landscape structure on native plant pollination', *Journal of Ecology*, vol 98, pp440–450

Beerling, D. J. and Perrins, J. M. (1993) '*Impatiens glandulifera* Royle (*Impatiens roylei* Walp.)', *Journal of Ecology*, vol 81, pp367–382

Beerling, D. J. and Woodward, F. I. (1994) 'Climate change and the British scene', *Journal of Ecology*, vol 82, pp391–397

Blatter, E. (1927) *Beautiful Flowers of Kashmir*, John Bale, Sons and Danielsson, Ltd., London

Bond, R. (1998) *Himalayan Flowers*, The Variety Book Depot, New Delhi

Brock, J. H. (1999) 'Ecological characteristics of invasive alien plants', in B. Tellman, D. M. Finch, C. Edminster and R. Hamre (eds) *The Future of Arid Grasslands: Identifying Issues, Seeking Solutions*, Diane Publishing, Darby, USA, pp137–143

Burton, R. M. (1983) *Flora of the London Area*, Natural History Society, London

CABI (2004) *Crop Protection Compendium*, 2004 edition, CAB International, Wallingford

Chittka, L. and Schürkens, S. (2001) 'Successful invasion of a floral market', *Nature*, vol 411, p653

Clements, D. R., Feenstra, K. R., Jones, K. and Staniforth, R. (2008) 'The biology of invasive alien plants in Canada. 9. *Impatiens glandulifera* Royle', *Canadian Journal of Plant Science*, vol 88, pp403–417

Cronk, Q. C. B. and Fuller, J. (2001) *Plant Invaders: The Threat to Natural Ecosystems*, Earthscan, London

DAISIE European Invasive Alien Species Gateway (2010) '100 of the Worst – Terrestrial Plants', www.europe-aliens.org/speciesTheWorst.do, accessed 31 December 2010

Dawson, F. H. and Holland, D. (1999) 'The distribution in bankside habitats of three alien invasive plants in the UK in relation to the development of control strategies', *Hydrobiologia*, vol 415, pp193–201

DeBach, P. (1964) *Biological Control of Insects, Pests and Weeds*, Chapman and Hall, London

Drescher, A. and Prots, B. (2003) 'Distribution patterns of Himalayan balsam (*Impatiens glandulifera* Royle) in Austria', in A. Zając, M. Zając and Z. Zemanek (eds) *Phytogeographical Problems of Synanthropic Plants*, Institute of Botany, Jagiellonian University, Kraków, pp85–96

Dukes, J. S. and Mooney, H. A. (1999) 'Does global change increase the success of biological invaders?', *Trends in Ecology and Evolution*, vol 14, pp135–39

EC (European Commission) (2000) 'Directive 2000/60/EC of the European Parliament and of the Council of 23 October 2000 establishing a framework for Community action in the field of water policy', European Parliament, Strasbourg

Environment Agency (2003) *Guidance for the Control of Invasive Weeds in or Near Fresh Water*, Environment Agency, Bristol

Gaudet, C. L. and Keddy, P. A. (1988) 'A comparative approach to predicted competitive ability from plant traits', *Nature*, vol 334, pp242–243

Grey-Wilson, C. (1980) Impatiens *of Africa*, A. A. Balkema, Rotterdam

Grime, J. P., Hodgson, J. G. and Hunt, R. (1988) *Comparative Plant Ecology: A Functional Approach to Common British Species*, Unwin Hyman, London

Hejda, M. (2009) '*Impatiens glandulifera* Royle, Himalayan balsam (Balsaminaceae, Magnoliophyta)', in J. A. Drake (ed) *Handbook of Alien Species in Europe*, Springer, Netherlands, p351

Hejda, M. and Pyšek, P. (2006) 'What is the impact of *Impatiens glandulifera* on species diversity of invaded riparian vegetation?', *Biological Conservation*, vol 132, pp143–152

Hopkins, G. W., Thacker, J. I., Dixon, A. F. G., Waring, P. and Telfer, M. G. (2002) 'Identifying rarity in insects: The importance of host plant range', *Biological Conservation*, vol 105, pp293–307

Hulme, P. E. (2007) 'Biological invasions in Europe: Drivers, pressures, states, impacts and responses', in R. E. Hester and R. M. Harrison (eds) *Biodiversity Under Threat*, Royal Society of Chemistry, Cambridge, pp56–80

Hulme, P. E, and Bremner, E. T. (2006) 'Assessing the impact of *Impatiens glandulifera* on riparian habitats: Partitioning diversity components following species removal', *Journal of Applied Ecology*, vol 43, pp43–50

Kollmann, J. and Bañuelos, M. J. (2004) 'Latitudinal trends in growth and phenology of the invasive alien plant *Impatiens glandulifera* (Balsaminaceae)', *Diversity and Distributions*, vol 10, pp377–385

Kollmann, J., Bañuelos, M. J. and Nielsen, S. L. (2007) 'Effects of virus infection on growth of the invasive alien *Impatiens glandulifera*', *Preslia*, vol 79, pp33–44

Kowarik, I. (2003) 'Human agency in biological invasions: Secondary releases foster naturalisation and population expansion of alien plant species', *Biological Invasions*, vol 5, pp293–312

Kurtto, A. (1996) 'Impatiens glandulifera (Balsaminaceae) as an ornamental and escape in Finland, with notes on the other Nordic countries', Symbolae Botaniccae Upsalienses, vol 31, pp221–228

Manchester, S. J. and Bullock, J. M. (2000) 'The impacts of non-native species on UK biodiversity and the effectiveness of control', Journal of Applied Ecology, vol 37, pp845–864

Markov, M. V., Ulanova, N. G and Čubatova, N. V. (1997) 'Rod nedotroga. [Genus Impatiens]', in V. N. Pavlov and V. N. Tichomirov (eds) Biologičeskaja flora moskovskoj oblasti [Biological flora of Moscov region], Vypusk, vol 13, pp128–168

Maskell, L. C., Bullock, J. M., Smart, S. M., Thompson, K. and Hulme, P. E. (2006) 'The distribution and habitat associations of non-native plant species in urban riparian habitats', Journal of Vegetation Science, vol 17, pp499–508

Maule, H., Andrews, M., Watson, C. and Cherrill, A. (2000) 'Distribution, biomass and effect on native species of Impatiens glandulifera in deciduous woodland in northeast England', Aspects of Applied Ecology, vol 58, pp31–38

Mumford, P. M. (1988) 'Alleviation and induction of dormancy by temperature in Impatiens glandulifera Royle', New Phytologist, vol 109, pp107–110

Nair, N. C. (1977) Flora of Bashahr Himalayas, International Bioscience Publishers, Hissar

Nasir, Y. J. (1980) 'Balsaminaceae', in E. Nasir and S. I. Ali (eds) Flora of Pakistan, Pakistan Agricultural Research Council, Islamabad, pp1–17

Navchoo, I. A. and Kachroo, P. (1995) Flora of Pulwama (Kashmir), Bishen Singh Mahendra Pal Singh, Dehra Dun

Nozzolillo, C. and Thie, I. (1983) 'Aspects of germination of Impatiens capensis Meerb., formae capensis and immaculata, and I. pallida Nutt', Bulletin of the Torrey Botanical Club, vol 110, pp335–344

Perrings, C., Williamson, M., Barbier, E., Delfino, D., Dalmazzone, S., Shogren, J., Simmons, P. and Watkinson, A. (2002) 'Biological invasion risks and the public good: An economic perspective', Conservation Biology, vol 6, p1

Provan, J., Love, H. M. and Maggs, C. A. (2007) 'Development of microsatellites for the invasive riparian plant Impatiens glandulifera (Himalayan balsam) using intersimple sequence repeat cloning', Molecular Ecology, vol 7, pp451–453

Prowse, A. J. (1998) 'Patterns of early growth and mortality in Impatiens glandulifera', in U. Starfinger, I. Kowarik, K. Edwards and M. Williamson (eds) Plant Invasions: Ecological Mechanisms and Human Responses, Backhuys, Leiden, pp245–252

Pyšek, P. (1998) 'Alien and native species in Central European urban floras: A quantitative comparison', Journal of Biogeography, vol 25, pp155–163

Pyšek, P. and Prach, K. (1993) 'Plant invasions and the role of riparian habitats: A comparison of four species alien to Central Europe', Journal of Biogeography, vol 20, no 4, pp413–420

Pyšek, P. and Prach, K. (1995) 'Invasion dynamic of Impatiens glandulifera: A century of spreading reconstructed', Biological Conservation, vol 74, pp41–48

Scottish Executive (2006) Consultation on Proposals to Amend Schedule 9 and the Use of an Order Made under Section 14A of the Wildlife and Countryside Act 1981, Environment and Rural Affairs Department Development, Edinburgh

Sharma, B. M. and Jamwal, P. S. (1988) Flora of Upper Liddar Valleys of Kashmir Himalaya, Scientific Publishers, Jodhpur

Shaw, R. and Tanner, R. (2008) 'Weed like to see less of them', Biologist, vol 55, pp208–214

Sheppard, A. W., Shaw, R. H. and Sforza, R. (2006) 'Top 20 environmental weeds for classical biological control in Europe: A review of opportunities, regulations and other barriers to adoption', Weed Research, vol 46, pp93–117

Showler, K. (1989) 'The Himalayan balsam [Impatiens glandulifera] in Britain: An undervalued source of nectar', Bee World, vol 70, pp130–131

Stamp, N. E. and Lucas, J. R. (1983) 'Ecological correlates of explosive seed dispersal', Oecologia, vol 59, pp272–278

Sykes, W. R. (1982) 'Checklist of dicotyledons naturalised in New Zealand 15', New Zealand Journal of Botany, vol 20, pp333–341

Tanner, R. (2008) 'A review on the potential for the biological control of the invasive weed, Impatiens glandulifera in Europe', in B. Tokarska-Guzik, J. H. Brock, G. Brundu, L. Child, C. C. Daehler and P. Pyšek (eds) Invasions: Human Perception, Ecological Impacts and Management, Backhuys Publishers, Leiden, pp343–354

Tanner, R., Ellison, C., Shaw, R. H., Evans, H. C. and Gange, A. C. (2008) 'Losing patience with Impatiens. Are natural enemies the solution?', Outlooks on Pest Management, vol 19, pp86–91

UK Wildlife and Countryside Act (1981) www.jncc.gov.uk/page-1377, accessed 9 September 2010

USDA (United States Department of Agriculture) (2010) 'Plants database', http://plants.usda.gov, accessed 29 August 2010

Valentine, D. H. (1978) 'The pollination of introduced species, with special reference to the British Isles and the genus Impatiens', in A. J. Richards (ed) The Pollination of Flowers by Insects, Academic Press, London, pp117–123

Williamson, M. (1996) Biological Invasions, Chapman and Hall, London

Williamson, M., Pyšek, P., Jarošík, V. and Prach, K. (2005) 'On the rates and patterns of spread of alien plants in the Czech Republic, Britain and Ireland', Ecoscience, vol 12, pp424–433

7

Lagarosiphon major (Ridley) Moss ex Wager (curly water weed)

Tobias O. Bickel

Origin and Current Distribution

Lagarosiphon major (Hydrocharitaceae), commonly known as curly water weed or oxygen weed, is a submerged aquatic macrophyte, the biology of which was first described from South Africa (Wager, 1928) (Figure 7.1). The plant was spread by the aquarium trade (sometimes referred to as *Elodea crispa* or *Anacharis crispa*) and currently the plant is introduced in several countries of Europe and regarded as a serious pest in New Zealand and more recently Ireland (see Figure 7.2). *Lagarosiphon major* (hereafter *Lagarosiphon*) is native to the Republic of South Africa and Zimbabwe (Symoens and Triest, 1983). *Lagarosiphon* has a habit of excessive growth in man-made habitats in its native range in South Africa and is seen as problematic there (Obermeyer, 1964).

Currently there are eight species recognized within the genus *Lagarosiphon* (Symoens and Triest, 1983), all of which are found in sub-Saharan Africa; one species is endemic to Madagascar. None of the other species of *Lagarosiphon* appear to be naturalized outside their native range or are seen as problematic weeds. However, *Lagarosiphon ilicifolius* (native to the system) replaced other vegetation in the artificial Lake Kariba in the Zambesi catchment (Zambia, Zimbabwe) and now dominates the submerged macrophyte community in terms of biomass and areal cover (Machena and Kautsky, 1988). Therefore, it is conceivable that this species could become weedy if introduced elsewhere.

New Zealand and Australia

Lagarosiphon was first reported in New Zealand in the 1950s in the Hutt Valley, Wellington. By 1957 it had already reached nuisance proportions in Lake Rotorua (Howard-Williams and Davies, 1988) and by 1980 the weed was naturalized in a further ten regions. Today, *Lagarosiphon* is established throughout New Zealand and is still spreading, with recent new infestations in the South Island (e.g. Oreti River, Lake Wakatipu). A 2009 survey established that *Lagarosiphon* is present in 7.3 per cent of 344 surveyed lakes in New Zealand (de Winton et al, 2009).

The speed of invasion can be appreciated with the example of Lake Taupo, a large lake (616km^2) in the centre of the North Island. *Lagarosiphon* was first found in Lake Taupo in 1966, incidentally in the Lake Taupo boat harbour on the north-eastern shoreline. Within two years, *Lagarosiphon* had practically replaced its competition (the invasive *Elodea canadensis* Michaux) in the area and was recorded on the opposite south side of the lake. By 1980, the plant occurred throughout the lake and dominated the submerged vegetation in terms of biomass and areal cover (Howard-Williams and Davies, 1988). In the early years, *Lagarosiphon* and other adventive submersed species (*Elodea canadensis* and *Egeria densa*) were often deliberately spread and planted for habitat enhancement or as an ornamental plant. For example, *Lagarosiphon* was introduced to Lake Rotorua to improve sport fish production (trout)

Source: John Clayton

Figure 7.1 *A single stem of* Lagarosiphon major

(Chapman, 1967). Until it was declared a noxious plant in 1982, *Lagarosiphon* was also distributed through the aquarium trade, which led to several unintentional infestations through disposal of aquarium contents (Howard-Williams et al, 1987).

There are three records of *Lagarosiphon* in Australia (Tasmania – 1983, Victoria – 1977, Queensland – 1990) but it never established in the wild and is considered currently not to be present in Australia.

Europe

Lagarosiphon is recorded from the majority of countries of western and central Europe and therefore appears widespread: Austria, Belgium, the UK (Channel Islands, England, Scotland, Northern Ireland, Wales), France, Germany, Ireland, Italy and Switzerland (Symoens and Triest, 1983; DAISY, no date). Apart from Ireland, the plant is not widely regarded as a nuisance in Europe, therefore information on its introduction history and distribution is scarce.

Lagarosiphon was introduced to Germany in 1966 and now occurs in five of its states. It is not a common macrophyte yet but is locally expanding (Hussner et al, 2010). Nevertheless, the plant is seen as troublesome and interferes with human activities in at least one instance in Germany (Schwanensee, Bavaria) (Hilt et al, 2006). Similar to this, *Lagarosiphon* is widespread in Belgium (first recorded in 1993), occurs only in

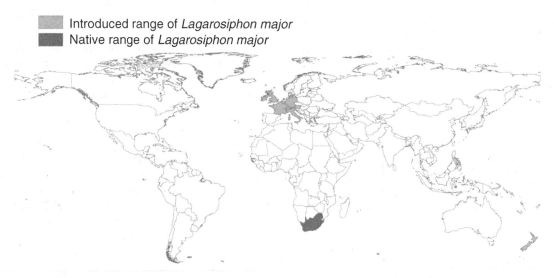

Source: Tobias O. Bickel

Figure 7.2 *Native and introduced distribution of* Lagarosiphon major

isolation, but is still expanding (BFIS, no date). *Lagarosiphon* was introduced fairly recently to the British Isles, where it is now displacing earlier established invasive species (*Elodea canadensis* and *E. nuttallii*), as it is competitively superior to the aforementioned species (James et al, 1999). Of concern in its introduced range in Britain is that even a small increase in average temperature due to climate change could favour *Lagarosiphon* and improve its competitive advantage (McKee et al, 2002; Moss, 2010).

Lagarosiphon is still in its early stage of invasion in Ireland (first recorded in 1966) (Symoens and Triest, 1983). Currently it is present in a number of locations throughout the island but is still rather rare (Minchin, 2007; Hackney, no date). However, in Lough Corrib it forms dense surface-reaching stands, is seen as a serious threat to native plants and wildlife, and is impacting on tourism and particularly the fisheries value of the lake (Caffrey et al, 2010). The rapid colonization of Lake Corrib is very similar to that seen in New Zealand. *Lagarosiphon* was first identified in Lake Corrib in 2005 (but was present presumably prior to this date) and quickly started to spread through the entire system. By 2008, the plants were already present in 113 sites and attained a high standing crop biomass in sheltered locations. At infested sites, *Lagarosiphon* completely replaced the diverse native macrophyte community (Caffrey et al, 2010).

The Ecology of *Lagarosiphon major* in its Native and Introduced Range

Lagarosiphon is a rooted submerged macrophyte that is found in a range of freshwater habitats. It is physiologically limited to water depths of less than 6.6m due to its inability to produce anchorage roots with increasing water pressure (Coffey and Wah, 1988). In New Zealand, the plant is found most commonly from 0.5 to 6.5m but usually dominates the submerged vegetation from 2 to 5m where it displaces the native tall vascular community of milfoils, pondweeds and characean meadows. Hence, in infested lakes the diverse native macrophyte assemblages are susceptible to displacement by the taller growing *Lagarosiphon*, which subsequently often forms monospecific stands (Clayton, 1996), or at least strongly impacts on the abundance of native macrophytes (Chapman et al,

1971; Clayton, 1982). Once a system is invaded, diverse native characean communities often only persist beyond the depth range of *Lagarosiphon* and other tall adventives (Howard-Williams et al, 1987). These impacts are greater in lakes with low water transparency and therefore a reduced overall depth limit for plant growth, i.e. if the depth limit of plant growth coincides with the depth limitations for *Lagarosiphon* there are no refuges for native charophytes (Howard-Williams et al, 1987). Therefore, the invasion of New Zealand freshwaters by tall growing macrophytes such as *Lagarosiphon* is one of the primary reasons for the decline of native macrophytes and is of special concern for the conservation of endangered aquatic plant species (Clayton, 1996). Beside the direct displacement of native vegetation, invasive Hydrocharitaceae also reduce the seed bank of native macrophytes in infested areas (de Winton and Clayton, 1996), therefore hindering future recruitment and re-establishment of native plant communities once *Lagarosiphon* is removed. However, there is a high variability between sites and there is not always a clear pattern between invasion by Hydrocharitaceae and a reduction of the native seed bank (de Winton et al, 2002). This is particularly the case in clear water lakes where native charophyte communities can still persist beyond the depth limit of *Lagarosiphon* (Clayton, J., pers. comm.).

Lagarosiphon usually grows in dense monospecific stands and can attain biomass well in excess of 1000g dry weight m^{-2}, as recorded in Lake Taupo (Howard-Williams and Davies, 1988) (Figure 7.3). Biomass usually is at its highest in the 4m depth zone. The biomass and height of *Lagarosiphon* is mainly determined by exposure to wave action (effective fetch), the slope and proportion of fine sediment (Howard-Williams and Davies, 1988; Riis and Biggs, 2001; Clayton and Champion, 2006). In Lake Taupo, dense *Lagarosiphon* stands with closed canopies above 1m were restricted to sites with a wind fetch of less than 2km and no *Lagarosiphon* was recorded when effective wind fetch was more than 10km (Howard-Williams and Davies, 1988). The number of native species was negatively related to *Lagarosiphon* biomass and height (Howard-Williams and Davies, 1988). In Lake Rotoma, *Lagarosiphon* reached a record biomass for submerged aquatic plants, with 3518g m^{-2} dry mass (Clayton, 1982). However, a later study of *Lagarosiphon* in the artificial hydro lake Lake Dunstan in the South Island of New Zealand found an average of 2472g m^{-2}

Lagarosiphon dry mass and a maximum of 8321g m^{-2} dry mass (Bickel, 2006).

Lagarosiphon grows best in clear water due to its preference of high light conditions, and has a light compensation point in the order of 10–15μmol m^{-2} s^{-1} and reaches light saturation at 90–170μmol m^{-2} s^{-1} (Coffey and Wah, 1988; Rattray, 1989; Schwarz and Howard-Williams, 1993). *Lagarosiphon* is known to decline with increasing turbidity (Coffey and Clayton, 1988; Wells and Clayton, 1990) caused by planktonic algae in increasingly eutrophic waters. Also, other invasive submerged weeds that are more tolerant of shading can outcompete *Lagarosiphon* once water clarity declines, for example *Egeria densa* or *Ceratophyllum demersum* (Coffey and Clayton, 1988; Tanner et al, 1990a; Wells and Clayton, 1990). Fine sandy sediments seem to favour its colonization (Chapman et al, 1971) and support the highest standing crop biomass. However, *Lagarosiphon* can also be found growing in small scattered stands in rocky shores where it takes hold in crevices. As *Lagarosiphon* effectively filters sediment out of the water column, it practically enhances its own habitat once dense beds are established.

Contrary to popular belief, *Lagarosiphon* can grow successfully in nutrient-poor waters. This can be seen in New Zealand lakes where the plant grows luxuriantly in oligotrophic environments (Howard-Williams et al, 1987) as the record biomasses from oligotrophic Lake Rotoma and Lake Dunstan illustrate. In fact, contrary

Source: John Clayton

Figure 7.3 *A monospecific stand of* Lagarosiphon *in Lake Tarawera*

to intuition, eutrophic waters might even be detrimental to growth of *Lagarosiphon* due to reduced water transparency (phytoplankton blooms) (Coffey and Clayton, 1988; Wells and Clayton, 1990) and an increase in epiphyton growth (Rattray et al, 1991). The reason for the success of *Lagarosiphon* in oligotrophic systems is that it can satisfy its nutrient requirements from the substrate (e.g. Rattray et al, 1991). In New Zealand, *Lagarosiphon* is often problematic in hydro lakes; fertile farmland or glacial deposits (loess soils) are flooded and provide ideal growing conditions for the plant even though nutrient concentrations in the water itself are very low. When growing in eutrophic water, nutrient concentrations in the substrate seem to be less important, showing that the plant can successfully utilize nutrients from the water column and the sediment (Rattray et al, 1991). *Lagarosiphon* shows a plastic morphological response to nutrient loadings in the water (trophic status): when growing in high nutrient concentrations *Lagarosiphon* has a lower root to shoot mass ratio and a reduced root biomass (James et al, 2006). *Lagarosiphon* also has the ability to take up and store nitrogen, and to some degree phosphorus, beyond needs if available at high concentrations (James et al, 2006); this might give the plant a competitive advantage if conditions change (depletion of nutrients). However, the tissue concentration of phosphorus and nitrogen in *Lagarosiphon* is lower than in *Elodea* spp. (Rattray et al, 1991; James et al, 2006) and there does not always seem to be a relationship between trophic status of water or sediment and nutrient concentrations in plant tissue of *Lagarosiphon* (Rattray et al, 1991).

Overall, excessive growth of *Lagarosiphon* seems to be more dependent on shelter, water temperature and a lack of herbivores (enemy release) (Howard-Williams et al, 1987; Riis and Biggs, 2001; Clayton and Champion, 2006). The comparably mild water temperatures and lack of ice cover in New Zealand allow *Lagarosiphon* to grow year round as there seems to be no variation in biomass between seasons (Schwarz and Howard-Williams, 1993; Riis et al, 2003). Given the broad environmental tolerance of *Lagarosiphon*, dispersal, combined with an empty ecological niche (lack of canopy-forming native species), seems to be a more important factor than habitat in determining the ultimate distribution of this species (e.g. Howard-Williams, 1993; de Winton et al, 2009). *Lagarosiphon* has been classified as an R-strategist, associated with traits such as fast colonizing abilities, fast growth and high disturbance resistance (Riis and Biggs, 2001). All of the plants classified as R-strategists in New Zealand are introduced submersed weeds, showing that one of the reasons for the success of these invaders in New Zealand is the vacant ecological niche they are occupying (Howard-Williams et al, 1987; Clayton, 1996).

Since increased nutrient concentration in the water (eutrophication) does not seem to explain the competitive advantage of *Lagarosiphon* towards other macrophyte species in Europe (James et al, 2006), it is more likely that the fast growth rates of *Lagarosiphon* enable it to outcompete other plants by shading them, similar to findings with other fast growing species (*Elodea nuttallii*) (Barrat-Segretain and Elger, 2004). When shoot fragments of *Lagarosiphon* are planted, the growth resources are primarily channelled into shoot development and plants rapidly increase in size after establishment. This fast development enables the plant to outgrow native macrophytes that invest more heavily in the development of an extensive root system (Rattray et al, 1994). Once *Lagarosiphon* outgrows other species it then forms a canopy that shades any potential competitors. Light reduction to below 1 per cent of surface light occurs at about 0.5–2.7m depending on conditions (Schwarz and Howard-Williams, 1993). *Lagarosiphon* also has a competitive advantage due to its ability to cope with adverse water conditions (high pH, low CO_2 and high oxygen stress) (James et al, 1999). *Lagarosiphon* is able to use bicarbonate as an alternative inorganic carbon source (Bain and Proctor, 1980). The high photosynthetic rate of *Lagarosiphon* results in a fast increase in pH, which could disadvantage pH-sensitive plants (James et al, 1999). Dense stands of submerged macrophytes, especially canopy-forming species, are known to significantly alter the water chemistry (DO, CO_2, pH) (e.g. Frodge et al, 1990). Therefore, dense stands of *Lagarosiphon* could create areas of high pH and low dissolved inorganic carbon that extend into surrounding water and negatively suppress photosynthetic performance of potential competitors, while *Lagarosiphon* itself can grow successfully in water with high pH and low concentrations of free CO_2 (James et al, 1999). However, rapid uptake of dissolved inorganic carbon in dense *Lagarosiphon* canopies can lead to the occasional cessation of photosynthesis if there is a lack of HCO_3^- as alternative carbon source (Schwarz and Howard-Williams, 1993).

All *Lagarosiphon* species are dioecious, i.e. they have separate male and female plants with morphologically different flowers. The pollination process is intricate (Wager, 1928; Symoens and Triest, 1983). The male flowers break off and freely float on the water surface; three of the stamens function as little sails. The female flowers sit on the water surface in a way that creates a small depression to attract the male flowers and to initiate fertilization. After ripening, the seeds are released and initially float before sinking to the sediments and are therefore effectively disseminated (Wager, 1928; Symoens and Triest, 1983).

Sexual reproduction of *Lagarosiphon* is restricted to its native range. In its introduced range it exclusively reproduces through vegetative propagation. In New Zealand for example, all plants are female and therefore sexual reproduction is not possible. However, the plant readily reproduces from stem fragments, which are spread between water bodies by human vectors (e.g. boating and fishing). *Lagarosiphon*, being practically monoclonal in New Zealand, shows a very low genetic variability there (Lambertini et al, 2010). However, even in its native range many *Lagarosiphon* populations are unisexual and monoclonal, and isozyme analysis reveals low genetic variability between populations (Triest, 1991).

Management of *Lagarosiphon*

As *Lagarosiphon* spreads only vegetatively in its introduced range, the control of dispersal between freshwaters systems is critical in managing this species. This is especially the case as upon establishment in a large system the species is virtually impossible to eradicate. In New Zealand, inter-lake dispersal of *Lagarosiphon* is directly related to boating activities. As the weed is primarily taken up at boat launches, control efforts in these areas are thought to greatly reduce the risk of spread to nearby unaffected water bodies and should be a prime focus for management. Much more difficult to manage is the long distance dispersal to unaffected catchments that happens mainly through new arrivals in the form of ornamental pond escapes or the thoughtless disposal of aquarium contents. There are several cases in New Zealand where escapes from ornamental ponds were responsible for new infestations, such as for example the escape of *Lagarosiphon* from a farm dam into the Ahuriri River and from there into Lake Benmore in the South Island.

Chemical control

Lagarosiphon is predominantly controlled with herbicides in New Zealand, mainly because of the low cost of control per area compared to other options. An early trial with sodium arsenite in Lake Rotoroa (Lake Hamilton) was described as 'spectacular' with all submerged aquatic vegetation destroyed due to the unusually high amount of arsenic applied, and macrophyte control lasting for a period of five years (Tanner and Clayton, 1990). However, subsequent applications of sodium arsenite in the Rotorua Lakes in 1961 had little success, either because of low water temperatures that prevented uptake by the plants or because *Lagarosiphon* was resistant to arsenic due to natural exposure in this geothermally active area (Fish, 1963). Further use of arsenic for control of *Lagarosiphon* in New Zealand was discouraged because of its toxicity and the availability of alternative herbicides such as diquat (Clayton, 1986).

There is now a long history of successful and environmentally sound use of diquat to control nuisance macrophyte growth in New Zealand, in particular *Lagarosiphon*. Diquat is currently being used on a large scale in New Zealand with annual and biannual applications in a large number of lakes and smaller water bodies. To minimize drift, improve target accuracy and reduce non-target impacts, diquat is used in the form of a gel or viscous formulation. Despite the successful use of diquat for *Lagarosiphon* control, there are still occasional limitations in areas of low water clarity, as diquat readily binds to clay or charged particles in the water column and becomes deactivated (Hofstra et al, 2001). Additionally, deposits on the plant surface (epiphyton and organic particulate material) can be a significant or total barrier to diquat uptake by the target plant (Clayton and Matheson, 2010). If application is carefully planned and environmental conditions are taken into account, diquat can be used very successfully to treat even large areas cost effectively (Clayton, 1996; Clayton and Matheson, 2010). As some of the native New Zealand macrophytes (particularly characeans) are less sensitive to diquat exposure, these desirable native plant communities can selectively be maintained (Clayton and Tanner, 1988; Tanner et al, 1990a).

For example, the use of diquat at 0.5ppm completely removed *Lagarosiphon* from a bay in Lake Rotoiti, with a subsequent recovery of native macrophytes (*Nitella* sp.) (Fish, 1966). However, as treatment of *Lagarosiphon*

with diquat rarely achieves eradication from an entire system, repeated treatments (often twice a year) are necessary to achieve required control. The recovery of very high *Lagarosiphon* biomass in Lake Rotoma, New Zealand, only two years after diquat control illustrates this point (Clayton, 1982). Also, the removal of *Lagarosiphon* by diquat (or any other means) can enable other invasive submerged species to take advantage. For example, management of introduced *Elodea canadensis* and *Lagarosiphon* stands in Lake Rotoroa (Lake Hamilton) increased the abundance of native charophytes, but the invasive species *Egeria densa* also became abundant and replaced *Lagarosiphon* as the predominant species in this lake in the long term (Tanner et al, 1990a). However, *Egeria* has a competitive advantage over *Lagarosiphon*, depending on trophic state, and would therefore have outcompeted *Lagarosiphon* in this mesotrophic water eventually.

Experiments carried out with herbicides other than diquat have shown mixed results in New Zealand. Fluridone was deemed unsuccessful in controlling *Lagarosiphon* in controlled outdoor conditions and in the field (Wells et al, 1986). A controlled test of endothal, triclopyr and dichlobenil revealed that only endothal was able to control *Lagarosiphon* (Hofstra and Clayton, 2001). As endothal is not deactivated in turbid water like diquat, it is a viable alternative if environmental conditions do not favour the use of diquat (Hofstra et al, 2001). After this encouraging research, endothal was registered for use in New Zealand in 2004 (Wells and Champion, 2010). Recent field trials in the South Island, New Zealand, showed that endothal can be highly effective to control *Lagarosiphon*. Even at very low concentrations (0.11ppm, below drinking water restrictions) *Lagarosiphon* was completely removed from small gravel extraction ponds (Wells and Champion, 2010). Encouragingly, native milfoils completely recovered within ten months post treatment. The successful field use of endothal shows that this is a viable alternative to control *Lagarosiphon* infestations if conditions are unsuitable for the use of diquat, provided long contact times (three–seven days) can be achieved (Hofstra and Clayton, 2001).

Biological control

There have been considerable efforts to find biocontrol agents to control *Lagarosiphon*, both in Europe (Baars et al, 2010) and New Zealand (Clayton, J., pers. comm.).

There are generally fewer success stories for biological control of submerged weeds, but there is limited knowledge about the natural arthropod or fungal enemies of the plant in its native range, so the prospect of effective biological control should not be ruled out (Gassmann et al, 2006). There are currently new programmes to identify biocontrol agents for *Lagarosiphon* to control infestations in Europe, and several promising natural enemies have been identified in the native range (Baars et al, 2010).

Grass carp (*Ctenopharyngodon idella*), although strictly speaking not a biological control agent as they are not host-specific, can be a cost effective and long term option to control excessive aquatic weed growth in both standing and flowing waters. Grass carp can be selective in their food choice and feeding rates vary with water temperatures, which have to be taken into account when choosing stocking rates. Food preference studies in New Zealand show that grass carp consume a variety of native macrophytes, and often prefer them to exotic species (Edwards, 1974; Rowe and Schipper, 1985). However, the feeding selectivity is usually not an issue from a management perspective as grass carp are often used to control monospecific stands of submerged weeds. The use of grass carp is climatically limited to water bodies that reach summer water temperatures above 15°C, and they have been successfully used in several South Island lakes in New Zealand (Clayton and Wells, 1999).

Determining the correct stocking density of grass carp can be difficult if partial macrophyte control (as opposed to total control) is the objective (Pípalová, 2006) as too few stocked fish can result in little control at all and too many may indiscriminately remove all the plants in a system. Once control is achieved, it can be difficult and costly to remove all fish from large water bodies (Clayton and Wells, 1999). However, if total vegetation control or eradication of invasive weeds is the anticipated outcome, stocking with grass carp is a rather straightforward and cost efficient management option. There has been some excellent success with the eradication of *Egeria densa* from Lake Parkinson after heavy stocking with grass carp (Tanner et al, 1990b). Once the fish were removed using netting and piscicides (rotenone), the lake returned to its natural vegetation by recruitment from the seed bank (Tanner et al, 1990b).

The use of grass carp may be limited due to biosecurity issues with stocking alien fish that are considered invasive in some regions, as in some US states. We also currently lack a thorough understanding of environmental impacts if escaped fish are able to establish in the wild (Dibble and Kovalenko, 2009). This problem can be circumvented by using sterile triploid grass carp; these are, however, more expensive. Nevertheless, unwanted reproduction of grass carp is only an issue in areas where conditions are suitable as grass carp have very strict habitat requirements for spawning (Pípalová, 2006). There is no evidence of natural grass carp reproduction in New Zealand, therefore, the use of triploid grass carp is not necessary there. As it is difficult to completely contain fish in an open system, there is a small risk associated with escaped grass carp impacting on native macrophytes in non-target sites. However, as densities of escaped fish are usually low, overall impact should be negligible in areas where fish can't reproduce (Clayton and Wells, 1999).

Physical control

Harvesting of aquatic weeds is a further option, especially where there are legislative or societal barriers to the use of herbicides. This can be a very efficient and long-lasting means to remove troublesome weeds, especially in smaller streams and canals, under the condition that the target weeds are already present in the entire water body to avoid further establishment by fragments created through mechanical control activities. Problems with harvesting in large water bodies arise with the disposal of harvested material, the spread of fragments, speed of regrowth and obstacles in the water (Clayton, 1996). Weed harvesting is used in New Zealand waters where there is local opposition to chemical control. As the area that can be treated is limited (slow process), harvesting is restricted to smaller infestations or to high use areas (boat ramps, swimming areas). Examples in New Zealand show that if *Lagarosiphon* is harvested repeatedly and close to the bottom, a change to more desirable vegetation can occur in oligotrophic systems (Howard-Williams, 1993).

Removal by suction dredging is a further frequently used mechanical weed control method that is less restricted by water depth, irregular bottom contours and obstacles in the water (as opposed to cutting with cutter bars) (Clayton, 1996). Hand weeding can be a very efficient follow-up method after suction dredging to remove remaining fragments and to remove small-scale infestations.

The disadvantage of mechanical control methods are the relatively high cost associated with manual labour. If weed is only partially or locally removed, *Lagarosiphon* readily re-establishes from leftover fragments and encroachment of remaining *Lagarosiphon* stands (Howard-Williams and Reid, 1989; Bickel and Closs, 2009). However, depending on the objectives of control, this is not necessarily a problem; for example if the removal of surface-reaching *Lagarosiphon* is required for the recreational period only.

A sometimes very effective way of controlling *Lagarosiphon*, and submerged weeds in general, is the drawdown of water levels and subsequent desiccation of the plants; obviously this is only possible in artificial lakes. Lakes that have a large annual water level fluctuation (natural or due to operation regime) usually do not support much macrophyte growth (Clayton, 1982; Clayton et al, 1986). Drawdown and the subsequent desiccation of weeds works both in summer and in winter (frost). However, drawdown also has serious ecological and economic issues due to the disruption of important littoral macrophyte beds and the loss of water capacity and electricity generation. Therefore this method was discontinued in New Zealand from the late 1970s (Clayton, 1996). Also, from experience in New Zealand hydro lakes, the outcome in terms of *Lagarosiphon* control can be very variable (Clayton, 1996). In addition, fluctuating water levels can aid the spread of the plant in the system due to the fragmentation of *Lagarosiphon* stands and the dispersal of propagules (Clayton, 1982).

Shading of small infestations with plastic sheets as bottom cover can be a very efficient and cost effective means of control (Clayton, 1996). If applied properly, sheets can provide control for many years. However, this technique is restricted to sheltered sites with little water movement. There is also a range of more durable, though more expensive, geotextile products available that are less sensitive to wave action or uplift due to gas accumulation. More recently, the fisheries authorities of Ireland conducted experiments with decomposable jute mats to shade out *Lagarosiphon* infestations in Lake Corrib (Caffrey et al, 2010). The jute mats gave a very high degree of control, with the majority of *Lagarosiphon* decayed after four months. Seven months after placement, the mats were colonized by native charophytes and, to some degree, by other macrophyte species. Overall, the jute matting has several advantages over plastic sheeting: it is easier to place due to its negative buoyancy, it is biodegradable and therefore cost effective (no removal necessary) and gas permeable (preventing the creation of anoxic conditions), stabilizes sediments and, lastly, assists the regeneration of native macrophytes from the seed bank (Caffrey et al, 2010).

The combination of bottom lining and hand weeding over several years following detection of an early *Lagarosiphon* outbreak resulted in successful eradication from the substantial Lake Waikaremoana, New Zealand (Department of Conservation, 2008).

Controversies Regarding the Management of *Lagarosiphon*

Even though the use of diquat to control macrophytes in freshwater environments is deemed safe (Emmet, 2002), there are sometimes issues with public resistance (chemophobia) or legislative barriers to the use of herbicides in freshwater systems. Chemophobia has successfully been addressed with educational processes in New Zealand, as diquat has a usage history of 50 years with a good environmental record of acceptable use and because it is by far the most cost effective large-scale method of submerged aquatic weed control.

However, if weed management is limited to non-herbicidal techniques due to public resistance or legislation, this can hinder control efforts and limit successful management. The initial lack of any *Lagarosiphon* management and subsequent restriction to non-chemical control in Lake Wanaka, New Zealand, compromised the successful containment of *Lagarosiphon* within a small area of the lake, resulting now in a significant financial commitment to manage a much larger infestation after substantial spread.

The use of herbicides for aquatic weed control, like all control options, can cause undesired side effects. For example, deoxygenation of water after treatment of large infestations may occur due to the decay of large amounts of weed biomass. However, this can be avoided through better timing (cooler months) to prevent a significant reduction in oxygen (Fish, 1966). Release of nutrients from decaying plants after chemical treatment can lead to the proliferation of algae, particularly in small static water bodies (Fish, 1966). As a consequence, lakes can switch to undesirable alternative stable states and become dominated by planktonic algae after the removal of large amounts of macrophytes.

Some of the most challenging issues of macrophyte management are the widely varying perceptions by different user groups of the necessary level of *Lagarosiphon* control and the fact that management goals can be mutually exclusive (Johnstone, 1986; Van Nes et al, 2002). For example, boating enthusiasts and swimmers would prefer a complete removal of all macrophytes, while fishermen see a certain amount of *Lagarosiphon* as beneficial to fish growth as it provides large quantities of invertebrate food (Kelly and Hawes, 2005; Bickel and Closs, 2008). Keeping this in mind, there is a realistic risk of intentional spread of *Lagarosiphon* to improve fisheries by an uneducated public (Tanner et al, 1986).

Contrary to public perception, the do-nothing approach might be a valid option for macrophyte control and can be fully justified in certain situations (Clayton, 1996). Aquatic plants are important components of freshwater ecosystems, so a complete removal of plants can have unwanted ecological consequences. The dense growth and high biomass of *Lagarosiphon* is thought to have profound ecological impacts on affected freshwater systems. For example, littoral food webs in infested lakes shift from a detrital basis to one dominated by epiphytal production (Kelly and Hawes, 2005; Bickel, 2006). The reason for this is the large surface area for epiphyton growth that exotic macrophytes offer, boosting the amount of colonizable substrate and increasing the abundance of grazing invertebrates compared to the abundance attained within the much smaller and less structurally complex native macrophytes. Both the number and diversity of epiphytic invertebrates is higher on *Lagarosiphon* as compared to native vegetation (Kelly and Hawes, 2005; Bickel, 2006). Even in the early days of invasion, the large number of invertebrates colonizing *Lagarosiphon* was noted (Fish, 1963). Therefore, the plant is seen as a valuable foraging ground for sport fish, expected to increase sport fish production, and is also a valuable foraging habitat for waterfowl. However, the do-nothing approach should incorporate a strategy to prevent spread of propagules. Considering the impacts on ecology and biodiversity of *Lagarosiphon* in invaded systems, containment and prevention of spread are paramount and should be included in any management plan.

References

Baars, J. R., Coetzee, J. A., Martin, G., Hill, M. P. and Caffrey, J. M. (2010) 'Natural enemies from South Africa for biological control of *Lagarosiphon major* (Ridl.) Moss ex Wager (Hydrocharitaceae) in Europe', *Hydrobiologia*, vol 656, no 1, pp149–158

Bain, J. T. and Proctor, M. C. F. (1980) 'The requirement of aquatic bryophytes for free CO_2 as an inorganic carbon source: Some experimental evidence', *New Phytologist*, vol 86, no 4, pp393–400

Barrat-Segretain, M. and Elger, A. (2004) 'Experiments on growth interactions between two invasive macrophyte species', *Journal of Vegetation Science*, vol 15, no 1, pp109–114

BFIS (Belgian Forum on Invasive Species) (no date) 'Invasive species in Belgium: *Lagarosiphon major*', http://ias.biodiversity.be/species/show/68, accessed 20 December 2010

Bickel, T. O. (2006) '*Lagarosiphon major*: An introduced macrophyte and its ecological role in the littoral of Lake Dunstan, New Zealand', PhD Thesis, Department of Zoology, University of Otago, Dunedin, New Zealand

Bickel, T. O. and Closs, G. P. (2008) 'Fish distribution and diet in relation to the invasive macrophyte Lagarosiphon major in the littoral zone of Lake Dunstan, New Zealand', *Ecology of Freshwater Fish*, vol 17, no 1, pp10–19

Bickel, T. O. and Closs, G. P. (2009) 'Impact of partial removal of the invasive macrophyte *Lagarosiphon major* (Hydrocharitaceae) on invertebrates and fish', *River Research and Applications*, vol 25, no 6, pp734–744

Caffrey, J. M., Millane, M., Evers, S., Moran, H. and Butler, M. (2010) 'A novel approach to aquatic weed control and habitat restoration using biodegradable jute matting', *Aquatic Invasions*, vol 5, no 2, pp123–129

Chapman, V. J. (1967) 'Conservation of maritime vegetation and the introduction of submerged freshwater aquatics', *Micronesica*, vol 3, pp31–35

Chapman, V. J., Brown, J. M. A., Dromgoole, F. I. and Coffey, B. T. (1971) 'Submerged vegetation of the Rotorua and Waikato Lakes. 1. Lake Rotoiti', *New Zealand Journal of Marine and Freshwater Research*, vol 5, no 2, pp259–279

Clayton, J. S. (1982) 'Effects of fluctuations in water level and growth of *Lagarosiphon major* on the vascular plants in Lake Rotoma, 1973–80', *New Zealand Journal of Marine and Freshwater Research*, vol 16, pp89–94

Clayton, J. S. (1986) 'Review of diquat use in New Zealand for submerged weed control', in *Proceedings of the EWRS/AAB 7th Symposium on Aquatic Weeds*, EWRS, Loughborough, UK, pp73–79

Clayton, J. S. (1996) 'Aquatic weeds and their control in New Zealand lakes', *Lake and Reservoir Management*, vol 12, no 4, pp477–486

Clayton, J. and Champion, P. (2006) 'Risk assessment method for submerged weeds in New Zealand hydroelectric lakes', *Hydrobiologia*, vol 570, no 1, pp183–188

Clayton, J. and Matheson, F. (2010) 'Optimising diquat use for submerged aquatic weed management', *Hydrobiologia*, vol 656, no 1, pp159–165

Clayton, J. S. and Tanner, C. C. (1988) 'Selective control of submerged aquatic plants to enhance recreational uses for water bodies', *Verhandlung Internationale Vereinigung für Theoretische und Angewandte Limnologie*, vol 23, pp1518–1521

Clayton, J. S. and Wells, R. D. S. (1999) 'Some issues in risk assessment reports on grass carp and silver carp', *Conservation Advisory Science Notes 257*, Department of Conservation, Wellington

Clayton, J. S., Schwarz, A. and Coffey, B. T. (1986) 'Notes on the submerged vegetation of Lake Hawea', *New Zealand Journal of Marine and Freshwater Research*, vol 20, pp185–189

Coffey, B. T. and Clayton, J. S. (1988) 'Changes in the submerged macrophyte vegetation of Lake Rotoiti, central North Island, New Zealand', *New Zealand Journal of Marine and Freshwater Research*, vol 22, no 2, pp215–223

Coffey, B. T. and Wah, C. K. (1988) 'Pressure inhibition of anchorage-root production in *Lagarosiphon major* (Ridl.) Moss: A possible determinant of its depth range', *Aquatic Botany*, vol 29, pp289–301

DAISY (no date) 'DAISY – species factsheet', www.europe-aliens.org/speciesFactsheet.do?speciesId=1142#, accessed 20 December 2010

Department of Conservation (2008) 'Rosie Bay re-opened to the public', www.doc.govt.nz/about-doc/news/media-releases/2008/rosie-bay-re-opened-to-the-public, accessed 20 December 2010

de Winton, M. D., Champion, P. D., Clayton, J. S. and Wells, R. D. S. (2009) 'Spread and status of seven submerged pest plants in New Zealand lakes', *New Zealand Journal of Marine and Freshwater Research*, vol 43, no 2, pp547–561

de Winton, M. D. and Clayton, J. S. (1996) 'The impact of invasive submerged weed species on seed banks in lake sediments', *Aquatic Botany*, vol 53, no 1–2, pp31–45

de Winton, M. D., Taumoepeau, A. T. and Clayton, J. S. (2002) 'Fish effects on charophyte establishment in a shallow, eutrophic New Zealand lake', *New Zealand Journal of Marine and Freshwater Research*, vol 36, pp815–823

Dibble, E. D. and Kovalenko, K. (2009) 'Ecological impact of grass carp: A review of the available data', *Journal of Aquatic Plant Management*, vol 47, pp1–15

Edwards, D. J. (1974) 'Weed preference and growth of young grass carp in New Zealand', *New Zealand Journal of Marine and Freshwater Research*, vol 8, no 2, pp341–350

Emmet, K. (2002) 'Appendix A: Final risk assessments for diquat dibromide', Washington State Department of Ecology, Olympia, Washington State

Fish, G. R. (1963) 'Observations on excessive weed growth in two lakes in New Zealand', *New Zealand Journal of Botany*, vol 1, pp410–418

Fish, G. R. (1966) 'Some effects of the destruction of aquatic weeds in Lake Rotoiti, New Zealand', *Weed Research*, vol 6, no 4, pp350–358

Frodge, J. D., Thomas, G. L. and Pauley, G. B. (1990) 'Effects of canopy formation by floating and submergent aquatic macrophytes on the water quality of two shallow Pacific Northwest lakes', *Aquatic Botany*, vol 38, pp231–248

Gassmann, A., Cock, M., Shaw, R. and Evans, H. (2006) 'The potential for biological control of invasive alien aquatic weeds in Europe: A review', *Hydrobiologia*, vol 570, no 1, pp217–222

Hackney, P. (no date) 'Invasive alien species in Northern Ireland', www.habitas.org.uk/invasive/species.asp?item=2117, accessed 20 December 2010

Hilt, S., Gross, E. M., Hupfer, M., Morscheid, H., Mählmann, J., Melzer, A., Poltz, J., Sandrock, S., Scharf, E.-M., Schneider, S. and Van De Weyer, K. (2006) 'Restoration of submerged vegetation in shallow eutrophic lakes – a guideline and state of the art in Germany', *Limnologica – Ecology and Management of Inland Waters*, vol 36, no 3, pp155–171

Hofstra, D. E. and Clayton, J. S. (2001) 'Evaluation of selected herbicides for the control of exotic submerged weeds in New Zealand: I. the use of endothall, triclopyr and dichlobenil', *Journal of Aquatic Plant Management*, vol 39, no 20–24, p20

Hofstra, D. E., Clayton, J. S. and Getsinger, K. D. (2001) 'Evaluation of selected herbicides for the control of exotic submerged weeds in New Zealand: II. the effects of turbidity on diquat and endothall efficacy', *Journal of Aquatic Plant Management*, vol 39, pp25–27

Howard-Williams, C. (1993) 'Processes of aquatic weed invasions – the New Zealand example – plenary address', *Journal of Aquatic Plant Management*, vol 31, pp17–23

Howard-Williams, C. and Davies, J. (1988) 'The invasion of Lake Taupo by the submerged water weed *Lagarosiphon major* and its impact on the native flora', *New Zealand Journal of Ecology*, vol 11, pp13–19

Howard Williams, C. and Reid, V. (1989) 'Aquatic weed survey of Lake Whakamarino following dredging', DSIR unpublished report to Electrocorp

Howard-Williams, C., Clayton, J. S., Coffey, B. T. and Johnstone, I. M. (1987) 'Macrophyte invasions', in A. B. Viner (ed) *Inland Waters of New Zealand*, DSIR Science Information Publishing Centre, Wellington, pp307–331

Hussner, A., Van De Weyer, K., Gross, E. M. and Hilt, S. (2010) 'Comments on increasing number and abundance of non-indigenous aquatic macrophyte species in Germany', *Weed Research*, vol 50, pp519–526

James, C. S., Eaton, J. W. and Hardwick, K. (1999) 'Competition between three submerged macrophytes, *Elodea canadensis* Michx, *Elodea nuttallii* (Planch.) St John and *Lagarosiphon major* (Ridl.) Moss', *Hydrobiologia*, vol 415, pp35–40

James, C. S., Eaton, J. W. and Hardwick, K. (2006) 'Responses of three invasive aquatic macrophytes to nutrient enrichment do not explain their observed field displacements', *Aquatic Botany*, vol 84, no 4, pp347–353

Johnstone, I. M. (1986) 'Macrophyte management: An integrated perspective', *New Zealand Journal of Marine and Freshwater Research*, vol 20, pp599–614

Kelly, D. J. and Hawes, I. (2005) 'Effects of invasive macrophytes on littoral zone productivity and foodweb dynamics in a New Zealand high-country lake', *Journal of the North American Benthological Society*, vol 24, no 2, pp300–320

Lambertini, C., Riis, T., Olesen, B., Clayton, J. S., Sorrell, B. K. and Brix, H. (2010) 'Genetic diversity in three invasive clonal aquatic species in New Zealand', *Bmc Genetics*, vol 11, p18

Machena, C. and Kautsky, N. (1988) 'A quantitative diving survey of benthic vegetation and fauna in Lake Kariba, a tropical man-made lake', *Freshwater Biology*, vol 19, no 1, pp1–14

McKee, D., Hatton, K., Eaton, J. W., Atkinson, D., Atherton, A., Harvey, I. and Moss, B. (2002) 'Effects of simulated climate warming on macrophytes in freshwater microcosm communities', *Aquatic Botany*, vol 74, no 1, pp71–83

Minchin, D. (2007) 'A checklist of alien and cryptogenic aquatic species in Ireland', *Aquatic Invasions*, vol 2, no 4, pp341–366

Moss, B. (2010) 'Climate change, nutrient pollution and the bargain of Dr Faustus', *Freshwater Biology*, vol 55, no s1, pp175–187

Obermeyer, A. A. (1964) 'The South African species of *Lagarosiphon*', *Bothalia*, vol 8, pp139–146

Pípalová, I. (2006) 'A review of grass carp use for aquatic weed control and its impact on water bodies', *Journal of Aquatic Plant Management*, vol 44, pp1–12

Rattray, M. R. (1989) 'An ecophysiological investigation of the growth and nutrition of three submerged macrophytes in relation to lake eutrophication', PhD Thesis, Botany Department, University of Auckland, Auckland

Rattray, M. R., Howard Williams, C. and Brown, J. M. A. (1991) 'Sediment and water as sources of nitrogen and phosphorus for submerged rooted aquatic macrophytes', *Aquatic Botany*, vol 40, no 3, pp225–237

Rattray, M. R., Howard-Williams, C. and Brown, J. M. A. (1994) 'Rates of early growth of propagules of *Lagarosiphon major* and *Myriophyllum triphyllum* in lakes of differing trophic status', *New Zealand Journal of Marine and Freshwater Research*, vol 28, pp235–241

Riis, T. and Biggs, B. J. F. (2001) 'Distribution of macrophytes in New Zealand streams and lakes in relation to disturbance frequency and resource supply – a synthesis and conceptual model', *New Zealand Journal of Marine and Freshwater Research*, vol 35, no 2, pp255–267

Riis, T., Biggs, B. J. F. and Flanagan, M. (2003) 'Seasonal changes in macrophyte biomass in South Island lowland streams, New Zealand', *New Zealand Journal of Marine and Freshwater Research*, vol 37, pp381–388

Rowe, D. K. and Schipper, C. M. (1985) 'An assessment of the impact of grass carp (*Ctenopharyngodon idella*) in New Zealand waters', New Zealand Ministry of Agriculture and Fisheries Environmental Report, Wellington

Schwarz, A.-M. and Howard-Williams, C. (1993) 'Aquatic weed-bed structure and photosynthesis in two New Zealand lakes', *Aquatic Botany*, vol 46, no 3–4, pp263–281

Symoens, J. J. and Triest, L. (1983) 'Monograph of the African genus *Lagarosiphon* Harvey (Hydrocharitaceae)', *Bulletin du Jardin botanique national de Belgique*, vol 53, no 3/4, pp441–488

Tanner, C. C. and Clayton, J. S. (1990) 'Persistence of arsenic 24 years after sodium arsenite herbicide application to Lake Rotoroa, Hamilton, New Zealand', *New Zealand Journal of Marine and Freshwater Research*, vol 24, no 2, pp173–179

Tanner, C. C., Clayton, J. S. and Harper, L. M. (1986) 'Observations on aquatic macrophytes in 26 northern New Zealand lakes', *New Zealand Journal of Botany*, vol 24, pp539–551

Tanner, C. C., Clayton, J. S. and Coffey, B. T. (1990a) 'Submerged-vegetation changes in Lake Rotoroa (Hamilton, New Zealand) related to herbicide treatment and invasion by *Egeria densa*', *New Zealand Journal of Marine and Freshwater Research*, vol 24, no 1, pp45–58

Tanner, C. C., Wells, R. D. S. and Mitchell, C. P. (1990b) 'Re-establishment of native macrophytes in Lake Parkinson following weed control by grass carp', *New Zealand Journal of Marine and Freshwater Research*, vol 24, pp181–186

Triest, L. (1991) 'Isozymes in *Lagarosiphon* (Hydrocharitaceae) populations from South Africa: The situation in dioecious, but mainly vegetatively propagating weeds', in L. Triest (ed) *Isozymes in Water Plants Volume 4*, Opera Botanica Belgica, Meise, pp71–85

Van Nes, E. H., Scheffer, M., Van Den Berg, M. S. and Coops, H. (2002) 'Aquatic macrophytes: Restore, eradicate or is there a compromise?', *Aquatic Botany*, vol 72, pp387–403

Wager, V. A. (1928) 'The structure and life history of the South African *Lagarosiphons*', *Transactions of the Royal Society of South Africa*, vol 16, no 2, pp191–212

Wells, R. D. S. and Champion, P. D. (2010) 'Endothall for aquatic weed control in New Zealand', in S. M. Zydenbos (ed) *17th Australasian Weeds Conference*, New Zealand Plant Protection Society, Christchurch, New Zealand, pp307–310

Wells, R. D. and Clayton, J. S. (1990) 'Submerged vegetation and spread of *Egeria densa* Planchon in Lake Rotorua, central North Island, New Zealand', *New Zealand Journal of Marine and Freshwater Research*, vol 25, pp63–70

Wells, R. D. S., Coffey, B. T. and Lauren, D. R. (1986) 'Evaluation of fluridone for weed control in New Zealand', *Journal of Aquatic Plant Management*, vol 24, pp39–42

8

Lythrum salicaria L. (purple loosestrife)

Keith R. Edwards

Introduction

Lythrum salicaria L. (purple loosestrife) is a native plant species of Eurasian freshwater wetlands, but is an aggressive invader of temperate North American wetlands (Stuckey, 1980; Thompson et al, 1987; Malecki et al, 1993; Edwards et al, 1998), and has successfully invaded Australia and New Zealand (Rockwell, 2001). The species has a broad natural range in Eurasia, being found in wetlands from 23 to 65°N, from northern Scandinavia to the northern reaches of the African Mediterranean countries (Morocco, Algeria, Tunisia, Libya and Egypt). There are scattered populations in Iraq and Iran as well as north-eastern China. *Lythrum salicaria* is also considered to be a native species in Japan (Thompson et al, 1987). It has almost the same latitudinal spread in its invasive range (from 30 to 56°N in North America) (Mal et al, 1992; Edwards et al, 1999; Bastlová et al, 2004), although it is found sporadically outside of the Wisconsin glaciation in North America.

Biology of *Lythrum salicaria*

Lythrum salicaria is an emergent, perennial wetland plant of the family Lythracea. Of the 35 species of the genus *Lythrum* worldwide, there are 12 reported in the US and Canada (Shinners, 1953), with *L. salicaria* being one of three non-native species of the genus in North America but the only one which is aggressively invasive (Thompson et al, 1987). A mature plant grows to an average height of 1.5m (though some can be over 3m) and can consist of 30 to 50 herbaceous stems emerging from a perennating rootstock. Schematics of a typical *L. salicaria* plant and stem are shown in Figure 8.1.

Stems are annual and square to six-sided. The stems emerge in late April or May. Leaves are sessile and typically opposite or in whorls, although some of the leaves near the top of the stem may be alternate. The inflorescence is a showy terminal spike (10–40cm in length) covered with numerous magenta-coloured flowers. Flowering occurs from early June into September. Flowers are tubular with up to 12 stamens. Styles are trimorphic, being short, mid or long (Eckert et al, 1996). Numerous bee and butterfly species pollinate the flowers, with the honey bee (*Apis mellifera*) being an important pollinator (Edwards et al, 1995). The fruit is a capsule that contains many small (1mm in length and 0.06mg in weight) seeds (Thompson et al, 1987). On average, each inflorescence of a mature plant can produce 900 capsules, with each capsule containing an average of 120 seeds (Shamsi and Whitehead, 1974), although there can be great variation in seed number per capsule due to environmental conditions (Edwards et al, 1995). The flat, thin-walled seeds contain little endosperm. Seed rain begins in late September to early October with the greatest amount of seed release occurring in late October to early November (Klips, 1990). Seed dispersal is usually through water or by being attached to animal fur or the feet of waterfowl (Hayes, 1979; Rawinski, 1982). The seeds can float for several days before sinking.

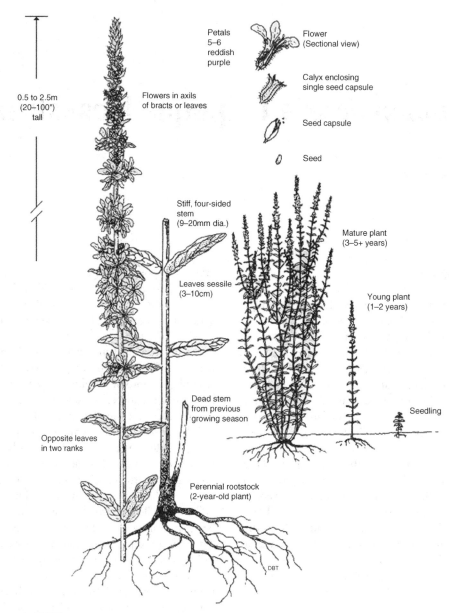

0.5 to 2.5m
(20–100")
tall

Petals
5–6
reddish
purple

Flower
(Sectional view)

Calyx enclosing
single seed capsule

Seed capsule

Seed

Flowers in axils
of bracts or leaves

Stiff, four-sided
stem
(9–20mm dia.)

Mature plant
(3–5+ years)

Leaves sessile
(3–10cm)

Young plant
(1–2 years)

Dead stem
from previous
growing season

Seedling

Opposite leaves
in two ranks

Perennial rootstock
(2-year-old plant)

DBT

Source: Thompson et al (1987)

Figure 8.1 *Schematic of a typical* Lythrum salicaria *plant*

Sexual reproduction is the most common form of generation, although the plant is capable of vegetative spread through regrowth from stem fragments (Stevens et al, 1997). Seed germination requires moist but not flooded (maximum of 2cm standing water) conditions, with temperatures of 15–20°C (Shamsi and Whitehead, 1974). New seeds mostly germinate in spring, in late April or May. By the third or fourth week after germination, seedlings can be fully established, having true leaves and a well-developed vascular system, although they may only be 10cm in height. After that, stem growth can be rapid, sometimes exceeding 1.0cm day^{-1}. Flowering occurs eight to ten weeks after germination, starting at the bottom of the inflorescence.

Thus, the capsules at the bottom of the inflorescence ripen and disperse their seeds while flowering is still occurring higher up the inflorescence (Rawinski, 1982). The above-ground shoots die back in autumn, with new shoots emerging from buds located on the rootstock the following spring.

History of Invasion and Spread

The history of the spread of *L. salicaria* in North America follows the general outline of Kowarik (1995). In this, there is an initial lag phase in which the newly established species spreads slowly in its new area. This is then followed by a sudden and rapid rate of spread. The first mention of *L. salicaria* in North America was in 1814 in the first edition of *Flora* by Torrey and Gray (Stuckey, 1980; Evans, 1982), noting scattered stands in the freshwater parts of the harbours of the major Atlantic coastal ports, including Philadelphia, New York, New Bedford and Boston. Scattered stands and single individuals were also noted growing along riverbanks and canals. Based on this citation, it has been estimated that *L. salicaria* first arrived in North America at the end of the 18th century or start of the 19th century. The first propagules were probably seeds transported from Europe either in sheep wool or ship ballast (Thompson et al, 1987).

Lythrum salicaria spread slowly over the next 130 years following this first introduction, usually along rivers, canals and ditches in the north-eastern part of the US and then slowly westward along the newly constructed Erie Canal (Thompson et al, 1987). Small and widely spaced groups of plants also became established in western Maryland and central Pennsylvania as new roads and turnpikes, such as the National Road, were constructed. By the 1850s and 1860s, *L. salicaria* was being recognized as a potential horticultural and landscape plant. It may be in this way that the plant was first introduced into the western Great Lakes region and the Pacific Northwest, with small populations establishing at this time in Muskegon, Michigan and Seattle, Washington, as well as Vancouver in British Columbia, Canada.

Different native European genotypes may have been introduced into North America in conjunction with the successive waves of European immigrants. *Lythrum salicaria* was used as a medicinal herb in Europe since the first century AD, when its uses as an astringent were noted by Dioscorides. Medieval herbals also mention that the leaves could be used to stop the flow of blood and that snakes and flies would leave a room in which it was burned. Therefore, *L. salicaria* was probably part of the traditional medicine practised by these immigrants, possibly making up part of their herbal gardens (Thompson et al, 1987). By 1900, the greatest number of established populations was still found in the north-eastern US. These populations were small and found mostly along riverbanks, canals and ditches. Smaller and more widely scattered populations occurred in the Midwest and northwest. Between 1850 and 1900, the species had also begun to spread southwards into Virginia and West Virginia as well as further westward along the Lake Erie shoreline, probably following the railways.

Lythrum salicaria began its phase of explosive spread in the 1930s, establishing rapidly in floodplain marshes along the St Lawrence, Merrimac, Hudson and Mohawk rivers. This was the first time that monocultural stands of the plant were recorded in these areas. This was followed by rapid spread westwards along the Great Lakes into the Midwest, as well as spread from the already established local populations (e.g., Muskegon, Michigan). Spread of the species also occurred from its loci in the Seattle/Vancouver areas, with *L. salicaria* moving rapidly south into northern California and east into Idaho, Utah and Wyoming. By 1985, *L. salicaria* was established in almost all states (except Montana) and Canadian provinces in a band between the 37th and 50th latitudes. Since then, it has continued spreading, with establishment in Montana and Alberta. Currently the species occurs in all states and provinces at 35–56°N, with some outlying populations such as around San Diego, California (Mal et al, 1992; Edwards et al, 1998, 2007).

Several hypotheses have been proposed to explain this rapid spread of *L. salicaria*. The 1930s was a time of numerous public works programmes initiated during the Great Depression (Thompson et al, 1987). Increased disturbance due to the large scale of some of these projects, such as the Tennessee River Valley Authority, construction of the Bonneville Dam along the Columbia River and construction of irrigation complexes and reservoirs in California and elsewhere in the western US, may have increased the available

(a)

(b)

Note: Native plants tend to be isolated or form small, sparse stands. Average plant height is 120cm but it depends on the environmental conditions of the site. Invasive plants reach an average height of 250cm and the populations may form monospecific stands.

Source: (a) Jan Květ, Třeboň Basin Biosphere Reserve; (b) King County, Washington Noxious Weed Control Program, near Seattle, Washington

Figure 8.2 *Populations of* Lythrum salicaria *in: (a) its native Eurasian range and (b) its invasive range*

habitat into which *L. salicaria* could establish and grow. Later, construction of the Interstate Highway System provided increased corridors by which the species could travel to invade previously inaccessible areas (Wilcox, 1989). In addition, beekeepers and honey producers began to realize the potential of *L. salicaria* as a honey plant, since honey bees (*Apis mellifera*) are a major pollinator of *L. salicaria* flowers (Levin, 1970). Beekeepers reportedly would deliberately spread *L. salicaria* seeds along streams and rivers in order to increase the density of plants (Thompson et al, 1987). Even as late as the 1990s, apiculturalists were still arguing against any control measures to reduce the size and number of *L. salicaria* populations (Wisconsin DATCP, 1993).

Native and Invasive Ecological Niches

Lythrum salicaria has many generalist traits that are thought to characterize many aggressive invasive species (Baker, 1974). It shows a high degree of phenotypic plasticity that allows it to grow in a wide range of different habitats and environmental conditions (Edwards et al, 1998).

In Eurasia, *L. salicaria* plants tend to grow singly or in small stands along the edges of lakes, rivers and in marshes (Polunin, 1969) (Figure 8.2a). Hejný (1960) found *L. salicaria* growing in tall sedge and degraded reed (*Phragmites australis*) communities, on drift material in the littoral of ponds and diffusely in non-flooded peat meadows. It is also a common species of floodplain fens and the littoral zone of lakes (Kask, 1982; Gavrilova, 1986) and is frequent in the floodplains near the mouths of large rivers flowing into the Black Sea (Dubina and Shelyag-Sosonko, 1989). It is a common species in roadside ditches (Edwards, pers. obs.).

Native plants in Eurasia may grow on average to 120cm in height, with occasional stems attaining heights of 2m (Clapham et al, 1959; Spencer-Jones and Wade, 1986). The stems have few or no branches. *Lythrum salicaria* seems to require some disturbance in order to become established in native communities (Hejný, 1960) and can become locally abundant following disturbances that have opened patches (Clapham et al, 1959). In such cases, *L. salicaria* may be one of the first colonizing species of the newly opened patch, being dominant for one to three years.

Afterwards, other species, such as *Juncus effusus*, expand into the patch, with *L. salicaria* taking on a more subordinate role, possibly due to herbivory by natural control agents, such as *Galerucella* spp. beetles (Ellis, 1963; Batra et al, 1986).

Invasive populations behave quite differently. Disturbance still seems to be a prerequisite for establishment in a wetland. However, once established, *L. salicaria* usually becomes the dominant species and remains so over time, resulting sometimes in the formation of monocultural stands (Rawinski, 1982) (Figure 8.2b). Invasive plants grow to an average height of 2.5m (Thompson et al, 1987), with particular stems reaching heights of 3m or more (Edwards, pers. obs.). The plants tend to be multi-stemmed, with no or several branches near the top of the plant. Large plants may have branches emerging in the upper third of the stems with inflorescences at the terminal end of each branch. By overgrowing native wetland flora, *L. salicaria* can become the top competitor in the invaded habitat (Gaudet and Keddy, 1988).

Bastlová-Hanzélyová (2001) found that native populations of *L. salicaria* grow in a wider range of environmental conditions than invasive populations in North America. The invasive populations were found almost exclusively in open areas that are either permanently or periodically flooded. In contrast, native European populations grow not only in typical wetland habitats but also in drier conditions, such as around the edges of agricultural fields. Recent disturbance seems to be a more important variable than moisture in describing the habitats associated with native *L. salicaria* populations. Thus, this species is more of a ruderal species in its native range.

Review of Management and Control Methods

Several methods have been used to try and control invasive *L. salicaria* populations in North America. These methods, and their advantages and disadvantages, are listed in Table 8.1. The type of method employed depends upon the size of the population, with no one method being best for all situations (Tu, 2009).

As with any invasive organism, the chances of successful control increase when dealing with small, new invasive populations. July and August, when the plants are in bloom and the colourful inflorescences are

Table 8.1 *Common methods used to control invasive* Lythrum salicaria *populations in North America, the population size for which each method is most useful and the advantages and disadvantages of each method*

Control method	Population size	Advantages	Disadvantages
Hand-pulling	Small	Eradicates population	Labour and time intensive; useful only for new, small populations
Cutting followed by flooding	Small, medium	Short term control possible; may work well in deep water habitats	High chance of reinvasion; labour and time intensive
Chemical	Small, medium and large	High levels of control possible; eradication possible	Adverse effects on desired native species; opens gaps that may allow for reinvasion or establishment of other invasive species
Biological	Large	Possible long term control; less adverse effects on native species	Introduces more non-native species; does not eradicate population; possible adverse effects on native species; poor results in deep water habitats

Note: Small = <100 plants; medium = 100–300 plants; large = >300 plants.

clearly visible, are the best times to search for new populations (Rendall, 1987 in Smith, 2009). These populations can be eradicated by hand-pulling before the plants have developed a large, tough rootstock. Pulling should occur in summer prior to seed release (Heidorn, 1990; Swearingen, 2005). The pulled plants need to be removed from the area and disposed of properly, usually by burning, to prevent regrowth from plant fragments (Stevens et al, 1997).

Other methods are required when faced with larger, more established populations, as time and labour restrictions would impair the use of hand-pulling. In this case, trying to contain the population rather than eradication may be the most realistic option. The methods available for medium to large populations are cutting/mowing, chemical control and biological control.

Physical control

Cutting of *L. salicaria* plants is usually conducted in association with flooding of the area to discourage regrowth. This technique works well with other wetland species, such as *Typha* (Beule, 1979). Edwards and Comas (2009) found that cutting stems flooded to at least a depth of 40cm resulted in good short term control. Stem densities in the next growing season, following cutting and followed by over-winter flooding,

had decreased by 93 per cent in plots in which the plants were cut compared to control, non-cut plots. Still, many managers discourage the use of this technique. There are important limitations to this method. First, the method is best used in situations with natural flooding regimes. Artificially flooding an area can result in great damage to native plants, with the end result being a greater infestation of *L. salicaria* (Heidorn, 1990). Thus, knowledge about the natural hydrological regime of the habitat is essential if cutting is to be used as a management technique. Second, care must be taken to remove all cut stems from the area followed by proper disposal, preferably burning. New stems can emerge from small stem fragments left in the field (Stevens et al, 1997) leading to reinfestation. Third, cutting is best done before seed maturation. Inflorescences should be bagged and removed from stems prior to stem cutting and the inflorescences disposed of properly. Fourth, as with any mechanical control measure, this method is labour and time intensive; thus, it is most appropriate for controlling small (<100 plants) to medium sized (100–300 plants) populations. This can be done by using a corps of well-trained volunteers (see Smith, 2009).

Lastly, long term monitoring of the control site will be necessary, with continued cutting of re-emergent shoots. Fluctuating water levels are a basic characteristic of wetlands (Mitsch and Gosselink, 2000), thus occasional drawdowns should be expected. An exposed, moist soil is the optimal condition for germination of *L. salicaria* seeds. With the possibility of high seed production (see above), it is likely that *L. salicaria* seeds are an important component of the seed bank. It is not known for how long *L. salicaria* seeds are viable, but Rawinski (1982) found that 80 per cent of seeds still germinated after four years. Monitoring is key to removing any new plants that emerge due to a drawdown.

Chemical control

Given the disadvantages associated with cutting, more managers use chemical control to combat *L. salicaria* invasions. Chemical control consists of using a herbicide approved for use over open water. For *L. salicaria* this is a glyphosate compound, which is known under the trade names of Roundup and Rodeo (Monsanto). Rodeo is the one approved for use over open water. Glyphosates are systemic in that they are translocated quickly from leaves to roots and rhizomes, where they are lethal (Malecki and Rawinski, 1985). Rodeo has a short retention time of one week in the soil, which reduces the risk of adverse environmental impacts (Thompson et al, 1987).

However, glyphosate is a non-specific, broad spectrum herbicide so care must be taken in how it is used. The recommended method is to first cut the stems 15cm above the soil or water surface and then spot treat the stems with a 20–30 per cent Rodeo solution. This method is most appropriate for medium sized invasions (>100 plants). The best time for this is at the beginning of the flowering season before seed maturation. In fact, as with cutting, it is better to first remove the inflorescences and dispose of them in an appropriate manner (Heidorn, 1990). Foliar spraying with a milder (2 per cent) glyphosate solution is better for controlling larger populations (>300 plants). The best results occur when *L. salicaria* plants are at their peak time of flowering, usually in August (Rawinski, 1982). Again, care must be taken to reduce the possibility of spraying desired, native species.

An alternative to this is to spray the plants twice during the growing season. The first time should be at the beginning of flowering using the cut and spot treat method and then several weeks later to treat plants that may have been missed the first time. This reduces the possibility of seed set and can lead to treating more plants (Heidorn, 1990). As with any control effort, spraying must be followed by monitoring of the treated site. This will allow for noticing plants that were missed during the first treatment application as well as treating any new plants. Monitoring needs to start in the same growing season as the initial treatment and then continue for several more years.

Biological control

Biological control may be the best method for long term control of large populations. A biological control programme, using three known control agents of *L. salicaria*, has been ongoing in North America since the 1990s (Hight and Drea, 1991; Malecki et al, 1993; Blossey et al, 2001). The three control agents, two folivorous beetles of the genus *Galerucella* (*G. calmariensis* and *G. pusilla*) and a root-boring weevil (*Hylobatis transversovitattus*) are considered to be monophagous for *L. salicaria* in its native Eurasian range. In preliminary trials, these herbivores were found to also feed on several native North American Lythraceae species (*L. alatum*, *Decodon verticillatus*), but the threat to North American wetland ecosystems by *L. salicaria* was considered to be a greater risk than the relatively low impact of the introduced control agents. Due to early positive results with controlled releases of the biocontrol agents, more managers began using this as the management technique of choice. Currently, insects have been released in at least 22 states in the US (Wisconsin DATCP, 1993; Malecki et al, 1993; Hight et al, 1995; van Driesche et al, 2002 in Smith, 2009).

As a result, significant reductions in plant height have been found at specific sites (see for example Blossey et al, 2001; Dech and Nosko, 2002; Denoth and Myers, 2005), but decreases in the abundance or size of invasive *L. salicaria* stands are less common (Grevstad, 2006). This has resulted in little or no change to *L. salicaria* invasions in some areas of North America, even ten years after release of the biological control agents (Grevstad, 2006). While *L. salicaria* may have declined at individual sites, no such reductions have been noted at regional scales.

This lack of broad-scale impacts may result from several factors. First, the length of time for observing impacts may not have been long enough. The longest monitoring study found (Grevstad, 2006) remeasured *L. salicaria* stands ten years after insect release, but it has been suggested that these programmes need 10–20 years to develop and for impacts to be seen (Blossey et al, 2001).

Second, the density of the initial releases of the insects may have been too small to produce the desired control. Grevstad (2006) in his resurvey of release sites found numerous areas that were unoccupied by the released *Galerucella* beetles, while areas containing them had lower than desired levels of feeding damage. Third, predation on the control agents by native North American species may reduce their densities and, thus, decrease their effectiveness (Hunt-Joshi et al, 2005).

Lastly, annual variation in the abundance of the control agents may reduce efficacy. In years with greater predation pressure and/or poorer environmental conditions, population levels of the control agents will decrease. At such times, *L. salicaria* plants will be able to store more energy, be more fertile and produce more seeds. Also, new seedlings will have a better chance of surviving to adulthood. These will then have a better chance of surviving at those times when control agent numbers increase. What this may mean is that, in some years, control agent numbers will be high enough to impact *L. salicaria* plant height and flowering ability, but there will be other years when insect numbers decrease and *L. salicaria* recovers. This would result in the little or no change seen at the population and regional scales. In addition, *Galerucella* have only one generation in temperate regions of North America and the adults diapause in early August. This allows *L. salicaria* plants to recover for almost two months, in which time they can store energy, flower and possibly even set seed (Grevstad, 2006).

While some consider that biological control offers the best method to date for long term control of *L. salicaria*, there are several general concerns about this particular programme and with biological control in general. First, biological control does not eradicate invasive species but only reduces them to manageable densities. Second, the success of this method for *L. salicaria* decreases greatly in deep water habitats (Hight and Drea, 1991). More studies are needed to determine what factors may be inhibiting the insects from successfully reducing *L. salicaria* populations in these communities. Third, as with chemical control, the death or weakening of individual plants due to the action of the control agents may result in the formation of open patches in the system. These may be colonized by desired native species, but also may allow other non-native or invasive species (e.g. *Phragmites australis*, *Phalaris arundinacea*) to establish. This possibility may be reduced by linking the control measures with a programme to seed the areas with native species (Heidorn, 1990). Last, there may be unpredictable adverse effects of the released biocontrol agents on native species and communities. While this does not seem to be the case for the insects used to control *L. salicaria*, adverse effects do seem to be a real concern in other cases (see for example the cases reviewed by Simberloff and Stiling, 1996; Williamson, 1996; Louda et al, 1997).

Challenges and Controversies

Lythrum salicaria is considered to be such an aggressive invasive species in North America that it has been placed on the noxious weeds lists of 32 states in the US. Such a designation means that it is illegal to sell or plant the species and, in cases such as Pennsylvania and Iowa, even cultivars are excluded (USDA, 2010). Given its large and showy inflorescences, some horticultural firms have tried to get around this by selling what they advertise as sterile cultivars (e.g. Morden Pink). It is questionable whether these cultivars are actually sterile (Anderson and Ascher, 1993).

However, several reviews of the scientific literature (e.g., Farnsworth and Ellis, 2001; Lavoie, 2010) have found that negative effects of invasive *L. salicaria* on native species and wetland ecosystem functions are inconclusive. It has been argued that invasion by *L. salicaria* leads to a decrease in plant diversity with possibly a decrease in wildlife use, as *L. salicaria* crowds out the native plants used as food by waterfowl and other wildlife (Thompson et al, 1987). In fact, one observational study (Bastlová-Hanzélyová, 2001) found much lower species richness in invaded North American sites (40 vascular plants species) compared to native European habitats (129 other vascular plant species). However, this study did not test whether these differences were due to the presence of *L. salicaria* or just characteristics of North American versus European habitats. However, others found no decrease in native North American species abundance following establishment of *L. salicaria* (Hager and McCoy, 1998; Treberg and Husband, 1999; Keller, 2000). In a review of the literature on *L. salicaria*, Anderson (1995) found that 29 native North American faunal species used invasive *L. salicaria* plants. All of these studies also found no clear evidence that *L. salicaria* outcompetes native North American wetland species, such as cattail (*Typha*). This finding appears to contradict the results of Mal et al (1997), who found that *L. salicaria* began to outcompete *Typha angustifolia* only after at least four years of co-occurrence. The lack of competitive advantage for *L. salicaria* in the former studies may be due to their being short term, observational studies and not experimental, as in the case of Mal et al (1997).

Several studies noted that invasions of *L. salicaria* could possibly negatively impact native community and ecosystem processes. For example, higher rates of litter decomposition were measured in areas dominated by *L. salicaria* compared to those dominated by native North American species (Bärlocher and Biddiscombe, 1996; Emery and Perry, 1996 for *T. latifolia*; Grout et al, 1997 for *Carex lyngbyei*), which could affect nutrient cycling and have longer term ecosystem effects, but these possible impacts have not yet been tested (Lavoie, 2010). In addition, Grabas and Laverty (1999) found that even a medium sized invasion of *L. salicaria* could negatively affect pollination and reproductive success of native flowering wetland plants growing in the same habitat.

What the reviews of the scientific literature show is that there is still a lot of uncertainty about the overall effect of *L. salicaria* on invaded wetland habitats. Such uncertainty makes it more difficult to understand, as well as predict, the impact of *L. salicaria* invasions. This also has consequences for management actions; all of the control methods listed above have certain costs in terms of time and money. A more effective and efficient use of limited resources requires that we have a better understanding of the ecology and possible effects of *L. salicaria* on native ecosystems (Luken, 1994; Edwards, 1998; Lavoie, 2010). For this to happen, management plans need to have a sound basis, integrating both ecological as well as cultural aspects (Edwards, 1998; Decocq, 2010). Management plans for many invasive species, including *L. salicaria*, focus only on the particular species without addressing larger-scale issues, such as possible underlying disturbance phenomena that allowed the species to invade originally and may help maintain the invasive populations. In this way, management actions mitigate the effect of environmental degradation but not the root causes (Hobbs and Humphries, 1995). What is required is a more integrated approach that would place the invasive species within a larger socioecological context (Edwards, 1998; Lavoie, 2010). This is not a new suggestion, but it is an objective that has rarely if ever been met.

Acknowledgements

Support for this chapter was provided by the Czech Ministry of Education (MSMT 6007665801). The comments of Claude Lavoie helped to greatly improve the text.

References

Anderson, M. G. (1995) 'Interactions between *Lythrum salicaria* and native organisms: A critical review', *Environmental Management*, vol 19, no 2, pp225–231

Anderson, N. O. and Ascher, P. D. (1993) 'Male and female fertility of loosestrife (*Lythrum*) cultivars', *Journal of the American Horticultural Society*, vol 118, no 6, pp851–858

Baker, H. G. (1974) 'The evolution of weeds', *Annual Review of Ecology and Systematics*, vol 5, pp1–24

Bärlocher, F. and Biddiscombe, N. R. (1996) 'Geratology and decomposition of *Typha latifolia* and *Lythrum salicaria* in a freshwater marsh', *Archiv für Hydrobiogie*, vol 136, no 3, pp309–325

Bastlová-Hanzélyová, D. (2001), 'Comparative study of native and invasive plants of *Lythrum salicaria* L.: Population characteristics, site and community relationships', in G. Brundu, J. Brock, I. Camarda, L. Child and M. Wade (eds) *Plant Invasions: Species Ecology and Ecosystem Management*, Backhuys Publishers, Leiden, pp33–41

Bastlová, D, Čížková, H., Bastl, M. and Květ, J. (2004) 'Growth of *Lythrum salicaria* and *Phragmites australis* plants originating from a wide geographical area: Response to nutrient and water supply', *Global Ecology and Biogeography*, vol 13, pp259–271

Batra, S. W. T., Schroeder, D., Boldt, P. E. and Mendl, W. (1986) 'Insects associated with purple loosestrife (*Lythrum salicaria* L.) in Europe', *Proceedings of the Entomological Society of Washington*, vol 88, pp748–759

Beule, J. D. (1979) 'Control and management of cattails in southeastern Wisconsin wetlands', *Technical Bulletin 112*, Wisconsin Department of Natural Resources, Madison, Wisconsin

Blossey, B., Skinner, L. C. and Taylor, J. (2001) 'Impact and management of purple loosestrife (*Lythrum salicaria*) in North America', *Biodiversity and Conservation*, vol 10, pp1787–1807

Clapham, A. R., Tutin, T. G. and Warburg, E. F. (1959) *Excursion Flora of the British Isles*, Cambridge University Press, London

Dech, J. P. and Nosko, P. (2002) 'Population establishment, dispersal and impact of *Galerucella pusilla* and *G. calmariensis*, introduced to control purple loosestrife in central Ontario', *Biological Control*, vol 23, pp228–236

Decocq, G. (2010) 'Invisibility promotes invasibility', *Frontiers in Ecology and the Environment*, vol 8, pp346–347

Denoth, M. and Myers, J. H. (2005) 'Variable success of biological control of *Lythrum salicaria* in British Columbia', *Biological Control*, vol 32, pp269–279

Dubina, D. V. and Shelyag-Sosonko, Yu. R. (1989) *Wetlands of the Black Sea Region*, Naukova Dumka, Kiev (in Russian)

Eckert, C. G., Manicacci, D. and Barrett, S. C. H. (1996) 'Genetic drift and founder effect in native versus introduced populations of an invading plant, *Lythrum salicaria* (Lythraceae)', *Evolution*, vol 50, pp1512–1519

Edwards, K. R. (1998) 'A critique of the general approach to invasive plant species', in U. Starfinger, K. Edwards, I. Kowarik and M. Williamson (eds) *Plant Invasions: Ecological Mechanisms and Human Responses*, Backhuys Publishers, Leiden, pp85–94

Edwards, K. R. and Comas, L. (2009) 'Evaluation of mechanical cutting to control littoral purple loosestrife stands', *Journal of Aquatic Plant Management*, vol 47, pp158–161

Edwards, K. R., Adams, M. S. and Květ, J. (1995) 'Invasion history and ecology of *Lythrum salicaria* in North America', in P. Pyšek, K. Prach, M. Rejmánek and M. Wade (eds) *Plant Invasions – General Aspects and Special Problems*, SPB Academic Publishing, Amsterdam, pp161–180

Edwards, K. R., Květ, J. and Adams, M. S. (1998) 'Comparison of the demographics of native and invasive populations of *Lythrum salicaria* L.', *Applied Vegetation Science*, vol 1, pp267–280

Edwards, K. R., Květ, J. and Adams, M. S. (1999) 'Comparison of *Lythrum salicaria* L. study sites in the midwest US and central Europe' *Ekologia (Bratislava)*, vol 18, pp113–124

Edwards, K. R., Květ, J. and Adams, M. S. (2007) 'Competitive abilities of native European and non-native North American populations of *Lythrum salicaria* L.' *Ekologia (Bratislava)*, vol 26, pp1–13

Ellis, E. A. (1963) 'Some effects of selective feeding by the coypu (*Myocastor coypus*) on the vegetation of Broadland', *Transactions of the Norfolk Norwich Naturalist Society*, vol 20, pp32–35

Emery, S. L. and Perry, J. A. (1996) 'Decomposition rates and phosphorus concentrations of purple loosestrife (*Lythrum salicaria*) and cattail (*Typha* spp.) in fourteen Minnesota wetlands', *Hydrobiologia*, vol 323, pp129–138

Evans, J. E. (1982) 'A literature review of management practices for purple loosestrife (*Lythrum salicaria*)', The Nature Conservancy, Midwest Regional Office, Minneapolis, Minnesota

Farnsworth, E. J. and Ellis, D. R. (2001) 'Is purple loosestrife (*Lythrum salicaria*) an invasive threat to freshwater wetlands? Conflicting evidence from several ecological metrics', *Wetlands*, vol 21, pp199–209

Gaudet, C. L. and Keddy, P. A. (1988) 'A comparative approach to predicting competitive ability from plant traits', *Nature*, vol 334, pp242–243

Gavrilova, G. (1986) 'Flora of vascular aquatic plants in the reserve Lake Cirisa (Latvian SSR)', in N. Ingerpuu, T. Ksenofontova and L. M. Laasimer (eds) *Vegetation Cover of Aquatic Macrophytes in Marshes Near the Coast of the Baltic Sea*, Estonian Academy of Sciences, Tallin, pp118–127

Grabas, G. P. and Laverty, T. M. (1999) 'The effect of purple loosestrife (*Lythrum salicaria* L.; Lythraceae) on the pollination and reproductive success of sympatric co-flowering wetland plants', *Ecoscience*, vol 6, no 2, pp230–242

Grevstad, F. S. (2006) 'Ten-year impacts of the biological control agents *Galerucella pusilla* and *G. calmariensis* (Coleopotera: Chrysomelidae) on purple loosestrife (*Lythrum salicaria*) in central New York State', *Biological Control*, vol 39, pp1–8

Grout, J. A., Levings, C. D. and Richardson, J. S. (1997) 'Decomposition rates of purple loosestrife (*Lythrum salicaria*) and Lyngbyei's sedge (*Carex lyngbyei*) in the Fraser River estuary', *Estuaries*, vol 20, no 1, pp96–102

Hager, H. A. and McCoy, K. D. (1998) 'The implications of accepting untested hypotheses: A review of the effects of purple loosestrife (*Lythrum salicaria*) in North America', *Biodiversity and Conservation*, vol 7, pp1069–1079

Hayes, B. (1979) 'Purple loosestrife – the wetlands honey plant', *American Bee Journal*, vol 119, pp382–383

Heidorn, R. (1990) 'Vegetation management guideline: Purple loosestrife (*Lythrum salicaria* L.)', *Illinois Natural Preserves Commission*, vol 1, p17

Hejný, S. (1960) *Ökologische Charakteristik der Wasser- und Sumpfpflanzen in den Slowakischen Tiefebenen (Donau- und Thiessgebeit)*, Slovak Academy of Sciences, Bratislava

Hight, S. D. and Drea, J. J. (1991) 'Prospects for a classical biological control project against purple loosestrife (*Lythrum salicaria* L.)', *Natural Areas Journal*, vol 11, pp151–157

Hight, S. D., Blossey, B., Laing, J. and Declerck-Floate, R. (1995) 'Establishment of insect biological control agents from Europe against *Lythrum salicaria* in North America', *Environmental Entomology*, vol 24, pp967–977

Hobbs, R. J. and Humphries, S. E. (1995) 'An integrated approach to the ecology and management of plant invasions', *Conservation Biology*, vol 9, pp761–770

Hunt-Joshi, T. R., Root, R. B. and Blossey, B. (2005) 'Disruption of weed biological control by an opportunistic predator', *Ecological Applications*, vol 15, pp861–870

Kask, M. (1982) 'A list of vascular plants of Estonian peatlands', in T. Frey, V. Masing and E. Roosaluste (eds) *Peatland Ecosystems*, Estonian Academy of Sciences, Tallin, pp39–49

Keller, B. E. M. (2000) 'Plant diversity in *Lythrum*, *Phragmites* and *Typha* marshes, Massachusetts, USA', *Wetlands Ecology and Management*, vol 8, pp391–401

Klips, R. A. (1990) 'Dissemination, germination, and early seedling establishment of purple loosestrife, *Lythrum salicaria* L.', MS Thesis, State University of New York at Buffalo, New York

Kowarik, I. (1995) 'Time lags in biological invasions with regard to the success and failure of alien species', in P. Pyšek, K. Prach, M. Rejmánek and M. Wade (eds) *Plant Invasions – General Aspects and Special Problems*, SPB Academic Publishing, Amsterdam

Lavoie, C. (2010) 'Should we care about purple loosestrife? The history of an invasive plant in North America', *Biological Invasions*, vol 12, pp1967–1999

Levin, D. A. (1970) 'Assortative pollination in *Lythrum*', *American Journal of Botany*, vol 57, pp1–5

Louda, S. M., Kendall, D., Connor, J. and Simberloff, D. (1997) 'Ecological effects of an insect introduced for the biological control of weeds', *Science*, vol 277, pp1088–1090

Luken, J. O. (1994) 'Valuing plants in natural areas', *Natural Areas Journal*, vol 14, pp295–299

Mal, T. K., Lovett-Doust, J., Lovett-Doust, L. and Mulligan, G. A. (1992) 'The biology of Canadian weeds 100: *Lythrum salicaria*', *Canadian Journal of Plant Science*, vol 72, pp1305–1330

Mal, T. K., Lovett-Doust, J. and Lovett-Doust, L. (1997) 'Time-dependent competitive displacement of *Typha angustifolia* by *Lythrum salicaria*', *Oikos*, vol 79, pp26–33

Malecki, R. A. and Rawinski, T. J. (1985) 'New methods for controlling purple loosestrife', *New York Fish and Game Journal*, vol 32, pp9–19

Malecki, R. A., Blossey, B., Hight, S., Schroeder, D., Kok, L. and Coulson, J. (1993) 'Biological control of purple loosestrife', *BioScience*, vol 43, pp680–686

Mitsch, W. J. and Gosselink, J. G. (2000) *Wetlands*, 3rd edition, John Wiley and Sons, New York

Polunin, O. (1969) *Flowers of Europe*, Oxford University Press, London

Rawinski, T. J. (1982) 'The ecology and management of purple loosestrife (*Lythrum salicaria*) in central New York', MS Thesis, New York Cooperative Wildlife Research Unit, Cornell University, Ithaca, New York

Rockwell, F. (2001) 'Introduced species summary project: Purple loosestrife (*Lythrum salicaria*)', www.columbia.edu/itc/cerc/danoff-burg/invasion_bio/inv_spp_summ/Lythrum_salicaria.html, accessed 25 August 2010

Shamsi, S. R. A. and Whitehead, F. H. (1974) 'Comparative eco-physiology of *Epilbium hirsutum* L. and *Lythrum salicaria* L. I. General biology, distribution and germination', *Journal of Ecology*, vol 62, pp279–290

Shinners, L. H. (1953) 'Synopsis of the United States species of *Lythrum* (Lythraceae)', *Field Laboratory*, vol 21, pp80–89

Simberloff, D. and Stiling, P. (1996) 'Risks of species introduced for biological control', *Biological Conservation*, vol 78, pp185–192

Smith, L. L. (2009) 'Invasive exotic plant management tutorial for natural lands managers', www.dcnr.state.pa.us/forestry/invasivetutorial/Purple_loosestrife, accessed 9 July 2009

Spencer-Jones, D. and Wade, M. (1986) *Aquatic Plants*, ICI Professional Products, Farnham, Surrey, UK

Stevens, K. J., Peterson, R. L. and Stephenson, G. R. (1997) 'Vegetative propagation and the tissues involved in lateral spread of *Lythrum salicaria*', *Aquatic Botany*, vol 56, pp11–24

Stuckey, R. L. (1980) 'The distributional history of *Lythrum salicaria* (purple loosestrife) in North America', *Bartonia*, vol 100, pp3–30

Swearingen, J. M. (2005) 'Plant conservation alliance, Alien plant working group, fact sheet – purple loosestrife', www.nps.gov/plants/alien/fact/lysa1.html, accessed 20 August 2010

Thompson, D. Q., Stuckey, R. L. and Thompson, E. B. (1987) *Spread, Impact, and Control of Purple Loosestrife* (Lythrum salicaria) *in North American Wetlands*, US Fish and Wildlife Service, Fish and Wildlife Research Report 2, US Department of Interior, Washington, DC

Treberg, M. A. and Husband, B. C. (1999) 'Relationship between the abundance of *Lythrum salicaria* (purple loosestrife) and plant species richness along the Bar River, Canada', *Wetlands*, vol 19, no 1, pp118–125

Tu, M. (2009) *Assessing and Managing Invasive Species Within Protected Areas*, The Nature Conservancy, Arlington, Virginia

USDA (United States Department of Agriculture) (2010) 'Noxious weeds list', http://plants.usda.gov/java/noxiousDriver, accessed 20 August 2010

Wilcox, D. A. (1989) 'Migration and control of purple loosestrife (*Lythrum salicaria* L.) along highway corridors', *Environmental Management*, vol 13, pp365–370

Williamson, M. (1996) *Biological Invasions*, Chapman and Hall, London

Wisconsin DATCP (Department of Agriculture, Trade and Consumer Protection) (1993) 'Environmental assessment for a permit to release biological control agents of purple loosestrife', Division of Agricultural Resource Management, Madison, Wisconsin

9

Myriophyllum aquaticum (Vell.) Verdcourt (parrot feather)

Andreas Hussner and Paul D. Champion

Introduction

Myriophyllum aquaticum (parrot feather) is a semi-aquatic or aquatic plant, able to form both submerged and emerged shoots. The species is indigenous to South America but now has a wide naturalized range, due to its popularity in the aquarium and ornamental pond plant trade. The emerged and floating-emerged form of this species is able to form dense beds, which can displace native vegetation (Moreira et al, 1999; Hussner, 2008), become a nuisance for recreational use of water bodies by humans such as fishing and boating, and affect drainage by clogging pumps and promoting flooding. Other impacts include obstruction of irrigation channels, river navigation and hydroelectric power production (Moreira et al, 1999; Shaw, 2003; Sheppard et al, 2006). *Myriophyllum aquaticum* can become a serious threat to native flora and alters water chemistry, resulting in decreasing oxygen concentration in the water and a general decrease of water pH values during the day caused by shading (van der Meijden, 1969; Ferreira et al, 1998; Ferreira and Moreira, 1999; Moreira et al, 1999; Bernez et al, 2006). Additionally, Orr and Resh (1992) observed a positive correlation between the egg and larvae abundance of the disease vector *Anopheles* mosquitoes and the stem density of *M. aquaticum* plants in waters in California.

Species Description

Stems of *M. aquaticum* are up to around 2m in length, and 4–5mm in diameter near the base. Emerged stems can extend up to 0.4m above the water surface (Figure 9.1). Adventitious roots are usually produced on the lower stem nodes, but creeping plants commonly root from nodes along the entire shoot. Leaves are in whorls of (4–)5–6, and there is a slight dimorphism between the submerged and emerged leaves. The submerged leaves are oblanceolate in outline, rounded at the apex and (17–)35–40mm long and (4–)8–12mm wide, pectinate with 25–30 linear and up to 7mm long pinnae. Emerged leaves are glaucous, erect near the apex, narrowly oblanceolate in outline, rounded at the apex and (15–)25–35mm long and (4–)7–8mm wide with (18–)24–36 pinnae, with a petiole approximately one fifth of the leaf length. The pinnae are linear-subulate, 4.5–5.5mm long, 0.3mm wide with very shortly apiculate tips (Orchard, 1979, 1981, 1985).

Outside of its native range in South America, only female plants of *M. aquaticum* are known, though even in its native habitats, male plants are not common and fruit production is very rare (Orchard, 1979).

Source: P. Champion

Figure 9.1 *Myriophyllum aquaticum*

Origin, History of Introduction and Current Distribution

The genus *Myriophyllum* is almost cosmopolitan, though it is absent from most of Africa. The genus is comprised of approximately 60 species, described in several revisions of the genus (van der Meijden, 1969; Orchard, 1979, 1981, 1985). There are three main centres of distribution of this genus: (1) Australasia with 38 different species (31 of which are endemic to Australia, 2 to New Zealand), (2) North America (13 species, 7 endemic) and (3) Asia (India/Indo-China, 10 species, 7 endemic) (Orchard, 1979, 1985). Beside the highly invasive *M. aquaticum*, two other *Myriophyllum* species, *M. spicatum* and *M. heterophyllum*, are problematic alien invasives in some parts of the world (Nelson and Couch, 1985; Hussner, 2008).

Myriophyllum aquaticum (synonyms: *Enydria aquatica* Vell., *Myriophyllum brasiliense* Cambess., *Myriophyllum proserpinacoides* Gillies ex. Hook and Arn.) is native to the lowlands of central South America,

but extends to altitudes of at least 3250m above sea level in Peru and 1900m in Brazil. *Myriophyllum aquaticum* is now almost cosmopolitan as an adventive in temperate and tropical regions (Orchard, 1979, 1985; EPPO, 2004). The species was introduced into Europe in 1880 (Sheppard et al, 2006), into the US in the late 1800s or early 1900s (Sutton, 1985), New Zealand in 1929 (Orchard, 1979), Australia in the 1960s (Aston, 1977), Japan in 1920 (Sutton, 1985) and Africa in 1918/1919 (Guillarmod, 1979). The species has also recently been documented in several other countries (Figure 9.2) (Mendes, 1978; Orchard, 1979, 1981, 1985; EPPO, 2004; Hussner, 2008).

The spread of *M. aquaticum* into new areas is almost exclusively caused by human dispersal, primarily because *M. aquaticum* plants are commonly sold as plants for aquaria (Hussner, 2008; Brunel, 2009; Hussner et al, 2010). In Java, *M. aquaticum* has been used as a protective cover for fish culture ponds and the tips of the shoots have been eaten as vegetables (Sutton, 1985). Once established in a new area, the plants can easily spread vegetatively, and *M. aquaticum* is known for its high regeneration capacity even from very small plant fragments (Hussner, 2009). Subsequent spread to new water bodies may occur as a contaminant of drainage machinery, fishing nets or boat trailers (Champion et al, 2002).

Ecology of *Myriophyllum aquaticum*

Myriophyllum aquaticum grows best in slow flowing or stagnant water with high nutrient conditions (Sytsma and Anderson, 1993a; Hussner, 2009). In general, depths less than 1m favour its establishment and growth (Moreira et al, 1999). In such shallow waters, the stems easily reach the water surface and the plants form floating mats, which shade the water beneath. Maximum biomass measured in the field can exceed 20kg of fresh weight m^{-2}, which has been reported from non-native stands in Portugal (Monteiro and Moreira, 1990). In California, biomass of 1.00 ± 0.08kg dry weight m^{-2} has been observed (Sytsma and Anderson, 1993c). In Germany, biomass of 2.061 ± 0.13kg dry weight m^{-2} of mainly floating and emerged plants has been found in shallow (<1m in depth) eutrophicated ponds in the second year after its introduction (Hussner, 2008). In fast flows, or in deep water, *M. aquaticum* only exists in its submerged form (Hussner and Lösch 2005). *Myriophyllum aquaticum* is also intolerant of saline conditions, and seawater was toxic to emerged shoots of *M. aquaticum* at concentrations

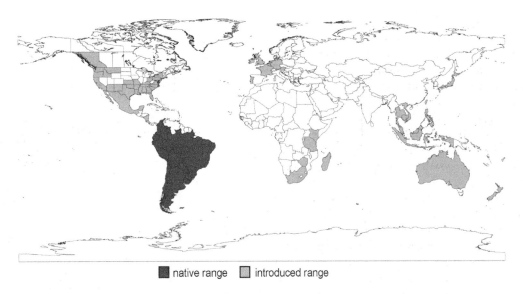

Source: Mendes (1978); Orchard (1979, 1981, 1985); EPPO (2004); Hussner (2008)

Figure 9.2 *Global distribution of* Myriophyllum aquaticum

between 10.0 and 13.2 parts per thousand (ppt), though lower salt concentrations (0.8 to 3.3ppt) stimulated root growth (Haller et al, 1974).

Myriophyllum aquaticum plants are known for their high regeneration capacity from plant fragments. This allows for rapid dispersal within a water body (Orchard, 1985; Kane et al, 1991; Hussner, 2009). Plant regeneration is even possible from a single emerged leaf within a few weeks, but is most likely to occur from shoot fragments containing nodes (Hussner, 2009) (Figure 9.3). The species over-winters

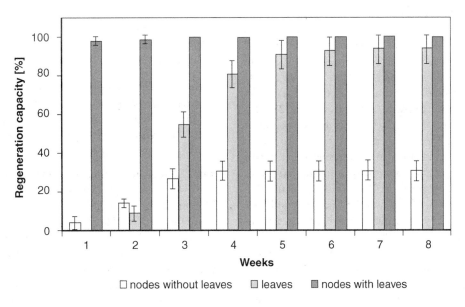

Source: Hussner (2008)

Figure 9.3 *Regeneration capacity of* Myriophyllum aquaticum *from 25 plant fragments (n = 4)*

either in its submerged form, or as rhizomes in the sediment and can survive at least six weeks under ice cover (Sytsma and Anderson, 1993d; Hussner, 2008). However, emerged shoots die when enclosed in ice (Hussner, 2008). The growth rate during winter is very low, with growth beginning at temperatures above 8°C (Moreira et al, 1999).

The relative growth rate and the root system development of *M. aquaticum* are influenced by nutrients and water level. Emerged plants of *M. aquaticum* showed increasing relative growth rates with increasing nutrient availabilities in the sediment (Figure 9.4c), and relative growth rates of emerged *M. aquaticum* plants were significantly higher under waterlogged (water level 10cm above soil surface) than under semi-drained (20cm below soil surface) and drained conditions (Figure 9.4a) (Hussner, 2009; Hussner et al, 2009). Plants performed best on waterlogged soils with high nutrient concentrations in the sediment, when decreasing water levels result in a faster (up to 10mm day⁻¹) and deeper root growth (Sutton, 1985; Hussner et al, 2009).

Water level fluctuations also influence the relative growth rate of *M. aquaticum*. In mesocosm experiments, *M. aquaticum* showed significantly higher relative growth rates under constant water levels (80cm water depth and waterlogged conditions), while changes in water level in two day/five day rhythms resulted in up to 40 per cent lower biomass production (Figure 9.4b).

Maximum relative growth rates of approximately 0.05g per gram dry weight per day have been reported from several studies of *M. aquaticum* (Figure 9.4) (Sytsma and Anderson, 1993d; Hussner, 2009). Root to shoot ratios of *M. aquaticum* plants are significantly influenced by nutrients, but not by water levels and water level fluctuation (Hussner, 2009; Hussner et al, 2009).

The critical, or growth limiting, phosphorus concentration for this species has been reported as 0.19 per cent and 0.1 per cent, expressed as percentage of dry weights of emerged leaves and stems of *M. aquaticum*, respectively. The critical concentration of nitrogen has been noted as 1.54 per cent and 0.42 per cent of the dry weight in emerged leaves and stems (Sytsma and Anderson, 1993b).

Plant porosity responses to habitat conditions include an increase in root porosity with increasing water level, and decreasing shoot porosity with increasing nutrient availabilities (Hussner et al, 2009). Photosynthetic measures of *M. aquaticum* show strong differences between submerged and emerged plants. In laboratory studies, submerged and emerged leaves

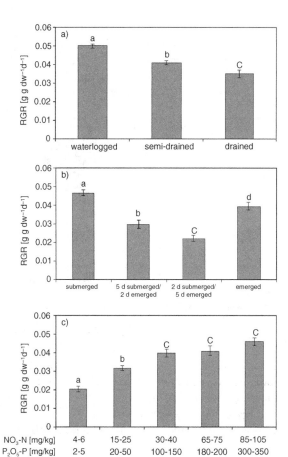

Note: Different letters mark significant differences (p<0.05). Mean ± standard error are shown.

Source: Hussner (2009); Hussner et al (2009)

Figure 9.4 *Relative growth rates of* M. aquaticum *under different soil water regimes under: (a) high nutrients, (b) water level fluctuations under high nutrients, and (c) different soil nutrients*

exhibit light compensation points of net photosynthesis of 42–45μE m⁻² s⁻¹ (Salvucci and Bowes, 1982). Submerged plants exhibit light saturation points of net photosynthesis of 250–300μE m⁻² s⁻¹ (Salvucci and Bowes, 1982). There are different reports about the light saturation point of emerged leaves of *M. aquaticum*. Salvucci and Bowes (1982) find a light saturation point of emerged *M. aquaticum* >2000μE m⁻² s⁻¹ under laboratory conditions, while Hussner (2009) reports light saturation of net photosynthesis at ~900μE m⁻² s⁻¹ and a temperature optimum at 27–37°C under field conditions (Figure 9.5).

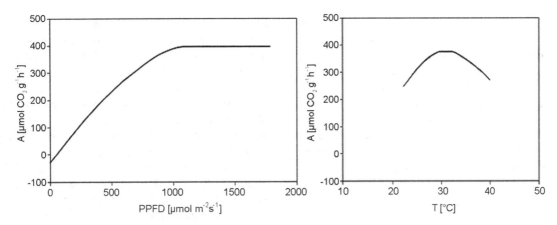

Note: PPFD = photosynthetic photon flux density.

Source: Hussner (2008)

Figure 9.5 *Light and temperature response curves of net CO$_2$ gas exchange of emerged* M. aquaticum

Management Techniques for *Myriophyllum aquaticum*

Biological control

In its native range in South America, several biological control agents have been found (Shaw, 2003; Gassmann et al, 2006). The most promising seems to be the leaf-feeding beetle *Lysathia* n. sp. (Chrysomelidae) and the stem-boring weevil *Listronotus marginicollis* (Curculionidae) (Cordo and DeLoach, 1982; Cilliers, 1999; Oberholzer et al, 2007). *Lysathia* n. sp. is a host-specific beetle, feeding and reproducing only on *M. aquaticum* (Cilliers, 1999). After its introduction into South Africa, *Lysathia* n. sp. caused a reduction in the abundance of *M. aquaticum* populations by up to 60 per cent over a three year period (Cilliers, 1999). However, *M. aquaticum* was able to recover from beetle damage when the beetle population declined after the migration of adults from heavily damaged plants.

Listronotus marginicollis is host-specific to *M. aquaticum* in the field, but in laboratory studies the species fed lightly on additional species (Cordo and DeLoach, 1982). The adult beetles fed on *M. aquaticum* foliage, whereas the larvae tunnelled in the stem. Oberholzer et al (2007) imported larvae and adults of this beetle into South Africa, and test trials showed that *L. marginicollis* did not attack non-target species but was a damaging agent against *M. aquaticum* in South Africa; particularly the larvae, which burrow down the

stem below the water surface, seem to be able to complement the work of the leaf-feeding *Lysathia* n. sp. (Oberholzer et al, 2007).

Although trialled, neither common carp (*Cyprinus carpio*) or grass carp (*Ctenopharyngodon idella*) appear to be effective biological control agents (Van Zon, 1977; Moreira et al, 1999), possibly due to the high tannin content of *M. aquaticum*.

Barreto et al (2000) have listed three different fungi species (*Chaetomella raphigera*, *Cercospora* sp. and *Mycosphaerella* sp.) associated with *M. aquaticum* and regard these as potential biological control agents for *M. aquaticum*. Joyner and Freeman (1973) reported that an isolate of *Rhizoctonia solani* was phytotoxic to tips of submersed *M. aquaticum*. An isolate of *Pythium carolinianum* has shown some promise as a potential biocontrol agent against *M. aquaticum*. Stems of *M. aquaticum* that were experimentally inoculated with this fungus showed significantly lower growth rates than control plants (Bernhardt and Duniway, 1984).

In North America, the native beaver (*Castor canadensis*) has been observed to reduce the abundance of invasive *M. aquaticum* by nearly 90 per cent (Parker et al, 2007), but beavers do not to seem to be a suitable biological control agent for this species as they are not species-specific.

Mechanical control

The mechanical control of *M. aquaticum* is difficult, due to the ease with which it breaks into fragments, which are not easily removed from the water (Guillarmod, 1977). Harvest of plant material is only effective if all plant material is removed; a partial harvest of the above-ground biomass only results in a reoccupation of the waters by *M. aquaticum* within several months (Ferreira and Moreira, 1990), with potentially increased spread via these shoot fragments (Machado and Rocha, 1998). In general, mechanical control measures are very time and cost intensive and rarely successful apart from at small scales.

Chemical control

Chemical control has generally relied on herbicides: although copper has been found to reduce growth of *M. aquaticum* in laboratory studies at concentrations of >1.0ppm in the root zone, this seems not to be practicable for application in the field (Sutton and Blackburn, 1971).

A range of herbicides has been tested to control *M. aquaticum*. Chemical applications have been more effective for longer term control than mechanical techniques. Excellent control has been reported when using endothall, diquat and 2,4-D in the US (Blackburn and Weldon, 1963; Westerdahl and Getsinger, 1988). Glyphosate control of *M. aquaticum* ranges from fair to excellent (Westerdahl and Getsinger, 1988; Machado and Rocha, 1998) to not recommended (Langeland, 1993; Moreira et al, 1999). Dichlobenil has shown some efficacy for *M. aquaticum* with fair control in the US (Westerdahl and Getsinger, 1988) and in Australia (Ripper and Milvain, 1989), but apart from 2,4-D, none of these products are regularly used to manage this plant.

More recently, triclopyr triethylamine has provided promising results in California (Anderson, 1999) and Hofstra et al (2006) found this product was the most effective (compared with clopyralid, fluridone, glyphosate, endothall and dichlobenil). Complete mortality of aerial shoots and no regrowth of *M. aquaticum* was noted in mesocosms treated with 0.5–8kg active ingredient (a.i.) ha^{-1} up to 20 weeks after treatment, with limited above- or below-ground biomass remaining after that time. Field trials confirmed the effectiveness of this product (Hofstra et al, 2006).

Wersal and Madsen (2007) evaluated the use of the herbicides imazapyr and imazamox for the control of *M. aquaticum*. Total mortality of plants was recorded eight weeks after treatment with 1.123 and 0.584kg a.i. ha^{-1} imazapyr. Lower rates of imazapyr and all tested rates of imazamox gave some reduction of *M. aquaticum* biomass, but were not as effective as imazapyr. Gray et al (2007) found carfentrozone did not effectively control *M. aquaticum*.

In summary, the three herbicides 2,4-D, triclopyr triethylamine and imazapyr can effectively control *M. aquaticum*. All are registered for aquatic use in the US, with triclopyr triethylamine also registered in New Zealand for this purpose.

Regulatory control

The introduction and spread of *M. aquaticum* outside of its indigenous range has occurred through its popularity as an aquarium and ornamental pond plant. A proactive strategy to prevent the further deliberate spread by this means was instigated in New Zealand in 2002, when the sale, propagation, distribution and commercial display of this and many other problematic aquatic weeds were banned nationally under the Biosecurity Act 1993 (Champion et al, 2002). Similar actions are planned for Australia (Petroeschevsky and Champion, 2008) and voluntary guidelines to prevent the spread of invasive non-native species such as *M. aquaticum* in Europe have been prepared (Heywood and Brunel, 2008). Such actions effectively reduce the rate of spread of such invasive species, as the volume of trade (or importation) of a species is a good indicator of the propagule pressure and introduction effort likely to result from the ornamental plant trade (Reaser et al, 2008; Simberloff, 2009; Champion et al, 2010).

References

Anderson, L. W. S. (1999) 'Foiling Watermilfoil', *Agricultural Research*, March

Aston, H. I. (1977) *Aquatic Plants of Australia*, Melbourne University Press, Carlton, Victoria

Barreto, R., Charudattan, R., Pomella, A. and Hanada, R. (2000) 'Biological control of neotropical aquatic weeds with fungi', *Crop Protection*, vol 19, pp697–703

Bernez, I., Aguiar, F., Violle, C. and Ferreira, T. (2006) 'Invasive river plants from Portugese floodplains: What can species attributes tell us?', *Hydrobiologia*, vol 570, pp3–9

Bernhardt, E. A. and Duniway, J. M. (1984) 'Root and stem rot of parrotfeather (*Myriophyllum brasiliense*) caused by *Pythium carolinianum*', *Plant Disease*, vol 68, pp999–1003

Blackburn, R. D. and Weldon, L. W. (1963) 'Suggested control measures of common aquatic weeds of Florida', *Hyacinth Control Journal*, vol 2, pp2–5

Brunel, S. (2009) 'Pathway analysis: Aquatic plants imported in 10 EPPO countries', *EPPO Bulletin*, vol 39, pp201–213

Champion, P. D., Clayton, J. S. and Rowe, D. (2002) *Lake Managers' Handbook: Alien Invaders*, Ministry for the Environment, Wellington

Champion, P. D., Clayton, J. S. and Hofstra, D. E. (2010) 'Nipping aquatic plant invasions in the bud – weed risk assessment and the trade', *Hydrobiologia*, vol 655, pp167–172

Cilliers, C. J. (1999) '*Lysathia* n. sp. (Coleotera: Crysomelidea), a host-specific beetle for the control of the aquatic weed *Myriophyllum aquaticum* (Haloragaceae) in South Africa', *Hydrobiologia*, vol 415, pp271–276

Cordo, H. A. and DeLoach, C. J. (1982) 'Weevils *Listronotus marginicollis* and *L. cinnamomeus* that feed on *Limnobium* and *Myriophyllum* in Argentina', *The Coleopterists' Bulletin*, vol 36, pp302–308

EPPO (European and Mediterranean Plant Protection Organization) (2004) '*Myriophyllum aquaticum*', EPPO data sheet on Invasive Plants *Myriophyllum aquaticum*', European and Mediterranean Plant Protection, www.eppo.org/QUARANTINE/Pest_Risk_Analysis/PRAdocs_plants/draftds/05-11833%20DS%20Myriophyllum%20aquaticum.doc, accessed 12 June 2010

Ferreira, T. and Moreira, I. (1990) 'Weed evolution and ecology in drainage canals of central Portugal', in *Proceedings EWRS 8th Symposium on Aquatic Weeds*, European Weed Research Society, Uppsala, pp97–102

Ferreira, M. T. and Moreira, I. S. (1999) 'River plants from an Iberian basin and environmental factors influencing their distribution', *Hydrobiologia*, vol 415, pp101–107

Ferreira, M. T., Catarino, L. and Moreira, I. (1998) 'Aquatic weed assemblages in an Iberian drainage channel system and related environmental factors', *Weed Research*, vol 38, no 4, pp291–300

Gassmann, A., Cock, M. J. W., Shaw, R. and Evans, C. R. (2006) 'The potential for biological control of invasive alien aquatic weeds in Europe: A review', *Hydrobiologia*, vol 570, pp217–222

Gray, C. J., Madsen, J. D., Wersal, R. M. and Getsinger, K. D. (2007) 'Eurasian watermilfoil and parrotfeather control using carfentrazone-ethyl', *Journal of Aquatic Plant Management* vol 45, pp43–46

Guillarmod, J. A. (1977) 'Some water weeds of the Eastern Cape Province – II. *Myriophyllum*', *Eastern Cape Naturalist*, vol 61, pp14–17

Guillarmod, J. A. (1979) 'Water weeds in Southern Africa', *Aquatic Botany*, vol 6, pp377–391

Haller, W. T., Sutton, D. L. and Barlowe, W. C. (1974) 'Effects of salinity on growth of several aquatic macrophytes', *Ecology*, vol 55, pp891–894

Heywood, V. and Brunel, S. (2008) 'Code of conduct on horticulture and invasive alien plants', Council of Europe (CoE) and the European and Mediterranean Plant Protection Organization (EPPO) paper

Hofstra, D. E., Champion, P. D. and Dugdale, T. M. (2006) 'Herbicide trials for the control of parrotsfeather', *Journal of Aquatic Plant Management*, vol 44, pp13–18

Hussner, A. (2008) 'Ökologische und ökophysiologische Charakteristika aquatischer Neophyten in Nordrhein-Westfalen', PhD Thesis, Universität Düsseldorf, Germany

Hussner, A. (2009) 'Growth and photosynthesis of four invasive aquatic plant species in Europe', *Weed Research*, vol 49, pp506–515

Hussner, A. and Lösch, R. (2005) 'Alien aquatic plants in a thermally abnormal river and their assembly to neophyte-dominated macrophyte stands (River Erft, Northrhine-Westphalia)', *Limnologica*, vol 35, pp18–30

Hussner, A., Meyer, C. and Busch, J. (2009) 'Influence of water level on growth and root system development of *Myriophyllum aquaticum* (Vell.) Verdcourt', *Weed Research*, vol 49, pp73–80

Hussner, A., van de Weyer, K., Gross, E. M. and Hilt, S. (2010) 'Comments on increasing number and abundance of non-indigenous aquatic macrophyte species in Germany', *Weed Research*, vol 50, pp519–526

Joyner, B. G. and Freeman, T. E. (1973) 'Pathogenicity of *Rhizoctonia solani* to aquatic plants', *Phytopathology*, vol 63, pp681–685

Kane, M. E., Gilman, E. F. and Jenks, M. A. (1991) 'Regenerative capacity of *Myriophyllum aquaticum* tissues cultured in vitro', *Journal of Aquatic Plant Management*, vol 29, pp102–109

Langeland, K. A. (1993) 'Hydrilla response to Mariner applied to lakes', *Journal of Aquatic Plant Management*, vol 31, pp175–178

Machado, C. and Rocha, F. (1998) 'Control of *Myriophyllum aquaticum* in drainage and irrigated channels of the Mondego river valley, Portugal', in A. Monteiro, T. Vasconcelos and L. Catarino (eds) *Management and Ecology of Aquatic Plants*, EWRS 10th Symposium on Aquatic Weeds, Lisbon, Portugal, pp373–376

Mendes, E. J. (1978) '*Myriophyllum aquaticum*', in E. Launert (ed) *Flora Zambesiaca: Volume 4*, Flora Zambesiaca Managing Committee, London, United Kingdom, p74

Monteiro, A. and Moreira, I. (1990) 'Chemical control of parrotfeather (*Myriophyllum aquaticum*)', in *Proceedings EWRS 8th Symposium on Aquatic Weeds*, European Weed Research Society, Uppsala, pp163–164

Moreira, I., Monteira, A. and Ferreira, T. (1999) 'Biology and control of parrotfeather (*Myriophyllum aquaticum*) in Portugal', *Ecology, Environment and Conservation*, vol 5, pp171–179

Nelson, E. and Couch, R. (1985) '*Myriophyllum spicatum* in North America', in L. W. J. Anderson (ed) *First International Symposium on Watermilfoil and Related Haloragaceae Species*, Aquatic Plant Management Society, Washington, pp19–26

Oberholzer, I. G., Mafokoane, D. L. and Hill, M. P. (2007) 'The biology and laboratory host range of the weevil, *Listronotus marginicollis* (Hustache) (Coleoptera: Curculionidae), a natural enemy of the invasive aquatic weed, parrot's feather, *Myriophyllum aquaticum* (Velloso) Verde (Haloragaceae)', *African Entomology*, vol 15, no 2, pp385–390

Orchard, A. E. (1979) '*Myriophyllum* (Haloragaceae) in Australasia. 1. New Zealand: A revision of the genus and a synopsis of the family', *Brunonia*, vol 2, pp247–287

Orchard, A. E. (1981) 'A revision of South American *Myriophyllum* (Haloragaceae), and its repercussions on some Australian and North American species', *Brunonia*, vol 4, pp27–65

Orchard, A. E. (1985) '*Myriophyllum* (Haloragaceae) in Australasia. 2. The Australian species', *Brunonia*, vol 8, pp173–291 + 58pp microfiche

Orr, B. K. and Resh, V. H. (1992) 'Influence of *Myriophyllum aquaticum* cover on *Anopheles* mosquito abundance, oviposition, and larval microhabitat', *Oecologia*, vol 90, pp474–482

Parker, J. D., Caudill, C. C. and Hay, M. E. (2007) 'Beaver herbivory on aquatic plants', *Oecologia*, vol 151, pp616–625

Petroeschevsky, A. and Champion, P. D. (2008) 'Preventing further introduction and spread of aquatic weeds through the ornamental plant trade', in *Proceedings of the 16th Australian Weed Conference, Cairns*, pp200–302

Reaser, L., Meyerson, A. and Von Holle, B. (2008) 'Saving camels from straws: How propagule pressure-based prevention policies can reduce the risk of biological invasion', *Biological Invasions*, vol 10, pp1085–1098

Ripper, C. and Milvain, H. (1989) *Aquatic Plant Control*, Department of Water Resources, NSW Agriculture and Fisheries, Griffith, New South Wales

Salvucci, M. E. and Bowes, G. (1982) 'Photosynthetic and photorespiratory responses of the aerial and submerged leaves of *Myriophyllum brasiliense* (*Myriophyllum aquaticum*)', *Aquatic Botany*, vol 13, pp147–164

Shaw, R. H. (2003) 'Biological control of invasive weeds in the UK: Opportunities and challenges', in L. E. Child, J. H. Brock, G. Brundu, K. Prach, P. Pyšek, P. M. Eade and M. Williamson (eds) *Plant Invasions: Ecological Threats and Management Sciences*, Backhuys Publishers, Leiden, The Netherlands, pp337–354

Sheppard, A. W., Shaw, R. H. and Sforza, R. (2006) 'Top 20 environmental weeds for classical biological control in Europe: A review of opportunities, regulations and other barriers to adoption', *Weed Research*, vol 46, pp93–117

Simberloff, D. (2009) 'The role of propagule pressure in biological invasions', *Annual Review of Ecology, Evolution, and Systematics*, vol 40, pp81–102

Sutton, D. L. (1985) 'Biology and ecology of *Myriophyllum aquaticum*', in L. W. J. Anderson (ed) *First International Symposium on Watermilfoil and Related Haloragaceae Species*, Aquatic Plant Management Society, Washington, DC, pp59–71

Sutton, D. L. and Blackburn, R. D. (1971) 'Uptake of copper by parrotfeather', *Weed Science*, vol 19, pp282–285

Sytsma, M. D. and Anderson, L. W. J. (1993a) 'Nutrient limitation in *Myriophyllum aquaticum*', *Journal of Freshwater Ecology*, vol 8, pp165–176

Sytsma, M. D. and Anderson, L. W. J. (1993b) 'Criteria for assessing nitrogen and phosphorus deficiency in *Myriophyllum aquaticum*', *Journal of Freshwater Ecology*, vol 8, pp155–163

Sytsma, M. D. and Anderson, L. W. J. (1993c) 'Biomass, nitrogen, and phosphorus allocation in parrotfeather (*Myriophyllum aquaticum*)', *Journal of Aquatic Plant Management*, vol 31, pp244–248

Sytsma, M. D. and Anderson, L. W. J. (1993d) 'Transpiration by an emergent macrophyte: Source of water and implications for nutrient supply', *Hydrobiologia*, vol 271, pp97–108

van der Meijden, R. (1969) 'An annotated key to the South-East Asiatic, Malesian, Mascarene, and African species of *Myriophyllum* (Haloragaceae)', *Blumea*, vol 17, pp303–311

Van Zon, J. C. J. (1977) 'Grass carp (*Ctenopharyngodon idella* Val.) in Europe', *Aquatic Botany*, vol 3, pp143–155

Wersal, R. M. and Madsen, J. D. (2007) 'Comparison of Imazapyr and Imazamox for control of parrotfeather (*Myriophyllum aquaticum* (Vell.) Verdc.)', *Journal of Aquatic Plant Management*, vol 45, pp132–136

Westerdahl, H. E. and Getsinger, K. D. (1988) *Aquatic Plant Identification and Herbicide Use Guide; Vol II: Aquatic Plants and Susceptibility to Herbicides*, Technical Report A-88-9, US Army Engineer Waterways Experiment Station, Vicksburg, MS

10

Spartina anglica C. E. Hubbard (English cord-grass)

Philip D. Roberts

History of the Species and its Introduction

Spartina anglica (Figure 10.1) is a vigorous perennial non-native grass species that occupies the lower intertidal zones of numerous coastlines and estuaries in temperate countries (Lacambra et al, 2004). The species is known by many common names and synonyms around the world (Table 10.1). *Spartina anglica* belongs to a relatively small genus consisting of approximately 14 species that are geographically centred along the east coast of North and South America, with outliers on the west coast of North America, Europe and Tristan da Cunha. All members of the genus occur primarily in wetlands, especially estuaries (Partridge, 1987).

The species originated at Hythe or Lymington, Hampshire (England) during the 19th century (Hubbard, 1957; Gray et al, 1991). *Spartina anglica* (2n = 122–124) was the result of chromosome doubling by *S. × townsendii* (2n = 61–62) *c*.1870, the sterile hybrid between the 'native' small cord-grass *S. maritima* (2n = 60) and the introduced North American smooth cord-grass *S. alterniflora* (2n = 62), which was introduced accidentally by shipping into Southampton *c*.1816 (Gray et al, 1991; Hammond and Cooper, 2003). This is a classic example of allopolypoid speciation; for a detailed description see Ainouche et al (2004).

In the UK, the early spread of *S. anglica* was largely natural, moving relatively slowly along the southern English coast, reaching Poole Harbour about 35 miles west around 30 years after the initial hybridization, and Pagham, 30 miles to the east, after almost 50 years. However, deliberate introductions of the species to the Beaulieu estuary, ten miles west of the site of origin, began in 1898 (Gray and Raybould, 1997). Despite the relatively slow spread along the English coast, *Spartina* sp. was discovered on the north coast of France by 1906, apparently having spread unaided. Further spread occurred on the English coast, to Rye, 90 miles along the south coast by 1925 (Gray and Raybould, 1997).

It was reported early in *S. anglica* expansion that the species could form extensive swards that could accrete tidal sediment in considerable volumes allowing for the substantial expansion in marsh elevation. It was this characteristic of *S. anglica* that was considered of value for coastal protection and reclamation projects, therefore causing the species to be intentionally introduced to coastal areas of the UK, Northern Europe, China, New Zealand and western US (Gray et al, 1991) (Figure 10.2).

The first assisted expansion of the species range was to the Norfolk coastline in 1907. During the 1920s and 1930s extensive plantings were made to most estuaries in southern Britain, mostly sourced as cuttings or seed from Arne Bay, Poole Harbour. By about 1965, more than 12,000ha of salt marsh in the UK was dominated by *Spartina* (an approximate quarter of the total). The populations of *S. anglica* on the UK south coast

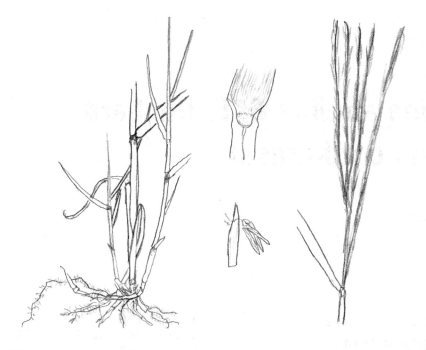

Figure 10.1 *Line drawing detailing* Spartina anglica

Table 10.1 *Synonyms and known common names for*
Spartina anglica

Synonyms
Spartina x townsendii sensu lato
Spartina x townsendii (fertile amphidiploid)
Spartina x townsendii agg.

Known common names
Preferred common name: English cord-grass
England/US: common cord-grass; English cord-grass; salt marsh grass
Australia: rice grass
Finland: englanninmarskiheinä
Germany: Englisches schlickgras: reisgras; salz-schlickgras
Netherlands: Engels slijkgras

Note: See text for detailed distribution notes.

Source: Philip D. Roberts

Figure 10.2 *Worldwide distribution of* Spartina anglica

harbours began to display 'die-back' by the 1920s (Doody, 1984). A typical pattern from Holes Bay, Poole Harbour, reports *S. anglica* arriving in 1899, expanding the sward to occupy around 208ha (60 per cent of the intertidal mudflats) by 1924, but retreating to less than half of that area by 1972, and to less than a third (63ha) by 1994. The die-back has been documented to have occurred at all English south and south-east coast estuaries, as well as in South Wales, northern France and south-west Netherlands (Gray and Raybould, 1997). The cause of the die-back is related to various phytotoxins produced under anaerobic soil conditions created by poor drainage, or even by rising relative sea levels. Gray et al (1991) report this expansion and subsequent retreat as a natural process in which a novel species, in exploiting an unoccupied niche, paved the way to its own destruction. Nehring and Hesse (2008) also add that die-back can be caused by prolonged cold periods, especially frost.

In Germany, *S. anglica* was planted at several sites along the coastline of the East and North Frisian Wadden Sea in 1927–1937 (Kolumbe, 1931; König, 1948). It is estimated that more than 70,000 shoots

were imported from the UK. The introduction of *S. anglica* to Germany was classed a success, to the extent that the species is apparently considered part of German coastal flora (Nehring and Adsersen, 2006). Within several decades *S. anglica* was frequently observed along the entire German Wadden Sea coastline (Nehring and Hesse, 2006). Nehring and Hesse (2008) mapped 20 introduced *S. anglica* plantations in German estuaries/Wadden Sea area between 1927 and 1952. Also mapped were seven plantations of *S. anglica* along The Netherlands Wadden Sea area between 1925 and 1938.

In Denmark, *S. anglica* was first planted in 1930 to 1931 in the Wadden Sea area. Approximately 6000 shoots were imported from the UK and planted at four sites on the island of Fanø and ten sites along a marshy part of the coastline from the German border to the peninsula of Skallingen (Nehring and Adsersen, 2006). Jørgensen (1934) reported that at 11 of the sites *S. anglica* stands were thriving. By 1943 *S. anglica* had spread to the island of Mandø and had been planted on Rømø in 1936, on Jordsand in 1944 and on Langi in 1946 (Pedersen, 1974). In the 1940s *S. anglica* had spread rapidly along the entire Danish Wadden Sea coastline, where it was reported as common between 0 to 40cm below sea level (Messenburg, 1972). Nehring and Hesse (2008) map 12 S. anglica plantations made in the Danish Wadden Sea area between 1931 and 1944.

During 1952 an unsuccessful experimental-scale planting was attempted in the Ringkøbing Fjord. In addition S. anglica was introduced to Randers Fjord and Mariager Fjord between 1948 and 1953 (approximately 70,000 plants, now reported as very common in the area and spreading along the coastline) and Limfjorden between 1950 and 1954 (several hundred plants, apparently only successful in the eastern part). It may have been introduced to Alrø in Horsens Fjord, from where it has spread to Vorsø (Adsersen, 1974) and the northern coast of the fjord. In the 1960s a few plants were experimentally planted in Stavns Fjord on Samsø. They were allegedly uprooted shortly after, but during the last 20 years S. anglica has infested approximately 4000m² of the Salicornia marsh (roughly half its extent) and is found scattered all around the bay (Randløv, 2006). Spartina anglica has recently been recorded at Bankel near Haderslev (Randløv, 2006), the first record along the Baltic coast. A few plants were transplanted to Norway but the plants died off after three years (Christiansen and Møller, 1983).

In China, descendants of only 21 individuals spread to cover more than 36,000ha by 1980 (Gray et al, 1991). Attempts to establish S. anglica in subtropical and tropical areas have failed (Ranwell, 1967).

In the US in Washington State, S. anglica is reported to occur along Skagit, Island, Snohomish, San Juan, Kitsap, Jefferson and King counties (Dethier and Hacker, 2004; NWCB 2005). It is also present in San Francisco Bay, California (Roberts, 2010).

In Australia, S. anglica was introduced during the 1920s although early attempts at establishment were unsuccessful and efforts continued throughout the 1940s and 1950s. Eventually colonization occurred and the species is now still spreading in three major sites: Tasmania, Victoria and South Australia (Hedge and Kriwoken 1997).

Spartina was first introduced into New Zealand in 1913, initially as S. townsendii. In 1928, Mr Bryce, a supplier of plants from Britain to New Zealand, supplied fertile seeds of S. anglica from Britain to plant at Napier, Kaipara Harour, Hokianga Harbour and Auckland. There are no records of the fate of the seed at the first two locations, and the third failed, but seed sent to Auckland had a 23 per cent germination rate (Bryce, 1936). Additional plants of S. anglica were also sent from the Essex marshes in 1955. Subsequent widespread and indiscriminate planting in estuaries around New Zealand has caused concern, particularly as native salt marsh is a sparse and valuable resource and encroachment of Spartina poses potential problems in relation to wildfowl feeding areas (Hubbard and Partridge, 1981). Lee and Partridge (1983) studied in detail the spread of S. anglica in the New River Estuary, Invercargill from 1973 to 1982, finding that the average spread was between 1.7 and 5.3m per annum.

Partridge (1987) summarizes the distribution of S. anglica as follows:

- North Island – Hokianga, Kaipara, Auckland, Hauraki Gulf, Tauranga, Gisborne, west coast north of Wellington.
- South Island – Westhaven, Farewell Spit, Tasman Bay, Linkwater, Havelock, Christchurch, Lyttelton Harbour, coast north of Dunedin, Taieri river, Catlins lock, Haldabe Bay, Invercargill.
- Stewart Island – Half Moon Bay.

Description, Physiology, Ecological Niche of *S. anglica* and its Impact

Spartina anglica is a vigorous, stout, rhizomatous salt marsh grass with round hollow stems approximately 5mm or more in diameter (Thompson, 1991). Individuals may grow 5–100cm tall. The leaves lack auricles and have ligules that have a fringe of hairs. The leaf blades, which may be flat or inrolled are approximately 35–45cm in length and 5–15mm broad, have a rough feel and have a green-grey complexion. The flowers occur in numerous, erect, contracted panicles, which consist of closely overlapping spikelets in two rows on either side of the rachis (Partridge, 1987). The species grows in roundish clumps of variable heights, depending on distance up shore; however, it commonly forms extensive meadows within years of introduction to an area. Other variable characteristics include shoot density, vegetative vigour, density of inflorescences, flowering times, seed production and seed germination (Lacambra et al, 2004; NWCB, 2005; Nehring and Adsersen, 2006; Roberts, 2010).

S. anglica is a C_4 photosynthesis strategy plant, like desert plants and those found in adverse environments where photosynthetic efficiency must be exploited, and physiological adaptations have evolved to tolerate high levels of stress (Matsuba et al, 1997). In general C_4 photosynthesis strategy plants are sensitive to low temperatures, C_4 grasses being generally more vigorous and productive in warm regions. However, *S. anglica* is an exception to this, being a cold-tolerant species found in cold temperate habitats, with net photosynthetic rates comparable to those of C_3 plants, yet able to maintain competitive growth rates at minimal temperatures (Matsuba et al, 1997; Lacambra et al, 2004).

The presence of cell glands in both surfaces of the leaf, the increased outflow of ions, the restriction of excess quantities of toxic ions entering the root system, and the evolution of root tissues with well-developed aerenchyma, have allowed the adaptation of *S. anglica* to difficult environmental conditions. The species can colonize low lying estuarine mudflat habitats that can remain immersed for at least six to nine hours daily and can adjust to extremely high external concentrations of sodium and chlorine ions (Thompson, 1991).

The leaves of *S. anglica* have a thick cuticle and stomata in grooves, reducing the potential uptake of salt by reducing the rate of transpiration. Waterlogging is counteracted by having an extensive root system in the surface layer of mud, and the roots cells contain large air spaces allowing oxygen to diffuse out and aerate the surrounding soil. The dense root system also allows resistance to the mechanical damage produced from waves (Lacambra et al, 2004). These additional characteristics of *S. anglica* allow it to be highly successful in salt marsh environments.

S. anglica is a rhizomatous perennial, spreading naturally via seed, rhizomes, tillering and rhizome fragments (Ranwell, 1964; Lacambra et al, 2004), with a range of natural dispersal mechanisms such as tidal, wind and animal. In addition, anthropogenic pathways for dispersal include shipping, packaging and intentional planting (Lacambra et al, 2004; Roberts, 2010).

Seed production is spatially and temporally variable, though seeds do not form part of the estuarine sediment seed bank (Gray et al, 1991). In Washington, *S. anglica* can flower as early as April, and flowering continues through the summer. In the UK, flowers emerge in July and August and seed ripens within approximately 12 weeks, while those flowering in September may not have time to mature (Mullins and Marks, 1987).

S. anglica seeds are relatively short lived. In the UK, seeds are only viable for one season under field conditions, with germination rates of 0.6–5 per cent. Laboratory studies have indicated that seeds stored at 4°C have remained viable for at least four years. Maximum germination occurred in the dark, with the germination rate rising as temperatures increased from 7 to 25°C (Hubbard, 1970). Seeds buried 1–3cm deep have the best chance of establishing. At shallower depths, seeds are subject to desiccation, while deeper burial may result in decreased viability due to anaerobic conditions (Groenendijk, 1986). On bare mud, *S. anglica* seedlings may grow densely, occurring at densities up to 13,000m^{-2}. Densities are lower in meadows (up to 9750m^{-2}), with many of the seedlings dying. In most cases, meadows are maintained by rhizome formation and tillering, rather than seedling establishment (Gray et al, 1991).

In the UK, the niche of *S. anglica* is between mean high water neap (MHWN) tides and mean high water spring (MHWS) tides (Gray et al, 1989). This comprises a range of low to high elevation estuarine habitat with varying degrees of tidal inundations. The species has been reported as withstanding submergence for up to nine hours (Ranwell, 1964). However, wave action has been suggested as a limiting factor for *S. anglica* establishment (Morley, 1973; Groenendijk, 1986). Upper limits of establishment are generally caused by a lack of immersion (Huckle et al, 2000) or by competition with other species. Successful establishment is more likely to occur in silt rather than sand sediments (Thompson et al, 1991; Huckle et al, 2000).

The *S. anglica* niche of low lying mudflat is below the growth of most other halophytes, allowing it to spread extensively. *Spartina anglica* spread occurs in two phases, initial invasion and establishment of seedlings or vegetational fragments on open mudflats, and then expansion of tussocks by radial clonal growth, for example over 30cm year^{-1} in organic mud in the Dovey Estuary (UK) (Charter and Jones, 1957). The tussocks spread and fuse to form clumps that expand into extensive meadows.

Marks and Truscott (1985) studied the growth pattern of *S. anglica* in salt marshes at Southport, Merseyside, UK. They reported distinct zones of *S. anglica* development within the sward that were characterized by shoot density and vegetative rigour. In the 'pioneer zone', all consisting of first year seedlings and circular clumps of up to 2m in diameter, a mean density of inflorescences of $4m^{-2}$ was reported. This then changed to more than $112m^{-2}$ in the adjacent 'transitional zone', where the clumps had merged to form continuous swards. They also reported two upper marsh zones, a 'mature zone' where *S. anglica* formed a single species sward with a continuous canopy of up to 1.3m high, and an 'invaded zone' in which other species such as *Puccinallia maritima* were found. In the last two zones the inflorescences were large, bearing twice as many spikelets as the lower zones; however, the proportion of mature seed per inflorescence was inversely related to the number of spikelets. Therefore the zone with the most vigorous vegetative growth and the largest inflorescences produced the smallest proportion of viable seed. The upper marsh colonization of *S. anglica* was also observed by Adam (1981) in UK marshes and Dethier and Hacker (2004) in Washington State (US).

The impact of *S. anglica* is on the whole negative. Although introduction for land reclamation was extremely successful (when local conditions suited establishment), the disadvantages to native biota were not generally realized during early reclamation projects (Doody, 1990; Lacambra et al, 2004). However, observations from UK naturalist, Stapf (1908, p35), include:

The immediate effect of the appearance of this pushful grass on the mudflats of the south coast has been to relieve their bareness and even to beautify them to some extent, and it has no doubt already affected animal life. Physical changes must follow, which, if the grass continues to flourish and spread, will react on the general conditions of the foreshore, resulting probably in the solidification and raising of the mud flat; but the process will take time. Whether the result will in the end be beneficial or to the contrary will depend greatly on local conditions. In any case it will be a change worth watching and studying.

S. anglica stems reduce wave energy (Knutson et al, 1982), which causes suspended sediment to accrete at stem bases (Gleason et al, 1979). The extensive rhizome system allows sediment to bind, leading to a rise in mudflat elevation (Dethier and Hacker, 2004). The sediment deposits cause the marsh surface to gradually become more elevated, quickly causing a gentle mudflat to transform into a raised marsh dissected by steep-sided channels (Figure 10.3). Chinese coastal regions experienced the benefits of *S. anglica* protection when hit by typhoons: *S. anglica* dominate marshes suffered little erosion whereas unvegetated areas were severely scoured (Chung, 1993;

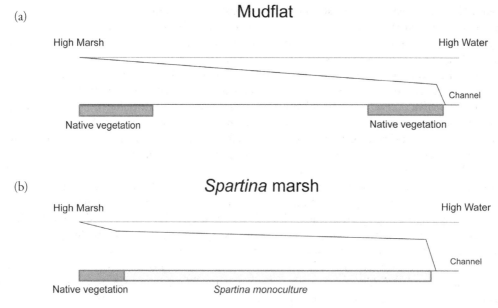

Note: Note the reduction of native vegetation and the expanse of *Spartina*.

Source: Adapted from Dethier and Hacker (2004)

Figure 10.3 *The profile of intertidal mudflats: (a) with native vegetation prior to* Spartina *invasion and (b) after invasion by* Spartina

Qin et al, 1997). However in European waters that experience high inputs of wave energy, it is generally considered that *S. anglica* does not assist effectively in coastal protection (König, 1948; Nehring and Hesse, 2008). Therefore *S. anglica* planting in the Wadden Sea area halted in the 1940s (Nehring and Hesse, 2008).

Spartina anglica causes amelioration of saline soils producing substrates suitable for the growth of crops such as *Triticum* spp. (Chung, 1993). Sediments under *S. anglica* stands are reported to have higher organic matter, porosity and pH than barren mudflat sediment (Hammond, 2001).

In general, it is reported that *S. anglica* has a negative effect on biota leading to partial or complete exclusion of native species (Hammond, 2001), including eelgrass (*Zostera*) beds, *Salicornia* spp., *Puccinellia maritima* and *Halimione portalacoides* (Oliver, 1925; Ranwell and Downing, 1959; Nairn, 1986; Scholten et al, 1987; Scholten and Rozema, 1990). In the UK, marshes dominated by *S. anglica* are reported to have fewer wading birds, such as dunlin, oystercatcher, ringed plover, sanderling, geese and widgeon (Davis and Moss, 1984; Nairn, 1986; Goss-Custard and Moser, 1990; Percival et al, 1998) due to the loss of feeding grounds. Finally, the overall effects for invertebrates are reported as variable (McCorry and Otte, 2000). Cobble beaches and mudflats with *S. anglica* are reported to have fewer worms, small bivalves and small crustaceans, due to the dense root mats. However, mobile surface invertebrates such as grasshoppers and spiders are more numerous as the thick sward produces new habitats for them to exploit (Cordell et al, 1998).

Management Efforts

Dykes can be used as a containment measure, as the dykes confine the lateral spread of *S. anglica* rhizomes. In addition, dykes remove tidal action, thereby inhibiting nutrient flow and oxygen exchange. Alternatively, dykes can be used to flood areas, eventually resulting in the death of *S. anglica*. However, the use of flooding also kills other species that cannot tolerate prolonged submersion, and therefore is not practical for large areas (Aberle, 1990).

Roberts and Pullin (2006, 2007) have, using systematic review and meta-analysis, extensively reviewed the efficacy of the control methods available for *S. anglica*. Within the appendix of their 2006 report they summarize the individual results of each disparate study and combine these within a meta-analysis to establish the most effective control method and attempt to obtain variables (e.g. inundation time, substrate) that might affect the outcomes of each control method. Table 10.2 shows the average *S. anglica* density reductions (or increases) achieved by various control methods.

Physical/mechanical control

On a small scale, seedlings can be pulled. Care must be taken to remove both the shoot and root for effective control. However, seedlings generally begin to tiller late in their first season. Once tillered, hand-pulling may break off portions of the root, allowing the plant to regrow. Repeated hand-pulling of small plants will eventually result in their death (Spartina Task Force, 1994).

Table 10.2 *The effectiveness of control methods at reducing the densities of* S. anglica

Control Method	Effectiveness (% reduction in densities, minus figures show increases of densities)
Cut and smother (mechanical)	97.9
Prokelisia spp. (biocontrol)	92.5
Fenuron (herbicide) spraying	88.2
Aminote-T (herbicide) spraying	75.8
Glyphosate (herbicide) spraying	42.8
Cut (mechanical) and glyphosate (spraying)	− 19.4
Cut only (mechanical)	− 42.8

Source: Data and analysis available in Roberts and Pullin (2006)

Cutting alone is not an effective control intervention of *S. anglica*, producing, on average, an overall increase in stem density of 42.8 per cent. However, when cutting is combined with a smothering element such as industrial black plastic sheeting then this control methodology can be highly significant, achieving a reported decline of 98 per cent stem density (Roberts and Pullin, 2006). In addition, Hammond and Cooper (2003) report that cutting and smothering was also the only management intervention that caused a decline in *S. anglica* dry root weight.

The amount of available evidence for mechanical control interventions against *S. anglica* is limited to personal accounts, with no numerical results available in the literature. However, for a closely related species (*S. alterniflora*), winter tilling produced the most effective control intervention, followed by discing and finally crushing. Crushing effectiveness was affected by the substrate type, with greatest control achieved on sand and soft silt, and least effective on firm silts or those areas with well-established *Spartina* meadows. However, tilling might be considered too costly for most *Spartina* management programmes, involving the purchase of an amphibious tiller, costing around £150,000; and is slow to implement, covering approximately 0.25ha hr^{-1} (Patten, K., unpublished data, 2004). Crushing is less expensive, costing around £50,000, and is quicker than tilling (0.5–1ha hr^{-1}), but for more effective control two or more treatments are required in one year (Roberts and Pullin, 2006, 2007).

Biological control

The potential use of *Prokelisia* spp. as a control method against *S. anglica* has been investigated by Wu et al (1999). Successful field trials have been undertaken at Puget Sound, Washington, US, resulting in a reduction of *S. anglica* densities by 92.5 per cent, but large numbers (>2000 0.5m^{-2}) of *Prokelisia* spp. were required to be applied to achieve this level of control, and multiple years of control are most likely needed.

Chemical control

Roberts and Pullin (2006) show that the herbicides most effective against *S. anglica* are fenuron (88.2 per cent reduction of original density of stands) and aminote-T (75.8 per cent reduction), but both had small datasets and require further trials prior to extensive use (though in some countries these herbicides are not licensed). One herbicide that is licensed for use against *S. anglica* in Washington state is Rodeo™ (glyphosate). However, most of the efficacy studies for this herbicide have been conducted on *S. alterniflora*.

The success of herbicidal control differs due to the concentration of active agent, the surfactants used, method of application and the natural conditions (tidal inundation and weather) (Hammond, 2001). The use of different surfactants with the herbicidal active agent affects the control achieved especially as *Spartina*'s leaves are covered by sediment and salt particles (Roberts and Pullin, 2006).

Challenges and Controversies

Attitudes to the value of *Spartina* have changed considerably since its original introduction to many countries. For example, Harbord (1949) advocated the use of *Spartina* plantations to add coastal stability and aid land reclamation. However, within the space of 20 years the attitude of most scientists and coastal land managers was that it posed a danger to the natural sedimentation of estuaries and the native flora and fauna (e.g. Ogle, 1982; Lee and Partridge, 1983; Hammond, 2001). There are still some conflicts over the use of *Spartina*, especially for land reclamation, due to the species' dense stands and ability to trap sediment. For example, in China, it has been reported that the annual value of crops on reclaimed areas is typically twice the total cost of reclamation (Chung, 1982). Other positive benefits of *Spartina* invasions are mainly economic. Livestock will eat *S. anglica* (Ranwell, 1967) and it is used as a green manure in China, with 50kg of *S. anglica* being equivalent to 0.5kg of urea. Chung (1993) reported that 3.8 million kg of fresh *S. anglica* was cut from a 76ha area in 1983 for use as fish food. *Spartina* biomass could also be used for biofuel production, sewage and pollution treatment, paper making or mushroom culture (e.g. Oliver, 1925; Ranwell, 1967; Scott et al, 1990; Chung, 2006). In addition, *S. anglica* has also been investigated as a health product: biomineral liquid from *S. anglica* culms has been put in sodawater, beer, milk, tea, wine and even bathing lotion (Hammond, 2001). *Spartina anglica* extracts are reported to assist the immune system, to have anti-inflammatory properties and to have a tonic effect on the heart (Qin et al, 1998). Total flavonoids of *S. anglica* may significantly prevent blood coagulation and encephalon thrombus (Qin et al, 1998). However, a commercial product for all of these applications has not been achieved. Overall, it appears that *S. anglica* may provide some economic benefits; however, this needs to be thoroughly compared with the risks of its uncontrolled propagation and detrimental alteration of native habitats.

References

Aberle, B. (1990) 'The biology, control and eradication of introduced *Spartina* (cordgrass) worldwide and recommendations for its control in Washington', Draft report to Washington Department of Natural Resources, Olympia

Adam, P. (1981) 'The vegetation of British salt marshes', *New Phytologist*, vol 88, pp143–196

Adsersen, H. (1974) '*Spartina* (Vadegraes) i Horsens Fjord)', *Flora og Fauna*, vol 80, pp37–42

Ainouche, M. L., Baumel, A. and Salmon, A. (2004) '*Spartina anglica* C. E. Hubbard: A natural model system for analysing early evolutionary changes that affect allopolyploid genomes', *Biological Journal of the Linnean Society*, vol 82, no 4, pp475–484

Bryce, J. (1936) '*Spartina townsendii* and *S. brasiliensis* in warm countries', *Kew Bulletin*, vol 1936, pp21–34

Charter, E. H. and Jones, H. (1957) 'Some observations on *Spartina townsendii* H. and J. Groves in the Dovey Estuary', *Journal of Ecology*, vol 45, no 1, pp157–67

Christiansen, C. and Møller, J. T. (1983) 'Rate of establishment and seasonal immersion of *Spartina* in Mariager Fjord, Denmark', *Holarctic Ecology*, vol 6, pp315–319

Chung, C. H. (1982) 'Low marshes, China', in R. R. Lewis III (ed) *Creation and Restoration of Coastal Plant Communities*, Boca Raton, Florida, CRC Press, pp131–145

Chung, C. H. (1993) 'Thirty years of ecological engineering with *Spartina* plantations in China', *Ecological Engineering*, vol 2, pp261–289

Chung, C. H. (2006) 'Forty years of ecological engineering with *Spartina* plantations in China', *Ecological Engineering*, vol 27, pp49–57

Cordell, J. R., Simenstad, C. A., Feist, B., Fresh, K. L., Thom, R. M., Stouder, D. J. and Luiting, V. (1998) 'Ecological effects of *Spartina alterniflora* invasion of the littoral flat community in Willapa Bay, Washington', *Eighth International Zebra Mussel and Other Nuisance Species Conference*, Sacramento, California

Davis, P. and Moss, D. (1984) '*Spartina* and waders – the Dyfi Estuary', in P. Doody (ed) Spartina anglica *in Great Britain*, Focus on Nature Conservation No 5, Nature Conservancy Council, Attingham, pp37–40

Dethier, M. N. and Hacker, S. D. (2004) 'Improving management practices for invasive cordgrass in the Pacific Northwest: A case study of *Spartina anglica*', Washington Sea Grant Program Publication, Seattle, WA

Doody, J. P. (1984) '*Spartina anglica* in Great Britain', A report of a meeting held at Liverpool University, 10 November 1982, Huntingdon, Nature Conservancy Council, (Focus on Nature Conservation Series), Nature Conservancy Council, Huntingdon

Doody, J. P. (1990) '*Spartina* – friend or foe? A conservation viewpoint', in A. J. Gray and P. E. M. Benham (eds) Spartina anglica: *A Research Review*, ITE Research Publication No 2, Natural Environment Research Council and HMSO, London, pp77–79

Gleason, M. L., Elmer, D. A., Pien, N. C. and Fisher, J. S. (1979) 'Effects of stem density upon sediment retention by salt marsh cordgrass, *Spartina alterniflora* Loisel', *Estuaries*, vol 2, pp271–273

Goss-Custard, J. D. and Moser, M. E. (1990) 'Changes in the numbers of dunlin (*Calidris alpina*) in British estuaries in relation to changes in the abundance of *Spartina*', in A. Gray and P. E. M. Benham (eds) Spartina anglica: *A Research Review*, ITE Research Publication No 2, Natural Environment Research Council and HMSO, London, pp39–47

Gray, A. J. and Raybould, A. F. (1997) 'The history and evolution of *Spartina anglica* in the British Isles', in K. Patten (ed) *Second International* Spartina *Conference Proceedings*, Washington State University, Olympia, pp13–16

Gray, A. J., Clarke, R.T., Warman, E. A. and Johnson, P. J. (1989) *Prediction of Marginal Vegetation in a Post-Barrage Environment*, Institute of Terrestrial Ecology (ITE) report, Wareham

Gray, A. J., Marshall, D. F. and Raybould, A. F. (1991) 'A century of evolution in *Spartina anglica*', *Advances in Ecological Research*, vol 21, pp1–62

Groenendijk, A. M. (1986) 'Establishment of a *Spartina anglica* population on a tidal mudflat: A field experiment', *Journal of Environmental Management*, vol 22, no 1, pp1–12

Hammond, M. E. R. (2001) 'The experimental control of *Spartina anglica* and *Spartina x townsendii* in estuarine salt marsh', PhD Thesis, University of Ulster

Hammond, M. E. R. and Cooper, A. (2003) '*Spartina anglica* eradication and inter-tidal recovery in Northern Ireland estuaries', in C. R. Veitch and M. N. Clout (eds) *Turning the Tide: The Eradication of Invasive Species*, Proceedings of the International Conference on Eradication of Island Invasives, IUCN, Gland, Switzerland, pp124–131

Harbord, W. L. (1949) '*Spartina townsendii* a valuable grass on tidal mud flats', *New Zealand Journal of Agriculture*, vol 78, pp507–508

Hedge, P. and Kriwoken, L. (1997) 'Managing *Spartina* in Victoria and Tasmania, Australia', in K. Patten (ed) *Second International* Spartina *Conference Proceedings*, Washington State University, Olympia, pp93–96

Hubbard, C. E. (1957) 'Report of the British Ecological Society Symposium on *Spartina*', *Journal of Ecology*, vol 45, pp612–616

Hubbard, J. C. E. (1970) 'Effects of cutting and seed production in *Spartina anglica*', *Journal of Ecology*, vol 58, no 2, pp329–334

Hubbard, J. C. E. and Partridge, T. R. (1981) 'Tidal immersion and the growth of *Spartina anglica* marshes in Waihopair River Estuary, New Zealand', *New Zealand Journal of Botany*, vol 19, pp115–121

Huckle, J. M., Potter, J. A. and Marrs, R. H. (2000) 'Influence of environmental factors on the growth and interactions between salt marsh plants: Effects of salinity, sediment and waterlogging', *Journal of Ecology*, vol 88, no 3, pp492–505

Jørgensen, C. A. (1934) 'Plantningsforsøg med *Spartina townsendii* i den Danske Vesterhavsmarsk', *Botanisk Tidsskrift*, vol 42, pp421–440

Knutson, P. L., Brochu, R. A., Seelig, W. N. and Inskeep, M. (1982) 'Wave damping in *Spartina alterniflora* marshes', *Wetlands*, vol 2, pp87–104

Kolumbe, E. (1931) *Spartina townsendii* – Anpflanzungen im Schleswig-holsteinischen Wattenmeer – Wissenschaftliche Meersuntersuchungen, *Abt. Kiel, N. F.*, pp66–73

König, D. (1948) '*Spartina townsendii* an der Westküste von Schlesig-Holstein', *Planta*, vol 36, pp34–70

Lacambra, C., Cutts, N., Allen, J., Burd, F. and Elliot, M. (2004) Spartina anglica: *A Review of its Status, Dynamics and Management*, English Nature Research Report No 527, Peterborough, UK

Lee, W. and Partridge, T. R. (1983) 'Rates of spread of *Spartina anglica* and sediment accretion in the New River Estuary, Invercargill, New Zealand', *New Zealand Journal of Botany*, vol 21, pp231–236

Marks, T. C. and Truscott, A. J. (1985) 'Variation in seed production and germination of *Spartina anglica* within a zoned saltmarsh', *Journal of Ecology*, vol 73, pp695–705

Matsuba, K., Imaizumi, N., Kaneko, S., Samejima, M. and Ohsugi, R. (1997) 'Photosynthetic responses to temperature of phosphoenolpyruvate carboxykinase type C_4 species differing in cold sensitivity', *Plant, Cell and Environment*, vol 20, no 2, pp268–274

McCorry, M. J. and Otte, M. L. (2000) 'Ecological effects of *Spartina anglica* on the macro-invertebrate infauna of the mud flats at Bull Islands, Dublin Bay, Ireland', *Web Ecology*, vol 2, pp71–73

Messenburg, H. (1972) 'Spartinas Kolonisation og Udbredelse Langs Ho Bugt', *Geografisk Tidsskrift*, vol 71, pp37–46

Morley, J. V. (1973) 'Tidal immersion of *Spartina* marsh at Bridgwater Bay, Somerset', *Journal of Ecology*, vol 61, pp383–386

Mullins, P. H. and Marks, T. C. (1987) 'Flowering phenology and seed production of *Spartina anglica*', *Journal of Ecology*, vol 75, no 4, pp1037–1048

Nairn, R. G. W. (1986) '*Spartina anglica* in Ireland and its potential impact on wildfowl and waders – a review', *Irish Birds*, vol 3, pp215–228

Nehring, S. and Adsersen, H. (2006) 'NOBANIS – invasive alien species fact sheet – *Spartina anglica*', Online Database of North European and Baltic Network on Invasive Alien Species – NOBANIS, www.nobanis.org, accessed 21 January 2011

Nehring, S. and Hesse, K.-J. (2006) 'The common cordgrass *Spartina anglica*: An invasive alien species in the Wadden Sea National Park', *Verhandlungen der Gesellschaft fur Okologie*, vol 36, pp333

Nehring, S. and Hesse, K.-J. (2008) 'Invasive alien plants in marine protected areas: The *Spartina anglica* affair in the European Wadden Sea', *Biological Invasions*, vol 10, pp937–950

NWCB (Noxious Weed Control Board) (2005) 'Common cordgrass (*Spartina anglica* C. Hubbard)', The State Noxious Weed Control Board, USA

Ogle, C. C. (1982) *Wildlife and Wildlife Values of Northland*, Fauna Survey Unit Report No 30. New Zealand Wildlife Service, Department of Internal Affairs, Wellington

Oliver, F. W. (1925) '*Spartina townsendii*: Its mode of establishment, economic uses and taxonomic status', *Journal of Ecology*, vol 13, pp74–91

Partridge, T. R. (1987) '*Spartina* in New Zealand', *New Zealand Journal of Botany*, vol 25, no 4, pp567–575

Pedersen, A. (1974) 'Gramineernes udbredelse i Danmark', *Botanisk Tidsskrift*, vol 68, pp178–343

Percival, S. M., Sutherland, W. J. and Evans, P. R. (1998) 'Intertidal habitat loss and wildfowl numbers: Applications of a spatial depletion model', *Journal of Applied Ecology*, vol 35, pp57–63

Qin, P., Xie, M., Jiang, Y. and Chung, C. H. (1997) 'Estimation of the ecological economic benefits of two *Spartina alterniflora* plantations in North Jiangsu, China', *Ecological Engineering*, vol 8, pp5–17

Qin, P., Xie, M. and Jiang, Y. (1998) '*Spartina* green food ecological engineering', *Ecological Engineering*, vol 11, pp147–156

Randløv, M. B. (2006) '*Spartina anglica* i Stavns Fjord pa Samso', Institute of Biology, University of Copenhagen

Ranwell, D. S. (1964) '*Spartina* salt marshes in southern England. II. Rate and seasonal pattern of sediment accretion. III. Rates of establishment, succession and nutrient supply at Bridgwater Bay, Somerset', *Journal of Ecology*, vol 52, pp79–105

Ranwell, D. S. (1967) 'World resources of *Spartina townsendii* (sensu lato) and economic use of *Spartina* marshland', *Journal of Applied Ecology*, vol 4, no 1, pp239–256

Ranwell, D. S. and Downing, B. M. (1959) 'Brent Goose (*Branta bernicla* (L.)) winter feeding pattern and *Zostera* resources at Scolt Head Island, Norfolk', *Animal Behaviour*, vol 7, pp42–56

Roberts, P. D. (2010) '*Spartina anglica* (common cordgrass) datasheet', CAB International, www.cabi.org/isc

Roberts, P. D. and Pullin, A. S. (2006) 'The effectiveness of management options used for the control of *Spartina* species', Systematic Review No 22, Centre for Evidence-Based Conservation, Birmingham, UK, www.environmentalevidence.org/Reviews.htm

Roberts, P. D. and Pullin, A. S. (2007) 'The effectiveness of management interventions for the control of *Spartina* species: A systematic review and meta-analysis', *Aquatic Conservation: Marine and Freshwater Ecosystems*, vol 18, no 5, pp592–618

Scholten, M. and Rozema, J. (1990) 'The competitive ability of *Spartina anglica* on Dutch salt marsh', in A. Gray and P. E. M. Benham (eds) Spartina anglica: *A Research Review*, ITE Research Publication No 2, NERC and HMSO, London, pp39–47

Scholten, M., Blaauw, P. A., Stroetenga, M. and Rozema, J. (1987) 'The impact of competitive interactions on the growth and distribution of plant species in salt marshes', in A. H. L. Huiskes, C. W. P. M. Blom and J. Rozema (eds) *Vegetation Between Land and Sea: Structure and Processes*, Dr W. Junk Publishers, Lancaster, pp270–281

Scott, R., Callaghan, T. V. and Lawson, G. J. (1990) '*Spartina* as a biofuel', in A. J. Gray and P. E. M. Benham (eds) Spartina anglica: *A Research Review*, ITE Research Publication No 2, NERC and HMSO, London, Chapter 8, pp48–51

Stapf, O. (1908) '*Spartina townendii*', *The Gardeners' Chronicle*, pp33–35

Spartina Task Force (1994) '*Spartina* management program: Integrated weed management for private lands in Willapa Bay', prepared for the Noxious Weed Board and County Commissioners, Pacific County, Washington

Thompson, J. D. (1991) 'The biology of an invasive plant: What makes *Spartina anglica* so successful?', *BioScience*, vol 41, pp393–401

Thompson, J. D., McNeilly, T. and Gray, A. J. (1991) 'Population variation in *Spartina anglica* C. Hubbard. III. Response to substrate variation in a greenhouse experiment', *New Phytologist*, vol 117, pp141–152

Wu, M.-Y., Hacker, S., Ayres, D. and Strong, D. R. (1999) 'Potential of *Prokelisia* spp. as biological control agents of English cordgrass, *Spartina anglica*', *Biological Control*, vol 16, no 3, pp267–273

11

Tamarix spp. (tamarisk, saltcedar)

Anna Sher

History of the Species

Tamarix is a genus in the family Tamaricaceae, which originates in Eurasia and North Africa (Baum, 1978). Several of these species have dispersed to other continents and regions by anthropogenic means for shade, erosion control and as an ornamental. Countries out of the native range where it has naturalized include Argentina (Gaskin and Schaal, 2003; Natale et al, 2008), Australia (Griffin et al, 1989), Canada (AARD, 2008), Mexico (Glenn and Nagler, 2005), South Africa (CARA, 2001) and the US (Robinson, 1965). The majority of research on the ecology of invasive tamarisk has been conducted on populations in the US; this chapter focuses on that work. In North America, 8–12 species have been introduced (Baum, 1967). Of these, *T. ramosissima*, *T. chinensis*, *T. parviflora*, *T. canariensis*/*T. gallica* (genetically indistinguishable morpho-species) and *T. aphylla* have all naturalized, with the majority of populations consisting of *T. ramosissima*, *T. chinensis* and their hybrid (Gaskin and Schaal, 2003). Given the difficulty in distinguishing these species from each other in the field, and the apparent similarity in their niche and behaviour, scientists and land managers have generally lumped this group together, although research is needed to test this assumption (Natale et al, 2010). Given that nearly all published research on the genus in North America is on these populations, *Tamarix* in this chapter refers to *T. ramosissima*, *T. chinensis* and their hybrid unless, otherwise specified.

There are no native species in the Tamaricaceae in North America, but genetic analysis supports Frankineaceae's status as a sister family, within which there are several species native to North America (Gaskin et al, 2004). As a group, members of the genus *Tamarix* are estimated to be the third most common woody species along rivers in this region, and are the second most dominant in terms of cover (Friedman et al, 2005). *Tamarix* has recently been estimated to cover several hundred thousand hectares in North America, from northern Mexico to Montana and from Kansas to California (Nagler et al, 2010). Although it has naturalized as far as the eastern coast and North Dakota (Figure 11.1), *Tamarix* has only achieved dominance in the warmer and drier regions of the US (NRCS, 2010). This may change under predicted scenarios of climate change, however (Bradley et al, 2009).

It is believed that *Tamarix* were first introduced to North America in the early 1800s, but may only have come to the Southwest many decades after this (Robinson, 1965; Chew, 2009). It was planted along stream sides as a bank stabilizer in the late 1800s by the Army Corps of Engineers, and by the early 1900s was well established (Graf, 1978). Between the 1920s and 1960s it is estimated to have increased from 4000ha to over 500,000ha (Robinson, 1965). By the late 1960s it began to be considered a pest, often in the context of its water use; however, it continued to be sold and promoted as a windbreak and an ornamental for many years.

At the time of writing, *Tamarix* is a 'listed species' on the state noxious weed lists of Colorado, Montana, Nebraska, Nevada, New Mexico, North Dakota, Oregon, South Dakota, Texas, Washington and Wyoming

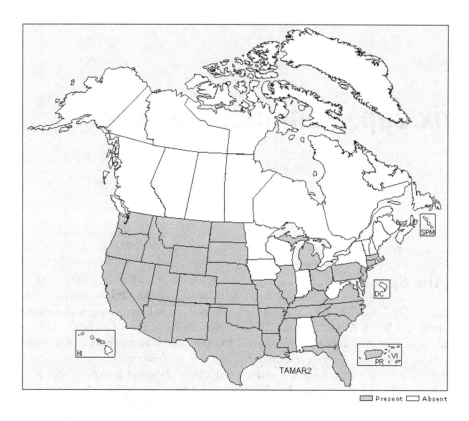

Source: Data and image from the United States Department of Agriculture Natural Resources Conservation Service (USDA-NRCS) PLANTS database

Figure 11.1 *Occurrence of species in the genus* Tamarix *within the US by state*

(NRCS, 2010). In most cases, it is not targeted for eradication (as this would probably be impossible), but is considered a significant problem and thus subject to localized control efforts (e.g. 'B' list status in Colorado). Economic losses due to *Tamarix* have been estimated to be $280–450 ha^{-1} (Zavaleta, 2000).

Ecological Niche of the Species

Tamarix is a deciduous tree with a shrub-like growth form that grows most thickly in the high moisture conditions of the riparian fringe, but which is highly drought tolerant once established and thus also occurs on upper terraces and far from perennial flows (Natale et al, 2010) (Figure 11.2). It can obtain heights of over 8m and will form dense thickets where conditions allow seedlings to establish. Conditions that promote

Source: Stephanie Gieck

Figure 11.2 *Mature* Tamarix ramosissima *tree growing in Florence, Colorado, US*

establishment from seed include alkaline soils, high moisture and shallow water tables for the first few weeks in a high light environment with little competition from other plants, and a gradually receding water table (Brotherson and Winkel, 1986; Everitt, 1980; Sher and Marshall, 2003). Near-monotypic populations of the tree can stretch for many kilometres along a river (Hink and Ohmart, 1984) (Figure 11.3). Such a vegetation community structure was unknown before the arrival of the species and as such is a dramatic departure from historic conditions along western rivers, which ranged from nearly empty (e.g. sand bars) to riparian forests with two and three canopy layers, fairly restricted to the riparian fringe (Webb et al, 2007).

Tamarix reproduction is primarily by seed; however, both shoot and root fragments have been known to reroot (Wilkinson, 1966). The root crown contains the meristomatic tissue, and as such is the source of new shoots each year, thus increasing the area occupied by an individual plant. There is some disagreement by *Tamarix* researchers as to the role of vegetative spread, but in an extensive survey in the Southwest, no excavated trees were found to be root sprouts (Shafroth, P., pers. comm.). However, the establishment of new seed-producing trees by asexual means, even if only minimal, could increase propagule pressure (Natale et al, 2010). *Tamarix* produces large sprays of tiny bisexual flowers that range from white to deep pink. Trees have been known to produce flowers as early as the second year

Figure 11.3 *Near-monotypic Tamarix population growing along the Colorado River in Utah, US*

after germination, and a single tree will produce many thousands of seeds each year. Each tiny seed is attached to many fine hairs that facilitate both wind and water dispersal. In a mature stand, an average of 17 seeds cm^{-2} per season have been recorded to reach the soil surface (Warren and Turner, 1975). These seeds tend to have high rates of viability, and although early publications reported lifespans for seeds of only a few weeks (Horton et al, 1960), the absence of a seed bank has yet to be verified in the literature. *Tamarix ramosissima* in Argentina was found to have a germination rate of approximately 80 per cent for fresh seed, but viability decreased to 30 per cent after 36 days of storage, decreasing to less than 5 per cent after 50 days (Natale, E., pers. comm.). *Tamarix* can disperse seed for the entire growing season (Warren and Turner, 1975), and a study of herbarium specimens suggests that trees at lower elevations in the Lower Colorado River Basin can flower all year long (Weisberg and Stevens, 2008). This long reproductive season allows the species to take advantage of monsoonal rains or over-bank flooding that occur at times that do not favour native species of *Populus* and *Salix*, which have evolved in a system of early spring flooding that occurred before anthropogenic disturbance to western hydrology (Warren and Turner, 1975). This is one of the features that is understood to have facilitated the spread and dominance of *Tamarix* in the New World (Sher et al, 2002).

Tamarix is considered shade intolerant and seedlings will be outcompeted by faster growing species such as native cottonwood (*Populus fremontii*) and box elder (*Acer negundo*) (Sher and Marshall, 2003; Sher et al, 2000, 2002; DeWine and Cooper, 2010). *Tamarix* trees have occasionally been observed as an understorey to *Populus*, but dense thickets in this context are unusual; the deeper the shade, the more stressed and sparse trees appear. However, research suggests that both *T. aphylla* and *T. ramosissima* may be able to outcompete the native willow, *Salix gooddingii* (Busch and Smith, 1995; Hayes et al, 2009). As a mature, well-established tree, *Tamarix* may be a strong competitor to other woody species, primarily by preventing establishment of other species in its understorey (Taylor et al, 1999). In addition to the shade created by a mature canopy, significant duff and detritus accumulates in the absence of over-bank flooding that can be highly saline (Rosel, 2006). Some have proposed that the species is allelopathic due to the resulting high salinity of the soil under these trees (e.g. Brock, 1994), although some plants can and do grow in the understorey (Lesica and DeLuca, 2004). It does not appear that *Tamarix* is mycorrhizal, and at least one study found that addition of arbuscular mycorrhizal spores reduced growth in *Tamarix* seedlings (Beauchamp et al, 2005).

Tamarix is a phreatophytic species, meaning that it is capable of putting down deep tap roots and utilizing groundwater; however, it will utilize moisture wherever it occurs (Nippert et al, 2010). The current consensus of researchers is that the range of water use of *Tamarix* is 0.7–1.4m per year, (ET_0F of 0.3–0.7, centring on a mean value of 0.5), although the extremes occur under very specific conditions (Zavaleta, 2009). The higher end has been observed for dense stands in mesic conditions, over a growing period of 300 days (ET_0F of 0.7) (Devitt et al, 1998). It has been observed in many areas that the presence of *Tamarix* is associated with the lowering of groundwater tables and drying up of streams; most likely this is in areas where other phreatophytic species (such as those of *Populus* and *Salix*) did not historically occur, or were sparse. Historic photographs suggest that such areas are not uncommon, and that *Tamarix* often took advantage of such habitat (Webb et al, 2007).

Both fire and flooding are sources of disturbance historically associated with Western rivers; however, the reduction in the latter has contributed to an increase in the former, and is probably exacerbated by *Tamarix* populations (Busch and Smith, 1993). Accumulation of *Tamarix* duff without cleansing over-bank flooding is a significant fire hazard. In addition, *Tamarix* wood burns hotter than other, native riparian species, thus increasing residence time and temperature of the soil. This has significant implications for the survival of species that have not evolved with such conditions. Given that *Tamarix* resprouts more readily after burning than some native species such as cottonwood, fire has been another mechanism that has probably contributed to the dominance of this species (Busch and Smith, 1995).

Tamarix is well adapted to high salinity, and has been found in soils as 'hot' as 30,000ppm total dissolved solids (equivalent to 30,000mg L^{-1} or 51dS m^{-1}), which is comparable to the salt content of ocean water (Glenn et al, 1998). Seedlings appear to favour water with less than 15dS m^{-1}, and cuttings prefer 6dS m^{-1} (Natale et al, 2010). As such, *Tamarix* is much more salt tolerant than native riparian species (Glenn et al, 1998). The leaves contain salt glands that isolate and exude the salt,

which accounts for the common name saltcedar. Roots also prevent uptake of some of the salts present in the soil, which results in elevated salinity of groundwater (Nagler et al, 2008). The degree to which *Tamarix* is responsible for altering environmental salinity through both exclusion and uptake is unknown, especially given that *Tamarix* is more salt tolerant than native tree species. Thus, the correlation between saline soils and *Tamarix* dominance may simply be a function of that species being able to survive where others cannot; saline soils are also a function of reduced over-bank flooding and high rates of surface evaporation (Stromberg et al, 2009). However, recent research conducted in New Mexico found a positive relationship between age and surface soil salinity until a threshold of approximately 15 years (Ohrtman, 2009). This suggests that the trees were contributing to soil salinity over time, made more compelling by the fact that sites that received cleansing over-bank flooding were less likely to show this relationship. Low soil salinity in the oldest, closed canopy stands is probably explained by the reduced surface evaporation in these sites, as well as the higher wood to leaf biomass ratio. *Tamarix* was also found to have a significantly lower normalized difference vegetation index (NDVI) than native cottonwood and willow, suggesting more light penetration (Nagler et al, 2004). Increased light and heat on the forest floor in *Tamarix* stands promotes more surface evaporation, also associated with elevated soil salinity, thus suggesting another mechanism whereby *Tamarix* may indirectly affect environmental salinity.

Given the changes in the physical structure of the forest, soil chemistry, fire regimes and hydrology associated with the species, it is not surprising that wildlife associated with *Tamarix* thickets tends to be markedly different from that in stands where native *Populus* (cottonwood) and *Salix* (willow) occur, either as a mixed stand or where *Tamarix* does not occur. Leaf tissue from *Tamarix* differs from native cottonwood and willow, which in one study was associated with a twofold decrease in macroinvertebrates relative to native species (Bailey et al, 2001). The only known invertebrate herbivores of *Tamarix* in North America are the introduced biocontrol *Diorhabda* spp., a tamarisk leafhopper *Opsius stactogalus,* a scale, *Chionespis* spp., and possibly a few generalist herbivores (Lewis et al, 2003). Research suggests that several predatory insect families may increase with the presence of the *Tamarix* biocontrol (Strudley, 2009).

Several studies on bird diversity in *Tamarix* stands have been conducted; although not entirely consistent, it is clear that there are distinct differences from native-dominated habitat. Several studies have measured lower bird diversity in *Tamarix* stands, owing in part to the complete absence of certain guilds (Hunter et al, 1988; Brand et al, 2008). In particular, raptors, cavity nesters, frugivores and nectivores tend to avoid the species, whereas some species of cup-nesting birds and insectivores sometimes appear to prefer it (Walker, 2006). As a result, while simple species richness is sometimes similar, bird species assemblages are found to differ greatly between pure native, mixed and *Tamarix*-dominated stands (Ellis, 1995; Brand et al, 2008; Walker, 2008). In one study, *Tamarix* and cottonwood were selected by birds with similar frequency but species differed so greatly that tree species was a more important predictor of bird species occurrence than any of the foliage density or distribution measures taken that might otherwise explain bird preference (Rice et al, 1984). Another study found that even for birds that used *Tamarix*, there appeared to be a threshold density, over which increase in the exotic tree did not increase bird abundance (van Riper et al, 2008). It should also be noted that bird use of *Tamarix* is highly variable and has been found to differ greatly between watersheds; for example, several species that were found to use *Tamarix* along the Pecos River do not use *Tamarix* growing along the Colorado River, despite their presence in the range (Hunter et al, 1988).

The south-western willow flycatcher (SWFL) (*Empidonax traillii extimus*) is the animal species for which the most literature exists on habitat value of *Tamarix*. SWFL is a listed species under the US Endangered Species Act, and its range overlaps with *Tamarix* infestations. There has been some disagreement as to whether *Tamarix* represents equivalent or inferior habitat for SWFL (Dudley and DeLoach, 2004; Sogge et al, 2008; van Riper et al, 2008), and nesting habitat of SWFL appears to be most strongly related to water resources, rather than vegetation type (Hatten et al, 2010). However, it is clear that this species does use *Tamarix*, and that any restoration plan must consider how it will mitigate habitat loss (Zavaleta et al, 2001).

Much less research exists on other animals. For mammals, observations suggest dramatically decreased diversity in *Tamarix* stands relative to other riparian areas (Hink and Ohmart, 1984). Although goats will browse young saplings, few other vertebrate herbivores appear to

eat it; for example, in one study of mule deer foraging, *Tamarix* was not among the 34 taxa listed as a food source, even though it was common in the study area (Marshal et al, 2004). Beavers will eat *Tamarix* but also strongly prefer native species (Kimball and Perry, 2008). Research on herpetofauna suggests that *Tamarix* control is likely to benefit or at least not harm native lizards (Bateman et al, 2008), and preliminary findings suggest that lizard diversity is greater in mixed stands over monotypic *Tamarix* stands (Bateman and Ostoja, 2010).

Management Efforts

Since the droughts of the late 1990s, *Tamarix* has been considered a high priority species for removal in many western states. Financial gains from *Tamarix* removal associated with the problems of lost water, increased flooding and loss of wildlife were computed for a 55 year period as ranging from $95 to nearly $7000 per acre net benefit, the lower end based on a discount rate of 6 per cent and the higher end on a 0 per cent discount rate (Zavaleta, 2000). In 2003 a not-for-profit called the Tamarisk Coalition was formed by river rafters and others frustrated by the observed change in their waterways, including river access. This group became important for joining together scientists, policy makers and stakeholder groups for the purpose of restoring the rivers to more functionality. In October 2006 President Bush signed into law the Salt Cedar and Russian Olive Control Demonstration Act, requiring the Department of Interior to assess the extent of *Tamarix* and Russian olive (*Eleagnus angustifolia*) infestation and to carry out demonstration projects to manage it. There was bipartisan support for the bill, and agreement about the necessity of controlling the species united many disparate groups, including commercial groups in the ranching and agriculture sector with non-profit conservation organizations such as The Nature Conservancy and Ducks Unlimited.

Management of the species was well understood by that time. Given that the root crown will vigorously resprout after any efforts to remove above-ground biomass, effective means of control all involve removing or killing this portion of the tree. This is most easily accomplished with pesticide or through a combination of mechanical and chemical control (Nissen et al, 2010). Approaches that have been successfully used on this species include foliar applications of a systemic herbicide (such as imazapyr), cut stump method (stumps are painted with herbicide after being cut), and combinations of fire and herbicide. Purely mechanical means include use of excavators and root rakes, taking care to mulch, burn or otherwise prevent root tissue from being reburied and therefore re-established (Wilkinson, 1966). Very large infestations have been successfully killed with aerial applications of herbicide. Given its poor competitive abilities, promotion of fast growing native trees has also been promoted as a means of natural control (Sher et al, 2002; DeWine and Cooper, 2010).

In the late 1990s, efforts to develop a biological control for *Tamarix* resulted in trials of species of *Diorhabda* (initially considered *D. elongata*, now reclassified as *D. carinulata* (Tracy and Robbins, 2009)), a host-specific beetle that feeds on the leaves of *Tamarix* in both the adult and larval stages (DeLoach et al, 2003). Cage trials were initiated at ten sites in 1999 with only mixed success, as there were some problems with beetles going into diapause (Lewis et al, 2003). In addition, trees were requiring repeated defoliations to be killed (Hudgeons et al, 2007). Nonetheless, it had promise and was deemed safe for the ecosystem, and so in 2001 was officially released in the wild. By 2010, *D. carinulata* could be found in *Tamarix* trees along the Colorado, Dolores, Green, Gunnison, San Juan, Virgin and White Rivers and their associated tributaries (Tamarisk Coalition, 2010).

Controversy

Although there was nearly universal support for its control through the late 1990s, as eradication programmes became more common, concern grew about the focus on the species rather than stewardship of the whole ecosystem, particularly with regard to bird habitat. Because *Tamarix* is able to exploit anthropogenic alterations of river systems, and because many of the environmental changes attributed to *Tamarix* could also be explained by these same changes, the focus on *Tamarix* removal (rather than hydrological management) began to be challenged (Shafroth et al, 2005, 2008; Stromberg et al, 2007). It was argued that human use of water resources and reduced over-bank flooding can also cause drought, lowering of water tables, increased

soil salinity, increased fire frequency and decreased establishment of native phreatophytes – all changes 'blamed' on *Tamarix* (Stromberg et al, 2009). Confounding the debate was frequent misuse and over-reaching interpretations of published field research (Stromberg et al, 2009).

Use of *Tamarix* by the endangered SWFL became the lightning rod for such concern, given that replacement vegetation would take years to become suitable habitat for the cup-nesting species (Sogge et al, 2003). Although chemical and mechanical control efforts could be easily regulated to provide strips of refuge for the birds, and made to avoid nesting areas completely, biological control was necessarily much more difficult to control. The US Fish and Wildlife Recovery Plan for the flycatcher in 2001 cited concern about *Tamarix* control efforts, and ultimately it was decided that no releases were to be made within 200 miles of SWFL habitat (USFWS, 2002). In addition, *D. carinulata* was not expected to tolerate latitudes lower than 38° due to day length requirements (Lewis et al, 2003), but the beetles were more mobile than expected and migrated south.

As a result, several groups became concerned about the spread of the saltcedar beetle, particularly after it came to light that there had been unauthorized distribution of the biocontrol in Utah (Hultine et al, 2010). In March of 2009, The Center for Biological Diversity and the Maricopa Audubon Society filed a notice of intent to sue the Animal and Plant Health Inspection Service (APHIS) and US Fish and Wildlife Service (USFWS) for violation of the Endangered Species Act by releasing the biocontrol beetle and thus presumably endangering habitat for the listed bird (Kenna, 2009). In June of that year the US Department of Agriculture formally ended the release programme of *D. carinulata* in 13 states (Colorado, Idaho, Iowa, Kansas, Missouri, Nebraska, Nevada, North Dakota, Oregon, South Dakota, Montana, Washington and Wyoming). This meant that APHIS no longer endorsed releasing the beetles, and that doing so could be punished with a fine of up to $250,000 per violation.

In May of 2010, the *Saltcedar and Russian Olive Control Demonstration Act Science Assessment* was published by the United States Geological Survey (USGS) (Shafroth et al, 2010). This exhaustive review publicized the current science revealing that *Tamarix* does have some ecological value and does not use more water than some native species (Shafroth et al, 2010). This report was then used in some media outlets to suggest that control efforts were misguided, although the review made no such claim and accurately reported the negative ecological impacts of the species as well (Barr, 2010).

Regardless of the controversy, there are currently many active *Tamarix* control programmes throughout the American West, with the primary goal of returning ecosystems to being native species dominated and recovering lost ecosystem services. Research, policy and land management have shifted in focus to some degree as a result of the proliferation of the biocontrol beetle. Given that *D. carinulata* is here to stay, it is clearly imperative that we understand the response of ecosystems to *Tamarix* defoliation. *Tamarix* was an ideal poster child to focus attention and funds on riparian restoration for over a decade; although this era may be ending, the partnerships formed as a result continue, to the benefit of ecosystem management in the West.

References

AARD (Alberta Agriculture and Rural Development) (2008) 'Weed alert – *Tamarix ramosissima*', Government of Alberta Agriculture and Rural Development, 26 June, http://www1.agric.gov.ab.ca/$department/deptdocs.nsf/all/prm12239, accessed 22 January 2011

Bailey, J. K., Schweitzer, J. A. and Whitham, T. G. (2001) 'Salt cedar negatively affects biodiversity of aquatic macroinvertebrates', *Wetlands*, vol 21, no 3, pp442–447

Barr, Z. (2010) 'Tamarisk owed an apology?', Radio interview with A. A. Sher, on *Colorado Matters*, Colorado Public Radio, June 10, 2010, www.cpr.org/article/Tamarisk_Owed_An_Apology

Bateman, H. L., Chung-MacCoubrey, A. and Snell, H. L. (2008) 'Impact of non-native plant removal on lizards in riparian habitats in the Southwestern United States', *Restoration Ecology*, vol 16, no 1, pp180–190

Bateman, H. L. and Ostoja, S. M. (2010) 'Saltcedar, herpetofauna, and small mammals: Evaluating the impacts of non-native plant biocontrol', *The Wildlife Society's 17th Annual Conference*, Snowbird, UT

Baum, R. (1967) 'Introduced and naturalized tamarisks in the United States and Canada [Tamaricaceae]', *Baileya*, vol 15, pp19–25

Baum, R. (1978) *The Genus Tamarix*, The Israel Academy of Sciences and Humanities

Beauchamp, V. B., Stromberg, J. C. and Stutz, J. C. (2005) 'Interactions between *Tamarix ramosissima* (saltcedar), *Populus fremontii* (cottonwood), and mycorrhizal fungi: Effects on seedling growth and plant species coexistence', *Plant and Soil*, vol 275, no 1–2, pp221–231

Bradley, B. A., Oppenheimer, M. and Wilcove, D. S. (2009) 'Climate change and plant invasions: Restoration opportunities ahead?', *Global Change Biology*, vol 15, no 6, pp1511–1521

Brand, L. A., White, G. C. and Noon, B. R. (2008) 'Factors influencing species richness and community composition of breeding birds in a desert riparian corridor', *Condor*, vol 110, no 2, pp199–210

Brock, J. H. (1994) '*Tamarix* spp. (salt cedar), an invasive exotic woody plant in arid and semi-arid riparian habitats of western USA', in L. C. De Waal, L. E. Child, P. M. Wade and J. H. Brock (eds) *Ecology and Management of Invasive Riverside Plants*, John Wiley and Sons, New York, USA, pp27–44

Brotherson, J. D. and Winkel, T. (1986) 'Habitat relationships of saltcedar (*Tamarix ramosissima*) in Central Utah', *Great Basin Naturalist*, vol 7, no 3, pp535–541

Busch, D. E. and Smith, S. D. (1993) 'Effects of fire on water and salinity relations of riparian woody taxa', *Oecologia*, vol 94, pp186–194

Busch, D. E. and Smith, S. D. (1995) 'Mechanisms associated with decline of woody species in riparian ecosystems of the southwestern US', *Ecological Monographs*, vol 65, no 3, pp347–370

CARA (Conservation of Agricultural Resources Act) (2001) 'Regulations 15 and 16 (regarding problem plants)', South Africa Conservation of Agricultural Resources Act, 1983 (Act No 43)

Chew, M. K. (2009) 'The monstering of tamarisk: How scientists made a plant into a problem', *Journal of the History of Biology*, vol 42, no 2, pp231–266

DeLoach, D. J., Lewis, P. A., Herr, J. C., Carruthers, R. I., Tracy, J. L. and Johnson, J. (2003) 'Host specificity of the leaf beetle, *Diorhabda elongata deserticola* (Coleoptera: Chrysomelidae) from Asia, a biological control agent for saltcedars (*Tamarix*: Tamaricaceae) in the western United States', *Biological Control*, vol 23, pp117–147

Devitt, D. A., Sala, A., Smith, D. B., Cleverly, J. R., Shaulis, L. and Hammett, R. (1998) 'Bowen ratio estimates of evapotranspiration for *Tamarix ramosissima* stands on the Virgin River in southern Nevada', *Water Resources Research*, vol 34, no 9, pp2407–2414

DeWine, J. M. and Cooper, D. J. (2010) 'Habitat overlap and facilitation in tamarisk and box elder stands: Implications for tamarisk control using native plants', *Restoration Ecology*, vol 18, no 3, pp349–358

Dudley, T. L. and DeLoach, C. J. (2004) 'Saltcedar (*Tamarix* spp.), endangered species, and biological weed control – Can they Mix?', *Weed Technology*, vol 18, no 5, pp1542–1551

Ellis, L. M. (1995) 'Bird use of saltcedar and cottonwood vegetation in the middle Rio Grande valley of New Mexico, USA', *Journal of Arid Environments,* vol 30, no 3, pp339–49

Everitt, B. L. (1980) 'Ecology of saltcedar – a plea for research', *Environmental Geology*, vol 3, no 2, pp77–84

Friedman, J. M., Auble, G. T., Shafroth, P. B., Scott, M. L., Merigliano, M. F., Freehling, M. D. and Griffin, E. R. (2005) 'Dominance of non-native riparian trees in western USA', *Biological Invasions*, vol 7, no 4, pp747–751

Gaskin, J. F. and Schaal, B. A. (2003) 'Molecular phylogenetic investigation of US invasive *Tamarix*', *Systematic Botany*, vol 28, no 1, pp86–95

Gaskin, J. F., Ghahremani-nejad, F., Zhang, D. and Londo, J. P. (2004) 'A systematic overview of Frankeniaceae and Tamaricaceae from nuclear rDNA and plastid sequence data', *Annals of the Missouri Botanical Garden*, vol 91, no 3, pp401–409

Glenn, E. P. and Nagler, P. L. (2005) 'Comparative ecophysiology of *Tamarix ramosissima* and native trees in western US riparian zones', *Journal of Arid Environments*, vol 61, no 3, pp419–446

Glenn, E., Tanner, R., Mendez, S., Kehret, T., Moore, D., Garcia, J. and Valdes, C. (1998) 'Growth rates, salt tolerance, and water use characteristics of native and invasive riparian plants from the delta of the Colorado River, Mexico', *Journal of Arid Environments*, vol 40, no 3, pp281–294

Graf, W. L. (1978) 'Fluvial adjustments to the spread of tamarisk in the Colorado Plateau region', *Geological Society of American Bulletin*, vol 89, no 10, pp1491–1501

Griffin, G. F., Stafford, D. M., Morton, S. R., Allan, G. E. and Masters, K. A. (1989) 'Status and implications of the invasion of tamarisk (*Tamarix aphylla*) on the Finke River, Northern Territory, Australia', *Journal of Environmental Management*, vol 29, pp297–315

Hatten, J. R., Paxton, E. H. and Sogge, M. K. (2010) 'Modeling the dynamic habitat and breeding population of Southwestern Willow Flycatcher', *Ecological Modelling*, vol 221, no 13/14, pp1674–1686

Hayes, W., Walker, L. and Powell, E. (2009) 'Competitive abilities of *Tamarix aphylla* in southern Nevada', *Plant Ecology*, vol 202, no 1, pp159–167

Hink, V. C. and Ohmart, R. D. (1984) 'Middle Rio Grande biological survey', US Army Corps of Engineers, Albuquerque, NM

Horton, J. S., Mounts, F. C. and Kraft, J. M. (1960) 'Seed germination and seedling establishment of phreatophyte species', US Department of Agriculture, Rocky Mountain Forest and Range Experiment Station, Paper No 48

Hudgeons, J. L., Knutson, A. E., Heinz, K. M., DeLoach, C. J., Dudley, T. L., Pattison, R. R. and Kiniry, J. R. (2007) 'Defoliation by introduced *Diorhabda elongata* leaf beetles (Coleoptera: Chrysomelidae) reduces carbohydrate reserves and regrowth of *Tamarix* (Tamaricaceae)', *Biological Control*, vol 43, no 2, pp213–221

Hultine, K. R., Belnap, J., van Riper III, C., Ehleringer, J. R., Dennison, P. E., Lee, M. E., Nagler, P. L., Snyder, K. A., Uselman, S. M. and West, J. B. (2010) 'Tamarisk biocontrol in the western United States: Ecological and societal implications', *Frontiers in Ecology and the Environment*, vol 8, no 9, pp467–474

Hunter, W. C., Ohmart, R. D. and Anderson, B. W. (1988) 'Use of exotic saltcedar (*Tamarix chinensis*) by birds in arid riparian systems', *The Condor*, vol 90, no 1, pp113–123

Kenna, M. (2009) Center for Biological Diversity and Maricopa Audubon Society vs. Animal and Plant Health Inspection Service and US Fish and Wildlife Service, www.biologicaldiversity.org/species/birds/southwestern_willow_flycatcher/pdfs/tamarisk_complaint.pdf, accessed 2 August 2011

Kimball, B. and Perry, K. (2008) 'Manipulating beaver (*Castor canadensis*) feeding responses to invasive tamarisk (*Tamarix* spp.)', *Journal of Chemical Ecology*, vol 34, no 8, pp1050–1056

Lesica, P. and DeLuca, T. (2004) 'Is tamarisk allelopathic?', *Plant and soil*, vol 267, no 1, pp357–365

Lewis, P. A., DeLoach, C. J., Knutson, A. E., Tracy, J. L. and Robbins, T. O. (2003) 'Biology of *Diorhabda elongata deserticola* (Coleoptera: Chrysomelidae), an Asian leaf beetle for biological control of saltcedars (*Tamarix* spp.) in the United States', *Biological Control*, vol 27, no 2, pp101–116

Marshal, J. P., Bleich, V. C., Andrew, N. G. and Krausman, P. R. (2004) 'Seasonal forage use by desert mule deer in southeastern California', *Southwestern Naturalist*, vol 49, no 4, pp501–505

Nagler, P. L., Glenn, E. P., Lewis Thompson, T. and Huete, A. (2004) 'Leaf area index and normalized difference vegetation index as predictors of canopy characteristics and light interception by riparian species on the Lower Colorado River', *Agricultural and Forest Meteorology*, vol 125, no 1/2, pp1–17

Nagler, P. L., Glenn, E. P., Didan, K., Osterberg, J., Jordan, F. and Cunningham, J. (2008) 'Wide-area estimates of stand structure and water use of *Tamarix* spp. on the lower Colorado River: Implications for restoration and water management projects', *Restoration Ecology*, vol 16, no 1, pp136–145

Nagler, P. L., Glenn, E., Jamevich, C. S. and Shafroth, P. B. (eds) (2010) *Distribution and Abundance of Saltcedar and Russian Olive in the Western United States*, US Geological Survey Scientific Investigations Report, Washington, DC

Natale, E., Gaskin, J., Zalba, S. M., Ceballow, M. and Reinoso, H. (2008) 'Species of the genus *Tamarix* (tamarisk) invading natural and semi-natural environments in Argentina', *Boletín de la Sociedad Argentina de Botanica*, vol 43, pp137–145

Natale, E., Zalba, S. M., Oggero, A. and Reinoso, H. (2010) 'Establishment of *Tamarix ramosissima* under different conditions of salinity and water availability: Implications for its management as an invasive species', *Journal of Arid Environments*, vol 74, no 11, pp1399–1407

Nippert, J. B., Butler Jr, J. J., Kluitenberg, G. J., Whittemore, D. O., Arnold, D., Spal, S. E. and Ward, J. K. (2010) 'Patterns of *Tamarix* water use during a record drought', *Oecologia*, vol 162, no 2, pp283–292

Nissen, S., Sher, A. A. and Norton, A. (eds) (2010) *Tamarisk Best Management Practices in Colorado Watersheds*, Colorado State University Press, Denver, CO

NRCS (Natural Resources Conservation Service) (2010) 'PLANTS Profile: *Tamarix ramosissima* Ledeb. saltcedar', PLANTS Database, http://plants.usda.gov/java/profile?symbol=TARA, accessed 20 January 2011

Ohrtman, M. K. (2009) 'Quantifying soil and groundwater chemistry in areas invaded by *Tamarix* spp. along the Middle Rio Grande, New Mexico', Doctoral Dissertation, University of Denver, Denver, CO

Rice, J., Anderson, B. W. and Ohmart, R. D. (1984) 'Comparison of the importance of different habitat attributes to avian community organization', *The Journal of Wildlife Management*, vol 48, no 3, pp895–911

Robinson, T. W. (1965) 'Introduction, spread and areal extent of saltcedar (*Tamarix*) in the western states', US Geological Survey Professional Paper 491-A

Rosel, C. E. (2006) 'Saltcedar (*Tamarix* spp.) leaf litter impacts on surface soil chemistry: Electrical conductivity and sodium adsorption ratio', MS Thesis, New Mexico State University, Las Cruces, NM

Shafroth, P. B., Cleverly, J. R., Dudley, T. L., Taylor, J. P., Van Riper, C., Weeks, E. P. and Stuart, J. N. (2005) 'Control of *Tamarix* in the Western United States: Implications for water salvage, wildlife use, and riparian restoration', *Environmental Management*, vol 35, no 3, pp231–246

Shafroth, P. B., Beauchamp, V. B., Briggs, M. K., Lair, K., Scott, M. L. and Sher, A. A. (2008) 'Planning riparian restoration in the context of *Tamarix* control in western North America', *Restoration Ecology*, vol 16, no 1, pp97–112

Shafroth, P. B., Brown, C. A. and Merritt, D. M. (eds) (2010) *Saltcedar and Russian Olive Control Demonstration Act Science Assessment*, US Geological Survey Scientific Investigations Report 2009-5247

Sher, A. A. and Marshall, D. L. (2003) 'Seedling competition between native *Populus deltoides* (Salicaceae) and exotic *Tamarix ramosissima* (Tamaricaceae) across water regimes and substrate types', *American Journal of Botany*, vol 90, no 3, pp413–422

Sher, A. A., Marshall, D. L. and Gilbert, S. A. (2000) 'Competition between native *Populus deltoides* and invasive *Tamarix ramosissima* and the implications for reestablishing flooding disturbance', *Conservation Biology*, vol 14, no 6, pp1744–1754

Sher, A. A., Marshall, D. L. and Taylor, J. P. (2002) 'Establishment patterns of native *Populus* and *Salix* in the presence of invasive nonnative *Tamarix*', *Ecological Applications*, vol 12, no 3, pp760–772

Sogge, M. K., Sferra, S. J., McCarthey, T. D., Williams, S. O. and Kus, B. E. (2003) 'Distribution and characteristics of southwestern willow flycatcher breeding sites and territories: 1993–2001', *Studies in Avian Biology*, vol 26, no, pp5–11

Sogge, M. K., Sferra, S. J. and Paxton, E. H. (2008) '*Tamarix* as habitat for birds: Implications for riparian restoration in the southwestern United States', *Restoration Ecology*, vol 16, no 1, pp146–154

Stromberg, J. C., Lite, S. J., Marler, R., Paradzick, C., Shafroth, P. B., Shorrock, D., White, J. M. and White, M. S. (2007) 'Altered stream-flow regimes and invasive plant species: The *Tamarix* case', *Global Ecology and Biogeography*, vol 16, no 3, pp381–393

Stromberg, J., Chew, M. K., Nagler, P. L. and Glenn, E. P. (2009) 'Changing perceptions of change: The role of scientists in *Tamarix* and river management', *Restoration Ecology*, vol 17, no 2, pp177–186

Strudley, S. (2009) 'Impacts of tamarisk biocontrol (*Diorhabda elongata*) on the trophic dynamics of terrestrial insects in monotypic tamarisk stands', University of Denver, Denver, CO

Tamarisk Coalition (2010) *2010 Tamarisk Leaf Beetle Monitoring Report*, Tamarisk Coalition, US

Taylor, J. P., Wester, D. B. and Smith, L. M. (1999) 'Soil disturbance, flood management, and riparian woody plant establishment in the Rio Grande floodplain', *Wetlands*, vol 19, no 2, pp372–338

Tracy, J. L. and Robbins, T. O. (2009) 'Taxonomic revision and biogeography of the *Tamarix*-feeding *Diorhabda elongata* (Brullé, 1832) species group (Coleoptera: Chrysomelidae: Galerucinae: Galerucini) and analysis of their potential in biological control of Tamarisk', *Zootaxa*, vol 2101, pp1–152

USFWS (US Fish and Wildlife Service) (2002) 'Southwestern willow flycatcher (*Empidonax traillii extimus*) final recovery plan', US Fish and Wildlife Service, Albuquerque, NM

van Riper III, C., Paxton, K. L., O'Brien, C., Shafroth, P. B. and McGrath, L. J. (2008) 'Rethinking avian response to *Tamarix* on the lower Colorado River: A threshold hypothesis', *Restoration Ecology*, vol 16, no 1, pp155–167

Walker, H. A. (ed) (2006) *Southwesertn Avian Community Organization in Exotic* Tamarix: *Current Patterns and Future Needs*, US Department of Agriculture, Forest Service, Rocky Mountain Research Station, Fort Collins, CO

Walker, H. A. (2008) 'Floristics and physiognomy determine migrant landbird response to tamarisk (*Tamarix ramosissima*) invasion in riparian areas', *The Auk*, vol 125, no 3, pp520–531

Warren, D. K. and Turner, R. M. (1975) 'Saltcedar (*Tamarix chinensis*) seed production, seedling establishment, and response to inundation', *Journal of the Arizona Academy of Science*, vol 10, no 3, pp135–144

Webb, R. H., Leake, S. A. and Turner, R. M. (2007) *The Ribbon of Green: Change in Riparian Vegetation in the Southwestern United States*, Arizona University Press, Tucson

Weisberg, P. and Stevens, L. E. (2008) 'Non-native *Tamarix ramosissima* recruitment along the Colorado River: Interactions among flow regimes and geomorphology', Project no NEV052PT, United States Department of Agriculture Research, Education and Economics Information System, University of Nevada

Wilkinson, R. E. (1966) 'Adventitious shoots on saltcedar roots', *Botanical Gazette*, vol 127, no 2–3, pp103–104

Zavaleta, E. (2000) 'Valuing ecosystem services lost to *Tamarix* invasion in the United States', in H. Mooney and R. J. Hobbs (eds) *Invasive Species in a Changing World*, Island Press, Washington, DC

Zavaleta, E. (2009) 'Independent peer review of tamarisk and Russian olive evapotranspiration', in *Colorado River Basin Tamarisk and Russian Olive Assessment*, Tamarisk Coalition, Grand Junction, CO, p128

Zavaleta, E. S., Hobbs, R. J. and Mooney, H. A. (2001) 'Viewing invasive species removal in a whole-ecosystem context', *Trends in Ecology and Evolution*, vol 16, no 8, pp454–459

Part II

Aquatic Invertebrates

12

Aedes albopictus Skuse (Asian tiger mosquito)

Paul Leisnham

Introduction

Aedes (Stegomyia) albopictus is a mosquito native to temperate and tropical parts of southeast Asia and India. It is thought to be one of the fastest spreading animal species and has been nominated as among 100 of the 'World's Worst' invaders (Global Invasive Species Database, 2005). *Aedes albopictus* is commonly called the 'Asian tiger mosquito' because it aggressively bites humans and has a distinctive black and white striped body (Figure 12.1). It can pose a serious public health risk by creating a severe biting nuisance and by transmitting pathogens that cause major human diseases, including dengue, chikungunya and West Nile encephalitis.

Invasion History of *Aedes albopictus*

Aedes albopictus has substantially extended its range during two periods in history. The first period was the 19th and first half of the 20th century, when it spread with human migration west to islands in the Indian Ocean (e.g. Mauritius, Seychelles and Madagascar) and east to Guam, Hawaii and other islands in the Pacific (Knudsen, 1995; Paupy et al, 2009). The second expansion is a dramatic spread that started in the late 1970s and is ongoing today. It has been facilitated by the intercontinental trade in used tyres and is widely considered the 'third wave' of human-aided global mosquito dispersal following the spread of *Ae. aegypti* and *Culex pipiens* in previous centuries (Lounibos, 2002). *Aedes albopictus* was first reported in Europe in 1979, North America in 1985, South America in 1986 and Africa in 1991 (Lounibos, 2002; Gratz, 2004; Eritja et al, 2005; Benedict et al, 2007; Paupy et al, 2009; Lambrechts et al, 2010). Over the last three decades, *Ae. albopictus* has been reported in over 40 countries and has become established in most of them (Table 12.1).

Ecology of *Aedes albopictus*

Container habitat

Originally a forest species, *Ae. albopictus* routinely uses small shaded natural containers that retain water as larval development sites, such as bamboo stumps, tree holes and plant axils (Hawley, 1988). Larvae (Figure 12.2) feed on decaying plant and animal matter (detritus) that fall into these containers and the associated microbial communities (Juliano, 2009). Its ecological preference for shaded containers allowed *Ae. albopictus* to utilize a range of artificial sites as it expanded into anthropogenic habitats, such as cemetery flower vases, bird baths, used tyres, buckets and guttering (Figure 12.3) (Hawley, 1988). In its native range today, *Ae. albopictus*

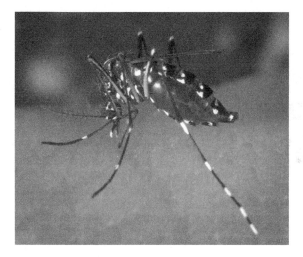

Source: James Gathway, CDC

Figure 12.1 *Adult female* Ae. albopictus *after taking a blood meal*

mainly occurs in rural and suburban areas but also densely populated cities areas such as Kuala Lumpur, Singapore and Tokyo where there is sufficient vegetation to provide food and shade for both adults and larvae (Hawley, 1988).

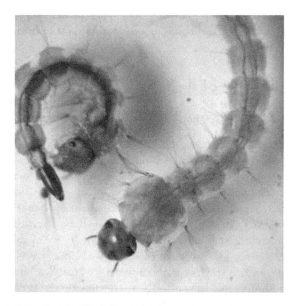

Source: Don Yee, Illinois State University

Figure 12.2 *Larvae of* Ae. albopictus

Source: Paul Leisnham

Figure 12.3 *Artificial containers at an industrial site that collect rainwater and provide common habitat for* Ae. albopictus *egg and larval development*

Aedes albopictus utilizes a variety of containers in its introduced range, although the specific type can vary between places. Occupying anthropogenic habitats and having cold-hardy long lived eggs are two traits commonly found in invasive mosquitoes and have allowed *Ae. albopictus* to survive its long distance dispersal around the world (Juliano and Lounibos, 2005). *Aedes albopictus* from temperate origins can enter diapause at the egg stage. Egg diapause increases survival in dry, cool conditions and can help reduce mortality during dispersal in used tyres and other containers. Diapause is expressed when female pupae and adults of the preceding generation encounter short day lengths (Hawley, 1988).

Table 12.1 *Countries by region in which* Ae. albopictus *was reported between 1979 and 2010*

First invasion year	Country	Current situation	Reference
Europe			
1979	Albania	Established	Adhami and Reiter, 1998
1990	Italy	Established	Eritja et al, 2005
	France	Established and spreading	Schaffner and Karch, 2000
2000	Belgium	Reported	Schaffner et al, 2004
2001	Montenegro	Established and spreading	Roiz et al, 2008
2003	Greece	Possibly established; scarce data	Schaffner et al, 2009
	Switzerland	Eradicated multiple incursions	Wymann et al, 2008
	Israel	Scarce data	L. Blaustein pers. comm. in Benedict et al, 2007
2004	Croatia	Established and spreading	Klobucar et al, 2006
	Spain	Established and spreading	Aranda et al, 2006
2005	Netherlands	Eradicated from nurseries	Scholte et al, 2008
	Slovenia	Established and spreading	Schaffner et al, 2009
	Bosnia	Possibly established; scarce data	Schaffner et al, 2009
2006	Monaco	Established	Schaffner et al, 2009
2007	Vatican City	Established	C. Venturelli pers. comm. in Schaffner et al, 2009
	Germany	Eradicated one incursion	Schaffner et al, 2009
	San Marino	Established	R. Mignani pers. comm. in Schaffner et al, 2009
Africa			
1991	Nigeria	Established	Savage et al, 1992
	South Africa	Reported	Cornel and Hunt, 1991
2000	Cameroon	Established	Fontenille and Toto, 2001
2001	Guinea	Established	Toto et al, 2003
2007	Gabon	Established	Coffinet et al, 2008
North America and Mexico			
1985	US	Established	Sprenger and Wuithiranyagool, 1986
1993	Mexico	Established	Benedict et al, 2007
Central America			
1983	Trinidad	Established	Lemaitre and Chadee, 1983
1993	Dominican Republic	Established	Peña, 1993
	Barbados	Established	Reiter, 1998
1995	Cuba	Established	Broche and Borja, 1999
	Honduras	Established	Ogata and Samayoa, 1996
	El Salvador	Established	Eritja et al, 2005
	Guatemala	Established	Ogata and Samayoa, 1996
1997	Bolivia	Established	Eritja et al, 2005
	Cayman Islands	Established	Lounibos et al, 2003
South America			
1986	Brazil	Established and spreading	Forattini, 1986
1998	Argentina	Established	Rossi et al, 1999
	Colombia	Established	Global Invasive Species Database, 2005

Table 12.1 *Countries by region in which* Ae. albopictus *was reported between 1979 and 2010* (Cont'd)

First invasion year	Country	Current situation	Reference
	Paraguay	Established	B. Cousinho pers. comm. in Benedict et al, 2007
2002	Panama	Established	Eritja et al, 2005
2003	Nicaragua	Established	Lugo et al, 2005
Pacific			
1982	New Zealand	Eradicated multiple incursions at ports	Derraik, 2004
1988	Fiji	Established	Kay et al, 1995
2005	Australia	Established on Torres Strait Islands; eradicated multiple incursions at mainland ports	Ritchie et al, 2006

Source: In addition to the tabled references, the current situation of some countries was compiled from Eritja et al, 2005; Global Invasive Species Database, 2005; Benedict et al, 2007; and Schaffner et al, 2009

Environmental conditions

Aedes albopictus is established in areas with a wide range of environmental conditions. Its distribution consists of areas where mean annual temperatures range from 5.0 to 28.5°C, mean annual minimum temperatures are as low as –1.8°C, and mean annual rainfall is between 29.2cm and 445.3cm (Benedict et al, 2007). Non-over-wintering expansions can extend into regions where the mean cold month temperature is as low as –5°C (Nawrocki and Hawley, 1987; Kobayashi et al, 2002). Studies have used field distribution information and laboratory experiments on the effects of environment on development and survival to help estimate the continuing spread of *Ae. albopictus* (e.g. Mitchell, 1995; Knudsen et al, 1996) and develop geographical information system (GIS)-based risk maps (e.g. Benedict et al, 2007; Schaffner et al, 2009). Benedict et al (2007) used Genetic Algorithm for Rule-Set Production (GARP) to describe *Ae. albopictus'* ecological niche based on climate and landscape features, and then predict its future spread. GARP simulates genetic processes of mutation, recombination and selection to non-randomly search and find rule-sets to describe the niches of species. Benedict et al (2007) also factored into models the risk of tyre imports from infested countries and the proximity of countries that have already been invaded to identify countries most at risk of future *Ae. albopictus* invasion. Mainland Australia, New Zealand, South Africa and numerous other African countries are at particularly high risk (Benedict

et al, 2007). *Aedes albopictus* expansion into the Caribbean, South America and most of Europe is also expected, regardless of whether tyres will be the mode of introduction. Under most of the climate change scenarios predicted by the Intergovernmental Panel on Climate Change (IPCC), *Ae. albopictus* is estimated to spread as far north as the Baltic states and into large parts of Sweden (Schaffner et al, 2009).

Ecological impacts on resident mosquitoes

Food within containers is limited and *Ae. albopictus* usually competes for resources with co-occurring mosquitoes (Juliano, 1998). Most studies of competition with *Ae. albopictus* have involved other non-native species in North America that are medically important, such as *Ae. aegypti* (dengue), *Ae. triseriatus* (LaCrosse encephalitis) and *Cx. pipiens* (West Nile virus) (Juliano, 2009). Numerous field and laboratory experiments have shown that *Ae. albopictus* is a superior competitor to these species under most conditions (Juliano and Lounibos, 2005).

Aedes albopictus spread in the US has been associated with the overall decline of *Ae. aegypti*, a pattern consistent with competitive exclusion (Juliano and Lounibos, 2005). However, at some places and during some seasons, *Ae. aegypti* persists and is even the dominant mosquito after the establishment of *Ae. albopictus* (e.g. Britch et al, 2008a, 2008b). Interestingly, the outcome of competition between these two species can vary across environmental gradients, which has prompted considerable interest in evolutionary and community ecology (Juliano, 2009). High resource levels from animal and some types of plant detritus tend to yield approximate competitive equality or even an advantage for *Ae. aegypti* (Daugherty et al, 2000; Murrell and Juliano, 2008). Additionally, habitat drying and resultant mortality of eggs differentially affects *Ae. albopictus* (Lounibos et al, 2010), and drier environments can reverse competitive advantage, favouring *Ae. aegypti* (Costanzo et al, 2005a). This finding highlights subtle ecological processes that can be involved when considering the invasion of mosquitoes and other organisms with complex lifecycles: the environment affects life stages that do not compete (e.g. eggs) but can still alter the population level consequences of competition in another lifecycle stage (e.g. larvae). In Brazil, there is also evidence of an overall decline of *Ae. aegypti* following *Ae. albopictus* invasion. *Aedes albopictus* from Brazil are also superior to *Ae. aegypti* in competition among larvae (Braks et al, 2004). Invasion of Europe by *Ae. albopictus* has been implicated as the cause of declines in abundance of *Ae. aegypti* (Simberloff and Gibbons, 2004), but the decline of *Ae. aegypti* in Spain and southern Europe preceded invasions by *Ae. albopictus* and probably results from eradication efforts (Eritja et al, 2005).

Although field and laboratory experiments show *Ae. triseriatus* and *Cx. pipiens* to be inferior competitors to *Ae. albopictus*, there is little evidence of a decline of either species after *Ae. albopictus* invasion (Juliano and Lounibos, 2005). *Culex pipiens* is thought to persist because it utilizes a much wider range of habitats, including ground pools, wastewater treatment wetlands and stormwater basins (Costanzo et al, 2005b). Persistence of *Ae. triseriatus* is likely due to differential vulnerability to predation. *Aedes albopictus* and *Ae. triseriatus* usually interact in forested areas where there are greater numbers of important predatory insects, such as the mosquito *Toxorhychites rutilus* and the midge *Corethrella appendiculata*. *Aedes albopictus* is more vulnerable to predation than *Ae. triseriatus* because of increased foraging behaviour and is therefore often suppressed in forested areas (e.g. Griswold and Lounibos, 2005).

Reciprocal niche-based distribution modelling compares models of occurrences in native vs. introduced ranges of invasive species to determine if native models accurately predict actual distributions in introduced ranges (e.g. Peterson and Robins, 2003). Reciprocal niche-based models of *Ae. albopictus* suggest that it has shifted niches in the Americas (North and South) and Europe compared to its native range in southeast Asia and India (Medley, 2010). Although the first location of establishment is predicted by niche models in all three continents, the subsequent spread has not been well predicted. The exclusion of *Ae. albopictus* from habitat patches due to competition or predation does not appear to constrain the extent of the US distribution of *Ae. albopictus*, suggesting that niche shifts and range expansion appear to be because of changes in the fundamental niche (Medley, 2010). *Aedes albopictus* has shown inter-population differentiation in a variety of life history traits subsequent to its invasion of the Americas, including larval development (Armbruster and Conn, 2006), survival and reproduction (Leisnham et al, 2008), larval competitive ability (Leisnham et al, 2009), and the expression of egg diapause (Lounibos et al, 2003). The evolution of increased incidence of egg diapause among populations along a north to south cline in Brazil and along a south to north cline in the US (Lounibos et al, 2003) is particularly noteworthy because this has probably helped the spread of *Ae. albopictus* at its latitudinal boundaries. *Aedes albopictus's* apparent ability to rapidly adapt to its local conditions suggests accurate predictions of future spread and management of the species may be very difficult.

Human health impacts

In most areas today, *Ae. albopictus* readily bites humans (anthropophagous) and a range of wild and domestic animals (zoophagous) (Niebylski et al, 1994), and can play a role in the epidemiology of naturally occurring or introduced arthropod-borne viruses (or arboviruses). *Aedes albopictus* is capable of transmitting almost all the viruses for which it has been experimentally tested (see reviews by Gratz, 2004; Paupy et al, 2009; Lambrechts et al, 2010). At least 11 viruses have been isolated from specimens in the field, including dengue, chikungunya, West Nile virus, Japanese encephalitis virus, Eastern equine encephalitis virus, Rift Valley virus and LaCrosse virus (Turell et al, 1988; Paupy et al, 2009 and references therein). Filarial parasites that cause dog heartworm (*Dirofilaria* spp.) have also been isolated from *Ae. albopictus* in Italy (Nayar and Knight, 1999).

Aedes albopictus ranks second only to *Ae. aegypti* in global importance as a dengue vector (Gratz, 2004). Dengue is the most prevalent human arboviral illness in the world and it is one of the leading causes of paediatric hospitalizations and mortality in southeast Asia (Kyle and Harris, 2008). Dengue epidemics have historically occurred when *Ae. aegypti* is present, and *Ae. albopictus* is usually considered an inefficient secondary vector. However, today, the importance of *Ae. albopictus* as a dengue vector is increasingly being recognized (e.g. Gratz, 2004; Paupy et al, 2010, but see Lambrechts et al, 2010). Most authors now implicate *Ae. albopictus* as the principal vector responsible for dengue epidemics in Japan, Guam, and Taiwan during World War II, when *Ae. aegypti* was absent or in low numbers (Hotta, 1994; Kobayashi et al, 2002; Paupy et al, 2009). *Aedes albopictus* is also thought to have been the main vector during a dengue outbreak in Réunion in 1977–1978 that infected 30 per cent of the population (Reiter et al, 2006) and a number of outbreaks on mainland China (Gratz, 2004). During recent outbreaks of dengue in central Africa, the dengue virus was isolated from *Ae. albopictus* but not *Ae. aegypti* (Paupy et al, 2010). *Aedes albopictus* transmits all four serotypes of the dengue virus, can often have similar rates of human feeding as *Ae. aegypti* in urban areas (Ponlawat and Harrington, 2005), and is thought to be at least as susceptible to dengue infection as *Ae. aegypti* (Lambrechts et al, 2010).

Similar to dengue, *Ae. albopictus* has historically been considered a secondary vector of chikungunya virus. Since 2004, however, *Ae. albopictus* has been implicated as the principal vector in outbreaks of chikungunya on islands in the Indian Ocean, in central Africa and in Italy, where *Ae. aegypti* was absent or in very low numbers (Reiter et al, 2006; Bonilauri et al, 2008; Pages et al, 2009). The 2007 chikungunya epidemic in Italy involved a new strain of chikungunya that has enhanced replication in *Ae. albopictus*, probably through independent viral exposure to and adaptation for *Ae. albopictus* via a single point mutation (de Lamballerie et al, 2008). The rapid emergence of this new chikungunya strain indicates the potential for *Ae. albopictus* to shape the evolution of resident vector virus systems and suggests that a similar event could occur for dengue or other arboviruses (Paupy et al, 2009; Lambrechts et al, 2010).

It is unclear if the spread of *Ae. albopictus* will result in a net gain or loss for public health. Because *Ae. albopictus* can displace *Ae. aegypti* and is usually a poorer vector of dengue, the spread of *Ae. albopictus* in areas with *Ae. aegypti* may lower dengue risk (Enserink, 2008). However, *Ae. albopictus* may act as a 'bridge vector' and increase animal to human transmission for

diseases that naturally circulate among animals by mainly zoophagous mosquitoes, such as West Nile virus and LaCrosse virus (Juliano and Lounibos, 2005; Juliano, 2009). LaCrosse virus has emerged in the Appalachia region, which has coincided with the invasion of *Ae. albopictus* and increased urban encroachment into forests in that region (Barker et al, 2003). Its native vector, *Ae. triseriatus*, is rarely found in anthropogenic areas, and it is possible that *Ae. albopictus* is acting as an efficient bridge between zoonotic cycles and humans (Barker et al, 2003).

Aedes albopictus may also affect disease cycles indirectly. Studies have shown that adult *Ae. triseriatus* develop infections of LaCrosse virus (Bevins, 2008) and adult *Ae. aegypti* develop infections of dengue (Alto et al, 2008) more often when they share containers at the larval stage with *Ae. albopictus*. Although the survival of both *Ae. triseriatus* and *Ae. aegypti* decreases from competition with *Ae. albopictus*, competition appears to reduce morphological and physiological barriers to dissemination of viruses within surviving mosquitoes and increase their vector competence (Alto et al, 2008).

Management of *Ae. albopictus* and Disease

Aedes albopictus control focuses on achieving the public health goals of reducing a biting nuisance and preventing human disease. Because there are no vaccines or drugs for dengue or chikungunya, preventing the spread of *Ae. albopictus* into new areas and suppressing existing populations are the most effective interventions against these diseases.

Preventing new colonizations

Considerable attention has been focused on controlling and regulating international used tyre imports. Several countries in South America (Venezuela, Chile, Bermuda, Costa Rica, Argentina and Brazil) have dictated embargoes on tyre imports in an attempt to prevent reintroductions of *Ae. albopictus*, *Ae. aegypti*, and dengue (Eritja et al, 2005). A comprehensive inspection and control programme was established in the US in 1986 aimed at preventing further introductions of *Ae. albopictus* after established populations were documented in 1985 (Table 12.1). Tyres that arrived from Asian countries that had *Ae. albopictus* had to be certified as being dry, clean and free from insects (Institute of Medicine, 2010). Non-compliant cargoes were fumigated with methyl bromide, or treated with a pressurized spray of detergent–water solution at 88°C, or by steam cleaning. Because of the considerable volume of container imports and labour intensiveness of these control actions, however, only about 10 per cent of all cargoes arriving at a few seaports could be practically inspected (Institute of Medicine, 2010). After several years the regulations were withdrawn and such inspection methods are now widely considered impractical (Institute of Medicine, 2010).

There have been examples of successful quarantine, inspection and eradication of *Ae. albopictus* in cargoes more amenable to control that arrive in less volume and at fewer ports. In June 2001, shipments of 'lucky bamboo' (*Dracaena sanderiana*) at the port in Los Angeles, US, were found to contain *Ae. albopictus*. Successful eradication measures included: (1) treating containers with insecticide at the port and at nurseries receiving bamboo, (2) an embargo prohibiting shipments of bamboo in standing water, and (3) extensive ongoing monitoring over successive years (Linthicum et al, 2003). *Aedes albopictus* has also been reported in bamboo shipments in The Netherlands and successfully eradicated at nurseries (Scholte et al, 2008). Australia and New Zealand have also eradicated introductions of *Ae. albopictus* arriving at coastal ports in used tyres, machinery and vehicles through comprehensive surveillance and insecticidal programmes at port locations and surrounding suburban areas (Derraik, 2004; Eritja et al, 2005).

Surveillance

Surveillance activities routinely involve placing oviposition traps (ovitraps) and adult traps in places where *Ae. albopictus* are likely to first colonize, such as shipping ports, airports and surrounding suburban areas (Silver, 2008). Ovitraps consist of black water containers with wooden paddles or paper lining that provides a textured surface on which female *Ae. albopictus* can lay eggs (Figure 12.4a). Because *Aedes* eggs are difficult to speciate, collected eggs are usually hatched and the resultant larvae are raised to adulthood and identified. Adult traps commonly consist of a light and are baited with CO_2 (e.g. dry ice or gas cylinder) to attract mosquitoes, and have a battery powered fan to blow specimens into a collection bag (Figure 12.4b)

(a)

(b)

Note: A battery is in the foreground beside the trap, and an igloo cooler holding dry ice CO_2 is hanging in the tree.

Source: Paul Leisnham

Figure 12.4 *Traps used to monitor populations of* Ae. albopictus *and other* Aedes *mosquitoes: (a) oviposition cup (ovitrap) used to collect eggs laid by females, and (b) adult trap (BG-Sentinel™) used to collect blood-seeking adult females*

(Silver, 2008). Because they need a power supply and source of CO_2, adult traps require daily maintenance thus the costs of employing them are higher than ovitraps. However, surveillance programmes that utilize adult traps allow collected specimens that have blood fed to be tested for viruses. Traditional trap designs, such as the CO_2-baited CDC light trap, have been unsuccessful in catching large numbers of adult *Ae. albopictus* (Silver, 2008). A relatively new type of low cost trap, the BG-Sentinel™, has been demonstrated to catch significantly higher numbers of *Aedes albopictus* (Farajollahi et al, 2009). This trap is usually baited with ammonia, fatty acids and lactic acids common in human skin secretions in addition to CO_2.

Suppressing established populations

Control of *Ae. albopictus*, like many container mosquitoes, usually involves the elimination of larval habitats and the application of larvicides (e.g. methoprene) or the toxic bacteria *Bacillus thuringiensis israelensis* (*BTI*) (Paupy et al, 2009). Due to the large number and cryptic location of containers on often limited access properties, regular control is usually labour intensive and often beyond the resources of public agency mosquito control programmes (Eritja et al, 2005). Even when thorough elimination campaigns have been undertaken, there has largely been limited success in suppressing *Ae. albopictus* and disease (Heintze et al, 2007; Richards et al, 2008). There has been some achievement in eliminating used tyres in the US that can produce extremely large numbers of *Ae. albopictus* by establishing strict guidelines on the disposal, storage and shredding of tyres (US EPA, 2006). But millions of tyres remain abandoned in a variety of habitats (e.g. woods, vacant lots, ditches) and continue to be discarded illegally, often in isolated areas (Yee, 2008). Other methods of *Aedes* control include adulticiding (usually using pyrethroids or organophosphates), biological controls (e.g. larvivorous invertebrates and fish), or genetic control by releasing sterile mosquitoes (e.g. Paupy et al, 2009; Bellini et al, 2010). However, few studies have yet to illustrate sustained success using these methods.

The most effective management of *Ae. albopictus* has been with strategies that integrate multiple methods based on a good understanding of *Ae. albopictus* biology and its invasion (e.g. Linthicum et al, 2003; Wymann et al, 2008; Holder et al, 2009). Examples of successful eradications have usually consisted of widespread surveillance programmes that detect an incursion in its earliest stages. These consist of strategically positioned ovitraps or adult traps near where *Ae. albopictus* is likely to colonize, such as along main traffic routes that transport containers, border crossings or seaports and airports. Once *Ae. albopictus* has been detected, larval and adult spraying is instigated in vegetated areas around incursion points or other important locations (e.g. nurseries receiving lucky bamboo). Wide ranging container removal is undertaken by public agency personnel and citizens, who are informed by thorough community-based education programmes. Increasingly sophisticated GIS and long term surveillance datasets are being used to map invasions of *Ae. albopictus*, associate environmental variables with *Ae. albopictus* spread, and predict areas of future risk of invasion (e.g. Benedict et al, 2007; Britch et al, 2008a, 2008b).

Personal protection

Given the difficulties in controlling *Ae. albopictus* populations, personal protection against biting is important in reducing the species' nuisance and vector potential. Mosquito control agencies frequently educate citizens to reduce sources of breeding habitats on their properties; use US Environmental Protection Agency (EPA) approved repellents, such as DEET on exposed skin and permethrin on clothing; and adjust personal behaviour to avoid the outdoors during peak biting times. However, few studies have evaluated the effectiveness of control and education programmes and then used this information to improve their implementation. Starting in 2008, a four year multi-million dollar programme is being funded by the United States Department of Agriculture in New Jersey that seeks to demonstrate practical and sustainable area-wide control of *Ae. albopictus*, provide a widely applicable model for *Ae. albopictus* control, reduce the application of insecticides, and enhance community-wide involvement in mosquito management (Healy et al, 2009; USDA, 2011).

References

Adhami, J. and Reiter, P. (1998) 'Introduction and establishment of *Aedes (Stegomyia) albopictus* Skuse (Diptera: Culicidae) in Albania', *Journal of the American Mosquito Control Association*, vol 14, no 3, pp340–343

Alto, B. W., Lounibos, L. P., Mores, C. N. and Reiskind, M. H. (2008) 'Larval competition alters susceptibility of adult *Aedes* mosquitoes to dengue infection', *Proceedings of the Royal Society B: Biological Sciences*, vol 275, no 1633, pp463–471

Aranda, C., Eritja, R. and Roiz, D. (2006) 'First record and establishment of the mosquito *Aedes albopictus* in Spain', *Medical and Veterinary Entomology*, vol 20, no 1, pp150–152

Armbruster, P. and Conn, J. E. (2006) 'Geographic variation of larval growth in North American *Aedes albopictus* (Diptera: Culicidae)', *Annals of the Entomological Society of America*, vol 99, no 6, pp1234–1243

Barker, C. M., Brewster, C. C. and Paulson, S. L. (2003) 'Spatiotemporal oviposition and habitat preferences of *Ochlerotatus triseriatus* and *Aedes albopictus* in an emerging focus of La Crosse virus', *Journal of the American Mosquito Control Association*, vol 19, no 4, pp382–391

Bellini, R., Albieri, A., Balestrino, F., Carrieri, M., Porretta, D., Urbanelli, S., Calvitti, M., Moretti, R. and Maini, S. (2010) 'Dispersal and survival of *Aedes albopictus* (Diptera: Culicidae) males in Italian urban areas and significance for sterile insect technique application', *Journal of Medical Entomology*, vol 47, no 6, pp1082–1091

Benedict, M. Q., Levine, R. S., Hawley, W. A. and Lounibos, L. P. (2007) 'Spread of the tiger: Global risk of invasion by the mosquito *Aedes albopictus*', *Vector-Borne and Zoonotic Diseases*, vol 7, no 1, pp76–85

Bevins, S. N. (2008) 'Invasive mosquitoes, larval competition, and indirect effects on the vector competence of native mosquito species (Diptera: Culicidae)', *Biological Invasions*, vol 10, no 7, pp1109–1117

Bonilauri, P., Bellini, R., Calzolari, M., Angeflni, R., Venturi, L., Fallacara, F., Cordioli, P., Angelini, P., Venturolli, C., Merialdi, G. and Dottori, M. (2008) 'Chikungunya virus in *Aedes albopictus*, Italy', *Emerging Infectious Diseases*, vol 14, no 5, pp852–854

Braks, M. A. H., Honório, N. A., Lounibos, L. P., Lourenco-De-Oliveira, R. and Juliano, S. A. (2004) 'Interspecific competition between two invasive species of container mosquitoes, *Aedes aegypti* and *Aedes albopictus* (Diptera: Culicidae), in Brazil', *Annals of the Enotomological Society of America*, vol 97, no 1, pp130–139

Britch, S. C., Linthicum, K. J., Anyamba, A., Tucker, C. J., Pak, E. W., Maloney, F. A., Cobb, K., Stanwix, E., Humphries, J., Spring, A., Pagac, B. and Miller, M. (2008a) 'Satellite vegetation index data as a tool to forecast population dynamics of medically important mosquitoes at military installations in the continental United States', *Military Medicine*, vol 173, no 7, pp677–683

Britch, S. C., Linthicum, K. J., Anyamba, A., Tucker, C. J., Pak, E. W. and Team, M. S. (2008b) 'Long term surveillance data and patterns of invasion by *Aedes albopictus* in Florida', *Journal of the American Mosquito Control Association*, vol 24, no 1, pp115–120

Broche, R. G. and Borja, E. M. (1999) '*Aedes albopictus* in Cuba', *Journal of the American Mosquito Control Association*, vol 15, no 4, pp569–570

Coffinet, T., Mourou, J. R., Pradines, B., Toto, J. C., Jarjaval, F., Amalvict, R., Kombila, M., Carnevale, P. and Pages, F. (2008) 'First record of *Aedes albopictus* in Gabon', *Journal of the American Mosquito Control Association*, vol 23, no 4, pp471–472

Cornel, A. J. and Hunt, R. H. (1991) '*Aedes albopictus* in Africa? First records of live specimens in imported tires in Cape Town', *Journal of the American Mosquito Control Association*, vol 7, no 1, pp107–108

Costanzo, K. S., Kesavaraju, B. and Juliano, S. A. (2005a) 'Condition specific competition in container mosquitoes: The role of non-competing life-history stages', *Ecology*, vol 86, no 12, pp3289–3295

Costanzo, K. S., Mormann, K. and Juliano, S. A. (2005b) 'Asymmetrical competition and patterns of abundance of *Aedes albopictus* and *Culex pipiens* (Diptera: Culicidae)', *Journal of Medical Entomology*, vol 42, no 4, pp559–570

Daugherty, M. P., Alto, B. W. and Juliano, S. A. (2000) 'Invertebrate carcasses as a resource for competing *Aedes albopictus* and *Aedes aegypti* (Diptera: Culicidae)', *Journal of Medical Entomology*, vol 37, no 3, pp364–372

de Lamballerie, X., Leroy, E., Charrel, R. N., Tsetsarkin, K., Higgs, S. and Gould, E. A. (2008) 'Chikungunya virus adapts to tiger mosquito via evolutionary convergence: A sign of things to come?', *Virology Journal*, vol 5, p33

Derraik, J. G. B. (2004) 'Exotic mosquitoes in New Zealand: A review of species intercepted, their pathways and ports of entry', *Australian and New Zealand Journal of Public Health*, vol 28, no 5, pp433–444

Enserink, M. (2008) 'A mosquito goes global', *Science*, vol 320, no 5878, pp864–866

Eritja, R., Escosa, R., Lucientes, J., Marques, E., Molina, R., Roiz, D. and Ruiz, S. (2005) 'Worldwide invasion of vector mosquitoes: Present European distribution and challenges for Spain', *Biological Invasions*, vol 7, no 1, pp87–97

Farajollahi, A., Kesavaraju, B., Price, D. C., Williams, G. M., Healy, S. P., Gaugler, R. and Nelder, M. P. (2009) 'Field efficacy of BG-Sentinel and industry-standard traps for *Aedes albopictus* (Diptera: Culicidae) and West Nile virus surveillance', *Journal of Medical Entomology*, vol 46, no 4, pp919–925

Fontenille, D. and Toto, J. C. (2001) '*Aedes (Stegomyia) albopictus* (Skuse), a potential new dengue vector in southern Cameroon', *Emerging Infectious Diseases*, vol 7, no 6, pp1066–1067

Forattini, O. P. (1986) '*Aedes (Stegomyia) albopictus* (Skuse) identification in Brazil', *Revista de Saude Publica*, vol 20, no 3, pp244–245

Global Invasive Species Database (2005) '*Aedes albopictus*', www.issg.org/database/species/ecology.asp?si=109&fr=1&sts=sss&lang=EN, accessed 1 August 2010

Gratz, N. G. (2004) 'Critical review of the vector status of *Aedes albopictus*', *Medical and Veterinary Entomology*, vol 18, no 3, pp215–227

Griswold, M. W. and Lounibos, L. P. (2005) 'Does differential predation permit invasive and native mosquito larvae to coexist in Florida?', *Ecological Entomology*, vol 30, no 1, pp122–127

Hawley, W. A. (1988) 'The biology of *Aedes albopictus*', *Journal of American Mosquito Control Association*, vol 4 (Supplement), no 1, pp1–40

Healy, S., Farajollahi, A., Fonseca, D., Gaugler, R., Hamilton, G., Worobey, J., Clark, G., Kline, D., Strickman, D. and Shepard, D. (2009) 'Area-wide management of the Asian tiger mosquito: 2008 project update', in *Mid-Atlantic Mosquito Control Association Annual Meeting*, 25–27 February, Virginia Beach, VA

Heintze, C., Garrido, M. V. and Kroeger, A. (2007) 'What do community-based dengue control programmes achieve? A systematic review of published evaluations', *Transactions of the Royal Society of Tropical Medicine and Hygiene*, vol 101, no 4, pp317–325

Holder, P., George, S., Disbury, M., Singe, M., Kean, J. M. and McFadden, A. (2009) 'A Biosecurity response to *Aedes albopictus* (Diptera: Culicidae) in Auckland, New Zealand', *Journal of Medical Entomology*, vol 47, no 4, pp600–609

Hotta, S. (1994) 'Dengue vector mosquitoes in Japan: The role of *Aedes albopictus* and *Aedes aegypti* in the 1942–44 dengue epidemics of Japanese main islands', *Medical Entomology and Zoology*, vol 49, no 4, pp267–274

Institute of Medicine (2010) *Infectious Disease Movement in a Borderless World*, The National Academies Press, Washington, DC

Juliano, S. A. (1998) 'Species introduction and replacement among mosquitoes: Interspecific resource competition or apparent competition?', *Ecology*, vol 79, no 1, pp255–268

Juliano, S. A. (2009) 'Species interactions among larval mosquitoes: Context dependence across habitat gradients', *Annual Review of Entomology*, vol 54, no 1, pp37–56

Juliano, S. A. and Lounibos, L. P. (2005) 'Ecology of invasive mosquitoes: Effects on resident species and on human health', *Ecology Letters*, vol 8, no 5, pp558–574

Kay, B. H., Prakash, G. and Andre, R. G. (1995) '*Aedes albopictus* and other *Aedes (Stegomyia)* in Fiji', *Journal of the American Mosquito Control Association*, vol 11, no 2, pp230–234

Klobucar, A., Merdic, E., Benic, N., Baklaic, Z. and Krcmar, S. (2006) 'First record of *Aedes albopictus* in Croatia', *Journal of the American Mosquito Control Association*, vol 22, no 1, pp147–148

Knudsen, A. B. (1995) 'Global distribution and continuing distribution of *Aedes albopictus*', *Parassitologia*, vol 37, no 1, pp91–97

Knudsen, A. B., Romi, R. and Majori, G. (1996) 'Occurrence and spread in Italy of *Aedes albopictus*, with implications for its introduction into other parts of Europe', *Journal of the American Mosquito Control Association*, vol 12, no 2, pp177–183

Kobayashi, M., Nihel, N. and Kurihara, T. (2002) 'Analysis of northern distribution of *Aedes albopictus* (Diptera: Culicidae) in Japan by geographical information system', *Journal of Medical Entomology*, vol 39, no 1, pp4–11

Kyle, J. L. and Harris, E. (2008) 'Global spread and persistence of dengue', *Annual Review of Microbiology*, vol 62, no 1, pp71–92

Lambrechts, L., Scott, T. W. and Gubler, D. (2010) 'Consequences of the expanding global distribution of *Aedes albopictus* for dengue virus transmission', *Plos Neglected Tropical Diseases*, vol 4, no 5, e646, doi:10.1371/journal.pntd.0000646

Leisnham, P. T., Sala, L. M. and Juliano, S. A. (2008) 'Geographic variation in adult survival and reproductive tactics of the mosquito *Aedes albopictus*', *Journal of Medical Entomology*, vol 45, no 2, pp210–221

Leisnham, P. T., Lounibos, L. P., O'Meara, G. F. and Juliano, S. A. (2009) 'Interpopulation divergence in competitive interactions of the mosquito *Aedes albopictus*', *Ecology*, vol 90, no 9, pp2405–2413

Lemaitre, A. and Chadee, D. D. (1983) 'Arthropods collected from aircraft at Piarco International Airport, Trinidad, West Indies', *Mosquito News*, vol 43, no 1, pp21–23

Linthicum, K. J., Kramer, V. L., Madon, M. B., Fujioka, M. and the Surveillance-Control Team (2003) 'Introduction and potential establishment of *Aedes albopictus* in California in 2001', *Journal of the American Mosquito Control Association*, vol 19, no 4, pp301–308

Lounibos, L. P. (2002) 'Invasions by insect vectors of human disease', *Annual Review of Entomology*, vol 47, no 1, pp233–266

Lounibos, L. P., Escher, R. L. and Lourenco-de-Oliveria, R. (2003) 'Asymmetric evolution of photoperiodic diapause in temperate and tropical invasive populations of *Aedes albopictus* (Diptera: Culicidae)', *Annals of the Entomological Society of America*, vol 96, no 4, pp512–518

Lounibos, L. P., O'Meara, G. F., Juliano, S. A., Nishimura, N., Escher, R. L., Reiskind, M. H., Cutwa, M. and Greene, K. (2010) 'Differential survivorship of invasive mosquito species in south Florida cemeteries: Do site-specific microclimates explain patterns of coexistence and exclusion?', *Annals of the Entomological Society of America*, vol 103, no 5, pp757–770

Lugo, E. D., Moreno, G., Zachariah, M. A., Lopez, M. M., Lopez, J. D., Delgado, M. A., Valle, S. I., Espinoza, P. M., Salgado, M. J., Perez, R., Hammond, S. N. and Harris, E. (2005) 'Identification of *Aedes albopictus* in urban Nicaragua', *Journal of the American Mosquito Control Association*, vol 21, no 3, pp325–327

Medley, K. A. (2010) 'Niche shifts during the global invasion of the Asian tiger mosquito, *Aedes albopictus* Skuse (Culicidae), revealed by reciprocal distribution models', *Global Ecology and Biogeography*, vol 19, no 1, pp122–133

Mitchell, C. J. (1995) 'Geographic spread of *Aedes albopictus* and potential for involvement in arbovirus cycles in the Mediterranean Basin', *Journal of Vector Ecology*, vol 20, no 1, pp44–58

Murrell, E. G. and Juliano, S. A. (2008) 'Detritus type alters the outcome of interspecific competition between *Aedes aegypti* and *Aedes albopictus* (Diptera: Culicidae)', *Journal of Medical Entomology*, vol 45, no 3, pp375–383

Nawrocki, S. J. and Hawley, W. A. (1987) 'Estimation of the northern limits of distribution of *Aedes albopictus* in North America', *Journal of American Mosquito Control Association*, vol 3, no 2, pp314–317

Nayar, J. K. and Knight, J. W. (1999) '*Aedes albopictus* (Diptera: Culicidae): An experimental and natural host of *Dirofilaria immitis* (Filarioidea: Onchocercidae) in Florida, USA', *Journal of Medical Entomology*, vol 36, no 4, pp441–448

Niebylski, M. L., Savage, H. M., Nasci, R. S. and Craig, G. B. (1994) 'Blood hosts of *Aedes albopictus* in the United States', *Journal of the American Mosquito Control Association*, vol 10, no 3, pp447–450

Ogata, K. and Samayoa, A. L. (1996) 'Discovery of *Aedes albopictus* in Guatemala', *Journal of the American Mosquito Control Association*, vol 12, no 3, pp503–506

Pages, F., Peyrefitte, C. N., Mve, M. T., Jarjaval, F., Brisse, S., Iteman, I., Gravier, P., Tolou, H., Nkoghe, D. and Grandadam, M. (2009) '*Aedes albopictus* mosquito: The main vector of the 2007 chikungunya outbreak in Gabon', *Plos One*, vol 4, no 3, e4691, doi:10.1371/journal.pone.0004691

Paupy, C., Delatte, H., Bagny, L., Corbel, V. and Fontenille, D. (2009) '*Aedes albopictus*, an arbovirus vector: From the darkness to the light', *Microbes and Infection*, vol 11, no 14–15, pp1177–1185

Paupy, C., Ollomo, B., Kamgang, B., Moutailler, S., Rousset, D., Demanou, M., Herve, J. P., Leroy, E. and Simard, F. (2010) 'Comparative role of *Aedes albopictus* and *Aedes aegypti* in the emergence of dengue and chikungunya in central Africa', *Vector Borne Zoonotic Diseases*, vol 10, no 3, pp259–266

Peña, C. J. (1993) 'First report of *Aedes (Stegomyia) albopictus* (Skuse) from the Dominican Republic', *Vector Ecology Newsletter*, vol 23, no 1, pp4–5

Peterson, A. T. and Robins, C. R. (2003) 'Using ecological-niche modeling to predict barred owl invasions with implications for spotted owl conservation', *Conservation Biology*, vol 17, no 4, pp1161–1165

Ponlawat, A. and Harrington, L. C. (2005) 'Blood feeding patterns of *Aedes aegypti* and *Aedes albopictus* in Thailand', *Journal of Medical Entomology*, vol 42, no 5, pp844–849

Reiter, P. (1998) '*Aedes albopictus* and the world trade in used tires, 1988–1995: The shape of things to come?', *Journal of the American Mosquito Control Association*, vol 14, no 1, pp83–94

Reiter, P., Fontenille, D. and Paupy, C. (2006) '*Aedes albopictus* as an epidemic vector of chikungunya virus: Another emerging problem?', *Lancet Infectious Diseases*, vol 6, no 8, pp463–464

Richards, S. L., Ghosh, S. K., Zeichner, B. C. and Apperson, C. S. (2008) 'Impact of source reduction on the spatial distribution of larvae and pupae of *Aedes albopictus* (Diptera: Culicidae) in suburban neighborhoods of a Piedmont community in North Carolina', *Journal of Medical Entomology*, vol 45, no 4, pp617–628

Ritchie, S. A., Moore, P., Carruthers, M., Williams, C., Montgomery, B., Foley, P., Ahboo, S., Van den Hurk, A. F., Lindsay, M. D., Cooper, B., Beebe, N. and Russell, R. C. (2006) 'Discovery of a widespread infestation of *Aedes albopictus* in the Torres Strait, Australia', *Journal of the American Mosquito Control Association*, vol 22, no 3, pp358–365

Roiz, D., Eritja, R., Molina, R., Melero-Alcibar, R. and Lucientes, J. (2008) 'Initial distribution assessment of *Aedes albopictus* (Diptera: Culicidae) in the Barcelona, Spain, area', *Journal of Medical Entomology*, vol 45, no 3, pp347–352

Rossi, G. C., Pascual, N. T. and Krsticevic, F. J. (1999) 'First record of *Aedes albopictus* (Skuse) from Argentina', *Journal of the American Mosquito Control Association*, vol 15, no 3, pp422–422

Savage, H. M., Ezike, V. I., Nwankwo, A. C. N., Spiegel, R. and Miller, B. R. (1992) 'First record of breeding populations of *Aedes albopictus* in continental Africa: Implications for arboviral transmission', *Journal of the American Mosquito Control Association*, vol 8, no 1, pp101–103

Schaffner, F. and Karch, S. (2000) 'First record of *Aedes albopictus* (Skuse, 1894) in metropolitan France', *Comptes Rendus de l'Academie des Sciences Series III–Sciences de la Vie–Life Sciences*, vol 323, no 4, pp373–375

Schaffner, F., Van Bortel, W. and Coosemans, M. (2004) 'First record of *Aedes (Stegomyia) albopictus* in Belgium', *Journal of the American Mosquito Control Association*, vol 20, no 2, pp201–203

Schaffner, F., Hendrickx, G., Scholte, E. J., Ducheyne, E., Medlock, J. M. and Avenell, D. (2009) *Development of Aedes albopictus Risk Maps*, European Centre of Disease Prevention and Control, Stockholm

Scholte, E. J., Dijkstra, E., Blok, H., De Vries, A., Takken, W., Hofhuis, A., Koopmans, M., De Boer, A. and Reusken, C. (2008) 'Accidental importation of the mosquito *Aedes albopictus* into the Netherlands: A survey of mosquito distribution and the presence of dengue virus', *Medical and Veterinary Entomology*, vol 22, no 4, pp352–358

Silver, J. B. (2008) *Mosquito Ecology: Field Sampling Methods*, Springer, New York

Simberloff, D. and Gibbons, L. (2004) 'Now you see them, now you don't! – population crashes of established introduced species', *Biological Invasions*, vol 6, no 1, pp161–172

Sprenger, D. and Wuithiranyagool, T. (1986) 'The discovery and distribution of *Aedes albopictus* in Harris County, Texas', *Journal of American Mosquito Control Association*, vol 2, no 2, pp217–219

Toto, J. C., Abaga, S., Carnevale, P. and Simard, F. (2003) 'First report of the oriental mosquito *Aedes albopictus* on the West African Island of Bioko, Equatorial Guinea', *Medical and Veterinary Entomology*, vol 17, no 3, pp343–346

Turell, M. J., Bailey, C. L. and Beaman, J. R. (1988) 'Vector competence of a Houston, Texas strain of *Aedes albopictus* for Rift Valley fever virus', *Journal of the American Mosquito Control Association*, vol 4, no 1, pp94–96

USDA (United States Department of Agriculture) (2011) 'Research project: Areawide pest management program for the Asian tiger mosquito in New Jersey', USDA, www.ars.usda.gov/research/projects/projects.htm?accn_no=412820

US EPA (United States Environment Protection Agency) (2006) 'Scrap tire cleanup guide', US Environmental Protection Agency, Office of Solid Waste Management, and the Illinois Environmental Protection Agency, www.epa.gov/reg5rcra/wptdiv/solidwaste/tires/guidance

Wymann, M. N., Flacio, E., Radczuweit, S., Patocchi, N. and Luthy, P. (2008) 'Asian tiger mosquito (*Aedes albopictus*) a threat for Switzerland?', *Eurosurveillance*, vol 13, no 1, pp1–3

Yee, D. A. (2008) 'Tires as habitats for mosquitoes: A review of studies within the eastern United States', *Journal of Medical Entomology*, vol 45, no 4, pp581–593

13

An overview of invasive freshwater cladocerans: *Bythotrephes longimanus* Leydig as a case study

Angela L. Strecker

Introduction

Invasive species are considered one of the greatest threats to freshwater biodiversity, and the spread of many non-native species has been accelerated by increasing globalization (Carlton and Geller, 1993). Cladoceran zooplankton are a key component of freshwater ecosystems, transferring energy from primary producers to higher trophic level consumers. Cladocerans are a relatively diverse group, with more than 600 documented species, which is probably an underestimate given that many cladocerans have flexible phenotypes; it has been estimated that their diversity may actually be two to four times greater (Forro et al, 2008). Although there have been comparatively few documented intercontinental cladoceran species invasions (Havel and Medley, 2006), this is probably the result of difficulty in detecting introduced species (e.g. cryptic invaders: Mergeay et al, 2005). Here, I present a case study on a well-documented cladoceran invasion, that of *Bythotrephes longimanus* (Onychopoda, Cercopagidae) (Figure 13.1a), whose primary introduction and secondary spread in the introduced range have been generally well described.

Despite a large degree of phenotypic plasticity, there is but one species in the genus *Bythotrephes*, *B. longimanus* (Martin and Cash-Clark, 1995). Henceforth,

it will be referred to as *Bythotrephes*. *Bythotrephes* is exemplary in that it has a unique morphology, is comparatively large bodied, and can reach relatively high densities shortly after establishment (Yan and Pawson, 1997), all factors that can make detection of invasion events more likely.

Invasion History and Current Distribution

Bythotrephes is native to Eurasia, where it is widespread, including lakes in Russia, Italy, Germany, Sweden, The Netherlands, Britain, Belgium, Switzerland, Norway and Finland (Grigorovich et al, 1998; MacIsaac et al, 2000; Colautti et al, 2005). Information on the extent of *Bythotrephes* distribution in its native range is incomplete, but there are indications that it is spread longitudinally across all of Russia and into China (Grigorovich et al, 1998). Additionally, there have been some intra-continental invasions of *Bythotrephes* into reservoirs in western Europe (Ketelaars and Gille, 1994).

Bythotrephes was first detected in Lake Ontario in the early 1980s (Johannsson et al, 1991) and subsequently spread to the rest of the Laurentian Great Lakes of North America in the next decade (Bur et al, 1986; Lehman, 1987; Jin and Sprules, 1990). It is

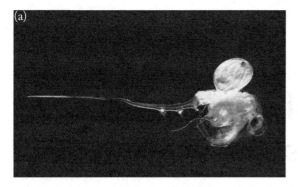

Source: (a) A. Strecker; (b) A. Jaeger-Miehls

Figure 13.1 *(a)* Bythotrephes longimanus, with late *stage embryos in its brood pouch, and (b) a fishing line from Lake Erie covered with* Bythotrephes

hypothesized that *Bythotrephes* was probably transported via ballast water of transoceanic ships moving from the Baltic Sea (Therriault et al, 2002a). Indeed, transoceanic shipping is responsible for 77 per cent of invasions to the Great Lakes since 1970 (Ricciardi, 2001). Secondary invasions of *Bythotrephes* into the Great Lakes from

other locations in Eurasia have been identified, which have reduced founder effects (Colautti et al, 2005).

Since its establishment in the Great Lakes, *Bythotrephes* has spread to more than 170 lakes in Ontario, Canada (Brinsmead, J., unpublished data), as well as lakes in Michigan, Minnesota, Ohio, Wisconsin and New York, US (Figure 13.2) (MacIsaac et al, 2000; Branstrator et al, 2006; Johnson et al, 2008; Brown, M., unpublished data; Papes, M., unpublished data).

The secondary spread of *Bythotrephes* has been associated with human vectors, such as live wells, bait buckets, bilge water, fishing lines, anchor lines and other recreational equipment (MacIsaac et al, 2004; Weisz and Yan, 2010a) (Figure 13.1b). The success of *Bythotrephes* as an invader via these pathways is probably attributable to the desiccation tolerance of certain life stages (see section below on life history strategies). These vectors are similar for other invasive cladocerans (Havel and Shurin, 2004; Havel and Medley, 2006). Hydrological connectivity and other natural vectors (e.g. flooding, waterfowl) have also been identified as substantial invasion pathways for some cladoceran species (e.g. *Daphnia lumholtzi*) (Havel and Shurin, 2004; Figuerola et al, 2005). Other anthropogenic vectors for invertebrates include canals, pipelines and construction equipment (Havel and Shurin, 2004), the aquarium trade (Duggan, 2010) and fish stocking (Sorensen and Sterner, 1992).

The closely related predatory cladoceran, *Cercopagis pengoi*, has followed a similar invasion pattern to *Bythotrephes*. *Cercopagis* is native to the Ponto-Caspian region, has colonized the Baltic Sea (Ojaveer et al, 1998) and has established in the lower Great Lakes (Ontario, Michigan, Erie) from the Black Sea (Critescu et al, 2001). *Cercopagis* has spread to inland lakes in New York and Michigan, including the Finger Lakes (Brown and Balk, 2008), Muskegon Lake (Therriault et al, 2002b) and Lake Champlain (Marsden and Hauser, 2009). Other intercontinental invasions of cladocerans include the taxa *Alona weinecki*, *Daphnia ambigua*, *D. curvirostris*, *D. galeata*, *D. obtusa*, *D. lumholtzi*, *D. parvula*, *D. pulex*, *D. pulicaria*, *Eubosmina coregoni* and *E. maritima* (Figure 13.3) (Bollens et al, 2002; Zanata et al, 2003; Mergeay et al, 2005; Havel and Medley, 2006 and references therein).

Notably, most intercontinental invasions have been between the Nearctic and Palearctic biogeographic provinces; however, this is likely to reflect a combination

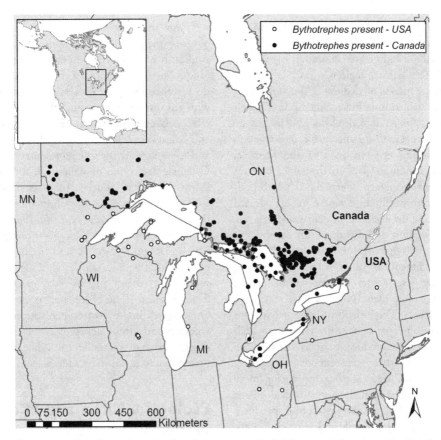

Note: In the upper left inset, map of North America and outline of current range. Verified detections in Canada and the US indicated by a black (n = 173) and white symbol (n = 31), respectively. State and province abbreviations: MI = Michigan, MN = Minnesota, NY = New York, OH = Ohio, ON = Ontario, WI = Wisconsin.

Source: based on MacIsaac et al (2000); Branstrator et al (2006); Johnson et al (2008) and on unpublished data by M. Brown, J. Brinsmead and M. Papes

Figure 13.2 *Map of current distribution of* Bythotrephes *in its introduced range*

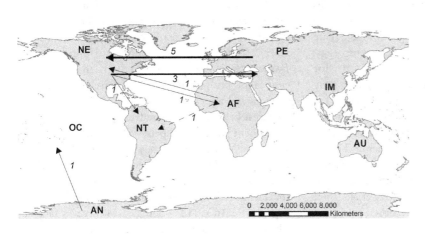

Note: AF = Afrotropic, AN = Antarctic, AU = Australasia, IM = Indo-Malaya, NE = Nearctic, NT = Neotropic, OC = Oceanic, PE = Palearctic. Thicker arrows represent a greater number of invasions, indicated by the number associated with each vector. The dashed arrow from the Afrotropic to Neotropic represents the invasion of *Daphnia lumholtzi* into Brazil (Zanata et al, 2003), for which the source is unknown. *D. lumholtzi* is native to the Afrotropic, Australasian and Indo-Malayan provinces.

Figure 13.3 *Intercontinental invasions of cladoceran zooplankton species by biogeographic province*

of a higher density of trade routes mediating ballast water introductions (Ricciardi and MacIsaac, 2000; Holeck et al, 2004) and greater sampling intensity in developed northern latitude regions.

Insight on the success of cladoceran zooplankton as invaders may be gained from examining aspects of their life history. Most temperate zone cladoceran species are cyclic parthenogens: sexual (or gametogenic) reproduction generates diapausing eggs in part of the lifecycle, generally when conditions are unfavourable, whereas asexual (or parthenogenic) reproduction produces broods of eggs that develop within the carapace and are released into the water column. Cladoceran diapausing eggs are tolerant of desiccation and freezing, can survive gut passage through fish (*Bythotrephes*: Jarnagin et al, 2000), can survive in sediments of ballast or live wells for long periods, and can over-winter in lake sediments and remain viable for more than 100 years (Cáceres, 1998; Kerfoot et al, 1999). The individuals that hatch from diapausing eggs are parthenogenic females, which generally continue to reproduce asexually through the summer. The size of parthenogenic broods varies greatly between species, with some *Daphnia* producing more than 100 eggs in a single brood (life history of cladocerans reviewed in Lynch, 1980). *Bythotrephes* typically produces two to ten eggs per brood (Straile and Halbich, 2000). Diapausing egg clutches tend to be less numerous, with *Bythotrephes* producing two to five diapausing eggs (Yan and Pawson, 1998) and *Daphnia* constrained to two eggs contained in a protective ephippium (Wetzel, 2001).

For most cladocerans, widespread establishment following an introduction event is probably constrained by Allee effects, i.e. finding a mate. The parthenogenic females are cued to begin producing males as conditions deteriorate, at which point small populations in a large three dimensional habitat are challenged to find a mate in order to produce over-wintering diapausing eggs. For *Bythotrephes*, windows of opportunity for establishment have been identified as being from early to mid-summer (Drake, 2004; Drake et al, 2006), as conditions are favourable for rapid reproduction and populations have sufficient time to grow to a size where mate limitation is not a constraining factor. This is probably also the case for other cladocerans. Hence, propagule pressure and timing are important aspects of invasion success for *Bythotrephes* (MacIsaac et al, 2004; Weisz and Yan, 2010a), as well as other successful invaders (Lockwood et al, 2005).

From an ecological theory perspective, the invasions of *Bythotrephes* and *Cercopagis* are interesting case studies. It has been hypothesized that phylogenetically unique taxa are more likely to become successful invaders because native taxa lack evolutionary experience with the invading taxa, and therefore have a greater effect in communities that lack similar species (Ricciardi and Atkinson, 2004; Strauss et al, 2006). *Bythotrephes* and *Cercopagis* are the sole representatives of the family Cercopagidae, none of which are found outside the Palearctic region. Indeed, there is only one species within the entire order Onychopoda that is native to the Nearctic, *Polyphemus pediculus* (Forro et al, 2008). Both *Bythotrephes* and *Cercopagis* are predators that have successfully invaded, and in some systems have displaced native taxa (Yan et al, 2001). Therefore, the success of these species may stem from their distinctiveness. There are no species in the order Onychopoda in the Neotropic, Afrotropic, Australasian, Oceanic and Antarctic regions (Forro et al, 2008), suggesting that a number of other regions may be susceptible to invasion by species from this order.

Ecological Niche and Effects of Invasion

In Eurasia, *Bythotrephes* is typically found in large, deep, low productivity lakes (MacIsaac et al, 2000). Although there has been some debate about the importance of productivity and lake clarity in determining *Bythotrephes* distribution, recent evidence has suggested that the greater prevalence of the invader in clear lakes reflects its strong reliance on visual feeding, contrary to many other invertebrate predators that detect prey via tactile cues (Pangle and Peacor, 2009). In its introduced range in North America, *Bythotrephes* is commonly found in large, deep and clear lakes, but also in lakes that have boat launches, cottages and are close to large donor populations (i.e. the Great Lakes), probably reflecting the anthropogenic component of their dispersal (Johnson et al, 2008; Weisz and Yan, 2010a).

Using the discriminant function model of MacIsaac et al (2000), I assessed the vulnerability of lakes in the continental US to the invasion of *Bythotrephes* using a large dataset (n = 1157) designed to survey a range of lake types. This approach has been used previously in smaller regions (Branstrator et al, 2006), but has not been utilized over broad spatial scales. The model is:

(1) lake score = $-1.765 + 0.4309 *$ Secchi $+ 0.1925 *$ ln(SA +1) $+ 0.0004 * Z_{max} - 0.0144 *$ chl a

where Secchi = Secchi depth (m), SA = surface area (km^2), Z_{max} = maximum depth (m), and chl a = chlorophyll a concentration (µg/L).

Following Branstrator et al (2006), a lake score ≥ -0.1546 indicates that a lake is susceptible to invasion, whereas a score < -0.1546 is not susceptible to invasion. Lakes were sampled as part of the National Lakes Assessment (US EPA, 2009), and spanned a range of size (0.04–1674km^2, median = 0.70km^2), maximum depth (0.5–97m, median = 6m), clarity (0.04–36m Secchi depth, median = 1.36m) and productivity (0.07–871µg/L chl a, median = 7.72µg/L). Despite the broad ranges represented, these data suggest that study lakes tended to be biased towards small and shallow water bodies; however, this is likely to mean that model predictions are conservative.

The model predicted that 19.9 per cent of lakes in the survey are vulnerable to invasion (n = 230) (Figure 13.4a) and are widely distributed across the continental US. As previous studies have demonstrated that the presence of boat launches is a significant factor predicting *Bythotrephes* invasions (Johnson et al, 2008; Weisz and Yan, 2010a), I assessed the number of lakes that are both vulnerable to invasion and have a boat launch, which reduced the predictions to 15.3 per cent of lakes being susceptible to *Bythotrephes* invasion (n = 178) (Figure 13.4a). I also summarized the model predictions by major hydrologic drainage units in the US (Seaber et al, 1987) (Figure 13.4b). The western (California, Pacific Northwest, Upper Colorado), Great Lakes, New England and Tennessee drainages had the highest percentage of lakes in the National Lakes Assessment database that are susceptible to invasion, with the model predicting 50 per cent of lakes in the Tennessee drainage to be susceptible to *Bythotrephes* (Figure 13.4b). Lakes in the southeast (Arkansas-White-Red, Lower Mississippi, South Atlantic-Gulf, Texas-Gulf), north central (Souris-Red-Rainy) and the Lower Colorado drainages were predicted to be less susceptible to invasion, probably because lakes in these regions have higher productivity and tend to be shallower and smaller (Figure 13.4b).

It is notable that *Bythotrephes* has established in a number of lakes that are uncharacteristic of early

habitat descriptions, which suggested that environmental suitability was constrained to large, deep, clear lakes (Grigorovich et al, 1998; MacIsaac et al, 2000). Weisz and Yan (2010a) suggest that although propagule pressure might be greatest to lakes that are large, deep and unproductive, these may not necessarily reflect the environmental conditions that actually restrict *Bythotrephes* distribution. Overall, the implications are that *Bythotrephes* has the potential to invade a large number of lakes across North America given sufficient propagule pressure, as studies in its native range have indicated broad tolerance of other environmental variables, such as salinity, pH and temperature (Grigorovich et al, 1998; MacIsaac et al, 2000).

The invasion of *Bythotrephes* has been implicated in significant changes in the native plankton communities of North American lakes. Native zooplankton assemblages in invaded lakes tend to have fewer species, lower abundance and lower biomass compared to uninvaded lakes (Yan et al, 2002; Boudreau and Yan, 2003; Barbiero and Tuchman, 2004; Strecker et al, 2006). Additionally, epilimnetic productivity is significantly reduced in invaded lakes compared to uninvaded lakes, probably the combined result of direct predation and indirect non-lethal effects of migration of zooplankton to deeper strata in order to avoid the visual predation of *Bythotrephes* (Pangle et al, 2007; Strecker and Arnott, 2008). Although laboratory feeding experiments have determined that *Bythotrephes* can consume large *Daphnia* (Schulz and Yurista, 1999), field studies have generally supported the hypothesis that the invader typically preys on smaller cladocerans (Yan et al, 2001; Strecker and Arnott, 2010). *Cercopagis* has similar prey preferences, consuming small cladocerans and copepods (Laxson et al, 2003). *Bythotrephes* and *Cercopagis* have very high consumption rates (Dumitru et al, 2001; Laxson et al, 2003; Strecker and Arnott, 2008), and the abundance and consumption of other native invertebrate predators, which compete for the same prey, is lower in *Bythotrephes*-invaded lakes compared to reference lakes (Foster and Sprules, 2009). Additionally, the native invertebrate predator, *Leptodora kindtii*, is being displaced by *Bythotrephes* in lakes across the Canadian Shield (Weisz and Yan, 2010b). Other indirect effects of the introduction of *Bythotrephes* include increased rotifer abundance and altered composition of the phytoplankton community (Hovius et al, 2006; Strecker et al, 2011).

(a)

(b)

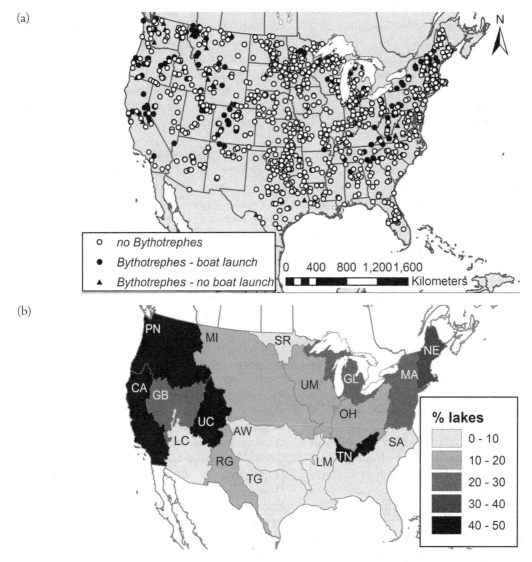

Note: In (a), white symbols indicate sites that are not suitable for *Bythotrephes* (n = 927), whereas black symbols represent sites that are suitable for *Bythotrephes* and have a boat launch (circle, n = 178) or no boat launch (triangle, n = 52). Abbreviations: AW = Arkansas-White-Red, CA = California, GB = Great Basin, GL = Great Lakes, LC = Lower Colorado, LM = Lower Mississippi, MA = Mid-Atlantic, MI = Missouri, NE = New England, OH = Ohio, PN = Pacific Northwest, RG = Rio Grande, SA = South Atlantic-Gulf, SR = Souris-Red-Rainy, TG = Texas-Gulf, TN = Tennessee, UC = Upper Colorado, UM = Upper Mississippi.

Source: Discriminant function model of MacIsaac et al (2000) and data from US EPA (2009)

Figure 13.4 *(a) Map of projected vulnerability of lakes to* Bythotrephes *invasion in the US and (b) percentage of lakes in dataset that are predicted to be suitable for* Bythotrephes *by hydrologic unit*

Management Strategies

It is generally understood that eradication of invasive species once they have become established is a significant challenge, and that management efforts to remove introduced species are rarely successful (Lodge et al, 2006). Hence, management of the vectors by which species are transported is considered a more effective strategy for preventing invasion (Ruiz and Carlton, 2003). Although *Bythotrephes* is now well established in North America, there is ample evidence that repeated introductions of propagules increases the probability of successful establishment and increases the genetic diversity of the invader (Lockwood et al, 2005). Thus, continued management of pathways is necessary and can inform the management of other potential invaders. For *Bythotrephes* and other freshwater invaders, integrated vector management should target the 'stepping stone' invasion model (Muirhead and MacIsaac, 2005): (1) long distance intercontinental introductions; and (2) secondary diffuse intracontinental introductions.

The first step in preventing long distance introductions involves the management of the most predominant pathway, transoceanic shipping. Dispersal in the ballast of ocean going ships has been implicated in the invasion of *Bythotrephes* and a number of other invaders (Ricciardi and MacIsaac, 2000). The exchange of freshwater ballast for saline waters mid-ocean can successfully reduce concentrations of freshwater invertebrates and some recruitment from diapausing eggs (Gray et al, 2007). In particular, ballast water exchange may be useful at reducing live animal survival: *Bythotrephes* and *Cercopagis* were both eliminated in salinity tolerance tests of short duration (<5 hours) (Ellis and MacIsaac, 2009). Although compliance with ballast water exchange regulations for the Great Lakes–St Lawrence Seaway is high (Buck, 2006), a large percentage of ships declare 'no ballast on board' (Holeck et al, 2004) and intracontinental or domestic-bound ballast water regulations are inconsistent (Lawrence and Cordell, 2010), indicating that there are many gaps in the management of this vector.

Ballast water exchange is generally considered far less effective for benthic or dormant life stages of species (MacIsaac et al, 2002). Diapausing eggs may survive in the sediments of ballast, which either do not get exchanged at sea in ships declared as no ballast on board or are residuals that remain after saltwater exchange (Bailey et al, 2003; Holeck et al, 2004). Bailey et al (2004) demonstrate that salinity reduces diapausing egg hatching success, but not completely. The poor efficacy of ballast water exchange is a management concern, particularly given that populations of invasive cladocerans can be started from a few individuals (e.g. *Daphnia pulex* in Kenya

originated from a single individual, probably introduced with non-native fish: Mergeay et al, 2005).

Other physical and chemical methods for reducing introductions of invasive freshwater invertebrates have been tested. Although chemical treatments can successfully reduce the survival of live cladocerans (Sano et al, 2004), gluteraldehyde and sodium hypochlorite induced variable mortality in diapausing eggs of *Daphnia mendotae*, often requiring high doses, suggesting that diapausing eggs are not as sensitive as live organisms to chemical control (Raikow et al, 2007a). Similar results were observed using SeaKleen™, a biocide (Raikow et al, 2006). In all cases, chemical treatments were significantly less effective when the eggs were buried in sediment. Additionally, hatching success of *Daphnia* diapausing eggs was significantly reduced by UV, deoxygenation and heat treatments, but these methods were not deemed viable because of the high doses required (Raikow et al, 2007b). Improved efficacy of seawater flushing was obtained for no ballast on board vessels when an additional treatment of concentrated sodium chloride was applied: both freshwater and brackish cladocerans and copepods were significantly reduced by the brine treatment (Santagata et al, 2009).

The second step of the vector management strategy is to prevent shorter distance diffuse invasions. For *Bythotrephes*, the likely pathway of short distance invasions is predominantly human vectors (see above). MacIsaac et al (2004) demonstrate a hub model invasion history for *Bythotrephes* in Ontario, Canada, whereby certain lakes with high recreational usage were identified as hubs, which were the sources for many subsequent invasions. Similarly, Buchan and Padilla (1999) show that models that incorporated recreational boater movements were better predictors of zebra mussel (*Dreissena polymorpha*) invasions compared to simple diffusion models. Educational campaigns that target boat launches, particularly at locations identified as hubs and with high outbound traffic, may be an effective management strategy. Some recommended actions to prevent the spread of *Bythotrephes* and other invaders include: (1) inspecting boat and equipment for any visible organisms; (2) drain water from live well, motor, bilge and transom while on land; (3) rinse equipment with hot tap water (>50°C) or spray with high pressure (250 psi) or dry in the sun for at least five days; and (4) empty bait buckets on land. Additionally, educational programmes can target broader audiences through distribution of materials and citizen-oriented science. The Invading Species Awareness Program in Ontario, Canada distributed more than 600,000 pieces of educational material and received more than 1900 sighting reports on an Invasive Species Hotline in 2009 (Invading Species Awareness Program for Ontario, 2010). Such methods, along with stronger legislative actions and regulations will be important in containing invasive cladoceran spread in the future.

Acknowledgements

Thanks to Meghan Brown, Monica Papes, Philip Shaw and Jeff Brinsmead for current *Bythotrephes* distribution data, and Eric Larson for constructive feedback.

References

Bailey, S. A., Duggan, I. C., van Overdijk, C. D. A., Jenkins, P. T. and MacIsaac, H. J. (2003) 'Viability of invertebrate diapausing eggs collected from residual ballast sediment', *Limnology and Oceanography*, vol 48, pp1701–1710

Bailey, S. A., Duggan, I. C., Van Overdijk, C. D. A., Johengen, T. H., Reid, D. F. and MacIsaac, H. J. (2004) 'Salinity tolerance of diapausing eggs of freshwater zooplankton', *Freshwater Biology*, vol 49, pp286–295

Barbiero, R. P. and Tuchman, M. L. (2004) 'Changes in the crustacean communities of Lakes Michigan, Huron, and Erie following the invasion of the predatory cladoceran *Bythotrephes longimanus*', *Canadian Journal of Fisheries and Aquatic Sciences*, vol 61, pp2111–2125

Bollens, S. M., Cordell, J. R., Avent, S. and Hooff, R. (2002) 'Zooplankton invasions: A brief review, plus two case studies from the northeast Pacific Ocean', *Hydrobiologia*, vol 480, pp87–110

Boudreau, S. A. and Yan, N. D. (2003) 'The differing crustacean zooplankton communities of Canadian Shield lakes with and without the nonindigenous zooplanktivore *Bythotrephes longimanus*', *Canadian Journal of Fisheries and Aquatic Sciences*, vol 60, pp1307–1313

Branstrator, D. K., Brown, M. E., Shannon, L. J., Thabes, M. and Heimgartner, K. (2006) 'Range expansion of *Bythotrephes longimanus* in North America: Evaluating habitat characteristics in the spread of an exotic zooplankter', *Biological Invasions*, vol 8, pp1367–1379

Brown, M. E. and Balk, M. A. (2008) 'The potential link between lake productivity and the invasive zooplankter *Cercopagis pengoi* in Owasco Lake (New York, USA)', *Aquatic Invasions*, vol 3, pp28–34

Buchan, L. A. and Padilla, D. K. (1999) 'Estimating the probability of long-distance overland dispersal of invading aquatic species', *Ecological Applications*, vol 9, pp254–265

Buck, E. H. (2006) 'Ballast water management to combat invasive species', Congressional Research Service, The Library of Congress, Order Code RL32344, Washington, DC

Bur, M. T., Klarer, D. M. and Krieger, K. A. (1986) 'First records of a European cladoceran, *Bythotrephes cederstroemi*, in Lake Erie and Huron', *Journal of Great Lakes Research*, vol 12, pp144–146

Cáceres, C. M. (1998) 'Interspecific variation in the abundance, production, and emergence of *Daphnia* diapausing eggs', *Ecology*, vol 79, pp1699–1710

Carlton, J. T. and Geller, J. B. (1993) 'Ecological roulette: The global transport of nonindigenous marine organisms', *Science*, vol 261, pp78–82

Colautti, R. I., Manca, M., Viljanen, M., Ketelaars, H. A. M., Bürgi, H., MacIsaac, H. J. and Heath, D. D. (2005) 'Invasion genetics of the Eurasian spiny waterflea: Evidence for bottlenecks and gene flow using microsatellites', *Molecular Ecology*, vol 14, pp1869–1879

Critescu, M. E. A., Hebert, P. D. N., Witt, J. D. S., MacIsaac, H. J. and Grigorovich, I. A. (2001) 'An invasion history for *Cercopagis pengoi* based on mitochondrial gene sequences', *Limnology and Oceanography*, vol 46, pp224–229

Drake, J. A. (2004) 'Allee effects and the risk of biological invasion', *Risk Analysis*, vol 24, pp795–802

Drake, J. A., Drury, K. L. S., Lodge, D. M., Blukacz, E. A., Yan, N. D. and Dwyer, G. (2006) 'Demographic stochasticity, environmental variability, and windows of invasion risk for *Bythotrephes longimanus* in North America', *Biological Invasions*, vol 8, pp843–861

Duggan, I. C. (2010) 'The freshwater aquarium trade as a vector for incidental invertebrate fauna', *Biological Invasions*, vol 12, pp3757–3770

Dumitru, C., Sprules, W. G. and Yan, N. D. (2001) 'Impact of *Bythotrephes longimanus* on zooplankton assemblages of Harp Lake, Canada: An assessment based on predator consumption and prey production', *Freshwater Biology*, vol 46, pp241–251

Ellis, S. and MacIsaac, H. J. (2009) 'Salinity tolerance of Great Lakes invaders', *Freshwater Biology*, vol 54, pp77–89

Figuerola, J., Green, A. J. and Michot, T. C. (2005) 'Invertebrate eggs can fly: Evidence of waterfowl-mediated gene flow in aquatic invertebrates', *American Naturalist*, vol 165, pp274–280

Forro, L., Korovchinsky, N. M., Kotov, A. A. and Petrusek, A. (2008) 'Global diversity of cladocerans (Cladocera; Crustacea) in freshwater', *Hydrobiologia*, vol 595, pp177–184

Foster, S. E. and Sprules, W. G. (2009) 'Effects of the *Bythotrephes* invasion on native predatory invertebrates', *Limnology and Oceanography*, vol 54, pp757–769

Gray, D. K., Johengen, T. H., Reid, D. F. and MacIsaac, H. J. (2007) 'Efficacy of open-ocean ballast water exchange as a means of preventing invertebrate invasions between freshwater ports', *Limnology and Oceanography*, vol 52, pp2386–2397

Grigorovich, I. A., Pashkova, O. V., Gromova, Y. F. and van Overdijk, C. D. A. (1998) '*Bythotrephes longimanus* in the Commonwealth of Independent States: Variability, distribution, and ecology', *Hydrobiologia*, vol 379, pp183–198

Havel, J. E. and Medley, K. A. (2006) 'Biological invasions across spatial scales: Intercontinental, regional, and local dispersal of cladoceran zooplankton', *Biological Invasions*, vol 8, pp459–473

Havel, J. E. and Shurin, J. B. (2004) 'Mechanisms, effects, and scales of dispersal in freshwater zooplankton', *Limnology and Oceanography*, vol 49, pp1229–1238

Holeck, K. T., Mills, E. L., MacIsaac, H. J., Dochoda, M. R., Colautti, R. I. and Ricciardi, A. (2004) 'Bridging troubled waters: Biological invasions, transoceanic shipping, and the Laurentian Great Lakes', *BioScience*, vol 54, pp919–929

Hovius, J. T., Beisner, B. E. and McCann, K. S. (2006) 'Epilimnetic rotifer community responses to *Bythotrephes longimanus* invasion in Canadian Shield lakes', *Limnology and Oceanography*, vol 51, pp1004–1012

Invading Species Awareness Program for Ontario (2010) 'Annual report for 2009/10', www.invadingspecies.com/indexen.cfm, accessed 10 December 2010

Jarnagin, S. T., Swan, B. K. and Kerfoot, W. C. (2000) 'Fish as vectors in the dispersal of *Bythotrephes cederstroemi*: Diapausing eggs survive passage through the gut', *Freshwater Biology*, vol 43, pp579–589

Jin, E. H. and Sprules, W. G. (1990) 'Distribution and abundance of *Bythotrephes cederstroemi* (Cladocera: Cercopagidae) in the St Lawrence Great Lakes', *Verhandlungen Internationale Vereinigung für theoretische und angewandte Limnologie*, vol 24, pp383–385

Johannsson, O. E., Mills, E. L. and O'Gorman, R. (1991) 'Changes in the nearshore and offshore zooplankton communities in Lake Ontario: 1981–1988', *Canadian Journal of Fisheries and Aquatic Sciences*, vol 48, pp1546–1557

Johnson, P. T. J., Olden, J. D. and Vander Zanden, M. J. (2008) 'Dam invaders: Impoundments facilitate biological invasions into freshwaters', *Frontiers in Ecology and the Environment*, vol 6, pp357–363

Kerfoot, W. C., Robbins, J. A. and Weider, L. J. (1999) 'A new approach to historical reconstruction: Combining descriptive and experimental paleolimnology', *Limnology and Oceanography*, vol 44, pp1232–1247

Ketelaars, H. A. M. and Gille, L. (1994) 'Range extension of the predatory cladoceran *Bythotrephes longimanus* Leydig 1860 (Crustaea, Onychopoda) in Western Europe', *Aquatic Ecology*, vol 28, pp175–180

Lawrence, D. J. and Cordell, J. R. (2010) 'Relative contributions of domestic and foreign sourced ballast water to propagule pressure in Puget Sound, Washington, USA', *Biological Conservation*, vol 143, pp700–709

Laxson, C. L., McPhedran, K. N., Makarewicz, J. C., Telesh, I. V. and MacIsaac, H. J. (2003) 'Effects of the non-indigenous cladoceran *Cercopagis pengoi* on the lower food web of Lake Ontario', *Freshwater Biology*, vol 48, pp2094–2106

Lehman, J. T. (1987) 'Palearctic predator invades North American Great Lakes', *Oecologia*, vol 74, pp478–480

Lockwood, J. L., Cassey, P. and Blackburn, T. (2005) 'The role of propagule pressure in explaining species invasions', *Trends in Ecology & Evolution*, vol 20, pp223–228

Lodge, D. M., Williams, S., MacIsaac, H. J., Hayes, K. R., Leung, B., Reichard, S., Mack, R. N., Moyle, P. B., Smith, M., Andow, D. A., Carlton, J. T. and McMichael, A. (2006) 'Biological invasions: Recommendations for US policy and management', *Ecological Applications*, vol 16, pp2035–2054

Lynch, M. (1980) 'The evolution of cladoceran life histories', *The Quarterly Review of Biology*, vol 55, pp23–42

MacIsaac, H. J., Ketelaars, H. A. M., Grigorovich, I. A., Ramcharan, C. W. and Yan, N. D. (2000) 'Modeling *Bythotrephes longimanus* invasions in the Great Lakes basin based on its European distribution', *Archiv für Hydrobiologie*, vol 149, pp1–21

MacIsaac, H. J., Robbins, T. C. and Lewis, M. A. (2002) 'Modeling ships' ballast water as invasion threats to the Great Lakes', *Canadian Journal of Fisheries and Aquatic Sciences*, vol 59, pp1245–1256

MacIsaac, H. J., Borbely, J. V. M., Muirhead, J. and Graniero, P. A. (2004) 'Backcasting and forecasting biological invasions of inland lakes', *Ecological Applications*, vol 14, pp773–783

Marsden, J. E. and Hauser, M. (2009) 'Exotic species in Lake Champlain', *Journal of Great Lakes Research*, vol 35, pp250–265

Martin, J. W. and Cash-Clark, C. E. (1995) 'The external morphology of the onychopod "cladoceran" genus *Bythotrephes* (Crustacea, Branchiopoda, Onychopoda, Cercopagididae), with notes on the morphology and phylogeny of the order Onychopoda', *Zoologica Scripta*, vol 24, pp61–90

Mergeay, J., Verschuren, D. and De Meester, L. (2005) 'Cryptic invasion and dispersal of an American *Daphnia* in East Africa', *Limnology and Oceanography*, vol 50, pp1278–1283

Muirhead, J. R. and MacIsaac, H. J. (2005) 'Development of inland lakes as hubs in an invasion network', *Journal of Applied Ecology*, vol 42, pp80–90

Ojaveer, E., Lumberg, A. and Ojaveer, H. (1998) 'Highlights of zooplankton dynamics in Estonian waters (Baltic Sea)', *ICES Journal of Marine Science*, vol 55, pp748–755

Pangle, K. L. and Peacor, S. D. (2009) 'Light-dependent predation by the invertebrate planktivore *Bythotrephes longimanus*', *Canadian Journal of Fisheries and Aquatic Sciences*, vol 66, pp1748–1757

Pangle, K. L., Peacor, S. D. and Johannsson, O. E. (2007) 'Large nonlethal effects of an invasive invertebrate predator on zooplankton population growth rate', *Ecology*, vol 88, pp402–412

Raikow, D. F., Reid, D. F., Maynard, E. E. and Landrum, P. E. (2006) 'Sensitivity of aquatic invertebrate resting eggs to SeaKleen® (menadione): A test of potential ballast tank treatment options', *Environmental Toxicology and Chemistry*, vol 25, pp552–559

Raikow, D. F., Landrum, P. F. and Reidt, D. F. (2007a) 'Aquatic invertebrate resting egg sensitivity to glutaraldehyde and sodium hypochlorite', *Environmental Toxicology and Chemistry*, vol 26, pp1770–1773

Raikow, D. F., Reid, D. F., Blatchley, E. R., Jacobs, G. and Landrum, P. F. (2007b) 'Effects of proposed physical ballast tank treatments on aquatic invertebrate resting eggs', *Environmental Toxicology and Chemistry*, vol 26, pp717–725

Ricciardi, A. (2001) 'Facilitative interactions among aquatic invaders: Is an "invasional meltdown" occurring in the Great Lakes?', *Canadian Journal of Fisheries and Aquatic Sciences*, vol 58, pp2513–2525

Ricciardi, A. and Atkinson, S. K. (2004) 'Distinctiveness magnifies the impact of biological invaders in aquatic ecosystems', *Ecology Letters*, vol 7, pp781–784

Ricciardi, A. and MacIsaac, H. J. (2000) 'Recent mass invasion of the North American Great Lakes by Ponto-Caspian species', *Trends in Ecology & Evolution*, vol 15, pp62–66

Ruiz, G. M. and Carlton, J. T. (2003) 'Invasion vectors: A conceptual framework for management', in G. M. Ruiz and J. T. Carlton (eds) *Invasive Species: Vectors and Management Strategies*, Island Press, Washington, DC, pp459–504

Sano, L. L., Mapili, M. A., Krueger, A., Garcia, E., Gossiaux, D., Phillips, K. and Landrum, P. F. (2004) 'Comparative efficacy of potential chemical disinfectants for treating unballasted vessels', *Journal of Great Lakes Research*, vol 30, pp201–216

Santagata, S., Bacela, K., Reid, D. F., McLean, K. A., Cohen, J. S., Cordell, J. R., Brown, C. W., Johengen, T. H. and Ruiz, G. M. (2009) 'Concentrated sodium chloride brine solutions as an additional treatment for preventing the introduction of nonindigenous species in the ballast tanks of ships declaring no ballast on board', *Environmental Toxicology and Chemistry*, vol 28, pp346–353

Schulz, K. L. and Yurista, P. M. (1999) 'Implications of an invertebrate predator's (*Bythotrephes cederstroemi*) atypical effects on a pelagic zooplankton community', *Hydrobiologia*, vol 380, pp179–193

Seaber, P. R., Kapinos, F. P. and Knapp, G. L. (1987) 'Hydrologic unit maps', US Geological Survey, Water Supply Paper 2294, http://water.usgs.gov/GIS/huc.html, accessed 13 December 2010

Sorensen, K. H. and Sterner, R. W. (1992) 'Extreme cyclomorphosis in *Daphnia lumholtzi*', *Freshwater Biology*, vol 28, pp257–262

Straile, D. and Halbich, A. (2000) 'Life history and multiple antipredator defenses of an invertebrate pelagic predator, *Bythotrephes longimanus*', *Ecology*, vol 81, pp150–163

Strauss, S. Y., Lau, J. A. and Carroll, S. P. (2006) 'Evolutionary responses of natives to introduced species: What do introductions tell us about natural communities?' *Ecology Letters*, vol 9, pp354–371

Strecker, A. L. and Arnott, S. E. (2008) 'Invasive predator, *Bythotrephes*, has varied effects on ecosystem function in freshwater lakes', *Ecosystems*, vol 11, pp490–503

Strecker, A. L. and Arnott, S. E. (2010) 'Complex interactions between regional dispersal of native taxa and an invasive species', *Ecology*, vol 91, pp1035–1047

Strecker, A. L., Arnott, S. E., Yan, N. D. and Girard, R. (2006) 'Variation in the response of crustacean zooplankton species richness and composition to the invasive predator *Bythotrephes longimanus*', *Canadian Journal of Fisheries and Aquatic Sciences*, vol 63, pp2126–2136

Strecker, A. L., Beisner, B. E., Arnott, S. E., Paterson, A. M., Winter, J. G., Johannsson, O. E. and Yan, N. D. (2011) 'Direct and indirect effects of an invasive planktonic predator on pelagic food webs', *Limnology and Oceanography*, vol 56, pp179–192

Therriault, T. W., Grigorovich, I. A., Critescu, M. E., Ketelaars, H. A. M., Viljanen, M., Heath, D. D. and MacIsaac, H. J. (2002a) 'Taxonomic resolution of the genus *Bythotrephes* Leydig using molecular markers and re-evaluation of its global distribution', *Diversity and Distributions*, vol 8, pp67–84

Therriault, T. W., Grigorovich, I. A., Kane, D. D., Haas, E. M., Culver, D. A. and MacIsaac, H. J. (2002b) 'Range expansion of the exotic zooplankter *Cercopagis pengoi* (Ostroumov) into western Lake Erie and Muskegon Lake', *Journal of Great Lakes Research*, vol 28, pp698–701

US EPA (United States Environmental Protection Agency) (2009) 'National lakes assessment: A collaborative survey of the nation's lakes', US Environmental Protection Agency, Office of Water and Office of Research and Development, EPA 841-R-09-001, Washington, DC

Weisz, E. J. and Yan, N. D. (2010a) 'Relative value of limnological, geographic and human use variables as predictors of the presence of *Bythotrephes longimanus* in Canadian Shield lakes', *Canadian Journal of Fisheries and Aquatic Sciences*, vol 67, pp462–472

Weisz, E. J. and Yan, N. D. (2010b) 'Shifting invertebrate zooplanktivores: watershed-level replacement of the native *Leptodora* by the non-indigenous *Bythotrephes* in Canadian Shield lakes', *Biological Invasions*, doi:10.1007/s10530-010-9794-8

Wetzel, R. G. (2001) *Limnology: Lake and River Ecosystems*, Academic Press, San Diego, CA

Yan, N. D. and Pawson, T. W. (1997) 'Changes in the crustacean zooplankton community of Harp Lake, Canada, following invasion by *Bythotrephes cederstroemi*', *Freshwater Biology*, vol 37, pp409–425

Yan, N. D. and Pawson, T. W. (1998) 'Seasonal variation in the size and abundance of the invading *Bythotrephes* in Harp Lake, Ontario, Canada', *Hydrobiologia*, vol 361, pp157–168

Yan, N. D., Blukacz, A., Sprules, W. G., Kindy, P. K., Hackett, D., Girard, R. E. and Clark, B. J. (2001) 'Changes in zooplankton and the phenology of the spiny water flea, *Bythotrephes*, following its invasion of Harp Lake, Ontario, Canada', *Canadian Journal of Fisheries and Aquatic Sciences*, vol 58, pp2341–2350

Yan, N. D., Girard, R. E. and Boudreau, S. (2002) 'An introduced invertebrate predator (*Bythotrephes*) reduces zooplankton species richness', *Ecology Letters*, vol 5, pp481–485

Zanata, L. H., Espíndola, E. L., Rocha, O. and Pereira, R. H. G. (2003) 'First record of *Daphnia lumholtzi* (Sars, 1885), exotic cladoceran, in São Paulo State (Brazil)', *Brazilian Journal of Biology*, vol 63, pp717–720

14

Invasive freshwater copepods of North America

Jeffery R. Cordell

Introduction

Copepods are small crustaceans, relatives of crabs and shrimp. They have been described as 'insects of the sea' because they are the most abundant metazoans in the sea and through their sheer abundance play a vital role in aquatic ecosystems (Huys and Boxshall, 1991). Copepods are usually the most abundant organisms found in plankton samples taken in the ballast waters of ships, where they can occur in densities of tens of thousands per cubic metre (Duggan et al, 2005; Cordell et al, 2009). It follows that millions of individual copepods are taken up, moved and discharged by ship ballasting operations, and it is not surprising that copepods have repeatedly been introduced to new habitats via this vector. Because of their high densities, any time water is moved from one place to another, copepods are moved as well, and activities such as fish hatchery operations, movement of aquatic plants and rice culture have also been implicated in the establishment of non-indigenous copepod species (Table 14.1). Even transport on the feet or feathers of migratory birds has been suggested as a vector for the spread of copepods (Reid and Reed, 1994). Another way in which copepods can invade new habitats is by coastal or estuarine species becoming adapted to fresh water and transported either by natural or human-mediated means through river systems into lakes and reservoirs (Lee, 1999; Lee and

Bell, 1999). The case histories presented below encompass several of the main vectors of spread of planktonic copepods.

Introductions of non-indigenous planktonic crustaceans into continental fresh waters have occurred in various locations throughout the world (Bollens et al, 2002). However, perhaps the most extensive invasion of lakes and rivers has occurred in North America, where at least nine planktonic copepods have invaded rivers and their estuaries (Cordell et al, 2008), and inland waters have seen the introduction of numerous other planktonic copepod and cladoceran species (Table 14.1) (USGS, 2009). Of these, most published studies have been conducted on the cladoceran *Bythotrephes longimanus* (Chapter 13). The ecology of non-indigenous copepods has been much less studied, but there are cases in which the introduced copepod has become so abundant that it dominates plankton abundance and must have similarly large ecological effects (Cordell et al, 2007; Bouley and Kimmerer, 2006; Cordell et al, 2008).

In the following case histories, three species or species groups are detailed that can be regarded as invasive, in that they have rapidly invaded new habitats in their introduced ranges and have become abundant and often dominant members of the zooplankton community. By necessity, I do not use a definition for 'invasive' that requires economic or environmental harm, because little or nothing is known about

Table 14.1 *Planktonic copepods introduced to North American fresh waters*

Species	Native range	Introduction	Possible vector(s)	References
Calanoida				
Arctodiaptomus dorsalis (Marsh)	Gulf of Mexico, Caribbean lowlands	Mexican and Colombian highlands, Venezuela, inland continental US	Fish hatcheries	Reid, 2007
Eurytemora affinis (Poppe)	Europe, Atlantic and Pacific North American coasts	Laurentian Great Lakes, Mississippi River system, Rio Bravo/Rio Grande system	Ballast water, passive dispersal	Saunders, 1993; Lee, 1999; Suárez-Morales et al, 2008
Pseudodiaptomus forbesi (Poppe and Richardson)	Yangtze River	Sacramento/San Joaquin and Columbia River systems[d]	Ballast water	Orsi and Walter, 1991; Cordell et al, 2008
Sinocalanus doerrii (Brehm)	Asia	Sacramento/San Joaquin and Columbia River systems[d]	Ballast water	Orsi et al, 1983; Cordell et al, 2008
Sinodiaptomus sarsi (Rylov)[a]	China	Western US	Tropical aquatic plants	Reid and Pinto-Coelho, 1994
Cyclopoida				
Apocyclops dengizicus Lepeshkin[b]	Old World	Eastern US	Unknown	Reid et al, 2002
Megacyclops viridis Jurine	Europe	Laurentian Great Lakes	Ballast water	Reid and Hudson, 2008
Mesocyclops pehpeiensis Hu	Eastern Asia	Southern US	Rice culture	Reid, 1993; Reid and Marten, 1995; Reid and Pinto-Coelho, 1994
Thermocyclops crassus Fischer	Europe, Asia, Australia	Northeast US (Lake Champlain), Mexico	Unknown	Duchovnay et al, 1992; Gutiérrez-Aguirre and Suárez-Morales, 2000
Mesocyclops longisetus curvatus Dussart[b]	South and Central America, the Antilles, Mexico, southern US	Yukon Territory, Canada	Migratory birds	Reid and Reed, 1994
Mesocyclops ogunnus Onabamiro	Africa, Asia	Florida Keys	Unknown	Hribar and Reid, 2008
Mesocyclops venezolanus Dussart[b]	South and Central America	Yukon Territory, Canada	Migratory birds	Reid and Reed, 1994
Limnoithona sinensis (Burckhardt)[c]	Yangtze River, China	Sacramento/San Joaquin and Columbia River systems	Ballast water	Ferrari and Orsi, 1984
Limnoithona tetraspina Zhang and Li	Yangtze River, China	Sacramento/San Joaquin and Columbia River systems[d]	Ballast water	Orsi and Ohtsuka, 1999

Note: [a] introduced population did not persist; [b] unknown whether or not introduced populations have persisted; [c] now rare where introduced; [d] may have been a secondary introduction to the Columbia River via ballast water from California (Cordell et al, 2008).

negative effects of the copepods. Instead, I refer to recent assertions by Valéry et al (2008) and Ricciardi and Cohen (2007) that the term 'invasive' does not have to connote negative impacts and encompasses a more mechanistic definition, as in the proposed definition given by Valéry et al (2008, p1349): 'A biological invasion consists of a species' acquiring a competitive advantage following the disappearance of natural obstacles to its proliferation, which allows it to spread rapidly and to conquer novel areas within recipient ecosystems in which it becomes a dominant population.'

Pseudodiaptomus forbesi

The calanoid copepod genus *Pseudodiaptomus* is distributed circumglobally in tropical and temperate nearshore waters. Prior to human-associated introductions, only two species occurred in North America north of Mexico: *P. pelagicus*, which is distributed on the Atlantic coast from the Gulf of Mexico north to Nova Scotia, Canada, and *P. euryhalinus*, which apparently has a limited distribution from San Francisco Bay, California, US, south to the northern coastline of Baja California, Mexico (Walter, 1989). However, since the late 1980s, three species of *Pseudodiaptomus* native to Asia have been introduced to the Pacific coast of the US. *Pseudodiaptomus marinus* was first found in southern California in 1986 (Fleminger and Kramer, 1988), and also appeared around this time in San Francisco Bay, where it is now abundant in higher salinity areas (Orsi and Walter, 1991; Bollens et al, 2011). *Pseudodiaptomus inopinus* was found in the Columbia River in the Pacific Northwest region of the US in 1990 (Cordell et al, 1992). Subsequently, it has become very abundant in brackish and tidal fresh waters of a number of other estuaries along the Pacific coast of the US (Cordell and Morrison, 1996; Cordell et al, 2007). The third species, *P. forbesi*, like *P. inopinus*, was initially introduced into estuaries, but has invaded truly fresh waters in its introduced range.

Pseudodiaptomus forbesi (Figure 14.1) is endemic to fresh waters of the Yangtze River in China and seems to be restricted to that region in its native distribution, with one questionable record from a lake in Japan (Mashiko, 1951; Orsi and Walter, 1991). It was most likely transported to North America in the ballast water

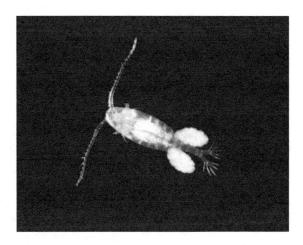

Figure 14.1 Pseudodiaptomus forbesi, *female with egg sacs*

of ships (Orsi and Walter, 1991; Cordell et al, 2008). The first record of this species outside of Asia was in the Sacramento-San Joaquin River delta, where it was first seen in plankton samples taken in 1987 by the California Department of Fish and Game, and was reported along with *P. marinus*, which was found in higher salinity waters (Orsi and Walter, 1991). By 1989, *P. forbesi* summer and autumn abundances were comparable to the previously dominant calanoid copepod *Eurytemora affinis*. Although *P. forbesi* abundance declined slightly after its introduction, it has remained relatively abundant in the fresh waters of the delta in summer and autumn, compared to other copepods that co-occur with it (Hennessy, 2010).

Pseudodiaptomus forbesi was next discovered approximately 900km north of the Sacramento-San Joaquin delta in the Columbia River, during a 2003 non-indigenous species survey (Sytsma et al, 2004). Subsequently, a plankton survey conducted in 2006 found that it had undergone an expansion from the lower, tidal portion of the river, into a number of reservoirs as far as 500km upstream in the Columbia River, and also occurred in small numbers in the lowest reservoir on the Snake River, the major tributary to the Columbia River (Cordell et al, 2008). As a follow-up to these surveys, plankton was sampled again in 2009, and the results are presented here (Figures 14.2 and 14.3, and Table 14.2). This latest survey shows that *P. forbesi* also now occurs in additional reservoirs in the Snake

■ Hydroelectric Dam
△ Stations sampled September 2009
○ Stations sampled monthly, July-December, 2009

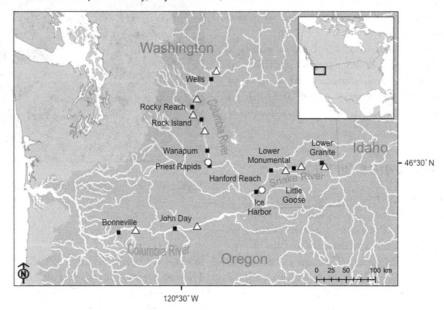

Note: Only dams associated with sampled reservoirs are shown. The Hanford Reach is the last free flowing section of the main stem of the Columbia River. Inset shows location of study area in North America. Samples were collected with 73μm mesh plankton nets and analysed as per Cordell et al (2008).

Figure 14.2 *Columbia-Snake River drainage (darker shade) and stations sampled in 2009 for the introduced copepod* Pseudodiaptomus forbesi *on the Columbia and Snake Rivers, US*

River, but only in low numbers (Table 14.2), and still has not penetrated upstream beyond the last free flowing section of the Columbia River (Hanford Reach) (Figure 14.3). One of the most obvious explanations for this distribution is that there is extensive shipping commerce in the lower Columbia and Snake Rivers, and none in the main stem Columbia River above the Hanford Reach. There are eight hydroelectric dams with navigational locks: four on the lower Snake and four on the Columbia. The locks were designed for the tug and barge trade, and can accommodate barges up to 198m long (Wallack, 2002). Thus, movement of water from the lower river through the locks, and also possibly water carried by vessels into the reservoirs could account for the presence of *P. forbesi* in the lower Columbia and Snake Rivers. Interestingly, native copepods in the family Diaptomidae were abundant only where *P. forbesi* did not occur

(Cordell et al, 2008) (Figure 14.3). It is not known if diaptomids were previously abundant where *P. forbesi* has invaded, or conversely if they were not naturally abundant and this helped to facilitate the invasion. In the reservoirs on the lower Columbia River where it has invaded, *P. forbesi* has become quite successful, comprising up to 98 per cent of the numbers in plankton net samples (Cordell et al, 2008). The reasons for its much lower relative abundances in the Snake River are unknown, but one possibility is that physico-chemical conditions favour the native cladocerans and cyclopoids that were dominant there. It may also be early in the invasion trajectory, and *P. forbesi* may become more abundant in the Snake River in the future.

The ecological effects of *Pseudodiaptomus forbesi* in invaded habitats are largely unknown. In the Sacramento-San Joaquin delta, two other copepods, *Eurytemora*

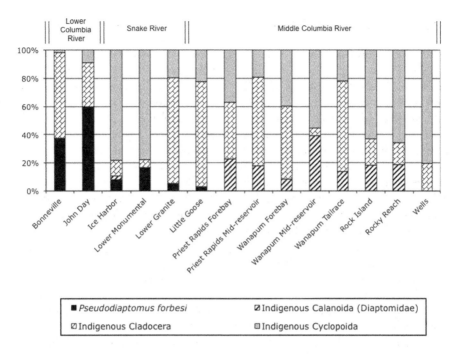

Note: Stations labelled Middle Columbia River are upstream of the Hanford Reach.

Figure 14.3 *Percentage composition of the introduced copepod* Pseudodiaptomus forbesi *and indigenous taxa groups in reservoirs on the Columbia and Snake Rivers, US, September 2009*

Table 14.2 *Densities of combined life history stages of* Pseudodiaptomus forbesi *in lower Columbia and Snake River reservoirs sampled in September 2009*

Reservoir	River system	*P. forbesi* density, no. m^{-3}
Bonneville	Lower Columbia	9472
John Day	Lower Columbia	25,295
Ice Harbor	Snake	749
Lower Monumental	Snake	231
Lower Granite	Snake	71
Little Goose	Snake	1077

affinis and *Acartia* spp. decreased greatly in abundance coincident with the introduction of *P. forbesi* (Kimmerer, 2004). However, in the case of *E. affinis* the decline was apparently caused mainly by another exotic species, the clam *Corbula amurensis*, consuming copepod larvae (Kimmerer et al, 1994). *Pseudodiaptomus forbesi* became abundant over much of the region formerly occupied by *E. affinis*, and the apparent coincidence in timing may be due to a competitive interaction between the copepods or to the ability of *P. forbesi* larvae to avoid predation by the introduced clam (Kimmerer and Orsi, 1996). *Pseudodiaptomus forbesi* is now an established and abundant member of the plankton assemblage in the tidal fresh regions of the delta, and comprises an important food source for several smelt species (e.g. Nobriga, 2002; Hobbs et al, 2006). Ecological effects of *P. forbesi* in the Columbia-Snake River system have not been studied.

Eurytemora affinis Species Complex

The calanoid copepod genus *Eurytemora* occurs in coastal subtropical to subarctic regions of the northern hemisphere in a broad range of habitats. Species diversity of the genus increases toward northern latitudes, with highest diversity in Alaska and northeastern Asia (Dodson et al, 2010 and references cited therein). Species of *Eurytemora* occur across a broad range of salinities, and this is especially true for the *E. affinis* species complex, which occupies habitats ranging from hypersaline salt marshes and brackish estuaries to completely fresh water (Lee, 1999).

Eurytemora affinis (Figure 14.4) is abundant in coastal waters of the northern hemisphere. It is mainly confined to coastal and estuarine habitats, but within the last century it has invaded many freshwater reservoirs and lakes in North America, Asia and Europe (Lee, 1999). However, the most extensive invasion of fresh waters by this species has occurred in North America, in such widely separated geographical settings as the Laurentian Great Lakes, the southern Great Plains of the US and the Chihuahuan Desert in Mexico (Saunders, 1993; Lee 1999; Lee et al, 2007; Suárez-Morales et al, 2008) (Figure 14.5a). These freshwater invasions have occurred largely in waters that have been significantly altered by human development (e.g. reservoirs and canals) or have intense water-based commerce (e.g. St Lawrence Seaway). For example, *E. affinis* is abundant in the St Lawrence estuary of

North America and invaded the St Lawrence River and Great Lakes after the opening of the St Lawrence Seaway *c.*1959 (Faber and Jermolajev, 1966; Lee et al, 2007; Farrell et al, 2010) (Figure 14.5b). There may be multiple mechanisms for the dispersal of *E. affinis* from coastal into fresh waters, and release of ballast water, the transplantation of water with live recreational fishes and transport by waterfowl have been identified as possible vectors (Lee, 1999).

The species '*E. affinis*' represents a number of closely related species or subspecies composed of genetically distinct clades that are for the most part morphologically indistinguishable (Lee and Frost, 2002; but see Alekseev and Souissi, 2011, for description of a new species from the complex based on morphology). There are large genetic distances and varying patterns of reproductive isolation among the clades (Lee, 2000), making understanding the invasion ecology of this 'species' complicated. For example, in the St Lawrence estuary, two reproductively isolated clades coexist in native saline habitats (Lee, 1999, 2000; Winkler et al, 2008). One of the clades is invasive, having extended its range into freshwater reservoirs and the Great Lakes, while the other has remained restricted to its native range (Lee, 1999, 2000; Winkler et al, 2008). Furthermore, the two clades show a striking geographic partitioning within their native range, with one clade dominating low salinity upstream regions and highly fluctuating salt marsh ponds, and the other dominating the central mesohaline portion of the middle estuary (Winkler et al, 2008).

The *E. affinis* species complex has proven to be a fertile subject for the study of the evolutionary genetics of invasions. For example, for more saline populations that invaded fresh water, the development of freshwater tolerance cannot be explained by either short term or developmental phenotypic plasticity alone, but rather the differences in tolerance between the populations are genetically based, and freshwater invasions are accompanied by evolutionary shifts in freshwater tolerance (Lee and Petersen, 2003; Lee et al, 2003). Both the saline and freshwater populations possess genetic variation in salinity tolerance, allowing selection to act rapidly when they invade new habitats (Lee et al, 2003, 2007).

The economic and ecological consequences of the invasion of continental fresh waters by the *E. affinis* species complex have not been studied. This topic

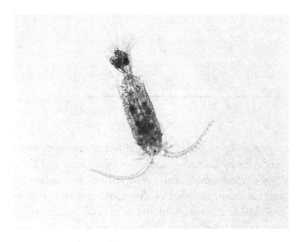

Source: Jeffery R. Cordell

Figure 14.4 Eurytemora affinis, *female with partial egg sac and attached spermatophore*

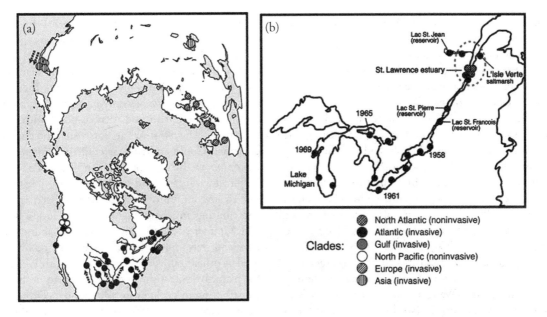

Note: Patterned circles represent genetically distinct clades. In (a) dashed arrows indicate pathways of independent freshwater invasions. The Atlantic clade (black dots) has extended its range into fresh water, while the North Atlantic clade has not (hatched dots). The dotted circle shows the native range for both clades; other populations are recent invasions into reservoirs and the Great Lakes. Dates are approximate timing of invasions.

Source: Lee et al, 2007

Figure 14.5 *Invasions by the* Eurytemora affinis *species complex from saline sources into freshwater lakes and reservoirs: (a) global pattern of freshwater invasions by* E. affinis *from genetically distinct saline source populations, and (b) populations of* E. affinis *in the St Lawrence drainage system*

deserves attention because *E. affinis* can be dominant in invaded habitats (e.g. the upper Ohio River, US) (Counahan et al, 2005), and it is very important in estuarine food webs, including tidal fresh waters (Dauvin and Desroy, 2005; Hoffman et al, 2008). In addition to possible impacts on food webs, *E. affinis* has been mentioned as a possible vector for disease because it is known to be a host of pathogenic species of *Vibrio*, including the causative agent of cholera (Rawlings et al, 2007).

Limnoithona sinensis and L. tetraspina

The cyclopoid copepod genus *Limnoithona* contains only two species, *L. sinensis* and *L. tetraspina* (Figure 14.6). Both species are endemic to the Yangtze River estuary,

Source: Jeffery R. Cordell

Figure 14.6 Limnoithona tetraspina, *female with egg sacs*

with *L. sinensis* occurring at least 300km upstream (Ferrari and Orsi, 1984). *Limnoithona sinensis* has also recently been found in estuaries on the south coast of Korea (Yoon and Chang, 2008). Both species have been introduced to the Sacramento-San Joaquin and Columbia River estuaries on the Pacific coast of the US.

The first species of *Limnoithona* to appear in North America was *L. sinensis*, found in samples taken by the California Department of Fish and Game during August 1979 in the San Joaquin River near Stockton, California, and soon thereafter found to be abundant throughout the Sacramento-San Joaquin delta (Ferrari and Orsi, 1984). This species also occurred in the Columbia River estuary in the early 1990s but is no longer found there (Cordell et al, 2008). Within its native range, *L. sinensis* also appears able to invade new habitats such as reservoirs, displacing previously dominant species (Wang et al, 2009). *Limnoithona tetraspina* was first seen in North America in 1993 in the San Francisco estuary near the confluence of the Sacramento and San Joaquin Rivers. At this time it was much more abundant in all seasons, particularly summer, than *L sinensis* was. Since the introduction of *L. tetraspina* in this system, densities of *L. sinensis* have remained very low (Bouley and Kimmerer, 2006). Orsi and Ohtsuka (1999) argue that *L. tetraspina* probably did not cause the decrease in *L. sinensis* because where they co-occur in their native habitat at the mouth of the Yangtze River their salinity distributions show little overlap: *L. tetraspina* is reported from brackish water and *L. sinensis* from fresh water. However, in both the Sacramento-San Joaquin and Columbia River systems, *L. tetraspina* occurs in completely fresh waters, and the disappearance or reductions in *L. sinensis* in those systems could have been the result of competition with *L. tetraspina*.

Limnoithona tetraspina is now the most abundant copepod in the low salinity parts of the San Francisco estuary (Bouley and Kimmerer, 2006). It is also established in the lower Columbia River to at least 75km upstream, possibly carried there in ballast water from California (Cordell et al, 2008). However, little is known about the ecology of *Limnoithona* species and their impacts on invaded ecosystems. The feeding behaviour of *L. tetraspina* suggests that they feed mainly on ciliates, with little grazing on diatoms (Bouley and Kimmerer, 2006; Gifford et al, 2007). Bouley and Kimmerer (2006) note that although

L. tetraspina comprised up to 80 per cent of the summer and autumn plankton numbers between 1993 and 1996, it rarely made up more than 10 per cent of the diet of planktivorous juvenile delta smelt over that same period. This may be due to their small size; one *L. tetraspina* has only one tenth the biomass of other calanoid copepods that co-occur with it, and it may not be worthwhile for visual predators to consume (Gould and Kimmerer, 2010). *Limnoithona tetraspina* is thriving in the San Francisco estuary despite low growth and fecundity rates, presumably due to low mortality, and this implies that it is not contributing much to higher trophic levels (Gould and Kimmerer, 2010). Because it is so abundant, *L. tetraspina* may impact microzooplankton food resources used by other omnivorous copepods that are consumed by planktivorous fish, and in doing so it may be contributing to declines in fish abundances that have been noted in the low salinity zone of the San Francisco estuary (Gould and Kimmerer, 2010).

Other Introduced Copepods

The copepod species listed in Table 14.1 are considered introduced in North America because they have expanded into the North American continent beyond their native ranges. In some cases, they have been introduced from other continents or from disjunct biogeographical provinces. These include the calanoid copepod *Sinocalanus doerrii*, which is native to Asia and has persisted in relatively low numbers in tidal fresh waters of the San Francisco and Columbia River estuaries, and a number of species of cyclopoid copepods in the genera *Apocyclops*, *Megacyclops*, *Mesocyclops* and *Thermocyclops* reported by Reid and several other authors (Table 14.1). However, these species are not considered invasive per se because they occur mainly in restricted distributions or are not abundant in the introduced habitats. One exception may be the case of the calanoid copepod *Arctodiaptomus dorsalis*, in which the introduction consists of a range expansion within North America, including several disjunct occurrences. From a core range in tropical and subtropical lowlands bordering the Gulf of Mexico and Caribbean Sea, this species has apparently been introduced to the Mexican and Colombian highlands, one location in Venezuela and a number of locations in the US, most successfully in the lower Mississippi basin (Texas, Oklahoma,

Arkansas), and north to Missouri, Indiana and Michigan (Reid, 2007). Reid (2007) speculates that several of the outlying populations were established via transport of fish or aquatic plants, and *A. dorsalis* often occurs in eutrophic impoundments and fish hatcheries, suggesting that it has the potential to extend its range farther north and south. As such it may represent the beginning stages of an invasion into continental fresh waters similar to that of *Eurytemora affinis*.

Secondary Spread

It is clear from the case histories and studies cited therein that once invasive copepods are introduced, they likely spread from that source to new regions. *Eurytemora affinis* and *Pseudodiaptomus forbesi* have expanded from estuaries into continental fresh waters, and *Limnoithona tetraspina* and *P. forbesi* have probably been transported from the San Francisco estuary to the Columbia River. For the last two species, ballast water is presumed to be the mode for secondary spread, by way of water taken up in California being discharged into the Columbia River (Cordell et al, 2008; Lawrence and Cordell, 2010). This is suggested by Cordell et al (2008), who sampled plankton in the ballast water of ships entering Puget Sound, Washington: for 44 ships sampled that had listed California as the ballast water source, *P. forbesi* occurred in 9 per cent of them and *L. tetraspina* occurred in 39 per cent.

In addition to coastwise spread, expansion of invasive copepod populations from coastal to inland waters is also associated with human activities. Lee (1999) found that phylogenetic relationships among *Eurytemora affinis* populations in the continental US indicated movement within river systems and also that recent freshwater populations almost always occurred in freshwater sites directly connected to river systems, suggesting that rivers provided the routes for invasion. Furthermore, patchy distribution of populations along rivers suggests that transport within river systems has been mediated by episodic events, such as water discharged from boats or transplantation of recreational fishes (Lee, 1999). One factor that the spread of both *E. affinis* and *Pseudodiaptomus forbesi* have in common is the presence of reservoirs in the invaded systems. Reservoirs appear to be particularly important as stepping stones for the spread of invasive aquatic species (Havel et al, 2005). Johnson et al (2008) analysed spatial distributions of five freshwater invaders in the US (including a planktonic species, the spiny water flea *Bythotrephes longimanus*), and their results suggest that the conversion of free flowing rivers into reservoirs facilitates the spread of invasive species. In their study, reservoirs were 2.5 to 7.8 times more likely than natural lakes to have established populations, and impoundment status was a significant predictor of occurrence for four of the five invaders. More than 90 per cent of the recent freshwater invasions by *E. affinis* have been in reservoirs (Lee, 1999). In the Columbia-Snake River system, *P. forbesi* has invaded where reservoirs are contiguously present, but not where a section of free flowing river interrupts the chain of reservoirs (Cordell et al, 2008).

Ecological Effects

The ecological effects of invasive copepods in North American fresh waters are not well understood. No native species have been eliminated by the invasive copepods, although some native copepods have experienced reduced abundances or changes in their distributions in time and space after the introductions (Orsi and Ohtsuka, 1999; Kimmerer, 2004; Hennessy, 2010). However, invasive copepods elsewhere have been associated with declines or extinctions of other copepods. For example, the calanoid *Eudiaptomus gracilis* invaded alpine and subalpine lakes in Italy during the late 1980s, and in one lake apparently caused or contributed to the extinction of its congener *E. padanus*, due to competitive interactions (Riccardi and Giussani, 2007). As mentioned earlier, other invasive freshwater planktonic crustaceans such as cladocerans can have large effects on invaded communities, and recent research on the effects of invasive copepods in North American estuarine brackish and marine habitats suggests that invading copepods might also have large ecological effects. Perhaps the best example of this is in the San Francisco estuary, where the copepod fauna now consists of mainly Asian species: the lower salinity parts of the estuary are dominated by species from China and Korea, while the community in the higher salinity regions is remarkably similar to that reported in Japanese estuaries (Bollens et al, 2011). One species that has invaded brackish waters of the estuary, the large calanoid *Tortanus dextrilobatus*, may have significant predatory impact on copepods including native *Acartia* species (Hooff and Bollens, 2004). Others may affect higher trophic levels because they have better avoidance behaviours than other copepods (e.g. *Sinocalanus doerrii*) (Meng and Orsi, 1991), or are too small to be eaten efficiently by plankton-feeding fish (e.g. *Limnoithona tetraspina*) (Bouley and Kimmerer, 2006). Whether or not similar effects accompany invasions of copepods into fresh waters is not known, and this represents a large and potentially important gap in understanding their impacts on native freshwater communities.

Acknowledgements

Results shown in Figures 14.2 and 14.3 and Table 14.2 were made possible by support from the Edward B. and Phyllis E. Reed Endowment, Smithsonian Institution. Samples from the lower Columbia River were provided courtesy of Dr S. Bollens, Washington State University, Vancouver, Washington; these and samples from Wanapum, Priest Rapids and Ice Harbor reservoirs were collected and data analysed in collaboration with T. Counihan, US Geological Survey, Cook, Washington.

References

Alekseev, V. R. and Souissi, A. (2011) 'A new species within the *Eurytemora affinis* complex (Copepoda: Calanoida) from the Atlantic Coast of USA, with observations on eight morphologically different European populations', *Zootaxa*, vol 2767, pp41–56

Bollens, S. M., Cordell, J. R., Avent, S. and Hooff, R. (2002) 'Zooplankton invasions: A brief review, plus two case studies from the northeast Pacific Ocean', *Hydrobiologia*, vol 480, pp87–110

Bollens, S. M., Breckenridge, J. K., Hooff, R. C. and Cordell, J. R. (2011) 'Mesozooplankton of the lower San Francisco Estuary: Spatio-temporal patterns, ENSO effects, and the prevalence of non-indigenous species', *Journal of Plankton Research*, doi:10.1093/plankt/FBR034

Bouley, P. and Kimmerer, W. J. (2006) 'Ecology of a highly abundant, introduced cyclopoid copepod in a temperate estuary', *Marine Ecology Progress Series*, vol 324, pp210–228

Cordell, J. R. and Morrison, S. M. (1996) 'The invasive Asian copepod *Pseudodiaptomus inopinus* in Oregon, Washington, and British Columbia estuaries', *Estuaries*, vol 19 pp629–638

Cordell, J. R., Morgan, C. A. and Simenstad, C. A. (1992) 'Occurrence of the Asian calanoid copepod *Pseudodiaptomus inopinus* in the zooplankton of the Columbia River estuary', *Journal of Crustacean Biology*, vol 12, pp260–269

Cordell, J. R., Rassmussen, M. and Bollens, S. M. (2007) 'Biology of the invasive copepod *Pseudodiaptomus inopinus* in a northeast Pacific estuary', *Marine Ecology Progress Series*, vol 333, pp213–227

Cordell, J. R., Bollens, S. M., Draheim, R. and Sytsma, M. (2008) 'Asian copepods on the move: Recent invasions in the Columbia-Snake River system, USA', *ICES Journal of Marine Science*, vol 65, pp753–758

Cordell, J. R., Lawrence, D. J., Ferm, N. C., Tear, L. M., Smith, S. S. and Herwig, R. P. (2009) 'Factors influencing densities of non-indigenous species in the ballast water of ships arriving at ports in Puget Sound, Washington, United States', *Aquatic Conservation: Marine and Freshwater Ecosystems*, vol 19 pp322–343

Counahan, D. F., Carline, R. F. and Reid, J. W. (2005) 'The occurrence of the calanoid copepod *Eurytemora affinis* (Poppe) in the upper Ohio River', *Northeastern Naturalist*, vol 12, no 4, pp541–545

Dauvin, J.-C. and Desroy, N. (2005) 'The food web in the lower part of the Seine estuary: A synthesis of existing knowledge', *Hydrobiologia*, vol 540, pp13–27

Dodson, S. I., Skelly, D. A. and Lee, C. E. (2010) 'Out of Alaska: Morphological diversity within the genus *Eurytemora* from its ancestral Alaskan range (Crustacea, Copepoda)', *Hydrobiologia*, vol 653, pp131–148

Duchovnay, A., Reid, J. W. and McIntosh, A. (1992) '*Thermocyclops crassus* (Crustacea: Copepoda) present in North America: A new record from Lake Champlain', *Journal of Great Lakes Research*, vol 18, pp415–419

Duggan, I. C., van Overdijk, C. D. A., Bailey, S. A., Jenkins, H. L. and MacIsaac, H. J. (2005) 'Invertebrates associated with residual ballast water and sediments of cargo-carrying ships entering the Great Lakes', *Canadian Journal of Fisheries and Aquatic Sciences*, vol 62, pp2463–2474

Faber D. J. and Jermolajev, E. G. (1966) 'A new copepod genus in the plankton of the Great Lakes', *Limnology and Oceanography*, vol 11, pp301–303

Farrell, J. M., Holeck, K. T., Mills, E. L., Hoffman, C. E. and Patil, V. J. (2010) 'Recent ecological trends in lower trophic levels of the international section of the St Lawrence River: A comparison of the 1970s to the 2000s', *Hydrobiologia*, vol 647, pp21–33

Ferrari, F. D. and Orsi, J. (1984) '*Oithona davisae*, new species, and *Limnoithona sinensis* (Burckhardt, 1912) (Copepoda: Oithonidae) from the Sacramento-San Joaquin estuary, California', *Journal of Crustacean Biology*, vol 4, pp106–126

Fleminger, A. and Kramer, S. H. (1988) 'Recent introduction of an Asian estuarine copepod, *Pseudodiaptomus marinus* (Copepoda: Calanoida), into southern California embayments', *Marine Biology*, vol 98, pp535–541

Gifford, S. M., Rollwagen-Bollens, G. and Bollens, S. M. (2007) 'Mesozooplankton omnivory in the upper San Francisco estuary', *Marine Ecology Progress Series*, vol 358, pp33–46

Gould, A. L. and Kimmerer, W. J. (2010) 'Development, growth, and reproduction of the cyclopoid copepod *Limnoithona tetraspina* in the upper San Francisco Estuary', *Marine Ecology Progress Series*, vol 412, pp163–177

Gutiérrez-Aguirre, M. and Suárez-Morales, E. (2000) 'The Eurasian *Thermocyclops crassus* (Fischer, 1853) (Copepoda, Cyclopoida) found in southeastern Mexico', *Crustaceana*, vol 73, no 6, pp705–713

Havel, J. E., Lee, C. E. and Vander Zanden, M. J. (2005) 'Do reservoirs facilitate passive invasions into landscapes?', *Bioscience*, vol 55, pp518–525

Hennessy, A. (2010) 'Zooplankon monitoring 2009', *IEP Newsletter*, vol 23, no 2, pp15–22

Hobbs, J. A., Bennett, W. A. and Burton, J. E. (2006) 'Assessing nursery habitat quality for native smelts (Osmeridae) in the low-salinity zone of the San Francisco estuary', *Journal of Fish Biology*, vol 69, pp907–922

Hoffman, J. C., Bronk, D. A. and Olney, J. E. (2008) 'Organic matter sources supporting lower food web production in the tidal freshwater portion of the York River estuary, Virginia', *Estuaries and Coasts*, vol 31, pp898–911

Hooff, R. C. and Bollens, S. M. (2004) 'Functional response and potential predatory impact of *Tortanus dextrilobatus*, a carnivorous copepod recently introduced to the San Francisco estuary', *Marine Ecology Progress Series*, vol 277, pp167–179

Hribar, L. J. and Reid, J. W. (2008) 'New records of copepods (Crustacea) from the Florida Keys', *Southeastern Naturalist*, vol 7, no 2, pp219–228

Huys, R. and Boxshall, G. (1991) *Copepod Evolution*, The Ray Society, London

Johnson, P. T. J., Olden, J. D. and Vander Zanden, M. J. (2008) 'Dam invaders: Impoundments facilitate biological invasions into freshwaters', *Frontiers in Ecology and the Environment*, vol 6, no 7, pp357–363

Kimmerer, W. (2004) 'Open water processes in the San Francisco estuary: From physical forcing to biological responses', *San Francisco Estuary and Watershed Science*, vol 2, no 1, pp1–142

Kimmerer, W. J. and Orsi, J. J. (1996) 'Causes of long-term declines in zooplankton in the San Francisco Bay estuary since 1987', in J. T. Hollibaugh (ed) *San Francisco Bay: The Ecosystem*, Pacific Division, American Association for the Advancement of Science, San Francisco, pp403–424

Kimmerer, W. J., Gartside, E. and Orsi, J. J. (1994) 'Predation by an introduced clam as the probable cause of substantial declines in zooplankton in San Francisco Bay', *Marine Ecology Progress Series*, vol 113, pp81–93

Lawrence, D. J. and Cordell, J. R. (2010) 'Relative contributions of domestic and foreign sourced ballast water to propagule pressure in Puget Sound, Washington, USA', *Biological Conservation*, vol 143, pp700–709

Lee, C. E. (1999) 'Rapid and repeated invasions of fresh water by the copepod *Eurytemora affinis*', *Evolution*, vol 53, no 5, pp1423–1434

Lee, C. E. (2000) 'Global phylogeography of a cryptic copepod species complex and reproductive isolation between genetically proximate "populations"', *Evolution*, vol 54, pp2014–2027

Lee, C. E. and Bell, M. A. (1999) 'Causes and consequences of recent freshwater invasions by saltwater animals', *Trends in Ecology and Evolution*, vol 14, no 7, pp284–288

Lee, C. E. and Frost, B. W. (2002) 'Morphological stasis in the *Eurytemora affinis* species complex (Copepoda: Temoridae)', *Hydrobiologia*, vol 480, pp111–128

Lee, C. E. and Petersen, C. H. (2003) 'Effects of developmental acclimation on adult salinity tolerance in the freshwater-invading copepod *Eurytemora affinis*', *Physiological and Biochemical Zoology*, vol 76, pp296–301

Lee, C. E., Remfert, J. L. and Gelembiuk, G. W. (2003) 'Evolution of physiological tolerance and performance during freshwater invasions', *Integrative & Comparative Biology*, vol 43, pp439–449

Lee C. E, Remfert J. L. and Chang, Y.-M. (2007) 'Response to selection and evolvability of invasive populations', *Genetica*, vol 129, pp179–192

Mashiko, K. (1951) 'Studies of the fresh-water plankton of central China, I', *The Science Reports of the Kanazawa University*, vol 1, no 1, pp17–31

Meng, L. and Orsi, J. J. (1991) 'Selective predation by larval striped bass on native and introduced copepods', *Transactions of the American Fisheries Society*, vol 120, pp187–192

Nobriga, M. L. (2002) 'Larval delta smelt diet composition and feeding incidence: Environmental and ontogenetic influences', *California Fish and Game*, vol 88, no 4, pp149–164

Orsi, J. J. and Ohtsuka, S. (1999) 'Introduction of the Asian copepods *Acartiella sinensis*, *Tortanus dextrilobatus*, (Copepoda: Calanoida), and *Limnoithona tetraspina* (Copepoda: Cyclopoida) to the San Francisco Estuary, California, USA', *Plankton Biology and Ecology*, vol 46, no 2, pp128–131

Orsi, J. J. and Walter, T. C. (1991) '*Pseudodiaptomus forbesi* and *P. marinus* (Copepoda: Calanoida), the latest copepod immigrants to California's Sacramento-San Joaquin Estuary', in S.-I. Uye, S. Nishida and J.-S. Ho (eds) *Proceedings of the Fourth International Conference on Copepoda, Bulletin of the Plankton Society of Japan, Special Volume*, Hiroshima, pp553–562

Orsi, J. J., Bowman, T. E., Marelli, D. C. and Hutchinson, A. (1983) 'Recent introduction of the planktonic calanoid copepod *Sinocalanus doerri* (Centropagidae) from mainland China to the Sacramento-San Joaquin estuary of California', *Journal of Plankton Research*, vol 5, pp357–375

Rawlings, T. K., Ruiz, R. R. and Colwell, R. R. (2007) 'Association of Vibrio cholera 01 El Tor and 0139 Bengal with the copepods *Acartia tonsa* and *Eurytemora affinis*', *Applied Environmental Microbiology*, vol 73, pp7926–7933

Reid, J. W. (1993) 'New records and redescriptions of American species of *Mesocyclops* and of *Diacyclops bernardi* (Petkovski, 1986) (Copepoda: Cyclopoida)', *Bijdragen tot de Dierkunde*, vol 63, pp173–191

Reid, J. W. (2007) '*Arctodiaptomus dorsalis* (Marsh): A case history of copepod dispersal', *Banisteria*, no 30, pp3–18

Reid, J. W. and Hudson, P. L. (2008) 'Comment on "Rate of species introductions in the Great Lakes via ships' ballast water and sediments"', *Canadian Journal of Fisheries and Aquatic Sciences*, vol 65, pp549–553

Reid, J. W. and Marten, G. G. (1995) 'The cyclopoid copepod (Crustacea) fauna of non-planktonic continental habitats in Louisiana and Mississippi', *Tulane Studies in Zoology and Botany*, vol 30, pp39–45

Reid, J. W. and Pinto-Coelho, R. M. (1994) 'An Afro-Asian continental copepod, *Mesocyclops ogunnus*, found in Brazil; with a new key to the species of *Mesocyclops* in South America and a review of intercontinental introductions of copepods', *Limnologica*, vol 24, pp359–368

Reid, J. W. and Reed, E. B. (1994) 'Firsts records of two neotropical species of *Mesocyclops* (Copepoda) from Yukon Territory: Cases of passive dispersal?', *Arctic*, vol 47, no 1, pp80–87

Reid, J. W., Hamilton IV, R. and Duffield, R. M. (2002) 'First confirmed New World record of *Apocyclops dengizicus* (Lepeshkin), with a key to the species of *Apocyclops* in North America and the Caribbean region (Crustacea: Copepoda)', *Jeffersoniana*, vol 10, pp1–25

Riccardi, N. and Giussani, G. (2007) 'The relevance of life-history traits in the establishment of the invader *Eudiaptomus gracilis* and the extinction of *Eudiaptomus padanus* in Lake Candia (Northern Italy): Evidence for competitive exclusion?', *Aquatic Ecology*, vol 41, pp243–254

Ricciardi, A. and Cohen, J. (2007) 'The invasiveness of an introduced species does not predict its impact', *Biological Invasions*, vol 9, pp309–315

Saunders, J. F. (1993) 'Distribution of *Eurytemora affinis* (Copepoda: Calanoida) in the southern Great Plains, with notes on zoogeography', *Journal of Crustacean Biology*, vol 13, pp564–570

Suárez-Morales, E., Rodrígues-Almaraz, G., Gutiérrez-Aguirre, M. A. and Walsh, E. (2008) 'The coastal-estuarine copepod, *Eurytemora affinis* (Poppe) (Calanoida, Temoridae) from arid inland areas of Mexico: An expected occurrence?', *Crustaceana*, vol 81, no 6, pp679–694

Sytsma, M. D., Cordell, J. R., Chapman, J. W. and Draheim, R. C. (2004) 'Lower Columbia River aquatic nonindigenous species survey 2001–2004', *Final Technical Report Prepared for the United States Coast Guard and the United States Fish and Wildlife Service*, www.clr.pdx.edu/projects, accessed 10 September, 2010

USGS (United States Geological Survey) (2009) 'Nonindigenous aquatic species', http://nas.er.usgs.gov, accessed 9 September, 2010

Valéry, L., Fritz, H., Lefeuvre, J.-C. and Simberloff, D. (2008) 'In search of a real definition of the biological invasion phenomenon itself', *Biological Invasions*, vol 10, pp1345–1351

Wang, T., Wang, X. and Han, B. (2009) 'Population dynamics of *Limnoithona sinensis* and its effects on crustaceans in a pumped storage reservoir, south China', *Hupo Kexue*, vol 21, no 1, pp110–116

Wallack, R. L. (2002) 'Columbia-Snake River barge system', *American Journal of Transportation*, no 164, pp1–2

Walter, T. C. (1989) 'Review of the New World species of *Pseudodiaptomus* (Copepoda: Calanoida), with a key to the species', *Bulletin of Marine Science*, vol 43, no 3, pp590–628

Winkler, G., Dodson, J. J. and Lee, C. E. (2008) 'Heterogeneity within the native range: Population genetic analyses of sympatric invasive and noninvasive clades of the freshwater invading copepod *Eurytemora affinis*', *Molecular Ecology*, vol 17, pp415–430

Yoon, H. J. and Chang, C. Y. (2008) 'Two brackish cyclopoid copepods from southern coast of Korea', *Korean Journal of Systematic Zoology*, vol 24, no 3, pp241–250

15

Corbicula fluminea Müller (Asian clam)

Martina I. Ilarri and Ronaldo Sousa

Introduction

Bivalves are one of the most invasive faunal groups in aquatic ecosystems and the modifications they can make to invaded areas are well recognized, such as high filtration rates changing phytoplankton, zooplankton and water clarity; high production of faeces and pseudofaeces altering biogeochemical cycles; and addition of shells changing the physical properties of bottom sediments (Strayer, 1999; Ruesink et al, 2005). Several invasive bivalves (e.g. *Crassostrea gigas, Dreissena polymorpha, Mytilus* spp., *Limnoperna fortunei, Potamocorbula amurensis*) can also attain very high densities and biomass (Sousa et al, 2009).

Corbicula fluminea (Asian clam) (Figure 15.1) is one of the most pervasive invasive species in freshwater ecosystems, mainly due to nuisance characteristics responsible for high ecological and economic impacts and great capacity for dispersion (Sousa et al, 2008a). It is consequently regarded as one of the 100 worst invasive species in Europe (DAISIE, 2009).

A Review of the *C. fluminea* Literature

Using a bibliometic survey of the Scopus database, we assessed the number of studies publish from 1972 to 2009 that have investigated *C. fluminea*. In this survey, each published work was classified according to its year of publication, country and area of origin (Asia, Central America, Europe, North America and South America) and theme (ecology and distribution, ecotoxicology, genetics and evolution, management, molecular biology, physiology and others that include for example parasitology).

A total of 349 publications focusing on *C. fluminea* were analysed. Through time it was possible to observe a clear increase in the number of publications, with the highest scientific production in 2005 (n = 32) (Figure 15.2a). Since the end of the 1990s we observed a rapid increase in the number of publications for Asia, Europe and South America, and a slight decrease for North America (Figure 15.2b).

Overall, North America and Europe had published 266 studies by 2009, accounting for about 77 per cent of all publications and showing a clear dominance of the introduced over the native *C. fluminea* range (Figure 15.3). The US is the leading country with a total of 156 publications, corresponding to 44.7 per cent, followed by France with 69 (19.8 per cent) and China with 35 (10.0 per cent). The higher scientific production by the US may be related to *Corbicula's* earlier colonization. The current decline in the number of published studies may be related to the recent introduction of the zebra mussel (*D. polymorpha*) leading to a change in the target species of investigation by many of the researchers interested in invasive bivalves. Conversely, after the establishment in several European countries, the number of studies started to rapidly increase, and hence, since the late 1990s Europe has become the most productive continent. This increasing interest in Europe may be related to the continued spread of this species to northern and eastern areas (see below).

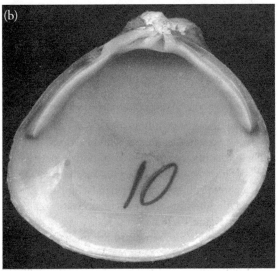

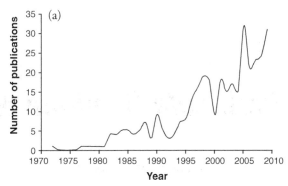

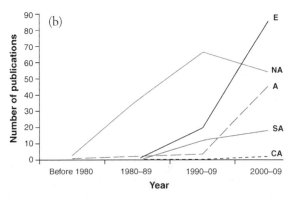

Note: A = Asia; CA = Central America; E = Europe; NA = North America; SA = South America.

Figure 15.2 *Total number of* C. fluminea *publications (a) per year and (b) per decade for each selected area*

Note: The shell is glossy and heavy with well-marked spaced striae (a). The cardinal teeth are tri-radiate with heavy protuberances and the lateral teeth are long and serrated (b).

Source: Ronaldo Sousa

Figure 15.1 C. fluminea *showing: (a) the external and (b) internal view of shells*

Among the themes, the most addressed was ecotoxicology with 178 papers (50.1 per cent), followed by ecology and distribution with 85 (23.9 per cent), physiology with 33 (9.3 per cent), while management was the least investigated with just 9 (2.5 per cent) (Figure 15.4). The US and China were the countries where management studies were conducted and most were published in the 1980s. The most recent management publication was from the mid-1990s. The small number of management-focused studies (9) may be indicative of a lack of interest in this topic, which is surprising given the generally recognized ecological and economic impacts caused by invasive populations of *C. fluminea*. In contrast, the great number of ecotoxicological studies may be indicative of the potential for using this species as a sentinel species in aquatic ecosystems.

Distribution

The native range of the *Corbicula* genus was confined, at the beginning of the last century, to Asia, Africa and Australia (McMahon, 1999). The precise distribution of *C. fluminea* in the native range is still controversial mainly because the accurate filiation of this species is impaired by taxonomical problems related to the high

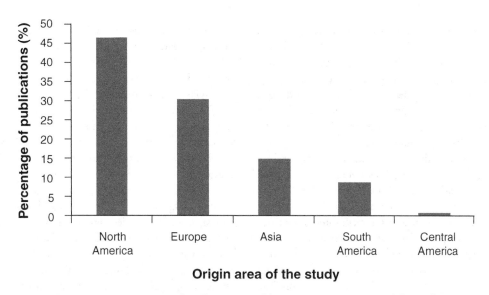

Figure 15.3 *Percentage of* C. fluminea *publications per origin area of the study*

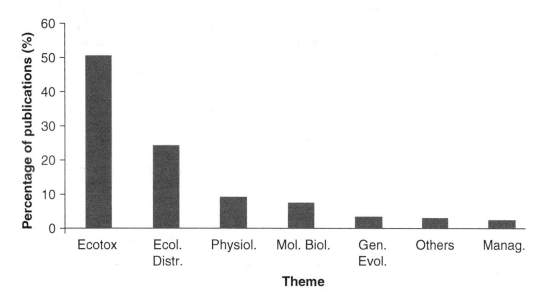

Note: Ecotox = ecotoxicology; Ecol. Distr. = ecology and distribution; Physiol. = physiology; Mol. Biol. = molecular biology; Gen. Evol. = genetics and evolution; Others; and Manag. = management.

Figure 15.4 *Percentage of* C. fluminea *publications per selected theme*

morphometric shell variability and by the relative paucity of genetic studies concluded so far (Renard et al, 2000; Pfenninger et al, 2002; Park and Kim, 2003; Sousa et al, 2007a).

According to McMahon (1982) the native range of *C. fluminea* includes southeast China, Korea and

south-eastern Russia. Since the beginning of the last century this species has dispersed worldwide, attaining a considerable geographic distribution (McMahon, 1999; Sousa et al, 2008a). Although somewhat controversial, since live specimens were never recorded after its original discovery, the first documented occurrence of this genus outside its original distribution was at Vancouver Island, British Columbia, in 1924 (Counts, 1986). Living specimens were only found in the 1930s, possibly introduced by Chinese immigrants as a food resource (Counts, 1986). In a few decades its distribution extended to the Atlantic coast, and is now widespread in 35 continental states of the US as well as in Hawaii (Mackie and Claudi, 2010). In South America, this genus was first recognized around the 1970s (Ituarte, 1994) and in Europe at the end of the 1970s (Mouthon, 1981). At present, the invasive range of *C. fluminea* extends from Patagonia in South America to Canada in North America, encompassing a great diversity of freshwater ecosystems. In Europe, the species has also attained a considerable distribution, being present in almost all European countries with the exception of northern countries where water temperatures are probably too low for winter survival. Most striking is that the species continues to expand its distribution to eastern European countries (Czech Republic, Hungary, Bulgaria, Serbia and Romania were recently invaded), mainly along the River Danube, and also northern countries such as Ireland, where this species was first detected during 2010 (Joe Caffrey, pers. comm.).

The introduction and subsequent dispersion of *C. fluminea* in aquatic ecosystems is clearly related to human activities (e.g. ballast water transport, food resource, utilization of specimens as fish bait, aquarium releases, transport of specimens as a tourist curiosity, juvenile byssal attachment to boat hulls and transport of sediments and/or water between different aquatic ecosystems with possible transfer of small juveniles or adults) (Darrigran, 2002; McMahon, 2002). In addition, *C. fluminea* has extensive capacities for natural dispersal since the juveniles can be transported by fluvial or tidal currents and also by birds or mammals (McMahon, 1999, 2002). This kind of natural transport may have a fundamental importance to the magnitude of secondary introductions and is possibly related to the spread of *C. fluminea* to remote areas with low or almost non-existent human pressure.

Information about dispersal mechanisms in addition to knowledge about the physiological and ecological tolerances of this species may be essential for predicting the future distribution and areas with higher risk of spread, with increased potential for the design of proper management plans. Fundamental abiotic variables for the growth, reproduction and survival of *C. fluminea* are summarized in Table 15.1, and are based on the review of Mackie and Claudi (2010). Obviously, this information is just a proxy of the basic ecological variables responsible for the success (or not) of *C. fluminea* in invaded areas since, for example, biotic interactions (e.g. predation, competition, parasites and diseases) may also have fundamental importance.

Lifecycle

Despite controversy about the reproductive mode, *C. fluminea* is generally described as a hermaphroditic species (Sousa et al, 2008a). Fertilization occurs inside

Table 15.1 *Fundamental abiotic variables determining the degree of invasion by* C. fluminea

Abiotic variables	No potential for adult survival	Little potential for larval development	Moderate potential for nuisance invasion	High potential for massive invasion
Calcium (mg Ca/L)	<1	1 to 2	2 to 5	>5
pH	<5	5–6 to >9.5	6 to 7	7 to 9
Hardness (mg CaCO$_3$/L)	<3	3 to 7	7 to 17	>17
Dissolved oxygen (mg/L)	<0.5	0.5 to 2	2 to 6	>6
Chrophyll a (µg/L)	<5 and >25	5 to 10	10 to 20	20 to 25
Secchi depth (m)	<0.1 and >8	0.1–0.3 to 6–8	0.3–0.5 to 3–6	0.5–3
Temperature (°C)	<2 and >36	2–14 to 30–36	15–18 to 25–30	18 to 25
Conductivity (µS/cm)	>12,600	11,000 to 12,600	8100 to 11,000	<8100
Total dissolved solids (mg/L)	>8400	7400 to 8400	5400 to 7400	<5400
Salinity	>8 *	7 to 8	5 to 7	<5

Note: * adults can support much higher salinities for small periods. For example, *C. fluminea* in the Minho estuary (Portugal) during high tides in the summer is subjected to salinities of 20.

Source: Based on Mackie and Claudi (2010)

the paleal cavity and the incubation of young occurs in the inner demibranches. The larvae pass through trocophore, veliger and pediveliger stages, being released as a D-shaped form with straight hinged shells (Mackie and Claudi, 2010). Juveniles at the time of release have small dimensions (around 250μm), being completely formed with a well-developed shell, adductor muscles, foot, statocysts, gills and digestive system (McMahon, 2002). After release to the water column, juveniles anchor to sediments or hard surfaces due to the presence of a mucilaginous byssal thread following a very brief period (a maximum of four days) in the plankton (Mackie and Claudi, 2010). Juveniles can also be resuspended by turbulent flows and dispersed for long distances, principally in the downstream direction (McMahon, 1999). The lifespan of this species is extremely variable, ranging from one to five years. The maturation period occurs within the first three to six months when the shell length reaches 6 to 10mm, and the number of annual reproductive periods can be highly variable (McMahon, 1999). Although the majority of studies advise that this species reproduces twice a year (e.g. once in spring/early summer corresponding to an increase in temperature, and again in late summer/autumn corresponding to a decrease in temperature), there have been some studies reporting just one reproductive period, while in others three were found, with differences among years even at the same site. This variability in the number of reproductive events may be related to water temperature and/or the quantity of food (Cataldo and Boltovskoy, 1999; Rajagopal et al, 2000; Mouthon, 2001).

Corbicula fluminea has a high fecundity (25,000 to 75,000 veligers produced during an individual's lifetime; McMahon, 2002), but has a low juvenile survivorship and a high mortality rate throughout adult life. The species grows rapidly, in part due to its high filtration and assimilation rates as much of its energy is allocated to growth and reproduction, typical for a species with an opportunistic lifecycle (McMahon, 2002). It has the highest net production efficiencies recorded for any freshwater bivalve, reflected by short turnover times (73–126 days) (McMahon, 2002; Sousa et al, 2008b).

Ecological Impacts

Usually, *C. fluminea* introductions contribute to abiotic changes that also influence biota, such as submerged vegetation, phytoplankton, zooplankton, macro-invertebrates and species at higher trophic levels (Phelps, 1994; Strayer, 1999; Sousa et al, 2008c, 2008d). In synthesis, four main processes can be emphasized for *C. fluminea*: (1) changes resulting from ecosystem engineering activities; (2) changes in biogeochemical cycles; (3) changes in trophic relationships; and (4) changes in biotic interactions, mainly with other bivalves (Figure 15.5).

Bivalves can be considered ecosystem engineers (i.e. organisms that can physically modify the environment) (Jones et al, 1994) and their importance has been recognized in shallow water habitats (Gutiérrez et al, 2003; Sousa et al, 2009). *Corbicula fluminea* possesses key attributes (e.g. shells, behaviour, size, abundance and distribution) that may affect the potential for engineering. The presence of live and dead *C. fluminea* shells may alter the substrate composition forming a more complex, sheltered and heterogeneous habitat that is attractive for several species (e.g. algae, freshwater sponges, crustaceans, insects, gastropods) (Sousa et al, 2008a). Substrate-based *C. fluminea* shells also contribute to reducing predation by higher trophic levels, can reduce physical and/or physiological stress, and can influence the transport of particles and solutes in the benthic environment, thus potentially affecting other species (Sousa et al, 2009). The species also has the capacity to bioturbate the top layer of the sediments, mainly by pedal movements, leading to significant changes in abiotic conditions (e.g. oxygen, redox potential, amount of organic matter, particle size) that could also affect other organisms. Moreover, its great filtration rate results in the removal of a wide range of suspended particles, with important repercussions for water clarity and light penetration that, among other alterations, may be advantageous for submerged plants (Phelps, 1994).

This large filtration capacity acts also as an important controlling element on phytoplankton and zooplankton, contributing to major changes in the flow of organic matter (Cohen et al, 1984; Phelps, 1994). Another consequence of this high filter feeding activity may be the increased deposition of ingested particles as faeces and pseudofaeces, which can lead to the addition of organic matter to bottom sediments, shifting primary production from planktonic to benthic communities (Vaughn and Hakenkamp, 2001). This situation can be responsible for significant changes in biogeochemical cycles, as invasive bivalves are well recognized for

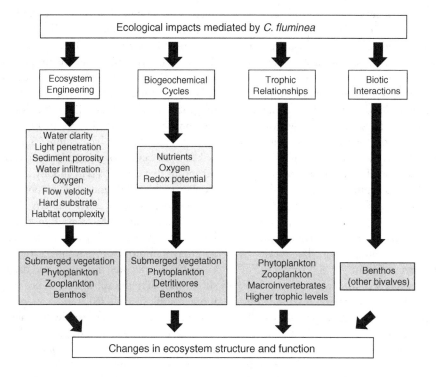

Figure 15.5 *General pathways of ecological impacts mediated by* C. fluminea

excreting large amounts of inorganic nutrients (Dame, 1996). This release of nutrients can stimulate primary production by submerged vegetation (which can also benefit from filtration rates that make the water clearer) and phytoplankton (which can also be negatively affected due to higher grazing pressure) (Phelps, 1994; Dame, 1996). In addition, *C. fluminea* is well recognized for suffering massive mortalities that usually occur when subjected to extreme conditions (severe winters or summers) (Werner and Rothhaupt, 2008; Vohmann et al, 2010; Ilarri et al, 2011). Recognizing that this species can attain great abundance and biomass, these massive die-offs can result in the release of huge amounts of organic material that might change biogeochemical cycles (Cherry et al, 2005; Cooper et al, 2005). These occurrences can abruptly increase nutrient concentrations causing massive mortalities in all the benthic fauna, and also affect the water quality (Sousa et al, 2007b, 2008b, 2008e). Typically, *C. fluminea* populations have a rapid recovery, reaching previous abundances and distributions while native species are still recovering (Ilarri et al, 2011). This phenomenon acts in favour of *C. fluminea* and may determine and/or accelerate the extirpation of some native species (Sousa et al, 2008b).

Corbicula fluminea can dominate the benthic biomass (Sousa et al, 2008c, 2008d) and thereby contribute substantially to trophic transfer through the benthic-pelagic food web, with significant impacts on lower and higher trophic levels. As stated earlier, phyto and zooplankton can be highly affected and the same is potentially true for detritivores and scavengers that can be highly subsidized, for example, by the massive die-offs. In addition, species at higher trophic levels are expected to consume *C. fluminea*, so it may act as an important food resource (Cantanhêde et al, 2008). Fish, birds and mammals are potential consumers, although this possibility has not been fully explored in ecological studies performed so far.

The rapid spread of *C. fluminea* has also raised much concern due to the possible effects on native bivalves from the Margaritiferidae, Unionidae and Sphaeriidae families (Strayer, 1999; McMahon, 2002; Sousa et al, 2008a, 2008c, 2008e). The arrival of

C. fluminea may affect native bivalves in several ways, such as:

- burrowing and bioturbation resulting in displacement and/or reduction of available habitats for juvenile unionoids and sphaeriids (Vaughn and Hakenkamp, 2001);
- suspension and deposit feeding that negatively impacts the recruitment of unionoid juveniles (Yeager et al, 1994);
- its larger filtration rates that make it a superior competitor for food resources compared to native bivalve species (Sousa et al, 2008b); and
- dense populations of *C. fluminea* being associated with notable ingestion of a number of unionoids' sperm, glochidia and newly metamorphosed juveniles (Strayer, 1999).

Moreover, *C. fluminea* can also be a vector for the introduction of new parasites and diseases (Darrigran, 2002). The majority of these effects remain speculative and further research is needed to clarify these interactions.

Economic Impacts

Corbicula fluminea frequently becomes the dominant species within an invaded habitat and when that habitat is responsible for important services, such as the supply of water for drinking and irrigation, provision of cooling water for a power plant, supporting aquaculture, commercial and sport fisheries or enabling recreational activities, it is not surprising that there can be substantial negative economic impacts. Most well-documented examples of economic impacts with invasive bivalves come from the industrial biofouling caused by the byssus-forming bivalves (e.g. *Dreissena* spp., *Mytilopsis* spp., *Mytilus* spp. and *L. fortunei*) (see Mackie and Claudi, 2010). *Corbicula fluminea* does not form physical attachments to solid structures, but the accumulation of shells can represent an important biofouling problem. Indeed, shells of *C. fluminea* have resulted in the occlusion of small diameter pipelines (<4cm), heat exchanger/condenser tubing, fire protection lines and course strainers in power plants, and irrigation pipes of small diameter (McMahon, 1999). In addition, large indirect impacts related to *C. fluminea* have resulted from the increased rates of sedimentation. These higher rates of sedimentation, mainly in managed systems such as irrigation canals or

waterways, require an increased frequency of dredging, which may have profound environmental implications (e.g. changes in the sediments, resuspension of nutrients and heavy metals, increase in turbidity and mortality of several organisms). Also important is the reduction or possible elimination of the recreational and aesthetic value of beaches and shorelines due to huge accumulations of live and empty shells. This situation may have negative impacts on tourism. The accumulation of empty shells can also result in a reduction in the efficiency of fishing nets, with consequent impacts on fisheries. Another possible impact of concern is the fouling of drinking water supplies since the accumulation of shells in water treatment plant can lead to incomplete filtration and consequent deterioration of water quality.

Management

Managing the invasion of *C. fluminea* requires the identification of major pathways or vectors, recognition of the species traits that confer invasive success and use of appropriate methods or strategies to prevent, eradicate or control this species (Clout and Williams, 2009). As with any invasive species, three main possibilities may exist to manage the impacts generated by the introduction of *C. fluminea*: (1) prevention; (2) eradication; and (3) containment and control.

Prevention

Prevention of introduction will always be the most effective tool (environmentally and economically) to combat the impacts of *C. fluminea*. Prevention can only be achieved through rigorous risk assessment, quarantine regulations and biosecurity activities (Maynard and Nowell, 2009). In the particular case of *C. fluminea*, information about possible vectors, mainly the vectors linked with human activities (being already well established), will have a fundamental role in the elaboration of a risk assessment.

Monitoring to detect the first appearance of young and adults, and their growth and abundance, will also help in determining the most effective management options. In the specific case of aquatic ecosystems, several programmes have already been implemented with the aim of assessing water quality. For example, the European Water Framework Directive that requires a great sampling effort could easily give important ecological information about aquatic invasive species such as *C. fluminea*. Such programmes, being already implemented, could be a fundamental help in the detection of invasive species or at least giving important details about their spatial distribution, so that managers can quickly and effectively combat this species. Also important, at this level, is to identify barriers to dispersal and habitats where *C. fluminea* cannot persist or cause harm.

It should be borne in mind that in general, if *C. fluminea* can be prevented from establishing in an area, the resources used in prevention are usually significantly lower than those needed for eradication, containment and control or even the consequences of doing nothing.

Eradication

The window of opportunity for a successful eradication may be quite narrow before *C. fluminea* becomes fully naturalized. Indeed, the total eradication of a species such as *C. fluminea* in aquatic ecosystems may be extremely difficult and in the most part almost impossible. Only in areas where the species has a very limited spatial distribution and where its abundance is low, can total eradication succeed. Therefore, eradication programmes may only be implemented with a thorough knowledge of the invaded area and when circumstances are appropriate, since the costs of promoting eradication when it is not possible can be huge and will lead to the abandonment of the problem (Parkes and Panetta, 2009). For example, a restricted area may be a viable option for manual eradication through a team of professionals and/or volunteers headed by a professional with knowledge about basic aspects of *C. fluminea* ecology and the study area. After this stage future monitoring will be necessary to detect possible individuals who were not eradicated in the first campaign. Another option for total eradication may follow the procedure described for the removal of *Mytilopsis* sp. in Australia (Willan et al, 2000). This drastic example relied on intensive chemical treatment with chlorine (sodium hypochlorite) and copper sulphate. This type of treatment can only be carried out in degraded systems within a small area (as the release of a quantity of chemicals in a system not degraded is neither environmentally sound nor publicly acceptable) and where the potential establishment of *C. fluminea* is likely to have great ecological and economic impacts.

There are significant economic and environmental benefits of eradication programmes, as opposed to perpetual control measures. However, the economic investment in eradication programmes can only be supported by a few countries that have sufficient financial and logistical capacity to detect these invasions at an early stage. However, given the widespread distribution of *C. fluminea*, it will always be possible that after a successful eradication programme the species could be introduced again from adjacent areas and initiate the problem once more.

Containment and control

When eradication is not possible, containment is necessary to prevent the species expanding and saturating all potential habitats. Full containment aims to limit the introduced range of a species to a fraction of the potential range (Grice, 2009); looking at the present distribution of *C. fluminea* it may be concluded that this battle has already been lost. Partial containment aims to slow the rate of spread of an invasive species (Grice, 2009), and in this case some efforts can be made to decrease human activities responsible for the spread of *C. fluminea*.

A control strategy is more appropriate for an advanced stage of invasion, where *C. fluminea* has reached a large abundance and has extended its spatial distribution. Control, as opposed to containment, does not have the aim of restricting the range of the invasive species but is an attempt to reduce the impacts through a reduction in abundance (and so the impact) to levels below those that would otherwise be attained (Grice, 2009). Hand-picking of adults is always a possible way of control that may have some success in sites recently invaded, where the species is restricted to a small area and has low densities. The use of chemicals, especially molluscicides, is commonly used against invasive snails and may be an option to control *C. fluminea*. However, since the most important impacts generated by Asian clams are in natural systems, chlorination or the use of other chemicals that impact other organisms may be highly controversial (Mackie and Claudi, 2010). Furthermore, the possible commercialization of the species in the invaded areas could also be a way to control its abundance.

Future Challenges

Although this species has been subjected to a great number of studies, there are numerous interesting issues that deserve future research, particularly addressing the management of impacts. There are only a few studies conducted in the native range (at least published in international journals) and thus this information is essential for the planning of management measures in the invaded systems. Future studies should also resolve some uncertainties in relation to *Corbicula* taxonomy, as well as the origin, sources and pathways of dispersion of the species, and the forms present in European and American continents. Therefore, it is fundamental to complete genetic and phylogenetic studies, and good cooperation between scientists and managers from the native and invaded range will be fruitful. Another field that should be developed is the use of this species to clean up polluted or highly modified environments; *C. fluminea* can be used in the biodeposition of suspended material, for clearing the water and also to remove pollutants from the water column. While the use of *C. fluminea* may have appeal for bioremediation programmes, the obvious risks of promoting the spread of the species means that such techniques have to be carefully evaluated. In the same vein, although *C. fluminea* has been extensively used in ecotoxicology, future studies may take this further, using its ubiquitous distribution for worldwide ecotoxicological comparisons in freshwater ecosystems. In addition, while many ecological effects have been documented following the arrival of *C. fluminea*, several aspects remain speculative and should receive more attention, such as the possible negative impact of the species on native bivalves, the changes resulting from massive die-offs and possible changes in food webs. Finally, educational programmes dealing with *C. fluminea* (and other invasive species) should be prioritized, since it could enhance the engagement of people in future management actions. Public participation is fundamental from ethical, legal and practical standpoints and as humans are the major vector of introduction and dispersal of *C. fluminea*, they necessarily must be the most powerful agents of future checks on the spread and impact of this species.

References

Cantanhêde, G., Hahn, N. S., Gubiani, É. A. and Fugi, R. (2008) 'Invasive molluscs in the diet of *Pterodoras granulosus* (Valenciennes, 1821) (Pisces, Doradidae) in the Upper Paraná River floodplain, Brazil', *Ecology of Freshwater Fish*, vol 17, pp47–53

Cataldo, D. and Boltovskoy, D. (1999) 'Population dynamics of *Corbicula fluminea* (Bivalvia) in the Paraná River delta (Argentina)', *Hydrobiologia*, vol 380, pp153–163

Cherry, D. S., Scheller, J. L., Cooper, N. L. and Bidwell, J. R. (2005) 'Potential effects of Asian clam (*Corbicula fluminea*) die-offs on native freshwater mussels (Unionidae) I: Water-column ammonia levels and ammonia toxicity', *Journal of the North American Benthological Society*, vol 24, pp369–380

Clout, M. N. and Williams, P. A. (2009) *Invasive Species Management: A Handbook of Principles and Techniques*, Oxford University Press, Oxford

Cohen, R. R. H., Dresler, P. V., Phillips, E. J. P. and Cory, R. L. (1984) 'The effect of the Asiatic clam, *Corbicula fluminea*, on phytoplankton on the Potomac River, Maryland', *Limnology and Oceanography*, vol 29, pp170–180

Cooper, N. L., Bidwell, J. R. and Cherry, D. S. (2005) 'Potential effects of Asian clam (*Corbicula fluminea*) die-offs on native freshwater mussels (Unionidae) II: Pore-water ammonia', *Journal of the North American Benthological Society*, vol 24, pp381–394

Counts, C. L. (1986) 'The zoogeography and history of invasion of the United States by *Corbicula fluminea* (Bivalvia: Corbiculidae)', *American Malacological Bulletin*, special edition, vol 2, pp7–39

DAISIE (2009) 'European Invasive Alien Species Gateway', www.europe-aliens.org/speciesTheWorst.do, accessed 20 January 2010

Dame, R. F. (1996) *Ecology of Marine Bivalves: An Ecosystem Approach*, CRS Press, New York

Darrigran, G. (2002) 'Potential impact of filter-feeding invaders on temperate inland freshwater environments', *Biological Invasions*, vol 4, pp145–156

Grice T. (2009) 'Principles of containment and control of invasive species', in M. N. Clout and P. A. Williams (eds) *Invasive Species Management: A Handbook of Principles and Techniques*, Oxford University Press, Oxford, pp61–76

Gutiérrez J. L., Jones, C. G., Strayer, D L. and Iribarne, O. O. (2003) 'Mollusks as ecosystem engineers: The role of shell production in aquatic habitats', *Oikos*, vol 101, pp79–90

Ilarri, M., Antunes, C., Guilhermino, L. and Sousa R. (2011) 'Massive mortality of the Asian clam *Corbicula fluminea* in a highly invaded area', *Biological Invasions*, vol 13, pp277–280

Ituarte, C. F. (1994) '*Corbicula* and *Neocorbicula* (Bivalvia: Corbiculidae) in the Paraná, Uruguay, and Rio de la Plata basins', *Nautilus*, vol 107, pp129–135

Jones, C. G., Lawton, J. H. and Shachak, M. (1994) 'Organisms as ecosystem engineers', *Oikos*, vol 69, pp373–386

Mackie, G. L. and Claudi, R. (2010) *Monitoring and Control of Macrofouling Mollusks in Fresh Water Systems*, CRC Press, Florida

Maynard, G. and Nowell, D. (2009) 'Biosecurity and quarantine for preventing invasive species', in M. N. Clout and P. A. Williams (eds) *Invasive Species Management: A Handbook of Principles and Techniques*, Oxford University Press, Oxford, pp1–18

McMahon, R. F. (1982) 'The occurrence and spread of the introduced Asiatic freshwater clam, *Corbicula fluminea* (Muller), in North America: 1924–1982', *Nautilus*, vol 96, pp134–141

McMahon R. F. (1999) 'Invasive characteristics of the freshwater bivalve, *Corbicula fluminea*', in R. Claudi and J. H. Leach (eds) *Nonindigenous Freshwater Organisms: Vectors, Biology and Impacts*, Lewis Publishers, CRC Press, London, pp315–343

McMahon, R. F. (2002) 'Evolutionary and physiological adaptations of aquatic invasive animals: r selection versus resistance', *Canadian Journal of Fisheries and Aquatic Sciences*, vol 59, pp1235–1244

Mouthon, J. (1981) 'Sur la présence en France et au Portugal de *Corbicula* (Bivalvia, Corbiculidae) originaire d'Asie', *Basteria*, vol 45, pp109–116

Mouthon, J. (2001) 'Life cycle and populations dynamics of the Asian clam *Corbicula fluminea* (Bivalvia: Corbiculidae) in the Saone River at Lyon (France)', *Hydrobiologia*, vol 452, pp109–119

Park, J.-K. and Kim, W. (2003) 'Two *Corbicula* (Corbiculidae: Bivalvia) mitochondrial lineages are widely distributed in Asian freshwater environment', *Molecular Phylogenetics and Evolution*, vol 29, pp529–539

Parkes, J. P. and Panetta, F. D. (2009) 'Eradication of invasive species: Process and emerging issues in the 21st century', in M. N. Clout and P. A. Williams (eds) *Invasive Species Management: A Handbook of Principles and Techniques*, Oxford University Press, Oxford, pp47–60

Pfenninger, M., Reinhardt, F. and Streit, B. (2002) 'Evidence for cryptic hybridization between different evolutionary lineages of the invasive clam genus *Corbicula* (Veneroida, Bivalvia) ', *Journal of Evolutionary Biology*, vol 15, pp818–829

Phelps, H. L. (1994) 'The Asiatic clam (*Corbicula fluminea*) invasion and system-level ecological change in the Potomac River estuary near Washington, D.C.', *Estuaries*, vol 17, pp614–621

Rajagopal, S., van der Velde, G. and bij de Vaate, A. (2000) 'Reproductive biology of the Asiatic clams *Corbicula fluminalis* and *Corbicula fluminea* in the river Rhine', *Archiv für Hydrobiologie*, vol 149, pp403–420

Renard, E., Bachmann, V., Cariou, M. L. and Moreteau, J. C. (2000) 'Morphological and molecular differentiation of invasive freshwater species of the genus *Corbicula* (Bivalvia, Corbiculidea) suggest the presence of three taxa in French rivers', *Molecular Ecology*, vol 9, pp2009–2016

Ruesink, J. L., Lenihan, H. S., Trimble, A. C., Heiman, K. W., Micheli, F., Byers, J. E. and Kay, M. C. (2005) 'Introduction of non-native oysters: Ecosystem effects and restoration implications', *Annual Review Ecology, Evolution, and Systematics*, vol 36, pp643–689

Sousa, R., Freire, R., Rufino, M., Méndez, J., Gaspar, M., Antunes, C. and Guilhermino, L. (2007a) 'Genetic and shell morphological variability of the invasive bivalve *Corbicula fluminea* (Müller, 1774) in two Portuguese estuaries', *Estuarine, Coastal and Shelf Science*, vol 74, pp166–174

Sousa, R., Antunes, C. and Guilhermino, L. (2007b) 'Species composition and monthly variation of the Molluscan fauna in the freshwater subtidal area of the River Minho estuary', *Estuarine, Coastal and Shelf Science*, vol 75, pp90–100

Sousa, R., Antunes, C. and Guilhermino, L. (2008a) 'Ecology of the invasive Asian clam *Corbicula fluminea* (Müller, 1774) in aquatic ecosystems: An overview', *Annales de Limnologie – International Journal of Limnology*, vol 44, pp85–94

Sousa, R., Nogueira, A. J. A., Gaspar, M., Antunes, C. and Guilhermino, L. (2008b) 'Growth and extremely high production of the non-indigenous invasive species *Corbicula fluminea* (Müller, 1774): Possible implications for ecosystem functioning', *Estuarine, Coastal and Shelf Science*, vol 80, pp289–295

Sousa, R., Rufino, M., Gaspar, M., Antunes, C. and Guilhermino, L. (2008c) 'Abiotic impacts on spatial and temporal distribution of *Corbicula fluminea* (Müller, 1774) in the River Minho Estuary, Portugal', *Aquatic Conservation: Marine and Freshwater Ecosystems*, vol 18, pp98–110

Sousa, R., Dias, S., Freitas, V. and Antunes, C. (2008d) 'Subtidal macrozoobenthic assemblages along the River Minho estuarine gradient (north-west Iberian Peninsula)', *Aquatic Conservation: Marine and Freshwater Ecosystems*, vol 18, pp1063–1077

Sousa, R., Dias, S., Guilhermino, L. and Antunes, C. (2008e) 'Minho River tidal freshwater wetlands: Threats to faunal biodiversity', *Aquatic Biology*, vol 3, pp237–250

Sousa, R., Gutiérrez, J. L. and Aldridge, D. C. (2009) 'Non-indigenous invasive bivalves as ecosystem engineers', *Biological Invasions*, vol 11, pp2367–2385

Strayer, D. L. (1999) 'Effects of alien species on freshwater mollusks in North America', *Journal of the North American Benthological Society*, vol 18, pp74–98

Vaughn, C. C. and Hakenkamp, C. C. (2001) 'The functional role of burrowing bivalves in freshwater ecosystems', *Freshwater Biology*, vol 46, pp1431–1446

Vohmann, A., Borcherding, J., Kureck, A., Vaate, A. B. D., Arndt, H. and Weitere, M. (2010) 'Strong body mass decrease of the invasive clam *Corbicula fluminea* during summer', *Biological Invasions*, vol 12, pp53–64

Werner, S. and Rothhaupt, K. O. (2008) 'Mass mortality of the invasive bivalve *Corbicula fluminea* induced by a severe low-water event and associated low water temperatures', *Hydrobiologia*, vol 613, pp143–150

Willan, R. C., Russell, B. C., Murfet, N. B., Moore, K. L., McEnnulty, F. R., Horner, S. K., Hewitt, C. L., Dally, G. M., Campbell, M. L. and Bourke, S. T. (2000) 'Outbreak of *Mytilopsis sallei* (Recluz, 1849)(Bivalvia: Dreissenidae) in Australia', *Molluscan Research*, vol 20, pp25–30

Yeager, M. M., Cherry, D. S. and Neves, R. J. (1994) 'Feeding and burrowing behaviours of juvenile rainbow mussels *Villoa iris* (Bivalvia: Unionidae)', *Journal of the North American Benthological Society*, vol 13, pp217–222

16

Eriocheir sinensis H. Milne-Edwards (Chinese mitten crab)

Matt G. Bentley

Introduction

Eriocheir sinensis (Chinese mitten crab) is a large crab with a distinctive square shaped carapace. It is one of only a few mitten crab species (Varunidae), all of which are native to the Far East. Mitten crabs were formerly assigned to the family Grapsidae (see Clark, 2006; Ng et al, 2008 for a discussion of the taxonomy) and much of the recent literature still refers to them as grapsid crabs. In addition to its characteristic carapace shape, *E. sinensis* (Figure 16.1) possesses setal mats covering the chelae, from which the crab gets its name. These 'mittens' are present in both sexes but form a more complete covering in males (females lack setae on part of the underside of the chela). These setal coverings make this crab easily identifiable in the field.

Of the mitten crab species, *E. sinensis* is the only species that has spread globally over the past century. Its present range extends to both coasts and river systems of North America, much of Europe from the Iberian Peninsula in the south to Scandinavia in the north, and eastwards into the Baltic Sea states and as far as Iran. It has not yet become established in the southern hemisphere. In addition to its invasive spread, the Chinese mitten crab is a commercially important species (Hymanson et al, 1999). Over the past two decades it has become an extremely important aquaculture species with a global production of 500,000 metric tonnes and a market value estimated at $2

billion in 2005 (Yang and Chang, 2005; Li et al, 2010). *Eriocheir sinensis* then, is important not only because of its invasive potential, the destructive nature of its burrowing habit and damage to fisheries but also because both wild caught and farmed crabs command a high price in the market place. This leads inevitably to conflict of interests among stakeholders.

Geographical Distribution

Native range

Eriocheir sinensis is one of only a small number of species belonging to the genus *Eriocheir* and closely related genera *Neoeriocheir* and *Platyeriocheir* that inhabit river systems of southeast Asia and Japan. The taxonomy is currently under revision and may lead to these species being assigned to fewer genera based on morphological characters (Tang et al, 2009) or molecular evidence (Tang et al, 2003). The native range of the Chinese mitten crab ranges from about 20°N in Hong Kong, through eastern China, to about 40°N in the Korean Peninsula (Hymanson et al, 1999). Its long distance migrations upstream from the estuaries mean that it is found in excess of 1400km from the coast (Dittel and Epifanio, 2009), in rivers and adjacent lakes. It is also stocked regularly into lakes, ponds and rice paddies for aquaculture purposes.

Note: Scale bar = 2cm.

Source: Courtesy of Alison Bentley

Figure 16.1 *The Chinese mitten crab* Eriocheir sinensis

European introduction

Eriocheir sinensis made its first appearance in Europe early in the 20th century, almost certainly through discharge of ballast water carrying larvae into European ports (Gollasch et al, 2002) after the adoption of seawater rather than solid ballast in ships. Introduction

into continental Europe was first reported in 1912 in the River Weser system, Germany (Peters, 1933) and into the River Thames, UK, in the 1930s (Harold, 1935). During the 20th century, further reports of populations throughout continental Europe were made (see Dittel and Epifanio, 2009), some of which arose because of coastal or inland spread and others through separate introductions (Herborg et al, 2003, 2005). The history of these reported introductions and spread is listed in Table 16.1. Herborg et al (2007) have predicted further spread in Europe.

European spread including the UK

Once mitten crabs had gained a foothold in Germany, The Netherlands, Belgium and France their spread was facilitated by the extensive network of canals that link the large river systems feeding the North Sea, English Channel and the Atlantic coast of France (Herborg et al, 2003). By the mid-20th century the crabs had reached the Baltic Sea and had been reported from Finland and Russia (Peters, 1938). The extension of the mitten crab's range in the UK was slow after the first report in the River Thames in the 1930s (Harold, 1935) and although it was reported again in the 1970s (Ingle and Andrews, 1976), it was not found in numbers until the 1990s (Clark and Rainbow, 1997). After this, the crab

Table 16.1 *History of Chinese mitten crab invasion of Europe during the 20th century*

Country	Location	Year	Source
Germany	River Aller (Weser)	1912	Peters (1933)
Germany	River Elbe	1914	Peters (1933)
Denmark	Baltic Sea	1927	Rasmussen (1987 cited in Herborg et al 2003)
Poland	Baltic coast	1928	Gollasch (2006)
Netherlands	Meuse/Maas River	1931	Kamps (1937 cited in Herborg et al 2003, 2007)
Russia	Vyborg	1933	Peters (1938)
Belgium		1933	Leloup (1937)
Lithuania	Curonian Lagoon and coast	1934	Bacevièius and Gasiûnaitè (2008)
Sweden	Gaevle	1934	Peters (1938)
UK	River Thames	1935	Harold (1935)
Northern France	Le Havre	1954	Hoestlandt (1959); Vincent (1996)
Western France	St Malo	1954	André (1954)
Southern France	Gironde	1954	André (1954)
UK	River Thames (2nd introduction)	1973	Ingle and Andrews (1976)
UK	River Humber	1976	Clark (1984)
Norway	Oslo	1977	Christiansen (1977)
Portugal	Tagus estuary	1994	Cabral and Costa (1999)

population spread rapidly through the River Thames catchment including into the Essex rivers that enter the north side of the Thames estuary (Clark et al, 1998). The Chinese mitten crab then became established in the Humber during the 1970s and the 1990s (Clark, 1984; Clark and Rainbow, 1996).

North America

To date the only established population in North America is that in the San Francisco Bay area of California. There exists some debate as to whether its introduction might have been a result of a deliberate attempt to establish a fishery or whether it followed the more normal route, following ballast water discharge of larvae into the Bay (Cohen and Carlton, 1997). Whatever the route of introduction, however, from the early 1990s the San Francisco Bay population spread rapidly during the next decade (Cohen and Carlton, 1995, 1997; Rudnick et al, 2003). Apart from the San Francisco Bay population, *E. sinensis* in the US and Canada has been recorded from only a few localities (Table 16.2). While there is no evidence that the other areas where mitten crabs have been found have yet established breeding populations, the environmental conditions and the estuaries would appear suitable for breeding to take place (Ruiz et al, 2006).

Limitations to further spread

There can be little doubt that the Chinese mitten crab will extend its spread further in Europe and North America. Physiological limitations are likely to restrict further extension of *Eriocheir* eastwards in the Baltic Sea, as the low salinity there will not meet the physiological requirements of the larvae (Dittel and Epifanio, 2009) and will therefore prevent successful reproduction. The population is likely to be maintained, however, by migration of adults from the North Sea–Baltic entrance area (Ojaveer et al, 2007). In the US and Canada, low temperatures, coupled with some of the physical characteristics of the estuaries, are likely to limit further extension northwards on the west coast (Hanson and Sytsma, 2008). The most recently reported extensions to the range of *Eriocheir sinensis* are listed in Table 16.3. While *Eriocheir sinensis* has yet to be reported from the southern hemisphere, there is likelihood that where appropriate estuarine and river systems are present, it could extend its invasive range to cover temperate areas of southern Africa, Australasia and South America. In the northern hemisphere, *Eriocheir* shares a coastal range with the shore/green crab *Carcinus maenas* both in its native range (Europe) and invasive range (North America). *C. maenas* has been introduced and become invasive in Australia, South Africa and South America (also *C. aestuarii* in Japan) (Carlton and Cohen, 2003) and continues to expand its range in these regions. Except for the additional requirement for access to appropriate river systems by *E. sinensis*, where temperature regimes are appropriate there is little reason not to expect mitten crabs to arrive in the southern hemisphere at some point in the future.

Ecology

Habitat requirements and preferences

Once metamorphosed and settled in an estuary, juvenile mitten crabs seek shelter under refuges created by pebbles, boulders, etc. (Gilbey et al, 2008). During the course of the next two years they migrate upstream and make use of various refugia available. They are most noted for their burrowing habit, where the crabs tunnel

Table 16.2 *History of Chinese mitten crab reports in North America*

Location	Year	Source
Laurentian Great Lakes (Lake Erie), US	1965	Nepszy and Leach (1973)
San Francisco Bay, California, US and extensive reports since 1992	1992–	Cohen and Carlton (1995) Rudnick et al (2003)
Mississippi River, US	1987	Horwath (1989)
Columbia River, US	2003	Rudnick et al (2003)
St Lawrence River, Quebec, Canada	2004	de Lafontaine (2005)
Chesapeake Bay, Maryland, US	2005	Ruiz et al (2006)

Table 16.3 *Recently recorded populations or new records of* Eriocheir sinensis

Country	Location	Year	Source
Iran	Caspian Sea	2002	Robbins et al, 2006
Iraq	Basra	2005	Clark et al, 2006
Ireland	Waterford Estuary	2006	Minchin, 2006
Italy	Venetian Lagoon	2005	Mizzan, 2005
Finland	Gulf of Finland	2002–2004	Ojaveer et al, 2007
Sweden	Lake Vänern	2007	Drotz et al, 2010
UK	River Tyne	2002	Herborg et al, 2002
UK	River Great Ouse	2008	Walker et al, 2010
UK	River Torridge	2009	Stu Higgs, Data Archive for Seabed Species – DASSH, pers comm
UK	Rover Conwy	2010	Stu Higgs, DASSH, pers comm

into soft banks of watercourses (Dittel and Epifanio, 2009; Rudnick et al, 2003). The burrows can be up to 1m long and give rise to substantial deposition of silt into the water channel (Rudnick et al, 2005). The increased siltation and bank erosion ultimately can have economic consequences, including increased flooding risk (Dutton and Conroy, 1998; Rudnick et al, 2000). Mitten crabs are able to negotiate many obstacles during their upstream migration although their passage may be blocked by impassable obstacles such as weirs (Panning, 1939) (Figure 16.2), and increased development of such obstacles has led to a population decline in some areas of China (Jin et al,

2001). Mitten crabs are able to traverse damp ground to move between watercourses.

Dietary requirements and food preferences

The Chinese mitten crab is a fairly indiscriminate feeder. During the course of its life it changes from a largely animal diet as an estuarine juvenile to an adult diet of aquatic vegetation, reflecting the relative type of food availability in freshwater vs. the estuarine and marine environments. Mitten crabs have been shown to exhibit preferences for fresh waters with abundant macrophytes (Jin et al, 2001) with plant materials comprising the greatest proportion of the diet of two-year-old crabs (Jin et al, 2003). Nutritional studies on juvenile *E. sinensis* have shown that the crabs have the ability to alter their digestive enzyme profiles to suit different diets and grow best on a high protein diet, which may explain their preference towards an animal-based diet (Lin et al, 2010). In the Oder estuary (northwest Poland) juveniles eat a mixed detritus, animal and vascular plant diet with copepods, chironomids and tubificids comprising the greatest element of the animal diet (Czerniejewski et al, 2010).

Source: Courtesy of Stephan Gollasch

Figure 16.2 *A large number of Chinese mitten crab* Eriocheir sinensis *amassed at an obstacle to upstream migration on the River Elbe, Germany*

Physiological tolerances, including to pollution

One of the key attributes to the success of the Chinese mitten crab as an invasive species is the ability to tolerate a wide range of salinities at various stages of its lifecycle. In common with many crab and lobster species it possesses planktonic larval stages (usually five zoea stages) and a megalopa settlement stage before becoming juveniles. Dittel and Epifanio (2009) have recently written a comprehensive review on the larval biology and physiology of *Eriocheir sinensis*. In essence, the larvae do not have the same osmoregulatory capacity as adults and must live at higher salinities in the outer regions of estuaries. Paradoxically, it is this requirement for a marine environment that facilitates the global transport of this species. The adults must descend the rivers to breed and release larvae, which are then easily transported across the globe from port to port in ships' ballast water. Estuaries and ports are frequently contaminated with heavy metals, such as cadmium. Mitten crabs have been shown to acclimate well to chronic exposure to cadmium ($50\mu g$ Cd L^{-1}) over 30 days, and indeed chronic exposure at low but environmentally realistic levels increases their ability to cope with higher acute (5mg Cd L^{-1}) toxic exposure (Silvestre et al, 2005). Such acclimation in osmoregulatory ability has not been found in other estuarine species such as *Carcinus maenas* (Thurberg et al, 1973; Bjerregaard and Vislie, 1985) and may be a further reason for the invasive success of the Chinese mitten crab.

Prey, predation and competition

Like many invasive species, Chinese mitten crabs may benefit from the 'enemy release hypothesis' (see Weis, 2010) where their success is in part attributable to an absence of pathogens, competitors or predators. Larvae and juveniles are likely to suffer predation from fish species in estuaries and juveniles still form fish prey as they migrate upstream. Suzuki et al (1998) show that fish predation of *Eriocheir japonica* juveniles increases further upstream as alternative prey sources decrease. The same situation is likely to exist for *E sinensis*. In fresh waters, adult Chinese mitten crabs have few predators; these are essentially large birds such as owls, cormorants and herons, as well as otters and foxes (Wu et al, 2006; Weber, 2008). The mitten crabs may though have a significant impact on other fauna through predation and exert significant competitive interactions. Their presence at high densities in fresh waters can have significant effects on zoobenthos and submerged macrophytes, as has been shown in stocking experiments in its native range (Yu and Jiang, 2005).

Control

Early efforts at control in Europe

As is so often the case with invasive species, efforts to control the spread of the Chinese mitten crab were not initiated until the crab had become well established in Europe. Control measures were felt necessary because of the damage that was being caused to fishing gear deployed in rivers (e.g. eel fyke nets), and also the fact that the crabs will damage and devour fish caught in such gear. In addition, it soon became recognized that mitten crabs cause riverbank instability and siltation of river channels. Trapping was employed on the rivers in Germany; efforts during the early 1930s in the River Weser removed hundreds of thousands of crabs but the numbers were so large (220 tonnes in 1936) that trapping could no longer be sustained. Panning (1939) reports that more than 30,000 crabs were removed from the river daily during March 1936 at Bremen.

Pre-emptive approaches to control in the US

The extensive spread of the Chinese mitten crab in the UK during the 20th century prompted the US Fish and Wildlife Service to seek the introduction of pre-emptive legislation in an attempt to prevent the further introduction and control the spread in the US, despite only a couple of reports of mitten crabs (Lake Erie; Mississippi River) (Table 16.2). State legislation was passed in 1987 in California and federal legislation superseded this in 1989. This prohibited the landing, transport or possession of live mitten crabs (Horwath, 1989). In the face of ineffective control of ballast water discharge, however, and the possibility of deliberate introduction to establish a fishery to serve the large Chinese resident community in San Francisco, mitten crab populations became well established in San Francisco Bay, in the Sacramento and San Joaquim drainages and numerous other tributaries by the late 1990s (Hui et al, 2005).

Management plans and legislation

Following the introduction of California legislation (Section 671(h)(2) Title 14 CCR) and federal legislation (50 CFR 16.13) (see Veldhuizen and Stannish, 1999), a draft management plan was developed by the US

Aquatic Nuisance Task Force. This plan essentially comprised recommendations for monitoring and risk assessment that could be employed in regions of the US, other than San Francisco Bay, that may be at risk from mitten crab invasion, and because no effective management tools exist, recommended an 'adaptive management' approach. In the UK a status report was commissioned into the mitten crab population in the Ouse washes (Walker et al, 2010) by Natural England, which suggested that heavy trapping effort might offer the only possible solution for control. This approach had previously failed in Germany, however, and any such trapping programme is likely to be prohibitively expensive. Development of any management plan will require the interests of all stakeholders to be considered and this would include those who advocate the establishment of a fishery for commercial gain (Stokes et al, 2006). An approach such as that being considered for the River Thames (Clark et al, 2009), with fishing of crabs for human consumption as a potential control measure, might represent a compromise.

Routes of introduction

It is now generally accepted that the principal route of introduction of the Chinese mitten crab to its invasive range has been through ships' ballast water (Carlton, 1985; Carlton and Geller, 1993). The timing of spread that began in Europe in the early 20th century coincided with the change to seawater ballast from the solid ballast that had been used by maritime shipping up until that time. This spread has possibly been facilitated by movement of adults from spawning areas in the outer estuaries or by intercoastal maritime traffic (Symkanin et al, 2009). Once introduced, *Eriocheir sinensis* has been shown to have rapid rates of both coastal and inland spread. In the UK and continental Europe, rates of coastal spread of over 400km per year and upriver spread of over 50km per year have been reported (Herborg et al, 2005). Predictions suggest that in Europe *E. sinensis* will continue to expand its range southwards and eastwards (Herborg et al, 2007).

Value for aquaculture

Since the early 1990s, aquaculture of the Chinese mitten crab has expanded rapidly in China to meet the demand for a highly sought after, nutritious and high value product (Chen et al, 2007) and this aquaculture has utilized lakes, ponds and paddy fields for the extensive grow-out of crabs after stocking with juveniles (Zhang and Li, 2002 cited in Li et al, 2010). More recently, considerable research effort has focused on the nutritional requirements of broodstock production and intensive production of larvae for stocking (e.g. Li et al, 2010; Lin et al, 2010). The rapid development of *Eriocheir sinensis* aquaculture in China raises the possibility of the extension of these activities to other areas in the Far East and possibly further afield. Associated with this, there must be the possibility of the extension of the invasive range or a reservoir of farm escapees that may facilitate transmission of this invasive crab to parts of the world it has not yet reached.

Value as a fishery

Apart from the *Eriocheir sinensis* fishery in China, there is an established fishery for the related *E. japonica* in Japan (Kobayashi et al, 1997) where market values even in 1997 were in excess of ¥2000/kg ($25). In some areas, release of seed (juveniles) to boost the fishery population occurs. This tactic remains a possibility elsewhere in the world where there are interests in creating or maintaining a mitten crab fishery. Recent investigations into the possible exploitation of the Chinese mitten crab population in the River Thames estuary has examined organic contaminants to ascertain the safety of mitten crabs obtained from there for human consumption, and results suggest that contaminant levels are low enough to permit limited safe consumption (Clark et al, 2009). Therefore the possibility of establishing a fishery there remains a distinct possibility. In San Francisco Bay, studies on mercury contamination suggest that levels would be within limits for human consumption (Hui et al, 2005). Mitten crabs harvested from The Netherlands are reported to be sold on the international market.

Proposed exploitation and lifting of regulations in the River Thames

Exploitation of a population of an introduced invasive species is always likely to stimulate debate. There are those who see this as likely to exacerbate the problems caused by the invasive in question and there are those who advocate that exploitation could be regarded as a form of control that might act to limit further spread. This debate has recently been explored by Clark (2011) in relation to the population of *Eriocheir sinensis* in the River Thames and Clark weighs up the pros and cons of their exploitation. There are clear arguments for and against, but no easy resolution.

Parasite and pathogen transmission

The Chinese mitten crab plays host to a parasitic lung fluke, the trematode *Paragonimus westermani* (Ingle, 1985; Clark et al, 1998) that can infect humans if the crab, is eaten raw or not well cooked. The completion of the parasite lifecycle requires a snail (*Semulcospira* sp.) as an intermediate host. The parasite infests adult crabs, and larvae transported via shipping would be unaffected. In addition the snail host genus is not found in the current invasive range of the mitten crab. While infection of *Eriocheir sinensis* by *P. westermani* in

its invasive range (at least in the US and Europe) seems unlikely, there has been recent evidence of the possibility of *Vibrio* infection. Wagley et al (2009) recently showed the presence of *Vibrio parahaemolyticus* pathogenicity markers in *E. sinensis* from the River Thames estuary, which raises the possibility of *P. haemolyticus*-related infection in humans if the crabs were to be consumed.

Acknowledgements

I am grateful to Alison Bentley for providing Figure 16.1 and Stephan Gollasch for providing Figure 16.2. I would like to acknowledge the excellent work of my PhD student, Leif-Matthias Herborg, who undertook his PhD with the support of a Newcastle University Swales Studentship and who continues to do excellent work in the Great Lakes region of the US and Canada. I am grateful to my wife, Alison, for her critical comments on the manuscript.

References

André, M. (1954) 'Présence du crabe chinois (*Eriocheir sinensis* H. M.-Edw.) dans la Loire', *Bulletin du Muséum*, vol 26, p581

Bacevièius, E. and Gasiûnaitë, Z. R. (2008) 'Two crab species (*Eriocheir sinensis* Edw.) and mud crab (*Rithropanopeus harissii* (Gould)) ssp. *tridentatus* (Maitland) in the Lithuanian coastal waters', *Transitional Waters Bulletin*, vol 2, no 2, pp63–68

Bjerregaard, P. and Vislie, T. (1985) 'Effects of cadmium on hemolymph composition in the shore crab *Carcinus maenas*', *Marine Ecology Progress Series*, vol 27, no 1–2, pp135–142

Cabral, H. N. and Costa, J. F. (1999) 'On the occurrence of the Chinese mitten crab *Eriocheir sinensis*, in Portugal (Decapoda: Brachyura)', *Crustaceana*, vol 72, no 1, pp55–58

Carlton, J. T. (1985) 'Transoceanic and inter-oceanic dispersal of coastal marine organisms: The biology of ballast water', *Oceanography Marine Biology Annual Review*, vol 23, pp313–371

Carlton, J. T. and Cohen, A. N. (2003) 'Episodic global dispersal in shallow water marine organisms: The case history of the European shore crabs *Carcinus maenas* and *C. aestuarii*', *Journal of Biogeography*, vol 30, no 12, pp1809–1820

Carlton, J. T. and Geller, J. B. (1993) 'Ecological roulette – the global transport of non-indigenous marine organisms', *Science*, vol 261, pp78–82

Chen, D. W., Zhang, M. and Shrestha, S. (2007) 'Compositional characteristics and nutritional quality of Chinese mitten crab (*Eriocheir sinensis*)', *Food Chemistry*, vol 103, no 4, pp1343–1349

Christiansen, M. E. (1977) '1st record of the Chinese mitten crab *Eriocheir sinensis* from Norway', *Flora (Oslo)*, pp30134–30138

Clark, P. (1984) 'Recent records of alien crabs in Britain', *Naturalist*, vol 109, pp111–112

Clark P. F. (2006) '*Eriocheir sinensis* H. Milne Edwards: 1853 or 1854 – Grapsidae or Varunidae?', *Aquatic Invasions*, vol 1, no 1, pp17–27

Clark P. F. (2011) 'The commercial exploitation of the Chinese mitten crab *Eriocheir sinensis* in the River Thames, London: Damned if we don't and damned if we do', in B. Gailil, P. F. Clark and J. T. Carlton (eds) *In the Wrong Place: Alien Marine Crustaceans: Distribution, Biology and Impacts, Invading Nature – Springer Series on Invasion Ecology Vol 6*, Springer Science and Business Media, BV, Netherlands, pp537–580

Clark, P. F. and Rainbow, P. S. (1997) 'The Chinese mitten crab in the Thames catchment', A Report for the Environment Agency, Part 1, Environment Agency, London

Clark, P. F., Rainbow, P. S., Robbins, R. S., Smith, B., Yeomans, W. E., Myles, T. and Dobson, G. (1998) 'The alien Chinese mitten crab *Eriocheir sinensis* (Crustacea: Decapoda: Brachyura) in the Thames catchment', *Journal of the Marine Biological Association of the UK*, vol 78, no 4, pp1215–1221

Clark, P. F., Abdul-Sahib, I. M. and Al-Asadi, M. S. (2006) 'The first record of *Eriocheir sinensis* H. Milne Edwards, 1853, from the Basra area of southern Iraq', *Aquatic Invasions*, vol 1, no 2, pp51–54

Clark, P. F., Mortimer, D. N., Law, R. J., Averns, J. M., Cohen, B. A., Wood, D., Rose, M. D., Fernandes, A. R. and Rainbow, P. S. (2009) 'Dioxin and PCB contamination in Chinese mitten crabs: Human consumption as a control mechanism for an invasive species', *Environmental Science and Technology*, vol 43, no 5, pp1624–1629

Cohen, A. N. and Carlton, J. T. (1995) 'Biological study. Non-indigenous aquatic species in a United States estuary: A case study of the biological invasions of San Francisco Bay and Delta', Connecticut Sea Grant NITS Report No PB96–166525, US Fish and Wildlife Service, Washington, DC and National Sea Grant College Program

Cohen, A. N. and Carlton, J. T. (1997) 'Transoceanic transport mechanisms: Introduction of the Chinese mitten crab, *Eriocheir sinensis*, to California', *Pacific Science*, vol 51, pp1–11

Czerniejewski, P., Rybczyk, A. and Wawrzyniak, W. (2010) 'Diet of the Chinese mitten crab, *Eriocheir sinensis* H. Milne Edwards, 1853, and potential effects of the crab on the aquatic community in the River Odra/Oder estuary (N.-W. Poland)', *Crustaceana*, vol 83, no 2, pp195–205

de Lafontaine, Y. (2005) 'First record of the Chinese mitten crab (*Eriocheir sinensis*) in the St Lawrence River, Canada', *Journal of Great Lakes Research*, vol 31, pp367–370

Dittel, A. I. and Epifanio, C. E. (2009) 'Invasion biology of the Chinese mitten crab *Eriochier sinensis*: A brief review', *Journal of Experimental Marine Biology and Ecology*, vol 374, pp79–92

Drotz, M. K., Berggren, M., Lundberg, S., Lundin, K. and von Proschwitz, T. (2010) 'Invasion routes, current and historical distribution of the Chinese mitten crab (*Eriocheir sinensis* H. Milne Edwards, 1853) in Sweden', *Aquatic Invasions*, vol 5, no 4, pp387–396

Dutton, C. and Conroy, C. (1998) *Effects of Burrowing Chinese Mitten Crabs (*Eriocheir sinensis*) on the Thames Tideway*, Environment Agency, London

Gilbey, V., Atrill, M. J. and Coleman, R. A. (2008) 'Juvenile mitten crabs (*Eriocheir sinensis*) in the Thames estuary: Distribution, movement and possible interactions with the native crab *Carcinus meanas*', *Biological Invasions*, vol 10 pp67–77

Gollasch, S. (2006) 'NOBANIS – Invasive Alien Species Fact Sheet – *Eriocheir sinensis*', Online Database of the North European and Baltic Network on Invasive Alien Species – NOBANIS, www.nobanis.org, accessed 14 January 2011

Gollasch, S., MacDonald, D. W., Beldon, S., Botnen, H., Christensen, J. T., Hamer, J. P. Houvenhagel, G., Jelmert, A., Lucas, I., Masson, D., McCollin, T., Olenin, S., Person, A., Walletius, I., Wetsteyn, L. P. M. J. and Wittling, T. (2002) 'Life in ballast tanks', in E. Leppäkoski, S. Gollasch and S. Olenin (eds) *Invasive Aquatic Species of Europe: Distribution, Impacts and Management*, Kluwer Academic Publishers, Dordrecht, London, pp217–231

Hanson, E. and Sytsma, M. (2008) 'The potential for mitten crab *Eriocheir sinensis* H. Milne Edwards, 1853 (Crustacea: Brachyura) invasion of Pacific Northwest and Alaskan estuaries', *Biological Invasions*, vol 10, no 5, pp603–614

Harold, C. H. H. (1935) *Thirteenth Annual Report on the Results of the Chemical and Bacteriological Investigations of the London Waters for the 12 Months Ending 31 December 1935*, Metropolitan Water Board, London

Herborg, L.-M., Bentley, M. G. and Clare, A. S. (2002) 'First confirmed record of the Chinese mitten Crab (*Eriocheir sinensis*) from the River Tyne, United Kingdom', *Journal of the Marine Biological Association of the United Kingdom*, vol 82, pp921–922

Herborg, L.-M., Rushton, S. P., Clare, A. S. and Bentley, M. G. (2003) 'Spread of the Chinese mitten crab (*Eriocheir sinensis* H. Milne Edwards) in continental Europe: Analysis of an historical data set', *Hydrobiologia*, vol 503, no 1–3, pp21–28

Herborg, L.-M., Rushton, S. P., Clare, A. S. and Bentley, M. G. (2005) 'The invasion of the Chinese mitten crab (*Eriocheir sinensis*) in the United Kingdom and its comparison to continental Europe', *Biological Invasions*, vol 7, no 6, pp959–968

Herborg, L. M., Rudnick D. A., Siliang, Y., Lodge, D. M. and McIsaac, H. (2007) 'Predicting the range of Chinese mitten crabs in Europe', *Conservation Biology*, vol 5, pp1316–1323

Hoestlandt, H. (1959) 'Répartition actuelle du crabe chinois (*Eriocheir sinensis* H. Milne Edwards) en France', *Bulletin Francais Pisciculture*, vol 194, pp5–13

Horwath, J. L. (1989) 'Importation or shipment of injurious wildlife: Chinese mitten crabs', *US Federal Register*, vol 54, no 98, pp22286–22289

Hui, C. A., Rudnick, D. and Williams, E. (2005) 'Mercury burdens in Chinese mitten crabs (*Eriocheir sinensis*) in three tributaries of southern San Francisco Bay, California, US', *Environmental Pollution*, vol 133, pp481–487

Hymanson, Z., Wang, J. and Sasaki, T. (1999) 'Lessons from the home of the Chinese mitten crab', *IEP Newsletter*, vol 12, pp25–32

Ingle, R. W. (1985) 'The Chinese mitten crab *Eriocheir sinensis* H. Milne-Edwards: A contentious immigrant', *London Naturalist*, vol 65, pp101–105

Ingle, R. W. and Andrews, M. J. (1976) 'Chinese mitten crab reappears in Britain', *Nature*, vol 263, p638

Jin, G., Li., Z. and Xie, P. (2001) 'The growth patterns of juvenile and precocious Chinese mitten crabs, *Eriocheir sinensis* (Decapoda, Grapsidae), stocked in freshwater lakes of China', *Crustaceana*, vol 74, no 3, pp261–273

Jin, G., Xie, P. and Li, Z. J. (2003) 'Food habits of two-year-old Chinese mitten crab (*Eriocheir sinensis*) stocked in Lake Bao'an, China', *Journal of Freshwater Ecology*, vol 18, no 3, pp369–375

Kamps, L. F. (1937) 'De Chineesche wolhand krab in Nederland', *Akademisch Proefschrift: Groningen*, pp1–112

Kobayashi, S., Kagehira, M., Yoneji, T. and Matsuura, S. (1997) 'Questionnaire research on the ecology and fishery of the Japanese mitten crab *Eriocheir japonica* (de Haan)', *Science Bulletin of the Faculty of Agriculture Kyushu University*, vol 52, no 1–2, pp89–104

Leloup, E. (1937) 'Contributions à l'étude de la faune belge. VII. – La propagation du crabe chinois en Belgique pendant l'année 1936', *Bulletin du Musée royal d'histoire naturelle de Belgique*, vol 13, pp1–7

Li, W. W., Gong, Y. N., Jin, X. K., He, L., Jiang, H., Ren, F. and Wang, Q. (2010) 'The effect of dietary zinc supplementation on the growth, hepatopancreas fatty acid composition and gene expression in the Chinese mitten crab, *Eriocheir sinensis* (H. Milne-Edwards) (Decapoda: Grapsidae)', *Aquaculture Research*, vol 41, pp828–837

Lin, S., Lou, Y. and Ye, Y. (2010) 'Effects of dietary protein level on growth, feed utilization and digestive enzyme activity of the Chinese mitten crab *Eriocheir sinensis*', *Aquaculture Nutrition*, vol 16, pp290–298

Minchin, D. (2006) 'First Irish record of the Chinese mitten crab *Eriocheir sinensis* (Milne-Edwards 1854) (Decapoda: Crustacea)', *Irish Naturalist Journal*, vol 28, pp303–304

Mizzan, L. (2005) '*Rhithropanopeus harrisii* (Gould, 1841) (Crustacea, Decapoda, Panopeidae) and *Eriocheir sinensis* H. Milne Edwards, 1854 (Crustacea, Decapoda, Grapsidae): Two new exotic crabs in the Venetian Lagoon', *Bollettino del Museo Civico di Storia Naturale di Venezia*, vol 56, pp89–95

Nepszy, S. J. and Leach, J. H. (1973) 'First records of the Chinese mitten crab *Eriocheir sinensis* (Crustacea: Brachyura) from North America', *Journal of the Fisheries Research Board of Canada*, vol 30, pp1909–1910

Ng, P. K. L., Guinot, D. and Davie, P. J. F. (2008) 'Systema Brachyurorum. Part 1. An annotated checklist of extant brachyuran crabs of the World', *Raffles Bulletin of Zoology*, vol 17, pp1–286

Ojaveer, H., Gollasch, S., Jaanus, A., Kotta, J., Laine, A. O., Minde, A., Normant, M. and Panov, V. E. (2007) 'Chinese mitten crab *Eriocheir sinensis* in the Baltic Sea – a supply side invader?', *Biological Invasions*, vol 9, pp409–418

Panning, A. (1939) 'The Chinese mitten crab', *Annual Report Smithsonian Institution*, *1938*, pp361–375

Peters, N. (1933) 'Lebenskundlicher Teil, Die chinesische Wollhandkrabbe (*Eriocheir sinensis* H. Milne-Edwards) in Deutschland', in N. Peters, A. Panning and W. Schnakenbeck (eds) *Zoologischer Anzeiger*, Akademische Verlagsgesellschaft Geest & Portig, Leipzig, pp59–155

Peters, N. (1938) 'Ausbreitung und Verbreitung der chinesischen Wollhandkrabbe (*Eriocheir sinensis* H. M.-Edw) in Europa im Jahre 1933 bis 1935', *Mitteilungen aus dem Hamburgischen Zoologischen Museum und Institut*, vol 47, pp1–31

Rasmussen, E. (1987) 'Status over uldhåndskrabbens (*Eriocheir sinensis*) udbredelse og forekomst i Danmark', *Flora og Fauna*, vol 93, pp51–58

Robbins, R. S., Sakari, M., Nezami, B. S. and Clark, P. F. (2006) 'The occurrence of *Eriocheir sinensis* H. Milne Edwards, 1853 (Crustacea: Brachyura: Varunidae) from the Caspian Sea region, Iran', *Aquatic Invasions*, vol 1, no 1, pp32–34

Rudnick, D., Halal, K. and Resh, V. (2000) *Distribution, Ecology and Potential Impacts of the Chinese Mitten Crab* (Eriocheir sinensis) *in San Francisco Bay*, University of California Water Resources Center, Riverside, CA

Rudnick, D., Veldhuizen, T., Tullis, R., Culver, C., Hieb, K. and Tsukimura, B. (2003) 'A life history model for the San Francisco estuary population of the Chinese mitten crab, *Eriocheir sinensis* (Decapoda: Grapsoidea)', *Biological Invasions*, vol 7, pp333–350

Rudnick, D., Chan, V. and Resh, V. H. (2005) 'Morphology and impacts of the burrows of the Chinese mitten crab, *Eriocheir sinensis* H. Milne Edwards (Decapoda, Grapsoidea), in south San Francisco Bay, California, USA', *Crustaceana*, vol 78, pp787–807

Ruiz, G. M., Fegley, L., Fofonoff, P., Cheng, Y. and Lemaitre, R. (2006) 'First records of *Eriocheir sinensis* H. Milne Edwards, 1853 (Crustacea: Brachyura: Varunidae) for Chesapeake Bay and the mid-Atlantic coast of North America', *Aquatic Invasions*, vol 1, no 3, pp137–142

Silvestre, F., Trausch, G. and Devos, P. (2005) 'Hyper-osmoregulatory capacity of the Chinese mitten crab (*Eriocheir sinensis*) exposed to cadmium; acclimation during chronic exposure', *Comparative Biochemistry and Physiology, Part C*, vol 140, pp29–37

Stokes, K. E., O'Neill, K. P., Montgomery, W. I., Dick, J. T. A., Maggs, C. A. and MacDonald, R. A. (2006) 'The importance of stakeholder engagement in invasive species management: A cross-jurisdictional perspective in Ireland', *Biodiversity and Conservation*, vol 15, pp2829–2852

Suzuki, T., Hamano, T., Araki, A., Hayashi, K., Fujimura, H. and Fujita, Y. (1998) 'Predation by fishes on released seedlings of the Japanese mitten crab *Eriocheir japonica*', *Crustacean Research*, vol 27, pp1–8

Symkanin, C., Davidson, I., Falkner, M., Sytsma, M. and Ruiz, G. M. (2009) 'Intra-coastal ballast water flux and the potential for secondary spread of non-native species on the US West Coast', *Marine Pollution Bulletin*, vol 58, pp366–374

Tang, B. P., Zhou, K. Y., Song, D. X., Yang, G. and Dai, A. Y. (2003) 'Molecular systematics of the Asian mitten crabs, genus *Eriocheir* (Crustacea: Brachyura)', *Molecular Phylogenetics and Evolution*, vol 29, no 2, pp309–316

Tang, B. P., Chen, L. Q., Zhou, K. Y. and Song, D. X. (2009) 'On the morphological characters of mitten crabs (Brachyura, Varunidae) antennae and its application in the classification', *Acta Zootaxonomica Sinica*, vol 34, no 1, pp79–86

Thurberg, F. P., Dawson, M. A. and Collier, R. S. (1973) 'Effects of copper and cadmium on osmoregulation and oxygen consumption in two species of estuarine crabs', *Marine Biology*, vol 23, pp171–175

Veldhuizen, T. and Stannish, S. (1999) 'Overview of the life history, distribution, abundance, and impacts of the Chinese mitten crab, *Eriocheir sinensis*', Report to California Department of Water Resources Environmental Studies Office, Interagency Program Sacramento, CA 95816, pp1–26

Vincent, T. (1996) 'Le crabe chinois *Eriocheir sinensis* H. Milne-Edwards, 1854 (Crustacea, Brachyura) en Seine-Maritime, France', *Annales Institut Océanographique*, vol 72, pp155–171

Wagley, S., Koofhethile, K. and Rangdale, R. (2009) 'Prevalence and potential pathogenicity of *Vibrio parahaemolyticus* in Chinese mitten crabs (*Eriocheir sinensis*) harvested from the River Thames estuary, England', *Journal of Food Protection*, vol 72, pp60–66

Walker, P., Fraser, D. and Clark, P. (2010) 'Status, distribution and impacts of mitten crab in the Nene and Ouse washes: Stage 1 Report', APEM Scientific Report 411010, APEM, Oxford, UK

Weber, A. (2008) 'Predation of invasive species Chinese mitten crab (*Eriocheir sinensis*) by Eurasian otter (*Lutra lutra*) in the Drömling Nature Reserve, Saxony-Anhalt, Germany', *IUCN Otter Specialist Group Bulletin*, vol 25, pp104–107

Weis, J. S. (2010) 'The role of behavior in the success of invasive crustaceans', *Marine and Freshwater Behaviour and Physiology*, vol 43, no 2, pp83–98

Wu, H., Sun, Y. H., Wang, Y. and Tsang, Y. S. (2006) 'Food habits of tawny fish owls in Sakatang Stream', *Taiwan Journal of Raptor Research*, vol 42, no 11, pp111–119

Yang, W. L. and Chang G. H. (2005) 'The current status and sustainable development of Chinese mitten crab farming', *Freshwater Fisheries*, vol 35, pp62–64

Yu, H. X. and Jiang, C. (2005) 'Effects of stocking Chinese mitten crab on the zoobenthos and aquatic vascular plant in the East Lake Reservoir, Heilongjiang, China', *Acta Hydrobiologica Sinica*, vol 29, no 4, pp430–434

Zhang, L. S. and Li, J. (2002) 'Culture methods of *Eriocheir sinensis*', in Zhang, L. S. (ed) *The Breeding and Culture of Chinese Mitten Crab*, Jing Dun Press, Beijing, pp247–320

17

Pacifastacus leniusculus Dana (North American signal crayfish)

Jenny C. Dunn

Introduction

Invasive crayfish are one of the greatest threats to freshwater ecosystems. Many non-native crayfish species have been introduced worldwide, where they often occupy a different ecological niche to the native species that they compete with and commonly replace. This niche may also differ to that occupied in their native range (Bondar and Richardson, 2009). As keystone species, they play a crucial role in the ecosystem (Dorn and Wojdak, 2004) and have the potential for dramatic impacts outside of their native territories. *Pacifastacus leniusculus* (North American signal crayfish) (Figure 17.1) is the largest and most widespread of any introduced crayfish species, and this chapter summarizes the history of its introduction along with its current distribution, ecological problems that its introduction has caused and attempts to control the species within its introduced range.

History of Introduction

The native range of the signal crayfish stretches from southern British Columbia to northern California, and east into Montana and Utah (Bondar et al, 2005). Ironically, it is considered to be potentially endangered within its home range (Hamr, 1998), although this is partially due to a lack of knowledge regarding its status.

The species has been widely introduced throughout Europe, as well as to Japan and other parts of North America, and is the most widespread of any introduced crayfish species (Holdich et al, 2009). It was largely introduced for aquaculture, as it is a large, fecund and fast growing species that can reach high densities (Hiley, 2002), but has also been introduced for the aquarium trade, fish markets and probably by anglers as supplemental food for fish stocks (Peay et al, 2010). On occasion, signal crayfish have also been introduced for weed clearance (Rogers and Loveridge, 2000; Howells and Slater, 2004) or discarded as unused bait or even as unwanted aquarium pets (Holdich et al, 2009). They may well have also been accidentally introduced through transport in association with fish, a common mode of unintentional invertebrate introductions (Gherardi et al, 2008). They are still sold live in fish markets for human consumption, providing continuing opportunities for the species to expand its range through further escapes into the wild (Holdich and Sibley, 2009).

In Japan, signal crayfish were introduced for aquaculture on five separate occasions between 1926 and 1930 (Kamita, 1970). While only two of these original populations are thought to have survived to the present day, crayfish from these surviving populations have more recently been introduced to other water bodies in north and east Hokkaido (Ohtaka et al, 2005) and have been linked with the decline of the

Note: The name of the signal crayfish originates from the white patch at the hinge of the chelae

Source: Courtesy of Emily Imhoff

Figure 17.1 *A signal crayfish in a British river*

endemic native Japanese crayfish *Cambaroides japonicus* (Hiruta, 1999; Usio et al, 2001).

Although Signal crayfish are native only to the Klamath River drainage basin within California, they have been widely introduced throughout the state, probably as bait by sport fisherman (Light et al, 1995). The species is associated with declines in populations of the Shasta crayfish (*P. fortis*) and is thought to be partially responsible for the extinction of the sooty crayfish (*P. nigrescens*) (Light et al, 1995).

In Europe, signal crayfish were first introduced to Sweden from California in 1959 for aquaculture, as the native noble crayfish (*Astacus astacus*) had been badly affected by crayfish plague and signal crayfish were considered to be an ecological homologue that could be farmed successfully (Holdich, 2002a). From 1960 onwards, the species was released into thousands of lakes and ponds in Sweden, and then into other European countries (Holdich, 2002a), where it has been linked with declines in the noble crayfish, the stone crayfish (*Austropotamobius torrentium*) and the white-clawed crayfish (*Austropotamobius pallipes*) (Westman et al, 2002; Huber and Schubart, 2005; Dunn et al, 2009).

Signal crayfish were introduced into England from Sweden in the 1970s for aquaculture trials, as they were larger and faster growing than the native white-clawed crayfish (*Austropotamobius pallipes*) (Holdich and Rogers, 1997; Holdich and Sibley, 2009). These trials were successful and led to the widespread introduction of the species into fish farm ponds and lakes, and also directly into the wild (Holdich and Rogers, 1997). During the 1970s and 1980s signal crayfish were farmed profitably in England, particularly as a sideline to fish farming (Hiley, 2002). However, few farms were sufficiently well fenced to prevent signal crayfish escaping (Hiley, 2002); escapes from fish farms led to further naturalized populations, and it became increasingly difficult to farm signal crayfish at a profit as feral populations became widespread (Hiley, 2002). However, crayfish farms were widespread in the early 1990s, with up to 100 premises registered, although in 1997 only 14 were still active (Holdich and Rogers, 1997).

Current distribution

Signal crayfish are still increasing their range throughout Europe, although their spread in Japan and North America is less well documented. In 2002, the species was established in the wild in 21 European countries (Holdich, 2002a). In 2006 this had increased to 22, with another 2 countries in which the species was thought to be present in the wild (Machino and Holdich, 2006), and in 2009 the signal crayfish was known to be established in 27 European territories (Holdich et al, 2009) (Table 17.1). One individual was reported from a 28th territory (Estonia) in 2008, although the presence of an established population has yet to be confirmed (Holdich et al, 2009). Table 17.1 summarizes the current distribution of signal crayfish in Europe.

Ecology and Impacts

Crayfish plague

Signal crayfish can carry the Oomycete *Aphanomyces astaci*, the causative agent of crayfish plague (Alderman et al, 1990). The plague can be devastating to naïve populations of other crayfish species, and plague outbreaks often cause 100 per cent mortality (e.g. Vorburger and Ribi, 1999); the introduction of crayfish

Table 17.1 *European territories from which signal crayfish have been recorded in the wild, along with the year of first record*

Country	Year first recorded	Country	Year first recorded
Austria	1970	Lithuania	1972
Belgium	1979	Luxembourg	1972
Croatia	2008	Netherlands	2004
Czech Republic	1980	Norway	2006
Denmark	1970s	Poland	1971
England	1970s	Portugal	1974
Finland	1967	Scotland	1995
France	1972	Slovakia	2005
Germany	1972	Slovenia	2003
Greece	1982	Spain	1974
Hungary	2000	Sweden	1960
Italy	1981	Switzerland	Pre-1987
Kaliningrad (Russia)	Pre-1995	Wales	Pre-1990
Latvia	1983		

Source: Compiled from Holdich et al (2009), Machino and Holdich (2006), Pöckl and Pekny (2002), Holdich (2002b), Puky et al (2005) and Hefti and Stucki (2006)

plague-infected signal crayfish by humans is the main cause of disappearance of native crayfish in some regions (e.g. Diéguez-Uribeondo, 2006). However, outbreaks of crayfish plague in native crayfish populations are not only caused directly by signal crayfish introductions, as plague spores can be carried on angling equipment, water, mud and fish (Holdich and Rogers, 1999; Hiley, 2002). The signal crayfish is generally only susceptible to the effects of the plague when put under stress (Cerenius and Söderhäll, 1992) with few records of mortality from plague in signal crayfish populations (Nylund and Westman, 2000). However, exceptions have recently been reported following the introduction of plague into established signal crayfish populations in Finland that had been historically free of the disease (Pursiainen and Rajala, 2010). While populations now appear to be recovering, crayfish catches decreased dramatically and reproduction declined for a number of years following plague introduction (Pursiainen and Rajala, 2010).

Not all signal crayfish populations carry crayfish plague, and plague-free populations are relatively widespread (Nylund and Westman, 2000; Hiley, 2002; Holdich and Sibley, 2009). Where the Oomycete is present within resistant signal crayfish populations, prevalence can vary from 0.8 per cent to 52 per cent (Nylund and Westman, 2000; Kozubíková et al, 2009). Indeed, mixed populations of native crayfish exist

alongside the invading species when crayfish plague is absent (Spink and Rowe, 2002), although the larger size, greater fecundity and faster growth of the signal crayfish inevitably leads to the local extinction of the native species (Holdich and Sibley, 2009), even after long periods of coexistence (Westman et al, 2002; Huber and Schubart, 2005).

Competition with native species

Where plague-free signal crayfish populations coexist alongside native crayfish species, mixed populations can occur (e.g. Peay and Rogers, 1999; Spink and Rowe, 2002). The invading species usually replaces native crayfish over a period of four to five years (Peay and Rogers, 1999; Huber and Schubart, 2005) and the replacement of native species through competition with signal crayfish is thought to have been underestimated (Peay and Rogers, 1999; Hiley, 2002).

Competition in crayfish tends to be size dependent, which favours the larger, faster growing and more aggressive signal crayfish (Vorburger and Ribi, 1999). The species has a relatively large chelae to body size ratio (Figure 17.2), allowing competitive dominance over other species with larger body size but smaller chelae (Usio et al, 2001). The direct mechanism by which signal crayfish replace native species in mixed populations is unknown, although several authors have suggested mechanisms by

Source: Courtesy of Neal Haddaway

Figure 17.2 *Signal crayfish have a large chelae to body size ratio, giving them an advantage in competitive interactions with other species*

Source: Courtesy of Emily Imhoff

Figure 17.3 *Female signal crayfish have a high fecundity and females can carry between 100 and 400 abdominal eggs*

which species displacement may occur. Stebbing et al (2006) show the native white-clawed crayfish to be repelled by water that had been conditioned by signal crayfish, suggesting that chemical signals may play a role in species displacement. Conversely, male signal crayfish were attracted to water conditioned by female white-clawed crayfish in breeding condition, supporting Westman et al's (2002) suggestion of reproductive interference of native species by signal crayfish.

It has been suggested that predation of native species by the aggressive signal crayfish is important in this process, and is thought, in combination with other factors, to be responsible for the decline in British white-clawed crayfish populations (Nakata and Goshima, 2006; Dunn et al, 2009). Signal crayfish rarely suffer predation by other crayfish species (Nakata and Goshima, 2006), although Dunn et al (2009) provide evidence for interspecific predation of similarly sized juvenile signal crayfish and adult white-clawed crayfish during moult (Dunn et al, 2009).

Signal crayfish tend to have a higher fecundity and earlier hatching than native species, giving them a significant competitive advantage over the species they replace (Huber and Schubart, 2005) (Figure 17.3). They grow much faster than many native species as they have a greater length increment per moult and more frequent moulting (Kirjavainen and Westman, 1994), can have juvenile recruitment of up to 70 juveniles m^{-2} (Guan and Wiles, 1999) and can reach densities of 20 crayfish m^{-2} (Hiley, 2002).

The dispersal of signal crayfish has proved very difficult to prevent. Few farms were sufficiently well fenced to prevent escape, and most have a through-flow of water that could carry eggs and juvenile crayfish into the nearest watercourse; unless the outflow was fitted with a 1mm mesh then prevention of signal crayfish escape was impractical (Hiley, 2002). Once established in a watercourse, signal crayfish are capable of spreading rapidly both upstream and downstream (Bubb et al, 2005).

The signal crayfish is a good disperser and can spread very rapidly. Dispersal rates of 120m per day and 240m per month have been recorded (Wright and Williams, 2000; Light, 2003), with movements of 345m in three weeks being recorded in one pit-tagged individual (Bubb et al, 2006a). Annual dispersal rates of 1.2km, 2.4km and up to 24.4km year^{-1} (Peay and Rogers, 1999; Bubb et al, 2005; Hudina et al, 2009) have been reported, although colonization rates tend to be slower in the earlier stages of establishment of a new population and tend to be faster in a downstream direction (Bubb et al, 2005). The ability of the signal crayfish to disperse further than the native species it replaces may allow it to utilize ephemeral and patchy resources more efficiently (Bubb et al, 2006b). While physical barriers such as dams and weirs may slow the spread of signal crayfish, they cannot prevent them spreading (Peay, 2001), as they will readily walk on land in order to go around in-stream obstacles (Peay, 2001; Holdich, 2002a), and can travel several hundred metres over land in one night (Hiley, 2002).

Signal crayfish are a very resilient species, and tolerate more extreme temperatures (Nakata et al, 2002), salinity (Holdich et al, 1997) and levels of pollutants such as sulphates (Rallo and García-Arberas, 2002) than native species. In Britain it is thought that there may be no situation in which the native white-clawed crayfish has a competitive advantage over the invading species (Hiley, 2002), although there are parts of Finland where the native noble crayfish survives and reproduces better than the invasive species (Heinemaa and Pursiainen, 2008).

Environmental impacts

The signal crayfish has negative impacts upon other aspects of river ecology besides competition with and replacement of native crayfish. It is an omnivorous species, and as such consumes both macrophytes (Nyström and Strand, 1996) and invertebrates (Crawford et al, 2006), and can modify the structure of invertebrate communities. It decreases invertebrate species richness and diversity, and reduces overall invertebrate biomass by as much as 40 per cent (Nyström et al, 2001; Crawford et al, 2006). Amphibian populations also suffer as signal crayfish prey on eggs and tadpoles, resulting in fewer froglets where invasive crayfish are present (Axelsson et al, 1997; Nyström et al, 2001). Effects may be even more serious in ecosystems where no crayfish are naturally present, such as in Scotland.

Fish species also suffer as a consequence of signal crayfish introductions: the invasive crayfish preys directly on fish and fish eggs as well as indirectly causing increased fish predation by ousting them from shelters (Guan and Wiles, 1997; Griffiths et al, 2004; Peay et al, 2009). The species competes with both benthic (Guan and Wiles, 1997) and salmonid (Griffiths et al, 2004) fish for shelter in laboratory experiments, resulting in a significant negative relationship between fish and signal crayfish abundance in wild populations (Peay et al, 2009). Peay et al (2009) also found that streams with a high abundance of the

native white-clawed crayfish also had a high abundance of juvenile trout, indicating that it is specifically the alien crayfish that impacts upon fish species. Guan and Wiles (1997) note that local extinctions of benthic fish are likely to result from the continued spread of signal crayfish, and Peay et al (2009) warn of a significant impact upon the fishing industry.

The signal crayfish is also a burrowing species in its invasive range, although burrowing has not been recorded in its native range (Guan, 1994). As the species reaches such high densities, burrowing can have considerable impacts upon riverbanks and may even cause bank collapse (Sibley, 2000).

Control Strategies

A major problem with controlling the spread of the species is that individuals may be present for a number of years before they reach densities at which they can be detected by standard methods. Minimum viable population density for signal crayfish may be lower than the detection limit for the species and territoriality may play a part in its spread (Sibley, 2000), allowing signal crayfish to colonize watercourses long before they can be detected (Hiley, 2002).

Once signal crayfish populations have become established, they are extremely difficult to eradicate (Rogers and Loveridge, 2000), and elimination other than by complete extermination is futile as the species will recolonize (Hiley, 2002). The importance of early eradication must be emphasized as attempting to eliminate populations that have already spread is usually financially and logistically unfeasible (Rogers and Loveridge, 2000). However, preventing further introductions is the only realistic way to prevent further spread: while legislation preventing the introduction of alien crayfish species is in place in many countries, enforcement and lack of education remain problematic. Thus, continuing emphasis needs to be placed upon education and information sharing (Peay, 2009a). A considerable amount of work has gone into attempting eradication strategies for signal crayfish, these falling into four main strategies: mechanical, physical, biological and chemical (Peay, 2001).

Mechanical removal

Trapping and netting using seine, fyke and drag nets can be effective at removing large numbers of signal crayfish;

however, they will never be effective at removing an entire population (Rogers and Loveridge, 2000), and may not even prevent a population from spreading (Sibley, 2000). Trapping tends to select for the capture of larger males, probably due to females becoming less active during the breeding season (Wright and Williams, 2000), and the large mesh size of conventional traps prevents the retention of smaller individuals (Byrne et al, 1999). Even when traps are modified by the use of smaller mesh, traps are still biased towards the capture of large crayfish, probably because smaller individuals avoid entering a trap where larger crayfish are already present in order to avoid the risk of cannibalism (Byrne et al, 1999). The removal of large males through trapping may lead to the remaining individuals becoming sexually mature at a younger age (Sibley, 2000; Smith and Wright, 2000), and decreasing density is known to be associated with increased reproductive success and decreased natural mortality (Smith and Wright, 2000), suggesting that a large proportion would need to be removed in order to achieve an overall decrease in abundance (Smith and Wright, 2000).

A trial of selective removal of only females and juveniles, returning adult males to the population, was undertaken in the River Wreake in Leicestershire, UK. The premise of this trial was that a superabundance of large males in a population would increase cannibalism of smaller individuals, and suppress the activity and thus productivity of younger crayfish, suppressing population growth (Sibley, 2000). However, this trial had no clear effects on population growth and no discernable effect upon recruitment within the population (Sibley, 2000).

Electrofishing can be effectively used to sample crayfish populations (Alonso, 2001), but is ineffective in deep or fast flowing water, or for crayfish in burrows, and thus can only be used in certain conditions (Sibley et al, 2009). Sinclair and Ribbens (1999) established using mark-recapture that three electrofishing runs would only catch 24–35 per cent of the population, thus making electrofishing unsuitable as an eradication method. Indeed, electrofishing may inadvertently assist the spread of crayfish populations through immobilized individuals being washed downstream (Freeman et al, 2010).

Physical removal

Habitat destruction combined with dewatering has been attempted in a bid to remove signal crayfish. However, despite the drastic nature of this treatment, it is unlikely to remove or kill signal crayfish unless the area remains dry for at least three months and all suitable habitat is excavated, as they can survive out of water for up to three months in burrows (Hiley, 2002). Indeed, one experimental trial involved dewatering and over-wintering of a pond in the Czech Republic, and signal crayfish were still found the following year, indicating that the species can survive in a drained pond with over-winter air temperatures of –20°C (Kozák and Policar, 2002).

Biological control

A method of signal crayfish control that has been considered but not trialled is that of predator introduction. A recent study with another invasive crayfish species, the red swamp crayfish (*Procambarus clarkia*), showed that the European eel (*Anguilla Anguilla*) both consumed crayfish and altered their behaviour, resulting in reduced trophic activity and potentially increased mortality due to starvation (Aquiloni et al, 2010). This effect is likely to be applicable to signal crayfish as eels predate heavily on this crayfish species in the wild (Furst, 1977; Blake and Hart, 1995) and elicit predator avoidance behaviour in the species (Blake and Hart, 1993). However, the main concern with the introduction of eels for signal crayfish control is that the eels would not stay within the introduction area (Peay, 2001). While predation by fish species is unlikely to eliminate invasive crayfish entirely, Aquiloni et al (2010) suggest that it may be used as a complement to trapping in order to depress populations.

The use of sterile males to upset the breeding pattern of signal crayfish has been considered in order to reduce productivity (Rogers and Watson, 2005). However, a feasibility study concluded that the necessity of releasing large numbers of sterile males would have a greater impact upon the environment than the original population (Rogers and Watson, 2005), and would at best only slow the growth of a signal crayfish population.

Potentially the best method for controlling signal crayfish would be using a species-specific pathogen, such as a fungus, virus or microbial organism, that would only affect signal crayfish without impacting upon the wider environment (Sibley et al, 2009). Great care would need to be taken that this pathogen would not affect other aquatic fauna or native crayfish species, just the invasive species. The likelihood of finding a pathogen fitting these criteria would seem slim, especially given the resilience of the alien species to pathogens that cause harm to native species. For example, *Thelohania contejeani*, the causative agent of porcelain disease, is known to cause mortality in the native white-clawed crayfish but has been found in signal crayfish with no signs of clinical impact in the invasive species (Dunn et al, 2009).

Chemical eradication

The use of pheromones has been widely used with problem species, with success in many cases (e.g. Agosta, 1992). Trials with signal crayfish were the first time that pheromones had been employed with an aquatic target species (Stebbing et al, 2003); however, results were unconvincing. While traps baited with female sex pheromones successfully caught males, these traps caught no more male crayfish than traditional food-baited traps and in fact caught fewer crayfish overall as they caught no females (Stebbing et al, 2003). Trials were also carried out using traps baited with alarm pheromones, in the hope that the deployment of alarm pheromones in certain stretches of water would repel crayfish from these stretches, thus limiting the spread of signal crayfish. Unfortunately these trials were also unsuccessful, as similar numbers of both male and female signal crayfish were caught in alarm pheromone-baited traps to those caught in food-baited traps (Stebbing et al, 2003).

Pyrethrin biocides have been used to attempt eradication of some populations of signal crayfish, with promising results thus far (Peay et al, 2006; Sandodden and Johnsen, 2010). Peay et al (2006) trialled the use of a natural pyrethrin at sites in Scotland, with some success; however, the high costs and legislative problems are thought likely to prevent widespread use. No biocides specific to crayfish are currently available, so those that are used are also toxic to non-target invertebrates as well as fish (Peay et al, 2006); thus the use of biocides is limited to ponds or stretches of water that can be isolated, or where outflow can be back-pumped to avoid leakage (Peay et al, 2006), and legal issues may prevent the use of biocides in some countries (e.g. Frutiger et al, 1999). Peay et al (2006) showed 100 per cent mortality in caged crayfish after five days of biocide treatment, and no crayfish were trapped during the summer following treatment; however, monitoring over a period of two to five years following treatment is deemed necessary to ensure that eradication has been successful (Peay et al, 2006). A more recent trial in Norway, using a synthetic pyrethroid in conjunction with pond drainage also showed promise in eradicating signal crayfish from two ponds (Sandodden and Johnsen, 2010). As of August 2010, no crayfish had been trapped since the treatment in 2008, although further monitoring is required before eradication can be confirmed (Sandodden, R., pers. comm.).

Challenges and Controversies

The resilient nature of signal crayfish, along with their ability to adapt to a wide range of environmental conditions makes them an extremely efficient invader, and potentially impossible to eradicate. The wide range of habitats occupied by the species means that integrated pest management (IPM) is likely to be most successful, by implementing a range of site-specific control and containment strategies depending on the nature of each site (Freeman et al, 2010). For example, biocides show promise for controlling isolated populations of signal crayfish (Sandodden and Johnsen, 2010), but the prospects of eliminating them in watercourses are currently remote. In Britain, research is under way into the selection of Ark (new refuge) sites for the native species, which in practice tend to be isolated sites with a low risk of invasion from signal crayfish (Sibley et al, 2009). This is very much a last resort, but is now considered a necessity for the conservation of native crayfish species in the wild where signal crayfish have been introduced (Peay, 2009b).

The emphasis in preventing the spread of signal crayfish needs to be on education and information sharing (Peay, 2009a): fish markets still sell live crayfish, allowing for their continued introduction (Holdich and Sibley, 2009), and while legislation exists to prevent further introductions, it is rarely enforced (Peay, 2009a). Improving education, alongside legislation and its enforcement, regarding the sale of non-native crayfish species in Europe through fish markets and the aquarium trade, is the best means of slowing the spread of the signal crayfish (Peay, 2009a).

Acknowledgements

Thanks to Neal Haddaway for providing Figure 17.2, and to Emily Imhoff for permission to use Figures 17.1 and 17.3. Thanks also to Katie Arundell for comments on an earlier draft and to an anonymous reviewer for comments that greatly improved the manuscript.

References

Agosta, W. C. (1992) *Chemical Communication, The Language of Pheromones*, Scientific American Library, New York

Alderman, D., Holdich, D. and Reeve, I. (1990) 'Signal crayfish as vectors of crayfish plague in Britain', *Aquaculture*, vol 86, no 1, p3

Alonso, F. (2001) 'Efficiency of electrofishing as a sampling method for freshwater crayfish populations in small creeks', *Limnetica*, vol 20, no 1, pp59–72

Aquiloni, L., Brusconi, S., Cecchinelli, E., Tricarico, E., Mazza, G., Paglianti, A. and Gherardi, F. (2010) 'Biological control of invasive populations of crayfish: The European eel (*Anguilla anguilla*) as a predator of *Procambarus clarkii*', *Biological Invasions*, vol 12, no 11, pp3817–3824

Axelsson, E., Nyström, P., Sidenmark, J. and Brönmark, C. (1997) 'Crayfish predation on amphibian eggs and larvae', *Amphibia Reptilia*, vol 18, no 3, pp217–228

Blake, M. A. and Hart, P. J. B. (1993) 'The behavioural responses of juvenile signal crayfish *Pacifastacus leniusculus* to stimuli from perch and eels', *Freshwater Biology*, vol 29, no 1, pp89–97

Blake, M. A. and Hart, P. J. B. (1995) 'The vulnerability of juvenile signal crayfish to perch and eel predation', *Freshwater Biology*, vol 33, no 2, pp233–244

Bondar, C. A. and Richardson, J. S. (2009) 'Effects of ontogenetic stage and density on the ecological role of the signal crayfish (*Pacifastacus leniusculus*) in a coastal Pacific stream', *Journal of the North American Benthological Society*, vol 28, no 2, pp294–304

Bondar, C. A., Zhang, Y., Richardson, J. S. and Jesson, D. (2005) *The Conservation Status of the Freshwater Crayfish*, Pacifastacus leniusculus, *in British Columbia*, Ministry of Water, Land and Air Protection, Province of British Columbia

Bubb, D. H., Thom, T. J. and Lucas, M. C. (2005) 'The within-catchment invasion of the non-indigenous signal crayfish *Pacifastacus leniusculus* (Dana), in upland rivers', *Bulletin Français de la Pêche et de la Pisciculture*, vol 376–377, pp665–673

Bubb, D. H., Thom, T. J. and Lucas, M. C. (2006a) 'Movement patterns of the invasive signal crayfish determined by PIT telemetry', *Canadian Journal of Zoology*, vol 84, no 8, pp1202–1209

Bubb, D. H., Thom, T. J. and Lucas, M. C. (2006b) 'Movement, dispersal and refuge use of co-occurring introduced and native crayfish', *Freshwater Biology*, vol 51, no 7, pp1359–1368

Byrne, C. F., Lynch, J. M. and Bracken, J. J. (1999) 'A sampling strategy for stream populations of white-clawed crayfish, *Austropotamobius pallipes* (Lereboullet) (Crustacea, Astacidae)', *Biology and Environment: Proceedings of the Royal Irish Academy*, vol 99B, no 2, pp89–94

Cerenius, L. and Söderhäll, K. (1992) 'Crayfish diseases and crayfish as vectors for important disease', *Finnish Fisheries Research*, vol 14, pp125–133

Crawford, L., Yeomans, W. E. and Adams, C. E. (2006) 'The impact of introduced signal crayfish *Pacifastacus leniusculus* on stream invertebrate communities', *Aquatic Conservation: Marine and Freshwater Ecosystems*, vol 16, no 6, pp611–621

Diéguez-Uribeondo, J. (2006) 'The dispersion of the *Aphanomyces astaci*-carrier *Pacifastacus leniusculus* by humans represents the main cause of disappearance of the indigenous crayfish *Austropotamobius pallipes* in Navarra', *Bulletin Français de la Pêche et de la Pisciculture*, vol 380–381, pp1303–1312

Dorn, N. J. and Wojdak, J. M. (2004) 'The role of omnivorous crayfish in littoral communities', *Oecologia*, vol 140, no 1, pp150–159

Dunn, J. C., McClymont, H. E., Christmas, M. and Dunn, A. M. (2009) 'Competition and parasitism in the native white clawed crayfish *Austropotamobius pallipes* and the invasive signal crayfish *Pacifastacus leniusculus* in the UK', *Biological Invasions*, vol 11, no 2, pp315–324

Freeman, M. A., Turnbull, J. F., Yeomans, W. E. and Bean, C. W. (2010) 'Prospects for management strategies of invasive crayfish populations with an emphasis on biological control', *Aquatic Conservation: Marine and Freshwater Ecosystems*, vol 20, pp211–223

Frutiger, A., Borner, S., Büsser, T., Eggen, R., Müller, R., Müller, S. and Wasmer, H. R. (1999) 'How to control unwanted populations of *Procambarus clarkii* in Central Europe?', *Freshwater Crayfish*, vol 12, pp714–726

Furst, M. (1977) 'Introduction of *Pacifastacus leniusculus* (Dana) into Sweden: Methods, results and management', *Freshwater Crayfish*, vol 3, pp229–247

Gherardi, F., Bertolino, S., Bodon, M., Casellato, S., Cianfanelli, S., Ferraguti, M., Lori, E., Mura, G., Nocita, A., Riccardi, N., Rossetti, G., Rota, E., Scalera, R., Zerunian, S. and Tricarico, E. (2008) 'Animal xenodiversity in Italian inland waters: Distribution, modes of arrival, and pathways', *Biological Invasions*, vol 10, pp435–454

Griffiths, S. W., Collen, P. and Armstrong, J. D. (2004) 'Competition for shelter among over-wintering signal crayfish and juvenile Atlantic salmon', *Journal of Fish Biology*, vol 65, no 2, pp436–447

Guan, R.-Z. (1994) 'Burrowing behaviour of signal crayfish, *Pacifastacus leniusculus* (Dana) in the River Great Ouse, England', *Freshwater Forum*, vol 4, no 3, pp155–168

Guan, R.-Z. and Wiles, P. R. (1997) 'Ecological impact of introduced crayfish on benthic fishes in a British lowland river', *Conservation Biology*, vol 11, no 3, pp641–647

Guan, R.-Z. and Wiles, P. R. (1999) 'Growth and reproduction of the introduced crayfish *Pacifastacus leniusculus* in a British lowland river', *Fisheries Research*, vol 42, no 3, pp245–259

Hamr, P. (1998) *Conservation Status of Canadian Freshwater Crayfishes*, World Wildlife Fund Canada and the Canadian Nature Federation, Toronto

Hefti, D. and Stucki, P. (2006) 'Crayfish management for Swiss waters', *Bulletin Français de la Pêche et de la Pisciculture*, vol 380–381, pp937–950

Heinemaa, S. and Pursiainen, M. (2008) 'signal crayfish *Pacifastacus leniusculus* at northerly latitudes: A search for the distribution limits', *Freshwater Crayfish*, vol 16, pp37–41

Hiley, P. D. (2002) 'The slow quiet invasion of signal crayfish (*Pacifastacus leniusculus*) in England – prospects for the white clawed crayfish (*Austropotamobius pallipes*)', in *Management & Conservation of Crayfish*, Environment Agency, Bristol, pp127–138

Hiruta, S. (1999) 'The present status of crayfish in Britain and the conservation of the native species in Britain and Japan', *Journal of Environmental Education*, vol 2, pp119–132

Holdich, D. M. (2002a) 'Crayfish in Europe – An overview of taxonomy, legislation, distribution, and crayfish plague outbreaks', in *Management & Conservation of Crayfish*, Environment Agency, Bristol, pp15–34

Holdich, D. M. (2002b) 'Distribution of crayfish in Europe and some adjoining countries', *Bulletin Français de la Pêche et de la Pisciculture*, vol 367, pp611–650

Holdich, D. M. and Rogers, W. D. (1997) 'The white-clawed crayfish, *Austropotamobius pallipes*, in Great Britain and Ireland with particular reference to its conservation in Great Britain', *Bulletin Français de la Pêche et de la Pisciculture*, vol 347, pp597–616

Holdich, D. and Rogers, D. (1999) *Freshwater Crayfish in Britain and Ireland*, Environment Agency, Bristol

Holdich, D. M. and Sibley, P. J. (2009) 'ICS and NICS in Britain in the 2000s', in J. Brickland, D. M. Holdich and E. M. Imhoff (eds) *Crayfish Conservation in the British Isles*, British Waterways Offices, Leeds, UK, pp13–34

Holdich, D. M., Harlioğlu, A. G. and Firkins, I. (1997) 'Salinity adaptations of crayfish in British Waters with particular reference to *Austropotamobius pallipes*, *Astacus leptodactylus* and *Pacifastacus leniusculus*', *Estuarine, Coastal and Shelf Science*, vol 44, no 2, pp147–154

Holdich, D. M., Reynolds, J. D., Souty-Grosset, C. and Sibley, P. J. (2009) 'A review of the ever increasing threat to European crayfish from non-indigenous crayfish species', *Knowledge and Management of Aquatic Ecosystems*, vol 394–395, art 11.

Howells, M. and Slater, F. (2004) 'Remnant populations of *Austropotamobius pallipes* in Wales, UK: Counts, causes, cures and consequences', *Freshwater Crayfish*, vol 14, pp140–146

Huber, M. G. J. and Schubart, C. D. (2005) 'Distribution and reproductive biology of *Austropotamobius torrentium* in Bavaria and documentation of a contact zone with the alien crayfish *Pacifastacus leniusculus*', *Bulletin Français de la Pêche et de la Pisciculture*, vol 376–377, pp759–776

Hudina, S., Faller, M., Lucić, A., Klobučar, G. and Maguire, I. (2009) 'Distribution and dispersal of two invasive crayfish species in the Drava River basin, Croatia', *Knowledge and Management of Aquatic Ecosystems*, vol 394–395, art 09

Kamita, T. (1970) *Studies on the Fresh-water Shrimps, Prawns and Crayfishes of Japan*, Sonoyama shoten, Matsue

Kirjavainen, J. and Westman, K. (1994) 'Comparative growth from length composition and mark-recapture experiments for noble crayfish (*Astacus astacus*) and signal crayfish (*Pacifastacus leniusculus*) in Finland', *Nordic Journal of Freshwater Research*, vol 69, pp153–161

Kozák, P. and Policar, T. (2002) 'Practical elimination of signal crayfish, Pacifastacus leniusculus (Dana), from a pond', in *Management & Conservation of Crayfish*, Environment Agency, Bristol, 200–208

Kozubíková, E., Filipová, L., Kozák, P., Duriš, Z., Martín, M. P., Diéguez-Uribeondo, J., Oidtmann, B. and Petrusek, A. (2009) 'Prevalence of the crayfish plague pathogen *Aphanomyces astaci* in invasive American crayfish in the Czech Republic', *Conservation Biology*, vol 23, no 5, pp1204–1213

Light, T. (2003) 'Success and failure in a lotic crayfish invasion: The roles of hydrologic variability and habitat alteration', *Freshwater Biology*, vol 48, no 10, pp1886–1897

Light, T., Erman, D. C., Myrick, C. and Clarke, J. (1995) 'Decline of the Shasta crayfish (*Pacifastacus fortis* Foxon) of Northeastern California', *Conservation Biology*, vol 9, no 6, pp1567–1577

Machino, Y. and Holdich, D. M. (2006) 'Distribution of crayfish in Europe and adjacent countries: Updates and comments', *Freshwater Crayfish*, vol 15, pp292–323

Nakata, K. and Goshima, S. (2006) 'Asymmetry in mutual predation between the endangered Japanese native crayfish *Cambaroides japonicus* and the North American invasive crayfish *Pacifastacus leniusculus*: A possible reason for species replacement', *Journal of Crustacean Biology*, vol 26, no 2, pp134–140

Nakata, K., Hamano, T., Hayashi, K.-I. and Kawai, T. (2002) 'Lethal limits of high temperature for two crayfishes, the native species *Cambaroides japonicus* and the alien species *Pacifastacus leniusculus* in Japan', *Fisheries Science*, vol 68, no 4, pp763–767

Nylund, V. and Westman, K. (2000) 'The prevalence of crayfish plague (*Aphanomyces astaci*) in two signal crayfish (*Pacifastacus leniusculus*) populations in Finland', *Journal of Crustacean Biology*, vol 20, no 4, pp777–785

Nyström, P. and Strand, J. A. (1996) 'Grazing by a native and an exotic crayfish on aquatic macrophytes', *Freshwater Biology*, vol 36, no 3, pp673–682

Nyström, P., Svensson, O., Lardner, B., Brönmark, C. and Granéli, W. (2001) 'The influence of multiple introduced predators on a littoral pond community', *Ecology*, vol 82, no 4, pp1023–1039

Ohtaka, A., Gelder, S. R., Kawai, T., Saito, K., Nakata, K. and Nishino, M. (2005) 'New records and distributions of two North American branchiobdellidan species (Annelida: Clitellata) from introduced signal crayfish, *Pacifastacus leniusculus*, in Japan', *Biological Invasions*, vol 7, no 2, pp149–156

Peay, S. (2001) *Eradication of Alien Crayfish Populations*, Environment Agency, Bristol

Peay, S. (2009a) 'Invasive non-indigenous crayfish species in Europe: Recommendations on managing them', *Knowledge and Management of Aquatic Ecosystems*, vol 394–395, art 03

Peay, S. (2009b) 'Selection criteria for "ark sites" for white-clawed crayfish', in J. Brickland, D. M. Holdich and E. M. Imhoff (eds) *Crayfish Conservation in the British Isles*, British Waterways Offices, Leeds, UK, pp63–70

Peay, S. and Rogers, D. (1999) 'The peristaltic spread of signal crayfish (*Pacifastacus leniusculus*) in the River Wharfe, Yorkshire, England', *Freshwater Crayfish*, vol 12, pp665–676

Peay, S., Hiley, P. D., Collen, P. and Martin, I. (2006) 'Biocide treatment of ponds in Scotland to eradicate signal crayfish', *Bulletin Français de la Pêche et de la Pisciculture*, vol 380–381, pp1363–1379

Peay, S., Guthrie, N., Spees, J., Nilsson, E. and Bradley, P. (2009) 'The impact of signal crayfish (*Pacifastacus leniusculus*) on the recruitment of salmonid fish in a headwater stream in Yorkshire, England', *Knowledge and Management of Aquatic Ecosystems*, vol 394–395, art 12

Peay, S., Holdich, D. and Brickland, J. (2010) 'Risk assessments of non-indigenous crayfish in Great Britain', *Freshwater Crayfish*, vol 17, pp109–122

Pöckl, M. and Pekny, R. (2002) 'Interaction between native and alien species of crayfish in Austria: Case studies', *Bulletin Français de la Pêche et de la Pisciculture*, vol 367, pp763–776

Puky, M., Reynolds, J. D. and Schád, P. (2005) 'Native and alien decapoda species in Hungary: Distribution, status, conservation importance', *Knowledge and Management of Aquatic Ecosystems*, vol 376–377, pp553–568

Pursiainen, M. and Rajala, J. E. (2010) *Crayfish Review 2009*, Riista- ja kalatalous – Selvityksiä 8/2010

Rallo, A. and García-Arberas, L. (2002) 'Differences in abiotic water conditions between fluvial reaches and crayfish fauna in some northern rivers of the Iberian Peninsula', *Aquatic Living Resources*, vol 15, no 2, pp119–128

Rogers, D. and Loveridge, S. (2000) *Feasibility of Control of Signal Crayfish – A Case Study, River Ure, North Yorkshire*, Crayfish Conference Leeds, Environment Agency, Leeds, Environment Agency, Leeds

Rogers, D. and Watson, E. (2005) 'A feasibility study of the potential application of sterile male introduction to control signal crayfish (*Pacifastacus leniusculus*)', English Nature, Peterborough

Sandodden, R. and Johnsen, S. I. (2010) 'Eradication of introduced signal crayfish *Pacifastacus leniusculus* using the pharmaceutical BETAMAX VET', *Aquatic Invasions*, vol 5, no 1, pp75–81

Sibley, P. J. (2000) *Signal Crayfish Management in the River Wreake Catchment*, Crayfish Conference Leeds, Environment Agency, Leeds, Environment Agency, Leeds

Sibley, P. J., Holdich, D. M. and Lane, M.-R. (2009) 'Invasive crayfish in Britain – management and mitigation', in I. D. Rotherham (ed) *Exotic and Invasive Plants and Animals: International Urban Ecology Review, No 4*, Wildtrack Publishing, Sheffield, pp105–118

Sinclair, C. A. and Ribbens, J. C. H. (1999) *The Distribution of American Signal Crayfish (*Pacifastacus leniusculus*) in the Kirkcudbrightshire Dee (Dumfries and Galloway) and an Assessment of the Use of Electrofishing as an Eradication Technique for Crayfish Populations*, Scottish Natural Heritage, Commissioned Report (F99AC601)

Smith, P. A. and Wright, R. (2000) *A Preliminary Consideration of Some Aspects Relating to the Population Dynamics of Signal Crayfish (*Pacifastacus leniusculus*) With a View to Assessing the Utility of Trapping as a Removal Method*, Crayfish Conference, Leeds, Environment Agency, Leeds

Spink, J. and Rowe, J. (2002) 'Signal and native crayfish in Broadmead Brook, Wiltshire', in *Management & Conservation of Crayfish*, Environment Agency, Bristol, pp148–150

Stebbing, P. D., Watson, G. J., Bentley, M. G., Fraser, D., Jennings, R., Rushton, S. P. and Sibley, P. J. (2003) 'Reducing the threat: The potential use of pheromones to control invasive signal crayfish', *Bulletin Français de la Pêche et de la Pisciculture*, vol 370–371, pp219–224

Stebbing, P. D., Elwis, A., Watson, G. J. and Bentley, M. G. (2006) 'A possible mechanism for the displacement of *Austropotamobius pallipes* by *Pacifastacus leniusculus*', *Freshwater Crayfish*, vol 15, pp130–138

Usio, N., Konishi, M. and Nakano, S. (2001) 'Species displacement between an introduced and a "vulnerable" crayfish: The role of aggressive interactions and shelter competition', *Biological Invasions*, vol 3, no 2, pp179–185

Vorburger, C. and Ribi, G. (1999) 'Aggression and competition for shelter between a native and an introduced crayfish in Europe', *Freshwater Biology*, vol 42, no 1, pp111–119

Westman, K., Savolainen, R. and Julkunen, M. (2002) 'Replacement of the native crayfish *Astacus astacus* by the introduced species *Pacifastacus leniusculus* in a small, enclosed Finnish lake: A 30-year study', *Ecography*, vol 25, no 1, pp53–73

Wright, R. and Williams, M. (2000) *Long Term Trapping of Signal Crayfish at Wixoe on the River Stour, Essex*, Crayfish Conference Leeds, Environment Agency, Leeds

18

Apple snails

Robert H. Cowie and Kenneth A. Hayes

Introduction

Ampullariidae are freshwater snails predominantly distributed in humid tropical and subtropical habitats of Africa, South and Central America, and Asia. The family name Pilidae is a junior synonym (ICZN, 1999) and should not be used. They include the largest of all freshwater snails (up to 17cm) and are a major component of the native freshwater mollusc faunas of many regions. Among the nine genera usually recognized (Berthold, 1991; Cowie and Thiengo, 2003), the two largest are *Pomacea* and *Pila*. Snails in these genera particularly are frequently known as 'apple snails' because many species bear large, round, often greenish shells.

Six of the nine genera contain fewer than six species each: *Afropomus* and *Saulea* are African; *Asolene*, *Felipponea*, *Pomella* and *Marisa* are South American. The genera *Lanistes* (African), *Pila* (African and Asian; *Ampullaria* and *Ampullarius* are junior synonyms) and *Pomacea* (South and Central American), contain 21, about 30, and probably 50 or more species, respectively, and comprise the largest part of the family. An additional genus, *Pseudoceratodes* (African, fossil only), is included in the family only tentatively.

Species of *Pomacea* especially, introduced to southern and eastern Asia and islands of the Pacific, have become major agricultural pests, notably in rice and taro but also other crops (Cowie, 2002; Joshi and Sebastian, 2006). *Pomacea* species have also been introduced to the continental US (Rawlings et al, 2007), Europe (López et al, 2010) and Australia (Hayes

et al, 2008), and between locations within South and Central America (Hayes et al, 2008). The name 'golden apple snail' is used widely in Asia for introduced *Pomacea* (Lai et al, 2005; Joshi and Sebastian, 2006), implying a single species, although it had been identified or misidentified as numerous different species (Cowie et al, 2006; Joshi and Sebastian, 2006). It had also been suggested, in some cases based on misidentifications, that more than one species was present in Asia (Keawjam and Upatham, 1990; Mochida, 1991; Yipp et al, 1991). In Hawaii, four species were initially recognized but in fact only three have been introduced (Cowie et al, 2007). In the continental US, introduced apple snails were identified primarily as *Pomacea canaliculata* and given the common name 'channeled apple snail', an anglicization of the scientific species name.

We now know, as a result primarily of analysis of DNA sequences, that several species are involved. In Asia, Hayes et al (2008) showed that the 'golden apple snail' is in fact two species, *Pomacea canaliculata* and *P. insularum*, and that *P. diffusa* and *P. scalaris* have also been introduced (Plate 18.1). Note that the snails from Cambodia, illustrated by Cowie (2002) as *P. canaliculata*, are in fact *P. insularum*, and much of the information given by Cowie (2002) for *P. canaliculata* is confounded with information for *P. insularum*, the two species not having been distinguished reliably at that time. Tran et al (2008) showed that in Hawaii all the 'golden' or 'channeled' apple snails were *P. canaliculata*, while *P. diffusa* and *Pila conica* had also been introduced. In the continental US, Rawlings et al

(2007) distinguished three species within what had been identified previously as 'channeled apple snails' (i.e. *P. canaliculata*): *P. canaliculata*, *P. insularum* (now referred to as the 'island apple snail') and *P. haustrum* ('titan apple snail'). They also confirmed the presence of *Pomacea diffusa* ('spike-topped apple snail') and *Marisa cornuarietis* ('giant ramshorn snail') in addition to the one native North American apple snail, *Pomacea paludosa* ('Florida apple snail').

The following sections on the distribution of introduced apple snails, their ecology, behaviour and physiology, importance as pests and management options are based largely on the review of Cowie (2002), which should be consulted for a more comprehensive treatment dealing with ampullariids in general, not just the invasive species. However, much additional information, particularly on the pest species, has become available since that publication, so in general, citations emphasize later publications.

History of Introduction and Current Distribution

The current known distributions of non-native species are given in Table 18.1. Other species recorded as introduced are based on misidentifications, e.g. *Pomacea bridgesii*, *P. cuprina*, *P. gigas*, *P. levior*, *P. lineata* and *P. paludosa* in Asia (e.g. Mochida, 1991; Yipp et al, 1991; Cuong, 2006; Hendarsih-Suharto et al, 2006), *P. paludosa* in Hawaii (Cowie, 1995) and *P. canaliculata* in Australia and Texas (Ranamukhaarachchi and Wikramasinghe, 2006).

The most widespread species is *Pomacea canaliculata*. *Pomacea insularum* is also present in Asia but less widely than *P. canaliculata*. In the US, *P. insularum* is present in the southeast, contrasting with the initial distribution of *P. canaliculata* in the west. *Pomacea haustrum* has also been reported. Within South America, *P. canaliculata* has been reported beyond its native range, in Chile. *Pomacea diffusa*, native to the Amazonian region, is established in Sri Lanka and Australia, as well as non-native regions of South and Central America. *Pomacea scalaris* has been introduced to Taiwan.

The native range of *Marisa cornuarietis* may only encompass Venezuela and Colombia, it having been introduced to other parts of northern South America (Table 18.1), although it is possible that it occurs in these areas naturally. It is widely introduced elsewhere,

especially in the Caribbean. The Asian *Pila conica* has been introduced to Guam, Palau and Hawaii in the Pacific, and the African *Pila leopoldvillensis* has been recorded in Taiwan.

Introduction for food

Some ampullariids are used as human food in their native ranges, mostly in Asia but also in South America and Africa. However, deliberate transport of apple snails from their native ranges to new regions as novel human food resources is probably the most important cause of their spread and establishment.

In the Pacific, *Pila conica* was introduced without authorization, either accidentally or deliberately as a food item to both Guam (first recorded 1984) and Hawaii (first recorded 1966) (Smith, 1992; Cowie, 1995), probably from the Philippines (Tran et al, 2008). It was also introduced to Palau in 1984 or 1985 but was eradicated by 1987 (Eldredge, 1994). But it is the South American *Pomacea* species that have attracted most attention, notably in southern and eastern Asia, where they have become major agricultural pests.

Between 1979 and 1981 a species of *Pomacea* was introduced to Asia, initially from Argentina to Taiwan (Mochida, 1991), although it may have been introduced earlier in the 1970s to the Philippines, China and Viet Nam (Wu and Xie, 2006). Undoubtedly this was *Pomacea canaliculata*, the only widespread species in Taiwan. *Pomacea scalaris*, also now recorded in Taiwan, may have been introduced accidentally with the original introduction(s) of *P. canaliculata* (Wu et al, 2011). The initial introduction to Taiwan was illegal, its purpose being to develop the species for both local consumption and export to the gourmet restaurant trade. The subsequent spread of these snails in Asia and the Pacific, distributed primarily for the same purposes, has been summarized by Cowie (2002), Halwart and Bartley (2006), Wu and Xie (2006) and others (generally not distinguishing *P. canaliculata* from *P. insularum*). Halwart and Bartley (2006) listed the origins of many of the Asian introductions as 'Amazon basin', which is incorrect except possibly for *Pomacea diffusa*. In 1981 snails were taken from Taiwan to Japan, Korea (Lee and Oh, 2006), China and Indonesia. By 1982 they had been introduced to the Philippines, and introductions to the Philippines continued from various sources as snail farming was promoted by governmental and

Table 18.1 *Native and non-native ranges of introduced ampullariids in the wild*

Species	Native range	Non-native range	Representative references for non-native range
Marisa cornuarietis	Colombia	Costa Rica	Nguma et al, 1982
	Venezuela	Cuba	Hunt, 1958
		Dominican Republic	Perera and Walls, 1996
		Egypt[a]	Brown, 1994
		French Guyana[b]	Massemin et al, 2009
		Guadeloupe	Pointier and David, 2004
		Guyana	Nguma et al, 1982; Massemin et al, 2009
		Panama	Nguma et al, 1982
		Puerto Rico	Hunt, 1958; Peebles et al, 1972; Nguma et al, 1982; Perera and Walls, 1996
		Sudan[a]	Brown, 1994
		Surinam	Nguma et al, 1982
		Tanzania[a]	Nguma et al, 1982; Brown, 1994
		US (Florida)	Hunt, 1958; Rawlings et al, 2007
		US (Texas)	Neck, 1984, Cowie, 2002
Pila conica	Southeast Asia	Guam	Smith, 1992; Cowie, 2002
		Hawaii	Cowie, 1995; Tran et al, 2008
		Palau[c]	Eldredge, 1994, Cowie, 2002
Pila leopoldvillensis	Africa	Taiwan	Wu and Lee, 2005
Pomacea canaliculata	Argentina	Bangladesh[d]	Ranamukhaarachchi and Wikramasinghe, 2006; Wu and Xie, 2006
	Uruguay	Cambodia[e]	Ranamukhaarachchi and Wikramasinghe, 2006
	Paraguay	Chile	Letelier and Soto-Acuña, 2008
	Southern Brazil	China	Hayes et al, 2008
		Dominican Republic	Rosario and Moquete, 2006
		Egypt[d]	Wu and Xie, 2006
		Guam	Hayes et al, 2008
		Hawaii	Hayes et al, 2008; Tran et al, 2008
		India[d]	Ranamukhaarachchi and Wikramasinghe, 2006; Wu and Xie, 2006
		Indonesia	Hayes et al, 2008
		Japan	Hayes et al, 2008
		Laos	Hayes et al, 2008
		Malaysia	Hayes et al, 2008
		Myanmar (Burma)	Hayes et al, 2008
		Papua New Guinea	Hayes et al, 2008
		Philippines	Hayes et al, 2008
		Singapore	Halwart and Bartley, 2006
		South Africa[f]	Berthold, 1991
		South Korea	Hayes et al, 2008

Table 18.1 *Native and non-native ranges of introduced ampullariids in the wild* (Cont'd)

Species	Native range	Non-native range	Representative references for non-native range
		Spain	López et al, 2010
		Taiwan	Hayes et al, 2008; Wu et al, 2010
		Thailand	Hayes et al, 2008
		Viet Nam	Hayes et al, 2008
		US (California)	Rawlings et al, 2007
		US (Arizona)	Rawlings et al, 2007
		US (Florida)	Rawlings et al, 2007; Hayes, unpublished
Pomacea diffusa	Amazonia	Australia	Hayes et al, 2008
		Brazil (Rio de Janeiro)	Hayes et al, 2008
		Brazil (Pará)	Hayes et al, 2008
		Brazil (Pernambuco)	Hayes, unpublished
		Colombia	Hayes, unpublished
		French Guiana	Massemin et al, 2009
		Hawaii	Cowie, 1995
		New Zealand[g]	K. J. Collier, pers. comm., 2010
		Panama	Hayes et al, 2008
		Sri Lanka	Hayes et al, 2008
		US (Florida)	Rawlings et al, 2007
		Venezuela	Hayes, unpublished
Pomacea haustrum	Amazonia	US (Florida)	Rawlings et al, 2007
Pomacea insularum	Argentina to Amazonia	Cambodia	Hayes et al, 2008
		Malaysia	Hayes et al, 2008
		Singapore	Hayes et al, 2008
		South Korea	Hayes et al, 2008
		Thailand	Hayes et al, 2008
		US (Alabama)	Hayes, unpublished
		US (Florida)	Rawlings et al, 2007
		US (Georgia)	Rawlings et al, 2007
		US (Louisiana)	Hayes, unpublished
		US (South Carolina)	R. T. Dillon, Jr, pers. comm., 2010
		US (Texas)	Rawlings et al, 2007
		Viet Nam	Hayes et al, 2008
Pomacea scalaris	Argentina	Taiwan	Hayes et al, 2008; Wu et al, 2010
	Southern Brazil		

Note: [a] not known whether widely established; [b] not explicitly considered introduced; [c] thought to have been eradicated; [d] unconfirmed; [e] unconfirmed; may refer to *P. insularum*; [f] identified as *Pomacea lineata* but probably *P. canaliculata*; [g] a single record, may not be established.

non-governmental organizations. By 1983 about 500 snail businesses had opened up in Japan; they were present in Okinawa by at least 1984. *Pomacea insularum* may have been first introduced around this time, from Argentina and southern Brazil (Hayes et al, 2008). Later, the snails were taken to parts of Malaysia (Sarawak and Peninsular Malaysia, 1987), Viet Nam (1988 or 1989), Thailand (1989 or 1990) and Laos (1992). They were present in Hong Kong and Singapore by 1991 and Cambodia by at least 1994. In the Pacific they were in Hawaii by 1989 or perhaps earlier (Cowie et al, 2007), Guam by 1989 (perhaps introduced from Taiwan or more likely the Philippines; Tran et al, 2008), and Papua New Guinea in 1990 (Orapa, 2006), probably introduced from the Philippines.

Most of these reports, prior to the clarification by Hayes et al (2008), assumed that a single species was involved, usually identified as *Pomacea canaliculata*. Hayes et al (2008) concluded, based on mitochondrial DNA (mtDNA) diversity, that the Asian populations of both *P. canaliculata* and *P. insularum* resulted from multiple introductions. Tran et al (2008) showed that only a single haplotype was present in Hawaiian *P. canaliculata*, suggesting a single introduction or multiple introductions from a single location, probably the Philippines.

Pomacea canaliculata was recorded in California in 1998, perhaps introduced from Hawaii for food (Rawlings et al, 2007). By 2007 it was in Florida, perhaps transported from the western US.

The snails' economic potential was overestimated and while many, mostly small aquaculture operations arose, relatively few persisted (Acosta and Pullin, 1991). In Taiwan, the local market failed because consumers disliked the snails' taste and texture (Yang et al, 2006). Stringent health regulations in developed nations largely precluded its importation (Naylor, 1996). Snails escaped or were deliberately released, becoming widespread and abundant in many countries. Expansion of their distribution has been assisted, among other things, by floods and typhoons, movement of infested soil, deliberate distribution of snails for weed control, and use for fishing bait.

The aquarium trade

Ampullariids are popular domestic aquarium snails (Perera and Walls, 1996; Wilstermann-Hildebrand, 2009; www.applesnail.net). Various species have therefore been introduced around the world, perhaps also accidentally with aquarium plants. *Marisa cornuarietis* has been introduced to several countries (e.g. the US) (Perera and Walls, 1996). *Pomacea diffusa* (usually referred to as *P. bridgesii* until their distinction was clarified by Hayes et al, 2008) was probably introduced to Florida in the early 1960s and is now also established in Alabama (Rawlings et al, 2007). It is grown commercially on a large scale in Florida (Perera and Walls, 1996). The market has expanded since the discovery and development of bright yellow, orange and other colour variants of *P. diffusa* and to some degree other *Pomacea* species (Perera and Walls, 1996). The most widespread mtDNA haplotype in *P. diffusa* is shared by snails from pet stores as far afield as Australia, Hawaii, Florida and Tehran (Hayes et al, 2008). *Pomacea diffusa* has been intercepted by customs officials in Singapore. It is established in Australia and Sri Lanka (Hayes et al, 2008) and was reported in the wild in Hawaii (Cowie, 1995) but has declined and was not recorded in more recent surveys (Cowie et al, 2007). *Pomacea canaliculata* (including brightly coloured forms) is in California and Arizona, and although probably introduced for food (Rawlings et al, 2007), the aquarium trade may also have been involved. *Pomacea insularum* has been detected in the trade in Belgium (Hayes et al, 2008) and its presence in Spain probably originated in the trade. *Pomacea diffusa* has also been sold for food in Belgium, as 'sea snails' (Thiengo, S. C., pers. comm.).

Keawjam and Upatham (1990) considered the *Pomacea* in Thailand to have been imported by the aquarium trade, but it is also probable that they were introduced for food, as elsewhere in southeast Asia. In Hawaii, *Pomacea canaliculata* has been available in aquarium stores, and purchase followed by release for culture as food items may have been one reason for its spread (Cowie, 2002), although the original source of the aquarium snails was probably local, following their initial introduction for food. *Asolene spixii* has been seen in pet stores in Hawaii but is not in the wild. *Pomacea lineata* (probably misidentified *P. canaliculata*) has been introduced to South Africa, and *Pila leopoldvillensis* has been recorded in Taiwan, both perhaps introduced via the aquarium trade.

Biological control

Ampullariids have been introduced in attempts to control the snail vectors of schistosomes. In Guadeloupe, introduced *Pomacea glauca* and *Marisa cornuarietis* caused the decline of *Biomphalaria glabrata* through competition (Pointier et al, 1991; Pointier and David, 2004). In Puerto Rico, *Marisa cornuarietis* caused declines in *B. glabrata* and *Lymnaea columella* through predation (Peebles et al, 1972). *Marisa cornuarietis* is said to have had a similar effect in the Dominican Republic (Perera and Walls, 1996) and in field experiments in Egypt and Tanzania (Nguma et al, 1982), although it seems not to have become established in the wild in Africa.

Many ampullariids feed voraciously on aquatic plants, this being one reason for their success in controlling other snail species: they reduce the available food. They have therefore been used or suggested for aquatic weed control in both natural wetlands and irrigated rice, e.g. *Marisa cornuarietis* in Florida and Puerto Rico (Simberloff and Stiling, 1996), *Pomacea canaliculata* in Asia (Joshi and Sebastian, 2006; Wada, 1997), although there are concerns in Asia that this might lead to farmers introducing snails to areas they have not yet reached (Wada, 2006).

Ecology

Habitat

Some species inhabit faster flowing streams and rivers. However, the invasive species, notably *Pomacea canaliculata* and *P. insularum* but also *P. diffusa*, *P. haustrum*, *P. scalaris*, *Pila conica* and *Marisa cornuarietis*, generally occur in their native ranges in slower moving or stagnant shallow water in lowland swamps, marshes, ditches, ponds and lakes, usually with muddy bottoms. They are thus well suited for living in rice paddies, taro patches and similar artificial habitats (Plate 18.2). They have become environmental pests in areas such as the Florida Everglades, also similar to their native habitats.

Few studies have assessed the chemical characteristics of the water in habitats favoured by ampullariids. In Hong Kong, Kwong et al (2008) were able to predict the distribution of *P. canaliculata* with some accuracy,

but the water chemistry differed considerably from that in its native range (Martín et al, 2001). Because they can breathe air, ampullariids are also tolerant of low levels of oxygen in the water. *Pomacea canaliculata* in Hawaii tolerates highly polluted water (Lach and Cowie, 1999).

Various species, including *Pomacea canaliculata* (Cowie, 2002), *P. insularum* (Ramakrishnan, 2007) and *P. diffusa* (Jordan and Deaton, 1999) exhibit some salinity tolerance. However, in general apple snails do not inhabit saline or brackish water.

Feeding, growth and reproduction

Feeding, growth and breeding of ampullariids in their native ranges may be seasonal and related to latitude, temperature and rainfall, and influenced by other factors such as vegetation type, food availability and water chemistry. In seasonally wet tropical regions, they may aestivate during the dry seasons as their habitat dries up, breeding in the rainy seasons. In subtropical or temperate regions, they may only breed during summer. Locally, variation in reproductive regime may be related to local climatic variation, especially availability of water. Thus in its natural range in southern Brazil, Uruguay, Paraguay and Argentina, *Pomacea canaliculata* is active only in summer, whereas *Pomacea insularum*, with a more northerly range, is not constrained by temperature but breeds only when water levels are optimal. When levels are at their lowest they bury into the mud to avoid desiccation, but when at their highest they are less able to access the shallow water preferred for feeding and reproduction. In their introduced humid tropical southeast Asian range and the controlled environment of a rice paddy, both species can grow and breed year round as long as sufficient water is present. In Hong Kong, *P. canaliculata* reaches full size in four to six months and reproduction occurs almost year round, although with some variation in snail biomass and density related to water temperature (Kwong et al, 2010). Under artificial conditions *P. canaliculata* can grow even faster. In cooler regions such as Japan, as paddies dry out and temperatures drop during winter, the snails bury themselves in the mud and become dormant, awaiting warmer temperatures and reflooding of the paddies in spring. Winter temperatures may limit the northern spread of

P. canaliculata in Japan (Ito, 2002), although it can alter its behaviour and acclimate to these cooler temperatures to some degree, permitting over-wintering further north than would otherwise be possible (Wada and Matsukura, 2007; Matsukura et al, 2009). The cold tolerance of the more tropical *P. insularum* may limit its northerly spread in the US (Ramakrishnan, 2007).

Maximum size varies among populations, perhaps related to environmental factors including habitat size, microclimate and water regime, and population density. *Pomacea canaliculata* in Hawaii reaches only about 30mm but in Asia can grow to over 65mm (Schnorbach, 1995) or even 90mm (Heidenreich et al, 1997), although these larger sizes may result from misidentifications of *P. insularum*.

Most ampullariids are generalist herbivores. The two most invasive pests, *Pomacea canaliculata* and *P. insularum*, grow rapidly when fed on numerous plant species (e.g. Lach et al, 2000; Fellerhoff, 2002 [*P. insularum* misidentified as *P. lineata*]; Boland et al, 2008; Qiu and Kwong, 2009; Baker et al, 2010; Wong et al, 2010). Growth rate generally correlates with feeding on the preferred plant(s). Some species will feed on insects, crustaceans, worms, bryozoans, small fish, frogs etc., mostly as carrion but not always, and some will attack other smaller snails and their eggs (e.g. Aditya and Raut, 2001, 2002; Pointier and David, 2004; Kwong et al, 2009; Wong et al, 2009). In Hong Kong, detritus occurred more frequently than macrophytes in the stomachs of *P. canaliculata* and they also ate cyanobacteria, green algae and diatoms (Kwong et al, 2010). The predominant habit, however, is macrophytophagous, which from a pest standpoint is also the most significant. *Pomacea canaliculata* and *P. insularum* seem particularly voracious and generalist compared to *P. haustrum* and *P. diffusa* (Morrison and Hay, 2011).

Ampullariids are dioecious (separate sexes), internally fertilizing and oviparous. Females tend to be larger than males, at least in some species, including *Pomacea canaliculata* and *P. insularum*. Copulation in *P. canaliculata* occurs up to about three times per week at any time of day or night and may take 10–18 hours. In *P. canaliculata*, and other species for which it has been reported, oviposition takes place predominantly at night, or in the early morning or evening. On each occasion, a single clutch is laid. The interval between successive ovipositions is 5–16 days for *P. canaliculata*.

Hatching generally takes place about two weeks after oviposition, but this period varies greatly and is highly dependent on temperature (Koch et al, 2009). Hatchlings immediately fall or crawl into the water. The estimated average annual output of *P. canaliculata* is about 4400 eggs (Barnes et al, 2008) or as many as 10,000 (Wu and Xie, 2006). *Pomacea insularum* is more fecund, with an average clutch containing 2064 eggs, a female laying at least one clutch per week over an extended active season, and about 70 per cent hatching success (Barnes et al, 2008).

Pomacea generally lay their eggs above water on the exposed parts of vegetation, rocks, etc., perhaps to avoid aquatic predators or in response to low oxygen tension in their often near-stagnant aquatic habitats. The eggs are enclosed in a calcium carbonate shell, which may or may not be used as a source of calcium for the developing embryos. Egg morphology differs among species, with the phylogenetically most derived group having spherical eggs that cluster relatively loosely in the egg mass (Hayes et al, 2009). This group includes the pests *Pomacea canaliculata* and *P. insularum*, both of which lay bright pink egg masses (Plate 18.3), with the latter laying larger clutches of more but smaller eggs than the former, although the number of eggs is highly variable (Barnes et al, 2008). These bright pink eggs are often the first visible signs of an infestation. The more basal taxa in this group lay eggs that are polygonal in shape, abutting tightly against one another within the egg mass (*P. diffusa*, *P. scalaris* and *P. haustrum* in Plate 18.3). A number of egg characteristics (shape, clutch structure, pigmentation) may be associated with increased hatching efficiency and adaptation to ephemeral habitats, which may contribute to these species' invasiveness (Hayes et al, 2009).

The white eggs of *Pila* spp. are also laid out of water, but in depressions made by the snails on banks or mudflats (Cowie, 2002) or at the bases of plants. They also have a calcareous coating. Those of *Marisa*, *Asolene* and *Felipponea* species (Plate 18.3) lack the calcareous coating and are deposited in gelatinous masses under water on submerged vegetation or other surfaces.

Few studies have addressed ampullariid population dynamics directly (e.g. Kwong et al, 2010). However, changes in water availability are important. In Florida, recruitment of the native *Pomacea paludosa* is enhanced in years when the water table remains high, allowing the snails to remain active instead of having to aestivate. In Venezuela, habitats with more permanent standing water (rice fields rather than natural wetlands) allow the native *Pomacea dolioides*, which here is a rice pest, to grow larger and achieve higher population densities, essentially because the time in aestivation (no growth) is shorter. *Pomacea canaliculata* densities of over 130m^{-2} have been recorded in taro patches in Hawaii, and up to 150m^{-2} in Philippine rice paddies. *Marisa cornuarietis* has been reported at over 200m^{-2} in Florida. Irrigated systems may contain water for longer periods than natural ones allowing apple snails to reach higher densities than in their natural environments.

Behaviour and Physiology

Many ampullariids are amphibious. The mantle cavity contains a ctenidium ('gill') and a portion modified as a pulmonary sac ('lung'), thereby allowing activity both in and out of the water. This is clearly of adaptive value for life in habitats that dry out periodically.

Many species also aestivate when their habitat dries out, burying themselves into the mud. *Pomacea canaliculata* is only reported to survive buried for up to three months (Schnorbach, 1995) but other species can survive much longer without water, e.g. 526 days in *Pomacea urceus* (Burky et al, 1972), 100–400 in *P. lineata* (Little, 1968) and 308 in *P. insularum* (Ramakrishnan, 2007). Therefore drying out a rice paddy or taro patch, especially if only for short periods, may not control the snails. Even species that do not lay their eggs above water may be able to survive significant periods out of water, e.g. 30–120 days in *Marisa cornuarietis*, depending on humidity. These species may therefore be able to travel at least short distances over land, and when introduced into new habitats may be difficult to contain within circumscribed areas (e.g. rice paddies).

Short term dispersal activity does not necessarily translate into long term, long distance dispersal. The natural spread of ampullariids from foci of introduction is poorly documented. In a Florida canal, a *Marisa cornuarietis* population expanded at least 1.5km downstream in six to eight months and soon was widely distributed in the canal system around Miami, dispersing predominantly by floating downstream on vegetation. Crawling upstream is also possible, unless the flow rate is too great (Ranamukhaarachchi and Wikramasinghe, 2006). Snail kites, predators of native *Pomacea paludosa* in Florida, may disperse introduced *Pomacea*. However, the rapid spread of *Pomacea canaliculata* and *P. insularum* within Asia, Hawaii and the continental US following introduction has been predominantly human mediated.

Most ampullariid species are tropical with upper lethal temperatures around 40°C, depending on exposure time. In *Marisa cornuarietis*, exposure for one to four hours at 40°C is lethal but the snails may be able to withstand temperatures up to 45°C for short periods, and while feeding is normal at 33.5–35.5°C, eggs do not develop normally at 35–37°C. For *Pomacea canaliculata*, a species of more temperate regions, mortality is high at water temperatures above 32–35°C, although in one study little reduction in activity levels occurred over five days at 35°C (Seuffert et al, 2010). The upper limit for long term exposure in *P. insularum* is 36–37°C (Ramakrishnan, 2007).

Regarding low temperatures, *M. cornuarietis* can survive over 24 hours at 11°C (although egg development ceased at this temperature), succumbs in five hours when exposed to 8°C, but may withstand 6°C for short periods. In contrast, *P. canaliculata* can survive 5–20 days at 0°C, two days at –3°C and six hours at –6°C (Cowie, 2002; Wada and Matsukura, 2007; Matsukura et al, 2009), although activity almost ceases below 10°C (Seuffert et al, 2010). However, Wu and Xie (2006) suggested that the snails introduced to China are less tolerant of cold temperatures. The lower limit for long term exposure in *P. insularum* is 15°C, although they can survive lower temperatures for short periods (Ramakrishnan, 2007). Various *Pila* species cannot survive at 20°C for extended lengths of time.

Comparability among such studies is poor, largely because experimental procedures, especially exposure time, differed. Nevertheless, differences among species in lethal limits probably reflect adaptation to their natural climatic environment. Lower limits are more variable than upper limits. Intra-specific variation probably results from differing experimental protocols, region of origin of the snails, prior acclimation to different temperatures, and possibly misidentification. These differences are probably important for the establishment of species introduced to regions differing in climate from their natural ranges.

Apple Snails as Pests

In general, ampullariids are not serious pests in their native ranges, although in rice-growing areas of Suriname, *Pomacea dolioides* (incorrectly referred to as *P. lineata*) has caused problems (Wiryareja and Tjoe-Awie, 2006). Damage has been reported (with few details) in other parts of the Caribbean; and in Africa and Asia, species of *Lanistes* and *Pila*, respectively, have occasionally been implicated as pests (Cowie, 2002).

However, when introduced outside their native ranges they have become more serious and widespread pests. In Puerto Rico, *Marisa cornuarietis* is a rice pest, and *Pila conica* is a taro pest in Hawaii (Cowie et al, 2007). But it is *Pomacea canaliculata* and *P. insularum* that have become by far the most serious pests, attacking numerous crops, most significantly rice in Asia (Lai et al, 2005; Joshi and Sebastian, 2006) (Plate 18.4). *Pomacea canaliculata* is also a rice pest in the Dominican Republic (Rosario and Moquete, 2006). The extent of their impact and the economic costs are huge and they are now considered the most important pests in irrigated rice agriculture (Greene, 2008).

In addition to agricultural problems, there are concerns for the natural environment. Their relatively indiscriminate feeding on weeds and desirable native plants as well as native snails and other fauna, may have significant negative impacts (Simberloff and Stiling, 1996; Carlsson and Lacoursière, 2005; Fang et al, 2010). Introduced *Pomacea* have been implicated in the decline of native *Pila* species in southeast Asia, in part because of extensive pesticide application against introduced *Pomacea*. *Pomacea insularum* has become a major environmental pest in parts of Florida, apparently outcompeting native *Pomacea paludosa* (Connor et al, 2008), with implications for various bird species that depend on the latter (Darby et al, 2007). Furthermore, because of their high levels of secondary production compared to the native faunas of the regions to which they have been introduced (Kwong et al, 2010), they may cause major ecosystem level changes (Carlsson et al, 2004).

Introduced ampullariids are also of medical concern as they act as vectors for a number of parasites that cause human diseases, including schistosomes that cause dermatitis and a fluke that causes intestinal problems (Hollingsworth and Cowie, 2006). Most notably, various species, including *Pomacea canaliculata*, can act as vectors of *Angiostrongylus cantonensis*, the rat lungworm, which can infect humans if ingested and cause potentially fatal eosinophilic meningoencephalitis. However, many snail species can act as vectors of *A. cantonensis*; there is little evidence of a relationship between presence specifically of apple snails and incidence of the disease (Smith, 1992; but see Lv et al, 2009), and levels of susceptibility of *P. canaliculata* are reported to be low (Tesana et al, 2008). Nonetheless, increasing contact with and consumption of snails could lead to increased incidence. Thorough cooking is essential.

Management

No single approach has proven effective and safe. An integrated pest management approach that tries to minimize pesticide use, maximize yields and lower costs has therefore generally been adopted (Greene, 2008), with the suite of practices used differing among and within countries (Joshi and Sebastian, 2006). Efforts have so far been almost exclusively targeted at apple snails in agricultural settings, primarily Asian rice.

Chemical control

Chemical control was practised extensively and often illegally in many Asian countries, especially the Philippines, but with significant human health problems. Efforts are being made to reduce the use of chemicals, and in Japan, for instance, few chemicals are now used for control in transplanted rice (Wada, 2006). Tin compounds and endosulfan are now widely banned or have fallen into disfavour because of human toxicological, phytotoxic or environmental concerns, and the main pesticides remaining in use are metaldehyde and niclosamide (Schnorbach et al, 2006; Joshi and Sebastian, 2006; Wu et al, 2010). Organic pesticides have received some attention (San Martín, 2006), and are considered more environmentally and toxicologically friendly, but they may also cause problems, especially at high concentrations.

Biological control

Yusa (2006) reviewed the predators of *Pomacea canaliculata* in areas where it has been introduced. Yusa et al (2006) tested 46 species in the laboratory; 26 of them fed on *P. canaliculata*. Few of these have any potential for control and few actually live in the rice paddies. Only ducks and fish have attracted serious consideration as potential control agents (Cowie, 2002; Wada, 2004).

In both rice and taro, significant reduction of snail numbers can be achieved using ducks, although they prefer or are only able to eat juvenile snails, so they are generally used in conjunction with other approaches, most often hand-picking of larger snails. In Hawaii, use of ducks is controversial because introduced domestic ducks can hybridize with the native Hawaiian duck, leading potentially to the genetic extinction of the latter (Rhymer and Simberloff, 1996).

Various fish species have been tested, both simply as predators and in a combined control/aquaculture scenario, with the common carp being the most widely evaluated (e.g. Ichinose et al, 2002). Although somewhat successful, the need for relatively deep water, susceptibility of the fish to predation and the negative impact of the fish on local ecosystems have precluded this approach from being widely implemented (Yusa, 2006).

Cultural and mechanical control

Various approaches have been adopted, often taking advantage of aspects of the snails' behaviour and physiology outlined above.

Hand-picking and destroying snails and their eggs, although labour intensive, is the most effective non-chemical control measure. Lowering the water level or draining the paddy will not kill the snails because of their ability to survive long periods without water. However, snail activity is reduced if the water is shallower than their shell height and periodic lowering of the water may make the snails congregate in slightly deeper areas, facilitating hand-picking. Destruction of eggs is facilitated by placing stakes on which the snails oviposit, the stakes with eggs being readily removed. Knocking eggs into the water is also effective, as they do not survive immersion (Horn et al, 2008). Bounties have been offered for snails or

eggs, but, as happened in Taiwan, people began rearing snails to claim the reward offered for egg masses (Cheng and Kao, 2006).

Collecting snails can control them and augment both human (Tamaru et al, 2006) and fish aquaculture food resources (Castillo and Casal, 2006). However, promoting use of a pest to control it is a controversial approach as it may lead to its further spread as people take it to previously uninfested areas.

Wire mesh grilles covering the inlets to the rice paddies, taro patches, etc. prevent dispersal at least of larger snails. Snails that collect in the grilles are easily destroyed.

Maintaining tidy edges of rice paddies, taro patches, etc. and weeding irrigation ditches reduces egg-laying sites, allows snails to be more easily seen and destroyed, and may decrease the chances of dispersal between paddies.

In rice, susceptibility to damage declines with seedling age, so transplanting older seedlings reduces damage (Wada, 2004). However, adoption of direct seeding (Wada, 2004) precludes this approach, although denuded areas can be replanted.

Baits (e.g. leaves, rotten fruits that are attractive to the snails) may divert the snails from eating rice or taro and facilitate hand-picking of snails congregating at the baits. This technique has also been used in non-agricultural settings (Van Dyke, 2010).

Tilling the earth during the off-season when the snails are buried crushes the snails (Wada, 2004) and in colder climates (Japan, Korea) exposes buried snails to cold, lethal temperatures. Growing wheat as an off-season crop (Wada, 1997) and crop rotation with soybean (Wada, 2004; Wada et al, 2004) may reduce numbers. Burning rice straw after harvest kills snails near the surface of the mud and the ash reportedly repels the snails.

Acknowledgements

We thank Jian-Wen Qiu, who commented constructively on a draft of this contribution.

Note: Top row, left to right: *Marisa cornuarietis*, *Pomacea canaliculata*, *P. insularum*; bottom row, left to right: *P. haustrum*, *P. diffusa*, *P. scalaris*. All to the same scale. Shell morphology and colour are for many species poor characters on which to base identification, as there is considerable intra-specific variation.

Source: K. A. Hayes

Plate 18.1 *Shells of introduced ampullariids*

Source: K. A. Hayes

Plate 18.2 *Introduced* Pomacea canaliculata *crawling on the muddy bottom of a taro patch in Hawaii*

Note: From left: *Marisa cornuarietis, Pomacea canaliculata, P. insularum, P. diffusa, P. scalaris, P. haustrum.* All to the same scale. Egg mass size is variable and not a good diagnostic character but egg morphology and colour are useful for distinguishing certain species.

Source: K. A. Hayes

Plate 18.3 *Egg masses of introduced ampullariids*

Note: Eggs deposited on taro plants and on a filtering device intended to prevent the spread of snails from one taro patch to another. Crawling snails are visible in the background on the edge of the now mostly denuded patch.

Source: K. A. Hayes

Plate 18.4 *Abundant egg masses of* Pomacea canaliculata *indicating a major infestation in Hawaii*

References

Acosta, B. O. and Pullin, R. S. V. (1991) *Environmental Impact of the Golden Snail (*Pomacea *sp.) on Rice Farming Systems in the Philippines*, Freshwater Aquaculture Center, Central Luzon State University, Munoz, Nueva Ecija; ICLARM, Manila

Aditya, G. and Raut, S. K. (2001) 'Food of the snail, *Pomacea bridgesi*, introduced to India', *Current Science*, vol 80, pp919–921

Aditya, G. and Raut, S. K. (2002) 'Destruction of *Indoplanorbis exustus* (Planorbidae) eggs by *Pomacea bridgesi* (Ampullariidae)', *Molluscan Research*, vol 22, pp87–90

Baker, P., Zimmanck, F. and Baker, S. M. (2010) 'Feeding rates of an introduced freshwater gastropod *Pomacea insularum* on native and nonindigenous aquatic plants in Florida', *Journal of Molluscan Studies*, vol 76, pp138–143

Barnes, M. A., Fordham, R. K. and Burks, R. L. (2008) 'Fecundity of the exotic applesnail, *Pomacea insularum*', *Journal of the North American Benthological Society*, vol 27, pp738–745

Berthold, T. (1991) *Vergleichende Anatomie, Phylogenie und historische Biogeographie der Ampullariidae (Mollusca, Gastropoda)*, Abhandlungen des Naturwissenschaftlichen Vereins in Hamburg Series, No 29, Verlag Paul Parey, Hamburg

Boland, B. B., Meerhoff, M., Fosalba, C., Mazzeo, N., Barnes, M. A. and Burks, R. L. (2008) 'Juvenile snails, adult appetites: Contrasting resource consumption between two species of applesnails (*Pomacea*)', *Journal of Molluscan Studies*, vol 74, pp47–54

Brown, D. S. (1994) *Freshwater Snails of Africa and their Medical Importance*, Taylor & Francis, London

Burky, A. J., Pacheco, J. and Pereyra, E. (1972) 'Temperature, water, and respiratory regimes of an amphibious snail, *Pomacea urceus* (Müller), from the Venezuelan savannah', *Biological Bulletin*, vol 143, pp304–316

Carlsson, N. O. L. and Lacoursière, J. O. (2005) 'Herbivory on aquatic vascular plants by the introduced golden apple snail (*Pomacea canaliculata*) in Lao PDR', *Biological Invasions*, vol 7, pp233–241

Carlsson, N. O. L., Brönmark, C. and Hansson, L.-A. (2004) 'Invading herbivory: The golden apple snail alters ecosystem functioning in Asian wetlands', *Ecology*, vol 85, pp1575–1580

Castillo, L. V. and Casal, C. M. V. (2006) 'Golden apple snail utilization in small-scale aquaculture in the Philippines', in R. C. Joshi and L. C. Sebastian (eds) *Global Advances in Ecology and Management of Golden Apple Snails*, Philippine Rice Research Institute, Muñoz, Nueva Ecija, pp475–482

Cheng, E. Y. and Kao, C.-H. (2006) 'Control of golden apple snail *Pomacea canaliculata* (Lamarck), in Taiwan', in R. C. Joshi and L. C. Sebastian (eds) *Global Advances in Ecology and Management of Golden Apple Snails*, Philippine Rice Research Institute, Muñoz, Nueva Ecija, pp155–167

Conner, S. L., Pomory, C. M. and Darby, P. C. (2008) 'Density effects of native and exotic snails on growth in juvenile apple snails *Pomacea paludosa* (Gastropoda: Ampullariidae): A laboratory experiment', *Journal of Molluscan Studies*, vol 74, pp355–362

Cowie, R. H. (1995) 'Identity, distribution and impacts of introduced Ampullariidae and Viviparidae in the Hawaiian Islands', *Journal of Medical and Applied Malacology*, vol 5 [for 1993], pp61–67

Cowie, R. H. (2002) 'Apple snails (Ampullariidae) as agricultural pests: Their biology, impacts and management', in G. M. Barker (ed) *Molluscs as Crop Pests*, CABI Publishing, Wallingford, pp145–192

Cowie, R. H. and Thiengo, S. C. (2003) 'The apple snails of the Americas (Mollusca: Gastropoda: Ampullariidae: *Asolene, Felipponea, Marisa, Pomacea, Pomella*): A nomenclatural and type catalog', *Malacologia*, vol 45, pp41–100

Cowie, R. H., Hayes, K. A. and Thiengo, S. C. (2006) 'What are apple snails? Confused taxonomy and some preliminary resolution' in R. C. Joshi and L. C. Sebastian (eds) *Global Advances in Ecology and Management of Golden Apple Snails*, Philippine Rice Research Institute, Muñoz, Nueva Ecija, pp3–23

Cowie, R. H., Hayes, K. A., Tran, C. T. and Levin, P. (2007) 'Distribution of the invasive apple snail *Pomacea canaliculata* (Lamarck) in the Hawaiian Islands (Gastropoda: Ampullariidae)', *Bishop Museum Occasional Papers*, 96, pp48–51

Cuong, D. C. (2006) 'The golden apple snail in Vietnam', in R. C. Joshi and L. S. Sebastian (eds) *Global Advances in Ecology and Management of Golden Apple Snails*, Philippine Rice Research Institute, Muñoz, Nueva Ecija, pp243–254

Darby, P. C., Mellow, D. J. and Watford, M. L. (2007) 'Food-handling difficulties for snail kites capturing non-native apple snails', *Florida Field Naturalist*, vol 35, pp79–85

Eldredge, L. G. (1994) *Perspectives in Aquatic Exotic Species Management in the Pacific islands. Volume 1. Introductions of Commercially Significant Aquatic Organisms to the Pacific Islands*, South Pacific Commission, Noumea

Fang, L., Wong, P. K., Lin, L., Lan, C. and Qiu, J.-W. (2010) 'Impact of invasive apple snails in Hong Kong on wetland macrophytes, nutrients, phytoplankton and filamentous algae', *Freshwater Biology*, vol 55, pp1191–1204

Fellerhoff, C. (2002) 'Feeding and growth of apple snail *Pomacea lineata* in the Pantanal wetland, Brazil – a stable isotope approach', *Isotopes in Environmental and Health Studies*, vol 38, pp227–243

Greene, S. D. (2008) 'Extending integrated pest management to the golden apple snail: Examining a community centre approach in northeast Thailand', *International Journal of Pest Management*, vol 54, pp95–102

Halwart, M. and Bartley, D. M. (2006) 'International mechanisms for the control and responsible use of alien species in aquatic ecosystems, with special reference to the golden apple snail', in R. C. Joshi and L. S. Sebastian (eds) *Global Advances in Ecology and Management of Golden Apple Snails*, Philippine Rice Research Institute, Muñoz, Nueva Ecija, pp449–458

Hayes, K. A., Joshi, R. C., Thiengo, S. C. and Cowie, R. H. (2008) 'Out of South America: Multiple origins of non-native apple snails in Asia', *Diversity and Distributions*, vol 14, pp701–712

Hayes, K. A., Cowie, R. H., Jørgensen, A., Schultheiß, R., Albrecht, C. and Thiengo, S. C. (2009) 'Molluscan models in evolutionary biology: Apple snails (Gastropoda: Ampullariidae) as a system for addressing fundamental questions', *American Malacological Bulletin*, vol 27, pp47–58

Heidenreich, A., Poethke, H.-J., Halwart, M. and Seitz, A. (1997) 'Simulation der Populationsdynamik von *Pomacea canaliculata* (Prosobranchia) zur Bewertung von Managementmaßnahmen', *Verhandlungen der Gesellschaft für Ökologie*, vol 27, pp441–446

Hendarsih-Suharto, Marwoto, R. M., Heryanto, Mulyadi and Siwi, S. S. (2006) 'The golden apple snail, *Pomacea* spp., in Indonesia', in R. C. Joshi and L. S. Sebastian (eds) *Global Advances in Ecology and Management of Golden Apple Snails*, Philippine Rice Research Institute, Muñoz, Nueva Ecija, pp231–242

Hollingsworth, R. G. and Cowie, R. H. (2006) 'Apple snails as disease vectors', in R. C. Joshi and L. S. Sebastian (eds) *Global Advances in Ecology and Management of Golden Apple Snails*, Philippine Rice Research Institute, Muñoz, Nueva Ecija, pp121–132

Horn, K. C., Johnson, S. D., Boles, K. M., Moore, A., Siemann, E. and Gabler, C. A. (2008) 'Factors affecting hatching success of golden apple snail eggs: Effects of water immersion and cannibalism', *Wetlands*, vol 28, pp544–549

Hunt, B. P. (1958) 'Introduction of *Marisa* into Florida', *Nautilus*, vol 72 pp53–55

Ichinose, K., Tochihara, M., Suguiura, N. and Yusa, Y. (2002) 'Influence of common carp on apple snail in a rice field evaluated by a predator-prey logistic model', *International Journal of Pest Management*, vol 48, pp133–138

ICZN (International Commission on Zoological Nomenclature) (1999) 'Opinion 1913. *Pila* Röding, 1798 and *Pomacea* Perry, 1810 (Mollusca, Gastropoda): Placed on the Official List, and AMPULLARIIDAE Gray, 1824: Confirmed as the nomenclaturally valid synonym of PILIDAE Preston, 1915', *Bulletin of Zoological Nomenclature*, vol 56, pp74–76

Ito, K. (2002) 'Environmental factors influencing overwintering success of the golden apple snail, *Pomacea canaliculata* (Gastropoda: Ampullariidae), in the northernmost population of Japan', *Applied Entomology and Zoology*, vol 37, pp655–661

Jordan, P. J. and Deaton, L. E. (1999) 'Osmotic regulation and salinity tolerance in the freshwater snail *Pomacea bridgesi* and the freshwater clam *Lampsilis teres*', *Comparative Biochemistry and Physiology A*, vol 122, pp199–205

Joshi, R. C. and Sebastian, L. S. (eds) (2006) *Global Advances in Ecology and Management of Golden Apple Snails*, Philippine Rice Research Institute, Muñoz, Nueva Ecija

Keawjam, R. S. and Upatham, E. S. (1990) 'Shell morphology, reproductive anatomy and genetic patterns of three species of apple snails of the genus *Pomacea* in Thailand', *Journal of Medical and Applied Malacology*, vol 2, pp49–62

Koch, E., Winik, B. C. and Castro-Vazquez, A. (2009) 'Development beyond the gastrula stage and the digestive organogenesis in the apple-snail *Pomacea canaliculata* (Architaenioglossa, Ampullariidae)', *Biocell*, vol 33, pp49–65

Kwong, K.-L., Wong, P.-K., Lau, S. S. S. and Qiu, J.-W. (2008) 'Determinants of the distribution of apple snails in Hong Kong two decades after their initial invasion', *Malacologia*, vol 50, pp293–302

Kwong, K.-L., Chan, R. K. Y. and Qiu, J.-W. (2009) 'The potential of the invasive snail *Pomacea canaliculata* as a predator of various life-stages of five species of freshwater snails', *Malacologia*, vol 51, pp343–356

Kwong, K. L., Dudgeon, D., Wong, P. K. and Qiu, J.-W. (2010) 'Secondary production and diet of an invasive snail in freshwater wetlands: Implications for resource utilization and competition', *Biological Invasions*, vol 12, pp1153–1164

Lach, L. and Cowie, R. H. (1999) 'The spread of the introduced freshwater apple snail *Pomacea canaliculata* (Lamarck) (Gastropoda: Ampullariidae) on O'ahu, Hawai'i', *Bishop Museum Occasional Papers*, vol 58, pp66–71

Lach, L., Britton, D. K., Rundell, R. J. and Cowie, R. H. (2000) 'Food preference and reproductive plasticity in an invasive freshwater snail', *Biological Invasions*, vol 2, pp279–288

Lai, P.-Y., Chang, Y. F. and Cowie, R. H. (eds) (2005) *Proceedings – APEC Symposium on the Management of the Golden Apple Snail, September 6–11, 2004*, National Pingtung University of Science and Technology, Pingtung, Chinese Taipei, Taiwan

Lee, T.-G. and Oh, K. C. (2006) 'Golden apple snails in Korea', in R. C. Joshi and L. C. Sebastian (eds) *Global Advances in Ecology and Management of Golden Apple Snails*, Philippine Rice Research Institute, Muñoz, Nueva Ecija, pp516–525

Letelier, S. and Soto-Acuña, S. (2008) 'Registro de *Pomacea* sp. (Gastropoda: Ampullaridae) en Chile', *Amici Molluscarum*, vol 16, pp6–13

Little, C. (1968) 'Aestivation and ionic regulation in two species of *Pomacea* (Gastropoda, Prosobranchia)', *Journal of Experimental Biology*, vol 48, pp569–585

López, M. A., Altaba, C. R., Andree, K. B. and López, V. (2010) 'First invasion of the apple snail *Pomacea insularum* in Europe', *Tentacle*, vol 18, pp27–29

Lv, S., Zhang, Y., Chen, S.-R., Wang, L.-B., Fang, W., Feng, C., Jiang, J.-Y., Li, Y.-L., Du, Z.-W. and Zhou, X.-N. (2009) 'Human angiostrongyliasis outbreak in Dali, China', *PLoS Neglected Tropical Diseases*, vol 3, number e520, doi:10.1371/journal.pntd.0000520

Martín, P. R., Estebenet, A. L. and Cazzaniga, N. J. (2001) 'Factors affecting the distribution of *Pomacea canaliculata* (Gastropoda: Ampullariidae) along its southernmost natural limit', *Malacologia*, vol 43, pp13–23

Massemin, D., Lamy, D., Pointier, J.-P. and Gargominy, O. (2009) *Coquillages et escargots de Guyane. Seashells and Snails from French Guiana*, Biotope, Mèze (Collection Parténope); Muséum national d'Histoire naturelle, Paris

Matsukura, K., Tsumuki, H., Izumi, Y. and Wada, T. (2009) 'Physiological response to low temperature in the freshwater apple snail, *Pomacea canaliculata* (Gastropoda: Ampullariidae)', *Journal of Experimental Biology*, vol 212, pp2558–2563

Mochida, O. (1991) Spread of freshwater *Pomacea* snails (Pilidae, Mollusca) from Argentina to Asia. *Micronesica*, Supplement, vol 3, pp51–62

Morrison, W. E. and Hay, M. E. (2011) 'Feeding and growth of native, invasive and non-invasive alien apple snails (Ampullariidae) in the United States: Invasives eat more and grow more' *Biological Invasions*, vol 13, pp945–955

Naylor, R. (1996) 'Invasions in agriculture: Assessing the cost of the golden apple snail in Asia', *Ambio*, vol 25, pp443–448

Neck, R. W. (1984) 'Occurrence of the striped ram's horn snail, *Marisa cornuarietis*, in central Texas (Ampullariidae)', *Nautilus*, vol 98, pp119–120

Nguma, J. F. M., McCullough, F. S. and Masha, E. (1982) 'Elimination of *Biomphalaria pfeifferi*, *Bulinus tropicus* and *Lymnaea natalensis* by the ampullariid snail, *Marisa cornuarietis*, in a man-made dam in northern Tanzania', *Acta Tropica*, vol 39, pp85–90

Orapa, W. (2006) 'Golden apple snail in Papua New Guinea', in R. C. Joshi and L. C. Sebastian (eds) *Global Advances in Ecology and Management of Golden Apple Snails*, Philippine Rice Research Institute, Muñoz, Nueva Ecija, pp528–529

Peebles, C. R., Oliver-González, J. and Ferguson, F. F. (1972) 'Apparent adverse effect of *Marisa cornuarietis* upon *Lymnaea columella* and *Biomphalaria glabrata* in an ornamental pond in Puerto Rico', *Proceedings, Hawaiian Entomological Society*, vol 21, pp247–256

Perera, G. and Walls, J. G. (1996) *Apple Snails in the Aquarium*, T. F. H. Publications, Inc., Neptune City, New Jersey

Pointier, J. P. and David, P. (2004) 'Biological control of *Biomphalaria glabrata*, the intermediate host of schistosomes, by *Marisa cornuarietis* in ponds of Guadeloupe: Long-term impact on the local snail fauna and aquatic flora', *Biological Control*, vol 29, pp81–89

Pointier, J. P., Théron, A., Imbert-Establet, D. and Borel, G. (1991) 'Eradication of a sylvatic focus of *Schistosoma mansoni* using biological control by competitor snails', *Biological Control*, vol 1, pp244–247

Qiu, J.-W. and Kwong, K.-L. (2009) 'Effects of macrophytes on feeding and life-history traits of the invasive apple snail *Pomacea canaliculata*', *Freshwater Biology*, vol 54, pp1720–1730

Ramakrishnan, V. (2007) 'Salinity, pH, temperature, dessication and hypoxia tolerance in the invasive freshwater snail *Pomacea insularum*', PhD dissertation, University of Texas at Arlington

Ranamukhaarachchi, S. L. and Wikramasinghe, S. (2006) 'Golden apple snails in the world: Introduction, impact, and control measures', in R. C. Joshi and L. C. Sebastian (eds) *Global Advances in Ecology and Management of Golden Apple Snails*, Philippine Rice Research Institute, Muñoz, Nueva Ecija, pp133–151

Rawlings, T. A., Hayes, K. A., Cowie, R. H. and Collins, T. M. (2007) 'The identity, distribution, and impacts of non-native apple snails in the continental United States', *BMC Evolutionary Biology*, vol 7, no 97, doi:10.1186/1471-2148-7-97

Rhymer, J. M. and Simberloff, D. (1996) 'Extinction by hybridization and introgression', *Annual Review of Ecology and Systematics*, vol 27, pp83–109

Rosario, J. and Moquete, C. (2006) 'The aquatic snail *Ampullaria canaliculata* L.-plague of irrigated lowland rice in the Dominican Republic', in R. C. Joshi and L. C. Sebastian (eds) *Global Advances in Ecology and Management of Golden Apple Snails*, Philippine Rice Research Institute, Muñoz, Nueva Ecija, pp514–515

San Martín, R. (2006) 'Recent developments in the use of botanical molluscicides against golden apple snails (*Pomacea canaliculata*)', in R. C. Joshi and L. C. Sebastian (eds) *Global Advances in Ecology and Management of Golden Apple Snails*, Philippine Rice Research Institute, Muñoz, Nueva Ecija, pp393–403

Schnorbach, H.-J. (1995) 'The golden apple snail (*Pomacea canaliculata* Lamarck), an increasingly important pest in rice, and methods of control with Bayluscid', *Pflanzenschutz-Nachrichten Bayer*, vol 48, pp313–346

Schnorbach, H.-J., Rauen, H.-W. and Bieri, M. (2006) 'Chemical control of the golden apple snail, *Pomacea canaliculata*', in R. C. Joshi and L. C. Sebastian (eds) *Global Advances in Ecology and Management of Golden Apple Snails*, Philippine Rice Research Institute, Muñoz, Nueva Ecija, pp419–438

Seuffert, M. E., Burela, S. and Martín, P. (2010) 'Influence of water temperature on the activity of the freshwater snail *Pomacea canaliculata* (Caenogastropoda: Ampullariidae) at its southernmost limit (Southern Pampas, Argentina)', *Journal of Thermal Biology*, vol 35, pp77–84

Simberloff, D. and Stiling, P. (1996) 'Risks of species introduced for biological control', *Biological Conservation*, vol 78, pp185–192

Smith, B. D. (1992) 'Introduction and dispersal of apple snails (Ampullariidae) on Guam', *Pacific Science Association Information Bulletin*, vol 44, pp12–14

Tamaru, C. S., Ako, H. and Tamaru, C. C.-T. (2006) 'Control of apple snail, *Pomacea canaliculata*, in Hawai'i: Challenge or opportunity?', in R. C. Joshi and L. C. Sebastian (eds) *Global Advances in Ecology and Management of Golden Apple Snails*, Philippine Rice Research Institute, Muñoz, Nueva Ecija, pp459–473

Tesana, S., Srisawangwong, T., Sithithaworn, P. and Laha, T. (2008) '*Angiostrongylus cantonensis*: Experimental study on the susceptibility of apple snails, *Pomacea canaliculata* compared to *Pila polita*', *Experimental Parasitology*, vol 118, pp531–535

Tran, C. T., Hayes, K. A. and Cowie, R. H. (2008) 'Lack of mitochondrial DNA diversity in invasive apple snails (Ampullariidae) in Hawaii', *Malacologia*, vol 50, pp351–357

Van Dyke, J. (2010) 'Trapping tons of exotic snails from Wellman Pond', http://snailbusters.wordpress.com/2010/01/26/trapping-tons-of-exotic-snails-from-wellman-pond, accessed 28 October 2010

Wada, T. (1997) 'Introduction of the apple snail *Pomacea canaliculata* and its impact on rice agriculture', in *Proceedings, International Workshop on Biological Invasions of Ecosystems by Pests and Beneficial Organisms*, National Institute of Agro-Environmental Sciences, Ministry of Agriculture, Forestry and Fisheries, Tsukuba, pp170–180

Wada, T. (2004) 'Strategies for controlling the apple snail *Pomacea canaliculata* (Lamarck) (Gastropoda: Ampullariidae) in Japanese direct-sown paddy fields', *Japan Agricultural Research Quarterly*, vol 38, pp75–80

Wada, T. (2006) 'Impact and control of introduced apple snail, *Pomacea canaliculata* (Lamarck), in Japan', in R. C. Joshi and L. C. Sebastian (eds) *Global Advances in Ecology and Management of Golden Apple Snails*, Philippine Rice Research Institute, Muñoz, Nueva Ecija, pp181–197

Wada, T. and Matsukura, K. (2007) 'Seasonal changes in cold hardiness of the invasive freshwater apple snail, *Pomacea canaliculata* (Lamarck) (Gastropoda: Ampullariidae)', *Malacologia*, vol 49, pp383–392

Wada, T., Ichinose, K., Yusa, Y. and Sugiura, N. (2004) 'Decrease in density of the apple snail *Pomacea canaliculata* (Lamarck) (Gastropoda: Ampullariidae) in paddy fields after crop rotation with soybean, and its population growth during the crop season', *Applied Entomology and Zoology*, vol 39, pp367–372

Wilstermann-Hildebrand, M. (2009) *Apfel-Schnecken. Die Familie der Ampullariidae*, Natur und Tier - Verlag GmbH, Münster

Wiyareja, S. and Tjoe-Awie, J. R. (2006) 'Golden apple snail: Its occurrence and importance in Suriname's rice ecosystem', in R. C. Joshi and L. C. Sebastian (eds) *Global Advances in Ecology and Management of Golden Apple Snails*, Philippine Rice Research Institute, Muñoz, Nueva Ecija, pp337–342

Wong, P. K., Kwong, K. L. and Qiu, J.-W. (2009) 'Complex interactions among fish, snails and macrophytes: Implications for biological control of an invasive snail', *Biological Invasions*, vol 11, pp2223–2232

Wong, P. K., Liang, Y., Liu, N. Y. and Qiu, J.-W. (2010) 'Palatability of macrophytes to the invasive freshwater snail *Pomacea canaliculata*: differential effects of multiple plant traits', *Freshwater Biology*, vol 55, no 10, pp2023–2031

Wu, J.-Y., Meng, P.-J., Liu, M.-Y., Chiu, Y.-W. and Liu, L.-L. (2010) 'A high incidence of imposex in *Pomacea* apple snails in Taiwan: A decade after triphenyltin was banned', *Zoological Studies*, vol 49, pp85–93

Wu, M. and Xie, Y. (2006) 'The golden apple snail (*Pomacea canaliculata*) in China', in R. C. Joshi and L. S. Sebastian (eds) *Global Advances in Ecology and Management of Golden Apple Snails*, Philippine Rice Research Institute, Muñoz, Nueva Ecija, pp285–298

Wu, W.-L. and Lee, Y.-C. (2005) 'The biology and population analysis of the golden apple snail in Taiwan', in P.-Y. Lai, Y.-F. Chang and R. H. Cowie (eds) *Proceedings – APEC Symposium on the Management of the Golden Apple Snail, September 6–11, 2004*, National Pingtung University of Science and Technology, Pingtung, Chinese Taipei, Taiwan, pp25–33

Wu, Y.-T, Wu, J., Li, M.-C., Chiu, Y.-W., Liu, M.-Y. and Liu, L.-L. (2011) 'Reproduction and juvenile growth of the invasive apple snails *Pomacea canaliculata* and *Pomacea scalaris* (Gastropoda: Ampullariidae) in Taiwan', *Zoological Studies*, vol 50, pp61–68

Yang, P.-S., Chen, Y.-H., Lee, W.-C. and Chen, Y.-H. (2006) 'Golden apple snail management and prevention in Taiwan', in R. C. Joshi and L. C. Sebastian (eds) *Global Advances in Ecology and Management of Golden Apple Snails*, Philippine Rice Research Institute, Muñoz, Nueva Ecija, pp169–179

Yipp, M. W., Cha, M. W. and Liang, X. Y. (1991) 'A preliminary impact assessment of the introduction of two species of *Ampullaria* (Gastropoda: Ampullariidae) into Hong Kong', in C. Meier-Brook (ed) *Proceedings of the Tenth International Malacological Congress, Tübingen, 27 August–2 September 1989*, UNITAS Malacologia, Tübingen, pp393–397

Yusa, Y. (2006) 'Predators of the introduced apple snail, *Pomacea canaliculata* (Gastropoda: Ampullariidae): Their effectiveness and utilization in biological control', in R. C. Joshi and L. C. Sebastian (eds) *Global Advances in Ecology and Management of Golden Apple Snails*, Philippine Rice Research Institute, Muñoz, Nueva Ecija, pp345–361

Yusa, Y., Sugiura, N. and Wada, T. (2006) 'Predatory potential of freshwater animals on an invasive agricultural pest, the apple snail *Pomacea canaliculata* (Gastropoda: Ampullariidae), in southern Japan', *Biological Invasions*, vol 8, pp137–147

19

Potamopyrgus antipodarum J. E. Gray (New Zealand mudsnail)

Sarina Loo

Description and Biology

Potamopyrgus antipodarum (synonymous with *P. jenkinsi* and *Hydrobia jenkinsi*) is a small, aquatic hydrobiid snail (Figures 19.1 and 19.2). In its native range the height of the shell can reach sizes of up to 12mm, but in the invaded range its adult size is generally 4–6mm (Ponder, 1988; Levri, et al, 2007). The shell is dextral and adults normally have five to six whorls. There can be great variation in the shell shape (from slender and elongate to ventricose) and shell ornamentation (with or without spines or keels) (Winterbourne, 1970). The shell colour ranges from light to dark brown, but encrusted shells may be other colours.

The New Zealand mudsnail reproduces mainly by parthenogenesis, and although sexual reproduction can occur, it is extremely rare in introduced populations (Wallace, 1992). Asexual females develop eggs that can grow without fertilization and produce cloned genetically identical offspring. The snail is highly fecund and ovoviviparous, which increases the likelihood of populations establishing when individuals are released into new areas (Wallace, 1992). One female is sufficient to initiate a new population.

In New Zealand many genetically different clones are present. In the invaded range much fewer clones exist. In the US three clones have been recorded, probably arriving in the US through different pathways. Europe has three clonal populations. A single population is known from Japan and genetic markers suggest that it represents an independently founded population. The widespread population in Australia is all from a single clone. Interestingly, the clone in Australia is also one of the clones found in the US, indicating that the US population may have come from either Australia or New Zealand (New Zealand Mudsnail Management and Control Plan Working Group, 2007).

The mudsnail can occur in high population densities. Densities of 800,000m^{-2} (Dorgelo, 1987) have been observed in Europe. In the US, >300,000m^{-2} is common and densities have been estimated to be as high as 750,000m^{-2} in rivers in Yellowstone National Park (Richards et al, 2004). The highest reported density in Australia is ~50,000m^{-2} (Schreiber et al, 1998). In the snail's native range in New Zealand, population densities of 180,000m^{-2} have been recorded (Michaelis, 1977). Densities are usually highest in systems with high primary productivity, constant temperatures and constant flow (Richards et al, 2001). Seasonal fluctuations in density occur, with peaks in summer.

The New Zealand mudsnail has a generalized diet of plant and animal detritus, algae and diatoms (Haynes and Taylor, 1984). In New Zealand, the snail appears to strongly favour pasture streams (due to the higher photosynthesis rates and higher algae and periphyton biomass) over streams in comparatively heavily shaded pine and native forest (Quinn et al, 1997).

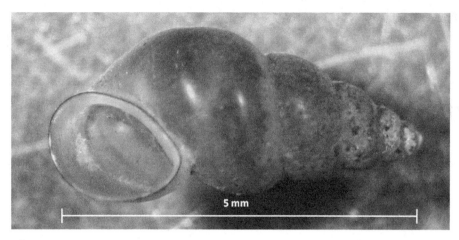

Figure 19.1 Potamopyrgus antipodarum *(New Zealand mudsnail)*

Trematode parasites are prevalent in New Zealand populations of the mudsnail. The parasite *Microphallus* sp. appears to be highly specific in the native range, infecting the most common genotypes (Dybdahl and Lively, 1998). The Australian population is also affected by *Microphallus* sp. trematodes (Schreiber et al, 1998). In France and Poland, native European trematode species have infected New Zealand mudsnail populations in very low frequencies (Gerard and Le Lannic, 2004; Zbikowski and Zbikowska, 2009).

Habitat and Environmental Tolerances

Potamopyrgus antipodarum can inhabit a wide range of ecosystems, including rivers, reservoirs, lakes and estuaries; can be found on a variety of substrates (for example, silt, sand, gravel, cobbles and vegetation), and in muddy to clear waters.

The species is more likely to be found in catchments that have suffered several types of anthropogenic disturbance (Zaranko et al, 1997; Schreiber et al, 2003; Alonso and Castro-Diez, 2008). The mudsnail has high competitive ability at early stages of succession, which explains its success in human-altered ecosystems (Quinn et al, 1998). Even in its native New Zealand range, the mudsnail was found to be more common in streams within agricultural catchments than streams with forested catchments (Quinn and Hickey, 1990; Quinn et al, 1997).

The snail has broad environmental tolerances and has been found in waters of high and low calcium content, on hard and soft substrates and in a wide variety of salinities, including in estuaries (Winterbourne, 1970; Jacobsen and Forbes, 1997). Population densities may decline in winter, but the snail can tolerate ice-covered lakes, so severe winter conditions with week-long periods at −4°C or less are needed to eliminate populations (Hylleberg and Siegismund, 1987). The snail has a high thermal tolerance and has been recorded alive in a laboratory at water temperatures up to 32°C (Quinn et al, 1994). However, the activity of

Figure 19.2 *New Zealand mudsnail populations can exhibit high densities*

the snail declines above 28°C (Winterbourne, 1969) and they are unlikely to reproduce above 24°C (Dybdahl and Kane, 2005). The greatest depth the snail has been found at is 25m (Zaranko et al, 1997). The mudsnails can resist desiccation for several days (Haynes et al, 1985).

Distribution

In its native range, the New Zealand mudsnail is found on both the North and South Islands of New Zealand and the surrounding smaller islands (Ponder, 1988). In its invaded range most of the available information is related to developed countries. There is a lack of distribution information for developing countries.

Europe

In Europe, the New Zealand mudsnail is very widespread and only a few countries have escaped invasion (Bank, 2007), although its presumed absence could be due to the lack of recent field observations (Cianfanelli et al, 2007). It has been in England since 1859, which was probably the first introduction in Europe (Boycott, 1936). It reached the European mainland about 1900 (Hubendick, 1950) and is now recorded in most countries, such as Germany (Boettger, 1963), France (Berner, 1963), Poland (Jackiewicz, 1973), Spain (Ibanez and Alonso, 1977), Portugal (Simoes 1988), Czech Republic (Kuchař, 1983), Turkey (Yildirim et al, 2006), Finland (Carlsson, 2000), Italy (Cianfanelli et al, 2007) Greece (Radea et al, 2008) and Russia (Filippenko and Son, 2008).

North America

In the US, the New Zealand mudsnail was first discovered in the Snake River in Idaho in 1987. In the first ten years of invasion it colonized 640km of the Snake River and its tributaries and continued its spread across the North American continental divide into the Madison River and Missouri River basin (Zaranko et al, 1997). The snail has now spread throughout the western states of Oregon, California, Washington, Arizona, Montana, Colorado, Nevada, Utah and Wyoming (Global Invasive Species Database, 2010). In the Great Lakes, the New Zealand mudsnail was first found established in Lake Ontario in 1991 (Zaranko et al, 1997) and in Lake Erie in 2005

(Levri et al, 2007). It may also be established in Lake Superior, where some individuals were found in 2001 (Grigorovich et al, 2003).

In Canada, mudsnails were found in north-eastern Lake Ontario, as well as Thunder Bay, Ontario, in 2001. The snail has been recorded to have expanded its range in the northern Pacific to Port Alberni in British Columbia, Canada (Davidson et al, 2008).

Ecological models of the potential distribution of the New Zealand mudsnail predict that the snail is likely to continue its spread in the western states of the US and across the Great Plains to the east coast (Loo et al, 2007a). The distribution in the Great Lakes is predicted to continue to expand and, if the snail reaches the rivers of the Mississippi basin, it will probably spread quickly through those states, as seen with the zebra mussel (*Dreissena polymorpha*). Southern Canada and parts of inland Mexico may also be suitable for invasion (Loo et al, 2007a).

Australia

The first records of the New Zealand mudsnail in Australia are from Hobart, Tasmania in 1872. By 1895 the snail had spread to mainland Australia where it was found in Melbourne, Victoria. It continued to move west to Adelaide, where it was found in 1926. By 1963 the snail had spread northwards to Sydney (Ponder, 1988).

Ecological models of the potential distribution of the New Zealand mudsnail in Australia predict that the snail is likely to continue its spread along the east coast. Given a suitable transport vector, it may also establish in south-western Australia (Loo et al, 2007a).

Asia

The New Zealand mudsnail has been found in Japan (Shimada and Urabe, 2003) and Iraq (Naser and Son, 2009).

Pattern and Vectors of Spread

The New Zealand mudsnail spreads both actively and passively. It has been estimated that populations can actively move upstream at a rate of 1km year^{-1} (Lassen, 1975). New Zealand mudsnails have also been found to float downstream independently or on aquatic vegetation (Vareille-Morel, 1983; Ribi and Arter, 1986).

Observations that the snail can pass live through the guts of several fish species, and then reproduce within an hour, indicate that the snail's spread can be aided by fish movement (Haynes et al, 1985). Because the New Zealand mudsnail is quite small it may be easily dispersed passively by birds (Haynes et al, 1985). It has been observed in the mud on the bill of ducks (Coates, 1922) and in continental Europe its first appearance was in the western Baltic on a migration flyway (Lassen, 1975). Given that mudsnails can live in moist areas along stream banks, the mudsnail could also be spread on the feet or fur of domestic livestock or native wildlife that are walking in streams or along the banks.

Commercial movement of aquaculture products, such as live trout or eggs for fish stocking, may also be an important vector of New Zealand mudsnail spread (Bowler, 1991); the species has been found in several fish hatcheries in the US and Australia. Additionally, given that the mudsnails are operculate and can resist desiccation for several days, they may easily be moved by angling gear or boats (Haynes et al, 1985). Loo et al (2007b) found that catchments in Victoria, Australia with high angling activity were more likely to be invaded by the mudsnail. Anecdotal evidence from the US suggests that the mudsnails have been spread extensively as 'hitchhikers' on angling equipment (Hosea and Finlayson, 2005). The spread of the New Zealand mudsnail in the US has been much more rapid then in Australia. This may be a result of greater average flow of human vectors. The US has a stronger angling culture and a much larger human population than Australia (Loo et al, 2007b).

The aquarium trade has potentially assisted the movement of the New Zealand mudsnail, especially via inadvertent movement on aquatic plants. Several authors have offered the hypothesis that initial introductions to Europe and Australia were a result of the transport of aquatic plants between Australia and botanical collections in Europe (Winterbourn, 1972; Ponder, 1988).

Natural resource management personnel involved in monitoring projects and restoration activities may transport New Zealand mudsnails to new water bodies via their gear, vehicles or clothing. Mudsnails can live in moist environments near the edges of streams, and therefore can be picked up and moved by people who are not wading in the water. Community and school monitoring groups are another potential vector for spread (New Zealand Mudsnail Management and Control Plan Working Group, 2007). Scientists undertaking research on water bodies may also transport the snail on waders and sampling nets.

Any waterway operations that remove and transport mud, sand and other bottom materials from areas with New Zealand mudsnails can serve as a vector for new introductions. Dredges that move frequently between rivers and estuaries are particularly likely sources of regional spread. Maintenance of canals and ditches by landowners, water and power agencies, and flood control personnel also have the potential to spread the snail. Introduction could also occur via firefighting machinery or equipment that is moved from one place to another across streams and rivers to fight backcountry or forest fires. Transporting large helicopter deployed water buckets between water bodies is a particular concern (New Zealand Mudsnail Management and Control Plan Working Group, 2007).

Impacts

Environmental impacts

The impacts of the New Zealand mudsnail are complex and variable. In an Australian stream, Schreiber et al (2002) found the colonization of native macroinvertebrates was positively correlated with densities of the mudsnail. A study from the US found the reverse pattern (Kerans et al, 2005). However, densities of the snail in the experimental plots differed markedly between the two studies (4500 vs. ~20,000m^{-2}, respectively). A possible shift from facilitation to competition as snail densities increase shows that the relationships between the mudsnail and other macroinvertebrates are complex (Kerans et al, 2005). At high densities, modification of natural benthic communities is likely.

In a highly productive stream in the US, the New Zealand mudsnail made up 97 per cent of the invertebrate biomass and consumed 75 per cent of the gross primary production, resulting in possible community level impacts (Hall et al, 2003). Hall et al (2006) measured extremely high rates of secondary production of the mudsnail when compared with native stream invertebrates.

In Mediterranean streams, the New Zealand mudsnail was found to have a relatively low impact on macroinvertebrate community structure, and the effect on chlorophyll *a* standing stocks was not significant (Murria et al, 2008). It is possible that harsh hydrologic conditions in the Mediterranean and Australia, such as seasonal droughts and floods, prevent mudsnail densities from becoming as large as in the US.

There are concerns about the snail's low nutritional value, especially for secondary consumers such as fish, because it provides little energy and may pass through the gut undigested (Haynes and Taylor, 1984; Haynes et al, 1985; McCarter, 1986). Vinson and Baker (2008) found that rainbow trout fed an exclusive and unlimited amount of New Zealand mudsnails lost 0.14–0.48 per cent of their initial body weight per day. They also found that the condition of wild caught brown and rainbow trout with New Zealand mudsnails in their guts was significantly lower than that of fish without mudsnails in their stomachs.

Social impacts

There is concern about the impacts of the New Zealand mudsnail invasion on the recreational fishing industry. The results from studies by Vinson and Baker (2008) confirm that North American trout fisheries face potential negative impacts from New Zealand mudsnail invasion. Management agencies may choose to close invaded water bodies to fishing or other water sports in attempts to prevent the spread of the New Zealand mudsnail. This may have localized impacts on recreational fisherman and other water sport enthusiasts.

Economic impacts

No studies into the economic impacts of the New Zealand mudsnail have been conducted, but the impacts are likely to vary regionally. Given the high densities that the mudsnail can occur in, the potential for biofouling (such as clogging water intake structures) is high. In Australia, the mudsnails have been distributed through water pipes to emerge from domestic taps and have blocked water pipes and meters (Ponder, 1988).

Indirect economic impacts include the resources that have been spent on research in numerous countries, monitoring programmes, decontamination activities

and public awareness raising campaigns (New Zealand Mudsnail Management and Control Plan Working Group, 2007).

Management Information

The management of aquatic invasive species is notoriously difficult. There can be limited options to control the spread of aquatic invasive species and prevention is the most effective and efficient means of management.

Prevention

Some prevention methods are applicable to a variety of pathways and invasive species. Increasing public awareness using education materials and programmes may be one of the best tools for reducing the spread of the New Zealand mudsnail. Education materials and programmes need to contain information on the potential harmful effects of the New Zealand mudsnail and methods to prevent spread. Promoting thorough drying of boats and trailers before launching at new locations would help to minimize introductions. The public (including natural resource management professionals and scientists) should all be careful to decontaminate equipment, clothing, vehicles and gear to avoid spreading existing populations or initiating new ones.

Several methods for decontaminating equipment have been trialled. Richards et al (2004) suggest that any heat drying treatment of potentially infected equipment at a temperature of no less than 29–30°C and a low humidity level for a minimum of 24 hours or, alternatively, at a temperature of at least 40°C and a low humidity level for at least two hours will ensure a high kill rate. Freezing equipment for 6–12 hours or placing gear in hot water maintained at 120°F (49°C) is also effective.

Equipment can be treated with a chemical solution by putting it in a bucket to soak, or placing it in a plastic bag with solution, shaking and allowing to stand for five to ten minutes. Laboratory tests have exposed New Zealand mudsnails to solutions of benzethonium chloride, chlorine bleach, Commercial Solutions Formula 409® Cleaner Degreaser Disinfectant, Pine-Sol®, ammonia, grapefruit seed extract, isopropyl alcohol, potassium permanganate and Copper sulphate. With the exception of grapefruit seed extract, potassium permanganate and isopropyl alcohol, these materials all killed mudsnails within five minutes (Hosea and Finlayson, 2005). According to this research, the most effective solutions for killing New Zealand mudsnails, which can be used in the field and not damage gear, are copper sulphate (252mg L^{-1} Cu), benzethonium chloride (1940mg L^{-1}) and 50 per cent Commercial Solutions Formula 409® Cleaner Degreaser Disinfectant. Schisler et al (2008) found that exposure to Formula 409® for five minutes was insufficient and ten minutes was required. They also found that the broad spectrum germicide Sparquat 256 at a concentration of 4mL L^{-1} was effective at providing a 100 per cent mortality rate after exposure for five minutes.

Non-chemical disinfection alternatives are advantageous in that they do not involve a chemical water bath (which requires proper disposal) and they have no potential to affect non-target organisms. However, freezing, hot water and drying at high temperatures may be difficult or impossible for many anglers or researchers who are moving from one water body to another in a short period of time and the impacts of repeated freezing or heating on angling gear are unknown. Dedicating a set of equipment to waters known to harbour New Zealand mudsnails and another to uninfected areas would reduce the risk of spread.

Where New Zealand mudsnails have been found in fish hatcheries or other aquaculture operations, releasing those fish that may have consumed live mudsnails only at sites already contaminated by mudsnails can avoid further spread. Voluntary or mandatory decontamination guidelines for waterway operations that remove and transport mud, sand and other bottom materials may help reduce the risk of spread. Government agencies may incorporate decontamination requirements when issuing permits for these activities (New Zealand Mudsnail Management and Control Plan Working Group, 2007).

Regulation and legislation can be used to control the movement of the New Zealand mudsnail. For example, some states in the western US specifically prohibit importation, possession and transport of the mudsnail. Other states use a permit system to control species that are transported into their waters. Quarantine requirements and inspection rules on the aquarium trade could also be used to restrict the spread of the mudsnail.

Ultimately the presence of aquatic invasive species often is symptomatic of underlying management problems that must be rectified to achieve long term improvements in overall stream condition. Given that the New Zealand mudsnail is more likely to be found in catchments that have suffered human disturbance, the best prevention technique may be to protect near-natural ecosystems from degradation.

Eradication

New populations of the New Zealand mudsnail can be eradicated with chemical or physical methods where it is feasible and practicable. Prior to enacting an eradication plan it must be determined:

- if total kill is likely, given that the survival of even one New Zealand mudsnail can negate an eradication attempt;
- if environmental damage will be caused and if so estimated recovery costs; and
- if there will be impacts to non-targeted and threatened/endangered species (New Zealand Mudsnail Management and Control Plan Working Group, 2007).

In natural water bodies, such as rivers or lakes, chemical eradication will often not be feasible and physical eradication difficult (New Zealand Mudsnail Management and Control Plan Working Group, 2007). Chemical treatment, such as the use of molluscide, would be hard to contain and not necessarily be selective for New Zealand mudsnails only. Native invertebrate populations may be at risk of being negatively affected by the treatment. Attempts at crushing or physical removal of the snails may only exacerbate the problem by spreading eggs to new sites (National Parks Service, 2003).

Areas where eradication may be possible are those where hydrological separation can occur (such as small lakes, ponds, irrigation canals and fish hatcheries). Chemicals may be applied to separated water bodies without causing downstream damage. In other cases draining and allowing the substrate to heat and dry in the summer or freeze in the winter would be effective (New Zealand Mudsnail Management and Control Plan Working Group, 2007).

Control and containment

To reduce the likelihood of spread of the New Zealand mudsnail to non-infested regions populations should be controlled or contained; for example, agencies may choose to close the invaded area to fishing or other water sports. Where possible, populations should be isolated and their numbers reduced. This can be achieved by periodic application of molluscicide or desiccation of a water body, where it is feasible and practicable.

Trematode parasites may also become useful to control population size by inhibiting reproduction. Studies of the efficacy and specificity of a New Zealand trematode parasite as a biological control agent have shown positive results so far (Dybdahl et al, 2005). However, biological control using another non-native species may have unintended consequences.

In addition, substantial research on specificity and effects of trematode introductions on vertebrates is required before the control method can be conducted to ensure that further harm to the environment does not result (New Zealand Mudsnail Management and Control Plan Working Group, 2007). In Poland, laboratory studies suggest that over time the capacity of native European parasites to inflict mudsnail mortality in invaded European ecosystems may increase (Zbikowski and Zbikowska, 2009). The use of trematodes that are native to the invaded region is a preferable control method to the introduction of a non-native trematode.

References

Alonso, A. and Castro-Diez, P. (2008) 'What explains the invading success of the aquatic mud snail *Potamopyrgus antipodarum* (Hydrobiidae, Mollusca)?', *Hydrobiologia*, vol 614, pp107–116

Bank, R. A. (2007) 'Mollusca: Gastropoda', Fauna Europea version 1.1, www.faunaeur.org, accessed 9 October 2010

Berner, L. (1963) 'Sur l'invasion de la France par *Potamopyrgus jenkinsi*', *Archiv fuer Molluskenkunde*, vol 92, pp19–29

Boettger, C. (1963) 'Die Herkunft und Verwandtschaftsbeziehungen der Wasserschnecke *Potamopyrgus jenkinsi*, nebst eine Angabe über ihr Auftreten im Mediterrangebiet', *Archiv fuer Molluskenkunde*, vol 80, pp57–84

Boycott, A. E. (1936) 'The habitats of freshwater mollusca in Britain', *Journal of Animal Ecology*, vol 5, pp111–186

Bowler, P. A. (1991) 'The rapid spread of the freshwater Hydrobiidae snail *Potamopyrgus antipodarum* (Gray) in the Middle Snake River, southern Idaho', *Proceedings of the Desert Fisheries Council*, vol 21, pp173–182

Carlsson, R. (2000) 'The distribution of the gastropods *Theodoxus fluviatilis* (L.) and *Potamopyrgus antipodarum* (Gray) in lakes on the Aland Islands, southwestern Finland', *Boreal Environment Research*, vol 5, pp187–195

Cianfanelli, S., Lori, E. and Bodon, M. (2007) 'Non-indigenous freshwater molluscs and their distribution in Italy', in F. Gherardi (ed) *Biological Invaders in Inland Waters: Profiles, Distribution, and Threats*, Springer, Dordrecht, The Netherlands, pp103–121

Coates, H. (1922) 'Invertebrate fauna of Perthshire: The land and freshwater Mollusca', *Transactions and Proceedings of the Perthshire Society of Natural Science*, vol 7, pp179–243

Davidson, T. M., Brenneis, V. E. F., de Rivera, C., Draheim, R. and Gillespie, G. E. (2008) 'Northern range expansion and coastal occurrences of the New Zealand mud snail *Potamopyrgus antipodarum* (Gray, 1843) in the northeast Pacific', *Aquatic Invasions*, vol 3, no 3, pp349–353

Dorgelo, J. (1987) 'Density fluctuations in populations (1982–1986) and biological observations of *Potamopyrgus jenkinsi* in two trophically differing lakes', *Hydrobiological Bulletin,* vol 21, pp95–110

Dybdahl, M. F. and Kane, S. L. (2005) 'Adaptation vs. phenotypic plasticity in the success of a clonal invader', *Ecology*, vol 86, no 6, pp1592–1601

Dybdahl, M. F. and Lively, C. M. (1998) 'Host-parasite coevolution: Evidence for rare advantage and time-lagged selection in a natural population', *Evolution*, vol 52, pp1057–1066

Dybdahl, M. F., Emblidge, A. and Drown, D. (2005) *Studies of a Trematode Parasite for the Biological Control of an Invasive Freshwater Snail*, Report to the Idaho Power Company

Filippenko, D. P. and Son, M. O. (2008) 'The New Zealand mud snail *Potamopyrgus antipodarum* (Gray, 1843) is colonising the artificial lakes of Kaliningrad City, Russia (Baltic Sea Coast)', *Aquatic Invasions*, vol 3, no 3, pp345–347

Gerard, C. and Le Lannic, J. (2004) 'Establishment of a new host–parasite association between the introduced invasive species *Potamopyrgus antipodarum* (Smith) (Gastropoda) and *Sanguinicola* sp. Plehn (Trematoda) in Europe', *Journal of Zoology*, vol 261, pp213–216

Global Invasive Species Database (2010) '*Potamopyrgus antipodarum*', www.issg.org/database/species/ecology.asp?si=19&fr=1&sts=sss, accessed 21 August 2010

Grigorovich, I. A., Korniushin, A. V., Gray, D. K., Duggan I. C., Colautti R. I. and. MacIsaac, H. J. (2003) 'Lake Superior: An invasion coldspot?', *Hydrobiologia*, vol 499, pp191–210

Hall, R. O., Tank, J. L. and Dybdahl, M. F. (2003) 'Exotic snails dominate nitrogen and carbon cycling in a highly productive stream', *Frontiers in Ecology and Environment*, vol 1, no 8, pp407–411

Hall, R. O., Dybdahl, M. F. and VanderLoop, M. C. (2006) 'Extremely high secondary production of introduced snails in rivers', *Ecological Applications*, vol 16, no 3, pp1121–1131

Haynes, A. and Taylor, B. J. R. (1984) 'Food finding and food preference in *Potamopyrgus antipodarum* (E. A. Smith) (Gastropoda: Prosobranchia)', *Archiv für Hydrobiologie*, vol 100, no 4, pp479–491

Haynes, A., Taylor, B. J. R. and Varley, M. E. (1985) 'The influence of the mobility of *Potamopyrgus jenkinsi* (Smith, E. A.) (Prosobranchia: Hydrobiidae) on its spread', *Archiv für Hydrobiologie*, vol 103, no 4, pp497–508

Hosea, R. C. and Finlayson, B. (2005) *Controlling the Spread of New Zealand Mudsnails on Wading Gear*, California Department of Fish and Game, Rancho Cordova, California

Hubendick, B. (1950) 'The effectiveness of passive dispersal of *Hydrobia jenkinsi*', *Zoologiska Bidrag från Uppsala*, vol 28, pp493–504

Hylleberg, J. and Siegismund, H. R. (1987) 'Niche overlap in mud snails (Hydrobiidae): Freezing tolerance', *Marine Biology*, vol 94, pp403–407

Ibanez, M. and Alonso, M. R. (1977) 'Geographical distribution of *Potamopyrgus jenkinsi* in Spain', *Journal of Conchology*, vol 29, pp141–146

Jackiewicz, M. (1973) 'New stations of the snail *Potamopyrgus jenkinsi* in Poland and some remarks about its distribution', *Przeglad Zoologiczny*, vol 17, pp364–366

Jacobsen, R. and Forbes, V. E. (1997) 'Clonal variation in life-history traits and feeding rates in the gastropod, *Potamopyrgus antipodarum*: performance across a salinity gradient', *Functional Ecology*, vol 11, pp260–267

Kerans, B., Dybdahl, M. F., Gangloff, M. M. and Jannot, J. E. (2005) '*Potamopyrgus antipodarum*: Distribution, density, and effects on native macroinvertebrate assemblages in the Greater Yellowstone Ecosystem', *Journal of the North American Benthological Society*, vol 24, no 1, pp123–138

Kuchař, P. (1983) '*Potamopyrgus jenkinsi* poprvé v Československu [First record of *Potamopyrgus jenkinsi* in Czechoslovakia]', *Živa*, vol 31, p23

Lassen, H. H. (1975) 'The migration potential of freshwater snails exemplified by the dispersal of *Potamopyrgus jenkinsi*', *Natura Jutlandica*, vol 20, pp237–242

Levri, E. P., Kelly, A. A. and Love, E. (2007) 'The invasive New Zealand mud snail (*Potamopyrgus antipodarum*) in Lake Erie', *Journal of Great Lakes Research*, vol 33, pp1–6

Loo, S. E., Mac Nally, R. and Lake, P. S. (2007a) 'Forecasting the range of invasion of the New Zealand mudsnail: A comparison of models built with native and invaded range data', *Ecological Applications*, vol 17, pp181–189

Loo, S. E., Keller, R. P. and Leung, B. (2007b) 'Freshwater invasions: Using historical data to analyse spread', *Diversity and Distributions*, vol 13, pp23–32

McCarter, N. H. (1986) 'Food and energy in the diet of brown and rainbow trout from Lake Benmore, New Zealand', *New Zealand Journal of Marine and Freshwater Research*, vol 20, pp551–559

Michaelis, F. B. (1977) 'Biological features of Pupu springs', *New Zealand Journal of Marine and Freshwater Research*, vol 11, pp357–373

Murria, C., Bonada, N. and Prat, N. (2008) 'Effects of the invasive species *Potamopyrgus antipodarum* (Hydrobiidae, Mollusca) on community structure in a small Mediterranean stream', *Fundamental and Applied Limnology*, vol 171/172, pp131–143

Naser, M. D. and Son, M. O. (2009) 'First record of the New Zealand mud snail *Potamopyrgus antipodarum* (Gray 1843) from Iraq: The start of expansion to Western Asia?', *Aquatic Invasions*, vol 4, no 2, pp369–372

National Park Service (2003) *New Zealand Mudsnail, Baseline Distribution and Monitoring Study*, Department of the Interior, Washington, DC

New Zealand Mudsnail Management and Control Plan Working Group (2007) *National Management and Control Plan for the New Zealand Mudsnail (Potamopyrgus antipodarum)*, Prepared for the Aquatic Nuisance Species Task Force, US

Ponder, W. F. (1988) '*Potamopyrgus antipodarum* – a molluscan coloniser of Europe and Australia', *Journal of Molluscan Studies*, vol 54, pp271–285

Quinn, G. P., Lake, P. S. and Schreiber, E. S. G. (1998) 'A comparative study of colonization by benthos in a lake and its outflowing stream', *Freshwater Biology*, vol 39, no 4, pp623–635

Quinn, J. M. and Hickey, C. W. (1990) 'Characterisation and classification of benthic invertebrate communities in 88 New Zealand rivers in relation to environmental factors', *New Zealand Journal of Marine and Freshwater Research*, vol 24, pp387–409

Quinn, J. M., Steele, G. L., Hickey, C. W. and Vickers, M. L. (1994) 'Upper thermal tolerances of twelve New Zealand stream invertebrate species', *New Zealand Journal of Marine and Freshwater Research*, vol 28, pp391–397

Quinn, J. M., Cooper, A. B., Davies-Colley, R. J., Rutherford, J. C. and Williamson, R. B. (1997) 'Land use effects on habitat, water quality, periphyton, and benthic invertebrates in Waikato, New Zealand, hill-country streams', *New Zealand Journal of Marine and Freshwater Research*, vol 31, pp579–597

Radea, C., Louvrou, I. and Economou-Amilli, A. (2008) 'First record of the New Zealand mud snail *Potamopyrgus antipodarum* J. E. Gray 1843 (Mollusca: Hydrobiidae) in Greece – notes on its population structure and associated microalgae', *Aquatic Invasions*, vol 3, no 3, pp341–344

Ribi, G. and Arter, H. (1986) 'Ausbreitung der Schneckenart *Potamopyrgus jenkinsi* im Zürichsee von 1980 bis 1984', *Vierteljahrsschrift der Naturfoschenden Gessllschaft in Zürich*, vol 131, no 1, pp52–57

Richards, D. C., Cazier, L. D. and Lester, G. T. (2001) 'Spatial distribution of three snail species, including the invader *Potamopyrgus antipodarum* in a freshwater spring', *Western North American Naturalist*, vol 6, pp375–380

Richards, D. C., O'Connell, P. and Cazier Shinn, D. (2004) 'Simple control method to limit the spread of the New Zealand mudsnail *Potamopyrgus antipodarum*', *North American Journal of Fisheries Management*, vol 24, pp114–117

Schisler, G. J., Vieira, N. K. M. and Walker, P. G. (2008) 'Application of household disinfectants to control New Zealand mudsnails', *North American Journal of Fisheries Management*, vol 28, pp1172–1176

Schreiber, E. S. G., Quinn, G. P. and Lake, P. S. (1998) 'Life history and population dynamics of the exotic snail *Potamopyrgus antipodarum* (Prosobranchia: Hydrobiidae) in Lake Purrumbete, Victoria, Australia', *Marine and Freshwater Research*, vol 49, pp73–78

Schreiber, E. S. G., Quinn, G. P. and Lake, P. S. (2002) 'Facilitation of native stream fauna by an invading species? Experimental investigations of the interaction of the snail, *Potamopyrgus antipodarum* (Hydrobiidae) with native benthic fauna', *Biological Invasions*, vol 4, pp317–325

Schreiber, E. S. G., Quinn, G. P. and Lake, P. S. (2003) 'Distribution of an alien aquatic snail in relation to flow variability, human activities and water quality', *Freshwater Biology*, vol 48, pp951–961

Shimada, K. and Urabe, M. (2003) 'Comparative ecology of the alien freshwater snail *Potamopyrgus antipodarum* and the indigenous snail *Semisulcospira* spp.', *Venus*, vol 62, pp39–53

Simoes, M. (1988) 'Distribución en Portugal de *Potamopyrgus jenkinsi* (Prosobranchia Hydrobiidae)', *Iberus*, vol 8, pp243–244

Vareille-Morel, C. (1983) 'Les mouvements journaliers du mollusque prosobranche *Potamopyrgus jenkinsi* Smith. Etude sur le terrain et en laboratoire', *Haliotis*, vol 13, pp31–34

Vinson, M. R. and Baker, M. A. (2008) 'Poor growth of rainbow trout fed New Zealand mud snails *Potamopyrgus antipodarum*', *North American Journal of Fisheries Management*, vol 28, pp701–709

Wallace, C. (1992) 'Parthenogenesis, sex and chromosomes in *Potamopyrgus*', *Journal of Molluscan Studies*, vol 58, pp93–107

Winterbourne, M. J. (1969) 'Water temperatures as a factor limiting the distribution of *Potamopyrgus antipodarum* (Gastropoda-Prosobranchia) in the New Zealand thermal region', *New Zealand Journal of Marine and Freshwater Research*, vol 3, pp453–458

Winterbourne, M. J. (1970) 'The New Zealand species of *Potamopyrgus* (Gastropoda: Hydrobiidae)', *Malacologia*, vol 10, no 2, pp283–321

Winterbourn, M. J. (1972) 'Morphological variation of *Potamopyrgus jenkinsi* (Smith) from England and a comparison with the New Zealand species, *Potamopyrgus antipodarum* (Gray)', *Proceedings of the Malacological Society of London*, vol 40, pp133–145

Yildirim, M. Z., Koca, S. B. and Kebapçi, U. (2006) 'Supplement to the prosobranchia (Mollusca: Gastropoda) fauna of fresh and brackish waters of Turkey', *Turkish Journal of Zoology*, vol 30, pp197–204

Zaranko, D. T., Farara, D. G. and Thompson, F. G. (1997) 'Another exotic mollusc in the Laurentian Great Lakes: The New Zealand native *Potamopyrgus antipodarum* (Gray 1843) (Gastropoda, Hydrobiidae)', *Canadian Journal of Fisheries and Aquatic Science*, vol 54, pp809–814

Zbikowski, J. and Zbikowska, E. (2009) 'Invaders of an invader – trematodes in *Potamopyrgus antipodarum* in Poland', *Journal of Invertebrate Pathology*, vol 101, pp67–70

Part III

Fish

20

Bigheaded carps of the genus *Hypophthalmichthys*

James E. Garvey

History of the Species and Introductions

Bighead (*Hypophthalmichthys nobilis*) and silver carp (*H. molitrix*) are phylogenetically similar fishes of the family Cyprinidae (Figure 20.1) (Howes, 1981). Bighead carp were previously placed in the genus *Aristichthys* but the species is now generally accepted to be in the genus *Hypophthalmichthys* (Howes, 1981). The characteristic large heads of both congeners led to the general moniker bigheaded carps (Kolar et al, 2007). Hybrids between congeners occur and appear to be more common outside of the parental species' native distribution (Schwartz, 1981; Kolar et al, 2007). Both species are native to the great rivers of eastern Asia including China, Russia and North Korea with much of their ranges overlapping (Figure 20.2). Their exact original distributions within Asia are difficult to determine because they have been widely introduced among drainages. In their native range, these species are harvested from wild fisheries and extensively produced in aquaculture (FAO, 2009). Individuals grow rapidly, typically reaching body lengths of nearly 1m in four to five years (Schrank and Guy, 2002; Neuvo et al, 2004; Williamson and Garvey, 2005). Fully grown adults can exceed 1m and 30kg. Individuals typically live between 5 and 10 years (Nuevo et al, 2004; Williamson and Garvey, 2005), although maximum ages of 20 years may occur (Kamilov and Salikhov, 1996). The flesh is palatable to many consumers, although bones are distributed throughout fillets, making the fish difficult to prepare and eat.

Populations of both species require free flowing rivers to complete their lifecycle, although a few exceptions may exist where populations are sustained in reservoirs (Kolar et al, 2007). Adult spawning and production of their semi-buoyant eggs and larvae appear to be initiated in part by rising water (Schrank et al, 2001; Chapman, 2006; DeGrandchamp et al, 2007). Larval Asian carp drift for several days in the swiftly flowing main channel and then settle into shallow slackwater areas such as river backwaters, side channels and island sloughs (Kolar et al, 2007; Lohmeyer and Garvey, 2009). Adults of both species use habitat similarly, preferring areas with slow velocities such as the borders of rivers behind natural and human-made structures, secondary channels and tributary mouths (Peters et al, 2006; Kolar et al, 2007; DeGrandchamp et al, 2008). Distance moved by adults increases with increasing flow velocity, reaching a maximum of 30km per week (DeGrandchamp et al, 2008). This ability to freely and rapidly move throughout rivers allows them to disperse and establish in novel areas.

Because bighead and silver carp are ecologically unique, they have garnered much attention by fisheries biologists. Both are planktivorous and perhaps detritivorous; they have been considered as ideal species

(a)

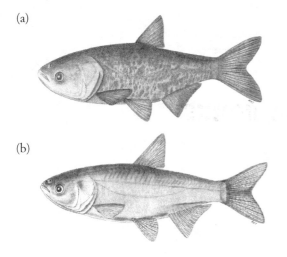

(b)

Note: Both are from the Southern Illinois University Carbondale fluid vertebrate collection.

Source: Matt R. Thomas

Figure 20.1 *(a) Bighead carp (b) and silver carp from Illinois, US*

for enhancing total fisheries production in areas outside their native range because many commercially important species do not directly occupy these trophic levels (Opuszynski, 1981). Bighead and silver carp are believed by some investigators to stimulate primary production and improve food availability for other fishes, particularly in polyculture (Yashouv, 1971).

Both species have been widely stocked to consume nuisance algae and organic matter in wastewater effluent or to improve water quality in aquaculture facilities (Dong et al, 1992). Although still open to debate (see below), these purported characteristics have led to worldwide introductions either through intentional releases or escape from aquaculture facilities, with bighead carp probably being established in the wild in 24 countries and silver carp in 34 countries (Kolar et al, 2007) (Figure 20.2). Introductions continue today. For example, bighead carp were recently found in the wild for the first time in England (Britton and Davies, 2007). An analysis of mitochondrial DNA variation from five large river systems in China, Hungary and the US showed that bighead carp probably originated in the Yangtze River, China (Li et al, 2010). Populations in the US perhaps originated from stocks in the Danube River rather than rivers in Asia.

The introduction and establishment of bighead and silver carp have been well documented in the US (Nico and Fuller, 1999). Both species were brought to the state of Arkansas in the US in the early 1970s (Freeze and Henderson, 1982) for human consumption and as potential biocontrol agents for aquaculture and wastewater treatment (Henderson, 1983). Fish were also transported to the states of Illinois (Malecha et al, 1981) and Alabama (Cremer and Smitherman, 1980) for research during this time. Accidental releases occurred for both species, resulting in their establishment and

■ Countries where introduced and likely established
■ Countries where native and introduced

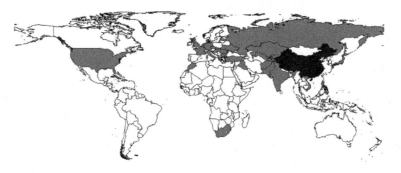

Source: James E. Garvey

Figure 20.2 *Countries in which bighead and/or silver carp are probably established*

eventual capture in the wild. Wild bighead carp were not discovered in the US until 1981. In contrast, silver carp were found in the White River, Arkansas in 1974, not long after they were brought into the US (Freeze and Henderson, 1982). Successful wild reproduction of bighead carp and silver carp was documented in US rivers by the early 1990s (Burr et al, 1996). Both species were widely distributed and reproducing throughout the Mississippi River and unimpounded Missouri River basins by the 2000s (Chick and Pegg, 2001).

Ecological Niche

In their native range, bighead and silver carp probably occupy separate niches by feeding on different sizes of particles and possibly also consuming different planktonic taxa. Both species have large mouths and buccal cavities (Figure 20.3), filtering water primarily by pumping large volumes of water past their gill rakers. Gill rakers of bighead carp are finely spaced, allowing them to sieve small plankton from the water. Silver carp can filter smaller particles than bighead carp (as small as 3–4μm; Kolar et al, 2007), probably because of their unique fused and sponge-like gill rakers (Smith, 1989). Also, silver carp possess an epibranchial organ that produces copious amounts of mucus to trap small items (Spataru, 1977). Bighead carp typically consume zooplankton and some phytoplankton (Dong and Li, 1994), whereas silver carp also consume small

Note: Note the wide mouth for filtering large quantities of water.

Source: Southern Illinois University, Fisheries and Illinois Aquaculture Center

Figure 20.3 *Bighead carp foraging*

phytoplankton (Williamson and Garvey, 2005) and bacteria on detritus (Schroeder, 1978). The relative contribution of these diet items to energy intake is not well known. A long, undifferentiated gut in these fishes probably facilitates digestion of low quality materials. Both species tolerate a range of environmental conditions (Bettoli et al, 1985; Garcia et al, 1999). The apparently broad fundamental niches of both species appear to make them ideal invaders in many rivers and lakes.

Although bigheaded carp may benefit some fisheries when introduced beyond their native range by elevating total fish production and landings by fishers (Xie and Yang, 2000), most reports suggest that the overall ecosystem effects are negative (Kolar et al, 2007). Bigheaded carp probably shift zooplankton communities to those dominated by zooplankters with small bodies (Spataru and Gophen, 1985; Wu et al, 1997; Sass, G., unpublished data). Both carps have high metabolic rates relative to other fishes with similar trophic requirements, meaning that demand for food is high and impact on plankton may be greater than that of many other filter feeders (Hogue and Pegg, 2009). In contrast to cladocerans, which are efficient herbivores, some species of copepods may be able to evade planktivory and thus may dominate when bigheaded carps are abundant (Xie and Yang, 2000; Williamson and Garvey, 2005). Loss of large bodied, efficient zooplankton grazers may induce a trophic cascade, causing nuisance phytoplankton blooms (Matyas et al, 2003). Carp excretion and egestion may be sufficiently high to elevate nutrients and further stimulate phytoplankton production (Starling, 1993). With the exception of experiments done by Schrank et al (2003) that suggest competitive interactions occur between young bighead carp and North American paddlefish, most researchers have reported broad changes in native fish community composition following bigheaded carp introductions without including control systems for comparison. Reductions in native fishes following invasion have been reported worldwide (Kolar et al, 2007), even within areas of China beyond the native range of bigheaded carps (Xie and Chen, 2001). Recently, Sampson et al (2009) documented that diets overlap between some native fishes – primarily, gizzard shad (*Dorosoma cepedianum*) and bigmouth buffalo (*Ictiobus cyprinellus*) – and bigheaded carps in the Mississippi and Illinois Rivers, US. In congruence with this finding, two planktivorous species declined following the invasion of the Illinois River by silver carp in 2000 (Irons et al, 2007). Given that bighead and silver carp overlap substantially in diet contents (Gu et al, 1996; Sampson et al, 2009), these congeners may compete with each other in non-native habitats, with silver carp perhaps being the superior competitor (Buck et al, 1978). The growing evidence for negative impacts is compelling; controlled experiments coupled with food web modelling would shed considerable light into the effects of the invasion on native species and ecosystems.

Predicting the range and impact of bigheaded carps in novel environments is challenging. A literature review of spatially explicit data within the US (Garvey, 2007) suggested that adults are concentrated in habitats of reduced flow and may be limited by slackwater habitat in some reaches. Ecological niche modelling by Chen et al (2007) accurately predicted current distributions of the species in North America and predicted that these species may invade much of eastern North America and parts of the continent's west coast. Cooke and Hill (2010), however, used bioenergetics models to demonstrate that plankton density within most of the Laurentian Great Lakes is insufficient to support bigheaded carp production. It is important to note that this study did not account for the impact of detritus and phytoplankton on the success of the species.

Most research has not considered the ability of both species to complete all aspects of their life history when becoming established. Reproductive success as indexed by the presence of larvae in the drift varied positively with seasonal discharge in the impounded reaches of the Mississippi River, but was consistently high in an unimpounded reach (Lohmeyer and Garvey, 2009). This is likely to be due to the way water is managed in the impounded reaches of the river, whereby gates of dams are closed during low flow, creating a series of disconnected pools (Lohmeyer and Garvey, 2009). This supports the long held idea that bigheaded carps require rising water levels and perhaps extended stretches of unimpeded river to successfully reproduce (Verigin et al, 1978). Although reduced flow and impoundment probably hinder dispersal and survival of eggs and larvae, low discharge also prevents reproduction by causing mature females to abandon spawning and reabsorb eggs (DeGrandchamp et al, 2007). Thus, if the species are to become abundant in lakes such as Lake Michigan where they currently threaten to invade, they will probably need to use long, free flowing tributary streams to successfully reproduce

(Cooke and Hill, 2010). Kolar et al (2007) and Mandrak and Cudmore (2004) identified tributaries in the North American Great Lakes that are likely candidates for successful reproduction.

Both food availability and water chemistry may limit the niches of bigheaded carps. Because of their rapid growth and large body size, both species have been assumed to require high primary productivity to successfully invade and persist. However, because both feed on a variety of particles and are very efficient at extracting food from the water (Schroeder, 1978; Smith, 1989; Dong et al, 1992), it is not unreasonable to expect that they can persist in unproductive environments by rapidly exploiting any resources available (e.g. plankton, particulate organic matter, microbes). In support of this idea, Calkins (2010) found that chlorophyll a concentrations (a surrogate for algal concentration) were three orders of magnitude greater in guts of adult silver carp than in Mississippi River backwaters in which these fish were collected (also see Pongruktham et al, 2010). Silver carp were found in areas of elevated chlorophyll a concentration in the river, suggesting individuals are able to seek out areas of elevated productivity even when overall primary productivity is quite low. Soft water may cause osmotic stress of fertilized eggs and reduce hatching success (Gonzal et al, 1987) thereby potentially limiting the ability for bigheaded carps to invade systems with low water hardness such as the Laurentian Great Lakes (Whittier and Aitkin, 2008). However, Rach et al (2010) and Chapman and Deters (2009) found that water hardness did not limit hatching success of silver carp and bighead carp, cautioning that the impact of water chemistry varies and may not limit the ability of these species to invade systems with soft water.

Management Efforts

Bigheaded carps continue to be important food fishes, primarily in Asia. Because of high demand, these species are overfished in their native range (Chapman, 2006). This problem has recently been exacerbated by the installation and operation of the Three Gorges Dam in the Yangtze River, which started in 2003 (Xie et al, 2007). Altered hydrology and temperature appear to have reduced reproductive output of these species (Xie et al, 2007). Simultaneously, commercial yield declined from 3.4 million metric tonnes in 2002 to 1.7 million metric tonnes in 2005 (Xie et al, 2007). A similar negative response is occurring in the impounded Pearl River (Tan et al, 2010). Because demand is high and they can be grown inexpensively on agricultural soy waste and manure (Buck et al, 1978), bighead and silver carp remain the most commonly aquacultured fishes in the world. Farmed bighead carp production was 15,306 tonnes in 1950 and rose to 2.3 million tonnes in 2008 (Figure 20.4) (FAO, 2009). Global production of farmed silver carp rose from 30,000 tonnes in 1950 to 3.8 million tonnes in 2008 (Figure 20.4) (FAO, 2009). As human population density rises in Asia and across the globe, the density of cultured bigheaded carp is sure to increase. Characteristics that make these species desirable, such as fast growth, appear to be highly heritable and are likely to be selected in hatcheries (Gheyas et al, 2009). These same characteristics may make them formidable invaders when they escape.

Although valuable in culture, some governments outside of Asia want to remove wild populations of bigheaded carp due to perceived and real threats to native resources. Silver carp also pose a risk because of their proclivity to porpoise from the water and injure boaters and skiers (Kolar et al, 2007). Reduced use of invaded rivers by the recreational community because of this perceived danger and its economic cost need to be quantified. In the US, a comprehensive management plan has been developed by the federal government to confront this ongoing invasion (Conover et al, 2007). This plan calls for government agencies within the US to prevent further introductions of bigheaded carps and to contain or extirpate wild populations. Transport of eggs and larvae by boats needs to be curbed largely through piscicide treatment of ballast and bilge water. In 2007, in a measure related to this effort, the US government made the live transport of all life stages of silver carp across state lines illegal, with several states following suit within their borders. No federal rule is in effect for bighead carp.

Outside of Asia, wild bigheaded carps are a minor component of fisheries. In the US, bigheaded carp were largely brought to domestic markets live from culture operations. These species are preferred by their target consumer group (typically Asian expatriates using city fish markets) to be consumed fresh and the flesh of these fishes has a very short shelf life. Little economic demand currently exists for wild caught bighead and silver carp in US rivers because carp cannot feasibly be harvested and transported live from capture fisheries. Given that bigheaded carps are overfished in Asia, it is possible that robust harvest may control them in US waters. However, a lack of a domestic market for processed bigheaded carps and reluctance of the federal or state governments to subsidize fishing for eradication has made fishing unprofitable to date. Public perception of bigheaded carps in the US may be improving and some states are attempting to enhance marketing as well as overseas export opportunities.

Major impediments to market development are contaminants and palatability issues. Organic contaminant concentrations are low in bigheaded carps in US waters (Rogowski et al, 2009). Heavy metal concentrations including mercury vary but are sufficiently elevated in some fish to merit consumption advisories (Rogowski et al, 2009). Barriers to the domestic market are the invasive nature of the fish, the name 'carp', which has a negative connotation to many US consumers, and the bones in the fillets. In an

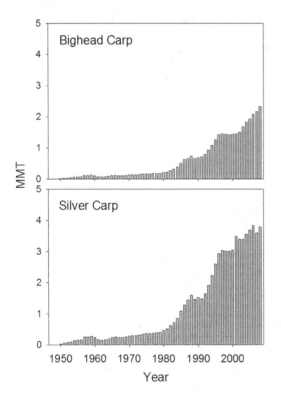

Source: FAO (2009)

Figure 20.4 *Global aquaculture production in million metric tonnes of bighead and silver carp, 1950–2008*

attempt to improve public perception, some agencies are marketing the bigheaded carps under the new name 'silverfin' and various methods for removing the bones from fillets are being developed.

Controversies about Control

No congruent worldwide effort to control feral bigheaded carps is in place. In the US, concerns about these species did not gain wide public attention until the late 1990s when silver carp densities apparently rose. The jumping ability of these large fish captured the attention of the media and sparked the concern of policy makers. The proximity of these species to the Laurentian Great Lakes also piqued public interest. The watersheds of the Mississippi River system including the Illinois River and the Laurentian Great Lakes are geologically separate (Figure 20.5). However, the two systems were connected by the construction of the Chicago Sanitary and Ship Canal in Chicago, Illinois in 1900, allowing aquatic invasive species to move among basins. In 2002, the US federal government installed an experimental electric barrier in the canal to dissuade the inter-basin movement of bigheaded carps and other invasive fishes (Figure 20.5); bigheaded carps were more than 80km below the barrier in the lower Illinois River at that time. A second set of electrical barriers was emplaced below the original barrier by 2010. Construction and maintenance costs of the barrier system exceeded $12 million by 2007 (USACE, 2010).

During summer 2009, researchers collected water samples from above and below the barriers and detected the environmental DNA (eDNA) of these species below the barrier (Lodge et al, 2009). In autumn 2009, eDNA was detected in Lake Michigan embayments and other locations in the canal above the barrier system. To date, no peer-reviewed publication exists to support the validity of the eDNA technique for bigheaded carps, although the method has been validated for other species (Ficetola et al, 2008). This eDNA may be attributed to the actual presence of the species in the river or from water contaminated with bigheaded carp cells (e.g. from boat bilge or other wastewater transported from a separate location where carp are present). Even though the validity of the technique and its relationship to carp densities remains questionable, the presence of eDNA in proximity to Lake Michigan caused an outcry from many environmental groups as well as Great Lakes states and

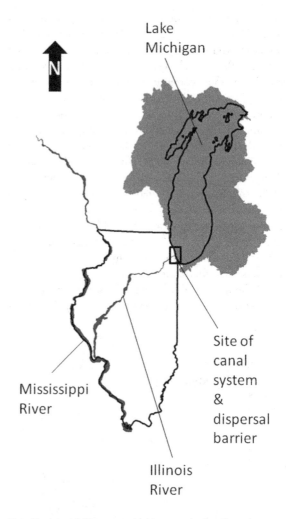

Note: The Lake Michigan watershed has grey shading. The only physical connection between the two basins is a canal system (boxed), where an electric barrier system is currently emplaced.

Source: James E. Garvey

Figure 20.5 *Intersection between the invaded Mississippi River System (including the Illinois River) and the Lake Michigan watershed*

Canada, prompting lawsuits demanding that Illinois close the canal. These groups argued that the multi-billion dollar economic value of the Great Lakes ecosystem was at risk and that the electrical barrier system was insufficient to keep carp from invading. Illinois depends on the canal for the movement of goods between the two basins via towed barges and for the transport of stormwater and wastewater downstream. The long term economic cost of closure to Chicago will be substantial. In spring 2010 this issue reached the US Supreme Court, which declined involvement. The canal remained open.

Following the detection of the eDNA, much sampling effort has been expended by US federal and state agencies and contracted fishers. This included releasing the piscicide rotenone in autumn 2009, which resulted in the capture of one bighead carp among thousands of other fishes below the barrier. Fishing revealed a single bighead carp above the barrier in June 2010. However, rotenone applied thereafter in the vicinity of the capture did not yield any additional bighead carp. The discovery in June 2010 prompted renewed criticism of the efficacy of the electrical barriers and caused many to question the decision to keep the canal in operation.

Prognosis

The contrasting cultural views on bigheaded carps make them a challenge for future management and control. As human demand for cheap and healthy protein sources grows, cultured bigheaded carp will be considered a viable option. The economic allure of aquaculture will continue to create incentives to transport them globally, even though their negative ecosystem impact outside of aquaculture is becoming better known. Biosecurity in culture is never 100 per cent foolproof and escapes are always possible. Thus, risk of establishing feral populations that will rapidly spread is always present.

Many questions about the biology and potential impact of these congeners remain. The reproductive cycle of these fishes is well understood in culture. Spawning of females is induced with injected hormones while survival of eggs and larvae is promoted in hatching jars (FAO, 2009). In contrast, behavioural and physiological cues influencing the onset, timing and duration of reproduction in the field are not well understood. Probably, a complex interplay among photocycle, flow, water level, temperature, habitat availability and body condition influences spawning. Hybridization and intersex conditions occur in feral bigheaded carps (Papoulias et al, 2006), although the causes and population consequences are not well understood. For most fishes, survival during the first year of life influences future cohort strength and thus population dynamics. Both species are highly fecund, with females potentially producing greater than one

million eggs per spawning bout (Figure 20.6) (Abdusamadov, 1987; DeGrandchamp et al, 2007). Populations probably build rapidly in new areas, but few investigators have quantified factors influencing growth and survival of new recruits. Also, the lateral movement of these fish between basins as eggs and larvae is a threat, but not well understood. In Illinois, several instances of adult bigheaded carp occurring in lakes were probably due to inundation and subsequent lateral transport of young carp from nearby streams during flooding (Garvey, unpublished data). Most fishes have planktivorous early life stages; however, the impact of planktivory by young and adult bigheaded carps on reproduction of native fishes has rarely been explored.

Critical to assessing risk of introducing these species into new systems such as the Laurentian Great Lakes is an understanding of their ability to exploit resources. Bigheaded carps lack cellulase (Bitterlich, 1985) and thus their ability to digest intact phytoplankton depends on grinding by the pharyngeal teeth and digestion of cell contents. Henebry et al (1988) suggest that bacteria in the guts of these fish may be critical for digestion. If fermentation within the guts of these fish is important, then this helps explain their ability to extract energy and nutrients from poor quality materials such as suspended particulate organic matter. Models predicting the population dynamics of bigheaded carps need to incorporate the influence of factors such as detritus, protozoans and suspended bacteria as well as plankton productivity. Accurate energy budgets need to be developed that weigh the cost of procuring energy

Source: Southern Illinois University, Fisheries and Illinois Aquaculture Center

Figure 20.6 *Ripe ovaries within an adult silver carp captured in the Illinois River*

through swimming and pump filter feeding vs. the available energy gain from the environment.

The most basic information about population dynamics of bigheaded carps is lacking. In Asia, the relative effects of overfishing and modified hydrology on populations are not understood. In the US, concerns about building carp populations are growing. However, only one attempt has been made to quantify population density and standing biomass, where Sass et al (2010) estimated total silver carp biomass to be 705 metric tonnes in a 124km reach of the Illinois River. The high reproductive rate and fast biomass growth of the species should allow populations to grow rapidly and resist environmental variation and harvest (Williamson and Garvey, 2005). A simple population model showed that adults withstand high mortality rates (Garvey et al, 2006). Populations will be more susceptible to reductions in reproductive output and survival of offspring than to removal of adults (Garvey et al, 2006).

Control of established bigheaded carps needs to be multifaceted. Even though populations may be resilient to adult mortality caused by harvest, overfishing in Asia confirms that intentional overfishing is possible. Developing a market and fishing down these species is feasible. If eradication by fishing does occur, an 'exit strategy' for the fishery must exist to ensure that fishers do not switch to native stocks for economic gain and deplete them. Reducing the ability for these fish to enter uninvaded river reaches and lakes will obviously curtail range expansions. Thus, developing effective barriers to movement is important. Electric barriers may be effective but are expensive to build and maintain and can threaten public safety. Cheaper curtains of fine bubbles combined with pulses of high frequency sound (up to 2000Hz) show promise in repelling the movement of bighead carp (Taylor et al, 2005). Barriers may be used downstream of areas where harvest is occurring to prevent new adults from recruiting to the target population. No selective piscicide has been developed for bigheaded carps. When rotenone was applied around the electric barrier in Illinois, public criticism arose because the toxin indiscriminately killed native species. Development of a selective toxicant similar to that used for larval sea lamprey in the Laurentian Great Lakes would be useful if applied to backwater areas where young bigheaded carp aggregate (Rach et al, 2009). Introducing transgenic individuals into bigheaded carp populations may be another form of control. Australia is exploring how introducing an aromatase blocker gene into invasive common carp (*Cyprinus carpio*) will induce production of only males in the population via introgression; this technology may be applied to bigheaded carps by stocking transgenic individuals into the wild (Conover et al, 2007). Transgenic males would mate with feral females, thereby increasing the number of males in the population and slowing population growth (see Davis et al, 1999). Introducing native piscivores at high densities to consume young bigheaded carps may also help reduce recruitment and control population growth (Sass, G., pers. comm.).

Bigheaded carps will create controversy and incur considerable environmental cost in many countries. In contrast, they will continue to be a desirable asset to many other countries, particularly China. Policy makers, managers and entrepreneurs must interact closely and devise ways to balance contrasting goals for these species on a global scale. Within non-native regions, one such way would be to determine how to successfully develop markets for these invasive fish domestically but also to reach existing consumers abroad. In contrast to managing a sustainable fishery, a host of alternative control measures need to be implemented to collapse the harvested populations and then use barriers to prevent recolonization. The effort to reduce populations will need to be sustained perhaps for decades before appreciable declines are seen. Well-designed, standardized and continuous monitoring must be conducted to quantify responses and make management changes when necessary.

References

Abdusamadov, A. S. (1987) 'Biology of white amur (*Ctenophyrngodon idella*), silver carp (*Hypophthalmichthys molitrix*), and bighead (*Aristichthys nobilis*) acclimatized in the Terek region of the Caspian Basin', *Journal of Ichthyology*, vol 26, no 4, pp41–49

Bettoli, P. W., Neill, W. H. and Kelsch S. W. (1985) 'Temperature preference and heat resistance of grass carp *Ctenopharyngodon idella* (Valenciennes), bighead carp *Hypophthalmichthys nobilis* (Gray), and their F1 hybrid', *Journal of Fish Biology*, vol 27, pp239–247

Bitterlich, G. (1985) 'Digestive enzyme pattern of 2 stomachless filter feeders, silver carp, *Hypophthalmichthys molitrix* Val, and bighead carp, *Aristichthys nobilis* Rich', *Journal of Fish Biology*, vol 27, pp103–112

Britton, J. R. and Davies, G. D. (2007) 'First UK recording in the wild of the bighead carp *Hypophthalmichthys nobilis*', *Journal of Fish Biology*, vol 70, pp1280–1282

Buck, D. H., Baur, R. J. and Rose, C. R. (1978) 'Utilization of swine manure in a polyculture of Asian and North American fishes', *Transactions of the American Fisheries Society*, vol 107, pp216–222

Burr, B. M., Eisenhour, D. J., Cook, K. M., Taylor, C. A., Seegert, G. L., Sauer, R. W. and Atwood, E. R. (1996) 'Nonnative fishes in Illinois waters: What do the records reveal?', *Transactions of the Illinois State Academy of Science*, vol 89, pp73–91

Calkins, H. A. (2010) 'Linking silver carp habitat selection to phytoplankton consumption in the Mississippi River', Master's Thesis, Department of Zoology, Southern Illinois University, Carbondale, Illinois

Chapman, D. C. (2006) 'Early development of four cyprinids native to the Yangtze River, China', *US Geological Survey Data Series*, 239, US Geological Survey

Chapman, D. C. and Deters, J. E. (2009) 'Effect of water hardness and dissolved-solid concentration on hatching success and egg size in bighead carp', *Transactions of the American Fisheries Society*, vol 138, pp1226–1231

Chen, P. F., Wiley, E. O. and McNyset, K. M. (2007) 'Ecological niche modeling as a predictive tool: Silver and bighead carps in North America', *Biological Invasions*, vol 9, pp43–51

Chick, J. H. and Pegg, M. A. (2001) 'Invasive carp in the Mississippi River Basin', *Science*, vol 292, pp2250–2251

Conover, G., Simmonds, R. and Whalen, M. (2007) *Management and Control Plan for Bighead, Black, Grass, and Silver Carps in the United States*, Asian Carp Working Group, Aquatic Nuisance Species Task Force, Washington, DC

Cooke, S. L. and Hill, W. R. (2010) 'Can filter-feeding Asian carp invade the Laurentian Great Lakes? A bioenergetic modelling exercise', *Freshwater Biology*, vol 55, pp2138–2152

Cremer, M. C. and Smitherman, R. O. (1980) 'Food habits and growth of silver and bighead carp in cages and ponds', *Aquaculture*, vol 20, pp57–64

Davis, S. A., Catchpole, E. A. and Pech, R. P. (1999) 'Models for the introgression of a transgene into a wild population within a stochastic environment, with applications to pest control', *Ecological Modelling*, vol 119, pp267–275

DeGrandchamp, K. L., Garvey, J. E. and Csoboth, L. A. (2007) 'Linking adult reproduction and larval density of invasive carp in a large river', *Transactions of the American Fisheries Society*, vol 136, pp1327–1334

DeGrandchamp, K. L., Garvey, J. E. and Colombo, R. E. (2008) 'Movement and habitat selection by invasive Asian carps in a large river', *Transactions of the American Fisheries Society*, vol 137, pp45–56

Dong, S. L. and Li, D. (1994) 'Comparative studies of the feeding selectivity of silver carp *Hypophthalmichthys molitrix*, and bighead carp, *Aristichthys nobilis* Rich', *Journal of Fish Biology*, vol 44, pp621–626

Dong, S. L., Li, D. S., Bing, X. W., Shi, Q. F. and Wang, F. (1992) 'Suction volume and filtering efficiency of silver carp (*Hypophthalmichthys-molitrix* Val) and bighead carp (*Aristichthys-nobilis* Rich)', *Journal of Fish Biology*, vol 41, pp833–840

FAO (Food and Agriculture Organization of the United Nations) (2009) *Fisheries Topics: Statistics. Fisheries Statistics and Information*, FAO Fisheries and Aquaculture Department [online], Rome

Ficetola, G. F., Miaud, C., Pompanon, F. and Taberlet, P. (2008) 'Species detection using environmental DNA from water samples', *Biology Letters*, vol 4, pp423–425

Freeze, M. and Henderson, S. (1982) 'Distribution and status of the bighead carp and silver carp in Arkansas', *North American Journal of Fisheries Management*, vol 2, pp197–200

Garcia, L. M. B., Garcia, C. M. H., Pineda, A. F. S., Gammad, E. A., Canta, J., Simon, S. P. D., Hilomen-Garcia, G. V., Gonzal, A. C. and Santiago, C. B. (1999) 'Survival and growth of bighead carp fry exposed to low salinities', *Aquaculture International*, vol 7, pp241–250

Garvey, J. E. (2007) 'Spatial assessment of Asian carp population dynamics: Development of a spatial query tool for predicting relative success of life stages', US Fish and Wildlife Service, http://fishdata.siu.edu/carptools

Garvey, J. E., DeGrandchamp, K. L. and Williamson, C. J. (2006) 'Growth, fecundity, and diets of Asian carps in the Upper Mississippi River system', US Army Corps of Engineers Technical Note, ERDC, Waterways Experimental Station, ERDC/TN ANSRP-06

Gheyas, A. A., Wooliams, J. A., Taggart, J. B., Sattar, M. A., Das, T. K., McAndrew, B. J. and Penman, D. J. (2009) 'Heritability estimation of silver carp (*Hypothalmichthys molitrix*) harvest traits using microsatellite parentage assignment', *Aquaculture*, vol 294, pp187–193

Gonzal, A. C., Aralar, E. V. and Pavico, J. M. F. (1987) 'The effects of water hardness on the hatching and viability of silver carp (*Hypophthalmichthys molitrix*) eggs', *Aquaculture*, vol 64, pp111–118

Gu, B. H., Schell, D. M., Huang, X. H. and Yie, F. L. (1996) 'Stable isotope evidence for dietary overlap between two planktivorous fishes in aquaculture ponds', *Canadian Journal of Fisheries and Aquatic Sciences*, vol 53, pp2814–2818

Henderson, S. (1983) *An Evaluation of Filter Feeding Fishes for Removing Excessive Nutrients and Algae from Wastewater*, US Environmental Protection Agency, EPA 600/2-83-019

Henebry, M. S., Gorden, R. W. and Buck, D. H. (1988) 'Bacterial-populations in the gut of the silver carp (*Hypophthalmichthys molitrix*)', *Progressive Fish Culturist*, vol 50, pp86–92

Hogue, J. L. and Pegg, M. A. (2009) 'Oxygen consumption rates for bighead and silver carp in relation to life-stage and water temperature', *Journal of Freshwater Ecology*, vol 24, pp535–543

Howes, G. (1981) 'Anatomy and phylogeny of the Chinese major carps *Ctenopharyngodon* Steind., 1866 and *Hypophthalmichthys* Blkr., 1860', *Bulletin of the British Museum (Natural History), Zoology*, vol 41, no 1, pp1–52

Irons, K. S., Sass, G. G., McClelland, M. A. and Stafford, J. D. (2007) 'Reduced condition factor of two native fish species coincident with invasion of non-native Asian carps in the Illinois River, USA – Is this evidence for competition and reduced fitness?', *Journal of Fish Biology*, vol 71, pp258–273

Kamilov, B. G. and Salikhov, T. V. (1996) 'Spawning and reproductive potential of the silver carp *Hypophthalmichthys molitrix* from the Syr Dar'ya River' *Journal of Ichthyology*, vol 36, no 5, pp600–606

Kolar, C. S., Chapman, D. C., Courtenay, W. R., Jr, Housel, C. M., Williams, J. D. and Jennings, D. P. (2007) 'Bigheaded carps: A biological synopsis and environmental risk assessment', *American Fisheries Society Special Publication 33*, Bethesda, Maryland, USA

Li, S., Yang, Q., Xu, J. and Wang, C. (2010) 'Genetic diversity and variation of mitochondrial DNA in native and introduced bighead carp', *Transactions of the American Fisheries Society*, vol 139, pp937–946

Lodge, D. M., Chadderton, W. L., Mahon, A. R. and Jerde, C. L. (2009) *Sampling Fact Sheet for eDNA Surveillance*, Center for Aquatic Conservation, Department of Biological Sciences, University of Notre Dame, http://aquacon.nd.edu

Lohmeyer, A. M. and Garvey, J. E. (2009) 'Placing the North American invasion of Asian carp in a spatially explicit context', *Biological Invasions*, vol 11, pp905–916

Malecha, R., Buck, H. and Bauer, R. J. (1981) 'Polyculture of the freshwater prawn (*Macrobrachium rosenbergii*) with two combinations of carps in manured ponds', *Journal of the World Mariculture Society*, vol 12, pp203–213

Mandrak, N. E. and Cudmore, B. (2004) *Risk Assessment for Asian Carps in Canada*, Fisheries and Oceans Canada, Research Document 2004/103

Matyas, K., Oldal, I., Korponai, J., Tatrai, I. and Paulovits, G. (2003) 'Indirect effect of different fish communities on nutrient chlorophyll relationship in shallow hypertrophic water quality reservoirs', *Hydrobiologia*, vol 504, pp231–239

Neuvo, M. R., Sheehan, R. J. and Wills, P. S. (2004) 'Age and growth of the bighead carp *Hypophthalmichthys nobilis* (Richardson 1845) in the middle Mississippi River', *Archiv für Hydrobiologie*, vol 160, pp215–230

Nico, L. G. and Fuller, P. L. (1999) 'Spatial and temporal patterns of nonindigenous fish introductions in the United States', *Fisheries*, vol 24, pp16–27

Opuszynski, K. (1981) 'Comparison of the usefulness of the silver carp and the bighead carp as additional fish in carp ponds', *Aquaculture*, vol 25, pp223–233

Papoulias, D. M., Chapman, D. and Tillitt, D. E. (2006) 'Reproductive condition and occurrence of intersex in bighead carp and silver carp in the Missouri River', *Hydrobiologia*, vol 571, pp355–360

Peters, L. M., Pegg, M. A. and Reinhardt, U. G. (2006) 'Movements of adult radio-tagged bighead carp in the Illinois River', *Transactions of the American Fisheries Society*, vol 135, pp1205–1212

Pongruktham, O., Ochs, C. and Hoover, J. J. (2010) 'Observations of silver carp (*Hypophthalmichthys molitrix*) planktivory in a floodplain lake of the lower Mississippi River Basin', *Journal of Freshwater Ecology*, vol 25, pp85–93

Rach, J. J., Boogaard, M. and Kolar, C. (2009) 'Toxicity of rotenone and antimycin to silver carp and bighead carp', *North American Journal of Fisheries Management*, vol 29, pp388–395

Rach, J. J., Sass, G. G., Luoma, J. A. and Gaikowski, M. P. (2010) 'Effects of water hardness on size and hatching success of silver carp eggs', *North American Journal of Fisheries Management*, vol 30, pp230–237

Rogowski, D. L., Soucek, D. J., Levengood, J. M., Johnson, S. R., Chick, J. H., Dettmers, J. M., Pegg, M. A. and Epifanio, J. M. (2009) 'Contaminant concentrations in Asian carps, invasive species in the Mississippi and Illinois Rivers', *Environmental Monitoring and Assessment*, vol 157, pp211–222

Sampson, S. J., Chick, J. H. and Pegg, M. A. (2009) 'Diet overlap among two Asian carp and three native fishes in backwater lakes on the Illinois and Mississippi rivers', *Biological Invasions*, vol 11, pp483–496

Sass, G. G., Cook, T. R., Irons, K. S., McClelland, M. A., Michaels, N. N., O'Hara, T. M. and Stroub, M. R. (2010) 'A mark-recapture population estimate for invasive silver carp (*Hypophthalmichthys molitrix*) in the La Grange Reach, Illinois River', *Biological Invasions*, vol 12, pp433–436

Schrank, S. J. and Guy, C. S. (2002) 'Age, growth, and gonadal characteristics of adult bighead carp, *Hypophthalmichthys nobilis*, in the Lower Missouri River', *Environmental Biology of Fishes*, vol 64, no 4, pp443–450

Schrank, S. J., Braaten, P. J. and Guy, C. S. (2001) 'Spatiotemporal variation in density of larval bighead carp in the lower Missouri River', *Transactions of the American Fisheries Society*, vol 130, pp809–814

Schrank, S. J., Guy, C. S. and Fairchild, J. F. (2003) 'Competitive interactions between age-0 bighead carp and paddlefish', *Transactions of the American Fisheries Society*, vol 132, pp1222–1228

Schroeder, G. L. (1978) 'Autotrophic and heterotrophic production of microorganisms in intensely-manured fish ponds, and related fish yields', *Aquaculture*, vol 14, pp303–325

Schwartz, F. J. (1981) *World Literature to Fish Hybrids with an Analysis by Family, Species, and Hybrid: Supplement 1*, NOAA Technical Report, National Marine Fisheries Service, SSRF-750

Smith, D. W. (1989) 'The feeding selectivity of the silver carp, *Hypophthalmichthys molitrix* Val.', *Journal of Fish Biology*, vol 34, no 6, pp819–828

Spataru, P. (1977) 'Gut contents of silver carp *Hypophthalmichthys molitrix* (Val.) – and some trophic relations to other fish in a polyculture system', *Aquaculture*, vol 11, pp137–146

Spataru, P. and Gophen, M. (1985) 'Feeding-behavior of silver carp *Hypophthalmichthys molitrix* Val and its impact on the food web in Lake Kinneret, Israel', *Hydrobiologia*, vol 120, pp53–61

Starling, F. (1993) 'Control of eutrophication by silver carp (*Hypophthalmichthys molitrix*) in the tropical Paranoa Reservoir (Brasilia, Brazil) – a mesocosm experiment', *Hydrobiologia*, vol 257, pp143–152

Tan, X., Li, X., Lek, S., Li, Y., Wang, C. and Luo J. (2010) 'Annual dynamics of the abundance of fish larvae and its relationship with hydrological variation in the Pearl River', *Environmental Biology of Fishes*, vol 88, pp217–225

Taylor, R. M., Pegg, M. A. and Chick, J. H. (2005) 'Response of bighead carp to a bioacoustic behavioural fish guidance system', *Fisheries Management and Ecology*, vol 12, pp283–286

USACE (US Army Corps of Engineers) (2010) *Chicago Sanitary and Ship Canal Aquatic Nuisance Species Dispersal Barriers*, US Army Corps of Engineers, Chicago District, www.lrc.usace.army.mil/projects/fish_barrier

Verigin, B. V., Makeyeva, A. P. and Zaki Mokhamed, M. I. (1978) 'Natural spawning of the silver carp, *Hypophthalmichthys molitrix*, the bighead carp, *Aristichthys nobilis*, and the grass carp, *Ctenopharyngodon idella*, in the Syr-Dar'ya River', *Journal of Ichthyology*, vol 18, no 1, pp143–147

Whittier, T. R. and Aitkin, J. K. (2008) 'Can soft water limit bighead carp and silver carp (*Hypophthalmichthys* spp.) invasions?', *Fisheries*, vol 33, pp122–128

Williamson, C. J. and Garvey, J. E. (2005) 'Growth, fecundity, and diets of newly established silver carp in the middle Mississippi River', *Transactions of the American Fisheries Society*, vol 134, pp1423–1430

Wu, L., Xie, P., Dai, M. and Wang, J. (1997) 'Effects of silver carp density on zooplankton and water quality: Implications for eutrophic lakes in China', *Journal of Freshwater Ecology*, vol 12, pp437–444

Xie, P. and Chen, Y. Y. (2001) 'Invasive carp in China's plateau lakes', *Science*, vol 294, pp999–1000

Xie, P. and Yang, Y. (2000) 'Long-term changes of copepoda community (1957–1996) in a subtropical Chinese lake stocked densely with planktivorous filter-feeding silver and bighead carp', *Journal of Plankton Research*, vol 22, pp1757–1778

Xie, S., Li, Z., Liu, J., Wang, H. and Murphy, B. R. (2007) 'Fisheries of the Yangtze River show immediate impacts of the Three Gorges Dam', *Fisheries*, vol 32, pp343–344

Yashouv, A. (1971) 'Interaction between the common carp (*Cyprinus carpio*) and silver carp (*Hypophthalmichthys molitrix*) in fish ponds', *Bamidgeh*, vol 12, pp85–92

21

Cyprinus carpio L. (common carp)

Brendan J. Hicks, Nicholas Ling and Adam J. Daniel

History of the Common Carp and its Introduction to Countries Outside its Native Range

Common carp (*Cyprinus carpio* L.) is a large bodied cyprinid that is now one of the world's most widespread and ecologically detrimental invasive freshwater fishes (Cambray, 2003). The species probably evolved in Central Asia in the area of the Caspian Sea at the end of the Pliocene and spread into the Black and Aral Seas, and subsequently into Europe as far west as the Danube River 8000 to 10,000 years ago (Balon, 1995). Spread to China, Japan and southeast Asia occurred later, but the extensive translocation and mixing of different genetic stocks to improve aquaculture and ornamental strains hinders our understanding of the species' biogeography. Carp has been introduced to more than 100 countries outside its native range (FishBase, 2010) (Figure 21.1).

Common carp is also the world's oldest aquaculture species, having been cultured for at least 2000 years in Europe (Balon, 1995). Carp were cultured throughout inland, post-Roman, Christian Europe to provide a ready source of fish for abstinence days with most monasteries maintaining carp ponds. The early extensive translocation and culture of carp throughout Europe and Asia was followed by a second extensive period of translocation outside Eurasia over the past two centuries to North, Central and South America, the Middle East, Africa, Australia and Oceania. Common carp is currently the third most cultured freshwater fish worldwide, exceeded only by the silver carp (*Hypophthalmichthys molitrix*) and the grass carp (*Ctenopharyngodon idella*). Carp aquaculture produces

nearly 3 million tonnes annually with a value exceeding $3 billion in 2007 (FAO, 2009) with most production occurring in China. Common carp are used for aquaculture, food and as ornamental pond fish with the highly coloured Japanese nishikigoi (koi) fetching prices in excess of $100,000 for individual trophy fish.

Common carp are closely related to goldfish (*Carassius auratus*) and both species are thought to be naturally tetraploid because they have around twice the number (2n = 100) of chromosomes compared to other cyprinid fishes. David et al (2003) estimates that genome duplication occurred around 12 million years ago and the two species subsequently diverged around 11 million years ago. Four modern subspecies of common carp were identified by Kirpitchnikov (1967): *C. c. carpio* (Europe), *C. c. aralensis* (central Asia), *C. c. haematopterus* (east Asia) and *C. c. viridiviolaceus* (southeast Asia). Recent genetic studies confirm the divergence of at least two subspecies of common carp (Gross et al, 2002; Kohlmann et al, 2003) presumably resulting from Pleistocene glacial separation. *Cyprinus carpio carpio* is now commonly regarded as central Eurasian/eastern European and *C. carpio haematopterus* is the main east Asian subspecies, although some evidence suggests that the southeast Asian subspecies (*C. c. viridiviolaceus*) is also genetically distinct (Chistiakov and Voronova, 2009). Although many authors discount the existence of *C. c. aralensis*, Murakaeva et al (2003) found genetic support for the distinctiveness of wild central Asian populations. The separation between European and Asian subspecies is regarded as ancient (Gross et al, 2002), and Mabuchi et al (2006) further found that carp from the ancient

Introduced range of the common carp

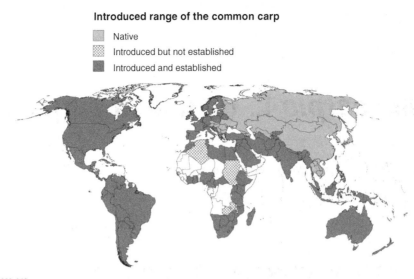

Native
Introduced but not established
Introduced and established

Source: www.fishbase.org

Figure 21.1 *The global distribution of common carp* (Cyprinus carpio L.)

Japanese Lake Biwa (4 million years ago) were also clearly distinct genetically. However, Kottelat (2001) regards the Asian coloured carp as an entirely separate species (*C. rubrofuscus*) and rejects *C. c. haematopterus*. A significant problem with studies of carp genetics is the considerable translocation and interbreeding that has occurred across the species' natural range over the last few hundred years (Matsuzaki et al, 2010). In addition, many genetically distinct varieties of carp have been bred

Source: Brendan Hicks

Figure 21.2 *Olive/bronze morph of wild common carp from the River Murray at Yarrawonga, Victoria, Australia (about 430mm fork length)*

for particular characteristics useful for their aquaculture, such as increased growth rate, greater cold tolerance or improved disease resistance (Hulata, 1995), and particular characteristics of the European carp such as the mirror scale form have been introduced into koi carp culture.

There are significant morphological and physiological differences between the subspecies. *C. c. carpio* (Figure 21.2) are predominantly a greyish/olive green colour whereas *C. c. haematopterus* produces highly coloured individuals. Although Balon (1995) believed that coloured (xanthoid and erythroid) carp originated in Asia by selective breeding from imported German stocks, recent genetic evidence tends to dispute this and the consensus viewpoint now regards the Asian subspecies as the origin of the coloured forms. Mabuchi et al (2006) confirm that ornamental coloured carp have probably been selectively bred from two independent sources. Chinese coloured strains can be traced back over 1200 years whereas modern Japanese koi culture is only around 180 years old. However, the inclusion of Eurasian haplotypes in modern Japanese koi is evidence of the introduction of Eurasian carp traits such as the mirror scale form. Outside their native range, feral populations of *C. c. carpio* tend to revert to a universally wild-type grey/olive colour whereas feral populations of pure strain *C. c. haematopterus* in Australia (Yanco strain) and New Zealand (koi strain) (Figure 21.3) remain highly coloured. European carp also possess

Source: Brendan Hicks

Figure 21.3 *Orange morph of wild koi carp from New Zealand (565mm fork length, 4.18kg)*

markedly serrated margins on the first rays of the dorsal and pectoral fins whereas those of the coloured Asian variety are smooth. The subspecies also differ in meristic (countable traits) features with highest counts of almost all meristic traits occurring in the European subspecies (Suzuki and Yamaguchi, 1980). *Cyprinus c. haematopterus* also seem untroubled by infections of the parasite *Thelohanellus nikolskii* that causes significant mortality among *C. carpio carpio* in central European fish farms (Molnar, 2002) and different genetic strains of carp also appear to confer some resistance to koi herpes virus (Shapira et al, 2005).

A further difficulty in defining carp species/subspecies boundaries is that hybridization occurs with other closely related cyprinids such as goldfish (*Carassius auratus*). Although it is still unclear whether the offspring of such pairings are fertile, Hänfling et al (2005) found some evidence of back-crossed hybrids. In New Zealand, koi carp-goldfish hybrids comprise about 1 per cent of the carp catch (e.g. Figure 21.4). In Australia, three separate carp strains have been identified, and one strain hybridized with goldfish (Shearer and Mulley, 1978) (Figure 21.5). Supporting

Source: Brendan Hicks

Figure 21.4 *Koi carp-goldfish hybrid from New Zealand (355mm fork length, 1.02kg)*

Source: Michael Lake, Kessels and Associates Ltd, New Zealand

Figure 21.5 *Common carp-goldfish hybrid from Australia (330mm fork length); note stout, highly serrated dorsal fin spine*

evidence of significant genetic differences between carp strains is further provided by experimental studies of hybridization: chromosomal number of experimental carp x rosy barb (*Barbus conchonius*) hybrids varied significantly depending on whether the carp strain was European or koi (Váradi et al, 1995).

Ecological Niche of Common Carp

Life history and habitat requirements

Common carp are naturally limnophilic, inhabiting slow flowing downstream reaches of large rivers, river deltas, backwaters, wetlands and floodplain lakes throughout their native range (Nikolski, 1933). They are euryhaline, tolerating 10ppt (parts per thousand) salinity for at least three months and surviving in water up to 14ppt salinity (Al-Hamed, 1971; Crivelli, 1981), and eurythermal, surviving from near-freezing to an upper lethal temperature of 43°C to 46°C (Opuszński et al, 1989). They also tolerate poor water quality and low dissolved oxygen (*Pcrit* ~15 per cent saturation at 15°C; Ultsch et al, 1980).

Carp have a complex life history that can be divided into spawning, foraging and over-wintering phases, each having different habitat associations. Habitat suitability curves have been developed for modelling key life history phases (Edwards and Twomey, 1982). In particular, carp move into interconnected shallow lakes, floodplains and wetlands to spawn (Stuart and Jones, 2006; Bajer and Sorensen, 2010) and juveniles develop in shallow floodplains and marshes (King, 2004). Adult carp tend to forage in warmer shallower waters in summer (Penne and Pierce, 2008), but over-winter in large aggregations in deeper water (Johnsen and Hasler, 1977).

Reproduction

Adult wild carp typically become sexually mature at two to three years old for females and one to two years for males (Crivelli, 1981). Relative fecundity ranged from 19,300 to 216,000 oocytes kg⁻¹ total body weight, with a mean of 97,200 oocytes kg⁻¹. Adults average around 400 to 500mm fork length and 2 to 3kg body weight but may grow much larger so adult females typically shed around 300,000 eggs per spawning. Tempero et al (2006) found that maximum gonad weight as a percentage of total body weight was about 20 per cent, and occurred in wild New Zealand koi in early spring, with females averaging 100,000 eggs kg⁻¹ female body mass. Common carp in Victoria, Australia, had higher fecundity (120,000 to 1,540,000 eggs per fish or 163,000 eggs kg⁻¹ total body weight; Sivakumaran et al, 2003) and in culture, fecundities may be up to 300,000 eggs kg⁻¹ total body weight (Billard, 1999).

Environmentally determined sex ratios of carp populations are typically skewed slightly towards females with ♂:♀ generally from 0.65 to 0.75. However, Crivelli (1981) observed a far higher ratio of males (1.66–1.76) in spawning aggregations in temporary marshes. These observations do not necessarily reflect genetic sex ratios. Carp in temperate latitudes spawn in spring as water temperatures rise above about 17°C, and depending on the duration of favourable conditions further spawning events may occur throughout summer (Tempero et al, 2006). In the tropical climate of West Bengal, Guha and Mukherjee (1991) observed two distinct reproductive cycles per year in winter and summer.

Despite the high fecundity of carp and the ability to spawn more than once per year under favourable conditions, recruitment appears to be highly variable and strongly dependent on climatic conditions. Carp spawn in warm shallow marginal zones of rivers and lakes. The eggs are adhesive and laid among submerged aquatic macrophytes or on inundated terrestrial plants. Ideal situations require high water temperatures (exceeding 17°C), high water levels to inundate marginal vegetated spawning habitat, and low wind conditions that can otherwise dislodge eggs, smother eggs with sediment or destroy the adhesive eggs laid on submerged vegetation

(Phelps et al, 2008). An analysis of 18 lakes in South Dakota, US, shows that carp population recruitment fluctuated synchronously throughout the region, apparently in response to climate (the Moran Effect; Phelps et al, 2008). A further demonstration of the influence of climate on carp recruitment is provided by Bajer and Sorensen (2010) who found that carp only recruit in shallow lakes following severe winter hypoxia, possibly in response to reduced predator pressure in shallow marginal nursery habitat. Spawning success is linked to water level, and strong year classes are associated with spring flooding (Inland Fisheries Service, 2010). It is likely that spawning aggregations are stimulated, at least in part, by the behavioural influence of pheromones. Sisler and Sorensen (2008) demonstrate that carp can discriminate between the odour of conspecifics and goldfish.

Age and growth

Although captive carp are reputed to be among the longest lived vertebrates, with some individuals exceeding 200 years, such cases appear unverified and probably exaggerated. Although captive fish may live considerably longer, wild carp typically average 4 to 5 years old, with few individuals surviving past age 15. The oldest individuals examined by Bajer and Sorensen (2010) were 17 and 34 in Lakes Echo and Susan, respectively. Carp exhibit a range of growth rates, with some evidence that males are smaller than females (Table 21.1), which is consistent with the earlier maturity of males.

New Zealand koi exceed the growth rates of carp from Europe and Chile after age four and achieve greater maximum size, although their growth rate is always less than that of fish in Australia (Tempero et al, 2006). Males rarely lived in excess of 8 years, whereas females lived to 12 years (Tempero et al, 2006). However, aging carp presents some problems because of the unsuitability of the larger saggital otoliths and most studies use scales instead. Phelps et al (2007) compared age estimations of carp using scales, fin rays, opercles, vertebrae and as1ericus otoliths and found that all structures other than otoliths consistently underestimated the ages of carp older than 10 to 13 years.

Diet

Analyses of carp diet are complicated by the maceration of food items by the pharyngeal teeth, which limits

Table 21.1 *von Bertalanffy growth curve parameters for* Cyprinus carpio, *including koi carp from New Zealand and common carp from Australia, France, Spain and Chile*

Country	L_∞	K	t_0	Source
Australia, Murray River, males	489	0.249	−0.519	Brown et al, 2005
Spain	500	0.300	0.200	Fernández-Delgado, 1990
New Zealand, Hikutaia Cut	500	0.215	0.000	Tempero, 2004
Chile	515	0.320	0.150	Prochelle and Campos, 1985
Australia, Murray River	515	0.236	−0.542	Brown et al, 2005
France	516	0.270	−0.400	Crivelli, 1981
Australia, Murray River, females	594	0.177	−0.609	Brown et al, 2005
Australia	655	0.260	−0.400	Vilizzi and Walker, 1999
New Zealand, Waikato River	675	0.210	0.150	Tempero et al, 2006
Croatia	820	0.122	0.811	Treer et al, 2003

Note: The equation for the von Bertalanffy growth is Lt = L∞(1− e−K(t−t°)), where Lt = length in cm at time t (age in years), L∞ = asymptotic maximum length, K = characteristic curvature. Both sexes combined unless otherwise stated.

Source: Tempero (2004); Tempero et al (2006)

visual identification of gut content. Adult common carp are benthivorous and omnivorous, consuming benthic invertebrates, detritus, plant seeds and macrophytes, although larvae and small juveniles are planktivorous (Vilizzi, 1998; Khan, 2003). Although the larval to juvenile transition occurs at around 25mm standard length, juveniles are not simply small adults (Villizi and Walker, 1999). Khan (2003) found that microcrustacea (copepods and cladocerans) dominate the diet of wild carp larvae and juveniles (<5cm) and that fish smaller than 15cm avoid benthic prey. Larval prey items increase in size as larvae grow and prey selectivity is density dependent with larvae selecting larger, more profitable prey as prey density increases (Khadka and Rao, 1986). Post-larval carp (<100g) will also alter their feeding niche between benthic or planktonic prey depending on prey availability (Rahman et al, 2010). Garcia-Berthou (2001) found that adult carp diet is also size dependent, with smaller adults tending to exploit meiobenthos (cladocerans, ostracods and small chironomid larvae) and larger carp consuming macrobenthos (larger chironomids). Fish of intermediate size (10 to 30cm) are able to retain small items (>250µm) with their branchial sieve but the efficiency of this diminishes with size (Sibbing, 1988).

Movements and migration

Although carp appear to show little habitat selectivity in Australasian rivers, they also show a remarkable degree of site fidelity. Jones and Stuart (2007) observe that small carp (284–328mm) occur equally among mainstream and off-stream areas, and among woody debris or aquatic vegetation, although they avoid high flows. The majority of carp move little (<5km), but occasionally undertake extensive migrations to access suitable spawning, feeding or over-wintering habitat. The home range of carp in Australia was a mean of 525m, with only 1 out 15 fish with a home range of 2.1km (Crook, 2004). Osborne et al (2009) found that 85 per cent of tagged wild koi in the Waikato River, New Zealand, were subsequently recaptured within 5km of their release site (mean time at liberty of 519 days). However, Daniel et al (2011), using a combination of acoustic and radio telemetry, found that koi in the same lowland river system undertook pre-spawning movements of hundreds of kilometres, associated with changes in water levels or temperature, although these fish often returned to near their site of origin. Temperature changes of as little as ~0.1°C per hour have been sufficient to alter activity patterns of common carp in heated effluent water

(Cooke and Schreer, 2003), and an increase of 2–3°C over one to two days triggered movements in the Waikato River, New Zealand (Daniel, 2009).

Almost identical results were obtained from movement studies of carp in the Murray-Darling river system in south-eastern Australia. 80 per cent of externally tagged carp were subsequently recaptured <5km from their release site (mean time at liberty 442 days; Stuart and Jones 2006). Jones and Stuart (2009) obtained similar results to Daniel (2009) from radio-tagged fish, with 65 per cent of fish showing long term site fidelity to within 100m, whereas two fish travelled more than 650km downstream. However, most fish moved into adjacent floodplain habitat upon flooding. A study of radio-tagged fish in a 40ha lake in Minnesota revealed that carp typically occupied a highly restricted home range (~100m × 70m) but quickly learned the location of a supplemental food supply and henceforth undertook substantial night-time movements (>300m) to the food source, returning to within their home range during daytime hours (Bajer et al, 2010). In conclusion, common carp appears to be a 'partial migrator', in which only a portion of the population moves; movement patterns do not fit classical definitions of 'true' migration in which all of a cohort migrate.

Driver et al (2005) examined differences in carp recruitment and residency in regulated vs. unregulated rivers in the Murray-Darling catchment and found that while floods enhanced opportunities for dispersal, regulated flows reduced the likelihood of high-flow mortality. Movements from rivers into lateral floodplain wetlands for spawning seem to be a common observation, with subsequent export of larvae and juveniles from wetlands. The possibility of excluding carp from entering wetland habitats has the potential to reduce spawning and the impacts of carp in such systems.

Carp also undergo seasonal movements in cold temperate lakes, aggregating in deeper, warmer water during winter. Johnsen and Hasler (1977) followed the movements of carp tagged with ultrasonic transmitters in Lake Mendota. Fish aggregated in two areas of deeper water in the lake from late autumn and were able to be successfully targeted by commercial fishers. Similar winter aggregations have been observed in Lake Sorrell, Tasmania, by following the movements of radio-tagged fish (Wisniewski, C., Inland Fisheries Service, pers. comm.). Adult carp over-winter in deep lakes that are not subject to winter-kill, but aggressively move into winter-kill-prone shallow regions in the spring to spawn. This accounts for recruitment peaks in years following severe winter hypoxia, which allows carp to exploit nursery habitat that is relatively free of predators (Bajer and Sorensen, 2010).

Population density

Few studies have examined carp population density within the species' native range. One study of the fishery of Lake Balaton in Hungary showed that carp did not dominate the fish biomass in this multispecies ecosystem within the species' natural range despite supplemental stocking for recreational angling (Biro, 1997). However, many studies have demonstrated that carp quickly dominate fish biomass and cause significant density-dependent ecological impacts following introductions outside the natural range. In a survey of 20 major river basins throughout the US, carp were the most commonly collected introduced fish species (Meador et al, 2003). Carp outside the natural range typically dominate fish populations and commonly exceed 500kg ha⁻¹ in biomass. Carp populations can expand rapidly to attain a biomass of 3144kg ha⁻¹ with densities exceeding 1000 individuals ha⁻¹ (Koehn, 2004). Driver et al (2005) obtained average carp biomass of up to 692kg ha⁻¹ in some regions of the Murray-Darling Basin, with localized densities exceeding 2000kg ha⁻¹. In New Zealand, localized spawning biomass, estimated from boat electrofishing, may reach 4030kg ha⁻¹ (Hicks et al, 2006). In the Camargue, France, biomasses of 8–335kg ha⁻¹ were estimated by sampling with rotenone, and carp was the dominant fish by biomass (mean 62 per cent; Crivelli, 1981). Introductions of carp in this area of the south of France (Etang de Mauguio) associated with dyke construction extend back as far as 1695.

Predators

Larval sibling cannibalism in carp is density and size dependent (van Damme et al, 1989); however, carp quickly outgrow the exploitable size range for most piscivorous fish. Even where carp constitute a major proportion of the fish biomass, they are rarely taken by piscivores such as pikeperch (*Stizostedion lucioperca*) and pike (*Esox lucius*) (Crivelli 1981; Liao et al, 2001). A review of published studies on nuisance fish control found that stocking with piscivores was the least successful method of control (Meronek et al, 1996). The ability of carp to degrade water quality in shallow lakes probably reduces the efficiency of visually feeding piscivorous fish, and an increase in carp abundance in Lake Naivasha, Kenya, coincided with a decline in an introduced piscivore (Britton et al, 2010). Adamek et al (2003) reported significant predation on adult carp by otters in fishponds in the Czech Republic, taking individuals up to 11.7kg in body weight and it is likely that otters and piscivorous birds prey on carp throughout their native range.

Ecological effects

Concerns about the environmental consequence of carp introduction date back to the 1850s in the US (McCrimmon, 1968). Weber and Brown (2009) provide a recent review. The detrimental effects of common carp on the water quality of shallow lakes was first reported in southern Wisconsin, US, by Cahn (1929) with a change from clear to turbid water, loss of aquatic vegetation and an implied loss of fish diversity. A large number of aquaria, exclosure and lake studies have since confirmed the ability of carp to resuspend sediments and nutrients due to their vigorous benthic feeding activity (e.g. Cahoon, 1953; Zambrano and Hinojosa 1999; Zambrano et al, 2001, Driver et al, 2005; Weber and Brown, 2009), which results in a characteristic pockmarked appearance (Figure 21.6). This results from the suctorial feeding mechanism in

common carp (Sibbing, 1988). Carp also reduce the abundance of aquatic macrophytes (Cahoon 1953; Zambrano and Hinojosa, 1999; Williams et al, 2002; Hinojosa-Garro and Zambrano, 2004) and benthic invertebrates (e.g. Zambrano and Hinojosa 1999; Hinojosa-Garro and Zambrano 2004), and increase abundance of cyanobacteria (Williams and Moss, 2003). Carp reduce abundance of submerged macrophytes indirectly by reduced light availability caused by increased turbidity and directly browsing and uprooting plants (Sidorkewicj et al, 1998; Miller and Crowl, 2006). Carp also maintain a turbid water state by resuspending algae (Roozen et al, 2007).

Biomanipulation studies using experimental ponds and lake enclosures have confirmed the relationship between carp biomass and reduced water quality (Lougheed et al, 1998), and between carp and reduced waterfowl abundance (Haas et al, 2007). It is clear that a soft mud bottom exacerbates the negative effects of carp compared to hard bottoms.

Carp are therefore capable of promoting an equilibrium shift in shallow lakes from a clear water, macrophyte-dominated system to one with highly turbid water dominated by excessive growth of phytoplankton (Scheffer et al, 1993).

Biomass thresholds for ecological effects have been the focus of considerable research, and in Australia a critical fish carp biomass of 450kg ha^{-1} has been

Source: Brendan Hicks

Figure 21.6 *Depressions from common carp feeding activity*

identified (Fletcher et al, 1985). However, loss of macrophytes in soft-bottomed water bodies can occur at biomasses of >200kg ha⁻¹ (Williams et al, 2002). Weber and Brown (2009) summarize a consensus threshold of 450kg ha⁻¹ from an extensive literature review, but Bajer et al (2009) found more recently that the ecological integrity of a shallow lake in North America was jeopardized at densities of ~100kg ha⁻¹ (Figure 21.7).

Carp have a reputation for increasing nutrient regeneration. Nuttall and Richardson (1991) suggest that minimal nitrogen and phosphorus are contributed from carp by excretion, but they used small, starved fish. However, when the effects of sediment resuspension were included, phosphorus concentrations and chorophyll *a* concentrations increased with increasing carp biomass between 300 and 800kg ha⁻¹, and with increasing fish size (Driver et al, 2005). Elevated nitrogen and phosphorus concentrations in the presence of carp are a consistent finding (King et al, 1997; Zambrano et al, 1999).

However, the introduction of common carp has not been universally associated with radical ecosystem changes. Common carp and silver carp (*Hypophthalmichthys molitrix*) were introduced in 1979 to Lake Heiliger See, northeast Germany. This is a 10.2ha lake (mean depth 6.5m) in degraded condition that has severe hypolimnetic oxygen depletion in midsummer. Accidental stocking resulted in a carp density of 482 carp ha⁻¹, which was regarded as overstocking (Barthelmes and Brämick, 2003). Native fish species persisted after the carp introduction at largely unchanged biomass. Addition of young-of-year common carp to mesocosms with native fish characteristic of Illinois lakes supported the biotic resistance hypothesis, i.e. that high species richness reduces the impact and success of carp (Carey and Wahl, 2010). Also, deeper lakes are predicted to have less severe effects from carp than shallow lakes (Zambrano et al, 2001).

Management Efforts Employed and Their Effectiveness

Population control methods

Most removal programmes are ineffective in the long term, and to date only small lakes have a high probability of successful removal (Koehn et al, 2000).

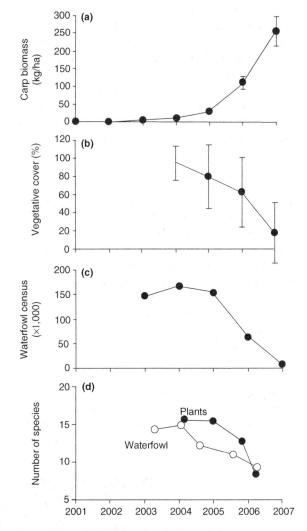

Source: Bajer et al (2009), reproduced with permission

Figure 21.7 *Reduction of aquatic macrophyte cover, number of wildfowl and species richness with increasing biomass of common carp in two shallow lakes, Hennepin and Hopper, Illinois: (a) Biomass of carp 2001–2007 (mean ± 1 SD); (b) per cent vegetative cover (mean ± 1 SD); (c) cumulative waterfowl (dabbling and diving ducks) count; and (d) number of aquatic plant and duck species*

Removal efforts generally fall into one of three categories: one-time removals, ongoing or annual removals, and eradication attempts. One-time removals of common carp can yield important scientific data but are of little long term value as water quality will return to the pre-removal state as fish biomass increases post removal (Meijer et al, 1999). Annual removals can be effective for reducing biomass of common carp and can potentially improve water quality, but are time consuming and costly. Population modelling using CARPSIM suggests that size selective removal might be useful to reduce biomass to <60 per cent of the virgin biomass (B_0), but that there is little prospect of reducing biomass to <10 per cent of B_0 unless fishing mortality F is <1.4, where F is a function of catchability, selectivity and fishing effort (Brown and Walker, 2004). Complete removal of common carp is often the goal of carp control programmes but is only feasible for relatively small water bodies that can be emptied or poisoned.

There are a host of potential removal methods for reducing or eradicating common carp including netting (Cahoon, 1953), trapping (Stuart et al, 2006), virus introduction (Matsui et al, 2008), exclusion (Lougheed et al, 1998), radio telemetry-assisted removal (Diggle et al, 2004), water level manipulation (Yamamoto et al, 2006), boat electrofishing (Hicks et al, 2006) and poisoning (Frederieke et al, 2005). Emerging technologies such as pheromone attraction (Sisler and Sorensen, 2008) and gene modification (i.e. daughterless carp; Grewe et al, 2005) show promise but may be decades away from being available as management tools. There is no single method for removing common carp that is effective in all situations.

Carp are caught commercially with a wide range of fishing gear, including large-meshed gill nets, fyke nets, or a combination of seine, trammel or fyke nets. Boat electrofishing is used for commercial carp fishing in Australia (Bell, K., K and C Fisheries, pers. comm.). In Ontario, Canada, 29,262 tonnes were caught between 1908 and 1966 (a mean of 496 tonnes year^{-1}; McCrimmon 1968). A total of 32 tonnes year^{-1} were caught commercially in the Camargue, but 84 per cent of the catch was released because of low commercial value (Crivelli, 1981). Removal of 13.6 tonnes annually from the artificial Lake Scucog (area 68km^2, mean depth 1.4m) in Ontario, Canada, had no apparent effect on the standing stock of carp (McCrimmon, 1968). Because of mercury contamination, adverse public reaction to fishing methods, an abundance of small bones and variable taste, carp has fallen from favour as a table fish in North America (Fritz, 1987), making commercial fishing unprofitable in some markets, and therefore unlikely to control carp abundance.

In water bodies that dry periodically, or have artificial water level controls, exclusion devices and traps have been used to prevent reinvasion of carp when reflooding occurs. Water level manipulation, where it is possible, offers significant advantages for carp control. Summer drawdown reduces recruitment of carp and goldfish (Yamamoto et al, 2006), and outflow manipulation can be used to attract carp to traps.

Winter mortalities can occur naturally in ice-covered lakes due to oxygen deficiencies, e.g. in Canada. Following the break-up of ice in Lake Scucog, Ontario, in 1959–1960, 80,000 dead carp were found (McCrimmon, 1968).

Effectiveness of control measures

Because carp can attain such high biomasses, reductions of carp abundance to below threshold biomasses can improve water quality even where eradication is not possible. A notable success has been carp removal from the Botany Wetlands, New South Wales, Australia, which comprises 11 interconnected ponds and adjacent land covering an area of 58ha. Ten tonnes of cyprinid biomass (4073 common carp and 261 goldfish) were removed over nine years, and during this time Secchi depth improved from 0.4m to 1.2m over the removal period, and cyanobacterial density decreased (Pinto et al, 2005). Deterioration of water clarity and macrophyte abundance occurred as carp proliferated in Hennepin and Hopper Lakes, Illinois, US, following initial near eradication of carp by rotenone treatment (Bajer et al, 2009).

Although there has been limited success eradicating common carp on a large scale, there has been progress that lends hope to future advancements in removal strategies. The most notable is the ongoing effort to remove common carp from two large lakes on the island of Tasmania in Australia. Common carp were discovered in Lake Sorell (4770ha) and Lake Crescent (2365ha) in 1995 (Diggle et al, 2004). These lakes now hold the last common carp in Tasmania after eradication of populations in farm dams on the northwest coast in the 1970s. The lakes also have an endemic galaxiid species, the endangered golden galaxias (Galaxias auratus) and state-renowned brown trout (Salmo trutta) fisheries. Since the discovery of the unwanted carp population, the state government has used a host of removal techniques including applying poison and fish removal guided by radio telemetry at a cost of AUS$300,000 to $400,000 per year (Diggle et al, 2004; Inland Fisheries Service, 2010). A total of 17,307 carp have been removed from Lake Sorell (including 14,517 juveniles from the 2009 spawning) and 7797 from Lake Crescent since 1995 (Inland Fisheries Service, 2010). Lake Crescent is now believed to be carp free. The key to the success achieved by the Inland Fisheries Service has been the ability to nearly eliminate recruitment by blocking littoral spawning habitat with barriers. Unfortunately, even physical exclusion is not 100 per cent effective as common carp reportedly spawned on the barrier netting used to prevent them from accessing more suitable spawning habitat.

Controversies around Presence and Control

In many countries where carp have been introduced there remains a tension between sport anglers and conservation and water quality management. For instance, in New Zealand carp have been spread illegally and deliberately by sport anglers. In many instances, people spreading the carp do not believe they have any detrimental effect. In some instances, carp appear to exist in the background, having little effect on water quality or native species. In many other instances, carp have acted as an invasive species, reproducing freely and spreading widely in warm river systems with extensive wetlands and shallow lakes attached. Common carp clearly have the fecundity, migratory instincts and feeding behaviour that predisposes them to be an invasive species.

There is also uncertainty around whether environmental degradation results from carp or not. In water bodies that once had relatively clear water and extensive native macrophyte communities, carp impacts are demonstrable; but other locations that support carp may have highly turbid water as a result not of the carp, but because of extensive areas of shallow water, soft sediments and frequent strong winds. Areas that are already degraded are unlikely to show further degradation as a result of carp introduction. Furthermore, since early introductions of carp, other factors that might degrade water quality and native biodiversity have been at work. For instance, in New Zealand increasing land use intensification for dairy farming and overharvest of freshwater eels (*Anguilla australis* and *A. dieffenbachii*) might be just as much to blame for degraded water quality and reductions in eel size as the proliferation of koi carp, as they happened contemporaneously. Only concerted and coordinated carp control efforts, adequate fishery management and water quality and biodiversity monitoring will clarify this.

Acknowledgements

This study was funded by contract UOWX0505 from the New Zealand Foundation for Research, Science and Technology.

References

Adamek, Z., Kortan, D., Lepic, P. and Andreji, J. (2003) 'Impacts of otter (*Lutra lutra* L.) predation on fishponds: A study of fish remains at ponds in the Czech Republic', *Aquaculture International*, vol 11, pp389–396

Al-Hamed, M. I. (1971) 'Salinity tolerance of common carp (*Cyprinus carpio* L.)', *Bulletin of the Iraq Natural History Museum*, vol 7, pp1–16

Bajer, P. G. and Sorensen, P. W. (2010) 'Recruitment and abundance of an invasive fish, the common carp, is driven by its propensity to invade and reproduce in basins that experience winter-time hypoxia in interconnected lakes', *Biological Invasions*, vol 12, pp1101–1112

Bajer, P. G., Sullivan, G. S. and Sorensen, P. W. (2009) 'Effects of a rapidly increasing population of common carp on vegetative cover and waterfowl in a recently restored Midwestern shallow lake', *Hydrobiologia*, vol 632, pp235–245

Bajer, P. G., Lim, H., Travaline, M. J., Miller, B. D. and Sorensen, P. W. (2010) 'Cognitive aspects of food searching behavior in free-ranging wild common carp', *Environmental Biology of Fish*, vol 88, pp295–300

Balon, E. K. (1995) 'Origin and domestication of the wild carp, *Cyprinus carpio*: From Roman gourmets to the swimming flowers', *Aquaculture*, vol 129, pp3–48

Barthelmes, D. and Brämick, U. (2003) 'Variability of a cyprinid lake ecosystem with special emphasis on the native fish fauna under intensive fisheries management including common carp (*Cyprinus carpio*) and silver carp (*Hypophthalmichthys molitrix*)', *Limnologica*, vol 33, pp10–28

Billard, R. (1999) *Carp Biology and Culture*, Springer-Verlag, Berlin

Biro, P. (1997) 'Temporal variation in Lake Balaton and its fish populations', *Ecology of Freshwater Fish*, vol 6, pp196–216

Britton, J., Harper, D., Oyugi, D. and Grey, J. (2010) 'The introduced *Micropterus salmoides* in an equatorial lake: A paradoxical loser in an invasion meltdown scenario?', *Biological Invasions*, vol 12, pp3439–3448

Brown, P. and Walker, T. I. (2004) 'CARPSIM: Stochastic simulation modelling of wild carp (*Cyprinus carpio* L.) population dynamics, with applications to pest control', *Ecological Modelling*, vol 176, pp83–97

Brown, P., Sivakumaran, K. P., Stoessel, D. and Giles, A. (2005) 'Population biology of carp (*Cyprinus carpio* L.) in the mid-Murray River and Barmah Forest Wetlands, Australia', *Marine and Freshwater Research*, vol 56, pp1151–1164

Cahn, A. R. (1929) 'The effect of carp on a small lake: The carp as a dominant', *Ecology*, vol 10, pp271–275

Cahoon, W. G. (1953) 'Commercial carp removal at Lake Mattamuskeet, North Carolina', *Journal of Wildlife Management*, vol 17, pp312–316

Cambray, J. A. (2003) 'Impact on indigenous species biodiversity caused by the globalisation of alien recreational freshwater fisheries', *Hydrobiologia*, vol 500, pp217–230

Carey, M. P. and Wahl, D. H. (2010) 'Native fish diversity alters the effects of an invasive species on food webs', *Ecology*, vol 91, pp2965–2974

Chistiakov, D. A. and Voronova, N. V. (2009) 'Genetic evolution and diversity of common carp *Cyprinus carpio* L.', *Central European Journal of Biology*, vol 4, pp304–312

Cooke, S. J. and Schreer, J. F. (2003) 'Environmental monitoring using physiological telemetry – a case study examining common carp responses to thermal pollution in a coal-fired generating station effluent', *Water Air and Soil Pollution*, vol 142, pp113–136

Crivelli, A. J. (1981), 'The biology of the common carp, *Cyprinus carpio* L. in the Camargue, southern France', *Journal of Fish Biology*, vol 18, pp271–290

Crook, D. A. (2004) 'Is the home range concept compatible with the movements of two species of lowland river fish?', *Journal of Animal Ecology*, vol 73, pp353–366

Daniel, A. J. (2009) 'Detecting exploitable stages in the life history of koi carp (*Cyprinus carpio*) in New Zealand', PhD thesis, University of Waikato, Hamilton, New Zealand

Daniel, A. J., Hicks, B. J., Ling, N. and David, B. O. (2011) 'Movements of radio- and acoustic-tagged adult common carp (*Cyprinus carpio* L.) in the Waikato River, New Zealand', *North American Journal of Fisheries Management*, vol 31, pp352–362

David, L., Blum, S., Feldman, M. W., Lavi, U. and Hillel, J. (2003) 'Recent duplication of the common carp (*Cyprinus carpio* L.) genome as revealed by analyses of microsatellite loci', *Molecular Biology and Evolution*, vol 20, pp1425–1434

Diggle, J., Day, J. and Bax, N. (2004) 'Eradicating European carp from Tasmania and implications for national European carp eradication', Inland Fisheries Service, Moonah, Report No 2000/182, Moonah, Australia

Driver, P. D., Harris, J. H., Closs, G. P. and Koen, T. B. (2005) 'Effects of flow regulation in the Murray-Darling Basin, Australia', *River Research and Applications*, vol 21, pp327–335

Edwards, E. A. and Twomey, K. A. (1982) 'Habitat suitability index models: Common carp', US Dept. Int. Fish Wildl. Serv. FWS/OBS-82/10.12, Fish and Wildlife Service, US Department of the Interior, Washington, DC

FAO (Food and Agriculture Organization of the United Nations) (2009) 'Fishery and aquaculture statistics', in *FAO Yearbook (2007)*, Food and Agriculture Organization of the United Nations, Rome

Fernández-Delgado, C. (1990) 'Life history patterns of the common carp, *Cyprinus carpio*, in the estuary of the Guadalquivir River in south-west Spain', *Hydrobiologia*, vol 206, pp19–28

FishBase (2010) www.fishbase.org/search.php, accessed 2 August 2011

Fletcher, A. R., Morison, A. K. and Hume, D. J. (1985) 'Effects of carp, *Cyprinus carpio* L., on communities of aquatic vegetation and turbidity of waterbodies in the lower Goulburn River basin', *Australian Journal of Marine and Freshwater Research*, vol 36, pp311–327

Frederieke, J., Kroon, P., Gehrke, C. and Kurwie, T. (2005) 'Palatability of rotenone and antimycin baits for carp control', *Ecological Management and Restoration*, vol 6, pp228–229

Fritz, A. (1987) 'Commercial fishing for carp', in E. L. Cooper (ed) *Carp in North America*, American Fisheries Society, Bethseda, Maryland, pp17–30

Garcia-Berthou, E. (2001) 'Size- and depth-dependent variation in habitat and diet of the common carp (*Cyprinus carpio*)', *Aquatic Sciences*, vol 63, pp466–476

Grewe, P., Botright, N., Beyer, J., Patil, J. and Thresher, R. (2005) 'Sex specific apoptosis for achieving daughterless fish', at 13th Australasian Vertebrate Pest Conference Wellington, New Zealand (Landcare Research, Wellington, New Zealand)

Gross, R., Kohlmann, K. and Kersten, P. (2002) 'PCR–RFLP analysis of the mitochondrial *ND-3r4* and *ND-5r6* gene polymorphisms in the European and East Asian subspecies of common carp (*Cyprinus carpio* L.)', *Aquaculture*, vol 204, pp507–516

Guha, D. and Mukherjee, D. (1991) Seasonal cyclical changes in the gonadal activity of common carp, *Cyprinus carpio* Linn.', *Indian Journal of Fisheries*, vol 38, pp218–223

Haas, K., Köhler, U., Diehl, S., Köhler, P., Dietrich, S., Holler, S., Jaensch, A., Niedermaier, M. and Vilsmeier. J. (2007) 'Influence of fish on habitat choice of water birds: A whole system experiment', *Ecology*, vol 88, pp2915–2925

Hänfling, B., Bolton, P., Harley, M. and Carvalho, G. R. (2005) 'A molecular approach to detect hybridisation between crucian carp (*Carassius carassius*) and non-indigenous carp species (*Carassius* spp. and *Cyprinus carpio*)', *Freshwater Biology*, vol 50, pp403–417

Hicks, B. J., Ling, N. and Osborne, M. W. (2006) 'Quantitative estimates of fish abundance from boat electrofishing', in M. J. Phelan and H. Bajhau (eds) *A Guide to Monitoring Fish Stocks and Aquatic Ecosystems, Darwin, 11–15 July 2005*, Australian Society for Fish Biology, Darwin, Australia, pp104–111

Hinojosa-Garro, D. and Zambrano, L. (2004) 'Interactions of common carp (*Cyprinus carpio*) with benthic crayfish decapods in shallow ponds', *Hydrobiologia*, vol 515, pp115–122

Hulata, G. (1995) 'A review of genetic improvement of the common carp (*Cyprinus carpio* L.) and other cyprinids by crossbreeding, hybridization and selection', *Aquaculture*, vol 129, pp143–155

Inland Fisheries Service (2010) 'Carp management program: Annual report 2009–2010', Inland Fisheries Service, New Norfolk, Tasmania, Australia

Johnsen, P. B. and Hasler, A. D. (1977) 'Winter aggregations of carp (*Cyprinus carpio*) as revealed by ultrasonic tracking', *Transactions of the American Fisheries Society*, vol 106, pp556–559

Jones, M. J. and Stuart, I. G. (2007) 'Movements and habitat use of common carp (*Cyprinus carpio*) and Murray cod (*Maccullochella peelii peelii*) juveniles in a large lowland Australian river', *Ecology of Freshwater Fish*, vol 16, pp210–220

Jones, M. J. and Stuart, I. G. (2009) 'Lateral movement of common carp (*Cyprinus carpio* L.) in a large lowland river and floodplain', *Ecology of Freshwater Fish*, vol 18, pp72–82

Khadka, R. B. and Rao, T. R. (1986) 'Prey size selection by common carp (*Cyprinus carpio var. communis*) larvae in relation to age and prey density', *Aquaculture*, vol 54, pp89–96

Khan, T. A. (2003) 'Dietary studies on exotic carp (*Cyprinus carpio* L.) from two lakes of western Victoria, Australia', *Aquatic Sciences*, vol 65, pp272–286

King, A. J. (2004) 'Ontogenetic patterns of habitat use by fishes within the main channel of an Australian floodplain river', *Journal of Fish Biology*, vol 65, pp1582–1603

King A. J., Robertson A. I. and Healey, M. R. (1997) 'Experimental manipulations of the biomass of introduced carp (*Cyprinus carpio*) in billabongs. I. Impacts on water-column properties', *Marine and Freshwater Research*, vol 48, pp435–443

Kirpichnikov, V. S. (1967) 'Homologous hereditary variation and evolution of wild common carp (*Cyprinus carpio* L.)', *Genetika*, vol 8, pp65–72 (in Russian)

Koehn, J. D. (2004) 'Carp (*Cyprinus carpio*) as a powerful invader in Australian waterways', *Freshwater Biology*, vol 49, pp882–894

Koehn, J. D., Brumley, A. and Gehrke, P. (2000) *Managing the Impacts of Carp*, Bureau of Rural Sciences, Canberra

Kohlmann, K., Gross, R., Murakaeva, A. and Kersten, P. (2003) 'Genetic variability and structure of common carp (*Cyprinus carpio*) populations throughout the distribution range inferred from allozyme, microsatellite and mitochondrial DNA markers', *Aquatic Living Resources*, vol 16, pp421–431

Kottelat, M. (2001) *Fishes of Laos*, Wildlife Heritage Trust Publications (Pty) Ltd, Colombo, Sri Lanka

Liao, H., Pierce, C. L. and Larscheid, J. G. (2001) 'Empirical assessment of indices of prey importance in the diets of fish', *Transactions of the American Fisheries Society*, vol 130, pp583–591

Lougheed, V. L., Crosbie, B. and Chow-Fraser, P. (1998) 'Predictions on the effect of common carp (*Cyprinus carpio*) exclusion on water quality, zooplankton, and submergent macrophytes in a Great Lakes wetland', *Canadian Journal of Fisheries and Aquatic Sciences*, vol 55, pp1189–1197

Mabuchi, K., Miya, M., Senou, H., Suzuki, T. and Nishida, M. (2006) 'Complete mitochondrial DNA sequence of the Lake Biwa wild strain of common carp (*Cyprinus carpio* L.): Further evidence for an ancient origin', *Aquaculture*, vol 257, pp68–77

Matsui, K., Honjo, M., Kohmatsu, Y., Uchii, K., Yonekura, R. and Kawabata, Z. (2008) 'Detection and significance of koi herpesvirus (KHV) in freshwater environments', *Freshwater Biology*, vol 53, pp1262–1272

Matsuzaki, S. S., Mabuchi, K., Takamura, N., Hicks, B. J., Nishida, M. and Washitani, I. (2010) 'Stable isotope and molecular analyses indicate that hybridization with non-native domesticated common carp influence habitat use of native carp', *Oikos*, vol 119, pp964–971

McCrimmon, H. R. (1968) 'Carp in Canada', *Fisheries Research Board of Canada Bulletin*, vol 165, pp1–89

Meador, M. R., Brown, L. R. and Short, T. (2003) 'Relations between introduced fish and environmental conditions at large geographic scales', *Ecological Indicators*, vol 3, pp81–92

Meijer, M.-L., De Boois, I., Scheffer, M., Portielje, R. and Hosper, H. (1999) 'Biomanipulation in shallow lakes in The Netherlands: An evaluation of 18 case studies', *Hydrobiologia*, vol 408–409, pp13–30

Meronek, T. G., Bouchard, P. M., Buckner, E. R., Burri, T. M., Demmerly, K. K., Hateli, D. C., Klumb, R. A., Schmidt, S. H. and Coble, D. W. (1996) 'A review of fish control projects', *North American Journal of Fisheries Management*, vol 16, pp63–74

Miller, S. A. and Crowl, T. A. (2006) 'Effects of common carp (*Cyprinus carpio*) on macrophytes and invertebrate communities in a shallow lake', *Freshwater Biology*, vol 51, pp85–94

Molnar, K. (2002) 'Differences between the European carp (*Cyprinus carpio carpio*) and the coloured carp (*Cyprinus carpio haematopterus*) in susceptibility to *Thelohanellus nikolskii* (Myxosporea) infection', *Acta Veterinaria Hungarica*, vol 50, pp51–57

Murakaeva, A., Kohlmann, K., Kersten, P., Kamilov, B. and Khabibullin, D. (2003) 'Genetic characterization of wild and domesticated common carp (*Cyprinus carpio* L.) populations from Uzbekistan', *Aquaculture*, vol 218, pp153–166

Nikolski, G. V. (1933) 'On the influence of the rate of flow on the fish fauna of the rivers of Central Asia', *Journal of Animal Ecology*, vol 2, pp266–281

Nuttall, P. M. and Richardson, B. J. (1991) 'Nitrogen and phosphorus excretion by European carp', *Chemosphere*, vol 23, pp671–676

Opuszyński, K., Lirski, A., Myszkowski, L. and Wolnicki, J. (1989) 'Upper lethal and rearing temperatures for juvenile common carp, *Cyprinus carpio* L., and silver carp, *Hypophthalmichthys molitrix* (Valenciennes)', *Aquaculture Research*, vol 20, pp287–294

Osborne, M. W., Ling, N., Hicks, B. J. and Tempero G. W. (2009) 'Movement, social cohesion, and site fidelity in adult koi carp, *Cyprinus carpio* L.', *Fisheries Management and Ecology*, vol 16, pp169–176

Penne, C. R. and Pierce, C. L. (2008) 'Seasonal distribution, aggregation, and habitat selection of common carp in Clear Lake, Iowa', *Transactions of the American Fisheries Society*, vol 137, pp1050–1062

Phelps, Q. E., Edwards, K. R. and Willis, D. W. (2007) 'Precision of five structures for estimating age of common carp', *North American Journal of Fisheries Management*, vol 27, pp103–105

Phelps, Q. E., Graeb, B. D. S. and Willis, D. W. (2008) 'Influence of the Moran Effect on spatiotemporal synchrony in common carp recruitment', *Transactions of the American Fisheries Society*, vol 137, pp1701–1708

Pinto, L., Chandrasena, N., Pera, J., Hawkins, P., Eccles, D. and Sim, R. (2005) 'Managing invasive carp (*Cyprinus carpio* L.) for habitat enhancement at Botany Wetlands, Australia', *Aquatic Conservation: Marine and Freshwater Ecosystems*, vol 15, pp447–462

Prochelle, O. and Campos, H. (1985) 'The biology of the introduced carp, *Cyprinus carpio* L., in the River Cayumpu, Valdivia, Chile', *Studies on Neotropical Fauna and Environment*, vol 20, pp65–82

Rahman, M. M., Kadowaki, S., Balcombe, S. R. and Wahab, M. A. (2010) 'Common carp (*Cyprinus carpio* L.) alters its feeding niche in response to changing food resources: Direct observations in simulated ponds', *Ecological Research*, vol 25, pp303–309

Roozen, F. C., Lürling, J. M., Vlek, H., Van der Pouw Kraan, E. A. J., Ibelings, B. W. and Scheffer, M. (2007) 'Resuspension of algal cells by benthivorous fish boosts phytoplankton biomass and alters community structure in shallow lakes', *Freshwater Biology*, vol 52, pp977–987

Scheffer, M., Hosper, S. H., Meijer, M-L., Moss, B. and Jeppesen, E. (1993) 'Alternative equilibria in shallow lakes', *Trends in Ecology and Evolution*, vol 8, pp275–279

Shapira, Y., Magen, Y., Zak, T., Kotler, M., Hulata, G. and Levavi-Sivan, B. (2005) 'Differential resistance to koi herpes virus (KHV)/carp interstitial virus and gill necrosis virus (CNGV) among common carp (*Cyprinus carpio* L.) strains and crossbreds', *Aquaculture*, vol 245, pp1–11

Shearer, K. D. and Mulley, J. C. (1978) 'The introduction and distribution of the carp, *Cyprinus carpio* Linnaeus, in Australia', *Australian Journal of Marine and Freshwater Research*, vol 29, pp551–563

Sibbing, F. A. (1988) 'Specializations and limitations in the utilization of food resources by the carp, *Cyprinus carpio*: A study of oral food processing', *Environmental Biology of Fishes*, vol 22, pp161–178

Sidorkewicj, N. S., López Cazorla, A. C., Murphy, K. J., Sabbatini, M. R. and Domaneiwski, J. C. J. (1998) 'Interaction of common carp with aquatic weeds in Argentine drainage channels', *Journal of Aquatic Plant Management*, vol 36, pp5–10

Sisler, S. P. and Sorensen, P. (2008) 'Common carp and goldfish discern conspecific identity using chemical cue', *Behaviour*, vol 145, pp1409–1425

Sivakumaran K. P., Brown P., Stoessel D. and Giles, A. (2003) 'Maturation and reproductive biology of female wild carp, *Cyprinus carpio*, in Victoria, Australia', *Environmental Biology of Fishes*, vol 68, pp321–332

Stuart, I. G. and Jones, M. (2006) 'Movement of common carp, *Cyprinus carpio*, in a regulated lowland Australian river: Implications for management', *Fisheries Management and Ecology*, vol 13, pp213–219

Stuart, I. G., Williams, A., McKenzie, J. and Holt, T. (2006) 'Managing a migratory pest species: A selective trap for common carp', *North American Journal of Fisheries Management*, vol 26, pp888–893

Suzuki, R. and Yamaguchi, M. (1980) 'Meristic and morphometric characters of five races of *Cyprinus carpio*', *Japanese Journal of Ichthyology*, vol 27, pp199–206

Tempero, G. W. (2004) 'Population biology of koi carp in the Waikato region', MSc thesis, University of Waikato, Hamilton

Tempero, G. W., Ling, N., Hicks, B. J. and Osborne M. W. (2006) 'Age composition, growth, and reproduction of koi carp (*Cyprinus carpio* L.) in the lower Waikato, New Zealand', *New Zealand Journal of Marine and Freshwater Research*, vol 40, pp571–583

Treer, T., Varga, B., Safner, R., Aničić, I., Piria, M. and Odak, T. (2003) 'Growth of the common carp (*Cyprinus carpio*) introduced into the Mediterranean Vransko Lake', *Journal of Applied Icthyology*, vol 19, pp383–386

Ultsch, G. R., Ott, M. E. and Heisler, N. (1980) 'Standard metabolic rate, critical oxygen tension, and aerobic scope for spontaneous activity of trout (*Salmo gairdneri*) and carp (*Cyprinus carpio*) in acidified water', *Comparative Biochemistry and Physiology A*, vol 67, pp329–335

van Damme, P., Appelbaum, S. and Hecht, T. (1989) 'Sibling cannibalism in koi carp, *Cyprinus carpio* L., larvae and juveniles reared under controlled conditions', *Journal of Fish Biology*, vol 34, pp855–863

Váradi, L., Hidas, A., Várkonyi, E. and Horváth, L. (1995) 'Interesting phenomena in hybridization of carp (*Cyprinus carpio*) and rosy barb (*Barbus conchonius*)', *Aquaculture*, vol 129, pp211–214

Vilizzi, L. (1998) 'Observations on ontogenetic shifts in the diet of 0+ carp, from the River Murray, Australia', *Folia Zoologica*, vol 47, pp225–229

Villizi, L. and Walker, K. F. (1999) 'The onset of the juvenile period in carp, *Cyprinus carpio*: A literature survey', *Environmental Biology of Fishes*, vol 56, pp93–102

Weber, M. J. and Brown, M. L. (2009) 'Effects of common carp on aquatic ecosystems 80 years after "Carp as a dominant": ecological insights for fisheries management', *Reviews in Fisheries Science*, vol 17, pp524–537

Williams, A. E. and Moss, B. (2003) 'Effects of different fish species and biomass on plankton interactions in a shallow lake', *Hydrobiologia*, vol 491, pp331–346

Williams, A. E., Moss, B. and Eaton, J. (2002) 'Fish induced macrophyte loss in shallow lakes: top–down and bottom–up processes in mesocosm experiments', *Freshwater Biology*, vol 47, pp2216–2232

Yamamoto, T., Kohmatsu, Y. and Yuma, M. (2006) 'Effects of summer drawdown on cyprinid fish larvae in Lake Biwa', *Japanese Journal of Limnology*, vol 7, pp75–82

Zambrano, L. and Hinojosa, D. (1999) 'Direct and indirect effects of carp (*Cyprinus carpio* L.) on macrophyte and benthic communities in experimental shallow ponds in central Mexico', *Hydrobiologia*, vol 408–409, pp131–138

Zambrano, L., Perrow, M. R., Macías-García, C. and Aguirre-Hidalgo, V. (1999) 'Impact of introduced carp (*Cyprinus carpio*) in subtropical shallow ponds in Central Mexico', *Journal of Aquatic Ecosystem Stress and Recovery*, vol 6, pp281–288

Zambrano, L., Scheffer, M. and Martínez-Ramos, M. (2001) 'Catastrophic response of lakes to benthivorous fish introduction', *Oikos*, vol 94, pp344–350

22

Gambusia affinis (Baird & Girard) and *Gambusia holbrooki* Girard (mosquitofish)

William E. Walton, Jennifer A. Henke and Adena M. Why

History of the Species and its Introduction

Two species of *Gambusia* are commonly referred to as mosquitofish and represent the most widespread freshwater fishes globally, occurring on all continents except Antarctica. *Gambusia affinis* was originally described in 1853. Placed within another genus by Baird and Girard, the species was reassigned within the genus *Gambusia* after the generic name was first assigned to a Cuban species (*G. punctata*) by Poey in 1854. The species name, *affinis*, denotes 'related' and is thought to refer to the similarity of the western form to an eastern species that had been characterized in an unpublished morphological description of a North American fish (Moyle, 2002). Girard formally described the eastern North American species of *Gambusia*, *G. holbrooki* in 1859 (Figure 22.1).

The western mosquitofish, *Gambusia affinis*, and the eastern mosquitofish, *Gambusia holbrooki*, are members of a genus of about 46 species (Moyle, 2002; Froese and Pauly, 2010) within the order Cyprinodontiformes and the family Poeciliidae, the top minnow live-bearers. Central America is the centre of poeciliid abundance (Moyle, 2002). Unlike the broad distributions of *G. affinis* and *G. holbrooki*, which are the result of introductions outside their native ranges for mosquito control, the 20 North American *Gambusia* species have comparatively restricted geographic ranges

and are found in rivers and spring systems in the south-central US and eastern Mexico (Page and Burr, 1991; Moyle, 2002).

Morphological similarity of *G. affinis* and *G. holbrooki* has caused historical changes in the taxonomic status of the two species that confound published scientific findings and stocking records. Both fish were considered subspecies of *G. affinis* for an extended period and publications did not always distinguish between the two forms (Gerberich and Laird, 1968; Moyle, 2002). After about 1990, species status was re-established based on morphological differences (Rauchenberger, 1989), genetic studies and geographic distribution (Wooten et al, 1988). Adults of the two species can be distinguished by the numbers of dorsal and anal fin rays and the morphology of the male anal fin or gonopodium: *G. holbrooki* usually has seven dorsal rays, ten anal rays and a gonopodium with a series of prominent teeth on ray three; whereas, *G. affinis* has six dorsal rays, nine anal rays and lacks prominent teeth on ray three of the gonopodium. More recently, molecular diagnostic tools have been developed to distinguish between the two species (Vidal et al, 2010).

Beginning in the early 1900s, the two mosquitofish species were introduced as biological control agents for mosquitoes in temperate and tropical countries (Figure 22.2). Based on published accounts of mosquitofish stocking and research focusing on

Source: © Chris Appleby; insert: © W. E. Walton

Figure 22.1 *A male (left) and female (right)* Gambusia holbrooki; *insert shows* gonopodium of the male

mosquito-eating fishes, Gerberich and Laird (1968) summarized the decadal trends of interest in biological control using fish. During the first half of the 20th century, these trends are associated with frequency of introductions of *Gambusia*, but they fail to reflect the rate of spread of *Gambusia* following introduction to locations outside the native ranges of the two species.

Organized mosquito control commenced in the late 1800s with an increased understanding of the role

that mosquitoes played in the transmission of diseases such as malaria and yellow fever. The first purposeful use of fish to control mosquitoes followed soon thereafter and was against container dwelling *Aedes aegypti* in Cuba at the turn of the century. The first long distance transplantation of mosquitofish from Seabrook, Texas to Hawaii occurred in 1905 (Figure 22.2). *Gambusia affinis* was introduced into Taiwan in 1911. During the period between 1911 and 1920 with the

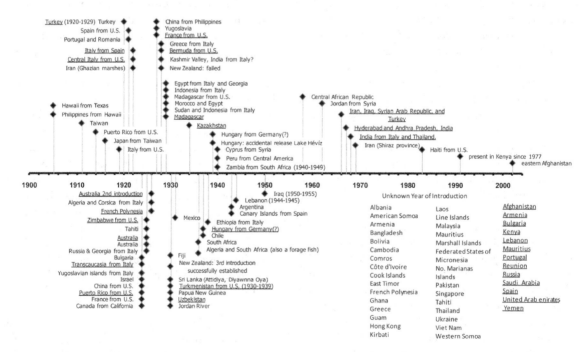

Note: The data represent a compilation of introductions listed in Gerberich and Laird (1968), Welcomme (1988), Haering (2005), ISSG (2006) and Froese and Pauly (2010). A question mark indicates uncertainty in a record.

Source: Timeline developed using Wittwer (2005)

Figure 22.2 *A timeline for the introduction of* Gambusia affinis *and* Gambusia holbrooki *(underlined) outside the US*

completion of the Panama Canal (in 1914) and during World War I, large numbers of men from North America were moved into the tropics and regions where mosquito-borne pathogens were present. At this time, *Gambusia* and guppies (*Poecilia reticulata*) were the primary species being transplanted for vector control.

During the following decade 1920–1930, interest in the use of larvivorous fish for vector control increased appreciably. During the 1920s, *Gambusia* were transplanted from eastern and south-central North America to Mexico, Central and South America (Figure 22.2) as part of a yellow fever reduction campaign sponsored primarily by the philanthropic efforts of the Rockefeller Foundation. The focus was still primarily on the control of mosquitoes inhabiting man-made containers such as rainwater storage jars. *Gambusia* also were transplanted to New Zealand, Australia and Europe. Recent genetic studies (Vidal et al, 2010) confirmed stocking records that indicated *G. holbrooki* was exported from North Carolina (US) to Spain in 1921. From Spain, the mosquitofish was introduced into Italy in 1922. The Italian population served as the primary source of *Gambusia* introductions to countries of Europe, western Asia, northern Africa and islands of the Mediterranean Sea between 1924 and 1930.

During this same time period, *Gambusia* were stocked outside (e.g. northern Illinois, California, Canada) its native range in North America. *Gambusia affinis* was introduced into California from Texas in 1922 (Dill and Cordone, 1997). By 1926, *G. affinis* had been introduced into 30 counties by the California State Board of Health and had spread rapidly from the introduction sites (Moyle, 2002). Pflieger (1975) notes that the distribution of *G. affinis* changed appreciably over 30 years. In survey collections from the 1940s *G. affinis* was restricted to the lowlands of south-eastern Missouri and waters adjacent to the Mississippi River. During a 30 year period the geographic distribution expanded to include central Missouri and two river systems in the south-western portion of the state as a consequence of widespread stocking for mosquito control.

Gerberich and Laird (1968) conclude that some of the *Gambusia* introductions during the 1920s had sound ecological bases, but others lacked consideration of the ecological conditions of the habitat relative to the physiological and ecological needs of the fish. The latter introductions either failed or gave equivocal results for mosquito control. During this time period,

Gambusia and other mosquito-eating fishes also were being used in abatement campaigns against vectors of malaria whose immature stages do not occur in container habitats; impact on non-target fauna was probably given little consideration. Mosquito-eating fish were naïvely thought to provide a long term solution for vector control that was a favourable alternative to the labour-intensive and ecologically damaging approaches that had been effective to date. Prior to this period the predominant approaches for mosquito control outside of residences and buildings were to drain and fill wetlands, to construct parallel ditching systems for draining standing water above the high tide line with no regard for natural drainage patterns and ecological interactions in coastal marshes, and to spread oil across the surface of large water bodies and slow moving sections of rivers and streams.

During the next two decades, interest in the use of biological control agents (based on the number of published papers) declined. However, in the 1930s, studies of the relationship between ecological conditions and the efficacy of biological control using mosquito-eating fish and of the evolutionary changes in fish stocks (i.e. cold-adapted and salinity-adapted strains of *Gambusia*) were carried out in regions of the former Soviet republics and North Africa (Gerberich and Laird, 1968). *Gambusia* were introduced into South Africa in 1936 and into various countries of South America during the late 1930s and early 1940s.

Interest in the use and the introductions of non-native fishes declined further during the 1940s after the realization in 1939 of the insecticidal properties of an organochlorine compound originally synthesized in 1874. Dichlorodiphenyltrichloroethane (DDT) was being used widely to reduce the populations of agricultural pest and public health insects. DDT provided the primary means of controlling malaria and other arthropod-borne diseases during World War II. During this time, interest in using fishes in biological control programmes against mosquitoes waned. However, *Gambusia* were introduced to Lebanon and Cyprus during the mid-1940s.

Interest in mosquito-eating fishes reached a nadir in the 1950s; nevertheless, *Gambusia* introductions occurred within western Asia. Following the discoveries of insecticide resistance to DDT and other synthetic insecticides in the mid-1950s, and the realization that these chemicals could not by themselves provide a viable long term strategy for mosquito control, interest in integrated mosquito control programmes that utilized a multifaceted approach to mosquito control increased during the 1960s and 1970s. The interest in use of larvivorous fish as components of integrated mosquito control programmes increased concomitantly. *Gambusia* were transplanted into additional countries in Eurasia and western Asia, the Middle East and India during the 1960s. Introductions slowed during the second half of the 20th century, but the end result was the spread of *G. affinis* and *G. holbrooki* across six continents (Figure 22.3).

Ecological Niche in Native and Introduced Ranges

The two *Gambusia* species that have been transplanted worldwide are native to the Atlantic and Gulf Coast drainages in eastern North America (Page and Burr, 1991; Nico and Fuller, 2010, Nico et al, 2010) (Figure 22.3). The eastern mosquitofish, *G. holbrooki*, occurs from southern New Jersey to Florida and to the eastern Mobile Bay, Alabama. The western mosquitofish, *G. affinis*, is native to the Gulf Slope drainage from central Indiana and southern Illinois to eastern Mexico and from the western Mobile Bay, Alabama to Texas and into eastern Mexico. The western extent of the native range in the south-central US was never adequately defined prior to the widespread movement of the fish by man. Intergrades of the two species can be found in the Mobile Bay basin.

The ecological niches of both species are similar. *Gambusia* spp. are common and locally abundant, often in vegetation along the periphery of lakes and ponds, and in backwaters and pools of streams within their native ranges. Mosquitofish are common in submerged and emergent plants, but tend not to penetrate dense plant beds preferring to reside near the vegetation-open water interface. They are occasionally found in brackish water. Both species are adapted for life in shallow, slow moving, warm water where piscivorous fish are absent or rare (Moyle, 2002).

Mosquitofish are omnivorous, opportunistic feeders. Diets typically include both plant and animal matter. *Gambusia* feed mostly at the water surface, but foraging is not restricted to the hypopneustic zone of the water column. Animal food includes insects, spiders, small crustaceans, rotifers and snails. Plant material is less important in the diet than is animal material but provides an important food when animal food is rare. It is common for diets to change during ontogeny from predominantly rotifers and micro-crustaceans in the diets of young individuals to mosquito larvae, other aquatic insects and organisms trapped in the surface film in the diets of adults. Cannibalism and predation on the eggs and immature stages of other co-occurring vertebrates are known to occur.

The two *Gambusia* species are hardy, capable of surviving broad ranges of environmental conditions, are comparatively tolerant of pesticides and exhibit high reproductive capacities. These characteristics, along with bearing live young (eliminating the need for nest building), omnivory, preference for habitats where predators are absent and mosquito larvae are present, and ease of culture, contribute to the success of *Gambusia* as mosquito control agents (Moyle, 2002). While these characteristics make them ideal candidates for mosquito control (Swanson et al, 1996), especially in the poor water quality and marginal habitats where mosquito production is typically greatest, many of these characteristics also are ideal for an invasive species (Lloyd et al, 1986; Moyle and Marchetti, 2006).

A detailed presentation of the environmental tolerances of *Gambusia* can be found in Swanson et al (1996); we provide only an overview of the broad environmental tolerances of mosquitofish. *Gambusia* spp. can survive low oxygen saturations (~0.2mg O_2 litre^{-1}) by breathing at the air–water interface. They can survive a broad temperature range (0.5–42°C) but persist in habitats where temperatures typically range annually between 10 and 35°C. Optimal temperatures for growth and reproduction are 25–30°C. Survival is greatly reduced during prolonged exposure to cold (<4°C) and this characteristic limits the geographic distribution of these species. Acclimation, however, is possible and cold hardy *Gambusia* strains exist. *Gambusia affinis* can survive in a pH range of 4.7–10.2 but typically occurs in waters with pH 7–9. Mosquitofish exhibit a broad salinity tolerance ranging from 0 to 58 parts per thousand (ppt), but survive best in fresh water and slightly brackish water with salinities <25ppt. Given these broad environmental tolerances, it should not be surprising that the ecological niche of *Gambusia* does not change appreciably in habitats outside its native range.

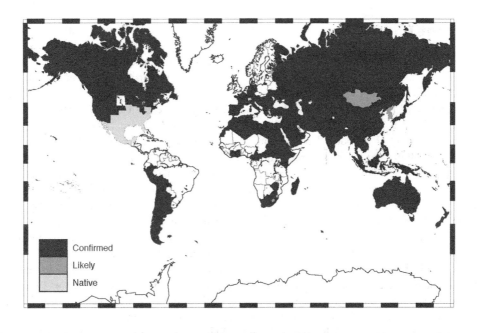

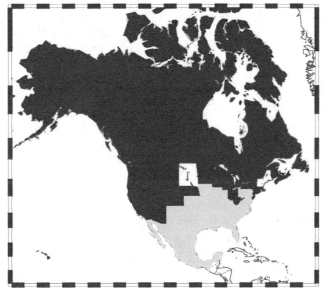

Note: Mosquitofish populations do not persist in regions where winter temperatures are below freezing for extended periods such as high latitude temperate zones unless geothermal or anthropogenic processes enhance water temperature during winter. Note that although entire countries are highlighted this does not mean that *Gambusia* spp. are present throughout all regions within the country.

Source: G. J. Funning created the distribution map using the Environmental Systems Research Institute, Inc,'s Digital Chart of the World®, available at www.maproom.psu.edu/dcw (accessed 10 August 2010) and generic mapping tools (Wessel and Smith, 1998)

Figure 22.3 *The native and introduced geographic distributions of* Gambusia *spp. worldwide (upper panel) and in North America (lower panel)*

Management Efforts

Attempts to eradicate introduced *Gambusia* populations have been largely unsuccessful and, even if possible, the cost of such efforts in established populations is prohibitively expensive. A primary focus of management efforts for *Gambusia* is to prevent new introductions and to limit future spread of introduced populations. This approach is recommended for the management of potentially widespread, persistent alien species because eradication after establishment is improbable (e.g. Lodge and Shrader-Frechette, 2003).

Given the rapid reproductive rate of mosquitofish populations following introduction, aggressive interactions of *Gambusia* with other freshwater biota, comparatively broad physical and chemical tolerances of mosquitofish, and the current widespread distribution of the species outside their native geographic ranges, management efforts that affect *Gambusia* yet have a minimum impact on non-target fish species have generally been unsuccessful. Control efforts applied against invasive species worldwide include:

- chemical control;
- cultural control and sanitary measures;
- physical removal/mechanical control;
- biological control;
- integrated pest management;
- control by utilization;
- ecosystem restoration; and
- containment (e.g. dispersal and movement control).

While some of these control strategies are currently not viable against mosquitofish and have not been attempted, most control strategies have a minimal long term impact on established populations and provide at best a temporary measure of population reduction in most natural habitats.

Chemical control

Gambusia spp. often have greater tolerances of chemical control agents than native fish and other aquatic biota (Lloyd and Arthington, 2010). *Gambusia* are more tolerant of the organophosphate pesticide chlorpyrifos (Dursban) than are several native fishes in Australia (Pyke, 2005) and are comparatively tolerant of a broad range of insecticides and herbicides (Walton, 2007).

The organic piscicide and insecticide rotenone has been used to attempt to control introduced mosquitofish, but because mosquitofish can often tolerate higher rotenone concentrations than native fishes, it may have detrimental effects on native fishes while concomitantly having little effect on mosquitofish. In Australia, most native fish are killed by a rotenone concentration of 0.5mg litre^{-1} but *G. holbrooki* can survive this concentration (Pyke, 2005). Impact on native fishes and other native fauna has been mitigated by releasing potassium permanganate downstream of the rotenone release point in flowing waterways (Lloyd and Arthington, 2010).

In addition to physiological tolerance of chemical control agents, *Gambusia* can use behavioural mechanisms to avoid toxic dosages. Rotenone applications failed to eliminate *Gambusia* from a hatchery spring pond and conveyance ditch (Gurtin, S., Arizona Game and Fish Department, pers. comm.). During an attempt to eliminate *Gambusia* from Scotty's Castle Creek at the north end of Death Valley National Park (California, US), mosquitofish were observed stranding themselves on algal mats at the water surface following the detection of rotenone applications and thereby survived chemical control methods by avoiding a toxic dosage of the piscicide. The rotenone application caused mortality but eradication by this chemical method was deemed infeasible because the numbers of surviving *Gambusia* were sufficient for re-establishment (Pister, P., pers. comm.). In another creek in the southwestern US (Corn Creek on the Desert National Wildlife Refuge north of Las Vegas, Nevada), a mosquitofish population was successfully eliminated using rotenone.

Cultural control and sanitary measures

Introduced fishes often thrive in disturbed habitats (Lloyd and Arthington, 2010) and the broad environmental tolerances of *Gambusia* provide a selective advantage over native fauna in such habitats. In hypereutrophic lentic habitats that are isolated from native habitats and that would be either stressful or potentially detrimental for native fishes (Walton et al, 2007; Peck and Walton, 2008), there is little reason not to utilize extant mosquitofish for vector control. As a component of integrated mosquito management programmes for such nutrient-enriched ecosystems, *Gambusia* can reduce larval and adult mosquito abundance by an order of magnitude or more as compared to the same control measures without larvivorous fish (Walton, W. E., unpublished data).

Arthington et al (1990) recommend that maintaining the natural flow regime and habitat characteristics of aquatic systems provides an effective management strategy to reduce the impact of introduced mosquitofish populations on native fauna. Non-native mosquitofish populations in the south-western US are detrimentally affected by disturbance from spates more than are native species with similar ecological requirements such as *Poeciliopsis* (Schoenherr, 1981; Minckley and Meffe, 1987). In addition to the potential negative effects of biotic interactions with *Gambusia* on native fauna, homogenization of flow regimes (Poff et al, 2007) probably also contributes significantly to the dominance of mosquitofish in some natural systems. In such systems, increasing the frequency and intensity of disturbance actually may be advantageous to native fishes (Bunn and Arthington, 2002).

Physical removal/mechanical control

Physical removal of mosquitofish requires a concerted effort to eliminate a substantial portion of the population, if not the entire population, in a water body or watercourse. The size (i.e. length, area) of the inundated habitats within a watershed, as well as the physical characteristics of the aquatic ecosystem under consideration, will determine the effort and cost of a well-planned and well-executed mosquitofish removal programme. *Gambusia* cannot be allowed to recolonize sites from which it was extirpated while control efforts are ongoing.

Physical removal programmes have required holders of scientific collection permits and persons carrying out research projects to sacrifice individuals of any exotic species taken during census and research activities. In Australia, *Gambusia* collected as bycatch in ecological research programs must be sacrificed immediately. Introduction to another water body or household aquaria is not permitted (Lloyd and Arthington, 2010). In the US, increasingly stringent regulations for the use of vertebrate animals in university-related research typically do not permit the elimination of bycatch in research programmes and, if euthanasia is permitted,

then the number of specimens to be killed must be specified beforehand. The practical outcome of compliance with such regulations necessitates that extermination of individuals of exotic species is left to agents of the natural resource agencies. Because collection and research activities are rarely carried out intensively across all habitats containing invasive species, such measures probably fail to cause a significant reduction in the populations of invasive species such as *Gambusia*.

The insertion of devices that macerate biological material within pipelines carrying reclaimed water appears to have been effective at eliminating mosquitofish from reclaimed water supplies. *Gambusia affinis* has been added for vector control in constructed wetlands treating municipal wastewater that provide a source of reclaimed water in southern California. Physicochemical conditions in wetlands treating highly enriched municipal effluent often are not conducive for the survival and proliferation of native fishes (Walton et al, 2007) and sometimes even *G. affinis* (Popko et al, 2009). The maceration devices eliminate clogging of small diameter water conveyance structures within reclaimed water systems by decaying vegetation and other biological material drawn into high volume pumps at the outflows of the wetlands.

Draining standing water bodies (e.g. outdoor ornamental ponds) and closing pathways for recolonization (e.g. dispersal control within conveyance structures between habitats containing mosquitofish and natural waters) have been somewhat successful for reducing *Gambusia* populations in Australia (Lloyd and Arthington, 2010). Barriers to spread must obstruct dispersal pathways as shallow as 3mm depth because *Gambusia* can move through shallow water half of their body depth (Alemadi and Jenkins, 2008).

Biological control and IPM

At the present time, biological control and integrated pest management (IPM) control strategies hold little promise for introduced *Gambusia* populations. Biological factors such as parasites, pathogens and predators that potentially regulate *Gambusia* populations occur in habitats within the native and introduced geographic ranges (Arthington and Lloyd, 1989; Courtenay and Meffe, 1989; Nagdali and Gupta, 2002); however, the majority of these mortality factors are not specific to *G. affinis* or *G. holbrooki*. Crandall and Bowser (1982) report a microsporidian infection (*Glugea*) that may be unique to mosquitofish but that has an unknown impact on *Gambusia*. Unlike efficacious biological control programmes in agricultural systems in which a parasitoid is intimately linked to a particular host pest insect, such narrowly focused interactions are not present for *Gambusia* and its potential biological control agents, at least for the important mortality factors that are known currently.

Moreover, the economic threshold of damage that can be tolerated within IPM programmes is comparatively easily calculated based on market conditions and the crop yield for agricultural systems. The concept of permissible crop losses does not translate readily to the ethics requisite in public health programmes and the level of control required for problematic invasive species. It is considerably more difficult to: (1) assign monetary values to a human life lost and to time lost while recouping from disease; and (2), as Service (1985) notes, achieve the high levels of reduction required for pathogens and vectors causing public health concerns. These same arguments are applicable to biological control and IPM programmes intended for invasive species. The inability of current global economic policies to assign full, credible dollar values to ecosystem services and biodiversity of natural environments (Ehrlich and Ehrlich, 2008) exacerbates the differences between IPM programmes for agriculture vs. invasive species. In many instances, the goal of invasive species control programmes is the elimination of the exotic species from its introduced range; this is not the goal of biological control and IPM programmes.

Lloyd and Arthington (2010) suggest that biomanipulation may afford some measure of *Gambusia* control through top-down predation. The proposed strategy using predatory galaxid (*Galaxias maculatus*) fishes in small water bodies of Australia does not, however, provide long term regulation of the target species. The relative cost–benefit of repeated introduction of predators vs. long term control by another control method(s) still needs to be evaluated.

Besides the direct benefit of consumption of *Gambusia* by piscivores, native piscivorous fishes provide an indirect benefit to vector control by forcing *Gambusia* to seek refuge in the vegetated habitats that are most likely to contribute significantly to mosquito production. The ability of predators to regulate *Gambusia* populations and the outcome of interactions between mosquitofish and fauna (e.g. juveniles of other fish species, macroinvertebrates) residing in the emergent and submerged vegetation will need to be evaluated carefully when determining the merits of this control strategy. The potential 'bio-synergistic' interactions of multiple biotic agents (Gerberich and Laird, 1968) functioning on different trophic levels warrants additional consideration.

Control by utilization

Utilization is not a control strategy that has been applicable to mosquitofish. If anything, utilization of mosquitofish for vector control has exacerbated the invasive species problem through the purposeful release of *Gambusia* into inappropriate habitats or unforeseen colonization of ecosystems outside the native geographical range. Prior to their use for vector control, these fish were not highly valued by humans, perhaps other than as a bait fish. Poey's name for the genus reflects this derisive point of view as it is a reference to a provincial Cuban term *gambusino* that refers to 'nothing' in the mocking sense that one says he was fishing for *gambusinos* when one catches nothing (Poey, 1861).

Ecosystem restoration

Depending on the foci of restoration activities, ecosystem restoration can either reduce or enhance the abundance of *Gambusia*. Maintenance of natural processes in aquatic ecosystems will promote native biodiversity and is the best way to suppress alien fishes (Arthington et al, 1990; Bunn and Arthington, 2002). Increasing the variation of discharge of flow regimes seems to reduce *Gambusia* populations associated with lotic environments (Minckley and Meffe, 1987). Re-establishing the natural variation in discharge that occurred historically in many drainage systems within suburban and urban areas is probably not possible because of human activities in and adjacent to the floodplain.

Restoration activities in riparian zones sometimes promote aquatic features that encourage human uses or attempt to satisfy multiple habitat goals. Depending on the morphometry and numerical occurrence of lentic type features within the restored riparian system, such features could enhance *Gambusia* populations. For example, the inclusion and increased frequency of large ponds and lakes in restoration projects that dissipate the impact of spates on the fish community may provide favourable habitat for *Gambusia* and non-native predators. A hands-off 'let nature take its course' approach to ecosystem management under homogeneous or reduced discharge regimes often leads to the development of thick stands of emergent vegetation. This approach to ecosystem restoration also will favour *Gambusia*.

Containment (e.g. dispersal and movement control)

Containment efforts include regulations and practices that prevent the spread of extant populations of *Gambusia* and limit its introduction into habitats that currently lack mosquitofish. Such measures include the quarantine and prohibition of stocking mosquitofish into waters of natural ecosystems. Categorization of *Gambusia* as a noxious species restricts the movement of mosquitofish between habitats and among watersheds in Australia (Lloyd and Arthington, 2010). Australia utilizes quarantine, certification and prohibition as measures to prevent the further introduction and spread of mosquitofish (Lloyd and Arthington, 2010).

The movement of any fish species, native or non-native, is typically limited by regulations in most US jurisdictions that require approval of the fish and wildlife agency, conservation district or other entity charged with jurisdictional oversight of natural resources. In many instances, this approval is contingent on the outcome of an examination of a representative sample of the stock population for disease, parasites, etc. Where stocking of mosquitofish for mosquito control outside the native geographic range of the fish is still permitted by law, containment is the primary strategy used for limiting the spread of *Gambusia*. In many jurisdictions, stocking of mosquitofish is limited to aquatic features (i.e. neglected residential swimming pools, ornamental ponds, etc.) that produce mosquitoes but have no direct connection to natural aquatic environments. These policies provide an effective means to halt subsequent introductions and the spread of mosquitofish distribution.

Whereas the management of mosquitofish populations in self-contained man-made aquatic environments is straightforward and generally effective, management of *Gambusia* in approved uses for semi-natural environments is potentially more difficult. Mosquitofish are also used as a component of integrated mosquito management programmes for semi-natural aquatic environments such as seasonal wetlands used for waterfowl hunting and inundated agriculture such as rice fields. Containment is also the primary strategy used in semi-natural aquatic environments to limit additional introductions of mosquitofish into natural aquatic ecosystems. Seasonal duck-hunting wetlands typically lack an outlet for water and are dried rather than drained. Water in rice fields may be drained overland or using procedures (e.g. pumps, water and fish barriers) that greatly reduce the survival and potential dispersal of mosquitofish.

Public awareness is an important component of mosquitofish management programmes in many places. Public education campaigns and outreach have been used to inform the public about the ecological consequences of alien fish introductions. Besides the two *Gambusia* species that have been spread as part of public health efforts, other live-bearing fishes and organisms (i.e., fish, amphibians, reptiles) in the aquarium trade have been introduced on several continents (Arthington et al, 1999; Saiki et al, 2010). Zoological parks are increasingly adding exhibits on the plight of native fauna, including fishes, as part of their educational activities. Public education should be requisite for vector control programmes in places where the public has access to mosquitofish. Boklund (1997) provides an example of the oversight required by the vector control agency, coordination with the natural resource agency and education/agreement with the recipient requisite for distribution of *Gambusia*. Lloyd and Arthington (2010) suggest that community involvement in monitoring habitats provides an early warning system for incipient population explosions following an introduction as well as a sense of ownership of the pest fish issue, and an improved understanding of the complexities of managing alien fish species. A concerned and informed public can provide important supplemental detection of species introductions for natural resource agencies dealing with ever-present budgetary constraints.

As a result of public education and stricter regulation of the movement of organisms between watersheds and outside native geographic ranges, mosquito control agencies are more aware of the ecological impact of non-native species than they were previously. In most places, mosquitofish are not currently planted into waters of the states of the US for the purpose of mosquito control.

Controversies

The two primary controversies related to *Gambusia* are its efficacy as a biological control agent for mosquitoes and the magnitude of its impact on aquatic ecosystems and biodiversity (Lloyd et al, 1986; Gratz et al, 1996; Rupp, 1996; Pyke, 2008). These differences of opinion stem in part from the evidence used by each side of the two controversies as regards the applicability of anecdotal accounts, laboratory and field experiments that differ in design and statistical power etc. to predict what happens in nature. In many cases the positive effect of *Gambusia* on reducing mosquito numbers or a negative effect of *Gambusia* on particular fauna is caused by more than one factor. Rarely has the relative importance of the factors causing a particular outcome been adequately quantified.

There are aquatic systems where *Gambusia* can exert significant levels of mortality on immature mosquitoes. Mosquitofish clearly have demonstrable negative effects on mosquitoes in environments that typically lack physical structure and have low predation pressure from consumers of small fishes. It could be argued justifiably that, in such situations, most native species could achieve the same result. The efficacy of *Gambusia* as an agent of mortality for mosquitoes declines as the physical structure of aquatic environments becomes more complex. This fact is not unique to *Gambusia* or any native fish species for that matter. Mosquitofish, as well as other species with high reproductive rates, can be effective mosquito control agents in isolated man-made aquatic environments such as man-made wetlands, ponds enriched by organic matter or with inputs of limiting nutrients, urban and agricultural drains and other systems where mosquito production is high enough to merit abatement and where mosquito predator populations need to increase rapidly to be large enough to provide persistent levels of control.

There is disagreement among vector ecologists, mosquito control practitioners, public health biologists and others involved in mosquito control as regards the efficacy of fish for vector control (Rupp, 1996; Gratz et al, 1996). Service (in Gratz et al, 1996) questions the utility of larvivorous fishes for vector control and argues that reduction of larval mosquito numbers is not the appropriate metric for assessing the effectiveness of a control agent: the biting densities of mosquito females and/or reduction of disease prevalence should be evaluated. Others claim that mosquito-eating fish are a useful component of integrated mosquito control strategies that include other control methods (see the opinions of some authors in Gratz et al, 1996). Because mosquitofish are opportunistic, generalist feeders, predation is not limited to mosquitoes. The lack of a tight predator–prey interaction between *Gambusia* and immature mosquitoes could slow control and impact non-target taxa.

There is ample observational and experimental evidence that non-native *Gambusia* have strong negative effects on the fauna and function of some ecosystems. Mosquitofish can be damaging to native fish (Schoenherr, 1981; Arthington and Lloyd, 1989; Courtenay and Meffe, 1989; Minckley and Marsh, 2009) and amphibian species (Gamradt and Kats, 1996; Goodell and Kats, 1999; Hamer et al, 2002). Mosquitofish are thought to eliminate some native species through predation on juveniles and/or eggs and interference competition (e.g. aggressive behaviour) that reduces reproductive success (Moyle, 2002). *Gambusia* can be detrimental to small fish with similar ecological requirements (Meffe and Snelson, 1989) and their greatest negative effects appear to be on fishes of similar size in small or isolated habitats where the mosquitofish can become dominant (Moyle, 2002).

Clearly, judicious use of mosquitofish and a greater appreciation of the roles that native mosquito-eating fish and invertebrates can play in integrated mosquito management programmes have merit. An all-out ban on the use of mosquitofish for mosquito control is probably not warranted, but use of *Gambusia* should be limited to the places where the fish are currently found and in habitats where the probability of dispersal into natural systems is very low. New introductions outside the native range should be prohibited. Efforts to manage and possibly extirpate *Gambusia* from natural habitats outside its native range should continue. A better understanding of the factors that facilitate dominance of *Gambusia* in some ecosystems and limit its numbers in native habitats is needed and should facilitate better management strategies for mosquitofish.

Acknowledgements

We benefited from discussions with members of the Southern California Native Freshwater Fauna Working Group, Desert Fishes Council, and Southern California Vector Control Environmental Taskforce. In particular, we thank the following individuals for collegiality and sharing of ideas and information: A. Arthington, W. Courtenay, J. Deacon, S. Gurtin, L. Lloyd, P. Moyle, P. Pister, R. Rader, M. Saba, A. Schoenherr, C. Swift and P. Unmack.

References

Alemadi, S. D. and Jenkins, D. G. (2008) 'Behavioral constraints for the spread of the eastern mosquitofish, *Gambusia holbrooki* (Poeciliidae)', *Biological Invasions*, vol 10, pp59–66

Arthington, A. H. and Lloyd, L. N. (1989) 'Introduced poeciliids in Australia and New Zealand', in G. K. Meffe and F. F. Snelson Jr (eds) *Ecology and Evolution of Livebearing Fishes (Poeciliidae)*, Prentice Hall, Englewood Cliffs, NJ, pp333–348

Arthington, A. H., Hamlet, S. and Blühdorn, D. R. (1990) 'The role of habitat disturbance in the establishment of introduced warm-water fishes in Australia', in D. A. Pollard (ed) *Introduced and Translocated Fishes and their Ecological Effects*, Bureau of Rural Resources Proceedings No 8, Australian Government Printing Service, Canberra, pp61–66

Arthington, A. H., Kailola, P. J., Woodland, D. J. and Zalucki, J. M. (1999) *Baseline Environmental Data Relevant to an Evaluation of Quarantine Risk Potentially Associated with the Importation to Australia of Ornamental Finfish*, Report to the Australian Quarantine and Inspection Service, Department of Agriculture, Fisheries and Forestry, Canberra

Boklund, R. J. (1997) 'Mosquitofish in control programs', *Journal of the American Mosquito Control Association*, vol 13, pp99–100

Bunn, S. E. and Arthington, A. H. (2002) 'Basic principles and ecological consequences of altered flow regimes for aquatic biodiversity', *Environmental Management*, vol 30, pp492–507

Crandall, T. A. and Bowser, P. R. (1982) 'A microsporidian infection in mosquitofish, *Gambusia affinis*, from Orange County, California', *California Fish and Game*, vol 68, no 1, pp59–61

Courtenay, W. R., Jr and Meffe, G. K. (1989) 'Small fishes in strange places: A review of introduced poeciliids', in G. K. Meffe and F. F. Snelson Jr (eds) *Ecology and Evolution of Livebearing Fishes (Poeciliidae)*, Prentice Hall, Englewood Cliffs, NJ, pp319–331

Dill, W. A. and Cordone, A. J. (1997) 'History and status of introduced fishes in California, 1871–1996', *Fish Bulletin 178*, California Dept. of Fish and Game, Sacramento, CA

Ehrlich, P. R. and Ehrlich, A. H. (2008) *The Dominant Animal*, Island Press, Washington, DC

Froese, R. and Pauly, D. (eds) (2010) 'FishBase', www.fishbase.org, version 05/2010, accessed 10 August 2010

Gamradt, S. C. and Kats, L. B. (1996) 'Effect of introduced crayfish and mosquitofish on California newts', *Conservation Biology*, vol 10, pp1155–1162

Gerberich, J. B. and Laird, M. (1968) 'Bibliography of papers relating to the control of mosquitoes by the use of fish: An annotated bibliography for the years 1901–1966', FAO Fisheries Technical Paper No 75 (FRs/T75), Rome, Italy

Goodell, J. A. and Kats, L. B. (1999) 'Effects of introduced mosquitofish on Pacific treefrogs and the role of alternative prey', *Conservation Biology*, vol 13, pp921–924

Gratz, N. S., Legner, E. F., Meffe, G. K., Bay, E. C., Service, M. W., Swanson, C., Cech Jr, J. J. and Laird, M. (1996) 'Comments on "Adverse assessments of *Gambusia affinis*"', *Journal of the American Mosquito Control Association*, vol 12, pp160–166

Hamer, A. J., Lane, S. J. and Mahony, M. J. (2002) 'The role of introduced mosquitofish (*Gambusia holbrooki*) in excluding the native green and golden bell frog (*Litoria aurea*) from original habitats in south-eastern Australia', *Oecologia (Berlin)*, vol 132, pp445–452

Haering, R. (2005) '*Gambusia holbrooki* (fish)', Global Invasive Species Database, www.issg.org/database/species/ecology. asp?si=617&fr=1&sts=sss, accessed 23 July 2010

ISSG (Invasive Species Specialist Group) (2006) '*Gambusia affinis* (fish)', Global Invasive Species Database, IUCN/SSC Invasive Species Specialist Group, www.issg.org/database/species/ecology.asp?si=126&fr=1&sts=sss, accessed 23 July 2010

Lloyd, L. and Arthington, A. H. (2010) '*Gambusia holbrooki*', Invasive Species Compendium. CABI, www.cabi.org/isc, accessed 10 August 2010

Lloyd, L. N., Arthington, A. H. and Milton, D. A. (1986) 'The mosquitofish – a valuable mosquito control agent or a pest?', in R. L. Kitching (ed) *The Ecology of Exotic Plants and Animals: Some Australian Case Studies*, John Wiley and Sons, Brisbane, pp6–25

Lodge, D. M. and Shrader-Frechette, K. (2003) 'Nonindigenous species: Ecological explanation, environmental ethics, and public policy', *Conservation Biology*, vol 17, pp31–37

Meffe, G. K. and Snelson Jr, F. F. (1989) 'An ecological overview of poeciliid fishes', in G. K. Meffe and F. F. Snelson Jr (eds) *Ecology and Evolution of Livebearing Fishes (Poeciliidae)*, Prentice Hall, Englewood Cliffs, NJ, pp13–31

Minckley, W. L. and Marsh, P. C. (2009) *Inland Fishes of the Greater Southwest: Chronicle of a Vanishing Fauna*, The University of Arizona Press, Tuscon, AZ

Minckley, W. L. and Meffe, G. K. (1987) 'Differential selection by flooding in stream-fish communities of the arid American Southwest', in W. J. Mathews and D. C. Heins (eds) *Community and Evolutionary Ecology of North American Stream Fishes*, University of Oklahoma Press, Norman, pp93–104

Moyle, P. B. (2002) *Inland Fishes of California*, University of California Press, Berkeley, CA

Moyle, P. B. and Marchetti, M. P. (2006) 'Predicting invasion success: Freshwater fishes in California as a model', *Bioscience*, vol 56, pp515–524

Nagdali, S. S. and Gupta, P. K. (2002) 'Impact of mass mortality of a mosquito fish, *Gambusia affinis* on the ecology of a freshwater eutrophic lake (Lake Naini Tal, India)', *Hydrobiologia*, vol 468, pp45–52

Nico, L. and Fuller, P. (2010) '*Gambusia holbrooki*. USGS Nonindigenous Aquatic Species Database', Gainesville, FL, http://nas.er.usgs.gov/queries/FactSheet.aspx?SpeciesID=849, accessed 23 July 2010

Nico, L., Fuller, P. and Jacobs, G. (2010) '*Gambusia affinis*. USGS Nonindigenous Aquatic Species Database', Gainesville, FL, http://nas.er.usgs.gov/queries/factSheet.aspx?SpeciesID=846, accessed 23 July 2010

Page, L. M. and Burr, B. M. (1991) *A Field Guide to Freshwater Fishes: North America North of Mexico*, Houghton Mifflin Co., Boston

Peck, G. W. and Walton, W. E. (2008) 'Effect of mosquitofish (*Gambusia affinis*) and sestonic food abundance on the invertebrate community within a constructed treatment wetland', *Freshwater Biology*, vol 53, pp2220–2233

Pflieger, W. L. (1975) *The Fishes of Missouri*, Missouri Department of Conservation, Jefferson City, MO

Poey, F. (1861 [1851–1858]) *Memorias Sobre la Historia Natural de la Isla de Cuba: Acompañadas de Sumarios Latinos y Extractos en Frances*, Impr. de Barcina, Habana, vol 1, pp382–386, www.biodiversitylibrary.org/item/20113#7

Poff, N. L., Olden, J. D., Merritt, D. M. and Pepin, D. M. (2007) 'Homogenization of regional river dynamics by dams and global biodiversity implications', *Proceedings of the National Academy of Sciences USA*, vol 104, pp5732–5737

Popko, D. A., Sanford, M. R. and Walton, W. E. (2009) 'The influence of water quality and vegetation on mosquitofish in mosquito control programs in wastewater wetlands', *Proceedings and Papers of the Mosquito and Vector Control Association of California*, vol 77, pp230–237

Pyke, G. H. (2005) 'A review of the biology of *Gambusia affinis* and *G. holbrooki*', *Reviews in Fish Biology and Fisheries*, vol 15, pp482–491

Pyke, G. H. (2008) 'Plague minnow or mosquito fish? A review of the biology and impacts of introduced *Gambusia* species', *Annual Review of Ecology, Evolution, and Systematics*, vol 39, pp171–191

Rauchenberger, M. (1989) 'Systematics and biogeography of the genus *Gambusia* (Cyprinodontiformes: Poecilidae)', *American Museum Novitates*, vol 2951, pp1–74

Rupp, H. R. (1996) 'Adverse assessments of *Gambusia affinis*: An alternative view for mosquito control practitioners', *Journal of the American Mosquito Control Association*, vol 12, pp155–166

Saiki, M., Martin, B. A. and May, T. W. (2010) *Final Report: Baseline Selenium Monitoring of Agricultural Drains Operated by the Imperial Irrigation District in the Salton Sea Basin, California*, US Geological Survey Open-File Report 2010-1064, Reston, VA

Schoenherr, A. A. (1981) 'The role of competition in the replacement of native fishes by introduced species', in R. J. Naiman and D. L. Soltz (eds) *Fishes in North American Deserts*, John Wiley and Sons Inc., New York, pp173–203

Service, M. W. (1985) 'Some ecological considerations basic to the biocontrol of Culicidae and other medically important insects', in M. Laird and J. W. Miles (eds) *Integrated Mosquito Control Methodologies, Volume 2*, Academic Press, London, pp9–30

Swanson C., Cech Jr, J. J. and Piedrahita, R. H. (1996) *Mosquitofish: Biology, Culture, and Use in Mosquito Control*, Mosquito and Vector Control Association of California and the University of California Mosquito Research Program, Sacramento

Vidal, O., Garciá-Berthou, E., Tedesco, P. A. and Garciá-Marín, J.-L. (2010) 'Origin and genetic diversity of mosquitofish (*Gambusia holbrooki*) introduced to Europe', *Biological Invasions*, vol 12, pp841–851

Walton, W. E. (2007) 'Larvivorous fish including *Gambusia*', in T. Floore (ed) *Biorational Control of Mosquitoes*, American Mosquito Control Association, Bulletin no 7, Mount Laurel, NJ, *Journal of the American Mosquito Control Association*, vol 23, no 2, Supplement, pp184–220

Walton, W. E., Wirth, M. C. and Workman, P. D. (2007) 'Environmental factors influencing survival of threespine stickleback (*Gasterosteus aculeatus*) in a multipurpose constructed treatment wetland in southern California', *Journal of Vector Ecology*, vol 32, pp90–105

Welcomme, R. L. (1988) 'International introductions of inland aquatic species', FAO Fisheries Technical Paper 294, Food and Agriculture Organization of the United Nations, Rome

Wessel, P. and Smith, W. H. G. (1998) 'New, improved version of generic mapping tools released', *Eos, Transactions, American Geophysical Union*, vol 79, pp579

Wittwer, J. W. (2005) 'How to create a timeline in Excel', Vertex42.com, www.vertex42.com/ExcelArticles/create-a-timeline. html, accessed 11 August 2010

Wooten M. C., Scribner, K. T. and Smith, M. H. (1988) 'Genetic variability and systematics of *Gambusia* in the southeastern United States', *Copeia*, vol 1988, pp283–289

23

Pseudorasbora parva Temminck & Schlegel (topmouth gudgeon)

Rudy E. Gozlan

History of *Pseudorasbora parva* and its Introduction to Host Countries

The history of the invasion of *Pseudorasbora parva* (topmouth gudgeon) has its origins in the lower part of Cháng Jiang (Yangzi River) at Wùhàn, a large county town on the east coast of China situated in Húběi Province. Early cooperation between China and ex-Soviet countries for the development of aquaculture and the fish farming of Chinese carp in particular have fuelled the fastest and widest fish invasion in the world, with 32 countries invaded from Central Asia to North Africa in less than 50 years (Gozlan et al, 2002).

This started in the 1950s with the first translocation of *P. parva* from the eastern part of the country into almost all natural lakes, reservoirs and the lower Upper Mekong River basin in Yunnan Province; the upper reach of Yellow River basin including Qinghai Province and Gansu Province; into inland waters of Inner Mongolia and inner waters including almost all natural lakes and rivers in lower altitude areas and reservoirs in Xinjiang (Anonymous, 1979; Li, 1982; Chu and Chen, 1989; Wu and Wu, 1991; Wang, 1995; Gao, 2005), and more recently in the City of Lhasa, Tibet (Gozlan et al, 2010a). The natural distribution of *P. parva* covers the eastern part of mainland China, Taiwan, North and South Korea, and Japan, with the exception of Hokkaido island where it was also translocated during the mid-1950s (Ishikari river; Hikita, 1961). This species is naturally abundant in the lower part of the Yellow River and the Yangtzi River but rare in the small southern coastal rivers of Fújàn Province and the lower part of the Pearl River, as well as in the northern part of the country in small coastal catchments of Liáoning Province and the lower part of the Amur basin, where abundance accounts for about 2 per cent of all catches.

The rise of the People's Republic of China led by Mao marked the post-World War II period, ending over 100 years of societal instability. This brought stability and prosperity to China that was characterized by rapid population increase. This resulted in a similarly rapid need for additional sources of animal protein, which was met by the development of fish farming in the west part of the country and the first translocations of cyprinid species such as black carp (*Mylopharyngodon piceus*; Richardson 1846), grass carp (*Ctenopharyngodon idella*; Valenciennes 1844), silver carp (*Hypophthalmichthys molitrix*; Valenciennes 1844) and bighead carp (*Hypophthalmichthys nobilis*; Richardson 1845). These fishes, from the east of China, especially from the middle and lower reaches of the Yangtze River basin, were taken west and north-westwards and introduced into many waters of Chinese provinces including Yunnan, Qinghai, Gansu and Xinjiang. Although in small localities *P. parva* is used for food, the accidental transfer and release of *P. parva* within translocations of key carp species for aquaculture characterize the primary pathway of *P. parva* introduction into its expanded range. Currently *P. parva*, although common in local markets, is often bought to feed other pet fish or reptiles. From the 32 markets surveyed across the native range, *P. parva* was never sold alive. However, in the west part of China, other secondary pathways, such as cultural or religious acts, have also been responsible for some secondary spread (Gozlan et al, 2010a). For example, people in Tibet traditionally buy small live fish in local fish markets and return them to the wild; in the last decade, *P. parva* has been reported from fish markets in big cities in Tibet such as Lhasa and Xigaze (Zhang, C., pers. comm.).

In Japan, after World War II, *P. parva* was also accidentally translocated into the northern part of Honshu island (Tohhoku district) and Hokkaido island (Nakamura, 1969) where it gradually hybridized with *P. pumila* (Takahashi, 1997). Today, the distribution of *P. parva* includes most rivers and lakes (Nakamura, 1969; Hikita, 1993), from Honshu island (central island), Shikoku island (southern main island) and Kyushu island (western main island).

The first introduction of *P. parva* has often been attributed to Romania with import of silver carp eggs and grass carp from Wùhàn, China (Nalbant, T., pers. comm.). However, similar cooperation programmes took place at the same time within countries of the Eastern Bloc and resulted in further introductions of *P. parva* to Hungary (Paks Fisheries Farm in 1963; Molnár, 1967), Lithuania (Dunojus Lake in 1963; Krotas, 1971), Romania (Nucet Fisheries Research Centre in 1961; Bănărescu, 1964) and the Ukraine (Kuchurganskoye reservoir in 1962; Chepurnov and Kubrak, 1965). Within its introduced range, 32 per cent of *P. parva* introductions originated from mainland China, 47 per cent from former USSR or socialist countries and 18 per cent from Russia (Gozlan et al, 2010a).

Prior to the 1990s, the primary mode of dispersal was accidental but human mediated through the translocation of fish stock from fish farm to fish farm (65 per cent of all introductions in the invasive range) and resulted in long distance dispersal (i.e. around 200km) (Gozlan et al, 2010a). Primary pathways of introduction also include recreational fishing (22 per cent), ornamental fish trade (9 per cent) and natural dispersal (1 per cent). Since the first introductions of *P. parva*, approximately five new countries have been invaded in each decade, with a mean detection time of 4.1 ± 5.2 years (mean ± SD, n = 19) between time of

first introduction into a country and first detection (Gozlan et al, 2010a). These long distance dispersals were then followed more recently by shorter dispersal distances (i.e. around 25km), with fish diffusing from their primary site of introduction (Gozlan et al, 2010a). According to Gozlan et al (2010a), natural dispersal represents the main secondary pathway (72 per cent) followed by angling (25 per cent) and the ornamental fish trade (3 per cent). The speed of the species' dispersal in the introduced range is rapid, with some early invaded countries such as, for example, The Netherlands and former Czechoslovakia already showing signs of saturation while other countries such as the UK have yet to reach their maximum capacity (Gozlan et al, 2010a). Genetic analysis of the introduced populations confirms the suspected pattern of colonization that was initiated by the introduction of *P. parva* to small geographic areas or a single location (Hanfling, B., pers. comm.) but associated with or preceded by the admixture of genetically diverse source populations that may have augmented its invasive potential (Falka et al, 2007).

All introductions have led to sustainable populations that spread into adjoining reservoirs and local catchments, with the exception of the Lithuanian introductions (1963, 2007) that both failed, highlighting the northern limit of *P. parva* distribution in Europe. The species spread west through Europe into countries within the Danube basin via early introductions around the Black Sea and east into countries such as Turkey and Iran. Following the earliest introductions, central Europe saw a more complicated pattern of intercountry spread. *P. parva* was introduced into former Czechoslovakia from Hungarian populations and then spread to Germany (Musil, J., pers. comm.) then to Holland, Belgium and the UK (Gozlan et al, 2002; Copp et al, 2009). Albanian populations (Allardi and Chancerel, 1988) were mainly responsible for the earliest introduction into France (1970s in the Sarthe region). In addition, it appears that the River Danube provided a dispersal pathway for natural colonization that is likely to be the case for Austria's earlier *P. parva* population, where the first records came from two relatively large rivers near the Danube confluence (Weber, 1984). For example, Ahnelt and Tiefenbach (1991) describe the rapid dispersal of the species along the River Raab (southeast Austria). From 1986 to 1990 the species was dispersing downstream, colonizing over 30km of river length in only four years (Figure 23.1).

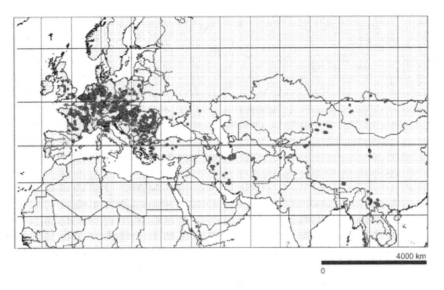

Source: Rudy E. Gozlan

Figure 23.1 *Current distribution of topmouth gudgeon* (Pseudorasbora parva) *in its invasive range*

Ecological Niche

Climatic niche

In its native range, *P. parva* is found in extremely contrasting climatic conditions and subject to the worst extremes of weather. Its natural distribution in China can be divided roughly into the following climatic regions; the north and northeast with temperatures fluctuating from –40°C in winter to +30°C in summer (i.e. Hēilóngjiāng and Jílín provinces); central with temperatures between 0°C and +30°C (Hénan and Ānhui provinces); and south with tropical conditions and temperatures of between 10°C and +30°C (i.e. Guangdōng and Guangxī provinces). Genetic studies indicate that *P. parva* populations associated with these three main climatic regions correspond to three separate lineages (Hanfling, B., pers. comm.). In fact, several subspecies have been described, such as *Pseudorasbora parva altipinna, P. p. depressirostris, P. p. parvula, P. p. tenuis, P. p. fowleri* and *P. p. monstrosa* (Nichols, 1925, 1929, 1943) but a recent visit to the collection of the Chinese Academy of Science, which includes thousands of specimens from all around China, did not highlight a great deal of variability in meristic characteristics. Further work is needed to confirm the relationship between genetic lineages, morphotypes and potential adaptation to specific climatic conditions. In addition, populations of *P. parva* in Japan and Taiwan are also genetically separated and represent two separate lineages.

At first sight, the climatic niche of *P. parva* seems extremely large and does not indicate any temperature limitations as it can withstand the cold winters of the north as well as the warm and humid summers of the south. However, the populations in the north and south part of the native distribution range are not very abundant, with *P. parva* representing less than 2 per cent of overall catches, while it fluctuates between 60 and 90 per cent of the catches in the lower part of the Yangtzi and Yellow Rivers.

In its introduced range, *P. parva* populations have also been subject to extreme climatic conditions with cold winters in northern Poland to warm summers in Italy or Algeria. However, populations have failed to establish in countries located north of Poland, such as Lithuania. This is probably due to the length of the summer period, which is too short and would not allow the establishment of *P. parva* populations in the early stage of colonization. As such, initial climatic analysis has shown that it is not so much the low winter temperatures that limit *P. parva* establishment, as it is found in regions with winter temperatures of –40°C, but rather the need for a period of approximately five months above 10°C, which is lacking in the northern European latitudes.

Of course, these thermal adaptations may be different among the different *P. parva* populations and further studies need to be done in experimental conditions to test the potential for adaptive plasticity to a thermal gradient from ancestral lineages. This could have direct consequences on the potential of this species to spread beyond its current invasive range.

Life history traits and habitat use

The invasive ability of *P. parva* is also facilitated by its life history traits that enhance its ability to colonize new waters rapidly (Rosecchi et al, 2001; Pinder et al, 2005; Beyer et al, 2007; Britton et al, 2007; Záhorská et al, 2009; Gozlan et al, 2010a). It is a small cyprinid species (Figure 23.2) with sexual maturity after one year and a limited lifespan (<5 years) (Rosecchi et al, 1993; Zhang et al, 1998a; Ye et al, 2006; Britton et al, 2007; Beyer, 2008; Kapusta et al, 2008; Yan and Chen, 2009, Gozlan et al, 2010a). In addition to early reproduction, multiple spawning with females producing a series of egg batches throughout a spawning season (Katano and Maekawa, 1997) ensures a high population growth that promotes colonization and establishment (Katano and Maekawa, 1997; Rosecchi

Note: Scale bar is 1cm. There are 38 scales in the lateral line and a total of 12 rows of scales, 5 of them being situated above the lateral line. The fin formula reads D II-III7; A II-III6; P I12-13; V I7.

Source: Rudy E. Gozlan

Figure 23.2 *(a) Female and (b) male topmouth gudgeon* (Pseudorasbora parva)

et al, 2001; Gozlan et al, 2002; Beyer et al, 2007; Gozlan et al, 2010a). Larvae survival rates are increased as batches laid between April and August decrease their susceptibility to mortality through changes in environmental conditions (Katano and Maekawa, 1997; Gozlan et al, 2003). Also, in contrast to most cyprinids, males establish and guard primitive nests that ensure a better protection of egg broods and thus a high hatching rate (Pinder and Gozlan, 2003; Gozlan et al, 2010a).

In line with its life history traits, *P. parva* demonstrates great plasticity in habitat use as it occupies a range of lotic and lentic waters including rivers, reservoirs, oxbows and canals, ponds and shallow lakes and other available water bodies (Arnold, 1990; Rosecchi et al, 1993; Adámek and Siddiqui, 1997; Hliwa et al, 2002). However, despite being able to establish populations under lotic conditions (Sunardi et al, 2005, 2007a, 2007b), *P. parva* establish larger populations under lentic conditions than under lotic conditions, supporting the assumption that main river channels serve as dispersal corridors (Muchacheva, 1950; Pollux and Korosi, 2006; Gozlan et al, 2010a). In mainland China, the large populations of *P. parva* found in the lower part of the Yellow and Yangzi Rivers are supported by large floodplains and enhanced by an extensive use of small-scale aquaculture ponds. In terms of microhabitat, *P. parva* does not have specific needs and can be found associated with pebbles, in-stream ligneous debris (Beyer et al, 2007; Gozlan et al, 2010a) or highly vegetated ponds (Trombitskiy and Kakhovskiy, 1987; Pollux and Korosi, 2006; Kapusta et al, 2008).

The diet of *P. parva* is varied and the species is generally described as an omnivore (Muchacheva, 1950; Weber, 1984; Xie et al, 2001), but has also been considered as planktivorous with a broad diet (Rosecchi et al, 1993; Zhang et al, 1998b; Sunardi et al, 2007a, 2007b). Due to small gape size, *P. parva* is limited to small food items (Arnold, 1990; Gozlan et al, 2010a) and generally feed on copepods, cladocerans, ostracods, molluscs, chironomid larvae, rotifers and detritus (Rosecchi et al, 1993; Adámek and Siddiqui, 1997; Zhang et al, 1998b; Hliwa et al, 2002; Nagata et al, 2005). The observed seasonal fluctuations in *P. parva* prey composition partly reflect the seasonal fluctuations of food resources in the water body (Xie et al, 2000; Beyer, 2008; Gozlan et al, 2010a, 2010b). Under specific conditions, such as in high density aquaculture ponds, *P. parva* has also been reported to be a facultative parasite causing injuries by nibbling other species, reaching the musculature in *H. molitrix*, *A. nobilis* and *C. idella* (Trombitskiy and Kakhovskiy, 1987; Libosvárský et al, 1990; Adámek et al, 1996).

Parasites and pathogens

A key ecological threat posed by *P. parva* is the introduction and spread of pathogens. The speed of its dispersal makes it particularly suited to introducing and spreading a wide range of pathogens. However, according to Gozlan et al (2010a), *P. parva* in its invasive range has a denuded parasitic fauna when compared to its native range, with parasites with complex lifecycles and using *P. parva* as a second intermediate host (i.e. Bucephalidae, Clinostomatidae, Cyathocotylidae, Heterophyidae and Opisthorchiidae parasites) being absent.

The two most pathogenic parasites associated with *P. parva* are *Anguillicola crassus* and the rosette agent *Sphaerothecum destruens* (Gozlan et al, 2005, 2009; Andreou et al, 2009). *Anguillicola crassus* is a parasitic nematode that infects the swimbladder of European eels (*Anguilla Anguilla*), causing high eel mortalities (Kennedy, 2007) and for which *P. parva* acts as an intermediate host. For example, 35 per cent of a French population of *P. parva* were infected with *A. crassus* (Cesco et al, 2001). In addition, the characterization of *P. parva* as a healthy carrier for the intracellular parasite *S. destruens* represents a major ecological risk as this pathogen has been responsible for mass mortality of salmonid fishes in the US (Arkush et al, 1998) and has also been associated with the decline of native European fish species including sunbleak (*Leucaspius delineates*; Gozlan et al, 2005, 2009). It has also been shown that many other species of cyprinids such as roach (*Rutilus rutilus*), bream (*Abramis brama*), carp (*Cyprinus carpio*) and fathead minnow (*Pimephales promelas*) are susceptible to this pathogen (Gozlan et al, 2005, 2006; Andreou, 2010). Although the origin of *S. destruens* in Europe remains unclear (Gozlan et al, 2009), it may have arrived with *P. parva* and consequently the dispersal of *P. parva* throughout Eurasia may have facilitated the spread of *S. destruens*, posing a threat to both salmonid and cyprinid populations (Gozlan et al, 2005, 2006, 2009). Further epidemiologic work is ongoing to characterize the true pattern of *S. destruens* dispersal.

Management, Control and Containment Efforts

Translocations of Chinese carp in China followed by introductions in Europe, North Africa and the Middle East constituted the main introduction pathway for *P. parva*, which hitchhiked along the target aquaculture species (Gozlan et al, 2010a, 2010b). *Pseudorasbora parva*'s introduction pathway was identified fairly early on in the invasion process (Bǎnǎrescu, 1964) and preventing further introductions should have been considered as a key initial step in its management. Effective management of *P. parva* should have started with risk communication and introduction prevention (Copp et al, 2005) using risk assessments such as the Fish Invasiveness Scoring Kit (FISK) (see www.cefas. co.uk/4200.aspx) as these tools maximize the potential for reducing adverse impacts and the associated costs (Gozlan et al, 2010b). Although it could be expected that managers would have intuitively used the 'precautionary principle' as a first option to prevent *P. parva* introductions and its spread, managers frequently wait until a non-native species is introduced before responding and limiting its impact (Britton et al, 2008). This is related to the fact that fish introductions often underpin important ecosystem services that are predicted to produce substantial economic returns and societal benefits (Gozlan, 2008; Gozlan and Newton, 2009) and that introductions still often occur even with strict and expensive prevention protocols in place. This is the typical pattern for the introduction of *P. parva* and its subsequent dispersal, as the introductions of Chinese carp were intentional and supported one of the largest aquaculture productions. In addition, at the time, the introduction of Chinese carp to former USSR countries was also seen as a political ideology, with species such as grass carp exemplified as a 'good communist fish', that grows large to feed the people, that is ecologically friendly as it is a macrophyte feeder, and that is native to communist countries. In other western countries such as England, grass carp was initially introduced as a biocontrol agent to reduce macrophyte growth in certain waters following a comprehensive assessment of the species (Stott, 1977). However, when accidental introductions occur, pre-introduction risk assessments are unlikely to have been carried out and non-target species such as *P. parva* exert the greatest ecological impact with little to no economic return (Gozlan and Beyer, 2006; Gozlan et al, 2005, 2010a, 2010b). In practice, the screening of fish consignments was not fully efficient, leading to secondary dispersal of *P. parva* (contaminant) via fish movements largely due to the absence of a 'rapid response strategy'.

Despite *P. parva* being listed as one of the top ten worst invasive species in Europe, there is, with the exception of England and Wales, no legislation or management procedures in place that specifically target this species. For example, Copp et al (2007) report a decrease in the rate of establishment success in the number of introduced non-native freshwater fishes to the UK after the 1970s. This coincided with the development and enactment of the Import of Live Fish Act (ILFA) in 1980 that was intended to be precautionary, but in practice is reactionary as a species must be demonstrated to be likely to cause harm in order to be placed on the list of controlled species. The progressive implementation of ILFA during the 1980s led to a strengthening of controls over which types of non-native species were imported to the UK and subject to a licensing procedure. *Pseudorasbora parva* was introduced for the first time in England in 1985 and the species was added to the ILFA list in 2005. During that period, *P. parva* went from one isolated population in an aquaculture farm in the south of the country to over 25 populations scattered all across England and Wales.

In 2005, the discovery in England of *P. parva* as a healthy carrier of *S. destruens* and subsequent parliamentary questions have led the Environment Agency to take novel measures in its fight against non-native species. For the first time, an effective lethal method designed to incur maximum mortality rates was used in closed water systems where *P. parva* populations were present and where there was either a risk of further dispersal in fluvial systems or where there was proximity to a site of specific conservation value. The decision to use rotenone (Ling, 2002), a naturally occurring ketone ($C_{23}H_{22}O_6$) that works by inhibiting oxygen utilization (Lockett, 1998), was driven by the high ecological risk posed by *P. parva* in spite of such techniques being viewed by many as controversial due to cost, difficulty of success and likelihood of damage to non-target species (Myers et al, 1998; Rayner and Creese, 2006; Britton et al, 2008; Gozlan et al, 2010a, 2010b). Toxicity of the rotenone depends on the target species and is influenced by many environmental

factors, such as temperature, light exposure, the degree of site enclosure, depth, pH, discharge and the binding to suspended matter, but also to many other specific aspects of the application method(s) used (Meadows, 1973; Lintermans, 2000; Willis and Ling, 2000). In England between 2005 and 2007, five populations of *P. parva* in fishing ponds/lakes (maximum 2ha and less than 5m deep) were successfully eradicated (Britton et al, 2008).

Although rotenone-based eradications have been effectively used against *P. parva* populations in England and Wales, the chance of success is proportional to the geographical spread of the populations (Anderson, 2005; Rayner and Creese, 2006), with eradication most effective in relatively small, closed, sparsely vegetated water bodies as is the case in the UK. However, at a European scale it would be unrealistic to imagine that *P. parva* populations could be eradicated, and control and containment efforts should aim to prevent the spread of *P. parva* into 'clean' catchments, using existing legislation and risk assessments (Hickley and Chare, 2004; Copp et al, 2005) and small-scale eradications whereby populations are eliminated from waters from which there is a high chance of dispersal into fluvial environments (Gozlan et al, 2010b).

Finally, *P. parva* has not yet reached the full extent of its invasive potential and countries such as the US, Canada, South Africa, Australia and New Zealand among others should learn from historical mismanagement in Europe that effective management of *P. parva* should start with risk communication and introduction prevention (Copp et al, 2005).

Challenges and Controversies

The history of *P. parva* invasion emphasizes the importance of screening and controlling human activity related to major ecosystem services such as aquaculture, recreational angling and the ornamental fish trade in order to avoid a non-native introduction becoming a pan-continental invasion (five new countries have reported *P. parva* introductions every decade since the initial introduction in 1960; Gozlan et al, 2010a). Introductions of non-native species are generally easier and cheaper to eradicate and/or control when their geographical distribution is still limited (Genovesi, 2005; Cacho et al, 2006) and is supported through the three step management concept of 'rapid detection, rapid assessment, rapid response' (Myers et al, 2000;

Zavaleta et al, 2001; Anderson, 2005). However, a key issue remains the ability to put rapid detection in place as detection efforts are inversely proportional to the non-native abundance and spread (Hayes et al, 2005). Although the first pathway of introduction for *P. parva* was long distance dispersal due to fish movements, the secondary pathway of dispersal (i.e. within countries) was in the vast majority through natural dispersal. So in the case of *P. parva* invasion, it would have been crucial to limit the time between introduction and detection to avoid their secondary spread. Gozlan et al (2010a) found that the detection time for *P. parva* within its invasive range was on average about four years, which appears to be too long to have prevented a pan-continental invasion. Hence, there is a need for networks designed to rapidly detect new introductions based on high risk locations, high value resources, important pathways and populations and species of most specific concern (Gozlan et al, 2010b).

One of the current key research goals related to *P. parva* invasion is to characterize the species' potential for further spread beyond its current distribution and for its introduction to continents where it is currently absent. Can we predict where *P. parva* is most likely to be introduced next and if so can we put in place rapid detection, rapid assessment and rapid response management? In the near future, climate models coupled with native population genetic analysis will allow predictive risk maps of introduction to be established based on the climate of the recipient country and the adaptability of the various genetic lineages to specific climatic conditions. It will then be the responsibility of policy makers to limit the risk of introduction through the use of adapted risk assessment on key introduction pathways, limiting the long term ecological risk for aquatic biodiversity and cost of associated ecosystem services (Copp et al, 2005; Gozlan and Newton, 2009; Takimoto, 2009; Gozlan et al, 2010b).

References

Adámek, Z. and Siddiqui, M. A. (1997) 'Reproduction parameters in a natural population of topmouth gudgeon *Pseudorasbora parva*, and its condition and food characteristics with respect to sex dissimilarities', *Polish Archives of Hydrobiology*, vol 44, pp145–152

Adámek, Z., Navrátil, S., Palíková, M. and Siddiqui, M. A. (1996) '*Pseudorasbora parva* Schlegel, 1842: Biology of non-native species in the Czech Republic', in M. Flašjshans (ed) *Proceedings of Scientific Papers to the 75th Anniversary of Foundation of the Research Institute of Fish Culture and Hydrobiology. Vodňany, Czech Republic*, Research Institute of Fish Culture and Hydrobiology, University of South Bohemia, pp143–152

Ahnelt, H. and Tiefenbach, O. (1991) 'Zum Auftreten des Blaubandbärblings (*Pseudorasbora parva*) (Teleostei: Gobioninae) in den Flüssen Raab und Lafnitz' ['Concerning the occurrence of the topmouth gudgen (*Pseudorasbora parva*) (Teleostei: Gobioninae) in the rivers Raab and Lafnitz', in German], *Österreichs Fischerei*, vol 44, pp19–26

Allardi, J. and Chancerel, F. (1988) 'Sur la presence en France de *Pseudorasbora parva* (Schegel, 1842)', *Bulletin Français de la Pêche et de la Pisciculture*, vol 308, pp35–37

Anderson, L. W. J. (2005) 'California's reaction to *Caulerpa taxifolia*: A model for invasive species rapid response', *Biological Invasions*, vol 7, pp1003–1016

Andreou, D. (2010) 'Life cycle and prevalence of *Sphaerothecum destruens* in native cyprinids', PhD Thesis, University of Cardiff, Cardiff, UK

Andreou, D., Paley, R. and Gozlan, R. E. (2009) 'Temperature influence on production and longevity of *Sphaerothecum destruens*' zoospores', *Parasitology*, vol 95, pp1539–1541

Anonymous (1979) 'Fishes of Xinjiang, Urumqi', Xinjiang Science and Technology Publishing House (ed C.A.o.S.a.F.B. Institute of Zoology), I, Xinjiang Uygur Autonomous Region

Arkush, K. D., Frasca, S. and Hedrick, R. P. (1998) 'Pathology associated with the rosette agent, a systemic protist infecting salmonid fishes', *Journal of Aquatic Animal Health*, vol 10, pp1–11

Arnold, A. (1990) *Eingebürgerte Fischarten: Zur Biologie und Verbreitung allochthoner Wildfische in Europa* [*Naturalised Fish Species: Biology and Distribution of Non-native Wild Fish in Europe*, in German], A. Ziemsen Verlag, Wittenberg, Lutherstadt

Bănărescu, P. (1964) *Pisces – Osteichthyes. Fauna Republicii Populare Romine* [In Romanian], vol 13, Ed. Acad. RPR, Bucuresti

Beyer, K. (2008) 'Ecological implications of introducing *Leucaspius delineatus* and *Pseudorasbora parva* into inland waters in England', PhD Thesis, University of Hull, UK

Beyer, K., Copp, G. H. and Gozlan, R. E. (2007) 'Microhabitat use of non-native topmouth gudgeon *Pseudorasbora parva* within a stream fish assemblage', *Journal of Fish Biology*, vol 71, pp224–238

Britton, J. R., Davies, G. D., Brazier, M. and Pinder, A. C. (2007) 'A case study on the population ecology of a topmouth gudgeon (*Pseudorasbora parva*) population in the UK and the implications for native fish communities', *Aquatic Conservation: Marine and Freshwater Ecosystems*, vol 17, pp749–759

Britton, J. R., Davies, G. D. and Brazier, M. (2008) 'Contrasting life history traits of invasive topmouth gudgeon (*Pseudorasbora parva*) in adjacent ponds in England', *Journal of Applied Ichthyology*, vol 24, pp694–698

Cacho, O. J., Spring, D., Pheloung, P. and Hester, S. (2006) 'Evaluating the feasibility of eradicating an invasion', *Biological Invasions*, vol 8, pp903–917

Cesco, H., Lambert, A. and Crivelli, A. J. (2001) 'Is *Pseudorasbora parva* an invasive fish species (Pisces, Cyprinidae) a new vector of the anguillicolosis in France?', *Parasite-Journal De La Société Française De Parasitologie*, vol 8, pp75–76

Chepurnov, V. S. and Kubrak, I. F. (1965) 'About the past, recent and future composition of the Kuchurgan lagoon ichthyofauna' [Materialy zoologicheskogo soveshaniya po probleme 'Biologicheskie osnovy rekonstruktsii, ratsional'nogo ispol'zovaniya i okhrany fauny yuzhnoi zony Evropeyskoi chasti SSSR'], *Kishinev*, pp284–288

Chu, X.-L. and Chen, Y.-R. (1989) *Fishes of Yunnan*, Science Press, Beijing

Copp, G. H., Garthwaite, R. and Gozlan, R. E. (2005) 'Risk identification and assessment of non-native freshwater fishes: A summary of concepts and perspectives on protocols for the UK', *Journal of Applied Ichthyology*, vol 21, pp371–373

Copp, G. H., Templeton, M. and Gozlan, R. E. (2007) 'Propagule pressure and the occurrence of non-native fishes in regions of England', *Journal of Fish Biology*, vol 71, Suppl. D, pp148–159

Copp, G. H., Villizi, L. and Gozlan, R. E. (2009) 'Fish movements: The introduction pathway for topmouth gudgeon *Pseudorasbora parva* and other non-native fishes in the UK', *Aquatic Conservation: Marine and Freshwater Ecosystems*, vol 20, pp269–273

Falka, I., Mérai, K. and Ferencz, B. (2007) 'Origin of introduced *Pseudorasbora parva* populations in Romania, based on genetic markers (16S rRNA)', *Studii şi Cercetări Biologie Series*, vol 13, pp1–5

Gao, X.-Y. (2005) *A Checklist on the Classification and Distribution of Vertebrate Species and Subspecies in Xinjiang*, Science and Technology Publishing House, Urumqi

Genovesi, P. (2005) 'Eradications of invasive alien species in Europe: A review', *Biological Invasions*, vol 7, pp127–133

Gozlan, R. E. (2008) 'Introduction of non-native freshwater fish: Is it all bad?', *Fish and Fisheries*, vol 9, pp106–115

Gozlan, R. E. and Beyer, K. (2006) 'Hybridisation between *Pseudorasbora parva* and *Leucaspius delineatus*', *Folia Zoologica*, vol 55, pp53–60

Gozlan, R. E and Newton A. C. (2009) 'Biological invasions: Benefits versus risks', *Science*, vol 324, p1015

Gozlan, R. E., Pinder, A. C. and Shelley, J. (2002) 'Occurrence of the Asiatic cyprinid *Pseudorasbora parva* in England', *Journal of Fish Biology*, vol 61, pp298–300

Gozlan, R. E., Pinder, A. C., Durand, S. and Bass, J. (2003) 'Could the small size of sunbleak, *Leucaspius delineatus* (Pisces, Cyprinidae) be an ecological advantage in invading British waterbodies?', *Folia Zoologica*, vol 52, pp99–108

Gozlan, R. E., St-Hilaire, S., Feist, S. W., Martin, P. and Kent, M. L. (2005) 'Biodiversity: Disease threat to European fish', *Nature*, vol 435, p1046

Gozlan, R. E., Peeler, E. J., Longshaw, M., St-Hilaire, S. and Feist, S. W. (2006) 'Effect of microbial pathogens on the diversity of aquatic populations, notably in Europe', *Microbes and Infection*, vol 8, pp1358–1364

Gozlan, R. E., Whipps, C., Andreou, D. and Arkush, K. (2009) 'Identification of the rosette-like agent as *Sphaerothecum destruens*, a multihost fish pathogen', *International Journal for Parasitology*, vol 39, pp1055–1058

Gozlan, R. E. et al (26 others) (2010a) 'Pan-continental invasion of *Pseudorasbora parva*: Towards a better understanding of freshwater fish invasions', *Fish and Fisheries*, vol 11, pp315–340

Gozlan, R. E., Britton, J. R., Cowx, I. and Copp, G. H. (2010b) 'Current knowledge on non-native freshwater fish introductions', *Journal of Fish Biology*, vol 76, pp751–786

Hayes, K. R., Cannon, R., Neil, K. and Inglis, G. (2005) 'Sensitivity and cost considerations for the detection and eradication of marine pests in ports', *Marine Pollution Bulletin*, vol 50, pp823–834

Hickley, P. and Chare, S. (2004) 'Fisheries for non-native species in England and Wales: Angling or the environment?', *Fisheries Management and Ecology*, vol 11, pp203–212

Hikita, T. (1961) 'On a new habitat of a cyprinid fish, *Pseudorasbora parva* (Temminck et Schlegel) in Hokkaido, Japan', *Scientific Reports of the Hokkaido Salmon Hatchery*, vol 16, pp91–92

Hikita, T. (1993) 'On the recent distribution of two small cyprinoid fishes, *Pseudorasbora parva pumila* (Miyadi) and *P. parva parva* (Temmink and Schlegel) in Hokkaido Island, Japan', *Scientific Reports of the Hokkaido Salmon Hatchery*, vol 18, pp113–116

Hliwa, P., Martyniak, A., Kucharczyk, D. and Sebestyén, A. (2002) 'Food preferences of juvenile stages of *Pseudorasbora parva* (Schlegel, 1842) in the Kis-Balaton Reservoir', *Archives of Polish Fisheries*, vol 10, pp121–127

Kapusta, A., Bogacka-Kapusta, E. and Czarnecki, B. (2008) 'The significance of stone moroko *Pseudorasbora parva* (Temminck and Schlegel) in the small-sized fish assemblages in the littoral zone of the heated Lake Lichenskie', *Archives of Polish Fisheries*, vol 16, pp49–62

Katano, O. and Maekawa, K. (1997) 'Reproductive regulation in the female Japanese minnow, *Pseudorasbora parva* (Cyprinidae)', *Environmental Biology of Fishes*, vol 49, pp197–205

Kennedy, C. R. (2007) 'The pathogenic helminth parasites of eels', *Journal of Fish Diseases*, vol 30, pp319–334

Krotas, R. (1971) *Freshwater Fish of Lithuania* [in Lithuanian], Mintis, Vilnius

Li, S. (1982) *Studies on Zoogeographical Divisions for Freshwater Fishes of China*, Science Press, Beijing

Libosvárský, J., Baruš, V. and Štěrba, O. (1990) 'Facultative parasitism of *Pseudorasbora parva* (Pisces)', *Folia Zoologica*, vol 39, pp355–360

Ling, N. (2002) 'Rotenone – a review of its toxicity and use for fisheries management', *Science for Conservation*, vol 211, p40

Lintermans, M. (2000) 'recolonization by the mountain galaxias *Galaxias olidus* of a montane stream after the eradication of rainbow trout *Oncorhynchus mykiss*', *Marine and Freshwater Research*, vol 51, pp799–804

Lockett, M. M. (1998) 'The effect of rotenone on fishes and its use as a sampling technique: A survey', *Zeitschrift für Fischkunde*, vol 5, pp13–45

Meadows, B. S. (1973) 'Toxicity of rotenone to some species of coarse fish and invertebrates', *Journal of Fish Biology*, vol 5, pp155–163

Molnár, K. (1967) 'Újabb kellemetlen vendég érkezett hazai vizeinkbe', *Halászat*, vol 13, pp171

Muchacheva, V. A. (1950) 'K biologii amurskogo Čebačka *Pseudorasbora parva* (Schlegel)', *Trudy Amurskoj Ichtiologičeskoj Ekspedicji, 1945–1949*, vol 1, pp365–374

Myers, J. H., Savoie, A. and van Randen, E. (1998) 'Eradication and pest management', *Annual Review of Entomology*, vol 43, pp471–491

Myers, J. H., Simberloff, D., Kuris, A. M. and Carey, J. R. (2000) 'Eradication revisited: Dealing with exotic species', *Trends in Ecology & Evolution*, vol 15, pp316–320

Nagata, T., Ha, J. Y. and Hanazato, T. (2005) 'The predation impact of larval *Pseudorasbora parva* (Cyprinidae) on zooplankton: A mesocosm experiment', *Journal of Freshwater Ecology*, vol 20, pp757–763

Nakamura, M. (1969) *Nihon no Koikagyorui, Shigenkagaku Series 4 [Japanese Cyprinids, Resource Science Series 4, in Japanese]*, Midori, Tokyo

Nichols, J. T. (1925) *Some Chinese Fresh-water Fishes. VII New Carps of the Genera Vacorhinus and Xenocypris. VIII. Carps Referred to the Genus Pseudorasbora. IX. Three New Abramidin Carps*, American Museum Novitates, New York

Nichols, J. T. (1929) *Some Chinese Freshwater Fishes*, American Museum Novitates, New York

Nichols, J. T. (1943). *The Fresh-Water Fishes of China*, American Museum Novitates, New York

Pinder, A. C. and Gozlan, R. E. (2003) 'Sunbleak and topmouth gudgeon – two new additions to Britain's freshwater fishes', *British Wildlife*, vol 15, pp77–83

Pinder, A. C., Gozlan, R. E. and Britton, J. R. (2005) 'Dispersal of the invasive topmouth gudgeon, *Pseudorasbora parva* in the UK: A vector for an emergent infectious disease', *Fisheries Management and Ecology*, vol 12, pp411–414

Pollux, B. J. A. and Korosi, A. (2006) 'On the occurrence of the Asiatic cyprinid *Pseudorasbora parva* in the Netherlands', *Journal of Fish Biology*, vol 69, pp1575–1580

Rayner, T. S. and Creese, R. G. (2006) 'A review of rotenone use for the control of non-indigenous fish in Australian fresh waters, and an attempted eradication of the noxious fish, *Phalloceros caudimaculatus*', *New Zealand Journal of Marine and Freshwater Research*, vol 40, pp477–486

Rosecchi, E., Crivelli, A. J. and Catsadorakis, G. (1993) 'The establishment and impact of *Pseudorasbora parva*, an exotic fish species introduced into Lake Mikri Prespa (North-Western Greece)', *Aquatic Conservation: Marine and Freshwater Ecosystems*, vol 3, pp223–231

Rosecchi, E., Thomas, F. and Crivelli, A. J. (2001) 'Can life-history traits predict the fate of introduced species? A case study on two cyprinid fish in southern France', *Freshwater Biology*, vol 46, pp845–853

Stott, B. (1977) 'On the question of the introduction of the grass carp (*Ctenopharyngodon ideila* Val.) into the United Kingdom', *Fisheries Management*, vol 8, pp63–71

Sunardi, Asaeda, T. and Manatunge, J. (2005) 'Foraging of a small planktivore (*Pseudorasbora parva*: Cyprinidae) and its behavioural flexibility in an artificial stream', *Hydrobiologia*, vol 549, pp155–166

Sunardi, Asaeda, T. and Manatunge, J. (2007a) 'Physiological responses of topmouth gudgeon, *Pseudorabora parva*, to predator cues and variation of current velocity', *Aquatic Ecology*, vol 41, pp111–118

Sunardi, Asaeda, T., Manatunge, J. and Fujino, T. (2007b) 'The effects of predation risk and current velocity stress on growth, condition and swimming energetics of Japanese minnow (*Pseudorasbora parva*)', *Ecological Research*, vol 22, pp32–40

Takahashi, K. (1997) *Present Status and Conservation of Japanese Endangered fish*, Midori, Tokyo

Takimoto, G. (2009) 'Early warning signals of demographic regime shifts in invading populations', *Population Ecology*, vol 51, pp419–426

Trombitskiy, I. D. and Kakhovskiy, A. E. (1987) 'On the facultative parasitism of the chebachok, *Pseudorasbora parva*, in fish ponds', *Journal of Ichthyology*, vol 27, pp180–182

Wang, D. (1995) 'The changes of fish fauna and protections of aboriginal fishes in the Tarim River', *Arid Zone Research*, vol 12, pp54–59

Weber, E. (1984) 'Die Ausbreitung der Pseudokeilfleckbarben im Donauraum' [in German], *Österreichs Fischerei*, vol 37, pp63–65

Willis, K. and Ling, N. (2000) 'Sensitivities of mosquitofish and black mudfish to a piscicide: Could rotenone be used to control mosquitofish in New Zealand wetlands?', *New Zealand Journal of Zoology*, vol 27, pp85–91

Wu, Y. and Wu, C. (1991) *The Fishes of the Qinghai-Xizang Plateau, Chengdu, Sichuan*, Publishing House of Science and Technology, Chengdu

Xie, S., Cui, Y., Zhang, T. and Li, Z. (2000) 'Seasonal patterns in feeding ecology of three small fishes in the Biandantang Lake, China', *Journal of Fish Biology*, vol 57, pp867–880

Xie, S., Zhu, X., Cui, Y., Wootton, R. J., Lei, W. and Yang, Y. (2001) 'Compensatory growth in the gibel carp following feed deprivation: Temporal patterns in growth, nutrient deposition, feed intake and body composition', *Journal of Fish Biology*, vol 58, pp999–1009

Yan, Y. and Chen, Y. (2009) 'Variations in reproductive strategies between one invasive population and two native populations of *Pseudorasbora parva*', *Current Zoology*, vol 55, pp56–60

Ye, S. W., Li, Z. J., Lek-Ang, S., Feng, G. P., Lek, S. and Cao, W. X. (2006) 'Community structure of small fishes in a shallow macrophytic lake (Niushan Lake) along the middle reach of the Yangtze River, China', *Aquatic Living Resources*, vol 19, pp349–359

Záhorská, E., Kováč, V., Falka, I., Beyer, K., Katina, S., Copp, G. H. and Gozlan, R. E. (2009) 'Morphological variability of the Asiatic cyprinid, topmouth gudgeon *Pseudorasbora parva*, in its introduced European range', *Journal of Fish Biology*, vol 74, pp167–185

Zavaleta, E. S., Hobbs, R. J. and Mooney, H. A. (2001) 'Viewing invasive species removal in a whole-ecosystem context', *Trends in Ecology & Evolution*, vol 16, pp454–459

Zhang, T., Cui, Y., Fang, R., Xie, S. and Li, Z. (1998a) 'Population biology of topmouth gudgeon (*Pseudorasbora parva*) in Bao'an Lake – I. Age and growth [In Chinese]', *Acta Hydrobiology Sinica*, vol 22 (Supplement), pp139–146

Zhang, T., Cui, Y., Li, Z., Xie, S. and Fang, R. (1998b) 'The ecology of topmouth gudgeon, *Pseudorasbora parva*, in Bao'an Lake – III. Food habits [In Chinese]', *Acta Hydrobiologica Sinica*, vol 22 (Supplement), pp157–167

24

Salmo trutta L. (brown trout)

Angus McIntosh, Peter McHugh and Phaedra Budy

Introduction

In comparison to most other freshwater invasive species, *Salmo trutta* (brown trout) (Plate 24.1) pose a paradox. They are regarded as one of 'world's worst invasive alien species' by international conservation authorities for their impact on native species (Lowe et al, 2000), but they are simultaneously loved by recreational anglers (Pascual et al, 2009). The complex social context surrounding brown trout necessitates that recreational angling and conservation values must be reconciled if non-native brown trout are to be managed effectively (Peterson et al, 2008a; Pascual et al, 2009; Cowx et al, 2010).

A History of Brown Trout Introductions and Invasions

Brown trout occur naturally throughout Europe, western Asia and North Africa, but can now be found on every continent excluding Antarctica, as well as many islands (MacCrimmon and Marshall, 1968; Figure 24.1). This range expansion was largely intentional and angler driven, and occurred in three distinct phases. From the 1860s to *c.*1925, brown trout embryos obtained from cultured European populations were transferred overseas to meet demand for fisheries in European colonies. Introductions during this early phase comprised both direct transfers (i.e. from native sources) and releases from naturalized, non-native populations. Releases occurred primarily in high latitude regions, and establishment success was high

(MacCrimmon and Marshall, 1968; MacCrimmon et al, 1970).

During the second phase (*c.*1925–1960s), introductions continued within countries with established populations, and also commenced in increasingly marginal and/or less populous regions. For instance, Venezuela and Papua New Guinea, two equatorial countries with potentially suitable highland areas, received multiple shipments of brown trout in the 1940s and 1950s (MacCrimmon et al, 1970). Similarly, brown trout were liberated in subtropical portions of the US (e.g. Hawaii) between 1930 and 1960. Not surprisingly, the establishment of naturally reproducing populations was less successful during this period, and the prevalence of 'put-and-take' fisheries (i.e. populations maintained only through release of cultured trout) increased markedly (Crawford and Muir, 2008). By the end of the 1960s, MacCrimmon and Marshall (1968, p2542) regarded 'the dissemination of European brown trout … to have been so complete … that most areas of the world capable of supporting significant natural populations have now received introductions'. The last phase of range expansion (1960s to present) is one characterized by the maintenance of some local populations through stocking programmes and a general cessation of broad-scale introduction efforts.

Available accounts (Figure 24.1) indicate that non-native brown trout populations are now established throughout Africa (Ethiopia, Kenya, Lesotho, Madagascar, Malawi, South Africa, Swaziland, Tanzania, Zimbabwe), Australasia (Australia, New Zealand, Papua New Guinea), the Americas (Argentina,

Bolivia, Canada, Chile, Peru, the US), the sub-Antarctic islands (e.g. the Falkland and Kerguelen islands), and previously uninhabited portions of Asia (Bhutan, China, Cyprus, India, Iran, Japan, Pakistan, areas of the Russian Federation, Sri Lanka). This list probably represents the minimum non-native range, as undocumented transfers have occurred in other countries (e.g. Mexico; Hendrickson et al, 2002) (see MacCrimmon and Marshall, 1968; MacCrimmon et al, 1970; Welcomme, 1988, and references therein for further distributional detail). That the list includes such far-flung places as the Falkland Islands is testament to the vigour with which brown trout introductions around the world have been pursued.

Brown trout range expansion has followed two main pathways: (1) initial release, followed by successful local establishment; and (2) secondary spread from naturalized populations (i.e. dispersal and colonization). At the broadest geographic scale, the first pathway is clearly of ultimate importance. At a finer scale, however, the fish's vagility combined with its tendency towards life history plasticity (see below) has allowed it to become regionally ubiquitous over relatively short periods. In the Kerguelen Islands, for example,

introductions of brown trout into 3 rivers facilitated the colonization of 16 coastal rivers in less than 40 years (Launey et al, 2010) and similar reports exist from Patagonian South America (e.g. Valiente et al, 2010). In addition to the overseas transfers summarized above, brown trout of cultured origin have been released liberally within the bounds of its native European range. By stocking these fish, non-native genetic material has been introduced throughout Europe (e.g. Spain; Almodovar et al, 2006), and has created a substantial conservation challenge.

In as much as the worldwide spread of brown trout is remarkable, places where they are not present are the most critical for the conservation of indigenous biodiversity. Quantitative estimates of the spatial extent of invader-free habitat are hard to come by, though available evidence suggests the extent is limited where brown trout are regionally abundant. In our Canterbury, New Zealand study sites for instance, <5 per cent of fish-inhabited streams are trout free (McIntosh et al, 2010). Further, there are no trout-free lakes bigger than 270ha in New Zealand (Chadderton, 2003), although Stewart Island, New Zealand's third largest island, has so far been spared brown trout invasion.

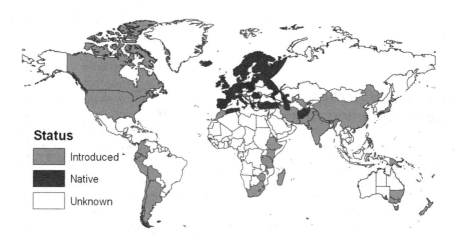

Note: Due to uncertainty associated with incomplete, conflicting, and/or anecdotal accounts for a number of locations, only records characterized by a high likelihood of occurrence (e.g. 'established' or 'probably established' in the FAO database) are depicted for the non-native range. This also means the distribution maps depict the 'minimum' introduced range.

Source: The non-native range is based on data from Fishbase (www.fishbase.org), the IUCN Invasive Species Specialist Group's Global Invasive Species Database (www.issg.org/database/welcome), and the FAO's database of introduced aquatic species (FAO, 2010), whereas the native range was delineated from MacCrimmon and Marshall (1968)

Figure 24.1 *Global distribution of brown trout within its native (dark grey) and introduced (light grey) range*

Niche Characteristics of Brown Trout

Brown trout have been presented with a genetic deck of cards that has allowed them to occupy both wide fundamental and realized niches relative to their cold-water relatives. They inhabit a diversity of habitats during their lifecycle, including estuaries, marine habitats, fjords, lakes, ponds, streams and rivers. Similarly, brown trout demonstrate an incredible array of life history expressions and plasticity (Klemetsen et al, 2003). As such, they can be diadromous, adfluvial, fluvial and all combinations of residential, migratory, sedentary and territorial within each of those broader life history categories. Further, at any given time and within the same population, several different life history expressions are often apparent, while across time and space, different expressions are readily turned on and off over relatively short evolutionary timescales (e.g. Ayllon et al, 2006). Clearly this plasticity and wide niche space has facilitated their successful invasion and establishment worldwide (Valiente et al, 2010). The basic niche requirements and preferences of the brown trout summarized below indicate the characteristics likely to have enhanced their successful circum-global invasion and establishment.

Although brown trout prefer cool water temperatures and moderately good water quality, compared to many cold-water fishes they can survive, grow and reproduce under a wide range of physical habitat conditions (i.e. fundamental niche). Their optimal temperature range for growth lies between 12 and 18°C; however, they can attain high densities where mean summer temperatures exceed 20°C (e.g. Rahel and Nibbelink, 1999), and their maximum thermal tolerance is near 30°C (Hari et al, 2006). Further, brown trout demonstrate considerable local adaption in thermal regulation of growth (e.g. Lobon-Cervia and Rincon, 1998). Their wide fundamental niche not only influences their success in novel habitats but may affect their competitive relations to other fishes (McHugh and Budy, 2005).

In lotic systems, suitable velocities, depths, substrata and temperatures for brown trout are highly dependent on life stage, season and diel period (Klemetsen et al, 2003). Habitat suitability criteria (Table 24.1) used in in-stream flow assessments give basic ranges for brown trout (Raleigh et al, 1986), and are assumed to apply in the absence of other abiotic or biotic factors (e.g. presence of a competitor). Brown trout are capable of utilizing a wide range of substrate types but generally prefer a minimum of 10 per cent cover (Hubert et al, 1996). In lentic systems, brown trout can grow well in nearly any standing water of suitable temperature and oxygen concentration (>5mg L^{-1}) (Elliott, 1994).

As autumn spawners, brown trout fry emerge in early spring. This emergence timing means winter and spring conditions determine early survival, and may ultimately limit their invasion success. In some high elevation streams, for instance, cold winter temperatures may delay fry emergence until during, or soon after the onset of peak snowmelt floods, making redds and emerging fry susceptible to scour, displacement and mechanical damage (e.g. Wood and Budy, 2009). Accordingly, strong relationships between peak stream flow and brown trout recruitment have been firmly established for many populations both within their native range (e.g. Lobon-Cervia, 2007) and for related species in their invaded range (e.g. Fausch et al, 2001).

Similar to their fundamental niche, brown trout also appear to have a wide realized niche and tolerate a broad range of biological conditions. Although brown trout production is greatest in alkaline, prey-rich streams (e.g. Almodovar et al, 2006), they also occupy and perform well in streams and lakes with low productivity (e.g. Dineen et al, 2007). Brown trout are primarily visual, diurnal feeders with an omnivorous diet, generally shifting towards larger prey as their body size increases (Klemetsen et al, 2003). They often switch to piscivory at 150–300mm of body length (e.g. Mittelbach and Persson, 1998), and when brown trout do include fish in their diet, their growth rate and optimum temperature for growth both increase markedly (e.g. Elliott and Hurley, 2000); an energetic optimization that may have facilitated their success as a non-native species.

As highly territorial, omnivorous and opportunistic feeders, brown trout are also known to be superior competitors in many situations. Intense agonistic interactions between territorial conspecifics – charges, bites and chases for example – lead to the establishment and maintenance of stable dominance hierarchies within native brown trout populations (e.g. Alanara et al, 2001). In its native Europe, brown trout are similarly effective at displacing other species, such as Atlantic salmon (Kennedy and Strange, 1986). This tendency

Table 24.1 *Habitat suitability criteria for four primary life stages of brown trout*

Life stage	Velocity (m/s)	Depth (m)	Substrate diameter (mm)	Temperature (°C)
Spawning and egg incubation	0.1–1.0	>0.1	25.4–81.3	3.3–12.8
Fry	0.1–0.5	0.3–0.8	<64	4.4–20.6
Juveniles	0–0.6	0.3–1.2	<250	3.9–26.7
Adults	0–0.6	0.2–1.3	<250	4.4–24.4

Note: Ranges shown for each factor are based on habitat utilization studies and effects on the growth, survival, or biomass of the species by life stage.

Source: Based on Raleigh et al (1986)

towards territoriality and aggression predisposes brown trout to competitive superiority outside of its native range.

The combination of niche characteristics outlined above not only allows brown trout to successfully invade new habitats, but also to become incredibly abundant under certain conditions. For example, we have observed brown trout densities in excess of one fish m^{-2} in our Utah, USA and South Island, New Zealand research sites, a number that rivals even the most productive populations inhabiting pristine European systems (e.g. Budy et al, 2008). Endogenous population regulation can limit abundance, however, particularly during early life stages (e.g. Lobon-Cervia, 2009). Exogenous factors that vary dramatically across space and time, such as hydrologic disturbance and extreme winter conditions, can also influence recruitment success (Milner et al, 2003). Nevertheless, high fecundity and strong recruitment in years of optimal conditions afford brown trout considerable resilience (e.g. Lobon-Cervia, 2008), an important population level trait for an invader.

Impacts of Brown Trout in Invaded Systems

Studies quantifying the impacts of brown trout in invaded habitats are a relatively recent phenomenon. This is largely due to a lack of information on the distribution and diversity of native fishes when brown trout were first liberated. Within the last 15 years, however, researchers have studied impacts for several taxonomic groups (fish; Table 24.2; and amphibians; Gillespie, 2001) at multiple levels of biological organization (Townsend and Simon, 2006), and in widely differing locales. Accordingly, mounting evidence for negative effects of brown trout on native fishes, the affected taxa receiving the most research attention, come in three forms:

1 distributional evidence suggesting historical displacement;
2 observational or experimental studies assessing mechanisms of impact; and
3 temporal datasets documenting decline following invasion.

First, catchment-level or broader-scale fish community datasets exhibit classic allopatric distributional patterns; natives are found where brown trout are not, and vice versa, with little overlap (e.g. Waters, 1983; Townsend and Crowl, 1991) (Plate 24.2a). Typically, brown trout are found in downstream areas, and native fish occur only above barriers (e.g. waterfalls) or in small headwater streams that are marginal (i.e. too cold) for brown trout occupancy (Hasegawa and Maekawa, 2008). Natives can also persist in habitats downstream of trout invaded reaches, particularly where warm temperatures and low flow extremes are the norm or streams are large (Leprieur et al, 2006; Habit et al, 2010; Woodford and McIntosh, 2011). Thus, disjunct native fish distributions, bounded by the presence of brown trout in habitats that would otherwise be suitable, strongly indicate historical displacement has occurred.

Experiments indicate that brown trout are capable of displacing native fishes through a combination of predator–prey, competitive and/or reproductive (i.e. genetic) interactions. These effects, alone or in combination, have been documented for more than 30

Note: In brown water streams (top line) brown trout are typically dark coloured with the traditional brown spots. However, in braided rivers (bottom line), influenced by glacial silt, they can be quite silvery in colouration (but still with spots), especially if the fish is marine migratory, as this 2.5kg fish probably was. In other locations (middle) they take on an intermediate appearance.

Source: Hamish Greig

This colour plate was funded by Ecology Centre, Utah State University.

Plate 24.1 *Colour variation associated with trout* (Salmo trutta) *from various freshwater situations in South Island, New Zealand, where they were caught*

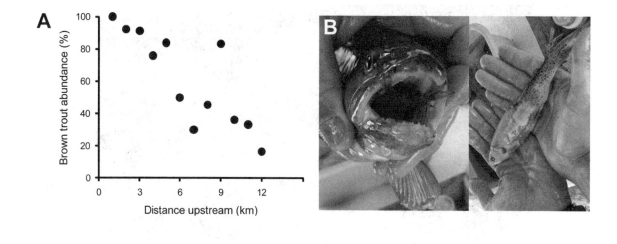

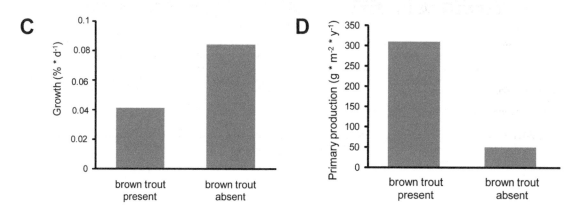

Note: (a) An example of a distribution pattern suggesting historic or ongoing invasion of brown trout associated with displacement of a native fish (white-spotted char, *Salvelinus leucomaenis*) from Monbetsu Stream, Japan. Individual points represent the abundance of brown trout relative to that of white-spotted char (expressed as per cent brown trout in total catch) from electrofishing within 200m sampling reaches along a 12km longitudinal gradient. (b) Evidence of brown trout predation on native fish (Bonneville cutthroat trout, *Oncorhynchus clarkii utah*) from a Utah, US, river. When sampled using gastric lavage, the 300mm brown trout pictured on the left regurgitated a 193mm cutthroat trout, as well as a snake. (c) Experimental evidence for negative competitive effects of brown trout on a native fish. In a field experiment that manipulated the local abundance of brown trout in 300m stream reaches, brown trout reduced cutthroat trout growth by half. (d) By suppressing invertebrate abundance and grazing behaviour, predation by brown trout indirectly influences primary producer (i.e. algae) standing biomass and production, relative to native fish. Data are for paired streams with and without brown trout from South Island, New Zealand.

Source: (a) recreated from Hasegawa and Maekawa (2008) with permission; (b) photograph courtesy of D. Weber; (c) reproduced from McHugh and Budy (2006) with permission; (d) redrawn from Huryn (1998)

Plate 24.2 *Impacts of non-native brown trout in invaded freshwater ecosystems*

Table 24.2 *Examples of native fishes affected by non-native brown trout invasions*

Family	Species	Mechanism(s) of impact	Type of evidence	Country	References
Balitoridae	*Barbatula barbatula*	predation	survey	Poland	Penczak, 1999
Catastomidae	*Catastomus platyrhynchus, C. microps*	predation	survey, experiment	US	Moyle and Marciochi, 1975; Olsen and Belk, 2005
Cottidae	*Cottus cognatus, C. gobio*	predation, competition	survey, experiment	Poland, US	Penczak, 1999; Zimmerman and Vondracek, 2006
Cyprinidae	*Gobio gobio, Lepidomeda aliciae, L. vittata, Phoxinus* sp., *Richardsonius balteatus*	predation	survey, experiment	Canada, US, Poland	Penzcak, 1999; Nannini and Belk, 2006; Nasmith et al, 2010
Galaxiidae	*Aplochiton taeniatus, A. zebra, Brachygalaxias bullocki,* several members of New Zealand's *Galaxias* sp. complex, *G. auratus, G. maculatus, G. olidus, G. platei, Neochanna burrowsius*	predation, competition, indirect effects	survey, experiment	Argentina, Australia, Chile, New Zealand, South Africa	McDowall, 2006; Stuart-Smith et al, 2008; McIntosh et al, 2010 and references therein; Habit et al, 2010
Petromyzontidae	*Lampetra planeri*	predation	survey	Poland	Penczak, 1999
Percichthydae	*Maccullochella macquariensis, Macquaria australasica*	predation, competition	survey	Argentina, Australia	Pascual et al, 2002
Salmonidae	*Salmo trutta*[1], *Salmo salar, Salmo marmoratus, Salvelinus fontinalis, Salvelinus leucomaenis, Salvelinus malma, Oncorhynchus aguabonita, O. apache, O. clarkii* subsp., *O. masou*	competition, predation, reproductive interactions	survey, experiment	Canada, Japan, Slovenia, US	Moyle, 1976; Waters, 1983; Mayama, 1999; Jug et al, 2005; McHugh and Budy, 2005, 2006; Shemai et al, 2007; Kitano et al, 2009

Note: [1] genetically distinct native brown trout stocks affected by the introduction of non-indigenous hatchery genotypes.

fish species belonging to at least eight families (Table 24.2). Predatory interactions are suspected to have caused some of the most dramatic declines (Jackson et al, 2004; McIntosh et al, 2010) (Plate 24.2b), especially in systems where brown trout attain a large size and/or native species lack evolutionary history with predators. Field and laboratory trials, for instance, illustrate that native fishes survive poorly and exhibit altered behaviours when exposed to predation by brown trout (Nannini and Belk, 2006). In other cases, competition is regarded a potent mechanism of brown trout impact, particularly where the affected native species belong to the family Salmonidae (e.g. US; McHugh and Budy, 2005). As in its native range, brown trout have been shown to dominate over several salmonids in behaviour, growth, and habitat use experiments (McHugh and Budy, 2005, 2006; Hasegawa and Maekawa, 2008) (Plate 24.2c); other experiments illustrate that brown trout can also displace more distantly related species (e.g. Galaxiidae; McIntosh et al, 1992). Although less studied, reproductive interactions with brown trout have also contributed to the decline of native fishes, especially for closely allied species such as Atlantic salmon (*Salmo salar*; Verspoor, 1988) and marble trout (*S. marmoratus*; Jug et al, 2005), but also from sterile hybrids arising from more distant matings (e.g. *Salvelinus leucomaenis* × *Salmo trutta*; Kitano et al, 2009). Yet, the most damaging genetic effects of brown trout introductions may exist within its native range, where repeated transfers of non-indigenous stocks have caused widespread introgression among formerly isolated brown trout populations (e.g. Almodovar et al, 2006).

Beyond the suite of direct interactions outlined above, brown trout probably affect native fishes through more complex and/or indirect pathways and have impacts that reach far beyond fishes. Most notably, trout have been shown to restructure the dynamics of ecosystems through strong cascading effects on algal communities (McIntosh and Townsend, 1996). As a consequence, non-native brown trout can modify primary and secondary production (Huryn, 1998) and biogeochemical processes (Simon et al, 2004), relative to invader-free systems (Plate 24.2d). Taken together, these studies indicate brown trout fundamentally alter the aquatic ecosystems they invade.

The most telling evidence implicating brown trout in the loss of biodiversity comes from 'natural experiments' or before–after studies that document the decline of native taxa following invasion. A study from Poland's Pilica River catchment (Penczak, 1999), for example, demonstrates that three species became locally extinct (or nearly so) within a decade of brown trout establishment. Studies also show that such trends can work in the reverse direction when brown trout are removed (Knapp et al, 2007).

Lastly, although available evidence illustrates that brown trout seriously affect many native species, researchers are increasingly recognizing the potential for variable and context-dependent invasion outcomes. Within streams there is considerable variation in the extent to which native species are excluded by

non-native brown trout. Landscape-level attributes (e.g. the spatial arrangement of native fish source populations within a catchment; Woodford and McIntosh, 2011), as well as hydrologic (i.e. flooding and river drying; McIntosh et al, 2010; Lepriuer et al, 2006) and thermal variability (McHugh and Budy, 2005), enable native species to co-occur with brown trout in many places, even though they are eliminated elsewhere. Such contingencies may provide avenues for conserving native biodiversity where it is threatened by brown trout invasion.

Management of Brown Trout Impacts

Efforts directed at the management of non-native brown trout need to be increased. There are few published accounts of attempted brown trout control, though isolated efforts have been undertaken and/or are ongoing in Australia, New Zealand and the US. However, given that two other salmonid invaders (rainbow trout, *Oncorhynchus mykiss*; brook trout, *Salvelinus fontinalis*; see Dunham et al, 2002 for a review) are adversely affecting many of the world's cold-water ecosystems, several active intervention and policy tools exist for managing non-native trout in novel environments.

Isolation management

Protecting native species through isolation management constitutes a powerful first order, 'protect what's left' intervention strategy. This is typically achieved by maintaining dispersal barriers downstream of habitats that have not yet been invaded by trout, or that have undergone recent trout removal (discussed below). Many native fish populations in the western US (e.g. Fausch et al, 2006) and New Zealand (e.g. Townsend and Crowl, 1991), for instance, exist mainly within invader-free habitats above waterfalls or unintended barriers (e.g. road culverts). Accordingly, these distributional observations have inspired the construction of artificial barriers and subsequent creation of refuge habitats (e.g. Thompson and Rahel, 1998). Thus, when applied in conjunction with eradication and population translocation, isolation management can also be used to expand the range of impacted native species (e.g. cutthroat trout, *Oncorhynchus clarkii*; Harig et al, 2000).

Though effective, there are also risks inherent to isolating populations for conservation purposes (Peterson et al, 2008b; Fausch et al, 2009). In particular, native fish can become locally extinct and/or suffer genetic effects when above-barrier refuge populations are too small (e.g. Harig and Fausch, 2002). Similarly, populations possessing both migratory and non-migratory components can experience reduced life history diversity if upstream passage is eliminated (Fausch et al, 2009). To this end, decision support tools that formalize these risks relative to those arising from trout invasion (Peterson et al, 2008b) have been produced, facilitating more systematic, catchment-scale management of non-native trout.

Chemical and mechanical removal methods

In conjunction with barrier construction or otherwise, there are two general classes of methods for eradicating or controlling the abundance of non-native trout: chemical and mechanical removal methods. The former includes piscicides such as rotenone and antimycin that are toxic to fish, whereas the latter includes methods such as electrofishing, trapping and gill netting. Piscicides have been used to eliminate non-native trout in both Australia and the US, and have proven effective when applied properly in a small stream setting (Lintermans, 2000). Indeed, post-treatment recovery by native fish can be quite rapid (e.g. within two to three years; Lintermans, 2000), particularly if colonist pools are adequate and barriers to reinvasion by trout are installed. Despite their promise, however, piscicides are underutilized in much of the brown trout's non-native range due to their high cost and toxicant status (Finlayson et al, 2005). Further, chemical methods have other shortcomings that render mechanical methods better suited to certain situations, such as impacts on non-target species (e.g. Hamilton et al, 2009) and limited social acceptability (Finlayson et al, 2005).

The suite of mechanical removal methods that has been used to control non-native trout to date includes electrofishing, trapping and gill netting. The first two techniques appear particularly suitable for reducing, but

not eliminating, non-native trout populations in small streams. In our own work in Utah, for example, brown trout removal from 5km of stream using 1-pass electrofishing significantly lowered adult brown trout numbers, and led to the first successful native trout recruitment event observed in over ten years. Passive trapping methods (e.g. hoop nets) are also useful and, while less effective than electrofishing, can be applied at a substantially lower cost (Lamansky et al, 2009). Beyond streams, gill nets have become a standard for the suppression of invasive trout in lake ecosystems. At the small extreme (lakes <0.1km^2), gill nets have been used to reduce trout abundance to near zero levels in alpine lakes in California, which has helped reverse declines of rare species (Knapp et al, 2007). Gill nets have also been used to reduce non-native trout abundance in larger systems (e.g. Yellowstone Lake, Wyoming, 352km^2), albeit with lower success (Gresswell, 2009). Although piscicides are superior when eradication is the goal, mechanical removal programmes have proven effective at reducing the local abundance of non-native trout (e.g. Thompson and Rahel, 1996; Lamansky et al, 2009). Thus, removals can greatly enhance the persistence of native fishes, when applied consistently and as part of a long term management strategy (Ruzycki et al, 2003; Peterson et al, 2008a).

Policy and education measures

A number of policy tools are available for reducing non-native trout impacts within invaded systems. These include public statutes, regulations associated with fishing licensure and voluntary agreements adopted by sports fish managers. Where used, the first policy tool has typically been applied to curtail new introductions. In particular, the congressional or parliamentary acts of several countries explicitly define certain non-native species as pests, and make their import, transfer and/or possession criminal acts (e.g. McGeoch et al, 2010). For example, the New Zealand Biosecurity Act (NZBA) of 1993, a model for aggressive invasive species legislation (e.g. Simberloff et al, 2005), imposes fines of up to NZ$100,000 (plus imprisonment) for the release of 'unwanted' species. At present, however, the NZBA register lacks mention of brown trout, and outside of South Africa, no other national invasive species policy classifies them as a nuisance to our knowledge. On the contrary, naturalized brown trout populations are often protected under environmental or fish and game legislation (McDowall, 2006). As a consequence, government policies can also foster interagency conflicts that reduce – or even prevent – action aimed at minimizing impacts of non-native brown trout (Jackson et al, 2004; Macchi et al, 2008). Broad national policies clearly have the potential to effect change in campaigns against non-native species, but at present they are an underutilized tool for brown trout management. Nevertheless, voluntary agreements to not extend the range of brown trout can be effective in preventing non-sanctioned sports fish spread where there is sufficient clout for managers to enforce restrictions. However, our experience is that policy tools are not effective in preventing continued salmonid stocking of areas where their populations appear to perform poorly or are more than likely not self-sustaining.

Beyond public statutes, recreational fishing regulations provide a means via which fishery managers can influence trends in non-native salmonid abundance. Overexploitation problems associated with recreational angling in the brown trout's native range (e.g. Almodovar and Nicola, 2004) in particular, suggest that fishing regulations facilitating greater harvest rates can impose population control. For this reason, bag limits for non-native trout have been liberalized in some places, and agencies have made it outright illegal for anglers to release non-natives alive in others (e.g. Martinez et al, 2009).

While each of the aforementioned policy and management tools has a role to play, any campaign to manage brown trout as a pest fish will benefit from a commensurate public education effort. Such efforts will provide the greatest benefits if user groups (i.e. anglers) are actively engaged and informed of relevant scientific, conservation and management issues, and given a stake in decision-making processes (e.g. Cowx et al, 2010). Anglers may otherwise perceive brown trout control efforts – no matter how localized – as the

beginning of a process that leads to wholesale eradication (Chadderton, 2003).

Controversies and uncertainties

To effectively minimize the impacts of non-native brown trout in the future, managers and scientists will have to confront head-on at least four controversies or uncertainties. First, the full scope of the brown trout's impact within invaded habitats, both in terms of mechanisms and geography, has not yet been fully characterized. For example, the majority of work published to date has come from only six countries (Argentina, Australia, Chile, Japan, New Zealand and the US). Second, the extent to which brown trout are actively invading new habitats needs to be clarified, since many perceive current distributional boundaries to be static phenomena, and this is very unlikely to be the case (McIntosh et al, 2010; Woodford and McIntosh, 2011). Third, sourcing money for the management of brown trout as a pest species, and allocating this resource optimally, is a challenge that will have to be squarely addressed. Finally, much controversy surrounds the concept of managing sought-after recreational species as a pest, and this may prove to be the greatest impediment to organized efforts aimed at their control. Meeting this challenge should prove easier in regions possessing one or more native species of similar recreational value (e.g. Japan, the US), compared to those lacking salmonid angling opportunity in the absence of brown trout (e.g. Australia and New Zealand). Further, control efforts will probably be more controversial in areas with non-native brown trout fisheries that comprise a non-trivial portion of local economies (Pascual et al, 2009). Despite these challenges the plight of indigenous aquatic ecosystems affected by brown trout demands solutions as the weight of evidence indicates that non-native populations of brown trout have adversely affected freshwater biodiversity in many regions.

References

Alanara, A., Burns, M. D. and Metcalfe, N. B. (2001) 'Intraspecific resource partitioning in brown trout: The temporal distribution of foraging is determined by social rank', *Journal of Animal Ecology*, vol 70, no 6, pp980–986

Almodovar, A. and Nicola, G. G. (2004) 'Angling impact on conservation of Spanish stream-dwelling brown trout *Salmo trutta* L.', *Fisheries Management and Ecology*, vol 11, pp173–182

Almodovar, A., Nicola, G. G., Elvira, B. and García-Marín, J. L. (2006) 'Introgression variability among Iberian brown trout evolutionary significant units: The influence of local management and environmental features', *Freshwater Biology*, vol 51, pp1175–1187

Ayllon, F., Moran, P. and Garcia-Vazquez, E. (2006) 'Maintenance of a small anadromous subpopulation of brown trout (*Salmo trutta* L.) by straying', *Freshwater Biology*, vol 51, pp351–358

Budy, P., Thiede, G. P. and McHugh, P., Hansen, E. S. and Wood, J. (2008) 'Exploring the relative influence of biotic interactions and environmental conditions on the abundance and distribution of exotic brown trout (*Salmo trutta*) in a high mountain stream', *Ecology of Freshwater Fish*, vol 17, pp554–566

Chadderton, W. L. (2003) 'Management of invasive freshwater fish: Striking the right balance', in *Managing Invasive Freshwater Fish in New Zealand: Proceedings of a Workshop Hosted by Department of Conservation*, Department of Conservation, Wellington, New Zealand, pp71–83

Cowx, I. G., Arlinghaus, R. and Cooke, S. J. (2010) 'Harmonizing recreational fisheries and conservation objectives for aquatic biodiversity in inland waters', *Journal of Fish Biology*, vol 76, no 9, pp2194–2215

Crawford, S. S. and Muir, A. M. (2008) 'Global introductions of salmon and trout in the genus *Oncorhynchus*: 1870–2007', *Reviews in Fish Biology and Fisheries*, vol 18, no 3, pp313–344

Dineen, G., Harrison, S. S. C. and Giller, P. S. (2007) 'Growth, production and bioenergetics of brown trout in upland streams with contrasting riparian vegetation', *Freshwater Biology*, vol 52, pp771–783

Dunham, J., Adams, S. B., Schroeter, R. and Novinger, D. C. (2002) 'Alien invasions in aquatic ecosystems: Toward an understanding of brook trout invasions and their potential impacts on inland cutthroat trout in western North America', *Reviews in Fish Biology and Fisheries*, vol 12, pp373–391

Elliott, J. M. (1994) *Quantitative Ecology and the Brown Trout*, Oxford University Press, Oxford, UK

Elliott, J. M. and Hurley, M. A. (2000) 'Daily energy intake and growth of piscivorous brown trout, *Salmo trutta*', *Freshwater Biology*, vol 44, pp237–245

FAO (Food and Agriculture Organization of the United Nations) (2010) 'Database on introductions of aquatic species', Fisheries and Aquaculture Department, www.fao.org/fishery/dias/en, accessed 20 July 2010

Fausch, F. D., Tanguchi, Y., Nakano, S., Grossman, G. D. and Townsend, C. R. (2001) 'Flood disturbance regimes influence rainbow trout invasion success among five holoarctic regions', *Ecological Applications*, vol 11, pp1438–1455

Fausch, K. D., Rieman, B. E., Young, M. K. and Dunham, J. B. (2006) 'Strategies for conserving native salmonid populations at risk from nonnative fish invasions: Tradeoffs in using barriers to upstream movement', USDA Forest Service Rocky Mountain Research Station, RMRS-GTR-174, Ft Collins, CO

Fausch, K. D., Rieman, B. E., Dunham, J. B., Young, M. K. and Peterson, D. P. (2009) 'Invasion versus isolation: Trade-offs in managing native salmonids with barriers to upstream movement', vol 23, no 4, pp859–870

Finlayson, B., Somer, W., Duffield, D., Propst, D., Mellison, C., Pettengill, T., Sexauer, H., Nesler, T., Gurtin, S., Elliot, J., Partridge, F. and Skaar, D. (2005) 'Native inland trout restoration on national forests in the western United States: Time for improvement?', *Fisheries*, vol 30, no 5, pp10–19

Gillespie, G. R. (2001) 'The role of introduced trout in the decline of the spotted tree frog (*Litoria spenceri*) in south-eastern Australia', *Biological Conservation*, vol 100, pp187–198

Gresswell, R. E. (2009) *Scientific Review Panel Evaluation of the National Park Service Lake Trout Suppression Program in Yellowstone Lake, August 25th–29th Final Report*, USGS Northern Rocky Mountain Science Center, Bozeman, Montana, YCR–2009–05

Habit, E., Piedra, P., Ruzzante, D. E., Walde, S. J., Belk, M. C., Cussac, V. E., Gonzalez, J. and Colin, N. (2010) 'Changes in the distribution of native fishes in response to introduced species and other anthropogenic effects', *Global Ecology and Biogeography*, vol 19, no 5, pp697–710

Hamilton, B. T., Moore, S. E., Williams, T. B., Darby, N. and Vinson, M. R. (2009) 'Comparative effects of rotenone and antimycin on macroinvertebrate diversity in two streams in Great Basin National Park, Nevada', *North American Journal of Fisheries Management*, vol 29, pp1620–1635

Hari, R. E., Livingstone, D. M., Siber, R., Burkhardt-Holm, P. and Guttinger, H. (2006) 'Consequences of climatic change for water temperature and brown trout populations in alpine rivers and streams', *Global Change Biology*, vol 12, pp10–26

Harig, A. L. and Fausch, K. D. (2002) 'Minimum habitat requirements for establishing translocated cutthroat trout populations', *Ecological Applications*, vol 12, pp535–551

Harig, A. L., Fausch, K. D. and Young, M. K. (2000) 'Factors influencing success of greenback cutthroat trout translocations', *North American Journal of Fisheries Management*, vol 20, pp994–1004

Hasegawa, K. and Maekawa, K. (2008) 'Different longitudinal distribution patterns of native white-spotted charr and nonnative brown trout in Monbetsu Stream, Hokkaido, northern Japan', *Ecology of Freshwater Fish*, vol 17, pp189–192

Hendrickson, D. A., Espinosa Perez, H., Findley, L. T., Forbes, W., Tomelleri, J. R., Mayden, R. L., Nielsen, J. L., Jensen, B., Campos, G. R., Romero, A. V., van der Heiden, A., Camarena, F. and de Leon, F. J. G. (2002) 'Mexican native trouts: A review of their history and current systematic and conservation status', *Reviews in Fish Biology and Fisheries*, vol 12, pp273–316

Hubert, W. A., Martin, T. D., Gerow, K. G., Binns, N. A. and Wiley, R. W. (1996) 'Estimation of potential biomass of trout in Wyoming streams to assist management decisions', *North American Journal of Fisheries Management*, vol 16, pp821–829

Huryn, A. D. (1998) 'Ecosystem-level evidence for top-down and bottom-up control of production in a grassland stream system', *Oecologia*, vol 115, pp173–183

Jackson, J. E., Raadik, T. A., Lintermans, M. and Hammer, M. (2004) 'Alien salmonids in Australia: Impediments to effective impact management, and future directions', *New Zealand Journal of Marine and Freshwater Research*, vol 38, no 3, pp447–455

Jug, T., Berrebi, P. and Snoj, A. (2005) 'Distribution of nonnative trout in Slovenia and their introgression with native trout populations as observed through microsatellite DNA analysis', *Biological Conservation*, vol 123, pp381–388

Kennedy, G. J. A. and Strange, C. D. (1986) 'The effects of intra-specific and inter-specific competition on the survival and growth of stocked juvenile Atlantic salmon, *Salmo salar*, and resident trout, *Salmo trutta*, in an upland stream', *Journal of Fish Biology*, vol 28, no 4, pp479–489

Kitano, S., Hasegawa, K. and Maekawa, K. (2009) 'Evidence for interspecific hybridization between native white-spotted charr *Salvelinus leucomaenis* and non-native brown trout *Salmo trutta* on Hokkaido Island, Japan', *Journal of Fish Biology*, vol 74, pp467–473

Klemetsen, A., Amundsen, P. A., Dempson, J. B., Jonsson, B., Jonsson, N., O'Connell, M. F. and Mortensen, E. (2003) 'Atlantic salmon *Salmo salar* L., brown trout *Salmo trutta* L. and Arctic charr *Salvelinus alpinus* (L.): A review of aspects of their life histories', *Ecology of Freshwater Fish*, vol 12, pp1–59

Knapp, R. A., Boiano, D. M. and Vredenburg, V. T. (2007) 'Removal of nonnative fish results in expansion of a declining amphibian (mountain yellow-legged frog, *Rana muscosa*)', *Biological Conservation*, vol 135, pp11–20

Lamansky, J. A., Keeley, E. R., Young, M. K. and Meyer, K. A. (2009) 'The use of hoop nets seeded with mature brook trout to capture conspecifics', *North American Journal of Fisheries Management*, vol 29, no 1, pp10–17

Launey, S., Brunet, G., Guyomard, R. and Davaine, P. (2010) 'Role of introduction history and landscape in the range expansion of brown trout (*Salmo trutta* L.) in the Kerguelen Islands', *Journal of Heredity*, vol 101, no 3, pp270–283

Leprieur, F., Hickey, M. A., Arbuckle, C. J., Closs, G. P., Brosse, S. and Townsend, C. R. (2006) 'Hydrological disturbance benefits a native fish at the expense of an exotic fish', *Journal of Applied Ecology*, vol 43, pp930–939

Lintermans, M. (2000) 'Recolonization by the mountain galaxias *Galaxias olidus* of a mountain stream after eradication of rainbow trout *Oncorhynchus mykiss*', *Marine and Freshwater Research*, vol 51, pp799–804

Lobon-Cervia, J. (2007) 'Numerical changes in stream-resident brown trout (*Salmo trutta*): Uncovering the roles of density independent factors across space and time', *Canadian Journal of Fisheries and Aquatic Sciences*, vol 61, pp1929–1939

Lobon-Cervia, J. (2008) 'Habitat quality enhances spatial variation in the self-thinning patterns of stream-resident brown trout (*Salmo trutta*)', *Canadian Journal of Fisheries and Aquatic Sciences*, vol 65, pp2006–2015

Lobon-Cervia, J. (2009) 'Recruitment as a driver of production dynamics in stream-resident brown trout (*Salmo trutta*)', *Freshwater Biology*, vol 54, no 8, pp1692–1704

Lobon-Cervia, J. and Rincon, P. A. (1998) 'Field assessment of the influence of temperature on growth rate in a brown trout population', *Transactions of the American Fisheries Society*, vol 127, pp718–728

Lowe, S. J., Browne, M. and Boudjelas, S. (2000) *100 of the World's Worst Invasive Alien Species*, IUCN/SSC Invasive Species Specialist Group (ISSG), Auckland, New Zealand

Macchi, P. J., Vigliano, P. H., Pascual, M. A., Alonso, M., Denegri, M. A., Milano, D., Asorey, M. G. and Lippolt, G. (2008) 'Historical policy goals for fish management in northern continental Patagonia Argentina: A structuring force of actual fish assemblages?', *American Fisheries Society Symposium*, vol 44, pp331–348

MacCrimmon, H. R. and Marshall, T. L. (1968) 'World distribution of brown trout, *Salmo trutta*', *Journal of the Fisheries Research Board of Canada*, vol 25, pp2527–2548

MacCrimmon, H. R., Marshall, T. L. and Gots, B. L. (1970) 'World distribution of brown trout, *Salmo trutta* – further observations', *Journal of the Fisheries Research Board of Canada*, vol 27, pp811–818

Martinez, P. J., Bigelow, P. E., Deleray, M. A., Fredenberg, W. A., Hansen, B. S., Horner, N. J., Lehr, S. K., Schneidervin, R. W., Tolentino, S. A. and Viola, A. E. (2009) 'Western lake trout woes', *Fisheries*, vol 34, pp424–442

Mayama, H. (1999) 'Predation of juvenile masu salmon (*Oncorhynchus masu*) and brown trout (*Salmo trutta*) on newly emerged masu salmon fry in the Chitose River', *Bulletin of the National Salmon Resources Center*, vol 2, pp21–27

McDowall, R. M. (2006) 'Crying wolf, crying foul, or crying shame: Alien salmonids and a biodiversity crisis in the southern cool-temperate galaxioid fishes?', *Reviews in Fish Biology and Fisheries*, vol 16, pp233–422

McGeoch, M. A., Butchart, S. H. M., Spear, D., Marais, E., Kleynhans, E. J., Symes, A., Chanson, J. and Hoffmann, M. (2010) 'Global indicators of biological invasion: Species numbers, biodiversity impact and policy responses', *Diversity and Distributions*, vol 16, pp95–108

McHugh, P. and Budy, P. (2005) 'An experimental evaluation of competitive and thermal effects on brown trout (*Salmo trutta*) and cutthroat trout (*Oncorhynchus clarkii utah*) performance along an altitudinal gradient', *Canadian Journal of Fisheries and Aquatic Sciences*, vol 62, pp2784–2795

McHugh, P. and Budy, P. (2006) 'Experimental effects of nonnative brown trout (*Salmo trutta*) on the individual- and population-level performance of native Bonneville cutthroat trout (*Oncorhynchus clarkii utah*)', *Transactions of the American Fisheries Society*, vol 135, pp1441–1455

McIntosh, A. R. and Townsend, C. R. (1996) 'Interactions between fish, grazing invertebrates and algae in a New Zealand stream: A trophic cascade mediated by fish-induced changes to grazer behaviour?' *Oecologia*, vol 108, pp174–181

McIntosh, A. R., Townsend C. R. and Crowl T. A. (1992) 'Competition for space between introduced brown trout (*Salmo trutta* L.) and common river galaxias (*Galaxias vulgaris* Stokell) in a New Zealand stream', *Journal of Fish Biology*, vol 41, pp63–81

McIntosh, A. R., McHugh, P. A., Dunn, N. R., Goodman, J. M., Howard, S. W., Jellyman, P. G., O'Brien, L. K., Nystrom, P. and Woodford, D. J. (2010) 'The impact of trout on galaxiid fishes in New Zealand', *New Zealand Journal of Ecology*, vol 34, pp195–206

Milner, N. J., Elliott, J. M., Armstrong, J. D., Gardiner, R., Welton, J. S. and Ladle, M. (2003) 'The natural control of salmon and trout populations in streams', *Fisheries Research*, vol 62, pp111–125

Mittelbach, G. G. and Persson, L. (1998) 'The ontogeny of piscivory and its ecological consequences', *Canadian Journal of Fisheries and Aquatic Sciences*, vol 55, pp1454–1465

Moyle, P. B. (1976) *Inland Fishes of California*, University of California Press, Berkeley, CA

Moyle, P. B. and Marciochi, A. (1975) 'Biology of the Modoc sucker, *Catostomus microps* (Pisces: Catostomidae) in northeastern California', *Copeia*, vol 1975, pp556–560

Nannini, M. A. and Belk, M. C. (2006) 'Antipredator responses of two native stream fishes to an introduced predator: Does similarity in morphology predict similarity in behavioural response?', *Ecology of Freshwater Fish*, vol 15, pp453–463

Nasmith, L. E., Tonn, W. M., Paszkowski, C. A. and Scrimgeour, G. J. (2010) 'Effects of stocked trout on native fish communities in boreal foothills lakes', *Ecology of Freshwater Fish*, vol 19, pp279–289

Olsen, D. G. and Belk, M. C. (2005) 'Relationship of diurnal habitat use of native stream fishes of the eastern Great Basin to presence of introduced salmonids', *Western North American Naturalist*, vol 65, no 5, pp501–506

Pascual, M., Macchi, P., Urbanski, J., Marcos, F., Rossi, C. R., Novara, M. and Dell'Arciprete, P. (2002) 'Evaluating potential effects of exotic freshwater fish from incomplete species presence-absence data', *Biological Invasions*, vol 4, pp101–113

Pascual, M. A., Lancelotti, J. L., Ernst, B., Ciancio, J. E., Aedo, E. and Garcia-Asorey, M. (2009) 'Scale, connectivity, and incentives in the introduction and management of non-native species: The case of exotic salmonids in Patagonia', *Frontiers in Ecology and the Environment*, vol 7, pp533–540

Penczak T. (1999) 'Fish production and food consumption in the Warta River (Poland): Continued post-impoundment study (1990–1994)', *Hydrobiologia*, vol 416, pp107–123

Peterson, D. P., Fausch, K. D., Watmough, J. and Cunjak, R. A. (2008a) 'When eradication is not an option: Modelling strategies for electrofishing suppression of non-native brook trout to foster persistence of sympatric native cutthroat trout in small streams', *North American Journal of Fisheries Management*, vol 28, pp1847–1867

Peterson, D. P., Rieman B. E., Dunham, J. B., Fausch, K. D. and Young, M. K. (2008b) 'Analysis of trade-offs between threats of invasion by non-native brook trout (*Salvelinus fontinalis*) and intentional isolation for native westslope cutthroat trout (*Oncorhynchus clarkii lewisi*)', *Canadian Journal of Fisheries and Aquatic Sciences*, vol 65, pp557–573

Rahel, F. J. and Nibbelink, N. P. (1999) 'Spatial patterns in relations among brown trout (*Salmo trutta*) distribution, summer air temperature, and stream size in Rocky Mountain streams', *Canadian Journal of Fisheries and Aquatic Sciences*, vol 56, pp43–51

Raleigh, R. F., Zuckerman, L. D. and Nelson, P. C. (1986) 'Habitat suitability index models and instream flow suitability curves: Brown trout, revised', US Fish and Wildlife Service Biological Report 82

Ruzycki, J. R., Beauchamp, D. A. and Yule, D. L. (2003) 'Effects of introduced lake trout on native cutthroat trout in Yellowstone Lake', *Ecological Applications*, vol 13, pp23–37

Shemai, B., Sallenave, R. and Cowley, D. E. (2007) 'Competition between hatchery-raised Rio Grande cutthroat trout and wild brown trout', *North American Journal of Fisheries Management*, vol 27, pp315–325

Simberloff, D., Parker, I. M. and Windle, P. N. (2005) 'Introduced species policy, management, and future research needs', *Frontiers in Ecology and Environment*, vol 3, pp12–20

Simon, K. S., Townsend, C. R., Biggs, B. J. F., Bowden, W. B. and Frew, R. D. (2004) 'Habitat-specific nitrogen dynamics in New Zealand streams containing native or invasive fish', *Ecosystems*, vol 7, pp777–792

Stuart-Smith, R. D., White, R. W. G. and Barmuta, L. A. (2008) 'A shift in the habitat use pattern of a lentic galaxiid fish: An acute behavioural response to an introduced predator', *Environmental Biology of Fishes*, vol 82, pp93–100

Thompson, P. D. and Rahel, F. J. (1996) 'Evaluation of depletion-removal electrofishing of brook trout in small Rocky Mountain streams', *North American Journal of Fisheries Management*, vol 16, pp332–339

Thompson, P. D. and Rahel, F. J. (1998) 'Evaluation of artificial barriers in small Rocky Mountain streams for preventing upstream movement by brook trout', *North American Journal of Fisheries Management*, vol 18, pp206–210

Townsend, C. R. and Crowl, T. A. (1991) 'Fragmented population structure in a native New Zealand fish: An effect of introduced brown trout', *Oikos*, vol 61, pp347–354

Townsend, C. R. and Simon, K. S. (2006) 'Consequences of brown trout invasion for stream ecosystems', in R. B. Allen and W. G. Lee (eds) *Biological Invasions in New Zealand*, Springer-Verlag, Berlin, pp213–225

Valiente, A. G., Juanes, F., Nunez, P. and Garcia-Vazquez, E. (2010) 'Brown trout (*Salmo trutta*) invasiveness: Plasticity in life-history is more important than genetic variability', *Biological Invasions*, vol 12, pp451–462

Verspoor, E. (1988) 'Widespread hybridization between native Atlantic salmon, *Salmo salar*, and introduced brown trout, *S. trutta*, in eastern Newfoundland', *Journal of Fish Biology*, vol 32, pp327–334

Waters, T. F. (1983) 'Replacement of brook trout by brown trout over 15 years in a Minnesota stream: Production and abundance', *Transactions of the American Fisheries Society*, vol 112, pp137–146

Welcomme, R. L. (compiler) (1988) 'International introductions of inland aquatic species', Fisheries and Aquaculture Organization of the United Nations Fisheries Technical Paper No 294

Wood, J. and Budy, P. (2009) 'The role of environmental factors in determining early survival and invasion success of exotic brown trout', *Transactions of the American Fisheries Society*, vol 138, pp756–767

Woodford, D. J. and McIntosh, A. R. (2011) 'Location of demographic sources affects distributions of a vulnerable native fish in invaded river networks', *Freshwater Biology*, vol 56, pp311–324

Zimmerman, J. K. H. and Vondracek, B. (2006) 'Interactions of slimy sculpin (*Cottus cognatus*) with native and nonnative trout: Consequences for growth', *Canadian Journal of Fisheries and Aquatic Sciences*, vol 63, pp1526–1535

Part IV

Amphibians and Reptiles

25

Rhinella marina L. (cane toad)

Richard Shine

The Species and Its History

Native to an extensive area of South and Central America, cane toads are large and highly toxic anurans (Figure 25.1), widely regarded as one of the most troublesome invasive anuran species (Kraus, 2009). Recent phylogenetic studies on toads (Bufonidae) have suggested significant nomenclatural changes, with the cane toad itself (traditionally '*Bufo marinus*') allocated to the genera *Chaunus* or *Rhinella* by some authors (e.g. Pramuk, 2006). Pending consensus, I use the name *Rhinella marina* in this chapter.

The history of cane toad introductions to more than 40 countries worldwide is summarized by Lever (2001) and Kraus (2009). Figure 25.2 shows the toad's native range, and the broad area of the globe to which it has been introduced by human activities. Those introductions largely comprise intentional translocations for the purpose of using toads as a biocontrol for invertebrate pests (Lever, 2001). The success or failure of those attempts has never been rigorously evaluated, but early claims of success in using cane toads to control beetles in commercial sugar cane plantations in Hawaii stimulated agricultural scientists to spread the toads widely (Lever, 2001). Cane toads were brought to many countries in the Pacific, the Caribbean, Asia and North America (Figure 25.2). Once having reached an area, cane toads proved adept not only at spreading in their own right, but also in stowing away on cars, trucks, boats and the like to reach other destinations (e.g. White and Shine, 2009).

Although cane toads have been blamed for causing environmental problems in several of the areas to which they were introduced, by far the most extensive information comes from the toads' invasion of Australia. Reflecting that continent's isolation from other landmasses through evolutionary time, many lineages of organisms widespread in Africa, Eurasia and the Americas did not reach Australia until they were assisted by anthropogenic translocation. A lack of co-evolution with bufonids has rendered many native Australian taxa unable to tolerate the powerful chemical defences (bufadienolides) produced by cane toads, with the result that cane toad invasion has caused catastrophic levels of mortality in populations of some native predators (Shine, 2010). For example, more than 70 per cent mortality within a year of toad arrival has been reported for some populations of varanid lizards (Griffiths and McKay, 2007; Doody et al, 2009; Ujvari and Madsen, 2009), bluetongue skinks (Price-Rees et al, 2010), freshwater crocodiles (Letnic et al, 2008) and marsupial quolls (Oakwood, 2003).

Reflecting public and scientific concern over the ecological impact of cane toads in Australia, the species' spread through that continent has been documented in more detail than is available for most other biological invasions (Urban et al, 2007, 2008). Originally, 101 adult toads were brought from Hawaii to north-eastern Queensland and bred in captivity. Genetic analyses suggest that the original stock used in these multiple translocations came from French Guiana or eastern Venezuela (Slade and Moritz, 1998). The progeny of

Source: David Nelson

Figure 25.1 *The cane toad,* Rhinella marina

the animals brought to Australia, plus some of the founding adults, were released in a range of sites along the cane-growing area of the Queensland coast (Lever, 2001). The toads thrived and began spreading southwards along the coast, and westwards through the tropics of Queensland and the Northern Territory. In 2010, the toad front crossed into Western Australia (Figure 25.3). The rate of westward invasion has accelerated through time, from about 10 to 15km per annum in the early years through to 50 or 60km per annum by 2000 (Phillips et al, 2006; Urban et al, 2007, 2008). In contrast, the invasion front in eastern Australia soon slowed down and has remained relatively stable in north-eastern New South Wales for several years, apparently reflecting thermal (cool climate)

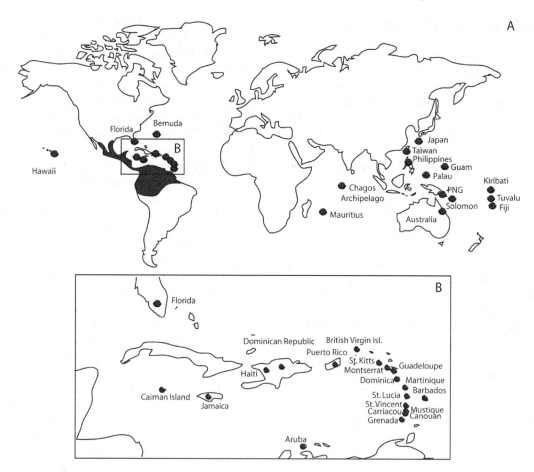

Note: Shading indicates the areas where cane toads occur naturally; symbols indicate introductions.

Source: Lever (2001)

Figure 25.2 *Areas in which cane toads occur naturally, and those to which the species has been deliberately or accidentally introduced*

Distribution of *Rhinella marina*:

Potential Habitat: – – – – – – –

Source: Modified from Figure 1 in Price-Rees et al (2010)

Figure 25.3 *Current and predicted distribution of the cane toad in Australia*

limits to the distribution of this tropical anuran (Kolbe et al, 2010).

The Ecological Niche of Cane Toads

Cane toads resemble many other ecologically generalized bufonid anurans in general body form (squat, with relatively short limbs and lacking the extreme morphological specializations seen in highly aquatic and arboreal anurans). However, cane toads are unusual in their large body sizes (to >200mm snout–urostyle length, 2kg, although typical adult size is far smaller; Lever, 2001) and (partly as a result of their large size) in their high toxicity. Many bufonids produce a complex mix of poisons (especially bufadienolides) that can be lethal to predators attacking and/or attempting to

ingest the toad (Lever, 2001; Shine, 2010). Toxin types and quantities show strong ontogenetic variation and thus toads are most highly toxic to aquatic predators during the egg stage, with toxicity then declining through tadpole life and reaching a minimum around the time of metamorphosis (Hayes et al, 2009). As soon as the young toad develops its own parotoid glands and begins to manufacture toxin, bufadienolide levels increase rapidly (Phillips and Shine, 2006a; Hayes et al, 2009).

More is known about the ecology of invasive populations of cane toads (in Australia) than about the species in its native range. In Australia, female cane toads probably mature at one to two years of age (Alford et al, 1995). Cane toads have a 'typical' anuran lifecycle, with females producing very large clutches (to >30,000 eggs per clutch) of small darkly pigmented eggs in long gelatinous strings. Breeding can occur over much of the year, but with clear peaks that may occur at different times of year in different areas. Males clasp the female behind her forearms (axillary amplexus) and fertilize the eggs externally, as they are extruded from the female's body. Males readily amplex even non-reproductive females, and sometimes other objects such as native frogs, or even human hands (Lever, 2001). Female toads are burdened by clasping males (e.g. move more slowly, and do not feed), but are unable to give the 'release call' that stimulates an amplexing male to abandon his post (Bowcock et al, 2008, 2009). However, the ability of female cane toads to inflate their bodies (a behaviour that typically functions as an antipredator tactic) provides some degree of female control over amplexus and partner choice; an inflated female is more difficult for a male to cling to (Bruning et al, 2010).

Cane toads frequently (but not exclusively) deposit their eggs in still water ponds rather than streams, and are highly selective about attributes of the spawning site. For example, they choose shallow water bodies with gently sloping rather than steep sides, and with open ground rather than dense vegetation around the water's edge (Hagman and Shine, 2006; Semeniuk et al, 2007). The eggs (approximately 2mm in diameter) are laid in shallow water, often twined around emergent vegetation, and typically hatch in less than 48 hours (depending on water temperature; Alford et al, 1995). The small black tadpoles develop rapidly, sometimes completing metamorphosis in less than 50 days (Alford et al, 1995). Toad tadpoles often aggregate into swarms, and are frequently seen in shallow water during daylight hours, perhaps reflecting their preference for high temperatures (Floyd, 1984). Tadpoles graze on a wide variety of food types, including dead animals (sometimes conspecifics) as well as vegetable material (Alford et al, 1995; Lever, 2001). Toad tadpoles produce alarm pheromones that repel conspecific tadpoles; frequent exposure to such chemical cues slows tadpole growth and reduces survival rates as well as body sizes at metamorphosis (Hagman et al, 2009).

Like many other bufonid species, cane toads metamorphose at a very small body size (approximately 10mm body length, <0.10g) – especially in relation to the very large size of adults. During the tropical dry season, hydric stress keeps metamorphs close to the edge of the natal pond for weeks or months (Child et al, 2008a, 2008b, 2009). Earlier hatched or faster growing metamorphs may reach body sizes large

enough to enable them to consume smaller conspecifics, creating a size class of specialist cannibals (Pizzatto and Shine, 2008). Adult toads are willing to consume smaller toads also, but an ontogenetic shift from diurnal to nocturnal habits reduces rates of encounter and hence cannibalism is mostly seen in sub-adult rather than adult toads (Pizzatto and Shine, 2008). A specialized toe luring behaviour appears to have evolved as an adaptation to facilitate such cannibalism (Hagman and Shine, 2008a). In their terrestrial phase, toads are carnivorous and primarily insectivorous (Zug and Zug, 1979). Although cases of predation on vertebrates have attracted considerable publicity, the overwhelming majority of the toads' diet consists of small insects such as beetles and ants (Zug and Zug, 1979).

Cane toads tend to avoid dense vegetation except as a diurnal retreat, and instead actively select the relatively open habitats created by human activity (Zug and Zug, 1979). Even at a micro scale, toads tend to select open areas for feeding and calling (Zug and Zug, 1979). Open areas may provide better visibility to detect edible prey for this ambush predator, and toads actively select foraging conditions (e.g. availability of artificial lights; substrate colour and rugosity) that facilitate the detection and capture of insect prey (González-Bernal et al, 2011). Disturbed habitats provide many of the resources that toads need, and a combination of shelter, water and food (insects attracted to lights) may explain the dense aggregations of toads often observed around built structures.

Movement patterns of invasive cane toads have been monitored in detail, using methods such as radiotelemetry and spool-and-line tracking, in several parts of Australia. Consistent with the acceleration of progress of the toad invasion through time (from 10km/year to >50km/year), toads at the invasion front move much further than do conspecifics from long established populations both in field studies (Alford et al, 2009) and in laboratory trials of locomotor endurance (Llewelyn et al, 2010). Remarkably, adult toads at the invasion front often travel more than 1km per night, and do so as long as the ground is wet enough to facilitate such movements (Phillips et al, 2007). These extensive movements are reflected in morphological features also, with invasion front toads exhibiting longer legs relative to body size, a trait strongly correlated with faster movements (Phillips et al, 2006). Theoretical models attribute the rapid evolution of dispersal-enhancing traits to spatial selection, whereby morphological and behavioural features that increase rates of toad displacement inevitably accumulate in invasion front populations. Because any toad without such features could not move quickly enough to remain in the fast moving vanguard of the invasion, there is a sorting out of traits across the landscape (Phillips et al, 2006). The end result has been the rapid evolution of an extreme morphotype – extreme to the point that many toads at the invasion front develop severe spinal arthritis (spondylosis) because of the pressures exerted by longer legs, continual long distance travel, and a compromised immune system (Brown et al, 2007). The ecology and movement patterns of cane toads within their native range are far less well known.

Management of Cane Toads

As for other topics on this species, most management information comes from one of the areas where toads have been introduced (Australia) rather than from other areas of translocation, or from the native range. Reflecting the high rates of toad-induced mortality of several species of large anuran-eating predators (lizards, snakes, crocodiles, marsupials; Shine, 2010), the invasion of cane toads has stimulated considerable efforts to manage this species in Australia. Governments have allocated at least AUS$20 million to research on this issue, most ($11 million) to a search for viral control by the CSIRO, a governmental research agency (Australian Government, 2010). An inability to find toad-specific pathogens within the toads' native range, or to create an effective or environmentally acceptable genetically manipulated viral control approach, resulted in termination of this research programme in 2009. Other suggestions about biocontrol methods such as 'daughterless genes' (manipulating toads genetically such that they produce only sons) or release of sterile males, have never been investigated in enough detail to overcome technical hurdles, let alone to implement in the field (e.g. Thresher and Bax, 2006).

Lacking any scientifically developed or tested means to control cane toads, local communities have thrown themselves enthusiastically into direct methods of removal such as hand-collecting (often called 'toad busting' or 'toad mustering'), trapping, or fencing of waterholes. These efforts have ranged from large-scale

well-organized 'Toad Day Out' activities, through to individual citizens waging their own vigilante campaigns by killing any toads that they encounter (Somaweera et al, 2010). A major advantage of the more highly organized activities is the ability to verify identification of any 'toads' collected. Surveys of the general public show high error rates in anuran identification, so many thousands of native anurans doubtless have been killed in the mistaken belief that they are cane toads (White, 2007; Somaweera et al, 2010). This killing often occurs well outside the known geographic range of cane toads in Australia, raising the rates of inadvertent mortality of frogs rather than toads (White and Shine, 2009; Somaweera et al, 2010).

As the toads have continued to spread at an unabated (indeed, increasing) rate across tropical Australia, public attitudes towards this invasive anuran have changed. Surveys suggest that concern by the general public about the ecological impact of cane toads is greatest in advance of the toad front, whereas people in long-colonized regions (i.e. who have had decades to adjust to toad presence) are less worried about the animals (Clarke et al, 2009). Nonetheless, toads remain unpopular and many people continue to kill them on sight (Somaweera et al, 2010).

Regrettably, there has been no quantitative analysis of the effectiveness of the major financial investment, and enormous volunteer effort, that has gone into toad control by local communities. Instead, the attitude has been simply to go out and kill as many toads as possible, and much confidence has been expressed by community leaders and politicians that direct removal by concerned citizens will be able to stop the cane toad invasion. As the failure of such activities to curtail the toads' advance has become obvious, the emphasis of such groups – and of federal governmental policy – has shifted towards attempts to reduce local toad abundance in order to protect endangered native species.

The primary methods used by local communities to reduce toad numbers include:

• Hand-collecting – adult toads select relatively open sites for many of their activities (see above) and can readily be approached and hand-collected at night by aid of flashlights. The male toads' call is distinctive, enabling easy identification of breeding sites. As noted above, collateral damage to native anurans through misidentification needs to be minimized if hand-collecting is to be employed by people without scientific training.

- Trapping – several trap designs have been popular, mostly involving artificial lights to attract insects (and thus toads) to enter through a one-way door. Recent research suggests that trap success could be increased by judicious selection of light sources and provision of recorded toad calls (Alford, R. A. and Schwarzkopf, L., pers. comm.), but few such improvements have been implemented, perhaps reflecting the challenge in communication between scientists and community groups. Regardless, trap success appears to be low relative to other methods (Brown, G. P. and Shine, R., unpublished data), and occasional captures of native species introduce the possibility of significant collateral damage and animal welfare concerns.

- Fencing of water bodies – especially in arid areas, cane toads need to rehydrate at water sources frequently (once every few nights; perhaps nightly in extreme conditions) and precluding access to water can kill a high proportion of toads in the local population. Unfortunately, domestic stock and native wildlife also need access to water, so any fencing has to exclude toads but not other species. The large body size and poor climbing ability of adult toads can be exploited for this purpose, but excluding sub-adult toads while allowing ingress and egress to similar-sized terrestrial frogs may prove to be impossible.

- Poisoning – the high fecundity of cane toads, combined with the reliance of metamorphs on moist conditions, can result in many thousands of newly metamorphosed individuals aggregating around water body margins. Hand-collecting such enormous numbers would be logistically difficult, so a common approach has been to spray a lethal poison (a household disinfectant) onto these hordes of juvenile toads. Although the method is claimed to be effective in killing toads, the likelihood of significant collateral damage to pond-side ecological systems has not been evaluated.

Another set of approaches, alternative to direct physical removal or poisoning, has emerged recently from intensive research on the ecology of cane toads in tropical Australia. Basic research on toad impact and biology has revealed a number of traits in which cane toads differ so consistently from native anurans that targeted control might be feasible. These traits include:

- Spawning site selection – cane toads select specific habitat features when choosing a water body in which to spawn (see above). In both the eastern and western edges of their current Australian range, those sites differ from those used by most of the local anuran species. Thus, planting dense vegetation cover around a pond might be enough to render it unsuitable for toad breeding while having little or no impact on its usage by local native frogs (Hagman and Shine, 2006; Semeniuk et al, 2007).

- Parasitism – cane toads in Australia frequently carry a nematode lungworm, *Rhabdias pseudosphaerocephala*, that is genetically identical to samples of the lungworm from cane toads (and other toad species) in South America. Presumably, the lungworm was brought to Australia with the toads. Sampling has yet to reveal this parasite in Australian native frogs, which instead have other species of *Rhabdias* (Dubey and Shine, 2008). Laboratory trials suggest that infection by the lungworm causes significant mortality of metamorph cane toads, and slows the growth of surviving individuals (Kelehear et al, 2009, 2011). The parasite is missing from the toad invasion front because infection dynamics during a biological invasion tend to leave pathogens some distance behind the front (Phillips et al, 2010). Hence, translocating the lungworm to the invasion front might help to control toad populations. Before any such method could be trialled, however, we would need to carefully assess the possibility that increased lungworm densities might negatively affect native frogs. Such trials are currently in progress.

- Vulnerability to local predators – although cane toads are highly toxic to many vertebrate predators, invertebrates tend to be less affected (Shine, 2010). Large carnivorous ants ('meat ants') are abundant in Australia, and kill many metamorphosing cane toads (Ward-Fear et al, 2010a, 2010b) (Figure 25.4). Presumably reflecting long co-evolution, metamorph native frogs are adept at avoiding and escaping meat ants, whereas cane toads are not (Ward-Fear et al, 2009, 2010a). Off-take rates of metamorph toads can be increased by laying out cat food baits to attract foraging ants to sites where cane toads are metamorphosing (Ward-Fear et al, 2010b). Collateral risks to native fauna appear to be low, especially compared to alternatives such as spraying poison onto the young toads (Ward-Fear et al, 2010b).

- Pheromonal communication – unlike the tadpoles of native frog species (or at least, those that have been tested to date), cane toad tadpoles produce specific chemicals that serve as communication mechanisms. For example, injured or stressed tadpoles produce an alarm pheromone that induces rapid retreat in other toad tadpoles (Hagman et al, 2009). The tadpoles of native frogs do not react to this substance, and probably do not even detect it (Hagman and Shine, 2008b, 2009). Long term daily exposure to the alarm pheromone reduces toad tadpole survival rates, and reduces body size at metamorphosis – a critical determinant of desiccation resistance (Child et al, 2008b, 2009) and vulnerability to both predators (Ward-Fear et al, 2010a) and parasites (Kelehear et al, 2009). Thus, application of the alarm pheromone to spawning sites might provide a toad-specific control method. Current work is attempting to clarify the chemical identity of this pheromone, as well as a toad-specific attractant pheromone that might be used to lure toad tadpoles into traps (Crossland and Shine, 2010).

Although such novel approaches have promise, much needs to be known before they can be implemented. Two critical issues are collateral damage (a biocontrol must target toads only, and not disadvantage native fauna) and feasibility. In particular, there is strong density dependence at several phases of the toads' life history: tadpoles eat newly laid eggs, tadpoles compete with each other, metamorphs compete with each other and consume each other, and so on (Alford et al, 1995). In such a system, even a major increment in mortality at one life history stage may have vanishingly small impact on recruitment to the next stage because of density-dependent compensation (Crossland et al, 2009). Field trials are needed to assess the effectiveness of alternative control methods not simply on the numbers of toads killed, but on the results of that additional mortality for recruitment into the adult population. Ultimately, the best criterion for success is whether or not application of the control method reduces the ecological impact of invasive toads on native wildlife; a parameter that is harder to quantify than changes in toad density, but that is more meaningful for conservation and management (Shine and Doody, 2011).

The critical issue thus is to reduce the ecological impact of toads, and reducing toad abundance is simply a means to that end. It may be an ineffective means if the magnitude of toad impact is only weakly related to toad densities. Most previous work has implicitly assumed that reducing the abundance of invasive toads will reduce their impact, but this may not be true. The major pathway of impact is lethal toxic ingestion of toads by native predators, and even a low density of toads may be enough for any local predator to locate at least one toad and be killed (Shine, 2010). If this is the case, we need alternative management approaches that do not rely on reducing toad numbers. Encouragingly, recent research has shown that it is possible to train native predators (marsupial quolls, *Dasyurus hallucatus*) to avoid toads as food (by adding nausea-inducing chemicals to small dead toads offered to captive raised quolls), and that such training enhanced rates of survival of these animals after they were released into toad-infested areas in the field (O'Donnell et al, 2010).

Source: Georgia Ward-Fear

Figure 25.4 *A native 'meat ant',* Iridomyrmex reburrus, *carrying away a metamorph cane toad beside a Northern Territory billabong*

Controversies about Cane Toad Impact and Control

Cane toads have achieved an iconic status in Australia and are a frequent subject of media commentary and debate (Australian Government, 2010). Some of those debates are between scientists (who disagree about issues such as the nature, severity and duration of the ecological impact of cane toads on native fauna); whereas other debates are between alternative community groups (about the most effective ways to control toads, or about the results of their own activities compared to those of other groups); or between scientists and community groups (about a wide range of topics).

There is broad scientific consensus that the primary ecological impact of cane toads is via lethal toxic ingestion of toads by native predators, but uncertainty continues about the exact levels of impact and how these vary through space and time (Shine, 2010). The level of toad impact on a given predator species can vary markedly through space, depending on how local habitats influence rates of encounter between toads and predators (Letnic et al, 2008). Field and laboratory studies also have shown that impact levels can fall rapidly (through aversion learning by predators) within a year or two after the toads invade (e.g. Webb et al, 2008; Greenlees et al, 2010; Nelson et al, 2010a, 2010b) and that over the next few decades, adaptive shifts in predator morphology, behaviour and physiology can enable previously vulnerable species to change in ways that permit coexistence with toads (Phillips and Shine, 2006b). Nonetheless, the degree to which predator populations have recovered in areas long colonized by toads remains unclear because of the lack of data on pre-invasion densities of such taxa (Shine and Doody, 2011).

Reflecting the focus of community groups on controlling toads, most disagreements between such groups and scientists involve issues such as the severity of impact. Local communities about to encounter toads for the first time often predict that toad impact will be devastating (a view less often expressed by communities with longer experience of toad presence; Clarke et al, 2009). So, scientists who point to a less extreme magnitude of toad impact can be unpopular with community leaders trying to amass support and funding in the face of an imminent invasion. Similarly, any comments from scientists about the lack of evidence for effectiveness of commonly employed methods of toad control such as traps and hand-collecting, or possible collateral damage from the use of poisons, are not likely to be well received by community groups who see no other options, and are keen to do what they can. Such disagreements probably reflect the differing motivations and backgrounds of the people involved, and emphasize the need for continued dialogue between scientists and community groups to maximize the effectiveness of their battles against the invasive toad (Shine and Doody, 2011).

Acknowledgements

I thank my research team for helping me to learn about cane toads and their impact, and the Australian Research Council for funding most of my work on this species. Special thanks go to Melanie Elphick and Sylvain Dubey, who assisted with the research as well as with preparation of this chapter and its figures.

References

Alford, R. A., Cohen, M. P., Crossland, M. R., Hearnden, M. N. and Schwarzkopf, L. (1995) 'Population biology of *Bufo marinus* in northern Australia', in C. M. Finlayson (ed) *Wetland Research in the Wet-Dry Tropics of Australia*, Supervising Scientist Report 101, Supervising Scientist, Canberra, pp173–181

Alford, R. A., Brown, G. P., Schwarzkopf, L., Phillips, B. and Shine, R. (2009) 'Comparisons through time and space suggest rapid evolution of dispersal behaviour in an invasive species', *Wildlife Research*, vol 36, pp23–28

Australian Government (2010) *Threat Abatement Plan for the Biological Effects, Including Lethal Toxic Ingestion, Caused by Cane Toads*, Department of the Environment, Water, Heritage, and the Arts, Australian Government, Canberra, ACT

Bowcock, H., Brown, G. P. and Shine, R. (2008) 'Sexual communication in cane toads (*Bufo marinus*): What cues influence the duration of amplexus?', *Animal Behaviour*, vol 75, pp1571–1579

Bowcock, H., Brown, G. P. and Shine, R. (2009) 'Beastly bondage: The costs of amplexus in cane toads (*Bufo marinus*)', *Copeia*, vol 2009, pp29–36

Brown, G. P., Shilton, C. M., Phillips, B. L. and Shine, R. (2007) 'Invasion, stress, and spinal arthritis in cane toads', *Proceedings of the National Academy of Sciences (USA)*, vol 104, pp17698–17700

Bruning, B., Phillips, B. L. and Shine, R. (2010) 'Turgid female toads give males the slip: A new mechanism of female mate choice in the Anura', *Biology Letters*, vol 6, pp322–324

Child, T., Phillips, B. L., Brown, G. P. and Shine, R. (2008a) 'The spatial ecology of cane toads (*Bufo marinus*) in tropical Australia: Why do metamorph toads stay near the water?', *Austral Ecology*, vol 33, pp630–640

Child, T., Phillips, B. L. and Shine, R. (2008b) 'Abiotic and biotic influences on the dispersal behaviour of metamorph cane toads (*Bufo marinus*) in tropical Australia', *Journal of Experimental Zoology*, vol 309A, pp215–224

Child, T., Phillips, B. L. and Shine, R. (2009) 'Does desiccation risk drive the distribution of metamorph cane toads (*Bufo marinus*) in tropical Australia?', *Journal of Tropical Ecology*, vol 25, pp193–200

Clarke, R., Carr, A., White, S., Raphael, B. and Baker, J. (2009) *Cane Toads in Communities*, Executive Report by the Australian Government Bureau of Rural Sciences, Canberra, ACT

Crossland, M. R. and Shine, R. (2010) 'Cues for cannibalism: Cane toad tadpoles use chemical signals to locate and consume conspecific eggs', *Oikos*, vol 120, pp327–332

Crossland, M. R., Alford, R. A. and Shine, R. (2009) 'Impact of the invasive cane toad (*Bufo marinus*) on an Australian frog (*Opisthodon ornatus*) depends on reproductive timing', *Oecologia*, vol 158, pp625–632

Doody, J. S., Green, B., Rhind, D., Castellano, C. M., Sims, R. and Robinson, T. (2009) 'Population-level declines in Australian predators caused by an invasive species', *Animal Conservation*, vol 12, pp46–53

Dubey, S. and Shine, R. (2008) 'Origin of the parasites of an invading species, the Australian cane toad (*Bufo marinus*): Are the lungworms Australian or American?', *Molecular Ecology*, vol 17, pp4418–4424

Floyd, R. B. (1984) 'Variation in temperature preference with stage of development in *Bufo marinus* larvae', *Journal of Herpetology*, vol 18, pp153–158

González-Bernal, E., Brown, G. P., Cabrera-Guzmán, E. and Shine, R. (2011) 'Foraging tactics of an ambush predator: The effects of substrate attributes on prey availability and predator feeding success', *Behavioral Ecology and Sociobiology*, vol 65, no 7, pp1367–1375

Greenlees, M., Phillips, B. L. and Shine, R. (2010) 'Adjusting to a toxic invader: Native Australian frog learns not to prey on cane toads', *Behavioral Ecology*, vol 21, pp966–971

Griffiths, A. D. and McKay, J. L. (2007) 'Cane toads reduce the abundance and site occupancy of freshwater goannas *Varanus mertensi*', *Wildlife Research*, vol 34, pp609–615

Hagman, M. and Shine, R. (2006) 'Spawning-site selection by feral cane toads (*Bufo marinus*) at an invasion front in tropical Australia', *Austral Ecology*, vol 31, pp551–558

Hagman, M. and Shine, R. (2008a) 'Deceptive digits: The functional significance of toe-waving by cannibalistic cane toads (*Chaunus marinus*)', *Animal Behaviour*, vol 75, pp123–131

Hagman, M. and Shine, R. (2008b) 'Australian tadpoles do not avoid chemical cues from invasive cane toads (*Bufo marinus*)', *Wildlife Research*, vol 35, pp59–64

Hagman, M. and Shine, R. (2009) 'Species specific communication systems in invasive toads *versus* Australian frogs', *Aquatic Conservation*, vol 19, pp724–728

Hagman, M., Hayes, R. A., Capon, R. J. and Shine, R. (2009) 'Alarm cues experienced by cane toad tadpoles affect post-metamorphic morphology and chemical defences', *Functional Ecology*, vol 23, pp126–132

Hayes, R. A., Crossland, M. R., Hagman, M., Capon, R. J. and Shine, R. (2009) 'Ontogenetic variation in the chemical defences of cane toads (*Bufo marinus*): Toxin profiles and effects on predators', *Journal of Chemical Ecology*, vol 35, pp391–399

Kelehear, C., Webb, J. K. and Shine, R. (2009) '*Rhabdias pseudosphaerocephala* infection in *Bufo marinus*: Lung nematodes reduce viability of metamorph cane toads', *Parasitology*, vol 136, pp919–927

Kelehear, C., Brown, G. P. and Shine, R. (2011) 'Influence of lung parasites on growth rates of free-ranging and captive adult cane toads', *Oecologia*, vol 165, pp585–592

Kolbe, J. J., Kearney, M. and Shine, R. (2010) 'Modeling the consequences of thermal trait variation for the cane toad invasion of Australia', *Ecological Applications*, vol 20, pp2273–2285

Kraus, F. (2009) *Alien Reptiles and Amphibians: A Scientific Compendium and Analysis*, Springer, Netherlands

Letnic, M., Webb, J. K. and Shine R. (2008) 'Invasive cane toads (*Bufo marinus*) cause mass mortality of freshwater crocodiles (*Crocodylus johnstoni*) in tropical Australia', *Biological Conservation*, vol 141, pp1773–1782

Lever, C. (2001) *The Cane Toad: The History and Ecology of a Successful Colonist*, Westbury Academic and Scientific Publishing, Otley, West Yorkshire

Llewelyn, J., Phillips, B. L., Alford, R. A., Schwarzkopf, L. and Shine, R. (2010) 'Locomotor performance in an invasive species: Cane toads from the invasion front have greater endurance, but not speed, compared to conspecifics from a long-colonised area', *Oecologia*, vol 162, pp343–348

Nelson, D., Crossland, M. R. and Shine, R. (2010a) 'Indirect ecological impacts of an invasive toad on predator-prey interactions among native species', *Biological Invasions*, vol 12, pp3363–3369

Nelson, D. W. M., Crossland, M. R. and Shine, R. (2010b) 'Foraging responses of predators to novel toxic prey: Effects of predator learning and relative prey abundance', *Oikos*, vol 120, pp152–158

Oakwood, M. (2003) 'The effect of cane toads on a marsupial carnivore, the northern quoll, *Dasyurus hallucatus*', Unpublished progress report, August 2003, Parks Australia North, Darwin

O'Donnell, S., Webb, J. K. and Shine, R. (2010) 'Conditioned taste aversion enhances the survival of an endangered predator imperiled by a toxic invader', *Journal of Applied Ecology*, vol 47, pp558–565

Phillips, B. L. and Shine, R. (2006a) 'Allometry and selection in a novel predator-prey system: Australian snakes and the invading cane toad', *Oikos*, vol 112, pp122–130

Phillips, B. L. and Shine, R. (2006b) 'An invasive species induces rapid adaptive change in a native predator: Cane toads and black snakes in Australia', *Proceedings of the Royal Society B*, vol 273, pp1545–1550

Phillips, B. L., Brown, G. P., Webb, J. K. and Shine, R. (2006) 'Invasion and the evolution of speed in toads', *Nature*, vol 439, p803

Phillips, B. L., Brown, G. P., Greenlees, M., Webb, J. K. and Shine, R. (2007) 'Rapid expansion of the cane toad (*Bufo marinus*) invasion front in tropical Australia', *Austral Ecology*, vol 32, pp169–176

Phillips, B. L., Kelehear, C., Pizzatto, L., Brown, G. P., Barton, D. and Shine, R. (2010) 'Parasites and pathogens lag behind their host during periods of host range-advance', *Ecology*, vol 91, pp872–881

Pizzatto, L. and Shine, R. (2008) 'The behavioral ecology of cannibalism in cane toads (*Bufo marinus*)', *Behavioral Ecology and Sociobiology*, vol 63, pp123–133

Pramuk, J. (2006) 'Phylogeny of South American *Bufo* (Anura: Bufonidae) inferred from combined evidence', *Zoological Journal of the Linnean Society*, vol 146, pp407–452

Price-Rees, S. J., Brown, G. P. and Shine, R. (2010) 'Are bluetongue lizards (*Tiliqua scincoides intermedia*, Scincidae) threatened by the invasion of toxic cane toads (*Bufo marinus*) through tropical Australia?', *Wildlife Research*, vol 37, pp166–173

Semeniuk, M., Lemckert, F. and Shine, R. (2007) 'Breeding-site selection by cane toads (*Bufo marinus*) and native frogs in northern New South Wales', *Wildlife Research*, vol 34, pp59–66

Shine, R. (2010) 'The ecological impact of invasive cane toads (*Bufo marinus*) in Australia', *Quarterly Review of Biology*, vol 85, pp235–291

Shine, R. and Doody, J. S. (2011) 'Invasive-species control: Understanding conflicts between researchers and the general community', *Frontiers in Ecology and the Environment*, in press, doi:10.1890/100090

Slade, R. W. and Moritz, C. (1998) 'Phylogeography of *Bufo marinus* from its natural and introduced ranges', *Proceedings of the Royal Society B*, vol 265, pp769–777

Somaweera, R., Somaweera, N. and Shine, R. (2010) 'Frogs under friendly fire: How accurately can the general public recognize invasive species?', *Biological Conservation*, vol 143, pp1477–1484

Thresher, R. E. and Bax, N. (2006) 'Comparative analysis of genetic options for controlling invasive populations of the cane toad *Bufo marinus*', in K. Molloy and W. Henderson (eds) *Science of Cane Toad Invasion and Control: Proceedings of the Invasive Animals CRC/CSIRO/Qld NRM&W Cane Toad Workshop, 5–6 June, Brisbane*, Invasive Animals CRC, Canberra, ACT, pp117–122

Ujvari, B. and Madsen, T. (2009) 'Increased mortality of naïve varanid lizards after the invasion of non-native cane toads (*Bufo marinus*)', *Herpetological Conservation and Biology*, vol 4, pp248–251

Urban, M., Phillips, B. L., Skelly, D. K. and Shine, R. (2007) 'The cane toad's (*Chaunus marinus*) increasing ability to invade Australia is revealed by a dynamically updated range model', *Proceedings of the Royal Society B*, vol 274, pp1413–1419

Urban, M., Phillips, B. L., Skelly, D. K. and Shine, R. (2008) 'A toad more traveled: The heterogeneous invasion dynamics of cane toads in Australia', *American Naturalist*, vol 171, ppE134–E148

Ward-Fear, G., Brown, G. P., Greenlees, M. and Shine, R. (2009) 'Maladaptive traits in invasive species: In Australia, cane toads are more vulnerable to predatory ants than are native frogs', *Functional Ecology*, vol 23, pp559–568

Ward-Fear, G., Brown, G. P. and Shine, R. (2010a) 'Factors affecting the vulnerability of cane toads (*Bufo marinus*) to predation by ants', *Biological Journal of the Linnean Society*, vol 99, pp738–751

Ward-Fear, G., Brown, G. P. and Shine, R. (2010b) 'Using a native predator (the meat ant, *Iridomyrmex reburrus*) to reduce the abundance of an invasive species (the cane toad, *Bufo marinus*) in tropical Australia', *Journal of Applied Ecology*, vol 47, pp273–280

Webb, J. K., Brown, G. P., Child, T., Greenlees, M. J., Phillips, B. L. and Shine, R. (2008) 'A native dasyurid predator (common planigale, *Planigale maculata*) rapidly learns to avoid toxic cane toads', *Austral Ecology*, vol 33, pp821–829

White, A. W. (2007) 'Living with *Bufo*', in D. Lunney, P. Eby, P. Hutchings and S. Burgin (eds) *Pest or Guest: The Zoology of Overabundance*, Royal Zoological Society of New South Wales, Mosman, pp16–29

White, A. W. and Shine, R. (2009) 'The extra-limital spread of an invasive species via "stowaway" dispersal: Toad to nowhere?', *Animal Conservation*, vol 12, pp38–45

Zug, G. R. and Zug, P. B. (1979) 'The marine toad, *Bufo marinus*: A natural history resumé of native populations', *Smithsonian Contributions to Zoology*, vol 284, pp1–54

26

Eleutherodactylus coqui Thomas (Caribbean tree frog)

Karen H. Beard and William C. Pitt

History of *Eleutherodactylus coqui* Introduction

Eleutherodactylus coqui (hereafter, the coqui) is a nocturnal, terrestrial frog endemic to the island of Puerto Rico (Figure 26.1). There are 16 *Eleutherodactylus* species endemic to the island, but the coqui is the most widespread and abundant. While larger than most other frogs in Puerto Rico, the coqui is a small frog (maximum snout–vent length (SVL) for males of 50mm and for females of 63mm; Joglar, 1998), that differentiates itself from other *Eleutherodactylus* species by using the full spectrum of vertical forest habitats and by its distinctive two note mating call, which sounds like 'ko-kee' and gave the frog its common name.

The coqui has established on a number of Caribbean islands to which it is not native, including Culebra and Vieques, Puerto Rico (Rivero and Joglar, 1979), St Thomas and St Croix, Virgin Islands (MacLean, 1982) and the Dominican Republic (Joglar, 1998). The coqui was also introduced to Florida in the early 1970s (Austin and Schwartz, 1975; Wilson and Porras, 1983), but has not been reported there since 2000 (Meshaka et al, 2004).

Most of the information on the coqui as an invasive has been obtained in Hawaii, and so Hawaii is the focus of this chapter. The coqui was introduced to Hawaii in the late 1980s via infested nursery plants (Kraus et al, 1999) (Figure 26.1), and, consistent with this, it first appeared in and around nurseries. There were two separate introductions: one to the island of Hawaii (Big Island) and one to Maui (Maliko Gulch), which, at least genetically, both originated near San Juan, Puerto Rico (Velo-Antón et al, 2007; Peacock et al, 2009). The coqui experienced a severe bottleneck when it was introduced, and all measures of genetic diversity are much higher in Puerto Rico than Hawaii (Peacock et al, 2009).

Since its initial introductions, the coqui has spread to the other two main islands: Kauai and Oahu; and two smaller islands: Molokai and Lanai (Kraus and Campbell, 2002; Anonymous, 2010). Subsequent spread originated from the Big Island, while the Maui introduction remains, for the most part, genetically isolated (Peacock et al, 2009). The coqui's spread was rapid. In 1998, there were only eight populations on the Big Island and Maui (Kraus et al, 1999). By 2001, there were over 200 populations on the Big Island, 36 on Maui, 14 on Oahu and 2 on Kauai (Kraus and Campbell, 2002).

Eradication efforts have been very successful on some islands. For example, on Kauai, there is now only one population, and control efforts have kept this from spreading and reduced it to a very small area (it remains on private property where the state does not have access). On Oahu, control efforts of infested nurseries, plant retailers and the one naturalized population were successful, such that Oahu has no known breeding

Figure 26.1 Eleutherodactylus coqui *in potted nursery plant*

populations (Anonymous, 2010). While there were reports of frogs on Molokai in 2001 and 2007 and Lanai in 2002, these individuals were eradicated and these islands are no longer thought to have frogs (Anonymous, 2010).

Naturalized populations still exist on Maui and Big Island. Maui had 14 naturalized population centres but now considers 7 of those eradicated, 6 to have very low numbers, and Maliko Gulch to be the last stronghold (Anonymous, 2010). There is a massive effort under way to eradicate the coqui in Maliko Gulch, which covers a 90ha area, and make Maui coqui free. The Big Island is a different story. On the Big Island, most of eastern part of the island is infested, and there are many established populations on the west side as well. Between 2006 and 2008, coqui-occupied areas expanded from 2800 to at least 25,000ha (Figure 26.2). Coquis are not believed to be eradicable on the Big

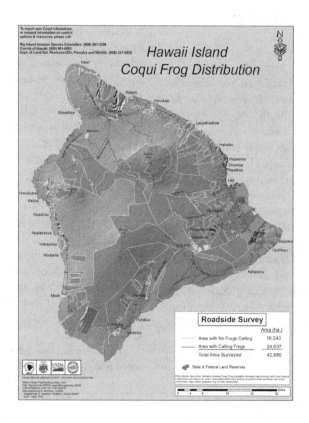

Figure 26.2 *Distribution of* Eleutherodactylus coqui *on the Big Island*
A colour version of this map is available at http://www.hawaiiinvasivespecies.org/iscs/biisc/pdfs/
bicoquiroadsidesurveymap2009.pdf

Island, but it is thought that areas may remain coqui free (Anonymous, 2010).

At first, coquis primarily spread through the sale and movement of nursery products (Kraus, 2003). For example, both populations that became naturalized on Kauai and Oahu originated from infested shipments sent to nurseries from the Big Island. However, several sites on the Big Island were established through intentional introductions conducted by those who wanted to encourage the coqui's presence in Hawaii, show that they were too widespread to eradicate, and as a misguided insect control effort. They were intentionally introduced to state and national parks and to private properties (Kraus and Campbell, 2002; Kraus, 2003). More recently, especially on the Big Island, coquis appear primarily to be spreading from existing populations and via vehicular traffic (Peacock et al, 2009).

The coqui has spread from Hawaii to other areas. Coquis in infested plant shipments have reached both California and Guam (Campbell and Kraus, 2002; Christy et al, 2007). In California, there are confirmed reports inside nurseries and unconfirmed reports outside nurseries (Beard et al, 2009). In Guam, coquis have been captured twice outside of nurseries; in both cases, individuals were eliminated and Guam is thought to be coqui free (Beard et al, 2009).

Ecological Niche

Coquis have direct development (eggs develop into froglets, not tadpoles), and therefore do not require water bodies for any life stage. However, coquis, like all anurans, have to balance thermoregulation and hydroregulation because of their permeable skin (Preest and Pough, 1989). This is most obviously observed in changes in behaviour and activity with changes in temperature and humidity. For example, frogs move, forage, call and breed more on warm and wet nights than on cold and dry nights (Woolbright, 1985; Townsend and Stewart, 1994; Fogarty and Vilella, 2002).

The need to balance thermoregulation and hydroregulation also determines their distribution (Rogowitz et al, 1999). Coquis inhabit almost anywhere in Puerto Rico from sea level to the highest peak (1200m) as long as there is high humidity and adequate cover (Schwartz and Henderson, 1991). Their densities are highest in forested habitats, typically reaching

around 20,000 frogs ha^{-1} (Stewart and Woolbright, 1996), but they also use other more marginal habitats, such as trees in urban areas and buildings (Joglar, 1998).

On the Big Island, coquis spread quickly at low elevations (<500m) of the eastern side, where mean annual precipitation is higher, but slower at high elevations (>1000m) and on the western side, where precipitation is lower (Chu and Chen, 2005). The highest elevation populations are found at 1200m, even though the highest peaks in Hawaii are around 4200m. Invasion into higher elevation forests is of concern because many endemic species are restricted to these habitats (Beard and Pitt, 2005). In Hawaii, coquis have primarily established in forests along roadsides, nurseries, residential gardens, resort areas, refuse areas and state parks.

In forests, the coqui prefers to forage and call on large-leafed tree species, such as *Cecropia*, *Heliconia* and palms, which are often found near streams (Figure 26.3) (Beard et al, 2003b). They prefer these species because they support their weight for calling and foraging, and they use large fallen leaves and leaf axils for nesting and diurnal retreat sites (Townsend, 1989; Beard et al, 2003b). Nesting and retreat sites are the primary factor limiting their populations (Stewart and Pough, 1983; Woolbright, 1991,1996). Thus, areas with more vegetation structure (i.e. more nesting and retreat sites) have more frogs (Fogarty and Vilella, 2001; Beard et al, 2008).

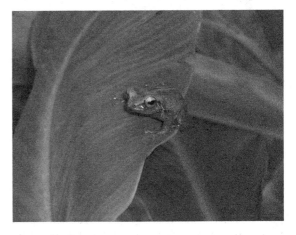

Source: William Pitt

Figure 26.3 Eleutherodactylus coqui *on a* Heliconia *leaf*

Economic Impacts

The main public concern regarding the coqui is the noise from their calls (80–90dB at 0.5m), which is greater than levels set to minimize interference with the enjoyment of life (Beard and Pitt, 2005). As a result, as people choose properties free or far from calls, property owners on the Big Island have felt the economic impacts of the invasion. If frogs are present before selling a property or home, there is a requirement to disclose this information. It has been determined that coquis cause an average of 0.16 per cent loss of real estate value per sale, which, when projected across the Big Island, is estimated to lower property values by $7.6 million (Kaiser and Burnett, 2006).

Because coquis are known to spread through the movement of plant products, the invasion has also affected Hawaii's nursery and floriculture industries, primary industries in Hawaii. In 2001, the Hawaii Department of Agriculture designated the coqui as a 'pest' and 'injurious wildlife', which makes it illegal to release, transport or export coquis. Because of this, these industries have had to pay to treat infestations (i.e. for added labour and treatment costs), they have lost time in shipping products, and they have lost products as ports of entry reject and destroy shipments (Anonymous, 2010). Nurseries with infestations have also experienced decreased sales (Beard et al, 2009). A collaborative agency certificate programme, Stop Coqui Hawaii, was initiated to educate nursery owners about protocols to reduce coqui and the public about which vendors are coqui free, but funding for the programme was discontinued (Anonymous, 2010).

County, state and federal governments also incur costs to control coqui. Costs for public agencies exceeded $4 million in 2006, but have declined in recent years. For example, the State of Hawaii Legislature spent $2 million for frog control in 2006, but only $800,000 in 2007, $400,000 in 2008, and $100,000 in 2009 (Anonymous, 2010). Current funding is not thought sufficient to keep Oahu and Kauai coqui free, eliminate frogs from Maui, and maintain levels of control on the Big Island. To do so is estimated to cost $150,000 per island each year for Oahu and Kauai; $800,000 year^{-1} for Maui; and $1.2 million year^{-1} for the Big Island (Anonymous, 2010).

Ecological Impacts

Because there are no native terrestrial amphibians or reptiles in Hawaii (Kraus, 2003), there were many concerns about the coqui's potential impacts on Hawaii's fragile native ecosystems (Kraus et al, 1999). The coqui has been described as one of the most abundant amphibians in the world, with densities approaching 50,000ha^{-1} at times in Puerto Rico (Stewart and Woolbright, 1996). Because of this and because the coqui is a generalist insectivore, it was thought that its most likely impacts would be through predation on invertebrate numbers (Beard and Pitt, 2005).

In areas in Hawaii where coquis consistently reach densities over 90,000 frogs ha^{-1}, they are thought to consume 690,000 invertebrates ha^{-1} night^{-1} (Beard et al, 2008) and reduce invertebrate populations (Sin et al, 2008). Fortunately, coquis have been found to consume primarily non-native leaf litter invertebrates in Hawaii: ants, amphipods and isopods (Beard, 2007). However, there are groups (including Acarina, Collembola, Gastropoda, Diptera and Coleoptera) that make up a significant portion of their diets and contain native species (Beard, 2007).

Coquis also may indirectly influence the ecosystem processes that invertebrates control. For example, invertebrates play key roles in breaking down plant and leaf litter material. In Puerto Rico, herbivory rates were lower, and plant growth and leaf litter decomposition rates were higher with than without coquis (Beard et al, 2003a). Similar patterns have been found in Hawaii (Sin et al, 2008). These results suggest that coquis could increase nutrient cycling rates in Hawaii and confer a competitive advantage to non-native plants in an ecosystem where natives evolved under nutrient-poor conditions (Beard and Pitt, 2005; Sin et al, 2008).

Other hypotheses regarding impacts include coquis competing with native insectivores, such as endemic birds, for prey (Kraus et al, 1999; Beard and Pitt, 2005). For example, the 'elepaio' (*Chasiempis* spp.), the 'i'iwi' (*Vestiaria coccinea*) and the endangered Hawaiian hoary bat (*Lasiurus cinereus semotus*) share prey and elevations with coquis (Beard and Pitt, 2005). Kraus et al (1999) suggest that coquis may increase native bird predators, such as the black rat (*Rattus rattus*) and small Indian mongoose (*Herpestes javanicus*). although coquis have been found to be a negligible part of their diets (Beard and Pitt, 2006). Finally, coquis may serve as a food source for other potentially devastating bird predators, such as the brown tree snake (*Boiga irregularis*) or other arboreal snakes, if introduced (Beard and Pitt, 2005).

Management Approaches

Since 1998, US Department of Agriculture (USDA) Wildlife Services has tested over 90 chemical agents (agricultural pesticides and pharmaceutical and household products) and 170 chemical formulations as potential frog toxicants. Only eight chemical products were highly effective (>80 per cent laboratory efficacy) and since 2001 only three (caffeine, hydrated lime and citric acid) were at various points in time approved for frog control.

While caffeine was very effective, it was only legal for use for registration and testing from 2001 to 2002, and never received government approval for more widespread use, primarily because of concerns regarding potential human health effects. Hydrated lime (3 to 6 per cent solutions) was also found to be highly effective, and legal for use from 2005 to 2008. Homeowners like it because it is inexpensive (~$0.02L^{-1}), but it leaves a white residue on plants, which makes it undesirable in nursery settings, and there are safety concerns because of its caustic effects (Pitt and Doratt, 2005).

At this time, citric acid, a minimum risk pesticide, is the only chemical that can be used legally for controlling coquis in Hawaii without restrictions. Citric acid (8 to 16 per cent solutions) is very effective (Pitt and Sin, 2004a; Pitt and Doratt, 2006; Tuttle et al, 2008; Doratt and Mautz, unpublished data). Its drawbacks include phytotoxic effects on plants, it can leave white to yellow dots on leaves, and it is relatively expensive (~$0.54L^{-1}) (Pitt and Sin, 2004b).

Hot water is also effective at killing frogs and eggs. Both sprayed hot water applied at 45°C for three minutes and vapour heat applied at 45°C, 90 per cent humidity will kill frogs (Hara et al, 2010). However, some plant species are sensitive to heat treatments (Hara et al, 2010).

Mechanical control has also been effective. Removing vegetation reduces the number of frogs in an area (Beard et al, 2008). Hand-capturing can effectively eliminate frogs if few are present (Beard, 2001). Traps providing retreat or nest sites capture frogs and eggs but must be monitored regularly to discourage breeding (Sugihara,

2000). Traps containing calling males can attract females but do not capture many frogs, and simple barriers can be used to contain frogs in small areas.

There have been suggestions to introduce a biocontrol agent to Hawaii, especially because there are no native frogs. However, no organism with the potential to reduce coquis has been identified. For example, chytrid fungus (*Batrachochytrium dendrobatidis* or Bd), which has been implicated in global amphibian declines, was proposed for introduction, but coquis are relatively resistant (Carey and Livo, 2008); Bd is already present in Hawaiian coqui (Beard and O'Neill, 2005); and the risk of spreading Bd to other areas outweighed the potential benefit of its introduction (Beard and O'Neill, 2005).

Investigations into potential parasites for biocontrol found eight species in coqui from Puerto Rico and two different species in coqui from Hawaii (Marr et al, 2008). Of the eight species found in Puerto Rico, one nematode species was identified as having potential as a safe and effective biocontrol agent. However, further testing suggested it only had limited potential as a biocontrol agent as it reduced coqui jumping performance but did not affect coqui growth or survivorship (Marr et al, 2010).

Control Effectiveness

Around 2005, the state of Hawaii began a major campaign to control the coqui. As mentioned previously, these efforts were very successful on Oahu and Kauai. For example on Oahu, control efforts on the one naturalized population were successful, with mostly ground operations (citric acid spraying, spot spraying operations and hand-capture), such that Oahu now has no naturalized populations. On Kauai, there was one naturalized population covering 6ha (Anonymous, 2010), but control efforts including large removals of vegetation and citric acid ground operations reduced this population to a very small area.

On Maui, control efforts led to the eradication of seven population centres and reduced another six populations, in addition to treating incipient populations (Anonymous, 2010). Eradication primarily occurred with ground operations of citric acid spraying, although hand-capturing was effective at removing incipient populations. On efforts in Maliko Gulch, the single, remaining large population has been problematic because of the terrain. Operations there include a variety of techniques: citric acid ground operations (citric acid fixed line delivery systems, trailer mounted storage tanks and spray systems, and spot spray operations), a high volume citric acid sprinkler system that can spray out over the gulch, aerial (helicopter) citric acid operations, and follow-up hand-capturing.

The Big Island has at least 25,000ha infested (Anonymous 2010). However, the area treated each year has been declining with reductions in funding. For example, over 415ha were treated in 2007, 340ha in 2008 and 147ha in 2009 (Figure 26.4). Over the years, treatments with citric acid, hydrated lime and mechanical techniques have been used to eradicate populations from isolated areas (such as greenhouses) and incipient populations. Aerial (helicopter) and

ground operations of citric acid were effective in reducing frog densities threefold in Manuka Natural Area Reserve (Tuttle et al, 2008). Traps have been effective where there are few frogs and natural retreat sites, such as in resort areas.

The main vector for the inter-island transportation of the coqui remains infested nursery products. Especially on the islands of Kauai and Oahu, which are coqui free, there needs to be effective inspection of shipments. Many shipments from the Big Island to these islands have been returned and destroyed (Anonymous, 2010). During quarantine, citric acid or, in limited areas, hot water treatments are used to eliminate frogs and their eggs from potted plants. However, these methods are not effective for large plant shipments and some growers are dissatisfied with the phytotoxic effects.

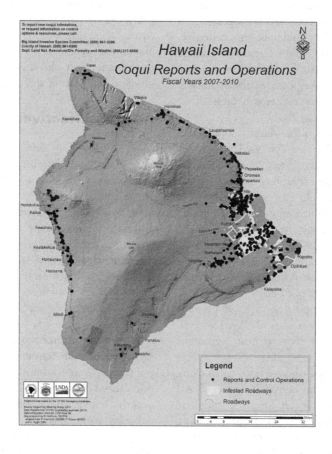

Source: McGuire et al (2010)

Figure 26.4 *Areas treated for control on the Big Island*

Opinions Regarding the Species

Coquis have been in the consciousness of Puerto Rican island dwellers for thousands of years as evidenced by the Taino Indians' (native to Puerto Rico) petroglyphs depicting coquis. Children in Puerto Rico grow up learning about these frogs, not only because of their ubiquity and conspicuous calls, but because the song of the coqui is the focus of Puerto Rican folk tales. One of these tales concludes that if coquis were ever to leave Puerto Rico, they would no longer sing. Furthermore, because Puerto Rico has no native ground mammals or other such charismatic fauna, this small frog became Puerto Rico's unofficial mascot. Thus, it probably comes as no surprise that during some of the initial control efforts, there was a campaign to have coquis shipped back to Puerto Rico. However, not everyone in Puerto Rico loves the frog, and there were individuals who called control operation managers in Hawaii to share methods for killing them.

There are also individuals in Hawaii that opposed control efforts (Kraus and Campbell, 2002). This resistance is best exemplified by the non-profit organization, the Coqui Hawaiian Integration and Reeducation Project (or CHIRP), which has a 30ha Coqui sanctuary in the south-eastern part of the Big Island. In addition to CHIRP, there are many individuals who opposed coqui control in their local community, and have been resistant to the community groups working to control coqui on their properties. Resistance to control probably has many roots, from those who:

- generally protest the control of any organism, but particularly vertebrates;
- enjoy the call and species;
- do not understand the problems associated with non-native species, especially when amphibians are declining globally;
- believe coquis might control unwanted pests; and

- do not approve of the funds and effort spent on control, especially when it involves placing chemicals in the environment or cutting down vegetation.

By contrast, there has been a lot of public support to control coquis in Hawaii. This is best exemplified by community groups such as the Kaloko Mauka Coqui Coalition, Kohala Coqui Coalition and Volcano Volunteer Coqui Patrol. These and other similar community associations organize themselves to control local infestations. The groups raise funds to rent or purchase equipment to control coquis, and have invested endless hours of volunteer time monitoring and controlling populations. For example, the groups received 80 awards up to $5000 from the County of Hawaii in 2006 and 2007 for chemicals, safety equipment and other expenses to control frogs (Anonymous, 2010). In fact, much of the control efforts on the Big Island have been conducted by these groups; in 2008, 43 per cent of land treated was done by community associations (Anonymous, 2010). Coqui control groups have many motivating factors, including keeping yards and forests near their homes quiet, improving quality of life (i.e. sleeping better) and maintaining property values, but some of these individuals also understand the coqui is non-native to Hawaii and believe it does not belong there.

As long as coquis and people have interacted there have been strong feelings about them. At this point in time, it is unlikely that the coqui will be eradicated from the Hawaiian Islands. With this invasive species, social issues will play a role in the final outcome.

Acknowledgements

Support for this research came from the Jack Berryman Institute at Utah State University, the US Fish and Wildlife Service, Hawaii Invasive Species Council and USDA/NWRC/APHIS/Wildlife Services.

References

Anonymous (2010) *Hawai'i's Coqui Frog Management, Research and Education Plan*, State of Hawaii

Austin, D. F. and Schwartz, A. (1975) 'Another exotic amphibian in Florida, *Eleutherodactylus coqui*', *Copeia*, vol 1975, p188

Beard, K. H. (2001) 'The ecological roles of a terrestrial frog, *Eleutherodactylus coqui* (Thomas), in the nutrient cycles of a subtropical wet forest in Puerto Rico', PhD Thesis, Yale University at New Haven, CT, USA

Beard, K. H. (2007) 'Diet of the invasive frog, *Eleutherodactylus coqui*, in Hawaii', *Copeia*, vol 2007, pp281–291

Beard, K. H. and O'Neill, E. M. (2005) 'Infection of an invasive frog *Eleutherodactylus coqui* by the chytrid fungus *Batrachochytrium dendrobatidis* in Hawaii', *Biological Conservation*, vol 126, pp591–595

Beard, K. H. and Pitt, W. C. (2005) 'Potential consequences of the coqui frog invasion in Hawaii', *Diversity and Distributions*, vol 11, pp427–433

Beard, K. H. and Pitt, W. C. (2006) 'Potential predators of an invasive frog (*Eleutherodactylus coqui*) in Hawaiian forests', *Journal of Tropical Ecology*, vol 22, pp1–3

Beard, K. H., Eschtruth, A. K., Vogt, K. A., Vogt, D. J. and Scatena, F. N. (2003a) 'The effects of the frog *Eleutherodactylus coqui* on invertebrates and ecosystem processes at two scales in the Luquillo Experimental Forest, Puerto Rico', *Journal of Tropical Ecology*, vol 19, pp607–617

Beard, K. H., McCullough, S. and Eschtruth, A. (2003b) 'A quantitative assessment of habitat preferences for the Puerto Rican terrestrial frog, *Eleutherodactylus coqui*', *Journal of Herpetology*, vol 10, pp1–17

Beard, K. H., Al-Chokhachy, R., Tuttle, N. C. and O'Neill, E. M. (2008) 'Population density and growth rates of *Eleutherodactylus coqui* in Hawaii', *Journal of Herpetology*, vol 42, pp626–636

Beard, K. H., Price, E. A. and Pitt, W. C. (2009) 'Biology and impacts of Pacific island invasive species: *Eleutherodactylus coqui*, the coqui frog (Anura: Leptodactylidae)', *Pacific Science*, vol 63, pp297–316

Campbell, E. W., III and Kraus, F. (2002) 'Neotropical frogs in Hawaii: Status and management options for an unusual introduced pest', in R. M. Timm and R. H. Schmitz (eds) *Proceedings of the 20th Vertebrate Pest Conference*, University of California, Davis, pp316–318

Carey, C. and Livo, L. (2008) 'To use or not to use the chytrid pathogen, *Batrachochytrium dendrobatidis*, to attempt to eradicate coqui frogs from Hawai'i', *First International Conference on the Coqui Frog, Feb 7–9, 2008*, Hilo, Hawaii, www.ctahr.hawaii.edu/coqui/WEBCCareyFICCF.pdf.pdf

Christy, M., Savidge, J. and Rodda, G. (2007) 'Multiple pathways for invasion of anurans on a Pacific island', *Diversity and Distributions*, vol 13, pp598–607

Chu, P. S. and Chen, H. Q. (2005) 'Interannual and interdecadal rainfall variations in the Hawaiian Islands', *Journal of Climate*, vol 18, pp4796–4813

Fogarty, J. H. and Vilella, F. J. (2001) 'Evaluating methodologies to survey *Eleutherodactylus* frogs in montane forests of Puerto Rico', *Wildlife Society Bulletin*, vol 29, pp948–955

Fogarty, J. H. and Vilella, F. J. (2002) 'Population dynamics of *Eleutherodactylus coqui* in Cordillera Forest reserves of Puerto Rico', *Journal of Herpetology*, vol 36, pp193–201

Hara, A. H., Jacobsen, C. M., Marr, S. R. and Niino-DuPonte, R. Y. (2010) 'Hot water as a potential disinfestation treatment for an invasive anuran amphibian, the coqui frog, *Eleutherodactylus coqui* Thomas (Leptodactylidae), on potted plants', *International Journal of Pest Management*, vol 56, pp255–263

Joglar, R. L. (1998) *Los coquíes de Puerto Rico: su história natural y conservation*, University of Puerto Rico, San Juan

Kaiser, B. and Burnett, K. (2006) 'Economic impacts of *E. coqui* frogs in Hawaii', *Interdisciplinary Environmental Review*, vol 8, pp1–11

Kraus, F. (2003) 'Invasion pathways of terrestrial vertebrates', in G. M. Ruiz and J. T. Carlton (eds) *Invasive Species: Vectors and Management Strategies*, Island Press, Washington, DC, pp68–92

Kraus, F. and Campbell, E. W. (2002) 'Human-mediated escalation of a formerly eradicable problem: The invasion of Caribbean frogs in the Hawaiian Islands', *Biological Invasions*, vol 4, pp327–332

Kraus, F., Campbell, E. W., Allison, A. and Pratt, T. (1999) '*Eleutherodactylus* frog introductions to Hawaii', *Herpetological Review*, vol 30, pp21–25

MacLean, W. P. (1982) *Reptiles and Amphibians of the Virgin Islands*, Macmillan Caribbean, London, Basingstoke

Marr, S. R., Mautz, W. J. and Hara, A. H. (2008) 'Parasite loss and introduced species: A comparison of the parasites of the Puerto Rican tree frog, (*Eleutherodactylus coqui*), in its native and introduced ranges', *Biological Invasions*, vol 10, pp1289–1298

Marr, S. R., Johnson, S. A., Hara, A. R. and McGarrity, M. E. (2010) 'Preliminary evaluation of the potential of the helminth parasite *Rhabdias elegans* as a biological control agent for invasive Puerto Rican coquís (*Eleutherodactylus coqui*) in Hawaii', *Biological Control*, vol 54, pp69–74

McGuire, R., Hamilton, R., Graves, P. and Rygh, C. (2010) *Coqui Frog Working Group*, DLNR-DOFAW, County of Hawaii, USDA/WS, BIISC, HDOA

Meshaka, W. E., Jr, Butterfield, B. P. and Hauge, J. B. (2004) *Exotic Amphibians and Reptiles of Florida*, Krieger Pub. Co., Melbourne, Florida

Peacock, M. M., Beard, K. H., O'Neill, E. M., Kirchoff, V. and Peters, M. B. (2009) 'Strong founder effects and low genetic diversity in introduced populations of coqui frogs', *Molecular Ecology*, vol 18, pp3603–3615

Pitt, W. C. and Doratt, R. E. (2005) 'Efficacy of hydrated lime on *Eleutherodactylus coqui* and an operational field-application assessment on the effects on non-target invertebrate organisms', USDA/APHIS/WS/NWRC, Hilo, Hawaii

Pitt, W. C. and Doratt, R. E. (2006) 'Screening for the evaluation of selected chemicals and pesticides to control *Eleutherodactylus* frogs in Hawaii', USDA/APHIS/WS/NWRC, Hilo, Hawaii

Pitt, W. C. and Sin, H. (2004a) 'Dermal toxicity of citric acid based pesticides to introduced *Eleutherodactylus* frogs in Hawaii', USDA/APHIS/WS/NWRC, Hilo, Hawaii

Pitt, W. C. and Sin, H. (2004b) 'Testing citric acid use on plants', *Landscape Hawaii*, July/August, pp5–12

Preest, M. R. and Pough, F. H. (1989) 'Interaction of temperature and hydration on locomotion of toads', *Functional Ecology*, vol 3, pp693–699

Rivero, J. A. and Joglar, R. L. (1979) '*Eleutherodactylus cochranae*', *Herpetological Review*, vol 10, p101

Rogowitz, G. L., Cortés-Rivera, M. and Nieves-Puigdoller, K. (1999) 'Water loss, cutaneous resistance, and effects of dehydration on locomotion of *Eleutherodactylus* frogs', *Journal of Comparative Physiology B: Biochemical Systemic and Environmental Physiology*, vol 169, pp179–186

Schwartz, A. and Henderson, R. W. (1991) *Amphibians and Reptiles of the West Indies: Description, Distributions, and Natural History*, University of Florida Press, Gainsville

Sin, H., Beard, K. H. and Pitt, W. C. (2008) 'An invasive frog, *Eleutherodactylus coqui*, has top-down effects on new leaf production and leaf litter decomposition rates through nutrient cycling in Hawaii', *Biological Invasions*, vol 10, pp335–345

Stewart, M. M. and Pough, F. H. (1983) 'Population density of tropical forest frogs: Relation to retreat sites', *Nature*, vol 221, pp570–572

Stewart, M. M. and Woolbright, L. L. (1996) 'Amphibians', in D. P. Reagan and R. B. Waide (eds) *The Food Web of a Tropical Rain Forest*, University of Chicago Press, Chicago, pp363–398

Sugihara, R. T. (2000) 'Coqui trap study summary', Report to: USDA/APHIS/WS/NWRC, Hilo, Hawaii

Townsend, D. S. (1989) 'The consequences of microhabitat choice for male reproductive success in a tropical frog (*Eleutherodactylus coqui*)', *Herpetologica*, vol 45, pp451–458

Townsend, D. S. and Stewart, M. M. (1994) 'Reproductive ecology of the Puerto Rican frog *Eleutherodactylus coqui*', *Journal of Herpetology*, vol 28, pp34–40

Tuttle, N. C., Beard, K. H. and Al-Chokhachy, R. (2008) 'Aerially applied citric acid reduces the density of an invasive frog', *Wildlife Research*, vol 35, pp676–683

Velo-Antón, G., Burrowes, P. A., Joglar, R., Martínez-Solano, I., Beard, K. H., Velo-Antón, G., Burrowes, P. A., Joglar, R., Martínez-Solano, I., Beard, K. H. and Parra-Olea, G. (2007) 'Phylogenetic study of *Eleutherodactylus coqui* (Anura: Leptodactylidae) reveals deep genetic fragmentation in Puerto Rico and pinpoints origins of Hawaiian populations', *Molecular Phylogenetics and Evolution*, vol 45, pp716–728

Wilson, L. D. and Porras, L. (1983) *The Ecological Impact of Man on the South Florida Herpetofauna*, University of Kansas, Museum of Natural History, Special Publication No 9

Woolbright, L. L. (1985) 'Patterns of nocturnal movement and calling by the tropical frog *Eleutherodactylus coqui*', *Herpetologica*, vol 41, pp1–9

Woolbright, L. L. (1991) 'The impact of Hurricane Hugo on forest frogs in Puerto Rico', *Biotropica*, vol 23, pp462–467

Woolbright, L. L. (1996) 'Disturbance influences long-term population patterns in the Puerto Rican frog, *Eleutherodactylus coqui* (Anura: Leptodactylidae)', *Biotropica*, vol 28, pp493–501

27

Rana [Lithobates] catesbeiana Shaw (American bullfrog)

Antonia D'Amore

Introduction

American bullfrogs, (Figure 27.1) (*Rana [Lithobates] catesbeiana* – henceforth 'bullfrogs'), are on the IUCN's list of 100 worst invasive species due to their broad global distribution and widespread effects on native communities where they have been introduced (Lowe et al, 2000). Bullfrogs are native to eastern North America, with the Rocky Mountains providing the western boundary of their natural range (Bury and Whelan, 1984). Native populations occur as far north as Canada and as far south as central Florida and north-eastern Mexico. The vast differences in climate across their native range probably play a contributing role in the success of bullfrog populations in a variety of different locales globally (Adams and Pearl, 2007). Their introduced range currently includes many countries in Europe, Asia, South America and the Caribbean islands and is thought to be expanding in many regions (Figure 27.2) (Adams and Pearl, 2007; Ficetola et al, 2010). Changes in land use will be relevant when trying to understand and predict expansion of their introduced range (Ficetola et al, 2010). Bullfrogs were introduced to the western US in the late 1800s in order to provide an additional source of frog legs after native stocks were depleted (Jennings and Hayes, 1985). Much of the research on bullfrog effects on native species has occurred in this region.

Natural History and Ecological Niche

Bullfrogs are the largest frog species in North America and among the largest frogs globally. Females reach slightly larger sizes than males, with adult females growing to average sizes of approximately 16cm SVL and males to approximately 15cm SVL (Howard, 1981). This sexual size dimorphism can vary with features of their introduced climate (Xuan et al, 2010). Due in part to their large size, female bullfrogs lay egg masses with up to 25,000 eggs, up to one quarter of their body mass, during the breeding season in late spring and early summer. The largest females have been recorded as laying more than 40,000 eggs (Bury and Whelan, 1984). These egg masses are attached to emergent vegetation in shallow water and will form large sheets on or just below the surface of the water. Choice oviposition sites are fiercely guarded by males, who fight to establish territories encompassing the best sites. Males are polygynous and mate with females that enter their territories and approach the calling male (Emlen, 1968; Wiewandt, 1969).

Bullfrogs can occupy a broad range of different wetland habitat types, including artificial and natural wetlands, streams, lakes and temporary pools, though they usually rely on permanent water sources for breeding habitat (Gahl et al, 2009). Within these

Source: K. D'Amore

Figure 27.1 *The American bullfrog* (Rana [Lithobates] catesbeiana)

broader habitat types, bullfrogs preferentially occupy shallow, lentic habitats with emergent vegetation (Bury and Whelan, 1984). While bullfrog populations are more likely to become invasive in areas with high annual precipitation (Ficetola et al, 2007) (which are more suitable for overland dispersal than arid regions), as long as suitable freshwater habitat is present, they are capable of invading arid areas as well (Rosen and Schwalbe, 1995). However, many of the best predictors of bullfrog invasions are directly tied to human behaviour, rather than habitat features (i.e. the presence of high density frog farms with simple enclosures (Liu and Li, 2009)).

As their native range spans a broad variety of climate types, the timing of breeding and other phenological events varies (Casper and Hendricks, 2005). In southern latitudes, for example, breeding begins in early spring and larvae are able to metamorphose in the same year. In cooler climates, bullfrogs may not begin breeding until the summer and primarily remain in the larval state for two to three years before transforming. Previously, it was commonly accepted that in all but the most southern bullfrog populations, the species requires a minimum of 12 months in its larval form before metamorphosis, thus requiring perennial water bodies for successful breeding (Bury and Whelan, 1984). However, recent research has determined that these animals can breed in more temporary habitat (Provenzano and Boone, 2009).

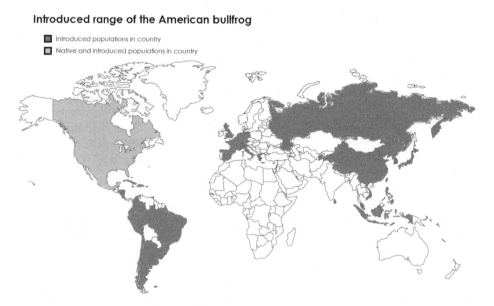

Source: Antonia D'Amore, based on open-access data

Figure 27.2 *Global distribution of the American bullfrog*

When managing hydroperiod in water bodies, there is the possibility for inadvertent selection of tadpoles that can metamorphose more quickly, further broadening the available habitat for this successful invader (Adams and Pearl, 2007).

Predation

Bullfrogs are generally the largest frog in regions where they are introduced and a large percentage of their diet is smaller frogs, both conspecifics and other species. One study found up to 80 per cent of their diet may be comprised of smaller conspecifics (Stuart and Painter, 1993) and their voracious appetites are implicated in the declines of more than a dozen North American amphibians (Casper and Hendricks, 2005). As a large, gape-limited predator, they are also known to consume small mammals, birds and snakes (Minton, 1949), though much of their diet is composed of invertebrates (Stewart and Sandison, 1972). One of the ways to measure the strong control that bullfrogs have on a freshwater community is to measure changes in community composition following bullfrog extinction or extirpation. Native green frogs (*Rana clamitans*), for example, showed a fourfold increase in numbers after bullfrogs went extinct from a site (Hecnar and M'Closkey, 1997). California red-legged frogs (*Rana draytonii*) similarly increased in numbers throughout the course of a bullfrog eradication campaign (D'Amore et al, 2009a). In both cases, the changes in number of native frogs encountered were probably due to both behavioural changes and increased population size.

Despite the bullfrogs' role as significant aquatic predators, there has been some debate as to whether bullfrogs are directly, solely responsible for the decline of native species, or if their presence is linked to other more detrimental factors, such as habitat change (Adams, 1999; Kiesecker et al, 2001) or introduced fish species, that greatly enhance their apparent effect (Hayes and Jennings, 1986; Adams et al, 2003). Certainly, the cumulative impacts of anthropogenic changes to habitat quality and community composition must be considered when evaluating causes for species decline.

Competition

Due to size alone, bullfrogs are likely to be the dominant anuran competitor in an aquatic community.

Experimental work in a California river found that bullfrog larvae exerted strong negative effects on survivorship rates and size at metamorphosis in two native species (Kupferberg, 1997). Another study found that competition from post-metamorphic bullfrogs was likely to be significant only for species that were also aquatic in their habitat preferences and similar in size. Otherwise, competitive interactions are dwarfed by predation (Werner et al, 1995).

Aside from their large size, there are a number of human-mediated factors that appear to give bullfrogs a competitive edge in their introduced range. Anthropogenic changes to freshwater habitats include destruction of native habitats, creation of artificial habitats, introduction of other invasive predators and alteration of the hydroperiod toward more perennial freshwater. Each of these changes generally favours bullfrogs over native competitors (Adams, 1999; Maret et al, 2006; D'Amore et al, 2010).

For example, artificial habitats tend to support highly simplified aquatic food webs with resources that are clumped in distribution, rather than scattered more uniformly. Kiesecker et al (2001) found that in habitats with clumped resources, bullfrog larvae reduced survivorship and size at metamorphosis in native red-legged frog tadpoles. In habitats with scattered resources, there was no significant effect of bullfrog larvae (Kiesecker et al, 2001). Artificial habitats are also often kept permanently wet in regions of the world where natural habitats dry down in the summer months. Pond permanence is strongly correlated with invasive bullfrog presence (Adams, 2000; D'Amore et al, 2010), and habitat change is a strong driver in the success of bullfrogs (Adams, 1999).

Another example of the correlation between human modification and bullfrog success is through the introduction of non-native fish species. Bullfrog eggs and tadpoles are generally unpalatable and are therefore able to coexist readily with many species of fish (Walters, 1975; Kruse and Francis, 1977). Indeed, in their invasive range, the presence of non-native fish species with which bullfrogs co-evolved, such as bluegill sunfish, may facilitate their invasional success (Adams et al, 2003). Reduced predation of bullfrog eggs and tadpoles, compared to native frogs, may then give bullfrogs a large numerical advantage over native frogs, which did not evolve in habitats with these fish and do not have the same defences. The numerical advantage, combined with their larger size and correspondingly larger metabolic needs, means that bullfrogs are formidable introduced competitors (Kiesecker et al, 2001; Wu et al, 2005; Wang et al, 2007).

Changes in habitat use

Some of the numerical reduction in native frog numbers in the presence of bullfrogs may be due to behavioural changes in the native species, rather than solely mortality. California red-legged frogs, for example, use shoreline habitat with more vegetation and hide a greater percentage of their body when bullfrogs are present (D'Amore et al, 2009a). More broadly, the presence of bullfrogs causes these native frogs to become more cryptic and more terrestrial. These changes in habitat use may make the native frogs more susceptible to terrestrial predators, disrupt their feeding patterns and/or increase predation during their attempts to attract and acquire mates. Bullfrog presence also appears to reduce ontogenetic resource partitioning among native frog size classes, causing adult frogs and smaller juveniles to occupy the same shallow shoreline habitat rather than spreading out across the habitat and segregating by size classes as they do in the absence of bullfrogs. This sharing of habitat by larger adults and smaller juveniles may increase rates of cannibalism in the native frogs (D'Amore et al, 2009a). Additionally, counts of California red-legged frog adults increase faster than can be attributed to population growth or immigration when numbers of bullfrog adults and juveniles are sharply reduced (D'Amore et al, 2009a), suggesting less cryptic behaviour in the absence of bullfrogs.

Changes in microhabitat use are not restricted to the post-metamorphic stages. In the aforementioned work by Kiesecker et al (2001), differences in habitat use among native northern red-legged frog larvae were noticed when bullfrog larvae were present. Substantially larger bullfrog larvae were able to sequester the majority of resources, resulting in slower growth and smaller size at metamorphosis for the red-legged frog (Kiesecker and Blaustein, 1998).

Disease

Bullfrogs are a carrier of the globally devastating amphibian chytrid fungus (*Batrachochytrium*

dendrobatidis) (Daszak et al, 2004). Amphibian chytrid has caused widespread decline of amphibian populations and appears to have driven some Neotropical and Australian species to extinction (Marca et al, 2005; Lips et al, 2006). Introduced bullfrogs in seven of eight countries were found to be infected by amphibian chytrid (Garner et al, 2006), despite not showing the characteristic lethargy and poor body condition that a sick animal would demonstrate. Bullfrog farms worldwide also appear to play an important role as a propagator of the disease and source of new invasions (Hanselmann et al, 2004; Schloegel et al, 2009).

The global trade in bullfrogs, some of which are wild caught and many of which are farmed, probably plays a significant role in the spread of chytrid and other amphibian pathogens to new locales. Not only does the transport of bullfrogs globally allow for their introduction into new areas, but the crowded conditions of farms increase the likelihood of culturing pathogens and discharging them into the wild via contaminated water or escaped bullfrogs (Gratwicke et al, 2009).

Reproductive interference

In most amphibian species, females are larger than males and the largest females are capable of laying the largest egg masses (Berven, 1981; Castellano et al, 2004). These large females are therefore highly desirable from an evolutionary perspective. In communities where introduced bullfrogs are larger than the largest native frog species, there is some indication that the presence of juvenile bullfrogs may cause reproductive interference for the native species. For example, prior to bullfrog introduction, 'mate with the largest frog possible' could be an adequate criterion for mate choice; upon bullfrog introduction, native frogs following this instinct may initiate amplexus with the wrong species. In California, it was found that native California red-legged frog males were preferentially initiating amplexus with juvenile bullfrogs that were considerably larger than California red-legged frog females (D'Amore et al, 2009b). Pearl et al (2005) highlight similar interspecific amplexus with juvenile bullfrogs; this and other work suggests that in populations with other stressors, this wasted reproductive effort could limit population growth (Orchard, 1999; Pearl et al, 2005; D'Amore et al, 2009b).

This behaviour is problematic for a number of reasons:

- attempting to mate with an animal that is too large could result in predation, rather than reproduction;
- amplectic pairs are more vulnerable to predation events;
- opportunity cost – these males are wasting time that could be spent mating with a conspecific female (juvenile bullfrogs do not give the species-specific release call that would alert a male frog to the mistake, resulting in prolonged attempts at amplexus); and
- potential risk of increased disease transmission, as bullfrogs are known carriers of amphibian chytrid.

Invasive Bullfrog Management and Control

Management of introductions and disease transmission

As stated above, the frog leg trade has high potential for introducing species, bullfrogs particularly, into new areas and increasing rates of disease transmission globally (Gratwicke et al, 2009). As there is mounting evidence that large differences exist in the virulence of different amphibian chytrid strains, transmission should be a major concern, even as amphibian chytrid infection becomes more broadly distributed around the world (Berger et al, 2005). Due to concern about new introductions of disease and animals, the California Department of Fish and Game recently voted to uphold a ban on the import of non-native frogs and turtles for food, while France has had a law banning both wild capture and farming of frogs since 1980, and the rest of the EU banned importation in the 1990s (Scalera, 2007). A complete ban on both harvesting native amphibians and importing non-natives is probably the only means of stopping the continued problem of pathogen contamination and overharvesting of native species.

Methods of control of existing invasive bullfrog populations

It is established that the best means of controlling invasive bullfrogs is in the preventative stage by not allowing new introductions (Adams and Pearl, 2007).

Once an invasion has occurred, however, there are two main pathways for control of invasive bullfrog populations – direct capture and removal of animals, or manipulation of the aquatic habitat to help prevent successful breeding.

Direct capture

Bullfrogs are active nocturnally and diurnally, as well as being highly aquatic. The best means of capturing adult and juvenile frogs is to carefully survey freshwater habitat at night using high powered headlamps to illuminate the ponds and help in spotting the eye-shine of each individual. In absence of a headlamp, a flashlight can be held next to the eyes. To aid in long distance identification of frogs, binoculars can also be used.

Each animal can then be carefully approached and caught by hand or net. Hawaiian slings (gigs) can also be used to capture and remove this species, with the advantage that the sling doesn't require approaching the animals as closely. In areas with high densities and large adults in deep water, pellet guns with scopes may be used to identify and remove animals. In pond systems with large amounts of structural complexity and thick vegetation, male bullfrog breeding calls may be used to identify male location throughout the breeding season (May–July) and increase the rate of adult capture. Most recently, there has been development of an electroshock device that allows animals to be stunned and removed at a distance of 7 feet with considerable accuracy (Orchard, S., unpublished data).

Habitat manipulation

Manipulating the habitat in order to reduce or eliminate the possibility of successful bullfrog breeding may be the preferable management tool for many conservation practitioners as it avoids direct physical contact with the animals. For this to be a feasible and effective method, there are several conditions that must be met: (1) the water body must be naturally temporary and/or fairly small; (2) the native frog species, if present, must have a breeding period that is shorter than that of bullfrogs; and (3) there must be human control of the hydroperiod. If a water body is naturally temporary, but human addition of water is causing it to be permanent and support breeding bullfrogs, simply cutting off the additional water source may be sufficient. Water manipulation should take into account the needs of the native species present. For freshwater sites where humans do not control the hydroperiod through water control structures or water addition, conservation practitioners have actively pumped water out of sites (costly, labour intensive and only feasible if no native tadpoles are present) or simply removed entire ponds through burying them when no native fauna are present. Routine habitat destruction, however, is not to be recommended as a conservation technique.

In addition to the difficulties mentioned above, the entire mosaic of available habitat should be considered. Eliminating the potential for breeding at one site may do little to reduce the total bullfrog population if there is other breeding habitat in the vicinity. Bullfrogs will flee a drying pond, moving long distances to relocate into new freshwater habitats and potentially compounding the problem. Predicted changes to available habitat should be considered when trying to identify areas for introduced range expansion (Ficetola et al, 2010).

Feasibility of control

There are currently three papers that use mathematical models to address the specific problem of how to control invasive bullfrog populations: Doubledee et al (2003), Govindarajulu et al (2005) and Grey (2009).

The first study deals specifically with methods of bullfrog control in areas that also have a threatened native species, the California red-legged frog (Doubledee et al, 2003). This model concludes that shooting adult frogs would deliver a positive benefit to California red-legged frog population growth, but also take extreme amounts of effort. Therefore, the authors suggest a method of bullfrog control that seeks a balance between effort and efficacy, draining ponds to remove the breeding stage and conducting lower intensity shooting of adults (Doubledee et al, 2003). A recent master's thesis takes the Doubledee model and expands it to consider management actions in several ponds, with juvenile dispersal among the ponds (Grey, 2009). The author concludes that management of bullfrogs is not necessary in temporary ponds and that removing 70 per cent of the larvae in permanent ponds every two years (through drying or seining) is frequent enough to keep a bullfrog population small.

The third model, based on demographic estimates out of British Columbia (Govindarajulu et al, 2005), addresses the question of which life stage bullfrog control efforts should focus on. Using a matrix model, the investigators conclude that removing larvae is labour intensive and therefore likely to only result in partial larval removal, as not all individual larvae can be caught. They conclude that this partial removal could result in larger larvae and higher overall survivorship rates. Similarly, removing adult individuals could reduce the effects of adult bullfrog cannibalism of juveniles, thereby promoting increased success of the juvenile age class. Due to the potential drawbacks to culling adults that their model identified, the authors suggest that the best means of controlling bullfrog populations is through the culling of metamorphosing young in the autumn.

As these modelling efforts reach varying conclusions, turning to case histories would be beneficial in resolving the considerable discrepancies. However, while there is much effort spent in the eradication of this, and other, invasive species (Zavaleta et al, 2001), there is a paucity of published literature on specific bullfrog control efforts, with the exception of a small handful of studies that describe ongoing efforts to control large invasions (Schwalbe and Rosen, 1988; Rosen and Schwalbe, 1995; Banks et al, 2000; Kahrs, 2006) or that describe eradication efforts briefly as part of a broader study (D'Amore et al, 2009a, 2009b). Better care needs to be taken in quantifying the numbers of invasive bullfrogs removed from a system, the hours of effort invested in the eradication and, if possible, the resulting changes in the native amphibian community. Once these data are available through careful, detailed relaying of case histories, a clear message on the utility and feasibility of control efforts will emerge.

Despite this caveat, a few general rules seem to emerge from the collective accounts:

- Control efforts should target all ponds within easy dispersal distance of each other in order to be effective. Otherwise, colonizing frogs from nearby ponds will quickly undo any progress gained.
- Smaller, isolated ponds with less vegetation will take less effort to control bullfrog numbers than larger wetlands with complex vegetation and connectivity to other sites.

• If water levels in the sites are under human control, drying down sites, while keeping in mind the needs of native species, can be an excellent means of removing the possibility of bullfrog breeding and fostering the continued presence of the native amphibian species. Long term widespread water management in this way could ultimately lead to bullfrog populations dying out.

Challenges and Controversies

A primary challenge in the effort to halt the continued introduction of bullfrogs and associated diseases is the global trade in amphibians. Moved internationally for both the pet trade and for human consumption, these practices create new bullfrog populations and thwart control efforts where bullfrogs are already established (Hanselmann et al, 2004; Fisher and Garner, 2007; Gratwicke et al, 2009). A direct conflict between human motivation for profit and conservation concerns is evident and will probably take considerable time to resolve.

There have been varying reports on the extent of impacts of bullfrogs on native amphibian populations (Bury and Whelan, 1984; Rosen and Schwalbe, 1995; Hecnar and M'Closkey, 1997; Banks et al, 2000; Boone et al, 2004; Daszak et al, 2004) and disagreement on the mechanism by which impacts occur (Kiesecker et al, 2001; Adams and Pearl, 2007). While bullfrogs are clearly voracious predators, capable of eating a broad range of taxa, some work suggests that amphibian declines in areas with bullfrogs are caused by factors associated with bullfrog presence (e.g. habitat modification, introduced fish species, altered hydrology,

disease), rather than predation or competition (Hayes and Jennings, 1986; Adams, 1999, 2000; Kiesecker et al, 2001). Despite these disputes, the cumulative direct and indirect effects of bullfrog presence presented previously add another, probably substantial, stressor to amphibian populations that are already affected by a suite of detrimental forces.

Control of non-native invasive species is often controversial, particularly when the species is an animal and control involves direct eradication. Some will claim that invasive species eradication is futile or overly labour intensive, while others maintain that eradication is and will remain an important conservation tool (Zavaleta et al, 2001; Veitch and Clout, 2002). However, in sites with declining, native amphibians, there are often multiple causes that can contribute to reduction in population sizes, from increased ultraviolet radiation, habitat destruction and degradation, chemical contamination, disease and many others. At the same time, there is an emerging understanding that effects of these threats and other stressors may be synergistic, such that the presence of an invasive predator makes an individual more susceptible to disease, to chemical exposure or other factors (Alford and Richards, 1999; Blaustein and Kiesecker, 2002). Ultimately, in the effort to restore communities and maintain biodiversity, there are factors that humans can change on a relatively short timescale (e.g. the size of an invasive population, direct chemical contamination) and those that we can't readily change (e.g. ultraviolet levels, disease). When confronted with the sixth mass extinction event (Wake and Vredenburg, 2008), conservation practitioners must tip the scales to benefit native biodiversity wherever possible.

References

Adams, M. J. (1999) 'Correlated factors in amphibian decline: Exotic species and habitat change in western Washington', *The Journal of Wildlife Management*, vol 63, pp1162–1171

Adams, M. J. (2000) 'Pond permanence and the effects of exotic vertebrates on anurans', *Ecological Applications*, vol 10, pp559–568

Adams, M. and Pearl, C. (2007) 'Problems and opportunities managing invasive bullfrogs: Is there any hope?', in F. Gherardi (ed) *Biological Invaders in Inland Waters – Profiles, Distribution, and Threats*, Springer, Amsterdam, pp679–693

Adams, M. J., Pearl, C. A. and Bury, R. B. (2003) 'Indirect facilitation of an anuran invasion by non-native fishes', *Ecology Letters*, vol 6, p343

Alford, R. A. and Richards, S. J. (1999) 'Global amphibian declines: A problem in applied ecology', *Annual Review of Ecology and Systematics*, vol 30, pp133–165

Banks, B. J., Foster, B., Langton, T. and Morgan, K. (2000) 'British bullfrogs?', *British Wildlife*, vol 11, pp327–330

Berger, L., Marantelli, G., Skerratt, L. F. and Speare, R. (2005) 'Virulence of the amphibian chytrid fungus *Batrachochytrium dendrobatidis* varies with the strain', *Diseases of Aquatic Organisms*, vol 68, pp47–50

Berven, K. A. (1981) 'Mate choice in the wood frog, *Rana sylvatica*', *Evolution*, vol 35, pp707–722

Blaustein, A. R. and Kiesecker, J. M. (2002) 'Complexity in conservation: Lessons from the global decline of amphibian populations', *Ecology Letters*, vol 5, pp597–608

Boone, M. D., Little, E. E. and Semlitsch, R. D. (2004) 'Overwintered bullfrog tadpoles negatively affect salamanders and anurans in native amphibian communities', *Copeia*, vol 2004, pp683–690

Bury, R. B. and Whelan, J. (1984) *Ecology and Management of the Bullfrog*, US Fish and Wildlife Service, Washington, DC

Casper, G. S. and Hendricks, R. (2005) '*Rana catesbeiana* Shaw, 1802, American bullfrog', in M. J. Lanoo (ed) *Amphibian Declines: The Conservation Status of United States Species*, University of California Press, Berkeley, pp540–546

Castellano, S., Cucco, M. and Giacoma, C. (2004) 'Reproductive investment of female green toads (*Bufo viridis*)', *Copeia*, vol 2004, pp659–664

D'Amore, A., Kirby, E. and McNicholas, M. (2009a) 'Invasive species shifts ontogenetic resource partitioning and microhabitat use of a threatened native amphibian', *Aquatic Conservation: Marine and Freshwater Ecosystems*, vol 19, pp534–541

D'Amore, A., Hemingway, V. and Kirby, E. (2009b) 'Reproductive interference by an invasive species: An evolutionary trap?', *Herpetological Conservation and Biology*, vol 4, pp325–330

D'Amore, A., Hemingway, V. and Wasson, K. (2010) 'Do a threatened native amphibian and its invasive congener differ in response to human alteration of the landscape?', *Biological Invasions*, vol 12, pp145–154

Daszak, P., Strieby, A., Cunningham, A. A., Longcore, J. E., Brown, C. C. and Porter, D. (2004) 'Experimental evidence that the bullfrog (*Rana catesbeiana*) is a potential carrier of chytridiomycosis, an emerging fungal disease of amphibians', *Herpetological Journal*, vol 14, pp201–207

Doubledee, R. A., Muller, E. B. and Nisbet, R. M. (2003) 'Bullfrogs, disturbance regimes, and the persistence of California red-legged frogs', *The Journal of Wildlife Management*, vol 67, pp424–438

Emlen, S. T. (1968) 'Territoriality in the bullfrog, *Rana catesbeiana*', *Copeia*, vol 1968, pp240–243

Ficetola, G. F., Thuiller, W. and Miaud, C. (2007) 'Prediction and validation of the potential global distribution of a problematic alien invasive species — the American bullfrog', *Diversity and Distributions*, vol 13, pp476–485

Ficetola, G. F., Maiorano, L., Falcucci, A., Dendoncker, N., Boitani, L., Padoa-Schioppa, E., Miaud, C. and Thuiller, W. (2010) 'Knowing the past to predict the future: Land-use change and the distribution of invasive bullfrogs', *Global Change Biology*, vol 16, pp528–537

Fisher, M. C. and Garner, T. W. J. (2007) 'The relationship between the emergence of *Batrachochytrium dendrobatidis*, the international trade in amphibians and introduced amphibian species', *Fungal Biology Reviews*, vol 21, pp2–9

Gahl, M., Calhoun, A. and Graves, R. (2009) 'Facultative use of seasonal pools by American bullfrogs (*Rana catesbeiana*)', *Wetlands*, vol 29, pp697–703

Garner, T. W. J., Perkins, M. W., Govindrarjulu, P., Seglie, D., Walker, S., Cunningham, A. A. and Fisher, M. C. (2006) 'The emerging amphibian pathogen *Batrachochytrium dendrobatidis* globally infects introduced populations of the North American bullfrog, *Rana catesbeiana*', *Biology Letters*, vol 2, pp455–459

Govindarajulu, P., Altwegg, R. and Anholt, B. R. (2005) 'Matrix model investigation of invasive species control: Bullfrogs on Vancouver Island', *Ecological Applications*, vol 15, pp2161–2170

Gratwicke, B., Evans, M. J., Jenkins, P. T., Kusrini, M. D., Moore, R. D., Sevin, J. and Wildt, D. E. (2009) 'Is the international frog legs trade a potential vector for deadly amphibian pathogens?', *Frontiers in Ecology and the Environment*, vol 8, pp438–442

Grey, I. (2009) 'Breeding pond dispersal of interacting California red-legged frogs (*Rana draytonii*) and American bullfrogs (*Lithobates catesbeianus*) of California: A mathematical model with management strategies', MS Thesis, California State University

Hanselmann, R., Rodríguez, A., Lampo, M., Fajardo-Ramos, L., Aguirre, A. A., Kilpatrick, A. M., Rodríguez, J. P. and Daszak, P. (2004) 'Presence of an emerging pathogen of amphibians in introduced bullfrogs, *Rana catesbeiana*, in Venezuela', *Biological Conservation*, vol 120, p115

Hayes, M. P. and Jennings, M. R. (1986) 'Decline of ranid frog species in western North America: Are bullfrogs (*Rana catesbeiana*) responsible?', *Journal of Herpetology*, vol 20, pp490–509

Hecnar, S. J. and M'Closkey, R. T. (1997) 'Changes in the composition of a ranid frog community following bullfrog extinction', *American Midland Naturalist*, vol 137, pp145–150

Howard, R. D. (1981) 'Sexual dimorphism in bullfrogs', *Ecology*, vol 62, pp303–310

Jennings, M. R. and Hayes, M. P. (1985) 'Pre-1900 overharvest of California red-legged frogs (*Rana aurora draytonii*): The inducement for bullfrog (*Rana catesbeiana*) introduction', *Herpetologica*, vol 41, pp94–103

Kahrs, D. A. (2006) 'American bullfrog eradication in Sycamore Canyon, Arizona, a natural open aquatic system', *Sonoran Herpetologist*, vol 19, pp74–77

Kiesecker, J. M. and Blaustein, A. R. (1998) 'Effects of introduced bullfrogs and smallmouth bass on microhabitat use, growth, and survival of native red-legged frogs (*Rana aurora*)', *Conservation Biology*, vol 12, pp776–787

Kiesecker, J. M., Blaustein, A. R. and Miller, C. L. (2001) 'Potential mechanisms underlying the displacement of native red-legged frogs by introduced bullfrogs', *Ecology*, vol 82, pp1964–1970

Kruse, K. C. and Francis, M. G. (1977) 'A predation deterrent in larvae of the bullfrog, *Rana catesbeiana*', *Transactions of the American Fisheries Society*, vol 106, pp248–252

Kupferberg, S. J. (1997) 'Bullfrog (*Rana catesbeiana*) invasion of a California river: The role of larval competition', *Ecology*, vol 78, pp1736–1751

Lips, K. R., Brem, F., Brenes, R., Reeve, J. D., Alford, R. A., Voyles, J., Carey, C., Livo, L., Pessier, A. P. and Collins, J. P. (2006) 'Emerging infectious disease and the loss of biodiversity in a Neotropical amphibian community', *Proceedings of the National Academy of Sciences of the United States of America*, vol 103, pp3165–3170

Liu, X. and Li, Y. (2009) 'Aquaculture enclosures relate to the establishment of feral populations of introduced species', *PLoS ONE*, vol 4, art e6199

Lowe, S., Browne, M. and Boudjelas, S. (2000) *100 of the World's Worst Invasive Alien Species*, IUCN/SSC Invasive Species Specialist Group, Auckland, New Zealand

Marca, E. L., Lips, K. R., Lötters, S., Puschendorf, R., Ibáñez, R., Rueda-Almonacid, J. V., Schulte, R., Marty, C., Castro, F., Manzanilla-Puppo, J., García-Pérez, J. E., Bolaños, F., Chaves, G., Pounds, J. A., Toral, E. and Young, B. E. (2005) 'Catastrophic population declines and extinctions in Neotropical harlequin frogs (Bufonidae: Atelopus)', *Biotropica*, vol 37, pp190–201

Maret, T. J., Snyder, J. D. and Collins, J. P. (2006) 'Altered drying regime controls distribution of endangered salamanders and introduced predators', *Biological Conservation*, vol 127, pp129–138

Minton, J. E. (1949) 'Coral snake preyed upon by the bullfrog', *Copeia*, vol 1949, p288

Orchard, S. A. (1999) 'The American bullfrog in British Columbia: The frog who came to dinner', in R. Claudi and J. H. Leach (eds) *Nonindigenous Freshwater Organisms: Vectors, Biology, and Impacts*, Lewis Publishers, Boca Raton, Florida, pp289–296

Pearl, C. A., Hayes, M. P., Haycock, R., Engler, J. D. and Bowerman, J. A. Y. (2005) 'Observations of interspecific amplexus between western North American ranid frogs and the introduced American bullfrog (*Rana catesbeiana*) and an hypothesis concerning breeding interference', *The American Midland Naturalist*, vol 154, pp126–134

Provenzano, S. E. and Boone, M. D. (2009) 'Effects of density on metamorphosis of bullfrogs in a single season', *Journal of Herpetology*, vol 43, pp49–54

Rosen, P. C. and Schwalbe, C. R. (eds) (1995) *Bullfrogs: Introduced Predators in Southwestern Wetlands*, US Department of the Interior, National Biological Service, Washington, DC

Scalera, R. (2007) 'Virtues and shortcomings of EU legal provisions for managing NIS: *Rana catesbeiana* and *Trachemys scripta elegans* as case studies', in F. Gherardi (ed) *Biological Invaders in Inland Waters: Profiles, Distribution, and Threats*, Springer, Amsterdam, pp669–678

Schloegel, L. M., Ferreira, C. M., James, T. Y., Hipolito, M., Longcore, J. E., Hyatt, A. D., Yabsley, M., Martins, A. M. C. R. P. F., Mazzoni, R., Davies, A. J. and Daszak, P. (2009) 'The North American bullfrog as a reservoir for the spread of *Batrachochytrium dendrobatidis* in Brazil', *Animal Conservation*, vol 13, pp53–61

Schwalbe, C. R. and Rosen, P. C. (eds) (1988) *Preliminary Report on Effects of Bullfrogs on Wetland Herpetofaunas in Southeastern Arizona*, US Department of Agriculture, Forest Service, Flagstaff, Arizona

Stewart, M. M. and Sandison, P. (1972) 'Comparative food habits of sympatric mink frogs, bullfrogs, and green frogs', *Journal of Herpetology*, vol 6, pp241–244

Stuart, J. N. and Painter, C. W. (1993) 'Life history notes: *Rana catesbeiana* (bullfrog), cannibalism', *Herpetological Review*, vol 24, p103

Veitch, C. R. and Clout, M. N. (2002) 'Turning the tide: The eradication of invasive species', in C. R. Veitch and M. N. Clout (eds) *The International Conference on Eradication of Island Invasives*, Occasional Paper of the IUCN Species Survival Commission No 27, pp4–12

Wake, D. B. and Vredenburg, V. T. (2008) 'Are we in the midst of the sixth mass extinction? A view from the world of amphibians', *Proceedings of the National Academy of Sciences*, vol 105, pp11466–11473

Walters, B. (1975) 'Studies of interspecific predation within an amphibian community', *Journal of Herpetology*, vol 9, pp267–279

Wang, Y., Guo, Z., Pearl, C. A. and Li, Y. (2007) 'Body size affects the predatory interactions between introduced American bullfrogs (*Rana Catesbeiana*) and native anurans in China: An experimental study', *Journal of Herpetology*, vol 41, pp514–520

Werner, E. E., Wellborn, G. A. and McPeek, M. A. (1995) 'Diet composition in postmetamorphic bullfrogs and green frogs: Implications for interspecific predation and competition', *Journal of Herpetology*, vol 29, pp600–607

Wiewandt, T. A. (1969) 'Vocalization, aggressive behavior, and territoriality in the bullfrog, *Rana catesbeiana*', *Copeia*, vol 1969, pp276–285

Wu, Z., Li, Y., Wang, Y. and Adams, M. J. (2005) 'Diet of introduced bullfrogs (*Rana catesbeiana*): Predation on and diet overlap with native frogs on Daishan Island, China', *Journal of Herpetology*, vol 39, pp668–674

Xuan, L., Yiming, L. and McGarrity, M. (2010) 'Geographical variation in body size and sexual size dimorphism of introduced American bullfrogs in southwestern China', *Biological Invasions*, vol 12, pp2037–2047

Zavaleta, E. S., Hobbs, R. J. and Mooney, H. A. (2001) 'Viewing invasive species removal in a whole-ecosystem context', *Trends in Ecology and Evolution*, vol 16, p454

28

Trachemys scripta (slider terrapin)

Gentile Francesco Ficetola, Dennis Rödder and Emilio Padoa-Schioppa

History of Introduction, Distribution and Impact on Native Species

Trachemys scripta, the slider terrapin, has been traded worldwide since at least the 1950s, and quickly became a very popular pet because of its cheap price and the reasonably simple husbandry. Sliders are probably the most commonly traded reptile: more than 52 million individuals were exported from the US during the period 1989–1997 (Telecky, 2001). Although sliders are mostly traded as pets, in some areas they are also imported or farmed for human consumption, particularly in Asia (Scalera, 2007). Three subspecies of *T. scripta* are currently recognized (Bonin et al, 2006): *Trachemys scripta scripta* (Thunberg in Schoepff, 1792), *T. s. elegans* (Wied, 1838) and *T. s. troostii* (Holbrook, 1836) (Figure 28.1). *Trachemys scripta elegans* (the red-eared slider terrapin) was the most widely traded subspecies until 1997. The European Union interrupted the import of *T. s. elegans* in 1997 (Regulation 338/1997; Regulation 349/2003) due to the high risk of biological invasion. However, these regulations considered only the subspecies *T. s. elegans* and, as a consequence, the trade in the other two subspecies (*T. s. scripta*, *T. s. troostii* and hybrids among subspecies) sharply increased after the ban (Scalera, 2007). Young sliders are sold at a size of just a few centimetres, but can grow quickly. As owners are rarely prepared to maintain large adults for many years, they often release terrapins into natural or semi-natural wetlands (Teillac-Deschamps et al, 2009).

Distribution

Trachemys scripta is native to the eastern US and northeast Mexico, but has been introduced worldwide (Figure 28.2). Feral individuals have been reported in at least 73 countries or overseas territories (Table 28.1) (Lever, 2003; Pupins, 2007; Kraus, 2009; Pendlebury, 2009; Scalera, 2009; Kikillus et al, 2010).

Reproduction has not been recorded in all areas with feral individuals; feral adults can survive long periods in suboptimal areas, where the climate is not suitable for reproduction because of low temperature or limited precipitation (Bringsøe, 2001; Ficetola et al, 2009). Also, ascertaining the reproduction of freshwater terrapins in natural wetlands can be challenging because juveniles are more difficult to spot than adults, and because it is difficult to ascertain whether juveniles originated from local reproduction or from recent release. Reproduction of non-native populations has been recorded in Mediterranean areas of Europe, Germany, in Japan and southeast Asia, Australia, New Zealand, in the West Indies and in the introduced range in the US (Lever, 2003; Cadi et al, 2004; Ramsay et al, 2007; Ficetola et al, 2009; Kikillus et al, 2010).

Impact on native species

Trachemys scripta can have notable impacts on native reptiles, amphibians, fish and invertebrates. There have been extensive studies on the interaction between sliders and two European species of terrapins: the European pond turtle, *Emys orbicularis*, and the Spanish terrapin, *Mauremys leprosa*. Freshwater terrapins often

(a)

(b)

Source: (a) D. Rödder; (b) G. F. Ficetola

Figure 28.1 *Feral individuals of (a)* Trachemys scripta elegans; *(b)* T. s. scripta

Note: Several presence localities, both in the native and in the invasive range, are depicted in Figure 28.3.

Source: Lever (2003); Pupins (2007); Kraus (2009); Pendlebury (2009); Scalera (2009); Kikillus et al (2010)

Figure 28.2 *Native range of* Trachemys scripta *(black) and countries where feral slider terrapins have been recorded (grey)*

Table 28.1 *Countries or territories where feral sliders have been reported*

Region	Countries/territories
Europe	Andorra, Austria, Belgium, Bulgaria, Croatia, Czech Republic, Denmark, Finland, France, Germany, UK, Greece, Hungary, Italy, Latvia, Lithuania, Malta, The Netherlands, Poland, Portugal, Romania, Russia, Serbia, Slovakia, Slovenia, Spain, Sweden, Switzerland
Africa	Egypt, Kenya, Namibia, Reunion (France), South Africa
Asia	Bahrain, Hong Kong, India, Indonesia, Israel, Japan, Malaysia, Saudi Arabia, Seychelles, Singapore, South Korea, Sri Lanka, Taiwan, Thailand, Turkey, Viet Nam
Central and North America	Aruba (Kingdom of The Netherlands), Bahamas, Bermuda (UK), British Virgin Islands (UK), Canada, Cayman Islands (UK), Guadeloupe (France), Martinique (France), Mexico, Netherlands Antilles, Nicaragua, Panama, Trinidad and Tobago, US Virgin Islands, US outside the native range
South America	Brazil, Chile, Guyana, Paraguay, Suriname
Oceania	Australia, Commonwealth of the Northern Mariana Islands (US), French Polynesia, Federated States of Micronesia, Guam (US), New Zealand

compete for basking sites, since basking is vital in temperate regions for thermoregulation and to activate metabolism (Meek and Avery, 1988). As sliders are larger and more aggressive than other species of terrapins, they can outcompete them both in the native and in the invaded range (Lindeman, 1999; Cadi and Joly, 2003; Spinks et al, 2003; Macchi, 2008). Cady and Joly (2003, 2004) used experimental ponds to assess competition and behavioural interactions between sliders and the threatened European pond turtle. Both turtle species preferred the same sites for basking but, when sliders where present, the European pond turtle shifted to suboptimal basking sites. Subsequent experiments performed by Macchi et al confirmed that competition for basking sites, and aggressive interactions between sliders and the European pond turtle, can threaten this native species (Macchi, 2008; Macchi et al, 2008). Similarly, in presence of sliders, the Spanish terrapin reduces basking activity and avoids basking sites where sliders are present (Polo-Cavia et al, 2010b). Sliders also have behavioural and physiological advantages compared to the Spanish terrapin, such as a higher tolerance to human disturbance, and a body shape determining better thermoregulatory abilities (Polo-Cavia et al, 2008, 2009). The competition with sliders for basking sites and perhaps other resources (e.g. food, nesting sites) may also be a cause of the decline of other freshwater terrapins, such as the western pond terrapin *Actinemys marmorata* in California (Spinks et al, 2003).

Interactions with native species can also occur during foraging. In European ponds, there is a wide overlap between the diet of sliders and the diet of native terrapins, suggesting that competition for food may occur (Pérez-Santigosa et al, 2011); competition with sliders can decrease foraging success and even increase mortality in European pond turtles (Cadi and Joly, 2004). Furthermore, tadpoles of several European amphibians can chemically detect the presence of native predatory terrapins and modify their behaviour to reduce predation risk, but they are unable to appropriately respond to the presence of sliders. Therefore, sliders might capture and consume tadpoles more easily than native terrapins, and thus have a competitive advantage during foraging (Polo-Cavia et al, 2010a). The release of sliders into natural ecosystems may also increase the risk of transmission of pathogens (such as nematodes and bacteria) to native terrapins (e.g. Spinks et al, 2003; Hidalgo-Vila et al, 2009).

The impact of sliders on other components of biota is less studied, but can be important. Sliders are omnivorous, and shift from a carnivorous to a more herbivorous diet during growth (Hart, 1983; Prévot-Julliard et al, 2007). Sliders can predate on crustaceans, aquatic insects, fish and amphibians, and their presence can therefore affect whole freshwater communities (Lever, 2003; Teillac-Deschamps and Prévot-Julliard, 2006; Prévot-Julliard et al, 2007; Pérez-Santigosa et al, 2011). For instance, tadpoles of some European species of anuran amphibians (*Pelophylax perezi*, *Pelobates cultripes* and *Hyla arborea*) reduce activity in the presence of native predatory terrapins, but they do not show such antipredatory behaviour when alien terrapins (such as sliders) are present. The lack of antipredatory behaviour is probably caused by the absence of a shared evolutionary history between prey and the alien predator, and may expose native amphibians to a high predatory pressure (Polo-Cavia et al, 2010a). Additionally, adult sliders feed on wetland vegetation, and can heavily damage it, particularly if they are at high density, or in small wetlands (Ficetola, G. F., unpublished data).

Ecological Niche and Potential Distribution

Successful establishment of alien invasive reptiles at a given site strongly depends on the availability of suitable habitats, therefore specific climate conditions can be a good predictor of invasion success (Bomford et al, 2010). As an aquatic species, the slider depends on continuous water availability throughout the year, whereby almost any kind of water body provides suitable habitats. Its breeding behaviour and digestive turnover rates are strongly temperature dependent, and the species does not feed at body temperatures lower than about 10°C (Parmenter, 1980). This makes the slider dependent on certain ambient temperature regimes to maintain a positive annual energetic balance.

Breeding is the most critical stage necessary for long term establishment of slider populations. In the native range, sliders usually lay eggs in subterranean nests from April to July (Gibbons et al, 1982; Aresco, 2004). Depending on incubation temperatures, time from egg deposition to hatching of the newborns ranges from 60 to 130 days, with lower incubation temperatures causing slower development. In Louisiana, eggs were reported to hatch in approximately 68–70 days (Dundee and

Rossman, 1989). In areas with a high seasonality hatchlings may hibernate inside the nest. This may ultimately limit the slider's distribution in northern parts of its native range as hatchlings may die at temperatures below –0.6 to –4.0°C (Packard et al, 1997; Tucker and Packard, 1998). Furthermore, successful egg development depends on sufficient moisture (Tucker and Packard, 1998) and warmth during incubation, about 26.0–32.5°C (Wibbels et al, 1991; Crews et al, 1994).

A further important requirement for long term persistence of the slider is a balanced sex ratio within populations. In slider embryos, as in most chelonians, sex determination is temperature dependent. During egg incubation, low temperature during a sensitive phase of approximately two weeks increases the number of males. By contrast, with warmer egg incubation temperatures, more females hatch (Wibbels et al, 1991; Crews et al, 1994; Ewert et al, 1994). Only within a transitional range between 28.3 and 30.6°C are both sexes differentiated (Morosovsky and Pieau, 1991; Cadi et al, 2004). These requirements for successful clutch development and balanced sex ratio strongly influence the native range of the slider (Rödder et al, 2009a), and probably have a strong effect also on its establishment success in other areas.

Several authors have assessed the potential distribution of the slider at both the regional and global scale (Ficetola et al, 2009; Rödder et al, 2009a, 2009b; Kikillus et al, 2010). We herein provide results of a mechanistic species distribution model (Kearney and Porter, 2009) based on physiological thresholds of the species within its native range as described in Rödder et al (2009b). These include

variables affecting both physiology and reproduction of sliders: the upper avoidance temperature of the species (about 37°C; Lamb et al, 1995) reflected by the maximum temperature of the warmest month; frost tolerance of neonates described by the minimum temperature of the coldest month (–12.6°C); annual mean temperature >8.3°C accounting for a positive energetic balance; annual precipitation >278mm, and precipitation of the driest quarter >22mm to account for water availability. Regions meeting these requirements comprise huge areas of North, Central and South America, Europe, West and Central Africa, the East African coast, eastern Asia, and eastern and western parts of Australia (Figure 28.3). In many of these areas the slider is actually distributed. We used the area under the curve (AUC) of the receiver operator characteristic plot (Manel et al, 2001) to assess the capability of this model to correctly identify areas where native and invasive populations of sliders are present. The model was tested using 375 native and 205 invasive records compiled through online databases (Global Biodiversity Information Facility, www.gbif.org; HerpNET, www.herpnet.org) and literature search (for a detailed list of sources see Rödder et al, 2009b). The discrimination performance of the resulting model was good in both the native (AUC = 0.85) and invasive range of the slider (AUC = 0.80).

The model presented in Figure 28.3 characterizes the requirements of the slider at the global scale. However, at a finer scale, suitable microhabitat features may become more important. Being a generalist, the species is able to occupy most wetlands. Nevertheless,

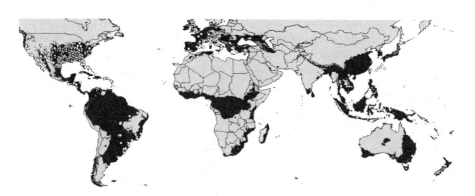

Note: Potential distribution of *Trachemys scripta* is indicated by black areas. Native populations are indicated as small white dots, and invasive populations as large white dots.

Source: Species records were compiled from online databases and a literature search (for details see Rödder et al, 2009b)

Figure 28.3 *Potential distribution of* Trachemys scripta *derived from physiological thresholds*

the slider prefers water bodies with certain features: still to slow running waters, eutrophic, 1 to 2m deep, and with dense vegetation providing cover from predators and supporting high densities of aquatic invertebrates on which it feeds (Morreale and Gibbons, 1986). The presence of suitable basking sites can be a further important feature for this species. Nevertheless, basking sites are rarely a limiting factor, because when no aquatic basking sites are available, the slider may be able to bask along the shore or at the water surface. The slider is a generalist omnivore (see above), and this probably contributes to its widespread distribution (Morreale and Gibbons, 1986).

Management Efforts

Prevention is certainly the most effective approach to limit the introduction of sliders into natural and semi-natural environments. Therefore, initial efforts should include the ban of import/trade of individuals, but regulations must be carefully planned to make them effective. The European regulation on the import of sliders (see above) is a clear example of the complexity of this task (Scalera, 2007). Furthermore, it is still legal to keep sliders and distribute them across the European countries (Scalera, 2007), including captive-bred juveniles hatched in the European Union. A more effective regulation should ban the trade of all slider subspecies, and of any other freshwater terrapin potentially capable of establishing naturalized populations. A first assessment of the probability of successful establishment may be performed through the use of bioclimatic models that evaluate suitability on the basis of climate similarity between the native range of the species and the areas where they are imported (see above) (Figure 28.3) (Jeschke and Strayer, 2008; Rödder et al, 2009b; Gallien et al, 2010; Kikillus et al, 2010).

Environmental education can also play an important role. Communication campaigns are essential to increase public awareness, and they should be a priority to avoid new introductions of terrapins by owners. Effective education campaigns should include communication targeted to explain the problems caused by introduced terrapins (e.g. exhibitions in public parks), but also more general information, encouraging people to change their perception toward nature and support biodiversity conservation. This can be achieved through enhanced outdoor activities and increasing personal contact with nature (Teillac-Deschamps et al, 2009).

Multiple techniques can be used to remove non-native sliders from wetlands. Traps are the most frequent approach to capture freshwater terrapins, including sliders, wherein basking traps and funnel hoop traps are the common techniques (Fowler and Avery, 1994; Savage, 2002). Basking traps (also called sink box traps) are floating boxes or barrels, with the rim just above the waterline. Terrapins crawl up onto the top of the box to bask in the sun and fall into the trap. The boxes can be made with hardware cloth with a mesh size of <2cm, allowing the capture of individuals of all sizes. The shape of the entrance must ensure that trapped individuals cannot climb the trap and escape.

Funnel hoop traps are barrel shaped traps that may be made by hardware cloth or cord/rope nets. They should have a funnel shaped entrance allowing terrapins to enter easily but that is reduced to a small slit entering the main trap space. The main section of the trap should be baited with meat, entrails or fish, wherein the bait should be suspended from the upper frame. The trap should be kept rigidly in place, and the upper part should be above the waterline to avoid the terrapins drowning. Traps of different sizes can be used to capture individuals of different body size (Fowler and Avery, 1994; Gianaroli et al, 2001; Savage, 2002; Gamble, 2006). Overall, basking traps seem to be a particularly effective approach. However, although both trapping techniques can be useful for adults, small juveniles are more difficult to spot and capture, particularly when using hoop traps (Gianaroli et al, 2001; Chen, 2006; Gamble, 2006). For this reason, it has been proposed that early removal of feral individuals, before they become naturalized and start to reproduce, would be desirable (Ficetola et al, 2009). Other approaches for the control of feral individuals include the use of nets, the complete draining of wetlands followed by removal of alien turtles, and the use of sniffer dogs to detect terrapins and their eggs, which can be removed after the identification of nesting areas (O'Keeffe, 2005; Scalera, 2009).

However, as individuals are still being released in natural and semi-natural wetlands, the capture of feral individuals is not sufficient to fully remove sliders, and can result only in short time effects. A combination of capture with environmental communication is probably the most effective approach to reduce the number of sliders present both in areas where they breed and in areas where reproduction is not successful (Teillac-Deschamps et al, 2009).

Controversies

Controversies on potential establishment/impact

Since the 1980s there have been remarkable controversies on the potential for establishment and impact of sliders in Europe. Despite the introduction of a very large number of individuals in multiple countries, some researchers suggested that the likelihood of slider establishment in the wild, and its potential impact on the native biota, would be limited. For instance, Bringsøe (2001) suggests that in northern Europe the climate is too cold, while in the Mediterranean region summers are too dry for successful egg development, and only small areas in southern Europe would have a suitable climate for this species. Similarly, field experiments performed in the 1990s suggested a limited reproductive capability and a very low survival of juveniles in the introduced range in central Italy (Luiselli et al, 1997). However, during the last decade the number of reproduction records in Mediterranean Europe has steadily increased (see data in Kikillus et al, 2010), indicating that sliders are establishing over larger areas. Projections of species distribution models onto future climate change scenarios suggest that the invasiveness of populations may increase in the near future (Ficetola et al, 2009; Kikillus et al, 2010).

Terrapin eradication and management: The social dimension does matter

When alien species are domesticated or used as ornamental animals, the decision to remove them may face the opposition of the public. Most people are familiar with slider terrapins, and their management should take into account social aspects. It has been suggested that the presence of alien terrapins in urban green spaces may be a reason for the general public to visit these areas. Once attracted to green spaces, people may encounter other aspects of nature, increase their receptivity toward environmental communication, and therefore become more willing to support biodiversity conservation (Teillac-Deschamps et al, 2009). This line of reasoning suggests focusing the management efforts (terrapin removal) on natural wetlands where sliders pose a serious threat to native biodiversity, while a complete removal might not be necessary in urban contexts (Teillac-Deschamps et al, 2009).

Controversies arise also after terrapins are captured. People and animal rights organizations may contest the killing of captured individuals, as has happened for other ornamental alien species (Bertolino and Genovesi, 2003). For this reason, in some European countries captured individuals are maintained in rescue centres. However, sliders are long lived animals. Maintaining thousands of terrapins for decades requires important resources and large dedicated areas. Hence, allocating resources for these animals can reduce the already limited funding available for biodiversity conservation. A possible solution to this issue is adding a charge to the price of sold terrapins. This money could be allocated for the management of issues caused by sliders, such as removing feral individuals and maintaining the captured ones, or for environmental education campaigns. Explaining the reasons for the increased price to the consumers may also increase their awareness of the issues that can be caused by invasive species, and perhaps discourage them from introducing terrapins in natural wetlands.

Acknowledgements

We thank N. Polo-Cavia for constructive comments. GFF was funded by a grant of the University of Milano-Bicocca on invasive species.

References

Aresco, M. J. (2004) 'Reproductive ecology of *Pseudemys floridana* and *Trachemys scripta* (Testudines: Emydidae) in northwestern Flordia', *Journal of Herpetology*, vol 38, pp249–256

Bertolino, S. and Genovesi, P. (2003) 'Spread and attempted eradication of the grey squirrel (*Sciurus carolinensis*) in Italy, and consequences for the red squirrel (*Sciurus vulgaris*) in Europe', *Biological Conservation*, vol 109, pp351–358

Bomford, N., Barry, S. C. and Lawrence, E. (2010) 'Predicting establishment success for introduced freshwater fishes: A role for climate matching', *Biological Invasions*, vol 12, pp2559–2571

Bonin, F., Devaux, B. and Dupré, A. (2006) *Toutes les tortues du monde*, Delachaux et Niestlé, Paris

Bringsøe, H. (2001) '*Trachemys scripta* (Schoepff, 1792) – Buchstaben-Schmuckschildkröte', in U. Fritz (ed) *Handbuch der Reptilien und Amphibien Europas. Schildkröten (Testudines)*, AULA, Wiebelsheim, pp525–583

Cadi, A. and Joly, P. (2003) 'Competition for basking places between the endangered European pond turtle (*Emys orbicularis galloitalica*) and the introduced red-eared slider (*Trachemys scripta elegans*)', *Canadian Journal of Zoology*, vol 81, pp1392–1398

Cadi, A. and Joly, P. (2004) 'Impact of the introduction of the read-eared slider (*Trachemys scripta elegans*) on survival rates of the European pond turtle (*Emys orbicularis*)', *Biodiversity and Conservation*, vol 13, pp2511–2518

Cadi, A., Delmas, V., Prévot-Julliard, A.-C., Joly, P. and Girondot, M. (2004) 'Successful reproduction of the introduced slider turtle (*Trachemys scripta elegans*) in the south of France', *Aquatic Conservation: Marine and Freshwater Ecosystems*, vol 14, pp237–246

Chen, T.-H. (2006) 'Distribution and status of the introduced red-eared slider (*Trachemys scripta elegans*) in Taiwan', in F. Koike, M. N. Clout, M. Kawamichi, M. De Poorter, and K. Iwatsuki (eds) *Assessment and Control of Biological Invasion Risks*, Shoukadoh Book Sellers and IUCN, Kyoto, Japan and Gland, Switzerland, pp187–195

Crews, D., Bergeron, J. M., Bull, J. J., Flores, D., Tousignant, A., Skipper, J. K. and Wibbels, T. (1994) 'Temperature-dependent sex determination in reptiles: Proximate mechanisms, ultimate outcomes, and practical applications', *Developmental Genetics*, vol 15, pp297–312

Dundee, H. A. and Rossman, D. A. (1989) *Amphibians and Reptiles of Louisiana*, State University Press, Baton Rouge

Ewert, M. A., Jackson, D. R. and Nelson, C. E. (1994) 'Patterns of temperature-dependent sex determination in turtles', *The Journal of Experimental Zoology*, vol 270, pp3–15

Ficetola, G. F., Thuiller, W. and Padoa-Schioppa, E. (2009) 'From introduction to the establishment of alien species: Bioclimatic differences between presence and reproduction localities in the slider turtle', *Diversity and Distributions*, vol 15, pp108–116

Fowler, J. F. and Avery, J. L. (1994) 'Turtles', in S. E. Hygnstrom and R. M. Timm (eds) *Prevention and Control of Wildlife Damage*, University of Nebraska, Lincoln, ppF27–F31

Gallien, L., Münkemüller, T., Albert, C., Boulangeat, I. and Thuiller, W. (2010) 'Predicting potential distributions of invasive species: Where to go from here?', *Diversity and Distributions*, vol 16, pp331–342

Gamble, T. (2006) 'The relative efficiency of basking and hoop traps for painted turtles (*Chrysemys picta*)', *Herpetological Review*, vol 37, pp308–312

Gianaroli, M., Lanzi, A. and Fontana, R. (2001) 'Utilizzo di trappole del tipo "bagno di sole artificiale" per la cattura di testuggini palustri', in *Atti 3° Congresso nazionale SHI, Pianura, (Pavia, 2000)*, pp153–155

Gibbons, J. W., Greene, J. L. and Patterson, K. K. (1982) 'Variation in reproductive characteristics of aquatic turtles', *Copeia*, vol 1982, pp776–784

Hart, D. R. (1983) 'Dietary and habitat shift with size of red-eared turtles (*Pseudemys scripta*) in a southern Louisiana population', *Herpetologica*, vol 39, pp285–290

Hidalgo-Vila, J., Díaz-Paniagua, C., Ribas, A., Florencio, M., Pérez-Santigosa, N. and Casanova, J. C. (2009) 'Helminth communities of the exotic introduced turtle, *Trachemys scripta elegans* in Southwestern Spain: Transmission from native turtles', *Research in Veterinary Science*, vol 83, pp463–465

Jeschke, J. M. and Strayer, D. L. (2008) 'Usefulness of bioclimatic models for studying climate change and invasive species', *Annals of the New York Academy of Sciences*, vol 1134, pp1–24

Kearney, M. and Porter, W. (2009) 'Mechanistic niche modelling: Combining physiological and spatial data to predict species' ranges', *Ecology Letters*, vol 12, pp334–350

Kikillus, K. H., Hare, K. H. and Hartley, S. (2010) 'Minimizing false-negatives when predicting the potential distribution of an invasive species: A bioclimatic envelope for the red-eared slider at global and regional scales', *Animal Conservation*, vol 13 (supplement 1), pp5–15

Kraus, F. (2009) *Alien Reptiles and Amphibians: A Scientific Compendium and Analysis*, Springer, Dordrecht

Lamb, T., Bickham, J. W., Lyne, T. B. and Gibbons, J. W. (1995) 'The slider turtle as an environmental sentinel: Multiple tissue assays using flow cytometric analysis', *Ecotoxicology*, vol 4, pp5–13

Lever, C. (2003) *Naturalized Amphibians and Reptiles of the World*, Oxford University Press, New York

Lindeman, P. V. (1999) 'Aggressive interactions during basking among four species of Emydid turtles', *Journal of Herpetology*, vol 33, pp214–219

Luiselli, L., Capula, M., Capizzi, D., Filippi, E., Jesus, V. T. and Anibaldi, C. (1997) 'Problems for conservation of pond turtles (*Emys orbicularis*) in central Italy: Is the introduced red-eared turtle (*Trachemys scripta*) a serious threat?', *Chelonian Conservation and Biology*, vol 2, pp417–419

Macchi, S. (2008) 'Eco-ethological characterization of the alien slider *Trachemys scripta* and evaluation of the effects of its introduction on the conservation of *Emys orbicularis*', PhD Thesis, Insubria University, Varese, Italy

Macchi, S., Balzarini, L. L. M., Scali, S., Martinoli, A. and Tosi, G. (2008) 'Spatial competition for basking sites between the exotic slider *Trachemys scripta* and the European pond turtle *Emys orbicularis*', in C. Corti (ed) *Herpetologia Sardiniae*, Belvedere, Latina, pp338–340

Manel, S., Williams, H. C. and Ormerod, S. J. (2001) 'Evaluating presence–absence models in ecology: The need to account for prevalence', *Journal of Applied Ecology*, vol 38, pp291–931

Meek, R. and Avery, R. A. (1988) 'Thermoregulation in chelonians', *Herpetological Journal*, vol 1, pp253–259

Morosovsky, N. and Pieau, C. (1991) 'Transitional range of temperature, pivotal temperatures and thermosensitive stages for sex determination in reptiles', *Amphibia-Reptilia*, vol 12, pp169–179

Morreale, S. J. and Gibbons, J. W. (1986) *Habitat Suitability Index Models: Slider Turtle*, US Fish and Wildlife Service Biological Report 82, Washington, DC

O'Keeffe, S. (2005) 'Investing in conjecture: Eradicating the red-eared slider in Queensland', in *Proceedings of the 13th Australasian Vertebrate Pest Conference*, The Museum of New Zealand (Te Papa), Wellington, New Zealand

Packard, G. C., Trucker, J. K., Nicholson, D. and Packard, M. J. (1997) 'Cold tolerance in hatchling slider turtles (*Trachemys scripta*)', *Copeia*, vol 1997, pp339–345

Parmenter, R. R. (1980) 'Effects of food availability and water temperature on the feeding ecology of pond sliders (*Chrysemys s. scripta*)', *Copeia*, vol 1980, pp503–514

Pendlebury, P. (2009) '*Trachemys scripta elegans*', Global Invasive Species Database, IUCN/SSC Invasive Species Specialist Group, www.issg.org/database/species/ecology.asp?si=71&fr=1&sts=

Pérez-Santigosa, N., Florencio, M., Hidalgo-Vila, J. and Díaz-Paniagua, C. (2011) 'Does the exotic invader turtle, *Trachemys scripta elegans*, compete for food with coexisting native turtles?', *Amphibia-Reptilia*, vol 32, no 2, pp167–175

Polo-Cavia, N., Lopez, P. and Martin, J. (2008) 'Interspecific differences in responses to predation risk may confer competitive advantages to invasive freshwater turtle species', *Ethology*, vol 114, pp115–123

Polo-Cavia, N., Lopez, P. and Martin, J. (2009) 'Interspecific differences in heat exchange rates may affect competition between introduced and native freshwater turtles', *Biological Invasions*, vol 11, pp1755–1765

Polo-Cavia, N., Gonzalo, A., Lopez, P. and Martin, J. (2010a) 'Predator recognition of native but not invasive turtle predators by naïve anuran tadpoles', *Animal Behaviour*, vol 80, pp461–466

Polo-Cavia, N., Lopez, P. and Martin, J. (2010b) 'Interspecific differences in heat exchange rates may affect competition between introduced and native freshwater turtles', *Biological Invasions*, vol 12, pp2141–2152

Prévot-Julliard, A. C., Gousset, E., Archinard, C., Cadi, A. and Girondot, M. (2007) 'Pets and invasion risks: Is the slider turtle strictly carnivorous?', *Amphibia-Reptilia*, vol 28, pp139–143

Pupins, M. (2007) 'First report on recording of the invasive species *Trachemys scripta elegans*, a potential competitor of *Emys orbicularis* in Latvia', *Acta Universitatis Latviensis – Biology*, vol 723, pp37–46

Ramsay, N. F., Ng, P. K. A., O'Riordan, R. M. and Chou, L. M. (2007) 'The read-eared slider (*Trachemys scripta elegans*) in Asia: A review', in F. Gherardi (ed) *Biological Invaders in Inland Waters: Profiles, Distribution, and Threats*, Springer, Dordrecht, pp161–174

Rödder, D., Kwet, A. and Lötters, S. (2009a) 'Translating natural history into geographic space: A macroecological perspective on the North American slider, *Trachemys scripta* (Reptilia, Cryptodira, Emydidae)', *Journal of Natural History*, vol 43, pp2525–2536

Rödder, D., Schmidtlein, S., Veith, M. and Lötters, S. (2009b) 'Alien invasive slider turtle in unpredicted habitat: A matter of niche shift or of predictors studied?', *PLoS ONE*, vol 4, ppe7843

Savage, J. M. (2002) *The Amphibians and Reptiles of Costa Rica*, University of Chicago Press, Chicago

Scalera, R. (2007) 'Virtues and shortcomings of EU legal provisions for managing NIS: *Rana catesbeiana* and *Trachemys scripta elegans* as case studies', in F. Gherardi (ed) *Biological Invaders in Inland Waters: Profiles, Distribution, and Threats*, Springer, Dordrecht, pp669–678

Scalera, R. (2009) '*Trachemys scripta* (Schoepff), common slider', in DAISIE (ed) *Handbook of Alien Species in Europe*, Springer, Dordrecht, p374

Spinks, P. Q., Pauly, G. B., Crayon, J. J. and Shaffer, H. B. (2003) 'Survival of the western pond turtle (*Emys marmorata*) in an urban California environment', *Biological Conservation*, vol 113, pp257–267

Teillac-Deschamps, P. and Prévot-Julliard, A. C. (2006) 'Impact of exotic slider turtles on freshwater communities: An experimental approach', in *First European Congress of Conservation Biology, Book of Abstracts*, Society for Conservation Biology, Heger, Hungary, pp162–163

Teillac-Deschamps, P., Lorrilliere, R., Servais, V., Delmas, V., Cadi, A. and Prévot-Julliard, A. C. (2009) 'Management strategies in urban green spaces: Models based on an introduced exotic pet turtle', *Biological Conservation*, vol 142, pp2258–2269

Telecky, T. M. (2001) 'US import and export of live turtles and tortoises', *Turtle and Tortoise Newsletter*, vol 4, pp8–13

Tucker, J. K. and Packard, G. C. (1998) 'Overwinter survival by hatchling sliders (*Trachemys scripta*) in West-Central Illinois', *Journal of Herpetology*, vol 32, pp431–434

Wibbels, T., Bull, J. J. and Crews, D. (1991) 'Chronology and morphology of temperature-dependent sex determination', *The Journal of Experimental Zoology*, vol 260, pp371–381

Part V

Aquatic and Riparian Mammals

29

Castor canadensis Kuhl (North American beaver)

Building effective alliances between research and management to mitigate the impacts of an invasive ecosystem engineer: Lessons from the study and control of Castor canadensis in the Fuegian archipelago

Christopher B. Anderson, Nicolás Soto, José Luis Cabello, Guillermo Martínez Pastur, María Vanessa Lencinas, Petra K. Wallem, Daniel Antúnez and Ernesto Davis

Castor canadensis in South America's Sub-Antarctic Ecoregion

The North American (or Canadian) beaver (*Castor canadensis*) is the largest rodent native to North America (Figure 29.1), being found in an extensive range from northern Canada to northern Mexico. It has only one other extant congener, *C. fiber*, distributed originally throughout western and northern Eurasia. Both species affect extensive areas by 'engineering' stream and riparian habitat through their habits of dam building and cutting streamside vegetation. They were also both prized for their pelts, leading to near extinction in their native ranges. *Castor canadensis* was hunted heavily and locally extirpated throughout North America by the late 1800s; subsequent conservation and restoration efforts succeeded in reintroducing the species in much of its native range by the mid- to late 20th century (Naiman et al, 1988). Simultaneously, in the 1940s and 1950s, efforts in various countries sought to introduce North American beavers outside of their native range for the perceived commercial value of its fur, including countries in Europe (e.g. Finland, Poland, Austria and Russia), where the native *C. fiber* had been decimated, and Chile and Argentina, where no native species occupied a similar niche (Anderson and Valenzuela, 2011). Due to the unique nature of the introduction of beavers to southern South America, this chapter will focus on the socioecological role of this invasive exotic species in sub-Antarctic forests and the policy-research response of local and international managers and scientists.

The sub-Antarctic archipelago, shared between Chile and Argentina (Figure 29.2), presents a paradox. On one hand, portions of this ecoregion are some of the least disturbed ecosystems on the planet (i.e. low human population density, highly intact native vegetation cover and over 50 per cent of its territory falling within the system of state protected areas; Mittermeier et al, 2003), while at the same time, it experiences pressing global environmental threats such as invasive alien species, climate change, the ozone hole and rapid commercial development, including tourism and salmon farming (Anderson et al, 2006a; Rozzi et al, 2006). The North American beaver has large impacts

Source: G. Martínez Pastur

Figure 29.1 *North American beaver* (Castor canadensis) *swimming in a waterway on Tierra del Fuego Island*

across the Tierra del Fuego and Cape Horn Archipelagos. The species was introduced in a single release of 25 pairs in 1946 by the Argentine government to Isla Grande in efforts to create a fur industry (Lizarralde, 1993). Curiously though the hunting of beavers was not legally sanctioned until 1981 in Argentina (Lizarralde, 1993) and during that interval the population expanded south and west into Chile. By the 1960s beavers had crossed the Beagle Channel, occupying what is today the Cape Horn Biosphere Reserve and progressively colonizing the neighbouring islands of Navarino, Hoste, Picton, Nueva and Lennox (Anderson et al, 2009). Additionally, the mainland was invaded by the mid-1990s, especially in the area of the Brunswick Peninsula (Wallem et al, 2007). To date, beavers have not been confirmed in the Wollaston

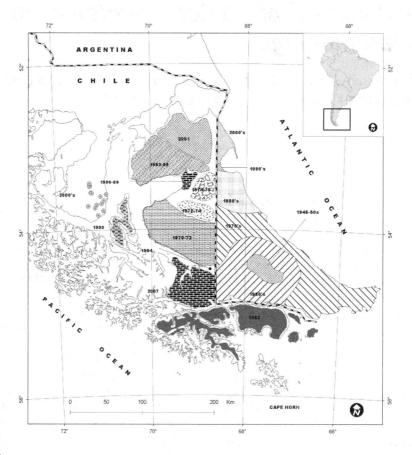

Source: SAG GIS Lab

Figure 29.2 *Map of the austral portions of Chile and Argentina, including the Tierra del Fuego and the Cape Horn Archipelagos, indicating the approximate dates of the range expansion for the North American beaver* (Castor canadensis) *since its initial introduction in 1946*

Islands (Cape Horn National Park) and the far western portion of the archipelago (Agostini National Park) (Anderson et al, 2006a; Moorman et al, 2006).

As a non-native ecosystem engineer, the beaver has large impacts on a range of taxa and levels of biological organization. These impacts have led the beaver to be considered a harmful invasive species, and governmental agencies in both Argentina and Chile, as well as various conservation organizations and research institutions, are interested in implementing and evaluating measures that will lead to its eradication in southern South America. To create eradication and management plans, we not only need to synthesize existing knowledge on sub-Antarctic ecosystems and the beaver's role as an invasive ecosystem engineer (Anderson et al, 2009), but also pay equal attention to learning the institutional and historic lessons that this species may provide on the successes and failures of linking researchers and managers to promote effective conservation action.

The Beaver as a Native and Non-native Ecosystem Engineer

Southern South American temperate forests are shared between Chile and Argentina (33°–56°S). They are the most extensive temperate forests in the southern hemisphere and have higher levels of endemism than forests at equivalent latitudes in North America (Armesto et al, 1995). Within this biome, the Tierra del Fuego and Cape Horn Archipelagos host the world's southernmost forested ecosystems, separated from the mainland by the Magellan Strait and numerous channels. Given its recent glacial history, high latitude and fragmented geography, the biotic communities found here are relatively species poor for most large taxa, such as terrestrial vertebrates and vascular plants. However, non-vascular plants (liverworts and mosses) are a notable exception with 5 per cent of the world's diversity of bryophytes occurring on this small landmass (Goffinet et al, 2006; Rozzi et al, 2008). Partially as a result of the low native diversity of terrestrial vertebrates, since the 1940s various government programmes in both Chile and Argentina have deliberately introduced potentially economically valuable species, particularly fur bearers, leading to a modern terrestrial mammal assemblage where non-natives outnumber native species (Table 29.1).

Overall, beavers behave similarly in both native and invaded ecosystems. Specifically, it has been shown that their ecological impacts generally are in the same direction and magnitude in both northern and southern hemisphere habitats (Anderson et al, 2009). However, in southern South America's extensive primary forests and bogs, the beaver has found an ideal habitat with abundant hydrologic and food resources, accompanied by low or non-existent predator pressure on many islands. This ecological context has led to high colony densities (Lizarralde, 1993; Skewes et al, 2006) and has provoked large-scale removal of riparian forests and conversion to beaver ponds and meadows (Anderson et al, 2006b; Martínez Pastur et al, 2006). Furthermore, in the southern hemisphere, the beaver's ecological niche has extended to suboptimal habitats, including treeline and steppe ecosystems. In effect, this invasion constitutes the largest landscape level alteration of the sub-Antarctic forest ecoregion in the Holocene (Anderson et al, 2009).

Yet estimates of abundance and area impacted by beavers are not without precedent in its native distribution, instead falling within the higher range of values expected for optimal native habitat (Anderson et al, 2009). Therefore, the differences detected between native and non-native beaver engineering effects lie not with the species itself, but rather the surrounding landscape. As such, the ultimate role of beavers in sub-Antarctic forests is predictable, based on understanding the natural history and evolutionary differences with North American ecosystems.

It should be noted that unlike North America's high latitude forests, the sub-Antarctic forest ecoregion is dominated by only three broadleaf tree species in the genus *Nothofagus* (Pisano, 1977). In stark contrast to boreal ecosystems, coniferous trees are not part of the forest community. In North America, the result of natural history characteristics, including regeneration strategies and defence mechanisms, means that beaver engineering activities create an unpalatable tree stand, particularly conifers, that remains and/or regrows (Naiman et al, 1988). *Nothofagus* forests naturally regenerate from direct seed deposition into gaps, since seed banks do not persist for long periods and vegetative reproduction is rare (Cuevas and Arroyo, 1999). Only one species (*N. antarctica*) is adapted for boggy soil conditions (Ramírez et al, 1985). As a result, the sub-Antarctic forest has less regeneration capacity and may become a long term stable meadow after beaver dam abandonment with modified succession based on *N. antarctica*, rather than the naturally co-dominant *N. pumilio* and *N. betuloides* (Wallem et al, 2010).

Table 29.1 *Native and exotic terrestrial mammals of the Tierra del Fuego and Cape Horn archipelagos (including both Argentina and Chile)*

Order	Scientific name	Common name	Site
Native species			
Artiodactyla	*Lama guanicoe*	Guanaco	IG & Nav
Carnivora	*Lontra provocax*	Large river otter	IG, Pic, Len, Gor & Wol
	Lontra felina	Sea otter	IG, Ho, Pic, Gor & Wol
	Pseudalopex culpaeus	Fuegian red fox	IG & Ho
Chiroptera	*Histiotus montanus*	Eared bat	IG, Nav & Wol
	Myotis chiloensis	Chiloé bat	IG, Nav & Gre
Rodentia	*Myocastor coypus*	Coypu	IG & Gor**
	Abrothrix xanthorhinus	Yellow-nosed mouse	IG, Nav & Ho
	Abrothrix longipilis	Long haired grass mouse	IG
	Akodon hershkovitzi	Cape Horn mouse	He, Ht & Hr
	Oligoryzomys longicaudatus	Long-tailed mouse	IG, Wol, Ht & Ho
	Euneomys chinchilloides	Patagonian chinchilla mouse	IG, Wol, Ht & Ho
	Ctenomys magellanicus	Magellanic tuco tuco	IG
Total	**13 species**		
Exotic species			
Artiodactyla	*Sus scrofa*	Feral pig	IG, Nav & Gor
	Bos tarus	Feral cow	IG & Nav
	Capra hircus	Feral goat	Sta
	Cervus elaphus	Red deer	Sta & IG¥
Carnivora	*Canis lupus familiaris*	Feral dog	IG, Nav & Hr*
	Felis domesticus	Feral cat	IG, Nav & Hr*
	Neovison vison	American mink	IG, Nav & Ho
	Pseudalopex griseus	Grey fox	IG
	Vulpes vulpes	Silver fox	IG***
Lagomorpha	*Oryctolagus cuniculus*	European rabbit	IG, Nav** & Len**
Perisodactyla	*Equus caballus*	Feral horse	IG & Nav
Rodentia	*Castor canadensis*	American beaver	IG, Nav, Ho, Pic, Nu & Len
	Ondatra zibethica	Muskrat	IG, Nav, Ho, Pic, Nu & Len
	Rattus rattus	Black rat	IG
	Rattus norvegicus	Norway rat	IG, Nav
	Mus musculus	House mouse	IG, Nav
Xenarthra	*Chaetophractus villosus*	Larger hairy armadillo	IG
Total	**17 species**		

Note: Site abbreviations (all names refer to islands in the archipelago): Gre = Grevy, Gor = Gordon, He = Herschel, Ht = Hermite, Ho = Hoste, Hr = Horn, IG = Isla Grande (the main island), Len = Lennox, Nav = Navarino, Nu = Nueva, Pic = Picton, Sta = Staten, Wol = Wollaston. * indicates domestic animals kept at remote Navy outposts. ** indicates unconfirmed reports. *** indicates sporadic sightings. ¥ indicates single population kept in captivity on IG.

Source: Allen (1905); Thomas (1916); Olrog (1950); Cabrera (1961); Peña and Barría (1972); Sielfeld (1977, 1984); Bridges (1978); Patterson et al (1984); Reise and Venegas (1987); Lizarralde and Escobar (2000); Anderson et al (2006b); Poljak et al (2007)

Furthermore, while in parts of North America beavers create a more diverse habitat mosaic and introduce a herbaceous species assemblage in an otherwise forested landscape (Wright et al, 2002), in South America this invasion has given an advantage to introduced plant species and other taxa already present in adjacent ecosystems in the landscape, such as grasslands and ñirre forests (Martínez Pastur et al, 2006; Anderson et al, 2006b).

The effects of North American beaver on physical, chemical and geomorphological conditions of streams are also similar for both an alien and native species (McDowell and Naiman, 1986; Anderson and Rosemond, 2007, 2010). The characteristics, persistence and magnitude of these effects depend on the beaver pond's placement in the catchment, including factors such as slope and precipitation. The ultimate result of beaver engineering is to increase the dependence of in-stream food webs on terrestrially derived organic material, which enhances allochthonous resource subsidies to stream fauna (McDowell and Naiman, 1986; Anderson and Rosemond, 2007, 2010). For sub-Antarctic streams, the trophic consequences of these alterations are to push benthic ecosystem secondary production from typical conditions of high latitude cold streams to those of temperate streams (Anderson and Rosemond, 2007).

Pioneering Efforts to Control Beaver in Chile's Magallanes Region

As a remote part of the Americas, until relatively recently research in most of the archipelago was based on short term projects, concentrated in summer months and conducted by researchers from other parts of the country and world (Rozzi et al, 2006). Only in 2000 did a permanent group of place-based academics consolidate around the creation of the Omora Ethnobotanical Park in Puerto Williams, Chile. Furthermore, in June 2008, a nascent Long-Term Socio-Ecological Research (LTSER) Network was financed by the Chilean National Scientific and Technological Commission's Basal Financing Program to better equip and provide infrastructure to conduct long term studies that contribute to the nation's development and well-being (Anderson et al, 2010). Such long term programmes provide potentially crucial support and opportunities

for researchers and decision makers to engage one another in a meaningful dialogue about the conservation, management and use of this biome.

The case of the North American beaver has become an emblematic example of both invasive species research and LTSER linked with policy making for southern South America. Specifically in the Chilean context, it is the catalyst behind a series of other invasive species control and management initiatives at the local and regional level that have also had impact at national and international scales (Choi, 2008). As such, the beaver has sparked a great deal of interest from scientists and also multinational conservation organizations, government officials and the general public. Successful eradication programmes, however, require a series of conditions to be met that include ecological, social, economic and political factors (Veitch and Clout, 2002). While significant attention is being paid to some of these points (see Parkes et al, 2008; Wallem et al, in press), the sociopolitical process that has led to the current situation and its lessons and implications for the future have not been taken into account. Here, we analyse the relationship between the historical timeline of research achievements and policy developments, focusing on the Chilean experience.

We analysed available bibliographic databases, considering two peer-reviewed, scientific sources: ISI® and Scielo indexed journals, to determine research trends on invasive beavers in Chile and Argentina. The Scielo journals were divided into Chilean and Argentine periodicals (Figure 29.3). We did not incorporate into this analysis broad review articles on invasive species for both countries (e.g. Jaksic et al, 2002), instead focusing on primary research about specific invasive species. Our search returned a total of 18 ISI, 3 Chilean non-ISI and 2 Argentine non-ISI publications. The topics addressed in the extant scientific literature have been as varied as ethical implications of invasive species management (Haider and Jax, 2006), population genetics (Lizarralde et al, 2008), properties of beaver meat (Hofbauer et al, 2005) and general community and ecosystem ecology for both terrestrial and aquatic habitats (review in Anderson et al, 2009). Overall, we found that more research has been conducted in Chile (71 per cent) than Argentina (29 per cent), and only beginning in 2009 have investigators from both countries begun to publish their work jointly (n = 2; 9 per cent; Anderson et al, 2009; Wallem et al, 2010).

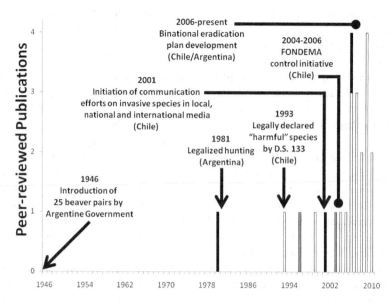

Note: ISI (no fill), Chilean non-ISI (black fill) and Argentine non-ISI (grey fill). Key sociopolitical decision-making events are also included in the timeline.

Figure 29.3 *Scientific literature citations regarding invasive beaver in Chile and Argentina per year in ISI, Chilean non-ISI and Argentine non-ISI peer-reviewed journals*

With regards to the relationship between the generation of information (expressed as peer-reviewed publications) and its use in policy initiatives, we found that the transfer of scientific knowledge to management via traditional academic publications is tenuous at best (Table 29.2). While beavers were introduced to Tierra del Fuego in 1946, the first scientific publication did not appear until 1980 in the *Anales del Instituto de la Patagonia*, a regional Chilean journal (Sielfeld and Venegas, 1980). In the same year, the Argentine government authorized the legal hunting of beavers for the first time (Lizarralde, 1993), which was doubtfully linked with a local Chilean scientific publication. More than a decade passed before the next scientific publication appeared, this time an Argentine researcher publishing in an international journal *Ambio* (Lizarralde, 1993), and again ironically in the same year the Chilean government declared the beaver 'harmful', thereby authorizing its hunting year round (Anderson et al, 2009), which is unlikely to be a causal result of the ISI research publication in English.

Instead, the link between research and decision making regarding invasive beavers in the early decades (1980s and 1990s) is apparently based on personal relationships between the few researchers and managers working on the subject and the managers' own field-based experiences, rather than academic information being generated and used on its own. This trend, however, began to change with the establishment of long term research initiatives and a critical mass of scientists and students coming to live in the Magallanes region as of 2000. From this point, personal relationships, direct knowledge and experience of managers, and academic information have had the chance to blend into the decision-making process. While the Omora Ethnobotanical Park was founded in 2000 with a priority line of research on invasive species in the remote Cape Horn Archipelago, due to the delay between initiating research and the publication of results, the first scientific papers began to appear in 2003 for the study of invasive American mink (Rozzi and Sherriffs, 2003) and 2006 for the study of invasive beavers (Anderson et al, 2006a, 2006b). The first concerted effort of the regional government to fund a beaver control programme began before these publications (2004–2006), and this effort clearly could not have been informed by peer-reviewed literature only. Instead, the pertinent role of researchers in this phase was the direct collaboration of scientists with managers

Table 29.2 *Historical timeline of benchmarks in research and application regarding the study and control of North American beaver in the Magallanes region of Chile and their implications for the sociopolitical process of linking academia with decision making*

Year	Benchmark	*Type of action*: Results and lessons
1946	Argentine government introduces 25 pairs of North American beavers to Tierra del Fuego Island to create a fur industry (Lizarralde, 1993).	*Application*: Due to a lack of understanding and/or appreciation of native biodiversity, various Argentine and Chilean government-sponsored programmes in the 1940s and 1950s attempted to 'improve' high latitude ecosystems that are species poor in large vertebrates by introducing ungulates (e.g. deer in northern Patagonia) and fur bearers (e.g. mink, muskrats and beaver) in Tierra del Fuego.
1960s	Introduced beaver population invades Chilean portions of the archipelago.	*Inaction*: An absence of researchers and policy makers in the Fuegian Archipelago led to the invasion taking place for decades without either a research or policy response.
1980	First academic publication on beaver invasion in Chile: Sielfeld and Venegas (1980). Authorization of legal hunting of beavers in Argentina.	*Application/Research*: While the first scientific publication on beavers occurred the same year as the first policy response, these two events occurred in separate countries, Chile and Argentina, respectively, and there is no indication that they were linked.
1993	First ISI publication by Argentine research group: Lizarralde (1993). First Chilean management response: *C. canadensis* listed as 'harmful' species, thus authorizing its capture year round (D.S. 133, Chilean Hunting Law).	*Application/Research*: As above, the link between the Argentine research published in *Ambio* and the policy result of the Chilean government declaring the beaver a 'harmful' species is untenable.
1997–1999	First extensive baseline study of invasive beavers in Chile (Skewes et al, 1999) conducted at request of Regional Agriculture and Livestock Service (SAG).	*Application/Research*: The first systematic study of beavers in Chile was funded by a management agency (SAG). The scientific results of this research were not available until 2005/2006. However, this and other grey literature were important sources of information for Chilean decision makers in the preparation of the first control efforts.
2000	Establishment of the Omora Ethnobotanical Park as a Long-Term Socio-Ecological Research (LTSER) programme in Cape Horn.	*Research*: Beginning of long term research presence in Cape Horn Archipelago with a priority line of study in invasive species, but also the park aspires to link research, education and conservation. Plus, its action plan (Rozzi et al, 2006) includes explicit and direct communication via the media to inform policy, public perceptions and decisions.
2001	First detected presence of invasive mink in Cape Horn Archipelago. SAG and Omora work together to alert the public and authorities to invasive species in Cape Horn in regional (*La Prensa Austral*) and national (*El Mercurio*) news outlets.	*Research*: Rozzi and Sherriffs (2003) report the presence of mink on Navarino Island. Previously, while this invasive species was known to have been liberated on Tierra del Fuego Island, researchers still reported its presence as 'possible or theoretical' as late as the 1990s (Massoia and Chébez, 1993). *Application*: A national media campaign in the leading Chilean newspaper led to the visit of the national director of wildlife for SAG to Navarino and the impetus behind the development of the first 'harmful' species control programme in the Magallanes region, which was funded by the regional government in 2004.
2002	Series of floods on Navarino Island provoke the failure of numerous beaver dams, creating a sediment flow that washes out bridges on the island.	*Application*: Recognition on a political level, particularly the provincial governor, of economic impacts associated with beavers, vis-à-vis the subsequent need to construct stronger concrete bridges on Navarino Island.

Table 29.2 *Historical timeline of benchmarks in research and application regarding the study and control of North American beaver in the Magallanes region of Chile and their implications for the sociopolitical process of linking academia with decision making* (Cont'd)

Year	Benchmark	*Type of action*: Results and lessons
2004–2006	Regional Fund for the Development of Magallanes (FONDEMA) finances the first large-scale Chilean beaver control programme, based on commercial uses of meat and pelt to create market incentives for hunting.	*Application*: As the first major effort to address the issue of beaver invasion in the Chilean portion of the archipelago, this project is of note for (i) pioneering the administrative structure necessary to carry out large-scale wildlife management projects; (ii) harvesting 11,700 beavers, 234 mink, 250 muskrats; (iii) training 276 trappers (of which 30 became proficient) and 45 artisans to use cured pelts; and (iv) demonstrating that the regional government is capable of assuming wildlife management issues by investing US$450,000. The project reduced beaver densities in targeted areas from 1.12 active colonies km^{-1} to 0.44 (Soto et al, 2007).
2005	Creation of Cape Horn Biosphere Reserve (CHBR).	*Application/Research*: The nomination of the CHBR involved a governmental partnership with researchers at the Omora Park to link research and policy for sustainable conservation and use of biodiversity in the archipelago. The implementation of the biosphere reserve involves the prioritization of invasive species research combined with regional decision makers and agencies, including SAG.
2006	Publication of 'Exotic vertebrates of the Cape Horn Archipelago' (Anderson et al, 2006a)	*Research*: While the sub-Antarctic ecoregion has generally been considered remote and 'pristine' (Mittermeier et al, 2003) and also in good conservation status, given its extensive national parks (51% of the Chilean portion), Anderson et al (2006a) also showed that it has not escaped the global impacts of introduced species with more non-native terrestrial vertebrates than native.
	First Argentine–Chilean binational meeting on beaver control, including researchers, authorities and managers.	*Application/Research*: Convened by the SAG and Wildlife Conservation Society (WCS), this first binational meeting was attended by more than 50 researchers and policy makers, representing the academic, public and private sectors, from Argentina, Chile, New Zealand and the US.
2007–2009	Communication by Omora Park scientists and SAG authorities of the beaver issue in local (*La Prensa Austral*), national (*El Mercurio*), scientific (*Nature*) and international (BBC, *Der Spiegel*) media outlets.	*Application*: Throughout the research projects on invasive species conducted in the CHBR, scientists and managers have worked actively as informants and also advisers for various programmes and reports, specifically relating to the beaver at the local/regional, national and international levels.
2008	The Chilean Agriculture and Livestock Service (SAG) finances feasibility study of beaver eradication in south Patagonia (Argentina and Chile).	*Research*: As a result of the 2006 binational meetings, SAG obtains institutional funds to conduct a feasibility study to determine the viability of eradication and possible other alternatives. An international team determined that eradication is technically feasible and costly (>US$30 million), but that more research would be needed to determine distribution, population dynamics and effective methods for control and detection.
2008	On 26 September 2008, Chile and Argentina signed a binational agreement on the Restoration of the Southern Ecosystems Affected by the American Beaver (*Castor canadensis*).	*Application*: (i) The fact that an issue of regional impact in a remote part of both countries arrived at the highest level of political decision making demonstrates that decision makers have recognized the seriousness of the impacts of this invader, which has been facilitated by nearly a decade of work between researchers and managers at the local and regional level. (ii) SAG financed a feasibility study to determine the cost benefits of eradication versus other control options.

Table 29.2 *Historical timeline of benchmarks in research and application regarding the study and control of North American beaver in the Magallanes region of Chile and their implications for the sociopolitical process of linking academia with decision making* (Cont'd)

Year	Benchmark	*Type of action*: Results and lessons
		(iii) Chile and Argentina develop a Contingency and Action Plan to eliminate beavers from the continent (Brunswick Peninsula), and currently SAG has obtained a regional government commitment of US$1.2 million for 2011–2013. (iv) Together both nations have worked to develop a Binational Strategy to prepare pilot projects and obtain international funding for a major binational, joint initiative.
2010	Establishment of memorandum of understanding (MoU) between University of Magallanes Master's of Science in Conservation and Management of Sub-Antarctic Ecosystems and SAG.	***Research/Application***: (i) Inclusion of SAG managers in master's thesis commissions; (ii) development of master's thesis projects in areas relevant to SAG; (iii) support of academics in the development and implementation of regional control plans; and (iv) joint publications written by scientists and managers.

in the development and carrying out of control plans (e.g. forming part of scientific advisory boards) and grey literature sources (Skewes et al, 1999). However, it must also be pointed out that from the management perspective, the academic ties proved to be crucial in the legitimization of control plans for politicians, who provided the resources, and also for broader communication with the media to create the necessary social conditions for such initiatives.

Since 2006 a new era for research is evident with an explosion of literature resulting from several projects and theses initiated on both sides of the Argentine/Chilean border. The year 2006 also coincides with the initiation of a binational process that aspires for eventual beaver eradication and has led to the signing of a binational treaty on the topic. Therefore, it is clear that the amount of scientific literature has increased, and it is concomitantly being better incorporated into the development of control plans with the formal involvement of researchers in the binational working groups. However, our analysis also detected gaps in key areas of investigation. If scientists truly intend to conduct relevant studies that are meaningful for managers, they must confront the existing bias in their agenda and publications towards general ecology and a clear lack of applied research into trapping methods, habitat selection and population dynamics models, and studies of the social perceptions of both alien species and eventual control programmes. To help remedy this problem, in 2010 the University of Magallanes and the Regional Office of the Chilean Agriculture and Livestock Service (SAG), which is charged with dealing with harmful species, signed a memorandum of understanding that links master's theses research with the application needs of SAG, specifically emphasizing projects (currently there are two under development) that would address issues related to invasive species management.

On the institutional level in Chile, these research advances have influenced a parallel improvement in the regulatory framework for prevention, control and/or eradication of invasive species. For example, there are now specific chapters about invasive species in the regional and national biodiversity strategies. However, these institutional advances have not been accompanied yet by an effective operational phase or translation into specific processes, largely due to a lack of funding. Furthermore, greater work on the legal framework is needed at the national level, as currently the laws that regulate invasive species are within Chile's 'Hunting Law' that is applied to the sustainable use of wild animals. In total, 5 articles (of 45) in the law and 7 (of 91) in the associated regulations make reference to 'harmful' species. Subsequently, SAG has made two resolutions (863/1999 and 5006/2004) that name more than 4900 species as potentially causing disturbance to the ecological balance. But, while the legal framework establishes the regulations for quarantine of potentially invasive species in the country as well as authorizes their hunting without restrictions, it does *not* impose obligatory control of these taxa. Furthermore, if we compare the regulations for different taxonomic categories of invasive species, administrative procedures and technical policies are much more advanced for the prevention, detection, control and eradication of invasive agents such as viruses, bacteria, parasites and insects, which affect the health of agricultural plants and animals and their derived products, while very little action can be taken towards species that affect 'ecosystem health'. This problem is largely due to the nature of the current administrative system that places SAG (a part of the Ministry of Agriculture) in this role of also working on invasive species, and it is hoped that the implementation of Chile's new Ministry of the Environment will resolve some of these institutional and regulatory bottlenecks and inconsistencies.

Final Comments on Linking Research, Decision Making and Eventual Restoration Initiatives

Putting the sub-Antarctic ecoregion on the map

As an invasive ecosystem engineer, the North American beaver has arguably created the largest landscape alteration in the sub-Antarctic forest biome since the end of the last ice age (Anderson et al, 2009). So, why did decision makers not notice this species for nearly 60 years? A comparison with the case of the European rabbit (*Oryctolagus cuniculus*) may prove illustrative. Unlike the beaver, various failed rabbit introductions took place in both the Argentine and Chilean sectors of Tierra del Fuego in 1874 and 1913, respectively, before a successful population was established in 1936. In only four years, this species had become a plague in the grassland biome due to low predator pressure and ample habitat. Its high numbers quickly put it into competition with the extensive and lucrative sheep ranching industry, causing degradation of rangelands that directly affected economic and political decision making. Between December 1953 and March 1954 the Ministry of Agriculture and private ranchers established a plan that effectively eliminated the population of invasive rabbits with the myxamatosis virus, making the time from introduction to invasion and finally to eradication only 17 years.

In contrast, the beaver largely went unnoticed for decades because it was mostly colonizing portions of

the archipelago that are uninhabited and its local ecological and economic effects did not include social sectors with power in regional and national decision making. It is clear that the establishment of long term research – in Chile dating to the seminal works financed by SAG and carried out by Skewes et al (1999) and documented most clearly on an archipelago level by Anderson et al (2006a) – put impacts into perspective and subsequently inserted this topic into the sociopolitical process. By late 2010, the documentation of the beaver's impacts had reached a critical point and the future implications on broader economic sectors were sufficiently clear to both countries to lead to the elevation of this topic to a binational treaty.

Addressing crucial knowledge gaps

The issues that must continue to be addressed in the agenda of research-management include: (1) how will scientists and managers work together to implement future control programmes to be more efficient and effective, and (2) in what ways can academic research and the application of these studies lead to ecosystem restoration of degraded riparian and stream habitats? Learning these lessons to better link research and management is an urgent matter, as in the Brunswick Peninsula (continental area) there is a planned project that is to be financed by the regional government from 2011–2013 with $1.2 million to eliminate beavers from the mainland. This proposal is a major administrative, logistical and technical challenge and will test the capacity of researchers and managers to eradicate beavers in a defined area and effectively integrate research and application in a real-time situation for the first time.

Building on strength

It is important to underline that in the various decades of work leading up to the present situation, the development of networks between individuals and institutions has been a key achievement in and of itself. The generation of scientific knowledge, positioning the 'problem' of invasive species in the public agenda and political debate, the contribution of designing regulations and policies between both countries, and the development of human capacity are all the product of multiple individuals and institutions working together. This alliance has led to concrete and substantial improvements in the link between research and management. Furthermore, it has allowed local scientists and managers to learn from previous experiences, including what is presently the world's largest successful island eradication project on the Galapagos Archipelago, where feral goats were removed from a total of approximately 512,000ha (Cruz et al, 2009; Lavoie et al, 2007). At the same time, it must be pointed out that Isabela Island is an order of magnitude smaller than Isla Grande alone, not to mention the entire Fuegian Archipelago, and the cost was approximately $9 million dollars over a ten year period, placing the estimate for Tierra del Fuego in the speculative realm of >$30 million (Parkes et al, 2008).

While eradication is a desirable goal, in terms of its costs, difficulty and uncertainty, there is a clear need to maintain a critical perspective and to continue evaluating the cost and benefits for a range of possible solutions that integrate the ultimate objective of restoration on the social, economic and ecological levels. Therefore, the establishment of a current dogma of 'eradication or nothing' is a potentially disturbing trend to some researchers and managers, as it does not take into account the dynamic nature of research and knowledge or adaptive management from not only the technical perspective, but also the political level for policy development and implementation. This paradigmatic shift, for example, has diverted attention from the original steps taken by SAG with the FONDEMA project, cutting short the three years of experience gained in developing market incentives to mitigate invasive species impacts.

References

Allen, J. D. (1905) *Mammalia of Southern Patagonia*, Princeton University, Princeton, New Jersey

Anderson, C. B. and Rosemond, A. D. (2007) 'Ecosystem engineering by invasive exotic beavers reduces in-stream diversity and enhances ecosystem function in Cape Horn Chile', *Oecologia*, vol 154, pp141–153

Anderson, C. B. and Rosemond, A. D. (2010) 'Beaver invasion alters terrestrial subsidies to subantarctic stream food webs', *Hydrobiologia*, vol 652, pp349–361

Anderson, C. B. and Valenzuela, A. E. J. (2011) 'Mammals, aquatic', in D. Simberloff and M. Rejmánek (eds) *Encyclopedia of Introduced Invasive Species*, University of California Press, pp445–449

Anderson C. B., Rozzi, R., Torres-Mura, J. C., McGehee, S. M., Sherriffs, M. F., Schüttler, E. and Rosemond, A. D. (2006a) 'Exotic vertebrate fauna in the remote and pristine sub-Antarctic Cape Horn Archipelago, Chile', *Biodiversity and Conservation*, vol 15, pp3295–3313

Anderson C. B., Griffith, C. R., Rosemond, A. D., Rozzi, R. and Dollenz, O. (2006b) 'The effects of invasive North American beavers on riparian plant communities in Cape Horn, Chile: Do exotic beavers engineer differently in sub-Antarctic ecosystems?', *Biological Conservation*, vol 128, pp467–474

Anderson, C. B., Martínez Pastur, G., Lencinas, M. V., Wallem, P. K., Moorman, M. C. and Rosemond, A. D. (2009) 'Do introduced North American beavers engineer differently in southern South America? – An overview with implications for restoration', *Mammal Review*, vol 39, pp33–52

Anderson, C. B., Rozzi, R., Armesto, J. J. and Gutierrez, J. (2010) 'Building a Chilean network for long-term socio-ecological research: Advances, perspectives and relevance', *Revista Chilena de Historia Natural*, vol 83, pp1–11

Armesto, J. J., Villagrán, C. and Arroyo, M. T. K. (eds) (1995) *Ecología de los Bosques Nativos de Chile*, Editorial Universitaria, Santiago

Bridges, L. (1978) *El Último Confín de la Tierra*, Marymar, Buenos Aires (1st edition, Hodder & Stoughton, 1947, London)

Cabrera, A. (1961) 'Catálogo de los mamíferos de América del Sur', Tomos I & II, *Revista del Museo Argentino de Ciencias Naturales Bernardino Rivadavia, Ciencias Zoológicas* IV, pp1–732

Choi, C. (2008) 'Tierra del Fuego: The beavers must die', *Nature,* vol 453, p968

Cruz, F., Carrion, V., Campbell, K. J., Lavoie, C. and Donlan, C. J. (2009) 'Bio-economics of large-scale eradication of feral goals from Santiago Island, Galapagos', *The Journal of Wildlife Management*, vol 73, pp191–200

Cuevas, J. G. and Arroyo, M. T. K. (1999) 'Ausencia de banco de semillas persistente en *Nothofagus pumilio* (Fagaceae) en Tierra del Fuego, Chile', *Revista Chilena de Historia Natural*, vol 72, pp73–82

Goffinet B., Buck, W., Rozzi, R. and Massardo, F. (2006) *The Miniature Forests of Cape Horn*, Ediciones de la Universidad de Magallanes – Fundación Omora, Punta Arenas

Haider, S. and Jax, K. (2006) 'The application of environmental ethics in biological conservation: A case study from the southernmost tip of the Americas', *Biodiversity and Conservation*, vol 16, pp2559–2573

Hofbauer, P., Schnake, F. G., Skewes, O., Lopez, A. J. L., Smulders, F. J. M., Bauer, F., Macher, R., Konig, H. E. and Paulsen, P. (2005) 'Studies on muscular topography and meat properties of beavers (*Castor canadensis*) caught in Tierra del Fuego, Chile', *Wiener Tierarztliche Monatsschrift*, vol 92, pp157–164

Jaksic, F. M., Iriarte, J. A., Jiménez, J. E. and Martínez, D. R. (2002) 'Invaders without frontiers: Cross-border invasions of exotic mammals', *Biological Invasions*, vol 4, pp157–173

Lavoie, C., Cruz, F., Carrion, G. V., Campbell, K., Donlan, C. J., Harcourt, S. and Moya, M. (2007) *The Thematic Atlas of Project Isabela: An Illustrative Document Describing, Step-by-step, the Biggest Successful Goat Eradication Project on the Galapagos Islands, 1998–2006*, Puerto Ayora, Galapagos, Charles Darwin Foundation

Lizarralde, M. (1993) 'Current status of the introduced beaver (*Castor canadensis*) population in Tierra del Fuego, Argentina', *Ambio*, vol 22, pp351–358

Lizarralde, M. S. and Escobar, J. M. (2000) 'Exotic mammals in Tierra del Fuego', *Ciencia Hoy*, vol 10, pp52–63

Lizarralde, M. S., Bailliet. G., Poljak, S., Fasanella, M. and Giulivi, C. (2008) 'Assessing genetic variation and population structure of invasive North American beaver (*Castor canadensis* Kuhl, 1820) in Tierra Del Fuego (Argentina)', *Biological Invasions*, vol 10, pp673–683

Martínez Pastur, G., Lencinas, V., Escobar, J., Quiroga, P., Malmierca, L. and Lizarralde, M. (2006) 'Understory succession in areas of *Nothofagus* forests in Tierra del Fuego affected by *Castor canadensis*', *Journal of Applied Vegetation Science*, vol 9, pp143–154

Massoia, E. and Chébez, J. C. (eds) (1993) *Mamíferos silvestres del archipiélago Fueguino*, Ediciones LOLA, Buenos Aires

McDowell, D. M. and Naiman, R. J. (1986) 'Structure and function of a benthic invertebrate stream community as influenced by beaver (*Castor canadensis*)', *Oecologia*, vol 68, pp481–489

Mittermeier, R., Mittermeier, C., Brooks, T. M., Pilgram, J. D., Konstant, W. R., da Fonseca, G. A. B. and Kormos, C. (2003) 'Wilderness and biodiversity conservation', *PNAS*, vol 100, no 18, pp10309–10313

Moorman, M. C., Anderson, C. B., Gutiérrez, A. G., Charlin, R. and Rozzi, R. (2006) 'Watershed conservation and aquatic benthic macroinvertebrate diversity in the Alberto D'Agostini National Park, Tierra del Fuego, Chile', *Anales del Instituto de la Patagonia*, vol 34, pp41–58

Naiman, R. J., Johnston, C. A. and Kelley, J. C. (1988) 'Alteration of North American streams by beaver: The structure and dynamics of streams are changing as beaver recolonize their historic habitat', *BioScience*, vol 38, pp753–762

Olrog, C. C. (1950) 'Notas sobre mamíferos y aves del archipiélago de Cabo de Hornos', *Acta Zoológica Lillonana* IX, pp505–532

Parkes, J. P., Paulson, J., Donlan, C. J., Campbell, K., Schiavini, A., Saavedra, B., Menvielle, M. F., Malmierca, L., Escobar, J., Muza, R., Briceno, C. and Silvia, C. (2008) *Estudio de Factibilidad de Erradicar el Castor Americano* (Castor canadensis) *en la Patagonia, Informe Final*, Encargado por el Servicio Agricola y Ganadero XII Reigon

Patterson, B. D., Gallardo, M. H. and Freas, K. E. (1984) 'Systematics of mice of the subgenus *Akodon* (Rodentia: Cricetidae) in southern South America, with the description of a new species', *Fieldiana Zoology*, vol 23, pp1–16

Peña, L. F. and Barría, N. D. (1972) 'Presencia de *Histiotus montanus magellanicus* Philp. y de *Myostis* ch. *chiloensis* Waterhouse (Chiroptera) al sur del Estrecho de Magallanes', *Anales del Museo de Historia Natural Valparaíso*, vol 5, pp202–202

Pisano, E. (1977) 'Fitogeografía de Fuelo-Patagonia Chilena. Comunidades vegetales entre las latitudes 52° y 56°S', *Anales del Instituto de la Patagonia*, vol 8, pp121–250

Poljak, S., Escobar, J., Deferrari, G. and Lizarralde, M. (2007) 'Un nuevo mamífero introducido en la Tierra del Fuego: el "peludo" *Chaetophractus villosus* (Mammalia, Dasypodidae) en Isla Grande', *Revista Chilena de Historia Natural*, vol 80, pp285–294

Ramírez, C., Correa, M., Figueroa, H. and San Martín, J. (1985) 'Variación del hábito y hábitat de *Nothofagus antarctica* en el centro sur de Chile', *Bosque*, vol 6, pp55–73

Reise, D. and Venegas, C. (1987) 'Catálogo de registros localidades y biotopos del trabajo de investigación acerca de los pequeños mamíferos de Chile y Argentina', *Gayana*, vol 51, pp103–130

Rozzi, R. and Sherriffs, M. F. (2003) 'El vison (*Mustela vison* Schreber, Carnivora: Mustelidae), un nuevo mamífero exotico para la isla Navarino', *Anales del Instituto de la Patagonia*, vol 31, pp97–104

Rozzi R., Massardo, F., Anderson, C. B., Berghoefer, A., Mansilla, A., Mansilla, M. and Plana, J. (2006) *Reserva de Biosfera Cabo de Hornos*, Ediciones de la Universidad de Magallanes, Punta Arenas

Rozzi, R., Armesto, J., Goffinet, B., Buck, W., Massardo, F., Silander, Jr, J., Arroyo, M. T. K., Russell, S., Anderson, C. B., Cavieres, L. and Callicott, J. B. (2008) 'Changing biodiversity conservation lenses: Insights from the subantarctic non-vascular flora of southern South America', *Frontiers in Ecology and the Environment*, vol 6, pp131–137

Sielfeld, W. (1977) 'Reconocimiento macrofaunístico terrestre en el área de Seno Ponsonby (Isla Hoste)', *Anales del Instituto de la Patagonia*, vol 8, pp275–296

Sielfeld, W. (1984) 'Alimentación de las nutrias *L. felina* y *L. provocax* en el medio marino al sur del Canal de Beagle', in *Primera Reunión de Trabajo de Experiencias en Mamíferos Acuáticos de América del Sur*, Buenos Aires

Sielfeld, W. and Venegas, C. (1980) 'Poblamiento e impacto ambiental de *Castor canadensis* Kuhl en Isla Navarino, Chile', *Anales del Instituto de la Patagonia*, vol 2, pp247–257

Skewes, O., Gonzalez, F., Rubilar, L. and Quezada, M. (1999) *Investigación, Aprovechamiento y Control del* Castor, *Islas Tierra del Fuego y Navarino*, Instituto Forestal – Universidad de Concepcion, Punta Arenas

Skewes, O., González, F., Olave, R., Ávila, A., Vargas, V., Paulsen, P. and Konig, H. E. (2006) 'Abundance and distribution of American beaver, *Castor canadensis* (Kuhl 1820), in Tierra del Fuego and Navarino Islands', *European Journal of Wildlife Research*, vol 52, pp292–296

Soto, N., Cabello, J. L. and Antunez, D. (2007) *Programa Control de Fauna Dañina en la Región de Magallanes, 2004–2007, Informe Final*, SAG-FONDEMA, Código BIP 0.027.043-0, Punta Arenas, Chile

Thomas, O. (1916) 'Notes on Argentina, Patagonia and Cape Horn Muridae', *Annals and Magazine Natural History*, vol 8, pp182–187

Veitch, C. and Clout, M. (2002) 'Turning the tide: The eradication of invasive species', in *Proceedings of the International Conference on Eradication of Island Invasives*, Occasional Paper Series of the IUCN Species Survival Commission No 27, Gland, Switzerland

Wallem, P. K., Jones, C. G., Marquet, P. A. and Jaksic, F. M. (2007) 'Identificación de los mecanismo subyacentes a la invasión de *Castor canadensis* (Kuhl 1820, Rodentia) en el archipiélago de Tierra del Fuego, Chile', *Revista Chilena de Historia Natural*, vol 80, pp309–325

Wallem, P. K., Anderson, C. B., Martínez Pastur, G. and Lencinas, M. V. (2010) 'Community re-assembly by an exotic herbivore, *Castor canadensis*, in subantarctic forests, Chile and Argentina', *Biological Invasions*, vol 12, pp325–335

Wallem, P. K., Soto, N., Cabello, J. L., Castro, S. A. and Jaksic, F. M. (in press) 'El caso de la invasión de *Castor canadensis* en Tierra del Fuego como oportunidad para analizar la valoración del impacto y manejo de vertebrados exóticos invasivos en Chile', *Invasiones Biológicas en Chile*

Wright, J. P., Jones, C. G. and Flecker, A. S. (2002) 'An ecosystem engineer, the beaver, increases species richness at the landscape scale', *Oecologia*, vol 132, pp96–101

30

Myocastor coypus Molina (coypu)

Sandro Bertolino, M. Laura Guichón and Jacoby Carter

Introduction

Myocastor coypus (Coypu) (Figure 30.1) is a large semi-aquatic rodent native to South America that is now present in all continents, except Oceania and Antarctica, after widespread introductions in the 1930s and 1940s (Carter and Leonard, 2002). There is a division in English-speaking countries as to common name usage. In England and former British colonies (e.g. Kenya) they are called 'coypus'. In North America and Asia they are generally referred to as 'nutria'. However, it should be noted that in Spanish speaking countries this name refers to otters (Lutrinae).

At a first glance the coypu resembles a very large rat, with short legs and a long cylindrical tail. Head and body length of adults ranges between 472 and 575mm, and the tail is 340–405mm (Woods et al, 1992). Mean weights are 4–6kg, with males larger than females, even though adults can be over 10kg (Gosling and Baker, 2008; Guichón et al, 2003a). Four out of five toes of the hind feet are webbed, indicating the coypu's adaptation to swimming. The fur colour ranges from yellow brown to dark brown, and the chin is covered by white hair.

In many regions the coypu is considered a pest because of its negative impact on ecosystems, crops and irrigation systems (Carter and Leonard, 2002; Bertolino and Genovesi, 2007). Coypus cause damage by overgrazing natural vegetation, damaging crops and undermining riverbanks and dykes by burrowing into them (Boorman and Fuller, 1981; Foote and Johnson, 1993; Panzacchi et al, 2007). For these reasons the species is considered one of the most problematic invasive species globally (Bertolino, 2008, 2009).

Native and Introduced Ranges

Coypus are hystricomorphic rodents endemic to South America (Nowak, 1991) (Figure 30.2). *Myocastor coypus* is the only species of the Myocastoridae family, and is native to southern South America (Woods et al, 1992). Four subspecies have been described for *M. coypus*:

1   *M. c. bonariensis* (Geoffroy St Hilaire, 1805) in Central and Northern Argentina, Bolivia, Southern Brazil, Paraguay and Uruguay;
2   *M. c. coypus* (Molina, 1782) in Central Chile;
3   *M. c. melanops* (Osgood, 1943) in Chiloé Island, Chile; and
4   *M. c. santacruzae* (Hollister, 1914) in Patagonia, Argentina.

Available data indicate that introduced coypus are from the subspecies *M. c. bonariensis* (Willner et al, 1979; Gosling and Skinner, 1984).

The coypu is listed as a species of least concern for conservation in view of its wide distribution and presumed large populations (Lessa et al, 2008). Hunting can control coypu populations and can, under certain circumstances, drive local populations to extinction (Deems and Pursley, 1978; Guichón and Cassini, 2005). In South America the coypu is intensively exploited because the fur constitutes an important economic resource for rural people and farmers. The interest for this fur diffused worldwide, and as a consequence from the late 1800s and early 20th century, coypu farms were started around the world to provide fur for the international market (Carter and

Source: Aurelio Perrone

Figure 30.1 Myocastor coypus *(coypu)*

Leonard, 2002). Coypu populations are now established in North America, many European countries, central and eastern Asia including Japan and Korea, Kenya in East Africa, and the Middle East (Carter and Leonard, 2002) (Figure 30.2). Repeated efforts to establish wild coypu populations in Ontario, Canada in the 1970s failed due to severe winters (Southerland, D., Ontario Ministry of Natural Resources, pers. comm.). In these countries, coypu were directly released into the wild to create populations that trappers could exploit, or were

maintained for breeding and production in fur farms from which they frequently escaped. Later, when the demand for coypu fur declined, many of these farms ceased to be profitable and farmers would often release the animals into the wild. Regardless of how they were introduced, where habitats and climatic conditions were favourable, coypu spread fast due to their adaptability to different type of aquatic habitats. Using streams and canals as pathways, coypu can rapidly occupy disjunct but suitable habitats. Females are able to reproduce throughout the year starting when less than a year old and the mean litter size is 4–6 young (range 1–12) (Gosling, 1981; Bounds et al, 2003; Guichón et al, 2003a). Where environmental conditions are not limiting, females can have 2.7 litters year^{-1} after a four month gestation period, with an average of 8–15 young year^{-1} (Brown, 1975; Willner et al, 1979; Reggiani et al, 1993).

Ecology

Coypus are found in a variety of aquatic habitats including: wetlands, ponds, lakes, rivers and streams. Habitat use at fine spatial scales differs according to habitat characteristics, food availability, predation risk

Note: Black: native range; dark grey: introduced range with good distribution data; light grey: introduced range with rough data available; arrows: small ranges.

Source: Woods et al (1992); Carter and Leonard (2002); Abe et al (2005); Korean Ministry of Environment (2009); DAISIE (2010)

Figure 30.2 *Global distribution of coypu*

and hunting pressure. In linear habitats, coypus build burrow systems in the bank and forage in the water or close to it (<10m) when vegetation is available (D'Adamo et al, 2000; Guichón et al, 2003b, 2003c). The animals rarely move more than 100m away from the banks, whereas they can cover kilometres of a river longitudinally (e.g. Kim, 1980; Reggiani et al, 1993). Each burrow system averages 4.5 openings (range 1–17) and may extend over 0.3–25m of shoreline (Guichón et al, 2003c). The same main burrow systems can be used for more than ten years, even when flooding and droughts occurred throughout this period (Guichón, M. L., pers. obs.). In Argentinean marshlands, coypus build nests in tall sedges and grasses growing in shallow water and forage mainly using a strip of approximately 4m width in the interface between areas with tall and short vegetation (Bó and Porini, 2005).

Coypus feed mainly on aquatic and hydrophilic plants and, to a lesser extent, on terrestrial plants, mainly grasses near the water's edge (Table 30.1). Diet selection is influenced by microhabitat use with animals avoiding foraging far from the water as a way of reducing predation risk and aiding thermoregulation. In a field experiment, Borgnia et al (2000) showed

Table 30.1 *Vegetation consumed by coypus in various habitats in introduced and native ranges*

Genera	Habitat[a]	Country[b]	Source[c]
Agrostis*	C, L	F, USA	8, 10
Alternanthera*	FM, P, R	A, USA	5, 6, 7
Azolla	C	F	8
Bidens*	FM	USA	5
Bromus	P, R	A	1, 7
Callitriche	FM	I	2
Carex	C, FM	F, UK	8, 9
Ceratophyllum	C	F	8
Cynodon	P, R	A	1, 7
Cyperus	FM	USA	5
Dichondra	P	A	1
Echinochloa	P, R	A	7
Eichornia	FM	A, USA	3, 6
Eleocharis*	FM, P, R	A, USA	1, 5, 7
Elodea*	FM	I	2
Glyceria	FM, P, R	A, I	2, 7
Hydrocotyle*	FM, P, R	A, USA	5, 7
Lemna*	C, FM, P, R	A, F, I, USA	1, 2, 6, 7, 8
Limnobium	FM	A	3
Lolium	C, P, R	A, F	1, 7, 8
Lysimachia	FM	I	2
Myriophyllum	FM	I	2
Najas	FM	USA	5
Nasturtium	FM	I	2
Nuphar	FM	I, UK	2, 9
Nymphoides	FM	I	2
Panicum	BM, FM	USA	4, 6

Table 30.1 *Vegetation consumed by coypus in various habitats in introduced and native ranges* (Cont'd)

Genera	Habitat[a]	Country[b]	Source[c]
Paspalum	P, R	A	7
*Phragmites**	BM, FM	I, UK, USA	2, 4, 9
Poa	R, C	A, F	7, 8
Pontederia	FM	USA	6
Ranunculus	FM	I	2
Robinia	FM	I	2
*Sacciolepis**	FM	USA	5
Sagittaria	FM	A, USA	3, 5
*Scirpus**	BM, FM	I, USA	2, 4
Sparganium	BM, FM	UK	9
*Spirodela**	C, FM	F, USA	6, 8
Typha	BM, FM	UK	9
Vallisneria	FM	I	2

Notes: Most consumed plants are denoted by* and those eaten only occasionally are not included.

[a]P: pond; FM: freshwater marsh; BM: brackish marsh; R: rivers and streams; C: canal network; L: urban and suburban lawns and golf courses

[b]A: Argentina; I: Italy; USA: United States; F: France; UK: England

[c]1: Guichón et al, 2003b; 2: Prigioni et al, 2005; 3: Bó and Porini, 2005; 4: Willner et al, 1979; 5: Shirley et al, 1981; 6: Wilsey et al, 1991; 7: Borgnia et al, 2000; 8: Abbas, 1991; 9: Gosling and Baker, 2008; 10: J. Carter, pers. obs.

that consumption of a plant species is dependent on its location, either near or far from the water. They offered patches at 1 or 5m from the water edge either of *Eleocharis bonariensis*, a preferred plant item in the study site, or *Lolium multiflorum*, from a pasture growing 30m from the water edge and scarcely consumed. *Lolium multiflorum* was significantly more consumed at 1m than at 5m from the water, while *E. bonariensis* was not consumed (Borgnia et al, 2000). Selection of foraging areas in or close to the water explains diet composition better than food preferences based on its quality (Guichón et al, 2003b). This behaviour may explain feeding on grass of lawns near residential developments and golf courses (Carter, J., pers. obs.).

Coypus are gregarious and have a polygynous mating system (Guichón et al, 2003c; Túnez et al,

2009). In South America, groups are typically composed by about ten individuals including several adult and sub-adult males and females, one dominant male and a variable number of juveniles (Guichón et al, 2003c). Spatial segregation among groups may depend on habitat characteristics, being more evident in a linear habitat than in a pond (Guichón et al, 2003c). Behavioural observations were supported by genetic studies showing large genetic variability among groups of coypus in linear habitats, though genetic differences were not important among groups living in the pond habitat (Túnez et al, 2009). In this pond, one dominant male could monopolize most paternities resulting in high variance in reproductive success among males as opposed to the low variability reported in linear habitats (Túnez et al, 2009). Different behaviour according to habitat characteristics suggest that in linear habitats social behaviour determines coypu distribution along watercourses (parallel to water edge) while foraging behaviour determines habitat use in relation to distance to water (perpendicular to water edge).

Home range size depends on habitat type, gender, reproductive condition and resource availability. Smaller home range sizes have been reported for females than males (Doncaster and Micol, 1989: 2.5ha and 670m long for females, 5.7ha and 1913m long for males; Gosling and Baker, 1989b: 3–5.5ha females, 6.8–8.4ha males). While family groups mainly use 200–400m along a stream (Guichón et al, 2003c), solitary large males may overlap the range of various female groups (Gosling and Baker, 1989b). Smaller home range sizes were reported in urban areas with abundant food (Meyer, 2001: 1.17ha females, 2.3ha males).

On average, individuals in introduced populations put on weight more quickly (Figure 30.3), reach sexual maturity at a younger age and frequently live at higher population densities than in their native range (Guichón et al, 2003a). This may be related to a high hunting pressure in the native range that selects for smaller adult size in respect to introduced areas (Purvis, 2001); though it may also be explained by harsh climatic conditions in introduced ranges that favour heavier animals. European populations at higher densities have relatively lower survival probabilities (e.g. Gosling et al, 1981; Doncaster and Micol, 1989; Reggiani et al, 1995), which is probably related to severe winters when consecutive freezing days cause population crashes (Gosling et al, 1983; Doncaster and Micol, 1990). In such a situation a larger body reduces the ratio of exposed skin to thermoregulating volume and can support more fat accumulation, while a young age of sexual maturity could increase the intrinsic growth rates of the population during its recovery phase (Guichón et al, 2003a).

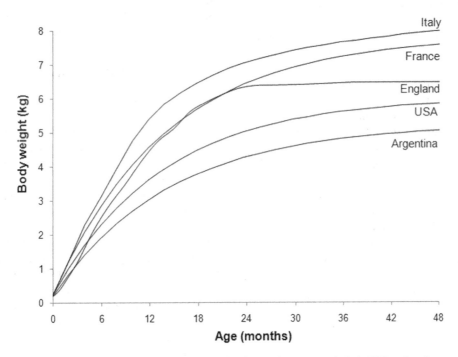

Note: Data for Argentina (Guichón et al, 2003a), USA (Dixon et al, 1979) and France (Doncaster and Micol, 1989) are from live trapping estimates; data for England (Gosling and Baker, 2008) and Italy (Bertolino, S., unpub. data) are from eye lens weight.

Source: Redrawn from Guichón et al (2003a)

Figure 30.3 *Relationship of body weight to age of coypus in the native range (Argentina) and countries of introductions*

Limiting Factors for Distribution and Abundance

Population density may range from a few animals per hectare in temperate and Mediterranean habitats (Norris, 1967; Doncaster and Micol, 1990; Reggiani et al, 1993) to 20–40 individuals ha⁻¹ in subtropical climates (Brown, 1975; LeBlanc, 1994); an exceptional peak of 138 individuals ha⁻¹ was reported in Oregon (LeBlanc, 1994). Densities of 6–8 individuals ha⁻¹ were reported in native range for non-hunted sites (Bó and Porini, 2001; Guichón and Cassini, 2005). Differences in food abundance, predation and climatic conditions explain high variability in coypu density, as well as hunting or control pressure where present.

Caimans in South America and alligators in North America are the most important predators of coypu (Woods et al, 1992). Other predators in the native and introduced ranges are felids and canids, other medium sized carnivores and some birds of prey (Woods et al, 1992; Bounds et al, 2003). Most of these species have dramatically declined in numbers both in wild and modified aquatic habitats, where domestic dogs and foxes are now the main predators.

Coypu populations are sensitive to climatic conditions and severe winters may be their most limiting factor (Doncaster and Micol, 1989). Cold weather can cause direct adult and juvenile mortality and can also cause reduction in fat reserves, which may increase abortion rates leading to reproductive failure (Willner et al, 1979; Gosling, 1981). During unusually cold winters, mortality rate can reach 80–90 per cent of the entire population (Doncaster and Micol, 1990). Severe winters have been credited with extirpating coypu populations in several regions including Scandinavian countries and in areas of the USA with more continental climates (Carter and Leonard, 2002). However, in those areas where small populations are able to survive coypu are able to quickly recover by increasing reproduction and survival rates (Doncaster and Micol, 1990; Reggiani et al, 1993). In response to a series of mild winters in recent years, coypu population ranges have been expanding northward in North America, in some cases re-establishing themselves in regions where they had previously died out.

Coypu are tolerant to human presence (Figure 30.4) and they can form stable populations, even in urban and recreational areas, if they are not hunted. Therefore, it is not expected that human presence alone can decrease coypu abundance, but hunting pressure has been postulated as the main determinant of coypu distribution at macrohabitat and landscape scales in the Argentinean Pampas (Guichón and Cassini, 1999, 2005; Leggieri et al, 2011).

Several diseases have been reported for coypus; however, prevalence was always low and clinical signs of infectious diseases are rarely observed in wild populations. The occurrence of *Toxoplasma gondii*, *Chlamydia psittaci* and *Leptospira* spp. was reported in Louisiana (Howerth et al, 1994). Leptospirosis was also observed in France (Michel et al, 2001) and England (Watkins et al, 1985). A skin infection in humans called 'nutria itch' is caused by the roundworm *Strongyloides myopotami* and it has been reported by people exposed to water with a large coypu population density (LeBlanc, 1994; Rossin et al, 2009). Coypus are potentially a source of zoonotic infections and caution should be taken when eating their meat and using it to feed other animals, when handling individuals or when in contact with water that might have been contaminated by coypus.

Ecological and Economic Impacts

Impact to ecosystems

Coypus are generalist herbivores that can feed on a large variety of plant materials (Table 30.1), including leaves, stems and roots. They generally select the parts of plants with the highest nutrient value, digging for below-ground energy-rich roots and tubers, leaving most of the plant material unconsumed. As a result of this feeding activity, large areas of *Nuphar lutea*, *Rumex* spp., *Sagittaria* spp., *Scirpus* spp., *Phragmites australis*, *Trapa natans* and *Typha* spp. may be eliminated (Ellis, 1963; Willner et al, 1979; Boorman and Fuller, 1981; Bertolino et al, 2005). Occasionally coypus might feed on crustaceans and freshwater mussels, but this has not been widely observed and may only happen a few localities. A possible impact on nesting birds may be due to the reduction of emergent vegetation used by water birds for resting and breeding, or by direct disturbance of nests.

In Louisiana (USA) the coypu has an important impact on the above-ground biomass of native marsh plant species, such as chairmaker's bulrush *Scirpus americanus* (Johnson and Foote, 1997) and arrowheads,

(a)

(b)

Source: Jacoby Carter, USGS

Figure 30.4 *(a) Coypu begging for food from one of the authors, and (b) sign warning the public not to feed animals (coypu are the target species), both in Germany*

Sagittaria latifolia and *S. platyphylla* (Llewellyn and Shaffer, 1993). In Louisiana and Maryland coypu feeding activity has been associated with the loss of brackish and freshwater marshes through a process known as eat-out (Foote and Johnson, 1993; Carter et al, 1999). In winter, as above-ground biomass goes into senescence, coypus switch to feeding on roots and rhizomes. The process of digging breaks up the vegetative mat that holds the fine marsh sediments together. These areas are regularly and frequently flooded and when they are, the newly exposed sediments are washed away leaving a 'hole' in the marsh that is difficult for plants to recolonize (Haramis, 1996; Colona et al, 2003). However, it is important to note that eat-outs occurred in Gulf of Mexico coastal marshes before coypus were introduced and were caused by the native muskrat *Ondatra zibethicus* (Dozier, 1952). Starting in the 1960s coypus began to

replace muskrats in the coastal and interior marshes of Louisiana and Maryland, US. As coypu populations increased, muskrat populations decreased. In 2007 estimates of coast-wide marsh damaged by coypu feeding activity ranged from 3400 to 41,500ha per year (Louisiana Department of Wildlife and Fisheries, 2010). In both Louisiana and Maryland, as coypu populations have been brought under control, muskrat populations have rebounded, indicating that competitive exclusion may be taking place between coypu and the native (for North America) muskrat.

Impacts on crops and irrigation systems

Coypus are known to eat crop plants, such as cereals, sugarcane, alfalfa, brassica, ryegrass, fruit and nut trees, and root crops, especially sugar beet (Schitoskey et al,

1972; Abbas, 1988; Gosling and Baker, 2008; Panzacchi et al, 2007). Impacts on crops are strongly dependent on food availability and proximity to water. If available, coypu prefer to forage on natural vegetation close to water rather than crops on more exposed land far from water (Borgnia et al, 2000; D'Adamo et al, 2000). However, they may impact on crops cultivated next to the water if natural vegetation is scarce, which may also cause damage in urban wetlands such as golf courses (Corriale et al, 2006). Hence, differences in habitat characteristics and management of riverbanks and irrigation canals may determine whether coypus behave as an agricultural pest (D'Adamo et al, 2000). Agricultural lands in many areas of the native range usually have a non-cultivated fringe of vegetation near the watercourse containing suitable food for coypus. This fringe of semi-natural riparian/border vegetation might not be available in other agro-systems where intensive land cultivation reaches the watercourse border, e.g. irrigation canals in European farmlands.

The most important economic damage is caused by the coypu's burrowing behaviour. Coypus dig extensive burrow systems into the riverbanks and ditches, disrupting drainage systems and posing a risk of flooding in low lying areas. In Italy, the cost of riverbank repair following damage by coypus was estimated at nearly €2 million year^{-1} (Panzacchi et al, 2007). Extensive burrowing makes dykes and levees susceptible to collapse due to other factors, such as flooding or vehicular traffic (Bounds et al, 2003). In North America, constructed wetlands are often used for secondary or tertiary sewage treatment. Coypu populations in these artificial wetlands can quickly grow to the point where they impair constructed marsh function.

Effectiveness of Management: Eradication and Control

The most effective strategy to reduce the negative consequences of biological invasions should be based on a hierarchical approach that comprises avoiding new introductions, prompt eradication of newly established species, and spatial containment and/or population control programmes (IUCN, 2000). In the case of the coypu, prevention failed almost everywhere due to direct introductions and the low security of farming. The coypu has been eradicated in two small areas in the USA (Carter and Leonard, 2002) and from a large area in England (Gosling and Baker, 1989a), and is controlled by trapping and shooting to reduce damage in several countries (Carter and Leonard, 2002; Bertolino and Genovesi, 2007).

In the USA, coypu control is managed at the state level. Most states with coypus regard them as another fur bearer and they are managed as such; control programmes are ongoing in Maryland and Louisiana. The goal of the Maryland programme is eradication of coypu in the Chesapeake Bay region. Most of the coypus have been successfully removed from the Delmarva Peninsula but access issues have prevented complete eradication (Linscomb, G., pers. comm.). The goal of the Louisiana programme is to reduce the populations enough to reverse their harmful effects on coastal marshes. In the 2009–2010 trapping season 445,963 coypu tails were turned in for bounty in Louisiana (Jordan and Mouton, 2010). Marsh damage attributed to coypu has seen a reduction since the programme began (Louisiana Department of Wildlife and Fisheries, 2010), but it is difficult to determine how much of the improvement was due to the control programme versus other factors such as marsh restoration, water diversion projects, or the impacts of hurricanes.

In Italy, during a six year period (1995–2000), despite the removal of 220,688 coypus with a cost of €2,614,408, the damage produced by the species amounted to €11,631,721 (Table 30.2). Damage to agriculture and riverbanks continued to increase during the six years, indicating an ineffective control both at national and local scales. According to previous experiences, non-intensive management operations may impact coypu populations with unexpected effects. The preferential capture of adult males in the first phases of control may create populations dominated by younger classes with a high potential for a subsequent population increase (Gosling and Baker, 1989a; Reggiani et al, 1993). The settlement in new locations of dispersing individuals escaping from areas disturbed by control actions may spread their impacts over a larger region.

Coypu populations have also been successfully managed in large-scale operations with successful containment and eradication programmes, leading to the reduction of economic losses and the preservation of biodiversity (Gosling and Baker, 1989a; Bertolino et al, 2005; Bertolino and Viterbi, 2010). An important feature of these projects was an adequate level of trapping effort that was maintained or even increased

Table 30.2 *Comparison between efforts and costs of the successful coypu eradication in England and of the permanent control campaign in Italy*

	East Anglia (England)		Italy	
Period	9 years (1981–1989)	6 years (1995–2000)	1 year (2000)	Predicted future efforts
Area	12,500 km²**	41,515 km²	41,515 km²	× 2.5–3.3
Coypus removed	34,822	220,688	64,338	160,845–212,315 year⁻¹
Total costs*	€5,000,000	€14,246,129	€3,773,786	€9,4–12,4 millions year⁻¹

Notes: Costs include damage to agriculture, irrigation systems and total costs of management. Predicted future costs consider the potential future range expansion of the coypu in Italy under a best case and a worst case scenario.

*updated to year 2000 currency values, **estimated from Figure 4 in Gosling and Baker 1989a, adding up each black dot that represents catches in a 10km by 10km grid square

Source: England data from Gosling and Baker (1989a); Italy data from Panzacchi et al (2007)

after first results were achieved (Baker, 2006; Bertolino and Viterbi, 2010). The relationship between local control efficacy and damage reduction is density and probably site dependent and it is not possible to point out a threshold for control to be effective. For example, a mean of 0.4–2.2 coypu km⁻² year⁻¹ removed reduced damage in two provinces in Piedmont, Italy, but not in another (Bertolino and Viterbi, 2010).

The eradication campaign against the coypus in England is considered one of the most successful eradication projects carried out on a mainland and should be used as a reference for future actions (Gosling and Baker, 1989a; Baker, 2006). Key points of the successful campaign were careful technical planning and thoughtful evaluation of the human dimension. During a six year trial, coypu were removed from a smaller area, allowing simulations of the efforts, costs and likely chances of success of the overall eradication under different scenarios (Gosling et al, 1988). Since trappers become unemployed if they are successful, it was considered important to consider how to ensure they remained motivated until their task was completed. This problem was overcome by restricting funds for eradication to a maximum of ten years and promising the trappers an extra bonus if they succeeded within this period (Gosling et al, 1988; Baker, 2006).

The costs associated with eradication campaigns may discourage authorities from starting new projects on mainland areas; however, permanent control to limit damage can be more expensive. Panzacchi et al (2007) compared the population control of coypu in Italy to eradication in England, pointing out a disparity of costs and efforts. The cost for the successful nine years of eradication in England was largely exceeded by the cost related to a few years of the permanent control campaign in Italy (Table 30.2). Likewise, the 2009–2010 trapping season in the state of Louisiana paid 306 hunters a collective $2,229,815 for 445,963 coypu tails. The payout to trappers has totalled $11,748,016 since 2002 (Louisiana Department of Wildlife and Fisheries, 2010).

The prompt eradication of isolated and newly colonized areas should be the basis for a proactive management strategy on coypu. Control plans must be conducted at an adequate, biologically sound spatial scale taking into account the potential counteracting effects of immigration. Whether the proposal is to attempt eradication or to initiate a permanent control plan, an adaptive resource management approach should be considered. This means collecting data during control operations, analysing, processing and feeding back results using a mix of experimentation and modelling to improve field operations (Roy et al, 2009).

Note

Disclaimer: Any use of trade, product, or firm names is for descriptive purposes only and does not imply endorsement by the US government.

References

Abbas, A. (1988) 'Impact du ragondin (*Myocastor coypus* Molina) sur une culture de mais (*Zea mays*) dans le Marais Poitevin?', *Acta Oecologica*, vol 9, pp173–189

Abbas, A. (1991) 'Feeding strategy of coypu (*Myocastor coypus*) in central western France', *Journal of Zoology*, vol 224, pp385–401

Abe, H., Ishii, N., Itoo, Y., Kaneko, Y., Maeda, K., Miura, S. and Yoneda, M. (2005) '*A Guide to the Mammals of Japan*', Tokai University Press, Kanagawa, Japan

Baker, S. J. (2006) 'The eradication of coypus (*Myocastor coypus*) from Britain: The elements of a successful campaign', in F. Koike, M. N. Clout, M. Kawamichi, M. De Poorter and K. Iwatsuki (eds) *Assessment and Control of Biological Invasion Risks*, Shoukadoh Book Sellers, Kyoto, Japan and International Union for Conservation of Nature, Gland, Switzerland, pp142–147

Bertolino, S. (2008) '*Myocastor coypus*, Global Invasive Species Database', Invasive Species Specialist Group, www.issg.org/database/welcome, accessed 24 August 2010

Bertolino, S. (2009) 'Species account of the 100 of the most invasive alien species in Europe: *Myocastor coypus* (Molina), coypu, nutria (Myocastoridae, Mammalia)', in DAISIE (ed) *Handbook of Alien Species in Europe*, Springer, Dordrecht, The Netherlands, p364

Bertolino, S. and Genovesi, P. (2007) 'Semiaquatic mammals introduced into Italy: Case studies in biological invasion', in F. Gherardi (ed) *Biological Invaders in Inland Waters: Profiles, Distribution, and Threats*, Springer, pp175–191

Bertolino, S. and Viterbi, R. (2010) 'Long-term cost-effectiveness of coypu (*Myocastor coypus*) control in Piedmont (Italy)', *Biological Invasions*, vol 12, pp2549–2558

Bertolino, S., Perrone, A. and Gola, L. (2005) 'Effectiveness of coypu control in small Italian wetland areas', *Wildlife Society Bulletin*, vol 33, pp714–720

Bó, R. F. and Porini, G. (2001) '*Proyecto "Nutria": Informe final de la primera etapa*', Executive report, Dirección Nacional de Fauna Silvestre, Argentina

Bó, R. F. and Porini, G. (2005) 'Estudios ecológicos básicos para el manejo sustentable de *Myocastor coypus* en Argentina', Executive report, Dirección Nacional de Fauna Silvestre, Argentina

Boorman, L. A. and Fuller, R. M. (1981) 'The changing status of reedswamp in the Norfolk broads', *Journal of Applied Ecology*, vol 18, pp241–269

Borgnia, M., Galante, M. L. and Cassini, M. H. (2000) 'Diet of the coypu (*Myocastor coypus*) in agro-systems of the Argentinean Pampas', *Journal of Wildlife Management*, vol 64, pp354–361

Bounds, D. L., Sherfy, M. H. and Mollett, T. A. (2003) 'Nutria', in G.A. Felhamer, B. C. Thompson and J. A. Chapman (eds) *Wild Mammals of North America*, Johns Hopkins University Press, Baltimore, Maryland, pp1119–1147

Brown, L. N. (1975) 'Ecological relationships and breeding biology of the nutria (*Myocastor coypus*) in the Tampa, Florida area', *Journal of Mammalogy*, vol 56, pp928–930

Carter, J. and Leonard, B. P., (2002) 'A review of the literature on the worldwide distribution, spread of, and efforts to eradicate the coypu (*Myocastor coypus*)', *Wildlife Society Bulletin*, vol 30, pp162–175

Carter, J., Foote, A. L. and Johnson-Randall, L. A. (1999) 'Modelling the effects of coypu (*Myocastor coypus*) on wetland loss', *Wetlands*, vol 19, pp209–219

Colona, R., Farrar, R., Kendrot, S., McKnight, J., Mollett, T., Murphy, D., Olsen, L. and Sullivan, K. (2003) 'Nutria (*Myocastor coypus*) in the Chesapeake Bay: A draft bay-wide management plan', The Chesapeake Bay Nutria Working Group

Corriale, M. J, Arias, S. M., Bó, R. F. and Porini, G. (2006) 'Habitat-use patterns of the coypu *Myocastor coypus* in an urban wetland of its original distribution', *Acta Theriologica*, vol 51, pp295–302

D'Adamo, P., Guichón, M. L., Bó, R. F. and Cassini, M. H. (2000) 'Habitat use of the coypu (*Myocastor coypus*) in agro-systems of the Argentinean Pampas', *Acta Theriologica*, vol 45, pp25–33

DAISIE (2010) 'Delivering Alien Invasive Species Inventories for Europe', www.europe-aliens.org, accessed 18 January 2010

Deems, E. F. and Pursley, D. (eds) (1978) 'North American furbearers: Their management, research and harvest status in 1976', International Association of Fish and Wildlife Agencies in cooperation with the Maryland Department of Natural Resources Wildlife Administration, University of Maryland, College Park, US

Dixon, K. R., Willner, G. R., Chapman, J. A., Lane W. C. and Pursley D. (1979) 'Effects of trapping and weather on body weights of feral nutria in Maryland', *Journal of Applied Ecology*, vol 16, pp69–76

Doncaster, C. P. and Micol, T. (1989) 'Annual cycle of a coypu (*Myocastor coypus*) population: Male and female strategies', *Journal of Zoology*, vol 217, pp227–240

Doncaster, C. P. and Micol, T. (1990) 'Response by coypus to catastrophic events of cold and flooding', *Holarctic Ecology*, vol 13, pp98–104

Dozier, H. L. (1952) 'The present status and future of nutria in the southeast states', *Proceedings of the Annual Conference of the Southeastern Association of Game and Fish Commissioners*, vol 6, pp368–373

Ellis, E. A. (1963) 'Some effects of selective feeding by the coypu (*Myocastor coypus*) on the vegetation of Broadland', *Transactions of the Norfolk and Norwich Naturalists' Society*, vol 20, pp32–35

Foote, A. L. and Johnson, L. A. (1993) 'Plant stand development in Louisiana coastal wetlands: Coypu grazing effects on plant biomass', in *Proceedings of the 13th Annual Conference of the Society of Wetland Scientists*, New Orleans, Louisiana, Society of Wetland Scientists, Utica, Mississippi, pp265–271

Gosling, L. M. (1981) 'Climatic determinants of spring littering by feral coypus (*Myocastor coypus*)', *Journal of Zoology*, vol 195, pp281–288

Gosling, L. M. and Baker, S. J. (1989a) 'The eradication of muskrats and coypus from Britain', *Biological Journal of the Linnean Society*, vol 38, pp39–51

Gosling, L. M. and Baker, S. J. (1989b) 'Demographic consequences of differences in the ranging behaviour of male and female coypus', in R. J. Putman (ed) *Mammals as Pests*, Chapman & Hall, London, pp155–167

Gosling, L. M. and Baker, S. J. (2008) 'Coypu *Myocastor coypus*', in S. Harris and D. W. Yalden (eds) *Mammals of the British Isles: Handbook*, 4th edition, The Mammal Society, Southampton, pp159–165

Gosling, L. M. and Skinner, R. J. (1984) 'Coypu', in I. L. Mason (ed) *Evolution of Domesticated Animals*, 2nd edition, Longman, Essex, pp246–251

Gosling, L. M., Watt, A. D. and Baker, S. J. (1981) 'Continuous retrospective census of the East Anglian coypu population between 1970 and 1979', *Journal of Animal Ecology*, vol 50, pp885–901

Gosling, L. M., Baker, S. J. and Skinner, J. R. (1983) 'A simulation approach to investigating the response of a coypu (*Myocastor coypus*) population to climatic variation', *EPPO Bulletin*, vol 13, pp183–192

Gosling, L. M., Baker, S. J. and Clarke, C. N. (1988) 'An attempt to remove coypus from a wetland habitat in East Anglia', *Journal of Applied Ecology*, vol 25, pp49–62

Guichón, M. L. and Cassini, M. H. (1999) 'Local determinants of coypu distribution along the Luján River, eastcentral Argentina', *Journal of Wildlife Management*, vol 63, pp895–900

Guichón, M. L. and Cassini, M. H. (2005) 'Population parameters of indigenous populations of *Myocastor coypus*: The effect of hunting pressure', *Acta Theriologica*, vol 50, pp125–132

Guichón, M. L., Doncaster, C. P. and Cassini, M. H. (2003a) 'Population structure of coypus (*Myocastor coypus*) in their region of origin and comparison with introduced populations', *Journal of Zoology*, vol 261, pp265–272

Guichón, M. L., Benitez., V., Abba., A., Borgnia., M. and Cassini, M. H. (2003b) 'Foraging behaviour of coypus *Myocastor coypus*: Why do coypus consume aquatic plants?', *Acta Oecologica*, vol 24, pp241–246

Guichón, M. L., Borgnia, M., Fernández Righi, C., Cassini, G. and Cassini, M. H. (2003c) 'Social behaviour and group formation in the coypus (*Myocastor coypus*) in the Argentinean pampas', *Journal of Mammalogy*, vol 84, pp254–262

Haramis, M. (1996) 'The effect of nutria (*Myocastor coypus*) on marsh loss in the lower eastern shore of Maryland: An enclosure study', United States Geological Survey website, www.pwrc.usgs.gov/resshow/nutria.htm, accessed 25 August 2010

Howerth, E. W., Reeves, A. J., McElveen, M. R. and Austin, F. W. (1994) 'Survey of selected diseases in nutria (*Myocastor coypus*) from Louisiana', *Journal of Wildlife Diseases*, vol 30, pp450–453

IUCN (International Union for the Conservation of Nature) (2000) 'IUCN guidelines for the prevention of biodiversity loss caused by alien invasive species', IUCN, Gland Switzerland, http://intranet.iucn.org/webfiles/doc/SSC/SSCwebsite/Policy_statements/IUCN_Guidelines_for_the_Prevention_of_Biodiversity_Loss_caused_by_Alien_Invasive_Species.pdf, accessed 30 November 2010

Johnson, L. A. and Foote, A. L. (1997) 'Vertebrate herbivory in managed coastal wetlands: A manipulative experiment', *Aquatic Botany*, vol 59, pp17–32

Jordan, J. and Mouton, E. (2010) 'Coastwide nutria control program 2009–2010', Louisiana Department of Wildlife and Fisheries, www.nutria.com/uploads/0910CNCPfinalreportB.pdf, accessed 19 August 2010

Kim, P. (1980) 'The coypu (*Myocastor coypus*) in the Netherlands: Reproduction, home range and manner of seeking food', *Lutra*, vol 23, pp55–64

Korean Ministry of Environment (2009) *Korean Red Data Book and Invasive Species in Korea*, UNJDP/GEF Korea Wetland Project

LeBlanc, D. J. (1994) 'Nutria', in S. E. Hygnstrom, R. M. Timm and G. E. Larsen (eds) *Prevention and Control of Wildlife Damage*, Nebraska Cooperative Extension Service, University of Nebraska-Lincoln, ppB71–B80

Leggieri, L. R., Guichón, M. L. and Cassini, M. H. (2011) 'Landscape correlates of the distribution of coypu (*Myocastor coypus*) in Argentinean Pampas', *Italian Journal of Zoology*, vol 78, pp124–129

Lessa, E., Ojeda, R., Bidau, C. and Emmons, L. (2008) '*Myocastor coypus*', in IUCN (2010) *IUCN Red List of Threatened Species*, Version 2010.2, www.iucnredlist.org, accessed 24 August 2010

Llewellyn, D. W. and Shaffer, G. P. (1993) 'Marsh restoration in the presence of intense herbivory: The role of *Justicia lanceolata* (Chapm) small', *Wetlands*, vol 13, pp176–184

Louisiana Department of Wildlife and Fisheries (2010) 'Nutria', www.nutria.com, accessed 25 August 2010

Meyer, J. (2001) 'Die Nutria *Myocastor coypus* (Molina, 1782) – eine anpassungsfhige Wildart', *Beitrage zur Jagd- und Wildforschung*, vol 26, pp339–347

Michel, V., Ruveon-Clouet, N., Menard, A., Sonrier, C., Fillonneau, C., Rakotovao, F., Ganière, J. P. and André-Fontaine, G. (2001) 'Role of the coypu (*Myocastor coypus*) in the epidemiology of leptospirosis in domestic animals and humans in France', *European Journal of Epidemiology*, vol 17, pp111–121

Norris, J. D. (1967) 'A campaign against feral coypus (*Myocastor coypus*) in Great Britain', *Journal of Applied Ecology*, vol 4, pp191–199

Nowak, R. M. (1991) *Walker's Mammals of the World*, 5th edition, The John Hopkins University Press, Baltimore

Panzacchi, M., Bertolino, S., Cocchi, R. and Genovesi, P. (2007) 'Cost/benefit analysis of two opposite approaches to pest species management: Permanent control of *Myocastor coypus* in Italy *versus* eradication in East Anglia (UK)', *Wildlife Biology*, vol 13, pp159–171

Prigioni, C., Balestrieri, A. and Remonti, L. (2005) 'Food habits of the coypu *Myocastor coypus*, and its impact on aquatic vegetation in a freshwater habitat of NW Italy', *Folia Zoologica*, vol 54, pp269–277

Purvis, A. (2001) 'Mammalian life histories and responses of populations to exploitation', in J. D. Reynolds, G. M. Mace, K. H. Redford and J. G. Robinson (eds) *Conservation of Exploited Species*, Cambridge University Press, Cambridge, pp169–181

Reggiani, G., Boitani, L., D'Antoni, S. and De Stefano, R. (1993) 'Biology and control of the coypu in the Mediterranean area', *Supplementi alle Ricerche di Biologia della Selvaggina*, vol 21, pp67–100

Reggiani, G., Boitani, L. and De Stefano, R. (1995) 'Population dynamics and regulation in the coypu *Myocastor coypus* in central Italy', *Ecography*, vol 18, pp138–146

Rossin, M. A., Varela, G. and Timi, J. T. (2009) '*Strongyloides myopotami* in ctenomyid rodents: Transition from semi-aquatic to subterranean life cycle', *Acta Parasitologica*, vol 54, pp257–262

Roy, S., Smith, G. C. and Russell. J. C. (2009) 'The eradication of invasive mammal species: Can adaptive resource management fill the gaps in our knowledge?', *Human–Wildlife Conflicts*, vol 3, pp30–40

Schitoskey, F., Evans, J. and Lavoie, G. K. (1972) 'Status and control of nutria in California', *Proceedings of the Vertebrate Pest Conference*, vol 5, pp15–17

Shirley, M. G., Chabreck, R. H. and Linscombe, G. (1981) 'Foods of nutria in fresh marshes of southeastern Louisiana', in J. A. Chapman and D. Pursley (eds) *Proceedings of the Worldwide Furbearer Conference*, Frotsburg, pp517–530

Túnez, J. I., Guichón, M. L., Centrón, D., Henderson, A., Callahan, C. and Cassini, M. H. (2009) 'Kinship and social organisation in coypus of Argentinean pampas', *Molecular Ecology*, vol 18, pp147–155

Watkins, S. A., Wanyangu, S. and Palmer, M. (1985) 'The coypu as a rodent reservoir of leptospira infection in Great Britain', *Journal of Hygiene*, vol 95, pp409–417

Willner, G. R., Chapman, J. A. and Pursley, D. (1979) 'Reproduction, physiological responses, food habitats, and abundance of nutria on Maryland marshes', *Wildlife Monographs*, vol 65, pp1–43

Wilsey, B. J., Chabreck, R. and Liscombe R. (1991) 'Variation in nutria diets in selected freshwater forested wetlands of Louisiana', *Wetlands*, vol 11, pp263–278

Woods, C. A., Contreras, L., Willner-Chapman, G. and Whidden, H. P. (1992) '*Myocastor coypus*', The American Society of Mammalogists, Mammalian Species No 398

31

Neovison vison Schreber (American mink)

Laura Bonesi and Michael Thom

Phylogeny and Distribution

Neovison vison (American mink) is a small, semi-aquatic carnivore of the weasel family Mustelidae (Figure 31.1). Although the American mink shares its common name with the European mink *Mustela lutreola*, molecular evidence suggests that the American species is in fact an outgroup of the *Mustela* species (see Harding and Smith, 2009). This is reflected in the recent reassignment of the species to the new genus *Neovison*, which it shares with the extinct sea mink *N. macrodon* (Wozencraft, 2005). American mink fossils are known from the Pleistocene fauna of North America, and although only around 20 samples have been found, it appears that the extent of the species' geographic range has contracted since the Pleistocene (Kurten and Anderson, 1980).

The native distribution of the American mink presently covers much of North America from the Arctic Circle to the southernmost zones of the US and into Mexico (e.g. Linscombe et al, 1982). The mink's valuable pelt has driven its spread by humans far beyond its native range and, as a consequence of escapes and intentional releases from fur farms, numerous wild populations have become established in many parts of Europe, South America, Russia and Asia (Figure 31.2).

American mink were introduced to Europe in the 1920s (Dunstone, 1993). Today mink farms are concentrated in northern Europe where the climate favours the growth of high quality fur. While the species is present throughout most of Europe, there is much variation in abundance both within and between countries (Bonesi and Palazon, 2007). For example, no reproductive populations are reported in Belgium or The Netherlands, despite the relatively large number of mink farms in these countries and their proximity to other countries, such as France or Germany, with widespread populations of mink (Bonesi and Palazon, 2007). Conversely, Poland has widespread populations that appear to have originated both from repeated fur farm escapes and invasion from neighbouring countries (Zalewski et al, 2010). Mink in Sweden and the UK are now apparently declining following an initial phase of increase, although the causes are still largely unknown (Bonesi and Palazon, 2007).

In the former Soviet Union, American mink were released into the wild to establish a harvestable population. According to some estimates, prior to 1971 up to 20,400 mink were released at more than 250 sites (Pavlov and Korasakova, 1973, in Macdonald and Harrington, 2003). Mink are now established as an invasive species in many former Soviet Union countries (Dunstone, 1993).

American mink feral populations are also present in Japan, in particular on the Islands of Hokkaido, Honsu and possibly Kyushu (Fukue et al, 2008; Shimatani et al,

Source: Mike Thom

Figure 31.1 *American mink*

2010a, 2010b), and in China where most farms are located in the northeast provinces of Heilongjiang, Jilin, Liaoning, Hebei and Shandong. It is likely that in these provinces there are some feral populations of mink (Zhao, J., pers. comm.).

In South America, the species was introduced in Argentina and Chile independently in the first half of the 20th century and it appears to be currently expanding its range (Jaksic et al, 2002). In Argentina, mink are presently distributed in Tierra del Fuego and along the western border with Chile throughout the Andean-Patagonian region (Lizarralde and Escobar, 2000). Feral populations are also present in Chile (e.g. Ibarra et al, 2009; Schüttler et al, 2010).

Note: Dashed: native range; dark grey: introduced range. Reliability of world distribution data varies greatly; best data are probably for North America and Europe.

Source: Modified from: Dunstone (1993); Bonesi and Palazon (2007); Genovesi and Scalera (2008); Global Invasive Species Database (www.issg. org); Reid and Helgen (2008) and with input from Yuko Fukue for Japan and Zhao Jaianjun for China.

Figure 31.2 *World distribution map of the American mink*

Ecology and Impacts

The American mink inhabits coastal as well as freshwater habitats, from sea level to the headwaters of hydrographical basins. Habitats in both its native and introduced ranges include rivers, lakes, wetlands, swamps and marshes (Linscombe et al, 1982;

Dunstone, 1993). At a macrohabitat scale, coastal areas represent superior habitat for the species, perhaps because of greater food availability compared to freshwater habitats (Hatler 1976; Dunstone and Birks 1985; Heylar 2005; Schüttler et al, 2008), and mink tend to be found at higher densities on the coast (Table 31.1).

Table 31.1 *Examples of the range of densities of American mink*

Country	Habitat	Density	References
Belarus	Riparian	3–22	Sidorovich and Macdonald, 2001
Canada	Coast	15–30	Hatler, 1976
Ireland	Riparian	3–10	Smal, 1991
Poland	Riparian	1–8	Sidorovich et al, 1996
Spain	Riparian	8–22	Melero et al, 2008
Patagonia	Coast	15	Previtali et al, 1998
UK	Riparian	1–7	Halliwell and Macdonald, 1996
UK	Coast	5–7	Dunstone and Birks, 1985

Note: Densities are expressed as number of mink per 10km of river or coastal shoreline.

Higher densities have also been observed in eutrophic water bodies when compared to oligotrophic water bodies (Dunstone and Birks, 1985). Habitat preferences of mink are linked to food availability, the presence of resting sites and breeding dens, and to a lesser extent to the distribution of competitors (e.g. Ben-David et al, 1995; Halliwell and Macdonald, 1996; Bonesi et al, 2000). As a general rule, mink usually prefer good vegetation cover with trees and shrubs, and avoid open areas (e.g. Previtali et al, 1998; Yamaguchi et al, 2003). This preference appears to be related more to prey availability (Burgess and Bider, 1980) than to the risk of predation, given that mink have few significant predators besides humans (Linscombe et al, 1982).

The American mink is a generalist and opportunist species with a variable diet that includes fish, crustaceans, birds (eggs, young and adults), small mammals, amphibians, insects and carrion. The trophic plasticity of mink allows them to specialize on certain prey species or generalize according to availability (Sidorovich, 1992) and their ability to swim and dive for fish and crustaceans, as well as hunt and scavenge on land, may insulate them against periods of food shortage (Dunstone, 1979; Macdonald et al, 1999).

The American mink has the ability to rapidly invade new territory (Gerell, 1970; Schüttler, et al 2010) and this, together with its impacts on native prey, makes American mink one of the most significant alien invaders threatening European biodiversity (Anonymous, 2007). Although the most severe impact of mink comes through predation on native species, additional threats may include competition with native riparian carnivores, disease transmission and hybridization (Kidd et al, 2009).

Mink can significantly deplete populations of birds (Nordström et al, 2003), especially ground nesting birds (e.g. Ferreras and Macdonald, 1999; Peris et al, 2009), rodents (Banks et al, 2004; Brzeziński et al, 2010), amphibians (Banks et al, 2005) and crayfish (Fischer et al, 2009). They have proved to be a particularly severe threat to the water vole *Arvicola terrestris*, causing widespread declines and local extirpations in the UK and in Belarus (Jefferies et al, 1989; Macdonald et al, 2002). Impacts cascading down through the trophic chain from mink to plant communities have also been observed (Fey et al, 2009). The impact of mink predation may be complicated by the presence of other, non-threatened prey. In the UK for example, the threat to water voles appears to be mediated by the presence of the abundant European rabbit – water vole populations that are isolated from rabbit habitat had a much higher likelihood of survival than populations that were connected to rabbit habitat (Oliver et al, 2009). This suggests that protection of rare prey species might partly be facilitated by habitat management or by removal of more common species that allow mink numbers to flourish.

Whether the presence of competitors can have a negative effect on mink populations by affecting distribution, spread and density has been subject of recent debate. Some large-scale studies using field signs and game bags to estimate population trends indicate that otters can slow the spread of mink, and cause a decline in existing populations (Ruiz-Olmo et al, 1997; Bonesi et al, 2006; McDonald et al, 2007; Bonesi and Macdonald, 2004a). However there is also evidence that while otters affect individual mink by negatively modifying their behaviour and body condition, these competitors have no effect at a population level (Harrington et al, 2009b). Discordant results also concern the diet of mink when sympatric with the otter. In the UK, the presence of otters causes mink to shift their diet from an aquatic-based one toward one that contains more terrestrial prey (e.g. Bonesi et al, 2004; Harrington et al, 2009b), while in Argentina the diet of sympatric mink converges more with that of the otter than does the diet of allopatric mink (Fasola et al, 2009). These results indicate that the effect of competitors is far from clear: habitat may play a role in shaping the different outcomes (Bonesi and Macdonald, 2004b), but more research is required. Recent studies also suggest indirect effects of foxes on the distribution of mink in Sweden (Carlsson et al, 2010) and indicate that sea eagles may restrict mink movements in Finland (Salo et al, 2008). In contrast, the presence of feral American mink has been proposed as a possible cause of the European mink's decline from its native range (Maran and Henttonen, 1995) although this is disputed (Lodé et al, 2001), partly because European mink have also declined where there are no American mink (Maran and Henttonen, 1995; Fournier and Maizeret, 2003).

Mink carry several diseases, such as distemper, Aujeszky's disease, rabies, mink enteritis virus and Aleutian disease (ADV) (Joergensen, 1985; McDonald and Lariviere, 2001) that can be transmitted to native mustelids. Distemper is potentially fatal in mink; however, tests for distemper virus antigens in mink from Denmark and England all proved negative (Yamaguchi and Macdonald, 2001; Hammershøj, 2003). ADV is commonly found in commercially farmed mink, and feral mink have been tested for ADV in four countries: Denmark, Iceland, Spain and England. The first three populations all showed a low incidence of antibodies to ADV (10–20 per cent) (Skírnisson et al, 1990; Mañas et al, 2001; Hammershøj, 2003), while in England 52 per cent of feral mink carried ADV antibodies (Yamaguchi and Macdonald, 2001). It is possible that captive and free ranging American mink may act as a source of the disease for other species of mustelids, but the likelihood of transmission is unknown.

The impact of feral mink on economic activities such as fish farming and poultry rearing is believed to be relatively small on a national scale (Harrison and

Symes, 1989), but it can be locally important (Moore et al, 2000).

Management Options

Management of introduced populations of American mink could in principle be achieved by several means, including:

- local eradication or control;
- restoration and manipulation of habitat;
- promotion of the natural recovery of native mink competitors;
- management of prey species; and/or
- prevention of escapes and rapid response after releases (Macdonald and Harrington, 2003; Bonesi and Palazon, 2007).

Of these, local control or eradication has been employed most frequently. This is sometimes followed by the reintroduction of impacted species such as the European mink (Maran, 2003) or the water vole (Moorhouse et al, 2009) and is sometimes guided by population modelling (Bonesi et al, 2007; Zabala et al, 2010).

Monitoring/surveying

Local control is generally preceded by surveying to establish the population size and extent. Surveying can either be carried out using the same methods that will then be used to catch the mink, i.e. bank-side traps or rafts (Reynolds et al, 2004), or through methods such as field sign searches (Macdonald and Harrington, 2003; Zuberogoitia et al, 2010) or camera trapping (Gonzalez-Esteban et al, 2004). One of the cheapest methods is searching for mink scats or tracks (e.g. Sidorovich et al, 1996). However, the relationship between sign surveys and trapping success tends to be very weak (Bonesi and Macdonald, 2004c; Zuberogoitia et al, 2006). Sign surveys are particularly problematic where similar signs are produced by co-occurring species, such as polecats, pine martens, otters and foxes (Harrington et al, 2010). To overcome these problems Harrington et al (2010) recommend undertaking DNA analyses for at least a subsample of collected scats. Surveys using tracking cartridges on floating rafts are perhaps a better alternative to field sign searches (Harrington et al, 2008a).

Catching methods

Barring methods that are either ineffective (shooting) or outlawed in many countries (hunting with dogs and leg-hold traps), there are two remaining effective methods for catching mink: bank-side traps and raft traps. Both involve live-trapping mink followed by humane dispatch, usually by shooting or lethal injection (Harrington et al, 2009a; Roy et al, 2009).

Bank-side traps are generally made from wire mesh and are set on dry land near a river or coastline (Figure 31.3). For ethical reasons, these need to be checked daily for mink and to release accidental captures of non-target species. They also need to be spaced relatively close together (300–500m) to maximize the encounter rate (King et al, 2009). For these reasons this method is relatively labour intensive, but in some areas remains the only practical option. In particular, these traps are useful on the coast or in fast flowing rivers where raft traps would be affected by water movement. Bank-side traps have been used successfully to trap mink in the large scale and long term project of mink eradication in the Outer Hebrides (Roy, 2006), where much of the trapping was in coastal habitat. Wooden bank-side traps were unsuccessfully used by local hunters on Hiiumaa Island, before eradication was achieved by an experienced field biologist using leg-hold traps adapted to capture mink without injury (Macdonald and Harrington, 2003). It is questionable whether the type of trap used in this case was as important for success as the experience of the researcher. In our opinion, expert knowledge of how the mink uses its habitat plays a crucial role in the successful placement of bank-side traps. King et al (2009) similarly emphasize the skill of those placing traps as important in the success of mustelid trapping operations.

Raft traps appear to be more independent of user experience and have a higher probability of trapping mink than bank-side traps (Reynolds et al, 2004; Bonesi, L., pers. obs.). This recently developed method (Reynolds et al, 2004) has been used successfully for mink control in a lowland British river (Harrington et al, 2009a). To reduce trapping workload, floating rafts can be fitted with a tracking cartridge to monitor mink and the trap activated only following detection of signs (Figure 31.4). Raft traps are also more time efficient as they intercept mink more successfully, and hence can be placed further apart (1–4km), than bank-side traps.

A number of studies have shown a strong seasonality in trapping success (e.g. Ireland, 1990; Smal, 1991; Roy, 2006). Recent field projects and modelling studies have shown that timing captures in the right seasons can improve trapping success (Bonesi et al, 2007; Harrington et al, 2009a). Mink have a very well-defined yearly cycle that is influenced by their physiology (Dunstone, 1993) as well as by prey availability (e.g. Ben-David et al, 1997; Ibarra et al, 2009). Mating, births, lactation and dispersal occur at specific times of year and are influenced by latitude but consistent within areas (Dunstone, 1993). These events influence the mobility and trap shyness of mink, two factors that are thought to be particularly important in determining the efficiency of trapping of wide ranging mustelids (King et al, 2009; Zuberogoitia et al, 2010). However, seasonality of trapping success may also be

Source: Laura Bonesi

Figure 31.3 *Bank-side trap*

Source: Laura Bonesi

Figure 31.4 *Floating raft trap*

dependent on the device used for trapping, with some methods apparently being less sensitive to seasonality of captures (Harrington et al, 2008a).

Traps can be baited with prey (fish or meat) or with lures (mink glands or artificially made lures). The efficacy of different baits has been tested in the Hebrides (Roy et al, 2006): lures (either natural or artificial) were significantly more effective than fish. An artificial lure was also tested against non-baited rafts by Reynolds et al (2004), who found no difference. With only two studies it is difficult to reach a conclusion, and artificially made lures can be rather different amongst themselves, but it is possible that rafts with their higher probability of intercepting mink are more independent of the effect of scent than bank-side traps.

Trapping success may also be improved with dog searches. Surveying to locate feeding dens and lie-up sites where traps can be placed can also be conducted with trained dogs (Roy et al, 2009). This has been used successfully in the Hebridean Mink Project to improve captures during May and June when mink were particularly difficult to trap (Roy, 2006). Dog searches may be essential in the last phase of an eradication campaign to detect and cull trap-shy mink (Zuberogoitia et al, 2006). Mustelids are cautious animals, and experimental work in New Zealand has demonstrated that wild ferrets (*Mustela furo*) are extremely resistant to capture, with up to a third of the individuals eluding recapture during the study (King et al, 2009). Similar difficulties exist with trapping the related American mink, and trap-shy individuals have being reported in field studies of this species (Hatler 1976; Zuberogoitia et al, 2006). Dog searches can also be used in mink control programmes to locate mink dens and then dispose of mink with methods other than trapping (shooting or air blasting), as has been done in Finland and Iceland (Nordström and Korpimäki, 2004; ISSG, 2007), but it is not known whether this is more or less efficient than trapping.

Bounty schemes

Cost is often one of the major constraints for mink control or eradication (Zabala et al, 2010) and mink pelts are commercially valuable. So, since the 1960s, strategies based on bounty schemes taken up by hunters or gamekeepers were attempted in the UK (King, 1983), in Lithuania (Bluzma, 1990, in Mickevicius and Baranauskas, 1992) and in Iceland (Hersteinsson, 1999). These have failed, since mink populations have increased in all these countries (Bonesi and Palazon, 2007). Mink have a relatively high reproductive rate, producing a litter of five to seven young per year and becoming reproductive in their first year of life (Dunstone, 1993). They are also able dispersers and highly mobile (Heylar, 2005) and being generalist and opportunist in their diet they have the flexibility to adapt to the local prey availability. Moreover, it has recently been demonstrated that mink exhibit a negative density-dependent survival following removal (Heylar, 2005). These schemes were also started when the population was already established, were not seasonally targeted, and were often of low intensity, a combination of factors that probably led to

Table 31.2 List of some of the projects for managing mink populations

Country	Year	Aim	Impacted species	Catching method	Season	Staff	Immigration	Scale	Outcome	Reference
Belarus	1992–2001	C	European mink	BST/ Hunting	All year	TF	YES	Catchment (240km²)	Mink reduced by 80% first year and then kept low	Sidorovich and Polozov, 2002
Estonia	1998–1999	E	European mink	LHT	All year	TF/V	NO	Regional (1000km²)	All mink removed (n = 50)	Maran et al, 2009
Finland	1993–2001	E	Birds	Dogs	All year	TF	YES	Small islands (130km²)	Nearly all mink removed, some reimmigration	Nordström et al, 2004; Nordström and Korpimäki, 2004
Iceland	1940s–	–	–	Hunting	–	BS	YES	Regional	Mink population overall increasing	Hersteinsson, 1999
Lithuania	1980s–	–	–	Hunting	–	BS	YES	Regional	Believed to have caused decline of mink	Bluzma, 1990, cited in Mickevicius and Baranauskas, 1992
Spain – Catalugna	1999–2006	C/E	–	BST	All year	TF	YES	Catchment (3 ×30km sections)	Two populations decreased one increased	Melero et al, 2010
Spain – northeast	2007	C	European mink	BST	Seasonal	TF	YES	Catchment	American mink reduced by ~60%	Zuberogoitia et al, 2010; Zabala et al, 2010
UK – England	1964–1970	E	–	Hunting/ BST	All year	BF	YES	Regional	Mink population overall increasing	Sheail, 2004
UK – Thames	2004–2005	C	Water voles	RT	Seasonal	TF	YES	Catchment (3 × 20km sections)	Mink reduced (20 to 67%) but reimmigration	Harrington et al, 2009a
UK – Western Isles	2001–2006	E	Ground nesting birds	BST	All year	TF	NO	Regional (1114km²)	Mink eradicated	Roy, 2006
UK – Scotland	2006–	C	Water voles	RT	All year	TF/V	YES	Large scale (~20,000km²)	Water voles have recolonized areas where mink removed	Water Vole Scotland, 2010

Note: Aim: C = control, E = eradication; Catching method: LHT = leg-hold traps; BST = bank-side traps; RT = raft traps; Staff: BS = bounty scheme, TF = task force, V = volunteers; Scale: catchment = up to 999km²; regional = between 1000 and 9999km²; large = over 10,000km²

failure. Such schemes also tend to be self-limiting by encouraging individuals to concentrate on high density populations, ignoring low density areas, to maximize their profits (Roy et al, 2009).

Importance of area and resources

Two fundamentally different strategies have been adopted successfully in mink management projects: (1) permanent eradication in areas where reimmigration by mink is highly unlikely; and (2) long term control in areas where reimmigration by mink is highly likely.

Permanent eradication has been carried out so far on islands that are sufficiently distant from sources of immigrants. Successful examples of this strategy are the Hebridean Mink Project (Roy, 2006) and the Hiiumaa project (Maran et al, 2009). However, recent work by Zabala et al (2010) suggests that the cost of eradicating mink in continental areas may not be as prohibitive as previously thought, and estimates the maximum cost for eradicating the American mink from the whole of Spain at €11 million. Recent research from the Outer Hebrides has shown that mink population behaviour changes substantially in response to continuous trapping effort (Bodey et al, 2010). Stable isotope analysis revealed that, as the trapping progressed, island living mink increased their reliance on marine food resources and moved towards the coast. The authors suggest that trapping may benefit from focusing on key areas with desirable resources, thereby drawing in nearby individuals. As some mink may nevertheless remain and breed in the suboptimal habitat, however, this approach alone may be more suited to long term control projects than to permanent eradication.

Perhaps surprisingly, permanent eradication projects are outnumbered by those attempting long term control, including some on a large scale (Table 31.2). The main problem facing the control strategy is that following removal, immigration from nearby areas must be prevented. There are, however, areas that can be defended more easily, including peninsulas and offshore islands (Nordström and Korpimäki, 2004; Roy et al, 2009). In long term control projects, culling must be followed by monitoring and trapping in the control area and in a buffer zone at least as large as the largest dispersal distance recorded, currently in the range of 30–40km (Gerell, 1970; Heylar, 2005).

Roy et al (2009) distinguish between 'targeted control in small areas' and 'seasonal or year round control over larger areas'. Targeted control in small areas could be used to reduce predation on species of conservation concern such as ground nesting birds (Craik, 1998) or salmonid stocks (Areal and Roy, 2009). This approach has been suggested for Navarino Island, in the Cape Horn Biosphere Reserve, Chile, where mink are relatively recent invaders that now threaten native species (Schüttler et al, 2010). Low level seasonal control could be used either to keep mink densities below a certain level or to defend native species impacted by mink at critical times of the year. In these cases, timing and length of trapping needs to be chosen accordingly, as supported both by field research and modelling (Bonesi et al, 2007; Harrington et al, 2009a). Year round control, especially over larger areas, is much more demanding (e.g. Melero et al, 2010). When traps can be used, this kind of project is best implemented by means of floating rafts set and checked by volunteers. This model is being used with success in southeast Scotland in the Cairngorms and northeast Scotland to protect the water vole (Water Vole Scotland, 2010); this project goes even further by trying to involve the whole community so that the control can be sustained long term (Evely et al, 2008). A further strategy, which has been applied in southwest England (Marshall-Ball, 2008) and evaluated but not yet applied in the Scottish Highlands (Harrington et al, 2008b), is to create a barrier to mink dispersal to protect areas that are already mink free.

Challenges and Controversies

In this chapter, we have discussed how the choice of the management strategy to reduce the damage caused by feral mink populations is very specific to the goals that are to be achieved, the resources available, and the area over which it needs to be implemented. With the exception of bounty schemes that have invariably proved unsuccessful, all the other strategies outlined here can be suitable depending on circumstances. Floating rafts, where they can be used, probably represent a better option than bank-side traps as they tend to intercept mink more frequently, are less labour intensive, and depend less on the experience of the trapper, but overall the specific strategy and equipment employed must be weighed against constraints. One of the challenges to the management of American mink is the development of a reliable and repeatable means of establishing presence and estimating numbers before engaging in eradication or control operations. Mink welfare and humane culling must also be considered in relation to potential benefits (Macdonald et al, 2006).

To minimize both the suffering to wild mink and the damage caused to native species, strategies to prevent escapes, together with immediate interventions after accidental escapes and intentional releases, should be put in place wherever mink are farmed. Most countries have been slow to regulate and enforce security measures for fur farms, and ensuring that such legislation is in place is one of the challenges for the future management of this species. While improving the security of fur farms against accidental escapes is possible, at least in principle, reducing the incidence of intentional releases by animal activists is proving more difficult. In any case, tightening the controls on mink farming can only ever be a partial strategy because established feral mink populations exist in many places where there are no operating mink farms. Indeed some countries, such as the UK, have tackled the problem by outlawing mink farming altogether (Fur Farming (Prohibition) Act 2000 and Fur Farming (Prohibition) (Scotland) Act 2002). Nevertheless, feral mink remain widespread there.

Our understanding of the complex interactions between invasive mink and their competitors, prey and habitat is continually improving. One of the challenges that we face is to better understand the role of competitors and prey on invasive mink populations. Some areas may benefit from management techniques that focus on increasing numbers of native competitors, or reducing numbers of non-threatened prey, as indirect mink control mechanisms to complement trapping strategies (Macdonald and Harrington, 2003; Bonesi and Macdonald, 2004b; Oliver et al, 2009). However, even in larger areas the long term control of mink through trapping may be sustainable on its own. For long term control it is not yet clear whether the community model of mink control, which is being experimented with in the Cairngorms Water Vole Conservation Project, is a more sustainable and successful approach than employing dedicated staff. Again, this may depend on circumstances and only time will tell whether both or either of these strategies is sustainable. As Zabala et al (2010) suggest, even eradication from entire countries may be possible where the resources are available, but tightening the control of fur farms and restricting the deployment of fur farms to less sensitive areas would be most beneficial to successful long term management.

Acknowledgements

We would like to thank Yuko Fukue, Piero Genovesi, Zhao Jianjun, Santiago Palazon and Iñigo Zuberogoitia, for providing some of the information reported. We would also like to thank an anonymous reviewer for their helpful comments.

References

Anonymous (2007) *Europe's Environment: The Fourth Assessment*, European Environment Agency, Copenhagen, Denmark

Areal, F. and Roy, S. (2009) 'A management decision tool for mink (*Mustela vison*) control in the western isles of Scotland (UK)', *International Journal of Design and Nature and Ecodynamics*, vol 4, pp16–31

Banks, P., Norrdahl, K., Nordström, M. and Korpimäki, E. (2004) 'Dynamic impacts of feral mink predation on vole metapopulations in the outer archipelago of the Baltic Sea', *Oikos*, vol 105, pp79–88

Banks, P., Nordström, M., Ahola, M. and Korpimäki, E. (2005) 'Variable impacts of alien mink predation on birds, mammals and amphibians of the Finnish archipelago: A long-term experimental study', in *IX International Mammalogical Congress*, Sapporo, Japan

Ben-David, M., Bowyer, R. and Faro, J. (1995) 'Niche separation by mink and river otters: Coexistence in a marine environment', *Oikos*, vol 75, pp41–48

Ben-David, M., Hanley, T. A., Klein, D. R. and Schell, D. M. (1997) 'Seasonal changes in diets of coastal and riverine mink: The role of spawning Pacific salmon', *Canadian Journal of Zoology*, vol 75, pp803–811

Bluzma, P. (1990) *Living Conditions and Situation of Mammal Populations in Lithuania: Mammals in Cultivated Landscape in Lithuania*, Mokslas Publisher, Vilnius, Lithuania (in Russian)

Bodey, T. W., Bearhop, S., Roy, S. S., Newton, J. and McDonald, R. A. (2010) 'Behavioural responses of invasive American mink *Neovison vison* to an eradication campaign, revealed by stable isotope analysis', *Journal of Applied Ecology*, vol 47, pp114–120

Bonesi, L. and Macdonald, D. (2004a) 'Impact of released Eurasian otters on a population of American mink: A test using an experimental approach', *Oikos*, vol 106, pp9–18

Bonesi, L. and Macdonald, D. (2004b) 'Differential habitat use promotes sustainable coexistence between the specialist otter and the generalist mink', *Oikos*, vol 106, pp509–519

Bonesi, L. and Macdonald, D. (2004c) 'Evaluation of sign surveys as a way to estimate the relative abundance of American mink (*Mustela vison*)', *Journal of Zoology*, vol 262, pp65–72

Bonesi, L. and Palazon, S. (2007) 'The American mink in Europe: Status, impacts, and control', *Biological Conservation*, vol 134, pp470–483

Bonesi, L., Dunstone, N., and O'Connell, M. (2000) 'Winter selection of habitats within intertidal foraging areas by mink (*Mustela vison*)', *Journal of Zoology*, vol 250, pp419–424

Bonesi, L., Chanin, P. and Macdonald, D. (2004) 'Competition between Eurasian otter *Lutra lutra* and American mink *Mustela vison* probed by niche shift', *Oikos*, vol 106, pp19–26

Bonesi, L., Strachan, R. and Macdonald, D. (2006) 'Why are there fewer signs of mink in England? Considering multiple hypotheses', *Biological Conservation*, vol 130, pp268–277

Bonesi, L., Rushton, S. and Macdonald, D. (2007) 'Trapping for mink control and water vole survival: Identifying key criteria using a spatially-explicit individual-based model', *Biological Conservation*, vol 136, pp636–650

Brzeziński, M., Romanowski, J., Zmihorski, M. and Karpowicz, K. (2010) 'Muskrat (*Ondatra zibethicus*) decline after the expansion of American mink (*Neovison vison*) in Poland', *European Journal of Wildlife Research*, vol 56, pp341–348

Burgess, S. A. and Bider, J. R. (1980) 'Effects of stream habitat improvements on invertebrates, trout populations, and mink activity', *Journal of Wildlife Management*, vol 44, pp871–880

Carlsson, N. O. L., Jeschke, J. M., Holmqvist, N. and Kindberg, J. (2010) 'Long-term data on invaders: When the fox is away, the mink will play', *Biological Invasions*, vol 12, pp633–641

Craik, J. C. A. (1998) 'Recent mink-related declines of gulls and terns in west Scotland and the beneficial effects of mink control', *Argyll Bird Report*, vol 14, pp98–110

Dunstone, N. (1979) 'Swimming and diving behaviour of the mink (*Mustela vison* Schreber)', *Carnivore*, vol 2, pp56–61

Dunstone, N. (1993) *The Mink*, Poyser, London

Dunstone, N. and Birks, J. D. S. (1985) 'The comparative ecology of coastal, riverine and lacustrine mink *Mustela vison* in Britain', *Zeitschrift für Angewardte Zoologie*, vol 72, pp59–70

Evely, A. C., Fazey, I. R. A., Pinard, M. and Lambin, X. (2008) 'The influence of philosophical perspectives in integrative research: A conservation case study in the Cairngorms National Park', *Ecology and Society*, vol 13, no 2, art52

Fasola, L., Chehébar, C., Macdonald, D. W., Porro, G. and Cassini, M. H. (2009) 'Do alien North American mink compete for resources with native South American river otter in Argentinean Patagonia?', *Journal of Zoology*, vol 277, pp187–195

Ferreras, P. and Macdonald, D. W. (1999) 'The impact of American mink *Mustela vison* on water birds in the upper Thames', *Journal of Applied Ecology*, vol 36, pp701–708

Fey, K., Banks, P. B., Oksanen, L. and Korpimäki, E. (2009) 'Does removal of an alien predator from small islands in the Baltic Sea induce a trophic cascade?', *Ecography*, vol 32, pp546–552

Fischer, D., Pavluvcik, P., Sedláček, F. and Šalek, M. (2009) 'Predation of the alien American mink, *Mustela vison* on native crayfish in middle-sized streams in central and western Bohemia', *Folia Zoologica*, vol 58, pp45–56

Fournier, P. and Maizeret, C. (2003) 'Status and conservation of the European mink (*Mustela lutreola*) in France', in *Proceedings of the International Conference on the Conservation of European Mink*, Gobierno de La Rioja, Logrono, La Rioja, Spain, pp95–100

Fukue, Y., Ashida, E. and Kishimoto, R. (2008) 'Distribution and population structure of the invasive species American mink *Neovison vison* in Nagano', *International Symposium CSIAM (Control Strategy of Invasive Alien Mammals)*, Okinawa, Japan

Genovesi, P. and Scalera, R. (2008) 'Distribution map of American mink', DAISIE, *Delivering Alien Invasive Species Inventories for Europe*, www.europe-aliens.org, accessed 11 January 2011

Gerell, R. (1970) 'Home ranges and movements of the mink *Mustela vison* in southern Sweden', *Oikos*, vol 21, pp160–173

Gonzalez-Esteban, J., Villate, I. and Irizar, I. (2004) 'Assessing camera traps for surveying the European mink, *Mustela lutreola* (Linnaeus, 1761), distribution', *European Journal of Wildlife Research*, vol 50, pp33–36

Halliwell, E. C. and Macdonald, D. W. (1996) 'American mink *Mustela vison* in the upper Thames catchment: Relationship between selected prey species and den availiability', *Biological Conservation*, vol 76, pp51–56

Hammershøj, M. (2003) 'Population ecology of free-ranging American mink *Mustela vison* in Denmark', PhD Thesis, National Environment Research Institute, Kalø, Denmark

Harding, L. E. and Smith, F. A. (2009) '*Mustela* or *Vison*? Evidence for the taxonomic status of the American mink and a distinct biogeographic radiation of American weasels', *Molecular Phylogenetics and Evolution*, vol 52, pp632–642

Harrington, L. A., Harrington, A. L. and Macdonald, D. W. (2008a) 'Estimating the relative abundance of American mink *Mustela vison* on lowland rivers: Evaluation and comparison of two techniques', *European Journal of Wildlife Research*, vol 54, pp79–87

Harrington, L. A., Huges, J. and Macdonald, D. W. (2008b) *Management of American Mink in the Northern Highlands: A Proposed Cordon Sanitaire Approach*, WildCRU, University of Oxford, Oxford

Harrington, L. A., Harrington, A. L., Moorhouse, T., Gelling, M., Bonesi, L. and Macdonald, D. W. (2009a) 'American mink control on inland rivers in southern England: An experimental test of a model strategy', *Biological Conservation*, vol 142, pp839–849

Harrington, L. A., Harrington, A. L., Yamaguchi, N., Thom, M. D., Ferreras, P., Windham, T. R. and Macdonald, D. W. (2009b) 'The impact of native competitors on an alien invasive: Temporal niche shifts to avoid interspecific aggression?', *Ecology*, vol 90, pp1207–1216

Harrington, L. A., Harrington, A. L., Hughes, J., Stirling, D. and Macdonald, D. W. (2010) 'The accuracy of scat identification in distribution surveys: American mink, *Neovison vison*, in the northern highlands of Scotland', *European Journal of Wildlife Research*, vol 56, pp377–384

Harrison, M. D. K. and Symes, R. G. (1989) 'Economic damage by feral American mink (*Mustela vison*) in England and Wales', in R. J. Putman (ed) *Mammals as Pests*, Chapman & Hall, London, pp242–250

Hatler, D. (1976) 'The coastal mink on Vancouver Island, British Columbia', PhD Thesis, University of British Columbia, Vancouver, Canada

Hersteinsson, P. (1999) 'Methods to eradicate the American mink (*Mustela vison*) in Iceland', in *Proceedings of the Workshop on the Control and Eradication of Non-native Terrestrial Vertebrates*, Environmental Encounters, Council of Europe Publishing, vol 41, pp25–29

Heylar, A. (2005) 'The ecology of American mink (*Mustela vison*): Response to control', PhD Thesis, University of York, York

Ibarra, J. T., Fasola, L., Macdonald, D. W., Rozzi, R. and Bonacic, C. (2009) 'Invasive American mink *Mustela vison* in wetlands of the Cape Horn Biosphere Reserve, southern Chile: What are they eating?', *Oryx*, vol 43, pp87–90

Ireland, M. (1990) 'The behaviour and ecology of the American mink *Mustela vison* Schreber in a coastal habitat', PhD Thesis, University of Durham, Durham

ISSG (Invasive Species Specialist Group) (2007) *Management: American Mink (*Mustela vison*)*, IUCN SSC, www.issg.org/database/species, accessed 19 August 2010

Jaksic, F. M., Iriarte, J. A., Jiménez, J. E. and Martínez, D. R. (2002) 'Invaders without frontiers: Cross-border invasions of exotic mammals', *Biological Invasions*, vol 4, pp157–173

Jefferies, D. J., Morris, P. A. and Mulleneux, J. E. (1989) 'An enquiry into the changing status of the water vole *Arvicola terrestris* in Britain', *Mammal Review*, vol 19, no 3, pp111–131

Joergensen, G. (1985) *Mink Production*, Scientifur, Hillerod, Denmark

Kidd, A. G., Bowman, J., Lesb, D. and Schulte-Hostedde, A. I. (2009) 'Hybridization between escaped domestic and wild American mink (*Neovison vison*)', *Molecular Ecology*, vol 18, pp1175–1186

King, C. M. (1983) 'Factors regulating mustelid populations', *Acta Zoologica Fennica*, vol 174, pp217–220

King, C. M., McDonald, R. M., Martin, R. D. and Dennis, T. I. (2009) 'Why is eradication of invasive mustelids so difficult?', *Biological Conservation*, vol 142, pp806–816

Kurten, B. and Anderson, E. (1980) *Pleistocene Mammals of North America*, Columbia University Press, New York

Linscombe, G., Kniler, N. and Aulerich, R. (1982) 'Mink', in J. A. Chapman and G. A. Feldhamer (eds) *Wild Mammals of North America: Biology, Management, and Economics*, John Hopkins University Press, Baltimore, pp629–643

Lizarralde, M. and Escobar, J. (2000) 'Exotic mammals in Tierra del Fuego', *Ciencia Hoy*, vol 10, pp52–63

Lodé, T., Cormier, J. P. and Le Jacques, D. (2001) 'Decline in endangered species as an indication of anthropic pressures: The case of European mink *Mustela lutreola* western population', *Environmental Management*, vol 28, pp727–735

Macdonald, D. W. and Harrington, L. A. (2003) 'The American mink: The triumph and tragedy of adaptation out of context', *New Zealand Journal of Zoology*, vol 30, pp421–441

Macdonald, D. W., Barreto, G. R., Ferreras, P., Kirk, B., Rushton, S., Yamaguchi, N. and Strachan, R. (1999) 'The impact of American mink, *Mustela vison*, as predators of native species in British freshwater systems', in D. P. Cowan and C. J. Fearem (eds) *Advances in Vertebrate Pest Management*, Filander-Verlag, Furth, pp5–24

Macdonald, D. W., Sidorovich, V. E., Anisomova, E. I., Sidorovich, N. V. and Johnson, P. J. (2002) 'The impact of American mink *Mustela vison* and European mink *Mustela lutreola* on water voles *Arvicola terrestris* in Belarus', *Ecography*, vol 25, no 3, pp295–302

Macdonald, D.W., King, C. M. and Strachan, R. (2006) 'Introduced species and the line between biodiversity conservation and naturalistic eugenics', in D. W. Macdonald and K. Service (eds) *Key Topics in Conservation Biology*, Wiley-Blackwell, Oxford, pp187–206

Mañas, S., Ceña, J. C., Ruiz-Olmo, J., Palazón, S., Domingo, M., Wolfinbarger, J. B. and Bloom, M. E. (2001) 'Aleutian mink disease parvovirus in wild riparian carnivores in Spain', *Journal of Wildlife Diseases*, vol 37, pp138–144

Maran, T. (2003) 'European mink: Setting of goal for conservation and the Estonian case study', *Galemys*, vol 15, pp1–11

Maran, T. and Henttonen, H. (1995) 'Why is the European mink (*Mustela lutreola*) disappearing? A review of the process and hypotheses' *Annales Zoologici Fennici*, vol 32, pp47–54

Maran, T., Podra, M., Polma, M. and Macdonald, D. W. (2009) 'The survival of captive-born animals in restoration programmes – Case study of the endangered European mink *Mustela lutreola*', *Biological Conservation*, vol 142, pp1685–1692

Marshall-Ball, R. (2008) *The Somerset Levels Green Shots Project*, Natural England, BASC, Environment Agency, Somerset Levels

McDonald, R. A. and Lariviere, S. (2001) 'Diseases and pathogens of *Mustela* spp., with special reference to the biological control of introduced stoat *Mustela erminea* populations in New Zealand', *Journal of the Royal Society of New Zealand*, vol 31, pp721–744

McDonald, R. A., O'Hara, K. and Morrish, D. (2007) 'Decline of invasive alien mink (*Mustela vison*) is concurrent with recovery of native otters (*Lutra lutra*)', *Diversity and Distributions*, vol 13, pp92–98

Melero, Y., Palazón, S., Revilla, E., Martelo, J. and Gosàlbez, J. (2008) 'Space use and habitat preferences of the invasive American mink (*Mustela vison*) in a Mediterranean area', *European Journal of Wildlife Research*, vol 54, pp1–9

Melero, Y., Palazón, S., Bonesi, L. and Gosàlbez, J. (2010) 'Relative abundance of culled and not culled American mink populations in Northeast Spain and their potential distribution: Are culling campaigns effective?', *Biological Invasions*, vol 12, pp3877–3885

Mickevicius, E. and Baranauskas, K. (1992) 'Status, abundance and distribution of mustelids in Lithuania', *Small Carnivore Conservation*, vol 6, pp11–14

Moore, N. P., Robertson, P. A. and Aegerter, J. (2000) *Feasibility Study into the Options for Management of Mink in the Western Isles*, CSL, MAFF, London

Moorhouse, T. P., Gelling, M. and Macdonald, D. W. (2009) 'Effects of habitat quality upon reintroduction success in water voles: Evidence from a replicated experiment', *Biological Conservation*, vol 142, pp53–60

Nordström, M. and Korpimäki, E. (2004) 'Effects of island isolation and feral mink removal on bird communities on small islands in the Baltic Sea', *Journal of Animal Ecology*, vol 73, pp424–433

Nordström, M., Hogmander, J., Laine, J., Nummelin, J., Laanetu, N. and Korpimäki, E. (2003) 'Effects of feral mink removal on seabirds, waders and passerines on small islands of the Baltic Sea', *Biological Conservation*, vol 109, pp359–368

Nordström, M., Ahola, M., Korpimäki, E. and Laine, J. (2004) 'Reduced nest defence intensity and improved breeding success in terns as responses to removal of non-native American mink', *Behavioral Ecology and Sociobiology*, vol 55, pp454–460

Oliver, M., Luque-Larena, J. J. and Lambin, X. (2009) 'Do rabbits eat voles? Apparent competition, habitat heterogeneity and large-scale coexistence under mink predation', *Ecology Letters*, vol 12, pp1201–1209

Pavlov, M. and Korasakova, I. (1973) 'American mink (*Mustela vison*)', in D. Kiris (ed) *Acclimatization of Game Animals in the Soviet Union*, Volgo-Vjatsk Book Publisher, Kirov, USSR, pp118–177

Peris, S. J., Sanguinetti, J. and Pescador, M. (2009) 'Have Patagonian waterfowl been affected by the introduction of the American mink *Mustela vison*?', *Oryx*, vol 43, pp648–654

Previtali, A., Cassini, M. H. and MacDonald, D. W. (1998) 'Habitat use and diet of the American mink (*Mustela vison*) in Argentinian Patagonia', *Journal of Zoology*, vol 246, pp482–486

Reid, F. and Helgen, K. (2008) '*Neovison vison*', in IUCN 2010, *IUCN Red List of Threatened Species, Version 2010.4*, www.iucnredlist.org

Reynolds, J. C., Short, M. J. and Leigh, R. J. (2004) 'Development of population control strategies for mink *Mustela vison*, using floating rafts as monitors and trap sites', *Biological Conservation*, vol 120, pp533–543

Roy, S. S. (2006) 'Mink control to protect important birds in SPAs in the Western Isles', Final Report to EU LIFE III – Nature, Scottish Natural Heritage, Edinburgh

Roy, S. S., Macleod, I. and Moore, N. P. (2006) 'The use of scent glands to improve the efficiency of mink (*Mustela vison*) captures in the Outer Hebrides', *New Zealand Journal of Zoology*, vol 33, pp267–271

Roy, S., Reid, N. and McDonald, R. A. (2009) *A Review of Mink Predation and Control in Ireland*, National Parks and Wildlife Service, Department of the Environment, Heritage and Local Government, Dublin

Ruiz-Olmo, J., Palazon, S., Bueno, F., Bravo, C., Munilla, I. and Romero, R. (1997) 'Distribution, status and colonization of the American mink *Mustela vison* in Spain', *Journal of Wildlife Research*, vol 2, pp30–36

Salo, P., Nordström, M., Thomson, R. L. and Korpimäki, E. (2008) 'Risk induced by a native top predator reduces alien mink movements', *Journal of Animal Ecology*, vol 77, pp1092–1098

Schüttler, E., Cárcamo, J. and Rozzi, R. (2008) Diet of the American mink *Mustela vison* and its potential impact on the native fauna of Navarino Island, Cape Horn Biosphere Reserve, Chile', *Revista Chilena de Historia Natural*, vol 81, pp585–598

Schüttler, E., Ibarra, J. T., Gruber, B., Rozzi, R. and Jax, K. (2010) 'Abundance and habitat preferences of the southernmost population of mink: Implications for managing a recent island invasion', *Biodiversity and Conservation*, vol 19, pp725–743

Sheail, J. (2004) 'The mink menace: The politics of vertebrate pest control', *Rural History*, vol 15, pp207–222

Shimatani, Y., Takeshita, T., Tatsuzawa, S., Ikeda, T. and Masuda, R. (2010a) 'Sex determination and individual identification of American minks (*Neovison vison*) on Hokkaido, Northern Japan, by fecal DNA analysis', *Zoological Science*, vol 27, pp243–247

Shimatani, Y., Fukue, Y., Kishimoto, R. and Masuda, R. (2010b) 'Genetic variation and population structure of the feral American mink (*Neovison vison*) in Nagano, Japan, revealed by microsatellite analysis', *Mammal Study*, vol 35, pp1–7

Sidorovich, V. E. (1992) 'Comparative analysis of the diets of European mink (*Mustela lutreola*) American mink (*M. vison*) and Polecat (*M. putorius*) in Byelorussia', *Small Carnivore Conservation*, vol 6, pp2–4

Sidorovich, N. V. and Macdonald, D. W. (2001) 'Density dynamics and changes in habitat use by the European mink and other native mustelids in connection with the American mink expansion in Belarus', *Netherland Journal of Zoology*, vol 51, pp107–126

Sidorovich, N. and Polozov, A. (2002) 'Partial eradication of the American mink *Mustela vison* as a way to maintain the declining population of the European mink *Mustela lutreola* in a continental area: A case study in the Lovat River head, NE Belarus', *Small Carnivore Conservation*, vol 26, pp12–14

Sidorovich, V., Jedrzejewska, B. and Jedrzejewski, W. (1996) 'Winter distribution and abundance of mustelids and beavers in the river valleys of Bialowieza Primeval Forest', *Acta Theriologica*, vol 41, pp155–170

Skírnisson, K., Gunnarsson, E. and Hjartardóttir, S. (1990) 'Plasmacytosis-sýking í villtum mink a Íslandi', *Icelandic Agricultural Science*, vol 3, pp113–122

Smal, C. M. (1991) 'Population studies on feral American mink *Mustela vison* in Ireland', *Journal of Zoology (London)*, vol 224, pp233–249

Water Vole Scotland (2010) 'Cairngorms and N.E. Scotland Water Vole Conservation Projects', www.watervolescotland.org

Wozencraft, W. C. (2005) 'Family Mustelidae', in D. E. Wilson and D. M. Reeder (eds) *Mammal Species of the World: A Taxonomic and Geographic Reference*, 3rd edition, The Johns Hopkins University Press, Baltimore, Maryland, pp601–605

Yamaguchi, N. and Macdonald, D. W. (2001) 'Detection of Aleutian disease antibodies in feral American mink in southern England', *Veterinary Record*, vol 149, pp485–488

Yamaguchi, N., Rushton, S. and Macdonald, D. W. (2003) 'Habitat preferences of feral American mink in the Upper Thames', *Journal of Mammalogy*, vol 84, pp1356–1373

Zabala, J., Zuberogoitia, I. and González-Oreja, J. A. (2010) 'Estimating costs and outcomes of invasive American mink (*Neovison vison*) management in continental areas: A framework for evidence based control and eradication', *Biological Invasions*, vol 12, pp2999–3012

Zalewski, A., Michalska-Parda, A., Bartoszewicz, M., Kozakiewicz, M. and Brzeziński, M. (2010) 'Multiple introductions determine the genetic structure of an invasive species population: American mink *Neovison vison* in Poland', *Biological Conservation*, vol 143, pp1355–1363

Zuberogoitia, I., Zabala, J. and Martinez, J. A. (2006) 'Evaluation of sign surveys and trappability of American mink: Management consequences', *Folia Zoologica*, vol 55, pp257–263

Zuberogoitia, I., González-Oreja, J. A., Zabala, J. and Rodriguez-Refojos, C. (2010) 'Assessing the control/eradication of an invasive species, the American mink, based on field data: How much would it cost?', *Biodiversity and Conservation*, vol 19, pp1455–1469

Part VI

Aquatic Pathogens

32

Bothriocephalus acheilognathi Yamaguti (Asian tapeworm)

Anindo Choudhury and Rebecca Cole

Introduction

The Asian tapeworm, also known as the Asian fish tapeworm, lives as an adult in the gastrointestinal tract of a wide variety of freshwater fishes across the globe. From its native east Asia, it hitched a ride in grass carp, and subsequently other fish hosts, to colonize every continent except Antarctica (Figure 32.1). It has even established on some oceanic islands such as Hawaii, Puerto Rico and Mauritius. It is now found in most countries in Asia, Europe and North America (Figures 32.1 and 32.2), and new reports continue to emerge. It remains by far the most successful invasive parasite species. The worm causes concern because it can be very pathogenic, especially in juvenile fish. It can also have more subtle but lasting negative effects. Its presence in fragile ecosystems, and where potential fish hosts are already at risk from other environmental factors, is therefore of special concern to fisheries biologists and natural resource managers.

History of the Species, Introductions and Host and Geographic Range

The Asian tapeworm was first described in 1934 by two names: Yamaguti (1934) described *Bothriocephalus acheilognathi* based on a single worm recovered from the cyprinid *Acheilognathus rhombeus* from Lake Ogura

(since drained), part of the Yodo River/Lake Biwa drainage in Japan. That same year, Yamaguti described a similar worm, *Bothriocephalus opsariichthidis*, from several specimens in another cyprinid, *Opsariichthys uncirostris* from Lake Biwa and the connected Yodo River. Subsequently, two other similar species were described from east Asia: *B. fluviatilis* from the cobitid (loach) *Hymenophysa curta* in Japan (Yamaguti, 1952) and *B. gowkongensis* from the grass carp *Ctenopharyngodon idellus* from Gowkong near Canton in southern China (Yeh, 1955).

Despite the original descriptions of this worm from native cyprinids in Japan, it is likely that the parasite was introduced into Japan with grass carp from China. Grass carp were introduced into Lake Biwa around 1916 (Naka, 1991 in Fausch and Nakano, 1998), well before Yamaguti described the tapeworms, and remain well established there (Fausch and Nakano, 1998). The historical biogeography of Japan's native fish fauna together with the history of fish translocation (Yuma et al, 1998), mostly unidirectional into Japan, also makes it likely that the tapeworm has its origins in mainland east Asia. Grass carp were also translocated to southern China from the more northern regions of the country, hence it is possible that Yeh's *Bothriocephalus gowkongensis* was/is also an introduced population of the Asian tapeworm. Following the work of Korting (1975), Molnar (1977), Pool and Chubb (1985) and others, a consensus emerged that the three nominal species, *B. acheilognathi*, *B. opsariichthydis* and *B. gowkongensis*

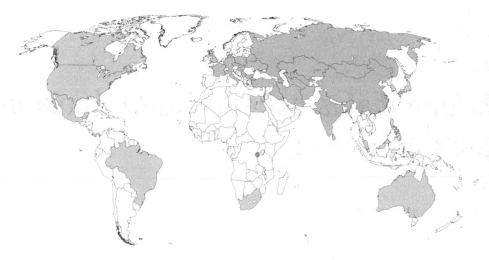

Note: Countries where Asian tapeworm has been reported are shaded grey. The '?' over Egypt refers to *'Bothriocephalus aegypticus'* and the grey dot over Lake Kivu in central Africa refers to *'Bothriocephalus kivuensis'* (see text for explanation).

Source: numerous sources (see References)

Figure 32.1 *World distribution of Asian fish tapeworm*

Note: States, provinces and territories where Asian tapeworm has been reported in the wild are in black. Areas in grey are states with reports from hatcheries only.

Source: numerous sources (see References)

Figure 32.2 *Distribution of Asian fish tapeworm in North America, Hawaii and the Caribbean islands*

are synonyms of one another; by the rules of precedence in taxonomy, *B. acheilognathi* stands as the sole Latin name for the Asian tapeworm (but see below for more on the taxonomic controversies surrounding this species).

Outside China and Japan, reports of the parasite, particularly in cultured carps, became common and widespread throughout the former Soviet Union in the 1950s and early 1960s (Bauer and Hoffman, 1976). Bauer and Hoffman (1976), Chubb (1981), Paperna (1996) and Hoffman (1999) reviewed the subsequent spread of this tapeworm. The current global distribution is shown in Figure 32.1, and it is probably present in more countries than has been documented in the literature.

While grass carp were responsible for the initial introduction of the Asian tapeworm into most countries, the parasite has subsequently expanded its range by colonizing other cyprinid and non-cyprinid hosts. It invaded Australia with introduced common carp *Cyprinus carpio* (Dove and Fletcher, 2000). In North America and Australia the parasite has also colonized a wide range of non-cyprinid hosts. García-Prieto and Osorio Sarabia (1991) and Salgado-Maldonado and Pineda-López (2003) describe its remarkable success in Mexico (see also Figure 32.2). According to a recent update from Mexico (Rojas-Sánchez and García-Prieto, 2008), the parasite has been reported from 72 fish species and 19 states in Mexico, and reports of new hosts and sites in Mexico continue to trickle in (Pérez-Ponce de Leon et al, 2009, from Durango; Méndez et al, 2010, from Baja California; Aguilar-Aguilar et al, 2010, from Puebla).

The parasite is spreading in the more northern regions of the North American continent as well (Figure 32.2); it has colonized the Lake Winnipeg drainage (Choudhury et al, 2006) where the parasite appears to be well established in native shiners (Patrick Nelson, pers. comm.), although it remains unclear how it entered that drainage (Choudhury et al, 2006). The tapeworm has also invaded the Lake Ontario/Erie drainages (Marcogliese, 2008). Elsewhere, it was recently reported from the Yampa River in Colorado (Ward, 2005a) and Río Grande/Río Bravo del Norte (Bean et al, 2007; Bean, 2008). It has spread to Hawaii and Mauritius with a very suitable host, the mosquitofish *Gambusia affinis*, a poeciliid (Paperna, 1996; Vincent and Font, 2003). In 2000, it was reported for the first time in the floodplain of the Yangtze River (Nie et al, 2000) and such range extensions will probably continue. In Eurasia and Africa, it has mainly remained confined

to cyprinids, as in the barbs in South Africa (Retief et al, 2007). This is not surprising given that relatively few of the more suitable non-cyprinid hosts, e.g. cyprinodontiforms and atherninforms, are present in the continental fresh waters there. In mainland Europe, however, at least one species of atherinid, *Atherina boyeri*, is infected (Giovinazzo et al, 2006). The tapeworm is well established in drainages that have been otherwise severely impacted, such as in reservoirs of the Aral Sea basin (Urazbaev and Kurbanova, 2006). There is also evidence that birds ingesting infected fish can seed waterways with faeces containing viable tapeworm eggs that survive the passage through the bird's gastrointestinal tract (Prigli, 1975).

The spread of Asian tapeworm in the US illustrates the interplay of various factors that results in the successful establishment of an invasive species. Grass carp first arrived in the US in 1963 as shipments of fingerlings from Malaysia and Taiwan to US Fish and Wildlife federal fish hatcheries in Stuttgart, Arkansas and Auburn, Alabama respectively (Fuller et al, 1999); grass carp were to be reared and further bred as biological control agents for aquatic weeds. It is likely that fish in at least one of these shipments were infected with Asian tapeworm. An unknown number of grass carp escaped their outdoor holding ponds/enclosures at the rearing facility in Stuttgart and took advantage of flooding to access the Mississippi drainage. Subsequent stockings of grass carp in lakes and reservoirs open to stream systems enabled the fish to colonize the Mississippi and Missouri Rivers by the early 1970s (Fuller et al, 1999). Presumably they carried their infections into these new locations. Early on, grass carp were also shipped to Florida to control aquatic macrophytes and as a result probably introduced Asian tapeworm into that state. While grass carp were being used to control aquatic weeds, mosquitofish, *Gambusia hollbrooki/G. affinis* were being used to control mosquito larvae. Overlap of grass carp and mosquitofish in places such as Florida and Louisiana, and a shared diet of copepods, allowed the tapeworm to effect one of its first major 'jumps' across host taxonomic boundaries, into a host of a distantly related order and family of fishes (Cyprinodontiformes: Poeciliidae). It appears that infected mosquitofish were translocated to distant places such as North Carolina, California and Hawaii. The Asian tapeworm also colonized other, cyprinid, hosts, such as the red shiner, golden shiner and common carp.

The movement of bait fish such as shiners is thought to have introduced the Asian fish tapeworm into Little Colorado River in Grand Canyon, and elsewhere (Choudhury et al, 2004). *Bothriocephalus acheilognathi* was first documented from the Grand Canyon in 1990 (Minckley, 1996). Infected hosts included the federally endangered humpback chub, *Gila cypha* and other native and non-native fish species (Brouder and Hoffnagle, 1997). These early surveys showed that the parasite was established in the seasonally warm Little Colorado River (LCR), a stream critical to the spawning and propagation of the endangered humpback chub in Grand Canyon (e.g. Stone and Gorman, 2006). Current conditions in Grand Canyon do not allow the propagation of humpback chub in the Colorado River mainstem, since the hydrology of the Colorado River in Grand Canyon has been drastically altered following the closure of Glen Canyon Dam (Stone et al, 2007). Amidst growing concerns about the impact of this potentially pathogenic tapeworm on native fish resources in Grand Canyon, a two year seasonal parasitological survey of native and non-native fishes in the LCR between June 1999 and April 2001 (Choudhury et al, 2004) found that humpback chub comprised only 8 per cent of fish sampled but harboured 54 per cent of the worms, and the tapeworm reached its highest abundance in this fish. The temperature profile of the LCR and its associated tributary creeks and backwaters provide environmental conditions ideal for the success of this parasite in this system. The study also showed that there may be two or more different circulation patterns of the parasite in two different host species (humpback chub and speckled dace) from different habitat types within the same drainage system and illustrates the complexity facing potential management strategies to control this parasite. Four years later, in 2005, Ward and Persons (2006) reported much lower intensities of infection in juvenile humpback chub in the LCR and suggested the winter and spring flooding in 2005 and its negative impact on copepod populations as a possible cause. It remains to be seen if this decline is long term and whether the parasite has reached an equlibrium with its host population whereby the initial high mortality rate in hosts supposedly abates (Hoffman, 1999). Another facet was added to this case with the recent finding of Asian tapeworm in carp (*C. carpio*) from the upper watershed of the LCR (above Grand Falls); these fish were thought to have acquired their infections even further upstream in headwater tributaries (Stone et al, 2007). Such upstream reservoirs of infection may ensure the continuous (re)population of the lower LCR.

Ecological Niche of the Species

The Asian tapeworm has been reported from large perennial rivers and small intermittent creeks, from lowland tropical ponds and floodplains to cold northern prairie waters and high altitude neovolcanic lakes, and from all sorts of waterways in between. The tapeworm is often referred to as thermophilic. Development of the tapeworm's eggs reportedly ceases below 12°C, but almost every body of water in colder climates has associated streams, creeks, backwaters and standing water that may warm up sufficiently during the summer months to allow development. For example, the cold and relatively constant temperature of the Colorado River in Grand Canyon prevents the parasite from developing in the mainstem of the river, but the parasite is common in its tributaries (Brouder and Hoffnagle, 1997; Choudhury et al, 2004). As a result, any drainage system that has fish and copepods should be considered potentially susceptible to this invasive species. The worm's name – 'Asian' fish tapeworm – may lull one into a sense of security about northern waters but it is worth remembering that the Asian tapeworm is also a northeast Asian tapeworm whose natural range includes the Amur River drainage.

The parasite clearly has a predilection for cyprinids but the long list of non-cyprinid hosts suggests that no species of fish can be considered 'safe' from this parasite a priori. Of the non-cyprinid hosts, the distantly related cyprinodontoids, such as poeciliids, seem to make for very suitable hosts (Scholz, 1997); fishes in this group are generally small bodied and commonly feed on copepods, so access to the right intermediate host may be a determining factor. Fishes more closely related to cyprinids, such as catostomids (suckers) and cobitids (loaches) seem to be less commonly reported as hosts and this may be due to their more benthic feeding habits. It is, however, access to intermediate hosts (copepods) in the diet together with suitable physiological (including immunological) conditions that enable a successful infection. It remains unclear what aspect of the worm's physiology makes it adapt so readily to different hosts. In its early development, the tapeworm also can infect a range of copepod species worldwide (e.g. Paperna, 1996; Scholz, 1997), which enhances its invasiveness.

Lifecycle: Normal and post-cyclic transmission

The Asian tapeworm lives as an adult in the gastrointestinal tract of its fish host (Figure 32.3). It is typically found in the intestine but can infect the poorly differentiated, less muscular, stomach of some cyprinids. Adult tapeworms typically release partially embryonated, tanned, operculate eggs in the gut of their fish hosts. Eggs pass out with the faeces, reach the water and settle down on the substrate where further development takes place. Within the egg, the embryo develops into a ciliated larva, the coracidium (pl. coracidia) that pushes its way out of an operculate (lidded) opening at one end of the egg. The rate of development is temperature dependent; at 28–30°C, development is rapid and the

majority hatch after a day, at 14–15°C, development and hatching are more protracted (10–28 days) (Hoffman, 1980; Granath and Esch, 1983a; Paperna, 1996). Egg development is reported to be seriously retarded below 12°C (Paperna, 1996).

The coracidium larva swims around with the help of the cilia that cover it. It swells as it swims around, but loses energy rapidly and dies within days if not eaten by a copepod. Coracidia survive two to three days at 20°C but five to six days at 15–16°C (Hanzelova and Žitňan, 1986). Granath and Esch (1983a) report low coracidia motility below 20°C.

Once consumed by a copepod, the larva loses its ciliated covering in the gut of the copepod, burrows through the gut wall and enters the body cavity (haemocoel) of the copepod where it develops into the first larval stage called a procercoid (Figure 32.3). Copepods may harbour several procercoids at the same time and heavily infected copepods become sluggish and eventually moribund. The procercoid is considered ready to infect the fish host once it has developed a tiny rounded tail called a 'cercomer' that is set off from the larval body by a visible constriction (Figure 32.3).

When copepods infected with fully developed procercoid larvae are consumed by fish, the procercoids are released from the confines of the copepod body, take refuge in the folds of the gut lumen, and begin developing into typical, externally segmented, 'strobilate' tapeworms. If recruitment of larval worms into fish occurs in the autumn – fairly typical in north temperate regions – then gravid (egg bearing) worms are commonly found by spring the following year. In lower latitudes, seasonality may not be as pronounced. Strobilation (the forming of the segmented body), which must precede maturation, is also temperature dependent. In grass carp, worms mature in 21–23 days at 28–29°C (Liao and Shih, 1956). Oškinis (1994) found that 89 per cent of the worms became segmented in experimentally infected common carp between 16 and 20 days at 25°C, whereas 80 per cent of the worms remained unsegmented four months later at 15°C. Davydov (1978) reported that at 15–22°C, maturation takes one and a half to two months in barbs and that temperatures below 15°C seriously delay development (six to eight months). Data reported by Granath and Esch (1983a) suggest slightly higher temperature requirements for the Asian tapeworm population in introduced mosquitofish in Belews Lake, a cooling

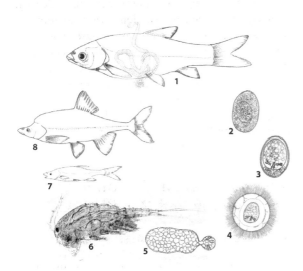

Note: 1 Adult tapeworm in definitive host; 2 partially embryonated egg; 3 fully embryonated egg; 4 coracidium – the first larval stage after swelling; 5 second larval stage – procercoid stage (fully developed, with cercomer); 6 cyclopoid opepod infected with procercoid stages. Final hosts (1, 7 and 8) become infected when they ingest infected copepods. When smaller fish (7) with tapeworms are ingested by larger piscivorous fish (8), the tapeworms can re-establish in the larger fish (post-cyclic transmission)

Figure 32.3 *Lifecycle of Asian tapeworm,* Bothriocephalus acheilognathi

reservoir subjected to warm water influxes. They report 'no apparent maturation' of worms at 20°C. Factors such as strain/population differences of the parasite and the species/population of host may cause variations in the lifecycle of this tapeworm.

Experimental infections have demonstrated that when an infected fish is consumed by another predatory fish, the worms can re-establish in the gut of that new host (post-cyclic transmission; Hansen et al, 2007). This results in an accumulation of worm biomass in predatory fishes and may account for the high numbers of tapeworms in larger humpback chub in the LCR in Grand Canyon.

Population dynamics

Seasonal differences in population structure are found in most populations of Asian tapeworm studied to date, but biotic and abiotic factors of the particular system influence these trends. The exceptional case in Hawaii (Vincent and Font, 2003) where no real seasonal trends were observed may, as the authors suggest, be due to the low prevalence and mean abundance of infection year round. Seasonal changes in temperate latitudes are correlated with seasonal changes in water temperature but the abundance of intermediate copepod hosts may also be a factor (Marcogliese and Esch, 1989). Studies from at least four systems, Belews Lake, LCR and Río Grande/Bravo in the US and Vaal Dam in South Africa, involving different hosts – mosquitofish, humpback chub (*Gila cypha*), red shiners (*Cyprinella lutrensis*) and yellowfish (*Labeobarbus kimberleyensis*) – have found lower abundance of infection during the summer when water temperatures are at their highest (Granath and Esch, 1983a, 1983b; Choudhury et al, 2004; Bertasso and Avenant-Oldewage, 2005; Bean, 2008). Granath and Esch (1983b) also found that infrapopulation densities (population densities in individual fish) decrease above 25°C. In Lake Beysehir in Turkey, infections in common carp peaked in April (Tekin-Özan et al, 2008). In contrast, the highest abundance of tapeworms in Mohave tui chub from Lake Tuende in California occurred during the highest water temperatures (Archdeacon, 2007).

The proportion of immature and gravid tapeworms also follows seasonal patterns. In LCR and Río Grande/Bravo del Norte, gravid worms were more abundant in the spring and early summer than in autumn while immature worms were more abundant in the autumn than in the summer (Choudhury, Hoffnagle and Cole, unpublished data; Bean, 2008). In Belews Lake, the prevalence of non-segmented worms in introduced mosquitofish showed an inverse relationship with both segmented and gravid worms; prevalence of non-segmented worms peaked during the autumn, winter and spring while those of segmented and gravid worms were highest during early summer to early autumn and decreased markedly at other times (Granath and Esch, 1983b). Riggs and Esch (1987) found two peaks of worm fecundity (measured as the number of gravid proglottids per gravid worm) in fathead minnows (*Pimephales notatus*) and red shiners in the spring and autumn, and proposed that continued feeding by these native cyprinids during the colder months of the year may contribute to such patterns as opposed to dynamics in the mosquitofish, an introduced warmer water species with different bioenergetic demands. Studies (Liao and Shih, 1956; Granath and Esch, 1983b) also report an inverse relationship between the abundance of gravid worms and prevalence and density. Temperature-dependent rejection related to immune responses has been proposed as a cause. Luo et al (2004) suggest that the up-regulation of certain genes during maturation may provide clues to this phenomenon but this remains speculative. The Belews Lake experience demonstrates that an interplay of temperature, water chemistry, fish temperature tolerance, as well as fish–copepod interactions influence seasonal population dynamics of the parasite.

Identification and Taxonomy: Morphology and Genetics

Tapeworms in cyprinids

The correct identification of parasites is important, particularly because appropriate management decisions depend on it. In the case of the Asian tapeworm, taxonomic issues and problems fall into two broad categories: (1) misidentification of species in other genera as Asian tapeworm, and (2) *Bothriocephalus* spp. in cyprinids that are very similar to the original trio of *B. acheilognathi*, *B. opsariichthydis* and *B. gowkongensis*.

The first category involves cases such as the misidentification of *Eubothrium tulipai*, a parasite of larger bodied western cyprinids, the pike-minnows

(*Ptychocheilus* spp.). The parasite was reported as *B. opsariichthydis* by Arai and Mudry (1983) from the northern pike-minnow, *Ptychocheilus oregonensis* in British Columbia, but Choudhury et al (2006) showed that the record was of *E. tulipai* (Ching and Andersen, 1983), a species with a superficially similar scolex. However, it is very likely that the Asian tapeworm would infect pike-minnows if it had access to them, so it cannot be assumed that tapeworms in pike-minnows will always be *E. tulipai*. Specimens should also be deposited in internationally known museum collections (e.g. Germany, Czech Republic, UK, Mexico, US, Russia, Japan, Brazil and Argentina).

The second category involves the various proposed synonyms of *B. acheilognathi* for worms described from cyprinids. The early controversy regarding the status of *B. acheilognathi*, *B. opsariichthydis* and *B. gowkongensis* seems to have been resolved by a broad consensus that they are names for a single species. Akhmerov's (1960) *Schizocoytle fluviatilis*, Molnar and Murai's (1973) *B. phoxini* from *Phoxinus phoxinus* in Hungary, Rysavy and Moravec's (1973) *B. aegypticus* from *Barbus bynni* in Egypt and *B. kivuensis* from *Barbus altianalis altianalis* in Lake Kivu (Baer and Fain, 1958) were also synonymized with *B. acheilognathi* (e.g. Korting, 1975; Molnar 1977; Pool and Chubb, 1985; Pool, 1987).

Dubinina (1982, 1987), however, concluded that there were two species of *Bothriocephalus* in cyprinids in the former Soviet Union, *B. acheilognathi* and the more common and widespread *B. opsariichthydis* (= *B. gowkongensis*, *B. phoxini*) that could be distinguished on the basis of scolex morphology. Pool and Chubb (1985), using scanning electron microscopy (SEM), conclude that the reported variation in scolex morphology is largely an artefact of the various methods used to fix and preserve these worms and upheld the synonymy of *B. acheilognathi*, *B. opsariichthydis*, *B. gowkongensis* and *B. phoxini*. Pool and Chubb (1985) suggest that Baer and Fain's (1958) *B. kivuensis*, while similar to *B. acheilognathi*, is a valid species based on certain morphological features. Paperna (1996) points out that *B. kivuensis* predates the introduction of non-native cyprinids, and hence *B. acheilognathi*, into the Lake Kivu system. Pool (1987) synonymizes both *B. kivuensis* and *B. aegypticus* with *B. acheilognathi* but Paperna (1996) lists them as valid species; it is unclear if Paperna (1996) was aware of Pool's (1987) work as it

is not cited. Grass carp were not introduced into the Nile drainage for aquaculture and stocking purposes until the late 1980s, although common carp were introduced in 1936 (FAO, 2003–2010). The taxonomy of the Indian species *B. teleostei* (Malhotra, 1984), which has been reported from cyprinids in hill streams of India (Chauhan and Malhotra, 1986) should be evaluated using standardized fixation, SEM and molecular methods, especially since Akhter et al (2008) report *B. acheilognathi* from *Schizothorax* spp. (cyprinids) in the Kashmir Valley.

More recently, molecular data have been used to address taxonomic issues and understand genetic diversity in *B. acheilognathi*. Liao and Lun (1998) suggest evidence for lineage diversification in *Bothriocephalus* from Chinese cyprinids, and for *B. opsariichthydis* being a valid species. Luo et al (2002), using samples from a wider host and geographical range, found evidence for lineage diversification based on ITS-1 and ITS-2 sequence datasets, although one of the three genotypes they identify was paraphyletic (an 'incomplete' group that does not include all the descendants of a common ancestor) and only one of the genotypes seemed to be associated with a particular type of cyprinid (*Culter* spp.). A follow-up study (Luo et al, 2003) using microsatellite markers of isolates from different cyprinids and localities within an interconnected drainage system (Yangtze River) indicates higher than expected genetic structuring. Four gene pools were identified, which suggests the possibility of a species complex (Luo et al, 2003). Although higher intraspecific variability is seen in certain regions of the genome, for example ITS-2 of the rRNA genome, low population variability in other regions of the rRNA genome has been used to argue that introduced Asian tapeworm populations all belong to the same species (Bean et al, 2007).

Accurate identification of *Bothriocephalus acheilognathi*

It is often stated that the Asian fish tapeworm is readily identified by its arrowhead-shaped or inverted heart-shaped attachment organ (scolex) when viewed in *lateral* profile (Figure 32.4a–f). While this remains a good starting point – and the scolex of *B. acheilognathi* is indeed unique within the genus – the

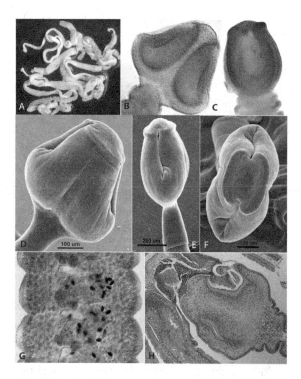

Note: (a) Several adult tapeworms; (b) stained scolex (lateral view); (c) stained scolex (dorso-ventral view); (d), (e) and (f) scanning electron micrographs of scolex of worms from humpback chub, LCR; (g) stained portion of gravid strobila showing rounded margins of segments; (h) histological section of attached scolex in the gut of humpback chub (LCR) showing necrotic material and damage to epithelia.

Source: (a) David Ward; other (b)–(h) Anindo Choudhury and Rebecca Cole

Figure 32.4 *Morphology of the Asian tapeworm*

final diagnosis should not depend on that single criterion when there may be other tapeworms with (superficially) similar scolices in the area (*Eubothrium tulipai* is a case in point) or when the scolex may not have been appropriately fixed or mounted (on slides). In addition to the morphology of the scolex, the *medial*, not lateral, position of the genital openings on the segments should be confirmed (Figure 32.4g). Samples of the worms should be stained and mounted after killing the worms with hot steaming fixative (preferably 5 per cent or 10 per cent buffered formalin, with adequate ventilation) to ensure that the scolex is not deformed but preserved in a relaxed state. If possible, a small portion of the mature

strobila (a few segments worth) should be cut from the worm *before* fixation and stored in 95 per cent or 100 per cent molecular grade ethanol for DNA extraction and verification using molecular analyses (such as in Bean et al, 2007).

In Eurasia, *Bathybothrium rectangulum*, a widespread parasite of certain cyprinids (e.g. *Barbus, Schizothorax, Oreoleuciscus* spp.) has a scolex that superficially resembles that of the Asian tapeworm, but like *Eubothrium tulipai*, *B. rectangulum* is an amphicotylid tapeworm and has lateral, not medial, genital openings (Dubinina, 1987).

Bothriocephalus spp. in non-cyprinid hosts

It appears that when *Bothriocephalus* is reported in cyprinids, it is *B. acheilognathi* or a species very similar to it (and probably closely related) such as *B. kivuensis* and perhaps *B. aegypticus*. The presence of other native *Bothriocephalus* species in non-cyprinid fishes in drainages where Asian tapeworm may be present makes discrimination between species of *Bothriocephalus* important, and biologists working in such systems need to be aware of these other species. This is especially true of North America where several native species of *Bothriocephalus* occur (Hoffman, 1999). One such species, *B. cuspidatus*, is common and widespread in a variety of North American freshwater fishes, including its most common and typical host, the walleye (*Zander vitreus*). The parasite is also often reported from yellow perch (*Perca flavescens*) and the mooneyes (*Hiodon* spp.). '*Bothriocephalus cuspidatus*' reported from sunfishes of the genus *Lepomis* is actually a separate species, based on morphological and molecular characteristics, and is described elsewhere (Choudhury, unpublished). *Bothriocephalus cuspidatus* has a less fleshy and more elongate arrowhead-shaped scolex than *B. acheilognathi*, while the species from *Lepomis* sunfishes has a scolex that is more rectangular in profile (Hoffman, 1980; Scholz, 1997). Segments of Asian tapeworm have rounded edges (Figure 32.4g) rather than more distinct and sharper projecting 'corners' as in *B. cuspidatus* (craspedote strobila). Since there is a higher diversity of *Bothriocephalus* spp. in non-cyprinids in North America than in the fresh waters of Eurasia, Africa or the neotropics (South and Central America) careful examination is advised. Ignorance of native

Bothriocephalus spp. in North American fishes resulted in some alarmist and misleading reporting in the local media (see below).

Impact

The detrimental effects of Asian tapeworm infections have been reviewed by several authors (e.g. Paperna, 1996; Hansen et al, 2006; Borucinska, 2008). Moderate to heavy infections of Asian tapeworm in small fish can be fatal. Mortality of fry and juveniles of cultured fish due to Asian tapeworm is well known in the farming of cyprinids such as grass carp, common carp and koi (Yeh, 1955; Bauer et al, 1969; Han et al, 2010) and has been reported in golden shiners (*Notemigonus chrysoleucas*) and mosquitofish (Hoffman, 1980; Granath and Esch, 1983c). The parasite has also been held partially responsible for the decline in humpback chub populations in the Grand Canyon ecosystem (Stone et al, 2007). Stone et al (2007) conclude that Asian tapeworm probably contributed to the estimated 30–60 per cent decline (Coggins et al, 2006) of humpback chub since the 1990s. Light infections do not show external signs, but moderate to heavy infections may cause abdominal distension. The parasite causes a decrease in body mass of carp (Bauer et al, 1969), retards growth of roundtail chub (Brouder, 1999) and Topeka shiner (Koehle and Adelman, 2007), and reduces condition factor in red shiners (Bean and Bonner, 2009). Experimental Asian tapeworm infections in bonytail chub (*Gila elegans*) resulted in reduced growth, negative changes in health condition indices, as well as accelerated mortality when food was reduced (Hansen et al, 2006). In smaller fish, mechanical obstruction of the gut and intestinal inflammation have severe effects. Pathogenic changes are probably initiated by the physical interaction between the worm's scolex and the gut mucosa at the attachment site (Figure 32.4h). Pathology includes inflammation, haemorrhaging, destruction and dysfunction of the intestinal mucosa, necrosis and even perforation (e.g. Hoole and Nisan, 1994). At the cellular level, separation and shedding of microvilli occurred at the interface between the gut and the tapeworm's bothridia (Hoole and Nisan, 1994). Inflammation is accompanied by migration and infiltration of lymphocytes, macrophages and eosinophils to the infected area and even out of the gut to the parasite surface. The disease caused by Asian tapeworm is called 'bothriocephalosis'.

Treatment

Early treatment of Asian tapeworm infections, typically in cultured fish, involved herbal extracts, which were replaced by niclosamide-based drugs (see Dick and Choudhury, 1995). In more recent years, praziquantel has become the drug of choice delivered either in a water bath or in medicated food (Borucinska, 2008). Ward (2007) found that both time and dosage were important factors in the efficacy of treatment; infections of bonytail chub were cleared at a dose of 1.5mg L^{-1} applied for 24 hours. Mitchell and Darwish (2009) found similar factors at play in treating grass carp; however, they also found that fish density during treatment played a role in its efficacy; only a 24 hour bath treatment of grass carp at a density of 60g L^{-1}, using 0.75mg L^{-1} of praziquantel or higher eliminated Asian fish tapeworm from the gut. All other conditions being similar, increasing the density of fish lowered the efficacy of the treatment. Thus, dosage must be adjusted to fish densities for the treatment to be effective. Drug delivery must also take into account the potential harm that the drug at higher dosages may do the fish (Mitchell and Hobbs, 2007). Ward (2005b) discusses whether strategic and targeted drug treatment of wild fish, such as adult humpback chub in the LCR, could be evaluated as a management strategy.

Management Efforts and Effectiveness

Management strategies in aquaculture

Since the main source of Asian tapeworm is infected fish, initial steps must be taken to prevent infected fish from entering aquaculture facilities. Fish farmers should enquire whether a 'disease free' certification includes testing for Asian tapeworm, and even if it does, they may wish to treat fish before introducing them to their aquaculture facilities. If infected fish are found in the aquaculture facility, they can be treated with drugs (see above). Praziquantel-treated fish cannot be sold commercially as food in the US since praziquantel is not approved for food fish (Merck, 2008). Praziquantel will kill and expel adult worms from fish but will not kill the eggs and free swimming larvae (Kline et al, 2009). This means that water baths used for deworming should be drained and disinfected at the end of a day's work, if the deworming is carried out over several days. Asian tapeworm, like other fish tapeworms, will release eggs when they come in contact with water, so treated fish should be passed through one or more changes of fresh water to minimize contamination by and carry-over of tapeworm eggs. Copepod densities may be controlled in indoor aquaculture operations but is difficult in outdoor ponds. In most countries, the stocking of fish in the wild is now controlled by laws but regulations vary. Irrespective of the variability of these laws, fish being translocated from hatcheries should also be checked for Asian tapeworm before they are stocked.

Management strategies in the wild

To our knowledge, no successful comprehensive or long term management strategies have been developed to eradicate or prevent the spread of Asian tapeworm in natural drainages (lakes and streams). In some systems such as the LCR that is circumscribed by defined and seasonally dry upstream regions and the cold Colorado River mainstem downstream, a management strategy for reducing Asian tapeworm abundance in targeted species such as humpback chub may be effective. Ward (2005b) discusses the pros and cons of trapping and deworming humpback chub with appropriate doses of praziquantel before releasing them back into the LCR. At appropriate doses, praziquantel is very effective in expelling worms from fish, but an additional precaution would be passing the fish through additional changes of fresh water to minimize contamination of fish nets with tapeworm eggs, especially since eggs and coracidia may survive the water bath treatment (Kline et al, 2009). Also, the water bath should be drained and disinfected at the end of the day. This deworming strategy would have to be sustained and repeated additional times (multiple times a year or every year) before long term benefits may be observed. The presence of reservoir hosts such as speckled dace and their abundance will probably ensure the continued presence of the parasite but targeted and strategic treatment of adult humpback chub at the right times of the year (preferably after recruitment and before maturation of the worms, i.e. late autumn) may reduce parasite abundance in the LCR. Undoubtedly, there are complicating factors, such as the movement of fish in and out of the LCR

from the Colorado mainstem, and these have to factored into any modelling and predictions that may be developed with a parasite management strategy (see below).

The abundance of Asian tapeworm in humpback chub in the LCR has also complicated the proposal to make the Colorado River in Grand Canyon warmer and more suitable for the propagation of native fishes by manipulating releases from Lake Powell. Hoffnagle et al (2006) suggest a re-evaluation of such proposals, given the presence of thermophilic parasites such as Asian tapeworm (and anchor worm, *Lernaea* sp.) in the warmer LCR and their potential spread should the Colorado River become warmer. Hoffnagle et al (2006) also found that humpback chub were healthier and in better condition in the colder Colorado River than in the more natural conditions of the LCR. Paradigms about the suitability of warmer waters for native fishes will need to accommodate the new reality of invasive thermophilic parasites as one of several interacting variables in developing effective management strategies in this ecosystem.

Asian tapeworm online and in the media

In countries such as Australia and the US, there is broad awareness of Asian tapeworm in fisheries and natural resources agencies, both at the state and central (federal) level. The US Geological Survey Nonindigenous Aquatic Species website maintains a useful factsheet on Asian tapeworm (USGS, 2009) that can be openly accessed. The USGS and USFWS also release reports and updates on the status of natural resources including those in Grand Canyon and mention is invariably made of Asian tapeworm (e.g. HT Media, 2006; Federal Information and News Dispatch Inc., 2009). The Food and Agricultural Organization of the United Nations (FAO) has reproduced Paperna's (1996) book online (FAO, 1996).

Asian tapeworm has also featured in the national and local media, for example Blakeslee (2002), Associated Press (2004) and O'Driscoll (2005). The name 'Asian tapeworm' has also been used in the European media for another species, *Nippotaenia mogurndae* (e.g. *The Slovak Spectator*, 2007), a parasite introduced into Slovak waters by the Amur sleeper (*Percottus glenii*), an eleotrid; this may cause confusion for non-specialists and the general public. The recent discovery of Asian tapeworm in the Great Lakes drainage (Marcogliese, 2008) prompted a series of stories in the local and regional media, often with misleading statements and implications. Marcogliese's publication coincided with anglers' reporting tapeworms in walleye, which led to much confused reporting. One story announced 'State issues sushi warning; Asian fish tapeworm found in Great Lakes; cook fish thoroughly' (*Grand Rapid Press*, 2008) and confused what is probably the common *Bothriocephalus* of walleye, *B. cuspidatus*, with Asian tapeworm, neither of which is found in the flesh of walleye or infective to humans. Marcogliese's report was also picked up by the *Muskegon Chronicle* and *Kalamazoo Gazette* (Alexander, 2008; *Kalamazoo Gazette*, 2008) that also mentioned 'tapeworms' in walleye. One article in the *Buffalo News* (Eliott, 2009) stated incorrectly that researchers from Environment Canada reported Asian tapeworm from walleye and hopelessly conflated Asian tapeworm with reports of species found in the gut and flesh of walleye.

Outlook

Although Asian tapeworm is no longer dependent upon grass carp, the spread of this original host is of as much relevance as the movement of baitfish and the stocking of other cyprinids. The spread of grass carp is being carefully followed and documented, especially in the US and Canada (Cudmore and Mandrak, 2004). Since first breaking out of relative confinement, the grass carp has gradually colonized a wide range of waterways in the US; it has been reported from 45 states (Nico et al, 2010) and has established breeding populations in Kentucky, Illinois, Missouri, Arkansas, Louisiana, Tennessee, Mississippi and Texas (Fuller et al, 1999; Nico et al, 2010). Grass carp have also been introduced into other areas, such as British Columbia and Alberta, Canada, through stocking (Cudmore and Mandrak, 2004). Herborg et al (2007) modelled the colonizing ability of grass carp and project that the species would be able to establish in Canada and throughout much of the US. Similarly, fisheries

biologists and agencies in Australia continue to study common carp, *Cyprinus carpio* (Koehn, 2004).

The parasite continues to draw attention, especially in the desert southwest of the US, where fragile ecosystems support populations, often fragmented and vulnerable, of several native fish species (Ward, 2005b). Ongoing long term parasite surveys at the Dominican University of California have targeted mosquitofish and other native fishes in the San Francisco Bay area where mosquitofish are stocked. As of now, the parasite seems to be confined to southern California (Warburton et al, 2002). Southern populations of *Gambusia* should not be used to stock new areas and any shipment of mosquitofish should be certified *Bothriocephalus* free. In the more northern regions of the continent, the situation may parallel that of Europe and northern Asia and the parasite may be mostly found in cyprinids. One of the main non-cyprinid host groups, the cyprinodontoids, is becoming increasingly rare in the northern reaches of the continent. It remains to be seen if the parasite encounters competition from native *Bothriocephalus* spp. as it attempts to move northwards. Strategic and focused surveys of fishes from susceptible areas will be crucial in monitoring the spread of this parasite and its impact on native fishes. Sensitive DNA probes to test faecal samples of cultured and wild fishes along with necropsy work may be the future of such parasite monitoring, especially in fish species that are vulnerable, threatened or endangered. It also remains to be seen if adverse effects on fish populations decrease over time as the host–parasite interaction supposedly stabilizes (Hoffman, 1999). At the same time, the Asian tapeworm lends itself to detailed studies of the population genetics of a model invasive species and may contribute to a deeper understanding of the nature of such colonizations.

Acknowledgements

We would like to thank David Ward, Arizona Game and Fish Department, Bubbling Ponds Research Facility, for Figure 32.4a, and are grateful for support from the USGS National Wildlife Health Center Madison and St Norbert College.

References

Aguilar-Aguilar, R., Jose-Abrego, A. and Pérez-Ponce de Leon, G. (2010) 'Cestoda, Bothriocephalidae, *Bothriocephalus acheilognathi* Yamaguti, 1934: Nematoda, Rhabdochonidae, *Rhabdochona canadensis* Moravec and Arai, 1971: New host records for the state of Puebla, Mexico, and a new fish host', *Checklist*, vol 6, no 3, pp437–438

Akhmerov, A. K. (1960) 'Fish cestodes of the Amur River', *Trudy Gel'mintologicheskoi Laboratorii Akademiya Nauk SSSR*, vol 10, pp15–21

Akhter, S., Fayaz, A., Chishti, M. Z. and Tariq, K. A. (2008) 'Seasonal dynamics of *Bothiocephalus* [*Bothriocephalus*] *acheilognathi* in *Schizothorax* spp. and cyprinid spp. from the Kashmir Valley', *Indian Journal of Applied and Pure Biology*, vol 23, no 1, pp67–72

Alexander, J. (2008) 'New tapeworm discovered in Great Lakes fish', *Muskegon Chronicle*, 23 December

Arai, H. P. and Mudry, D. R. (1983) 'Protozoan and metazoan parasites of fishes from the headwaters of the Parsnip and McGregor rivers, British Columbia: A study of possible parasite transfaunations', *Canadian Journal of Fisheries and Aquatic Sciences*, vol 40, pp1676–1684

Archdeacon, T. (2007) 'Effects of Asian tapeworm, mosquitofish, and food ration on Mohave tui chub growth and survival', MS Thesis, University of Arizona, Tucson, Arizona

Associated Press (2004) 'Panel tackles ailing Grand Canyon's woes', Associated Press Online, 8 June, LexisNexis Academic

Baer, J. C. and Fain, A. (1958) '*Bothriocephalus* (Clestobothrium) *kivuensis* n. sp. cestode parasite d'un barbeau du lac Kivu', *Annales de la Societe Royale Zoologique de Belgique*, vol 88, pp287–302

Bauer, O. N. and Hoffman, G. L. (1976) 'Helminth range extension by translocation of fish', in L. A. Page (ed) *Wildlife Diseases*, Plenum Publishing Corp, NY, pp163–172

Bauer, O. N., Musselius, V. A. and Strelkov, Y. A. (1969) *Diseases of Pond Fishes*, 'Kolos', Moscow, in English, Israel Program for Scientific Translations, Jerusalem, 1973

Bean, M. G. (2008) 'Occurrence and impact of the Asian fish tapeworm *Bothriocephalus acheilognathi* in the Rio Grande (Rio Bravo del Norte)', MS Thesis, Texas State University, San Marcos

Bean, M. G. and Bonner, T. H. (2009) 'Impact of *Bothriocephalus acheilognathi* (Cestoda: Pseudophyllidea) on *Cyprinella lutrensis* Condition and Reproduction', *Journal of Freshwater Ecology*, vol 24, no 3, pp383–391

Bean, M. G., Skerikova, A., Bonner, T. H., Scholz, T. and Huffman, D. G. (2007) 'First record of *Bothriocephalus acheilognathi* in the Rio Grande with comparative analysis of ITS2 and V4-18S rRNA gene sequences', *Journal of Aquatic Animal Health*, vol 19, pp71–76

Bertasso, A. and Avenant-Oldewage, A. (2005) 'Aspects of the ecology of the Asian tapeworm, *Bothriocephalus acheilognathi* Yamaguti, 1934 in yellowfish in the Vaal Dam, South Africa', *Onderstepoort Journal of Veterinary Research*, vol 72, no 3, pp207–217

Blakeslee, S. (2002) 'Restoring an ecosystem torn asunder by a dam', *The New York Times*, 11 June

Borucinska, J. D. (2008) 'Diseases caused by Cestoda', in J. C. Eiras, H. Segner, T. Wahli and B. J. Kapoor (eds) *Fish Diseases, Volume 2*, Science Publishers, Enfield, New Hampshire, US, pp977–1024

Brouder, M. J. (1999) 'Relationship between length of roundtail chub and infection intensity of Asian fish tapeworm *Bothriocephalus acheilognathi*', *Journal of Aquatic Animal Health*, vol 11, pp302–304

Brouder, M. J. and Hoffnagle, T. L. (1997) 'Distribution and prevalence of the Asian fish tapeworm, *Bothriocephalus acheilognathi*, in the Colorado River and tributaries, Grand Canyon, Arizona, including two new host records', *Journal of the Helminthological Society of Washington*, vol 64, no 2, pp219–226

Chauhan, R. S. and Malhotra, S. K. (1986) 'Population biology of the pseudophyllidean cestode *Bothriocephalus teleostei* (Malhotra, 1984) in the Indian Hill-Stream teleosts. I. Influence of season and temperature', *Boletín Chileno de Parasitología*, vol 41, nos 3–4, pp51–61

Ching, H. L. and Andersen, K. (1983) 'Description of *Eubothrium tulipai* sp. n. (Pseudophyllidea: Amphicotylidae) from northern squawfish in British Columbia', *Canadian Journal of Zoology*, vol 61, pp981–986

Choudhury, A., Hoffnagle, T. L. and Cole, R. A. (2004) 'Parasites of native and nonnative fishes in the Little Colorado River, Grand Canyon, Arizona', *Journal of Parasitology*, vol 90, pp1042–1052

Choudhury, A., Charipar, E., Nelson, P., Hodgson, J. R., Bonar, S. and Cole, R. A. (2006) 'Update on the distribution of the invasive Asian fish tapeworm, *Bothriocephalus acheilognathi*, in the US and Canada', *Comparative Parasitology*, vol 73, no 2, pp269–273

Chubb, J. C. (1981) 'The Chinese tapeworm *Bothriocephalus acheilognathi* Yamaguti, 1934 (synonym *B. gowkongensis* Yeh, 1955) in Britain', *Proceedings of the 2nd British Freshwater Fisheries Conference*, University of Liverpool, Liverpool, pp40–51

Coggins, L. G., Jr, Pine, III, W. E., Walters, C. J., Van Haverbeke, D. R., Ward, D. and Johnstone, H. C. (2006) 'Abundance trends and status of the Little Colorado River population of humpback chub', *North American Journal of Fisheries Management*, vol 26, pp233–245

Cudmore, B. and Mandrak, N. E. (2004) 'Biological synopsis of grass carp (*Ctenopharyngodon idella*)', *Canadian Manuscript Report of Fisheries and Aquatic Sciences*, vol 2705

Davydov, O. N. (1978) 'Growth development and fecundity of *Bothriocephalus gowkongensis* (Jeh, 1955), a cyprinid parasite', *Hydrobiologicheski Zournal*, vol 14, pp70–77 (in Russian, English translation: *Hydrobiological Journal*, vol 14, pp60–66, 1979)

Dick, T. A. and Choudhury, A. (1995) 'Cestoidea (Phylum Platyhelminthes)', in P. T. K. Woo (ed) *Fish Diseases and Disorders, Vol 1, Protozoan and Metazoan Infections*, CAB International, Wallingford, UK, pp391–414

Dove, A. D. M. and Fletcher, A. S. (2000) 'The distribution of the introduced tapeworm *Bothriocephalus acheilognathi* in Australian freshwater fishes', *Journal of Helminthology*, vol 74, pp121–127

Dubinina, M. N. (1982) 'On the synonymy of species in the genus *Bothriocephalus* (Cestoda: Bothriocephalidae), parasites of Cyprinidae of the USSR', *Parazitologiya*, vol 16, pp41–45

Dubinina, M. N. (1987) 'Class Cestoda Rudolphi, 1808', in O. N. Bauer (ed) *Key to the Parasites of Freshwater Fish of the USSR*, 2nd edition, vol 3, Nauka, Leningrad, pp5–76

Elliot, W. (2009) 'Tapeworms invade the Great Lakes; Agencies warn anglers of walleyes from Lake Huron', *Buffalo News*, 4 January

FAO (Food and Agriculture Organization of the United Nations) (1996) 'CIFA Technical Paper 31', FAO, Rome, www.fao.org/docrep/008/v9551e/v9551e15.htm

FAO (2003–2010) 'National aquaculture sector overview. Egypt. National aquaculture sector overview fact sheets', text by Salem, A. M. and Saleh, M. A., in *FAO Fisheries and Aquaculture Department*, FAO, Rome, (updated 16 November 2010), www.fao.org/fishery/countrysector/naso_egypt/en

Fausch, K. D. and Nakano, S. (1998) 'Research on fish ecology in Japan: A brief history and selected review', *Environmental Biology of Fishes*, vol 52, pp75–95

Federal Information and News Dispatch Inc. (2009) 'USGS news release: Endangered humpback chub population increases 50 percent from 2001 to 2008', Federal Information and News Dispatch Inc, 27 April 2009

Fuller, P. L., Nico, L. G. and Williams, J. D. (1999) 'Nonindigenous fishes introduced into the inland waters of the US', *American Fisheries Society Special Publication 27*, American Fisheries Society, Bethesda, US

García-Prieto, L. and Osorio Sarabia, D. (1991) 'Distribucion actual de *Bothriocephalus acheilognathi* en México', *Anales del Instituto de Biología Universidad Nacional Autónoma de México, Serie Zoologia*, vol 62, no 3, pp523–526

Giovinazzo, G., Antegiovanni, P., Dörr, A. J. M. and Elia, A. C. (2006) 'Presenza di *Bothriocephalus acheilognathi* (Cestoda: Pseudophyllidea) in *Atherina boyeri* del Lago Trasimeno. (Presence of *Bothriocephalus acheilognathi* (Cestoda: Pseudophyllidea) in Atherina boyeri of Lake Trasimeno)', *Ittliopatologia*, vol 3, pp61–67

Granath Jr, W. O. and Esch, G. W. (1983a) 'Temperature and other factors that regulate the composition and intrapopulation densities of *Bothriocephalus acheilognathi* (Cestoda) in *Gambusia affinis* (Pisces)', *Journal of Parasitology*, vol 69, pp1116–1124

Granath Jr, W. O. and Esch, G. W. (1983b) 'Seasonal dynamics of *Bothriocephalus acheilognathi* in ambient and thermally altered areas of a North Carolina cooling reservoir', *Proceedings of the Helminthologhical Society of Washington*, vol 50, no 2, pp205–218

Granath Jr, W. O. and Esch, G. W. (1983c) 'Survivorship and parasite-induced host mortality among mosquito fish in a predator-free, North Carolina cooling reservoir', *American Midland Naturalist*, vol 110, pp314–323

Grand Rapid Press (2008) 'State issues sushi warning; Asian fish tapeworms found in Great Lakes; cook fish thoroughly', *Grand Rapid Press*, Michigan, 28 December

Han, J. E., Shin, S. P., Kim, J. H., Choresca, C. H., Jr, Jun, J. W., Gomez, D. K. and Park, S. C. (2010) 'Mortality of cultured koi *Cyprinus carpio* in Korea caused by *Bothriocephalus acheilognathi*', *African Journal of Microbiology Research*, vol 4, no 7, pp543–546

Hansen, S. P., Choudhury, A., Heisey, D. M., Ahumada, J. A., Hoffnagle, T. L. and Cole, R. A. (2006) 'Experimental infection of the endangered bonytail chub (*Gila elegans*) with the Asian fish tapeworm (*Bothriocephalus acheilognathi*): Impacts on survival, growth and condition', *Canadian Journal of Zoology*, vol 84, pp1383–1394

Hansen, S. P., Choudhury, A. and Cole, R. A. (2007) 'Evidence of experimental postcyclic transmission of *Bothriocephalus acheilognathi* in bonytail chub (*Gila elegans*)', *Journal of Parasitology*, vol 93, no 1, pp202–204

Hanzelova, V. and Žitňan, R. (1986) 'Embryogenesis and development of *Bothriocephalus acheilognathi* Yamaguti, 1934 (Cestoda) in the intermediate host under experimental conditions', *Helminthologia*, vol 23, pp145–155

Herborg, L.-M., Mandrak, N. E., Cudmore, B. C. and MacIsaac, H. J. (2007) 'Comparative distribution and invasion risk of snakehead (Channidae) and Asian carp (Cyprinidae) species in North America', *Canadian Journal of Fisheries and Aquatic Science*, vol 64, pp1723–1735

Hoffman, G. L. (1980) 'Asian tapeworm, *Bothriocephalus acheilognathi* Yamaguti, 1934, in North America', *Fisch und Umwelt, Special Volume in Honor of Prof. Dr H.-H. Reichenbach-Klinke's 65th Birthday*, vol 8, pp69–75

Hoffman, G. L. (1999) *Parasites of North American Freshwater Fishes*, 2nd edition, Cornell University Press, Ithaca

Hoffnagle, T. L., Choudhury, A. and Cole, R. A. (2006) 'Parasitism and body condition in humpback chub from the Colorado and Little Colorado Rivers, Grand Canyon, Arizona', *Journal of Aquatic Animal Health*, vol 18, pp184–193

Hoole, D. and Nisan, H. (1994) 'Ultrastructural studies on intestinal response of carp, *Cyprinus carpio* L., to the pseudophyllidean tapeworm, *Bothriocephalus acheilognathi* Yamaguti, 1934', *Journal of Fish Diseases*, vol 17, pp623–629

HT Media (2006) 'US Fed News. Endangered humpback chub population in Grand Canyon stabilizing', HT Media, 3 August

Kalamazoo Gazette (2008) 'Love sushi? Avoid fish caught in the Great Lakes', *Kalamazoo Gazette*, 25 December

Kline, S. J., Archdeacon, T. P. and Bonar, S. A. (2009) 'Effects of praziquantel on eggs of the Asian tapeworm *Bothriocephalus acheilognathi*', *North American Journal of Aquaculture*, vol 71, no 4, pp380–383

Koehle, J. J. and Adelman, I. R. (2007) 'The effects of temperature, dissolved oxygen, and Asian tapeworm infection on growth and survival of the Topeka shiner', *Transactions of the American Fisheries Society*, vol 136, no 6, pp1607–1613

Koehn, J. D. (2004) 'Carp (*Cyprinus carpio*) as a powerful invader in Australian waterways', *Freshwater Biology*, vol 49, no 7, pp882–294

Korting, W. (1975) 'Larval development of *Bothriocephalus* sp. (Cestoda: Pseudophyllidea) from carp (*Cyprinus carpio* L.) in Germany', *Journal of Fish Biology*, vol 7, pp727–733

Liao, H. and Shih, L. (1956) 'On the biology and control of *Bothriocephalus gowkongensis* Yeh, A tapeworm parasitic in the young grass carp (*Ctenpharyngodon idellus* C. & V.)', *Acta Hydrobiologica Sinica*, vol 2, pp129–185

Liao, X.-H. and Lun, Z. (1998) 'Evolutionary relationship among *Bothriocephalus* parasitized in grass carp *Ctenopharhyngodon idellus* (C. et V.), common carp *Cyprinus carpio* L. and Ma Kou Yu *Opsariichthys bidens* Günther in China', *Chinese Science Bulletin*, vol 43, no 13, pp1115–1119

Luo, H. Y., Nie, P., Zhang, Y. A., Wang, G. T. and Yao, W. J. (2002) 'Molecular variation of *Bothriocephalus acheilognathi* Yamaguti, 1934 (Cestoda: Pseudophyllidea) in different fish host species based on ITS rDNA sequences', *Systematic Parasitology*, vol 52, no 3, pp159–166

Luo, H. Y., Nie, P., Zhang, Y. A., Yao, W. J. and Wang, G. T. (2003) 'Genetic differentiation in populations of the cestode *Bothriocephalus acheilognathi* (Cestoda, Pseudophyllidea) as revealed by eight microsatellite markers', *Parasitology*, vol 126, no 5, pp493–501

Luo, H. Y., Nie, P., Chang, M. X., Song, Y. and Yao, W. (2004) 'Characterization of development-related genes for the cestode *Bothriocephalus acheilognathi*', *Parasitology Research*, vol 94, no 4, pp265–274

Malhotra, S. K. (1984) 'Cestode fauna of hill-stream fishes in Garhwal Himalayas, India. II. *Bothriocephalus teleostei* n. sp. from *Barilius bola* and *Schizothorax richardsonii*', *Boletín Chileno de Parasitología*, vol 39, nos 1&2, pp6–9

Marcogliese, D. J. (2008) 'First report of the Asian fish tapeworm in the Great Lakes', *Journal of Great Lakes Research*, vol 34, pp566–569

Marcogliese, D. J. and Esch, G. W. (1989) 'Experimental and natural infection of planktonic and benthic copepods by the Asian tapeworm, *Bothriocephalus acheilognathi*', *Proceedings of the Helminthological Society of Washington*, vol 56, no 2, pp151–155

Méndez, O., Salgado-Maldonado, G., Caspeta-Mandujano, J. M. and Cabañas-Carranza, G. (2010) 'Helminth parasites of some freshwater fishes from Baja California Sur, Mexico', *Zootaxa*, vol 2327, pp44–50

Merck (2008) 'The Merck Veterinary Manual', (online), Merck & Co. Inc., Whitehouse Station, New Jersey, www.merckvetmanual.com/mvm/index.jsp?cfile=htm/bc/170404.htm

Minckley, C. O. (1996) 'Observations on the biology of the humpback chub in the Colorado River basin, 1908–1990', PhD Thesis, Northern Arizona University, Flagstaff, US

Mitchell, A. and Darwish, A. (2009) 'Efficacy of 6-, 12-, and 24-h praziquantel bath treatments against Asian tapeworms *Bothriocephalus acheilognathi* in grass carp', *North American Journal of Aquaculture*, vol 71, no 1, pp30–34

Mitchell, A. J. and Hobbs, M. S. (2007) 'The acute toxicity of praziquantel to grass carp and golden shiners', *North American Journal of Aquaculture*, vol 69, no 3, pp203–206

Molnar, K. (1977) 'On the synonyms of *Bothriocephalus acheilognathi* Yamaguti, 1934', *Parasitologica Hungarica*, vol 10, pp61–62

Molnar, K. and Murai, E. (1973) 'Morphological studies on *Bothriocephalus gowkongensis* Yeh, 1955 and *B. phoxini* Molnar, 1968 (Cestoda, Pseudophyllidea)', *Parasitologica Hungarica*, vol 6, pp99–110

Nico, L. G., Fuller, P. L. and Schofield, P. J. (2010) '*Ctenopharyngodon idella*', USGS Nonindigenous Aquatic Species Database, Gainesville, FL, http://nas.er.usgs.gov/queries/FactSheet.aspx?speciesID=514

Nie, P., Wang, G. T., Yao, W. J., Zhang, Y. A. and Gao, Q. (2000) 'Occurrence of *Bothriocephalus acheilognathi* in cyprinid fish from three lakes in the flood plain of the Yangtze River, China', *Diseases of Aquatic Organisms*, vol 41, no 1, pp81–82

O'Driscoll, P. (2005) 'River flush washes up new data', *US Today*, 27 October

Oškinis, V. (1994) 'Temperature and development of *Bothriocephalus acheilognathi* Yamaguti, 1934 (Cestoda) in laboratory conditions', *Ekologia*, vol 4, pp40–42

Paperna, I. (1996) 'Parasites, infections and diseases of fishes in Africa, an update', CIFA Technical Paper No 31, FAO, Rome

Pérez-Ponce de Leon, G., Rosas-Valdez, R., Mendoza-Garfias, B., Aguilar-Aguilar, R., Falcón-Ordaz, J., Garrido-Olvera, L. and Pérez-Rodriguez, R. (2009) 'Survey of the endohelminth parasites of freshwater fishes in the upper Mezquital River Basin, Durango State, Mexico', *Zootaxa*, vol 2164, pp1–20

Pool, D. W. (1987) 'A note on the synonomy of *Bothriocephalus acheilognathi* Yamaguti, 1934, *B. aegyptiacus* Rysavy and Moravec, 1975 and *B. kivuensis* Baer and Fain, 1958', *Parasitology Research*, vol 73, pp146–150

Pool, D. W. and Chubb, J. C (1985) 'A critical scanning electron microscope study of the scolex of *Bothriocephalus acheilognathi* Yamaguti, 1934, with a review of the taxonomic history of the genus *Bothriocephalus* parasitizing cyprinid fishes', *Systematic Parasitology*, vol 7, pp199–211

Prigli, M. (1975) 'Über die Rolle der Wasservögel bei der Verbreitung des *Bothriocephalus gowkongensis* Yeh, 1955 (Cestoda)', *Parasitologia Hungarica*, vol 8, pp61–62

Retief, N.-R., Avenant-Oldewage, A. and du Preez, H. H. (2007) 'Ecological aspects of the occurrence of Asian tapeworm, *Bothriocephalus acheilognathi* Yamaguti, 1934 infection in the largemouth yellowfish, *Labeobarbus kimberleyensis* (Gilchrist and Thompson, 1913) in the Vaal Dam, South Africa', *Physics and Chemistry of the Earth*, vol 32, pp1384–1390

Riggs, M. R. and Esch, G. W (1987) 'The suprapopulation dynamics of *Bothriocephalus acheilognathi* in a North Carolina reservoir: Abundance, dispersion, and prevalence', *Journal of Parasitology*, vol 73, no 5, pp877–892

Rojas-Sánchez, A. and García-Prieto, L. (2008) 'Distribución actual del céstodo *Bothriocephalus acheilognathi* en México', *Memorias XXV Simposio sobre Fauna Silvestre*, Universidad Nacional Autónoma de México, México, pp89–93

Rysavy, B. and Moravec, F. (1973) '*Bothriocephalus aegypticus* sp. n. (Cestoda: Pseudophyllidea) from *Barbus bynni*, and its life cycle', *Vestnik Ceskoslovenske Spolecnosti Zoologicke*, vol 39, pp68–75

Salgado-Maldonado, G. and Pineda-López, R. (2003) 'The Asian fish tapeworm *Bothriocephalus acheilognathi*: A potential threat to native freshwater fish species in Mexico', *Biological Invasions*, vol 5, pp261–268

Scholz, T. (1997) 'A revision of the species of *Bothriocephalus* Rudolphi, 1808 (Cestoda: Pseudophyllidea) parasitic in American freshwater fishes', *Systematic Parasitology*, vol 36, pp85–107

The Slovak Spectator (2007) 'Around Slovakia', *The Slovak Spectator*, 14 April

Stone, D. M. and Gorman, O. T. (2006) 'Ontogenesis of endangered humpback chub (*Gila cypha*) in the Little Colorado River, Arizona', *American Midland Naturalist*, vol 155, pp123–135

Stone, D. M., Van Haverbeke, D. R., Ward, D. L. and Hunt, T. A. (2007) 'Dispersal of nonnative fishes and parasites in the intermittent Little Colorado River, Arizona', *Southwestern Naturalist*, vol 52, no 1, pp130–137

Tekin-Özan, S., Kir, I. and Barlas, M. (2008) 'Helminth parasites of common carp (*Cyprinus carpio* L., 1758) in Beyşehir Lake and population dynamics related to month and host size', *Turkish Journal of Fisheries and Aquatic Sciences*, vol 8, pp201–205

Urazbaev, A. N. and Kurbanova, A. I. (2006) 'Parasitofauna of fish of the far east complex established in reservoirs of the southern Aral Sea', *Vestnik Zoologii*, vol 40, no 6, pp535–540

USGS (United States Geological Survey) (2009) '*Bothriocephalus acheilognathi*', http://nas.er.usgs.gov/queries/FactSheet. aspx?speciesID=2798

Vincent, A. G. and Font, W. F. (2003) 'Host specificity and population structure of two exotic helminths, *Camallanus cotti* (Nematoda) and *Bothriocephalus acheilognathi* (Cestoda), parasitizing exotic fishes in Waianu Stream, O'ahu, Hawai'i', *Journal of Parasitology*, vol 89, no 3, pp540–544

Warburton, M., Kuperman, B., Matey, V. and Fisher, R. (2002) 'Parasite analysis of native and non-native fish in the Angeles National Forest', 2001 Final Report, Prepared for US Forest Service, Angeles National Forest, US Geological Survey, Western Ecological Research Center, San Diego, CA

Ward, D. L. (2005a) 'Collection of Asian tapeworm (*Bothriocephalus acheilognathi*) from the Yampa river, Colorado', *Western North American Naturalist*, vol 65, no 3, pp403–404

Ward, D. L. (2005b) 'Removal and quantification of Asian tapeworm from humpback chub using praziquantel', archived presentation (pdf), Technical Work Group Public Meeting, 21–22 June 2005, Bureau of Indian Affairs, Phoenix, Arizona, www.usbr.gov/uc/rm/amp/twg/mtgs/05jun21/Attach_05.pdf

Ward, D. L. (2007) 'Removal and quantification of Asian tapeworm from humpback chub using praziquantel', *North American Journal of Aquaculture*, vol 69, no 3, pp207–210

Ward, D. L. and Persons, W. (2006) 'Little Colorado River fish monitoring: 2005 annual report', Arizona Game and Fish Department, Research Branch, Phoenix, AZ

Yamaguti, S. (1934) 'Studies on the helminth fauna of Japan. Part 4. Cestodes of Fishes', *Japanese Journal of Zoology*, vol 6, pp1–112

Yamaguti, S (1952) 'Studies on the helminth fauna of Japan. Part 49', *Acta Medica Okayama*, vol 8, no 1, pp1–98

Yeh, I. S. (1955) 'On a tapeworm *Bothriocephalus gowkongensis* n. sp. (Cestoda: Bothriocephalidae) from freshwater fish in China', *Acta Zoologica Sinica*, vol 7, pp69–74 (in Chinese, English summary)

Yuma, M., Hosoya, K. and Nagata, Y. (1998) 'Distribution of the freshwater fishes of Japan: An historical overview', *Environmental Biology of Fishes*, vol 52, pp97–124

33

Centrocestus formosanus Nishigori (Asian gill-trematode)

Lori R. Tolley-Jordan and Michael A. Chadwick

Introduction

The transport of species outside their native ranges to new locations is one of the greatest threats to global biodiversity (Clavero and García-Berthou, 2005). The present movement of invasive species into new habitats is unprecedented when compared to other geologic times (Ricciardi, 2007) and is expected to continue as a direct consequence of global trade (Soule, 1990). The impact of invasive species in novel communities, and success of invader establishment, can be dictated not only by the invader's life history traits, but also the parasites carried by the invader (Prenter et al, 2004). Invasive parasites can infect a wide variety of naïve hosts with intensive pathogenic effects that reduce fitness and cause mortality. Thus, by causing declines in native populations, these exotic parasites can enhance the ability of their invasive hosts to become established (Prenter et al, 2004).

Biodiversity losses due to invasive species are particularly exacerbated in freshwater systems that comprise around 1 per cent of the Earth's surface, but contain about 10 per cent of global biodiversity (Strayer and Dudgeon, 2010). Molluscs constitute a significant number of freshwater invasive species (Strayer, 2010) and many, particularly snails, are hosts to a variety of parasites, particularly digeneans (Trematoda: Digenea; Esch and Fernandez, 1994).

There are about 40,000 species of these digeneans (Dillon, 2000), endoparasites that require an intermediate host (a snail) for asexual larval development and a definitive host (a vertebrate) for sexual adult development. Many of these have secondary intermediate hosts, generally fishes (Hoffman, 1999), that serve as an additional host of larval metamorphosis. The ability of digeneans to use a wide variety of fish hosts, including novel fish hosts, is of particular interest to invasive species biologists, as invasive parasites can be more pathogenic to native fishes than some endemic fish diseases (Font, 2003). As fishes in novel habitats lack co-evolved defence mechanisms to minimize densities of exotic parasites in tissues (Taraschewski, 2006), these high parasite loads can result in decreased fitness or even mortality in the infected fishes (Poulin, 2006).

The digenean *Centrocestus formosanus* has harmful and sometimes lethal effects on fishes (Blazer and Gratzek, 1985; Hoffmann, 1999; Mitchell et al, 2005). Originally from southeast Asia (Scholz and Salgado-Maldanado, 2000), the species now occurs in freshwater systems on all continents except Africa and Antarctica. This chapter outlines the *C. formosanus* lifecycle, pathogenic effects on its first intermediate, second intermediate and definitive hosts, parasite zoogeography (based on published records), and efficacy for control and eradication.

Lifecycle of *Centrocestus formosanus*

Centrocestus formosanus is a digenetic trematode (Digenea: Heterophyidae) with a complex lifecycle involving snails, fishes, birds and mammals as hosts (Chen, 1948) (Figure 33.1). The digeneans often are hosted by a single species of snail (Wright, 1973; Sapp and Loker, 2000a, 2000b; Cribb et al, 2001), but there is little host specificity among their vertebrate hosts (Cribb et al, 2001). *Centrocestus formosanus* requires all hosts to live in perennial aquatic systems in order to complete the lifecycle and obtain successful establishment (Esch and Fernandez, 1994). The obligate aquatic nature of these parasites is reflected in the particular snail, fish, bird and mammal species used as hosts.

The lifecycle (Figure 33.1) begins when *C. formosanus* eggs, associated with a definitive host's

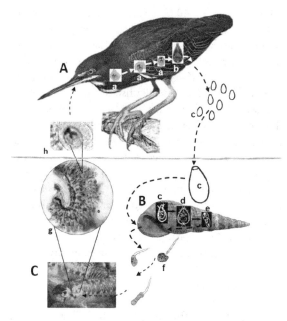

Note: (a) Metacercaria from fish eaten by bird; (b) eggs released from intestinal tract of definitive host; (c) miracidium; (d) germinal cells from miracidium develop into redia; (e) redia; (f) shed cercaria; (g) encysted metacercaria; (h) metacercaria exhibiting cartilaginous encapsulation

Source: Figure reproduced with permission from Mitchell et al (2005)

Figure 33.1 *Complex lifecycle of the gill trematode* Centrocestus formosanus *involving (A) a definitive bird (or mammal) host, (B) the first intermediate snail host and (C) an intermediate fish host*

faeces, are deposited into an aquatic habitat. Within 24 hours, the first larval stage, miricidiae, hatch and infect a snail, the first intermediate host. Miricidae probably enter snails through mantle tissue (Martin, 1958) and develop into rediae. Rediae of *C. formosanus* migrate and develop in digestive and gonadal tissue of their snail host. These rediae produce large numbers of cercariae that infect the next host (Martin, 1958). One snail host, *Melanoides tuberculata*, has been shown to shed on average 1600 but up to 63,400 cercariae d^{-1} (Lo and Lee, 1996). The free swimming cercariae shed into the water column attach to the gills of a fish (Chen, 1948) and within one hour of exposure (Blazer and Gratzek, 1985), the cercariae burrow into gill tissue and morph into metacercariae that will infect the definitive host. Infections usually range from 1 (light infection) to 200 (heavy infection) metacercariae per gill arch or a total of less than 10 to over 1000 metacercariae per fish (Blazer and Gratzek, 1985; Madhavi, 1986; Mitchell et al, 2000). Metacercariae reach infective maturity in 18–20 days (Chen, 1948) and are transmitted to the definitive hosts (birds and/or mammals) via ingestion of an infected fish. As fishes heavily infected with *C. formosanus* have slowed swimming they are believed to be more easily predated upon by birds (Balasuriya, 1988). The final transformation of the parasite into a hermaphroditic, sexually mature, adult worm occurs within 72 hours in the digestive tract of the vertebrate definitive host. Sexual reproduction occurs within the definitive host and fertilized eggs are shed in the faeces, starting about 11 days post-infection (Chen, 1942).

Pathogenic Effects of *Centrocestus formosanus*

First intermediate hosts: Pathology to snails

The reallocation of host resources toward parasite development has been shown to have severe consequences on the reproductive potential and fitness of snail hosts (Esch and Fernandez, 1994). *Centrocestus formosanus* has been shown to sterilize *Melanoides tuberculata* (Ben-Ami and Heller, 2005) via the complete destruction of gonadal tissue of the infected snail. The gonads of infected snails are filled with cercaria producing rediae instead of ova (Berry and

Kadri, 1973; Tolley-Jordan, pers. obs.). Similarly, brood pouches in the mantle of infected snails have been found to contain no embryos (Ben-Ami and Heller, 2005). Although no studies have demonstrated lethal effects of infection by *C. formosanus*, the reduction in fecundity of snails in areas of high parasite prevalence may induce changes in snail population structure.

Second intermediate hosts: Pathology to fishes

Centrocestus formosanus is unique among digeneans in that it can have severe effects on fish fitness and induce mortality through deformation of gill tissues (Blazer and Gratzek, 1985; Vélez-Hernández et al, 1998; Mitchell et al, 2000, 2005) (Figure 33.2). Infected fishes exhibit an inflammatory response to the presence of metacercariae on gills by forming cartilaginous encasements around each metacercariae (Blazer and Gratzek, 1985). These inflammations alter or destroy gill arches (Mitchell et al, 2000), reducing gas exchange (Blazer and Gratzek, 1985; Alcaraz et al, 1999) and inducing behavioural

changes in fishes such as lethargy and gasping (Balasuriya, 1988). These infected fishes incur a greater risk for predation by birds than non-infected fishes (Balasuriya, 1988). As the inflammatory response can kill the metacercariae shortly after infection (Mitchell et al, 2000), success of *C. formosanus* transmission in these definitive hosts requires relatively constant, low numbers of encystment through time. Thus, as a fish ages it accumulates more encapsulated cysts until a threshold is reached that causes death (Balasuriya, 1988; Mitchell et al, 2000). However, adults of large-bodied fishes (e.g. greater than 15cm in length) tend to have low levels of cercarial infections, which suggests susceptibility to infection occurs only in juveniles (Balasuriya, 1988).

Definitive hosts: Pathology to birds and mammals

No descriptions of pathology of this parasite to definitive hosts have been reported suggesting that the presence of digeneans has benign effects on their definitive hosts.

Note: Countries where *C. formosanus* occurs (in snails or birds or fishes) within the parasite's native range are shown in medium grey, and countries with invasive populations of *C. formosanus* are shown in black. Countries with no occurrence of *M. tuberculata* are shown in white, countries with native populations of *M. tuberculata* (Africa, Middle East and South Asia) are shown in light grey, and countries with invasive populations of *M. tuberculata* are shown in dark grey.

Source: Lori R. Tolley-Jordan and Michael A. Chadwick

Figure 33.2 *The global distribution, by country, of* Centrocestus formosanus *and its first intermediate snail host,* Melanoides tuberculata

Zoogeography of *Centrocestus formosanus*

First intermediate snail hosts

The native range for the parasite (within all hosts) is southeast Asia (Scholz and Salgado-Maldonado, 2000). *Centrocestus formosanus* was originally described from Formosa (Taiwan) by Nishigori (1924) in natural infections of an operculate snail, *Melanoides tuberculata*. *Melanoides tuberculata*, the most common first intermediate host used by the parasite (Yanohara, 1987), is considered native to Asia and Africa (Brown, 1980) with type locality from Coramandel (southeast coast of India). Although both *M. tuberculata* and *M. tuberculatus* appear in the literature, the proper use of the species name is the feminine form based on the International Code of Zoological Nomenclature (ICZN, 1999; Article 30.1.4.4), so the correct version is *M. tuberculata* (Cowie, R., pers. comm.).

Although *C. formosanus* is reported from three other thiarid snails of tropical origin: *Stenomelania newcombi* from Oahu, Hawaii (Martin, 1958), *S. libertina* from Korea (Cho et al, 1983) and Japan (Kagei and Yanohara, 1995) and *Thiara scabra* in Japan (Kagei and Yanohara, 1995), these snail species are not reported as hosts elsewhere. It is likely that *S. newcombi* from Hawaii was incorrectly identified and specimens are actually *M. tuberculata*, as *S. newcombi* has not been reported in other gastropod surveys and historical records on the island (Cowie, 1997, 1998) and there is no valid species of *Stenomelania* known as *S. newcombi*. Despite the fact that *C. formosanus* now occur in freshwater habitats throughout the world (Figure 33.2), we found no published records of parasite infection in any additional snail species confirming the strong evolutionary host specificity of *C. formosanus* for *M. tuberculata*.

Native range of *M. tuberculata*

Brown (1980) lists the range of *M. tuberculata* to be widespread in southeast Asia and Africa, though it is uncommon in western Africa and missing from the Zaire basin. In general, published accounts reported in this review supported this distribution (Table 33.1). However, phylogenetic analyses have shown differences between *M. tuberculata* of Asian versus African origin, and that *M. tuberculata* found in the Congo Basin

(West Africa) are invaders from Asia (Facon et al, 2003) as historical collections do not show *M. tuberculata* present in that region (McCullough, 1964; Brown, 1980). Due to the extreme variability in shell morphology of *M. tuberculata*, invasive lineages within the native range of the snail are often undetected without the aid of molecular analyses (Genner et al, 2004). Thus, populations of *M. tuberculata* can be invasive at regional scales within the globally recognized native range of the snail (Figure 33.2).

Invasive range of *M. tuberculata*

Melanoides tuberculata tolerates a wide range of temperatures (Mitchell and Brandt, 2005), salinities (Roessler et al, 1977), pH (Duggan et al, 2007), and can survive after several days of desiccation (Mitchell and Brandt, 2005). Further, individuals reproduce primarily through parthenogenesis, have long lifespans and slow intrinsic growth rates that yield high population densities in tropical streams (Pointier et al, 1993b) and warm water springs (in subtropical and temperate regions; Rader et al, 2003; Duggan et al, 2007; Tolley-Jordan and Owen, 2008). In combination, these traits have allowed *M. tuberculata* to invade a wide variety of freshwater systems throughout the world, potentially at the expense of native snails. For example, extirpations of pulmonate snails following the introduction *M. tuberculata* have been observed (Pointier, 1993b; Pointier et al, 1994) and declines of native prosobranchs are attributed to the invasion of this snail (Murray, 1971).

Populations of the snail are found on all continents (except Antartica), and distributions are not limited to tropical systems. These snails have also successfully invaded numerous countries in temperate regions (above latitudes of 40°N and 40°S). Fifty-four per cent of the global distribution (by country) of *M. tuberculata* is exotic (Table 33.1). Many of the introductions were probably unintentional and directly linked to the transport of aquatic plants for the aquarium (and ornamental pond) trade (Madsen and Frandsen, 1989). In temperate zones, all *M. tuberculata* populations are found in warm water springs (or artificially warmed ponds) as permanent populations are not supported in water temperatures less than 18°C (Mitchell and Brandt, 2005). The risk for establishment of *M. tuberculata* in other warm water spring systems is high as the transport of tropical aquatic plants and fishes into Europe and

Table 33.1 *The distribution of the first intermediate* (Melanoides tuberculata), *second intermediate (fishes) and definitive hosts (birds) of the gill trematode* Centrocestus formosanus *(CF)*

Continent	Country	First	Second	Definitive	Source
North America	Mexico	I-1973, +RE	+RE	BS	Abbott, 1973; Arizmendi, 1992; Amaya-Huerta and Almeyda-Artigas, 1994; Contreras-Arquieta, 1998; Vélez-Hernández et al, 1998; Scholz and Salgado-Maldonado, 2000
	US	I-1969, +RE	+RE	BV, AA	Murray, 1964, 1971; Dundee and Paine, 1977; Blazer and Gratzek, 1985; Vogelbein and Overstreet, 1988; Wu, 1989; Olson and Pierce, 1997; McDermott, 2000; Flemming, 2002; Anderson, 2004; Mitchell et al, 2005
Central America	Columbia	I-1969, +RE	+RE		Escobar et al, 2009
Caribbean	Cuba	I-1983			Perera et al, 1987
	Dominican Republic	I-1997			Giovanelli et al, 2001
	Guadeloupe	I-1979			Pointier et al, 1993b
	Martinique	I-1979			Pointier et al, 1993a
	St Lucia	I-1978			Pointier and Augustin, 1999
South America	Argentina	I			Gregoric et al, 2006
	Brazil	I-1984			DeMarco, 1999; Guimaraes et al, 2001; Giovanelli et al, 2005
	Peru	I-1990			Vivar et al, 1990
	Venezuela	I-1972			Pointier et al, 1994
Western Europe	Austria	I-1960			Horsák et al, 2004
	France	I			Glaubrecht et al, 2009
	Germany	I-1962			Gasull, 1974
	Italy	I-1984	+RE		Manfrin et al, 2002; Cianfanelli et al, 2007; Gherardi et al, 2007
	Netherlands	I			Cianfanelli et al, 2007
	Spain	I-1974			Gasull, 1974
	Sweden	I			Discovery Database, 2010
Eastern Europe	Czech Republic	I-1970s			Horsák et al, 2004
	Croatia		+RE		Gjurčevič et al, 2007
	Malta	I-1862			Gasull, 1974
	Slovakia				Horsák et al, 2004
	Ukraine	I-2000	+RE		Alexandrov et al, 2007
Western Asia	Afghanistan	N			Discovery Database, 2010
	Iran	N, + RE			Farahnak et al, 2004
	Iraq	N			Abdel-Azim and Gismann, 1956; Lazim et al, 1988
	India	N, + RE	+	EG	Rekharani and Madhavi, 1985; Gogoi and Sarma, 1986; Madhavi, 1986

Table 33.1 *The distribution of the first intermediate* (Melanoides tuberculata), *second intermediate (fishes) and definitive hosts (birds) of the gill trematode* Centrocestus formosanus *(CF)* (Cont'd)

Continent	Country	First	Second	Definitive	Source
	Israel	N, + RE	+RE		Livshits and Fishelson, 1983; Heller and Farstay,1990; Dzikowski et al, 2004
	Jordan	N			Abdel-Azim and Gismann, 1956
	Oman	N			Samadi et al, 1999
	Saudi Arabia	N, + RE	+RE		Ismail, 1990; Abdel-Azim and Gismann, 1956
	Syria	N			Abdel-Azim and Gismann, 1956
	Turkey	N	+RE		Yildrim, 1999
	United Arab Emirates	N, + RE	+RE		Ismail and Arif, 1991, 1993
	Yemen	N			Al-safadi, 1991
South Asia	Pakistan	N			Nazneen and Begum, 1994
	Sri Lanka	N, + RE	+		Balasuriya, 1988; Thilakaratne et al, 2003
Southeast Asia	China	N, +	+		Dudgeon and Yipp, 1983; Zeng and Liao, 2000; Sohn et al, 2009
	Hong Kong	N, +			Chen, 1942, 1948; Dudgeon,1986
	Japan	N, +	+	NN, LC	Yanohara et al, 1987; Kagei and Yanohara, 1995
	Kinmen Islands	N, +			Chao et al, 1993
	Korea				Cho et al, 1983
	Laos	N	+		Han et al, 2008
	Taiwan	N, +	+	NN	Nishigori, 1924
	Thailand	N, +	+		Sukontason et al, 1999; Dechruksa et al, 2007
	Viet Nam	N, +	+		Thien et al, 2007; Chi et al, 2008
Africa	Benin	N			Ibikounlé et al, 2009
	Burundi	N			Gryseels et al, 1987
	Cameroon	N			Ngonseu et al, 1992
	Comoros Islands	N			Genner et al, 2004
	Cote d'Ivoire	N			Samadi et al, 1999
	Dem. Rep. Congo	N			Samadi et al, 1999
	Egypt	N			Martin, 1959; Yaseen, 1995
	Ethiopia	N			Kloos, 1985
	Gabon	N			de Clercq, 1987
	Ghana	N			Blay and Dongdem, 1996
	Kenya	N			Muli and Mavuti, 2001
	Libya	N			Sandford, 1936

Table 33.1 *The distribution of the first intermediate* (Melanoides tuberculata), *second intermediate (fishes) and definitive hosts (birds) of the gill trematode* Centrocestus formosanus *(CF)* (Cont'd)

Continent	Country	First	Second	Definitive	Source
	Malawi	N			Genner et al, 2004
	Morocco	N			Laamrani et al, 1997
	Mozambique	N			Genner et al, 2004
	Namibia	N			Brown et al, 1992
	Nigeria	N			Ndifon and Ukoli, 1989
	Rwanda	N			Gryseels et al, 1987
	Somalia	N			Genner et al, 2004
	South Africa	N			de Kock and Wolmarans, 2009
	Sudan	N			Madsen et al, 1988
	Tanzania	N			Loker et al, 1981
	Tunisia	N			Discovery Database, 2010
	Uganda	N			Darlington, 1977
	Zambia	N			Hira, 1970
	Zimbabwe	N			Mukaratirwa et al, 2005
Indo-Pacific	Borneo	N			Supian and Ikhwanuddin, 2002
	Indonesia	N			Carney et al, 1973
	Java	N			Thornton and New, 1988
	Malaysia	N			Berry and Kadri, 1973
	Philippines	N	+		Martin, 1958; Ito, 1977
	Singapore	N	+		Genner et al, 2004
	Sumatra	N			Thornton and New, 1988
Oceania	Australia	I	+		Evans and Lester, 2001; Glaubrecht et al, 2009
	Fiji	I			Haynes, 2000
	French Polynesia	I-1980			Pointier and Marquet, 1990
	Guam	I			Smith, 2003
	Hawaii (US)	I	+		Martin, 1958; Cowie, 1997
	New Caledonia	I			Haynes, 2000
	New Guinea	I			Ball and Glucksman, 1978
	New Zealand	I-2001			Duggan, 2002
	Solomon Islands	I			Olivier, 1947
	Vanuatu Islands	I			Haynes, 2000

Note: The first intermediate host is listed as either Native (N) or Introduced (I) with date of introduction given (when available). + denotes presence of snails and fishes infected with CF and RE is range extension of the parasite. Known species of definitive bird hosts infected with CF are *Ardea alba*, (AA), *Butorides virescens* (BV), *B. striatus* (BS), *Egretta garzetta* (EG), *Lxobrychus cinnamomeus* (LC) and *Nycticorax nycticorax* (NC).

North America is a major economic industry (Madsen and Frandsen, 1989). Thus, continued range extensions in temperate zones may be anticipated. Monitoring is necessary as many spring systems harbour endemic snails at risk from competitive displacement by *M. tuberculata* (e.g. *Elimia comalensis* and *Cincinnatia comalensis* in springs in Texas, US; Murray, 1971; Tolley-Jordan and Owen, 2008), which could cause population declines of native snails and possible risk of extinction (Lydeard et al, 2004).

Dispersal of *Centrocestus formosanus* via the first intermediate snail host

The establishment of *C. formosanus* depends on the presence of the most common snail host, *M. tuberculata*, due to the high degree of host specificity (Wright, 1973; Sapp and Loker, 2000a, 2000b). Although parasite range extensions occur through natural mechanisms of dispersal, most modern exotic distributions are due to human interventions (Prenter et al, 2004). *Melanoides tuberculata* is host to over 50 species of parasites, several of which are human pathogens (Ismail, 1990), which makes the successful establishment of additional pathogens into novel ranges via transmission from *M. tuberculata* of strong concern. Currently, *M. tuberculata* infected with *C. formosanus* are reported from 16 countries (12 of which are range extensions), with the notable exceptions of snails from African waters (Table 33.1). The reason for this is not well understood; however, the presence of other digeneans such as *Paragonimus westermanni* (Hira, 1970) and *Philophthalmus gralli* (Mukaratirwa et al, 2005) in African *M. tuberculata* provides evidence that African strains do serve as hosts to digeneans. Further sampling efforts are necessary to determine if *C. formosanus* occurs in endemic *Melanoides* found in Africa. Recent introductions of *M. tuberculata* strains of Asian origin into Lake Malawi (Genner et al, 2004) and the Congo Basin (Facon et al, 2003) should provide a suitable host for *C. formosanus*. Currently, no reports of *M. tuberculata* infected with *C. formosanus* have been found in the Caribbean, Central America or Oceania. It is likely that if any host infected with *C. formosanus* is introduced into these areas the parasite will become established. Thus, through time, these regions will more than likely become range extensions for the parasite.

Dispersal of *Centrocestus formosanus* via second intermediate snail hosts: Fishes

The affinity of *C. formosanus* for a wide variety of fishes to serve as the second intermediate host is apparent in the 128 fish species in 83 genera of 26 families reported from 19 countries around the world (Table 33.2, Figure 33.2). Fourteen countries from North America, Europe, west Asia and Australia are range extensions of *C. formosanus* found in fishes. The earliest record of range extension comes from Hawaii (Martin, 1958) and the most recent record occurs in Croatia in 2007 (Gjurčević et al, 2007). Studies that listed dates for the introduction of *C. formosanus*-infected fishes showed that most range extensions occurred following the importations of aquarium plants and fishes and some aquaculture species of Asian origin. Fishes of Asian origin were found to be infected with *C. formosanus* in Central America, Europe and the Middle East, and all reports were from fish farms (Table 33.2). However, in North America in systems where *M. tuberculata* infected with *C. formosanus* occur, 67 of the 85 species of infected fishes were native to the region, 53 of which were regional endemics (Table 33.2). Interestingly, reintroductions of the parasite into southeast Asian countries (native range of *C. formosanus*) via exotic fishes (non-native to southeast Asia) transported from aquarium and aquaculture facilities were a common occurrence (Table 33.2). Although *C. formosanus* requires all three hosts (first and second intermediate and definitive host) to have an established population, several countries report only fishes infected with the parasite (generally of Asian origin that were imported) from aquarium facilities. Thus, establishment of the parasite in these regions in native fishes is not likely.

Native and invasive ranges: Definitive bird and mammal hosts

Six species of fish-eating wading birds in Ardeidae (Table 33.1) from seven countries were reported as definitive hosts. Birds from Mexico, Hawaii, the US, Israel and Iran are considered range extensions for the parasite; albeit many of these bird species are globally widespread within its native range. In addition to birds, mammals such as rodents, cats and humans have been reported as definitive hosts in the native range of the parasite (Chen, 1942; Yu et al, 1994; Kagei and

Table 33.2 *Families and species of fishes that serve as second intermediate hosts (gills infected) of* Centrocestus formosanus

Taxa	Location	Source
Anabantidae		
Anabas testudineus	Viet Nam (N, F), Philippines (N)	Martin, 1958; Thien et al, 2007
Anguillidae		
Anguilla japonica	Japan (I)	Kagei and Yanohara, 1995
Atherinidae		
Chirostoma humboldtianum	Mexico (N, RE, F)	Scholz and Salgado-Maldonado, 2000
Atherinopsidae		
Atherinella crystallina	Mexico (N, RE)	Salgado-Maldonado, 2006
A. ammophila	Mexico (N, RE)	Salgado-Maldonado, 2006
Belontiidae		
Trichogaster trichopterus	Mexico (N, RE, F)	Salgado-Maldonado, 2006
Polyacanthus operculatus	Mexico (N, RE)	Chen, 1942
Centrarchidae		
Lepomis macrochirus	US (N, RE)	Flemming, 2002
L. microlophus	US (N, RE)	McDermott, 2000
Micropterus dolomieu	US (N, RE)	Flemming, 2002
M. salmoides	US (N, RE)	McDermott, 2000; Flemming, 2002
M. treculi	Mexico (N, RE) US (N, RE)	Flemming, 2002; Salgado-Maldonado, 2006
Channidae		
Channa formosana	Taiwan (N)	Chen, 1942
Rhodeus ocellatus	Taiwan (N)	Chen, 1942
R. sinensis	China (N)	Youzhu and Wang, 2000
Characidae		
Astyanax sp.	Mexico (N, RE)	Salgado-Maldonado, 2006
A. fasciatus	Mexico (N, RE)	Salgado-Maldonado, 2006
A. mexicanus	US (I, RE)	Flemming, 2002; Mitchell et al, 2005
Bramocharax caballeroi	Mexico (N, RE)	Salgado-Maldonado, 2006
Hyphessobrycon sp.	US (I, RE, A)	Blazer and Gratzek, 1985
Cichlidae		
Aequidens pulcher	Columbia (I, RE)	Escobar et al, 2009
Archocentrus nigrofasciatum	Mexico (N, RE)	Salgado-Maldonado, 2006
Cichlasoma cyanoguttatum	US (N, RE)	McDermott, 2000
C. geddesi	Mexico (N, RE)	Salgado-Maldonado, 2006
C. gadovii	Mexico (N, RE)	Scholz and Salgado-Maldonado, 2000
C. helleri	Mexico (N, RE)	Scholz and Salgado-Maldonado, 2000

Table 33.2 *Families and species of fishes that serve as second intermediate hosts (gills infected) of*
Centrocestus formosanus *(Cont'd)*

Taxa	Location	Source
C. nigrofasciatum	Mexico (N, RE)	Scholz and Salgado-Maldonado, 2000
C. pasionis	Mexico (N, RE)	Scholz and Salgado-Maldonado, 2000
C. salvini	Mexico (N, RE)	Scholz and Salgado-Maldonado, 2000
C. urophthalamus	Mexico (N, RE)	Salgado-Maldonado, 2006
Herichthys cyanoguttatus	Mexico (N, RE)	Salgado-Maldonado, 2006
Oreochromis aureus	Mexico (I, RE, A)	Salgado-Maldonado, 2006
O. mossambicus	Mexico (I, RE, A)	Salgado-Maldonado, 2006
O. nobilis	Viet Nam (I, F)	Chi et al, 2008
O. urolepis	Mexico (N, RE)	Salgado-Maldonado, 2006
Parachromis freidrichsthalii	Mexico (N, RE)	Salgado-Maldonado, 2006
P. managuensis	Mexico (N, RE)	Salgado-Maldonado, 2006
Pterophyllum scalare	Turkey (I, RE, A)	Yildiz, 2005
Thorichthys helleri	Mexico (N, RE)	Salgado-Maldonado, 2006
Tilapia zillii	Israel (I, RE, F)	Dzikowski et al, 2004
Veija fenestra	Mexico (N, RE)	Salgado-Maldonado, 2006
Cobitidae		
Misgurnus anguillicaudatu	Taiwan (N)	Chen, 1942
M. asguiliaudatus	China (N)	Youzhu and Wang, 2000
Crangolandidae		
Clarias fuscus	Taiwan (N)	Chen, 1942
Cyprinidae		
Algansea tincella	Mexico (N, RE)	Salgado-Maldonado, 2006
Aplocheilus panchax	India (N)	Madhavi, 1986
Aristichtheys nobilis	Viet Nam (N, F)	Chi et al, 2008
Brachydanio retrio	US (I, RE, A)	Blazer and Gratzek, 1985
Carassius auratus	Croatia (I, RE, A), Japan, (N), Mexico (I, RE), Singapore (N, A), Sri Lanka (N, F), Taiwan (N) Turkey (I, RE, A)	Chen, 1942; Kagei and Yanohara, 1995; Thilakaratne et al, 2003; Yildiz 2005; Salgado-Maldonado, 2006; Gjurčevič et al, 2007
C. repasson	Laos (N)	Chi et al, 2008; Han et al, 2008
Cirrhinus molitorella	Laos (N), Viet Nam (N, F)	Han et al, 2008
Cyprinella lutrensis	US (N, RE)	Flemming, 2002
C. venusta	US (N, RE)	Flemming, 2002
Cyprinus carpio	Mexico (I, RE), Taiwan (N), Viet Nam (N, F)	Vélez-Hernández et al, 1998; Thien et al, 2007; Sohn et al, 2009; Chi et al, 2008

Table 33.2 *Families and species of fishes that serve as second intermediate hosts (gills infected) of* Centrocestus formosanus (Cont'd)

Taxa	Location	Source
Ctenopharngodon idella	Mexico (I, RE), China (N), Taiwan (N), Viet Nam (N, F)	Chen, 1942; Zeng and Liao, 2000; Salgado-Maldonado, 2006; Thien et al, 2007; Chi et al, 2008
Danio rerio	Mexico (I, RE, F)	Ortega et al, 2009
Dionda argentosa	US (N, RE)	McDermott, 2000
D. diaboli	US (N, RE)	McDermott, 2000
D. episcopa	US (N, RE)	Flemming, 2002
Hypophthalmichthys molitrix	Mexico (I, RE, A), Viet Nam (N, F)	Scholz and Salgado-Maldonado, 2000; Chi et al, 2008; Ortega et al, 2009
Labeo rohita	Viet Nam (N, F)	Thien et al, 2007; Chi et al, 2008
Megalobrama amblycephala	Mexico (I, RE, F)	Scholz and Salgado-Maldonado, 2000; Ortega et al, 2009
Mylopharyngodon piceus	Mexico (I, RE, F)	Scholz and Salgado-Maldonado, 2000; Ortega et al, 2009
Notropis amabilis	US (N, RE)	Mitchell et al, 2005
N. volucellus	US (N, RE)	Flemming, 2002
Osteochilus hasseltii	Laos (N)	Han et al, 2008
Pseudorasbora parval	Taiwan (N)	Chen, 1942
Puntius brevis	Laos (N)	Han et al, 2008
Xenotaca variata	Mexico (I, RE)	Salgado-Maldonado, 2006
Yuriria alta	Mexico (N, RE)	Salgado-Maldonado, 2006
Zacco platypus	Taiwan (N)	Chen, 1942
Elotridae		
Dormitator latifrons	Mexico (N, RE)	Salgado-Maldonado, 2006
Gobiomorus dormitor	Mexico (N, RE)	Salgado-Maldonado, 2006
G. maculatus	Mexico (N, RE)	Salgado-Maldonado, 2006
G. polylepis	Mexico (N, RE)	Scholz and Salgado-Maldonado, 2000
Gobiidae		
Glossogobius giurnus	Philippines (N)	Martin, 1958
Sicydium multipunctatum	Mexico (I, RE)	Salgado-Maldonado, 2006
Rhinogobius giurinus	China (N)	Youzhu and Wang, 2000
Sicydium multipunctatum	Mexico (I, RE)	Salgado-Maldonado, 2006
Goodeidae		
Goodea atripinnis	Mexico (N, RE)	Salgado-Maldonado, 2006
Ilyodon furcidens	Mexico (N, RE)	Salgado-Maldonado, 2006
I. whitei	Mexico (N, RE)	Salgado-Maldonado, 2006

Table 33.2 *Families and species of fishes that serve as second intermediate hosts (gills infected) of* Centrocestus formosanus (Cont'd)

Taxa	Location	Source
Hemiramphidae		
Hemiamphus dussumieri	Philippines (N)	Martin, 1958
Heptapteridae		
Rhamdia guatamalensis	Mexico (N, RE)	Salgado-Maldonado, 2006
Ictaluridae		
Ictalurus punctatus	Mexico (I, RE, F), USA (N, RE)	McDermott, 2000; Scholz and Salgado-Maldonado, 2000; Ortega et al, 2009
Loricariidae		
Hypostomus plecostomus	Mexico (I, RE, F)	Ortega et al, 2009
Mugilidae		
Agonostomus monticola	Mexico (I, RE)	Salgado-Maldonado, 2006
Mugil cephalus	India (N), Japan (N)	Rekharani and Madhavi, 1985; Kagei and Yanohara, 1995
Liza macrolepis	India (N)	Rekharani and Madhavi, 1985
Valamugil cunnesius	India (N)	Rekharani and Madhavi, 1985
Ophiocephalidae		
Ophiocephalus striatus	Philippines (N)	Martin, 1958
Percidae		
Etheostoma fonticola	US (N, RE)	Mitchell et al, 2005
E. grahami	US (N, RE)	McDermott, 2000
E. lepidum	US (N, RE)	Mitchell et al, 2005
Percina caproides	US (N, RE)	Flemming, 2002
Poeciliidae		
Gambusia affinis	Taiwan (I), China (I), Hawaii (I, RE)	Chen, 1942; Martin, 1958; Zeng and Liao, 2000
G. yucatana	Mexico (N, RE)	Salgado-Maldonado, 2006
Heterandria spp.	Mexico (N, RE)	Salgado-Maldonado, 2006
H. bimaculata	Mexico (N, RE)	Salgado-Maldonado, 2006
Poecilia sp.	Mexico (N, RE, F)	Salgado-Maldonado, 2006
P. butleri	Mexico (N, RE)	Salgado-Maldonado, 2006
P. mexicana	Mexico (N, RE)	Salgado-Maldonado, 2006
P. petenensis	Mexico (N, RE)	Salgado-Maldonado, 2006
P. reticulata	Mexico (N, RE, F), Turkey (I, RE, A), Singapore (I, A), Australia (I, RE, A)	Evans and Lester, 2001; Yildiz, 2005; Salgado-Maldonado, 2006
P. sphenops	Mexico (N, RE, F)	Salgado-Maldonado, 2006
Poeciliopsis baenschi	Mexico (N, RE, F)	Salgado-Maldonado, 2006
P. gracilis	Mexico (N, RE, F)	Salgado-Maldonado, 2006

Table 33.2 *Families and species of fishes that serve as second intermediate hosts (gills infected) of*
Centrocestus formosanus (Cont'd)

Taxa	Location	Source
P. infans	Columbia (I, RE)	Escobar et al, 2009
Xiphophorus hellerii	US, (I, RE, A), Columbia (I, RE), Hawaii (I, RE)	Martin, 1958; Blazer and Gratzek, 1985; Escobar et al, 2009
X. maculatus	Columbia (I, RE), Turkey (I, RE, A), Australia (I, RE, A)	Evans and Lester, 2001; Yildiz, 2005; Escobar et al, 2009
Salmonidae		
Oncorhynchus mykiss	US (N, RE)	Olson and Pierce, 1997
Siluridae		
Parasilurus asotus	Taiwan (N)	Chen, 1942
Theraponidae		
Therapon plumbeus	Philippines (N)	Martin, 1958

Note: Localities by country are given along with if fish is Native (N) or Introduced (I; according to Nelson, 1976); if the infected fish is a range extension (RE) of Centrocestus formosanus, and if fish was collected from a fish farm (F) or aquaria (A) used in ornamental fish pet trade.

Yanohara, 1995) and are probably definitive hosts in other regions of the world. Although the list of definitive hosts for birds and mammals is not as extensive as fishes, this is probably due to limited investigation. We anticipate many species of mammals that frequently use aquatic habitats with *C. formosanus* could serve as definitive hosts and ensure persistence of this parasite in invaded aquatic habitats.

Efficacy for Control of *Centrocestus formosanus*

Compromised fish health is of particular to concern to fish hatcheries that raise juvenile fish and ornamental fish producers that can suffer substantial economic losses by the invasion of this parasite (Francis-Floyd et al, 1997). In addition, the parasite has an indirect negative impact on endemic and rare fishes. Thus, finding a means to eradicate this parasite can be particularly important to fisheries managers and conservation agencies. However, eliminating snails and subsequently their concomitant parasites where they have been established generally are difficult. Molluscicides, such as latex from *Euphorbia splendens*, used to control *Biomphaliara glabrata*, require far higher

doses for control of *M. tuberculata* (Giovanelli et al, 2001). In addition, chemically based molluscicides can often be expensive and harmful to potable water supplies (Thomas, 1987) and to fishes (Hoffman, 1970).

Thus, means of control of *M. tuberculata*, and indirectly *C. formosanus*, should be focused on preventing dispersal to new habitats. Chemical treatments such as disinfectants (e.g. Roccal®-D Plus, Pfizer), and salt mixed with ice water have been found to be effective at killing *M. tuberculata* on fisheries equipment, a proposed mechanism of dispersal of this snail within the US (Mitchell et al, 2007; Mitchell and Brandt, 2009). Heat treatments at 50°C for about five minutes will also kill this snail host (Mitchell and Brandt, 2005). Preventative measures, such as these, are probably the most effective way of ensuring that snails and parasites do not spread and become established in novel habitats or regions. Yet, these will only be effective if education and training are included in efforts to reduce introduction and secondary spread. The biggest risk of dispersal of these snails and their concomitant parasites probably comes from the aquarium trade where hobby aquarists can easily purchase *M. tuberculata*, commonly referred to as Malaysian trumpet snails, from a variety of outlets including the internet.

References

Abbott, R. T. (1973) 'Spread of *Melanoides tuberculata*', *The Nautilus Monograph Review*, vol 87, no 1, p29

Abdel-Azim, A. and Gismann, A. (1956) 'Bilharziasis survey in south-western Asia covering Iraq, Israel, Jordan, Lebanon, Sa'udi Arabia, and Syria: 1950–51', *Bulletin of the World Health Organization*, vol 14, pp403–456

Alcaraz, G., Pérez-Ponce de Leon, G., Garcia, L., Leon-Regagnon, V. and Vanegas, C. (1999) 'Respiratory responses of grass carp *Ctenopharyngodon idella* (cyprinidae) to parasitic infection by *Centrocestus formosanus* (Digenea)', *The Southwestern Naturalist*, vol 44, no 2, pp222–226

Alexandrov, B., Boltachev, A., Kharchenko, T., Lyashenko, A., Son, M., Tsarenko, P. and Zhukinsky, V. (2007) 'Trends of aquatic alien species invasions in Ukraine', *Aquatic Invasions*, vol 2, no 3, pp215–242

Al-safadi, M. M. (1991) 'Freshwater molluscs of Yemen Arab Republic', *Hydrobiologia*, vol 208, no 3, pp245–251

Amaya-Huerta, D. and Almeyda-Artigas, R. J. (1994) 'Confirmation of *Centrocestus formosanus* (Nishigori, 1924) Price, 1932 (Trematoda: Heterophyidae) in Mexico', *Research and Reviews in Parasitology*, vol 54, no 2, pp99–103

Anderson, T. K. (2004) 'A review of the United States distribution of *Melanoides tuberculatus* (Müller, 1774), an exotic freshwater snail', *Ellipsaria*, vol 6, no 2, pp15–17

Arizmendi, M. A. (1992) 'Descripción de algunas etapas larvarias y de la fase adulta de *Centrocestus formosanus* de Tezontepec de Aldama, Hidalgo', *Anales. Inst. Biol. Univ. Nac. Auton. Mexico, Ser. Zool*, vol 63, no 1, pp1–11

Balasuriya, L. (1988) 'A study on the metacercarial cysts of a *Centrocestus* species (Digenea: Heterophyidae) occurring on the gills of cultured cyprinid fishes in Sri Lanka', *Journal of Inland Fisheries*, vol 4, pp3–10

Ball, E. and Glucksman, J. (1978) 'Limnological studies of Lake Wisdom, a large New Guinea caldera lake with a simple fauna', *Freshwater Biology*, vol 8 no 5, pp455–468

Ben-Ami, F. and Heller, J. (2005) 'Spatial and temporal patterns of parthenogenesis and parasitism in the freshwater snail *Melanoides tuberculata*', *Journal of Evolutionary Biology*, vol 18, pp138–146

Berry, A. J. and Kadri, A. H. (1973) 'Reproduction in the Malayan freshwater cerithiacean gastropod *Melanoides tuberculata*', *Journal of Zoology*, vol 172, pp369–381

Blay, J. and Dongdem, F. (1996) 'Preliminary observations on the benthic macrofauna of a polluted coastal lagoon in Ghana (West Africa)', *Tropical Ecology*, vol 37, no 1, pp127–133

Blazer, V. S. and Gratzek, J. B. (1985) 'Cartilage proliferation in response to metacercarial infections of fish gills', *Journal of Comparative Pathology* vol 95, pp273–280

Brown, D. S. (1980) *Freshwater Snails of Africa and their Medical Importance*, Taylor and Francis, London

Brown, D. S., Curtis, B. A., Bethune, S. and Appleton, C. C. (1992) 'Freshwater snails of East Caprivi and the lower Okavango River basin in Namibia and Botswana', *Hydrobiologia*, vol 246, no 1, pp9–40

Carney, W. P., Hadidjaja, P., Davis, G. M., Clarke, M. D., Djajasamita, M. and Nalim, S. (1973). '*Oncomelania hupensis* from the Schistosomiasis focus in central Sulawesi (Celebes), Indonesia', *Journal of Parasitology*, vol 59, no 1, pp210–211

Chao, D., Wang, L. and Huang, T. (1993) 'Prevalence of larval helminths in freshwater snails of the Kinmen Islands', *Journal of Helminthology*, vol 67, pp259–264

Chen, H. T. (1942) 'The metacercariae and adult of *Centrocestus formosanus* (Nishigori, 1924) with notes on the natural infection of rats and cats with *C. Armatus* (Tanabe, 1922)', *Japanese Journal of Parasitology*, vol 28, pp285–298

Chen, H. T. (1948) 'Some early larval stages of *Centrocestus formosanus* (Nishigori, 1924)', *Lingnan Science Journal*, vol 22, pp93–105

Chi, T. T. K., Dalsgaard, A., Turnbull, J. F., Tuan, P. A. and Murrell, K. D. (2008) 'Prevalence of zoonotic trematodes in fish from a Vietnamese fish-farming community', *Journal of Parasitology*, vol 94, no 2, pp423–428

Cho, H. C., Chung, P. R. and Lee, K. T. (1983) 'Distribution of medically important freshwater snails and larval trematodes from *Parafossarulus manchouricus* and *Semisulcospira libertina* around the Jinyang Lake in Kyongsang-Nam-Do, Korea', *Kisaengchunghak Chapchi*, vol 21, no 2, pp193–204

Cianfanelli, S., Lori, E. and Bodon. M. (2007) 'Non-indigenous freshwater mollusks and their distribution in Italy', in F. Gerhardi (ed) *Biological Invaders of Inland Waters: Profiles, Distributions, and Threats. Invading Nature*, Springer Series in Invasion Ecology, New York, pp103–121

Clavero, M. and García-Berthou, E. (2005) 'Invasive species are a leading cause of animal extinctions', *Trends in Ecology and Evolution*, vol 20, no 3, p110

Contreras-Arquieta, A. (1998) 'New records of the snail *Melanoides tuberculata* (Müller, 1774) (Gastropoda: Thiaridae) in the Cuatro Cienegas Basin and its distribution in the state of Coahuila, Mexico', *The Southwestern Naturalist*, vol 45, no 2, pp283–286

Cowie, R. H. (1997) 'Catalog and bibliography of the non-indigenous non marine snails and slugs of the Hawaiian islands', *Bishop Museum Occasional Papers*, vol 50, pp1–66

Cowie, R. H. (1998) 'Patterns of introduction of non-indigenous non-marine snails and slugs in the Hawaiian Islands', *Biodiversity and Conservation*, vol 7, pp349–368

Cribb, T. H., Bray, R. A. and Littlewood, D. T. J. (2001) 'The nature and evolution of the association among digeneans, molluscs and fishes', *International Journal for Parasitology*, vol 31, pp997–1011

Darlington, J. P. E. C. (1977) 'Temporal and spatial variation in the benthic invertebrate fauna of Lake George, Uganda', *Journal of Zoology*, vol 181, no 1, pp95–111

Dechruksa, W., Krailas, D., Ukong, S., Inkapatanakul, W. and Koonchornboon, T. (2007) 'Trematode infections of the freshwater snail family Thiaridae in Khek River, Thailand', *Southeast Asia Journal of Medicine and Tropical Health*, vol 38, no 6, pp1016–1028

de Clercq, D. (1987) 'La situation malacologique a Kinshasa et description d'un foyer autochtone de schistosomiase a *Schistosoma intercalatum*', *Annales de la Societe Belgue de Medecine Tropicale*, vol 67, pp345–352

de Kock, K. N. and Wolmarans, C. T. (2009) 'Distribution and habitats of *Melanoides tuberculata* (Müller, 1774) and *M. victoriae* (Dohrn, 1865) (Mollusca: Prosobranchia: Thiaridae) in South Africa', *Water South Africa*, vol 35, no 5, pp713–720

DeMarco Jr, P. (1999) 'Invasion by the introduced aquatic snail *Melanoides tuberculata* (Müller 1774) (Gastropoda: Prosobranchia: Thiaridae) of the Rio Doce State Park, Minas Gerais, Brazil', *Studies on Neotropical Fauna and the Environment*, vol 34, no 3, pp186–189

Dillon, R. T. (2000) *The Ecology of Freshwater Molluscs*, Cambridge University Press, Cambridge

Discovery Database (2010) www.discoverlife.org

Dudgeon, D. (1986) 'The lifecycle, population dynamics and productivity of *Melanoides tuberculata* (Müller 1774) (Gastropoda: Prosobranchia: Thiaridae) in Hong Kong', *Journal of Zoology, London*, vol 208, pp37–53

Dudgeon, D. and Yipp, M. W. (1983) 'The diets of Hong Kong freshwater gastropods' in B. Morton and D. Dudgeon (eds) *Proceedings of the Second International Workshop on the Malacofauna of Hong Kong and Southern China*, Hong Kong University Press, Hong Kong, pp491–509

Duggan, I. C. (2002) 'First record of a wild population of the tropical snail *Melanoides tuberculata* in New Zealand natural waters', *New Zealand Journal of Marine and Freshwater Research*, vol 36, pp825–829

Duggan, I. C., Boothroyd, I. K. G. and Speirs, D. A. (2007) 'Factors affecting the distribution of stream macroinvertebrates in geothermal areas: Taupo Volcanic Zone, New Zealand', *Hydrobiologia*, vol 592, pp235–247

Dundee, D. S. and Paine, A. (1977) 'Ecology of the snail, *Melanoides tuberculata* (Müller), intermediate host of the human liver digenean (*Opisthorchis sinensis*) in New Orleans, Louisiana', *Nautilus*, vol 91, no 1, pp17–20

Dzikowski, R., Levy, M. G., Poore, M. F., Flowers, J. R. and Paperna, I. (2004) 'Use of rDNA polymorphism for identification of Heterophyidae infecting freshwater fishes', *Diseases of Aquatic Organisms*, vol 59, pp35–41

Esch, G. W. and Fernandez, J. C. (1994) 'Snail-trematode interactions and parasite community dynamics in aquatic systems: A review', *The American Midland Naturalist*, vol 131, pp209–237

Escobar, J. S., Correa, A. C. and David, P. (2009) 'Did life history evolve in response to parasites in invasive populations of *Melanoides tuberculata*?', *Acta Oecologia*, vol 35, pp639–644

Evans, B. B. and Lester, R. J. G. (2001) 'Parasites of ornamental fish imported into Australia', *Bulletin of the European Association of Fish Pathologists*, vol 21, no 2, pp51–55

Facon, B., Pointier, J. P., Glaubrecht, M., Poux, C., Jarne, P. and David, P. (2003) 'A molecular phylogeography approach to biological invasions of the New World by parthenogenetic thiarid snails', *Molecular Ecology*, vol 12, pp3027–3039

Farahnak, A., Shiekhian, R. and Mobedi, I. (2004) 'A faunistic survey on the bird helminth parasites and their medical importance', *Iranian Journal of Public Health*, vol 33, no 3, pp40–46

Flemming, B. (2002) 'Downstream spread of the digenetic trematode, *Centrocestus formosanus*, into the Guadalupe River, Texas', MSc Thesis, Southwest Texas State University, Texas

Font, W. F. (2003) 'The global spread of parasites: What do Hawaiian streams tell us?', *Bioscience*, vol 53, pp1061–1067

Francis-Floyd, R., Gildea, J., Reed, P. and Klinger, R. (1997) 'Use of Bayluscide (Bayer 73) for snail control in fish ponds', *Journal of Aquatic Animal Health*, vol 9, pp41–48

Gasull, L. (1974) 'Una interesante localidad con *Melanoides tuberculata* (Müller) en la província de Castellón de la Plana (Mollusca, Prosobranquia)', *Bolletín de la Societat d'Història Natural de les Balears*, vol 19, pp148–150

Genner, M. J., Michel, E., Erpenbeck, D., De Voogd, N., Witte, F. and Pointier, J. P. (2004) 'Camouflaged invasion of Lake Malawi by an oriental gastropod', *Molecular Ecology*, vol 13, pp2135–2141

Gherardi, F., Bertolino, S., Bodon, M., Casellato, S., Cianfanelli, S., Ferraguti, M., Lori, E., Mura, G., Nocita, A., Riccardi, N., Rossetti, G., Rota, E., Scalera, R., Zerunian, S. and Tricarico, E. (2007) 'Animal xenodiversity in Italian inland waters: Distribution, modes of arrival, and pathways', *Biological Invasions*, vol 10, no 4, pp435–454

Giovanelli, A., da Silva, C. L. P. A. C., Medeiros, L. and Carvalho, de Vasconcellos, M. (2001) 'The molluscicidal activity of the latex of *Euphorbia splendens* var. *hislopii* on *Melanoides tuberculata* (Thiaridae), a snail associated with habitats of *Biomphalaria glabrata* (Planorbidae)', *Memórias do Instituto Oswaldo Cruz*, vol 96, no 1, pp123–125

Giovanelli, A., de Silva, C. L. P. A. C., Leal, G. B. E. and Baptista, D. F. (2005) 'Habitat preference of freshwater snails in relation to environmental factors and the presence of the competitor snail *Melanoides tuberculatus* (Müller, 1774)', *Memórias do Instituto Oswaldo Cruz*, vol 100, no 2, pp169–176

Gjurčević, E., Petrinec, Z., Kozarič, Z., Kužir, S., Kantura, V., Vučemilo, M. and Džaja, P. (2007) 'Metacercariae of *Centrocestus formosanus* in goldfish (*Carassius auratus* L.) imported into Croatia', *Helminthologia*, vol 44, no 4, pp214–216

Glaubrecht, M., Brinkmann, N. and Pöppe, J. (2009) 'Diversity and disparity "down under": Systematics, biogeography and reproductive modes of the "marsupial" freshwater Thiaridae (Caenogastropoda, Cerithioidea) in Australia', *Zoosystematics and Evolution*, vol 85, pp199–275

Gogoi, A. R. and Sharma, B. N. D. (1986) 'Aquatic snails and their infection with larval trematodes in Kamrup district of Assam', *Indian Journal of Animal Sciences*, vol 56, no 6, pp663–666

Gregoric, D. E. G., Núez-Rumi, V. and Roche, M. A. (2006) 'Freshwater gastropods from Del Plata basin, Argentina: Checklist and new locality records', *Communicaciones de la Sociedad Malacológica del Uruguay*, vol 89, no 9, pp51–60

Gryseels, B., Nkulikyinka, L., Kabahizi, E. and Margegeya, E. (1987) 'A new focus of *Schistosoma mansoni* in the highlands of Burundi', *Annales de la Societe Belgue de Medecine Tropicale*, vol 67, pp247–257

Guimaraes, C. T., Pereira de Souza, C. and de Moura Soares, D. (2001) 'Possible competitive displacement of planorbids by *Melanoides tuberculata* in Minas Gerais, Brazil', *Memórias do Instituto Oswaldo Cruz, Suppl.*, vol 96, no 1, pp173–176

Han, E., Shin, E., Phommakorn, S., Sengvilaykham, B., Kim, J., Rim, H. and Chai, J. (2008) '*Centrocestus formosanus* (Digenea: Heterophyidae) encysted in the freshwater fish, *Puntius brevis*, from Lao PDR', *Korean Journal of Parasitology*, vol 46, no 1, pp49–53

Haynes, A. (2000) 'The distribution of freshwater gastropods on four Vanuatu islands: Espíritu Santo, Pentecost, Éfate and Tanna (South Pacific)', *Annals de Limnologie*, vol 36, no 2, pp101–111

Heller, J. and Farstey, V. (1990) 'Sexual and parthenogenetic populations of the freshwater snail *Melanoides tuburculata* in Israel', *Israel Journal of Zoology*, vol 37, pp75–87

Hira, P. (1970) 'Schistosomiasis at Lake Kariba, Zambia. 1. Prevalence and potential intermediate snail hosts at Siavonga', *Tropical Geographic Medicine*, vol 22, no 3, pp323–334

Hoffman, G. L. (1970) 'Control methods for snail-borne zoonozes', *Journal of Wildlife Diseases*, vol 6, pp262–265

Hoffman, G. L. (1999) *Parasites of North American Freshwater Fishes*, 2nd edition, Cornell University Press, Ithaca, New York

Horsák, M., Dvořák, L. and Juřičková, L. (2004) 'Greenhouse gastropods of the Czech Republic: Current stage of research', *Malacological Newsletter*, vol 22, pp141–147

Ibikounlé, I., Mouahid, G., Sakiti, N. G., Massougbodji, A. and Moné, H. (2009) 'Freshwater snail diversity in Benin (West Africa) with a focus on human schistosomiasis', *Acta Tropica*, vol 111, pp29–34

ICZN (International Commission on Zoological Nomenclature) (1999) *International Code of Zoological Nomenclature*, 4th edition, The International Trust for Zoological Nomenclature, The Natural History Museum, London

Ismail, N. S. (1990) 'A new cercariae from the freshwater snail *Melanoides tuberculatus*, Müller, 1774, in Asir Province, Saudi Arabia', *Japanese Journal of Parasitology*, vol 39, no 2, pp172–175

Ismail, N. S. and Arif, A. M. S. (1991) 'Larval trematodes of *Melanoides tuberculatus* (Müller, 1774) (Gastropoda: Prosobranchia) in a brackish spring, United Arab Emirites', *Japanese Journal of Parasitology*, vol 40, no 2 pp157–169

Ismail, N. S. and Arif, A. M. S. (1993) 'Population dynamics of *Melanoides tuberculata* (Thiaridae) snails in a desert spring, United Arab Emirites and infection with larval trematodes', *Hydrobiologia*, vol 257, pp57–64

Ito, J. (1977) 'Studies on the fresh water cercariae in Leyte Island, Philippines', *Japanese Journal of Experimental Medicine*, vol 47, no 4, pp223–230

Kagei, N. and Yanohara, Y. (1995) 'Epidemiological study on *Centrocestus formosanus* (Nishigori, 1924) – surveys of its infection in Tanegashima, Kagoshima Prefecture, Japan', *Japanese Journal of Parasitology*, vol 44, no 2, pp154–160

Kloos, H. (1985) 'Water resources development and schistosomiasis ecology in the Awash Valley, Ethiopia', *Social Science and Medicine*, vol 20, no 6, pp609–625

Laamrani, H., Khallayoune, K., Delay, B. and Pointier, J. P. (1997) 'Factors affecting the distribution and abundance of two prosobranch snails in a thermal spring', *Journal of Freshwater Ecology*, vol 12, no 1, pp75–79

Lazim, M. N., Shuker-Khan, A. O. and Saeed, I. S. (1988) 'Some of the macrobenthic invertebrates from springs and streams of northern parts of Iraq', *Journal of Biological Science Research*, vol 19, no 2, pp425–427

Livshits, G. and Fishelson, L. (1983) 'Biology and reproduction of the freshwater snail *Melanoides tuberculata* (Gastropoda: Prosobranchia) in Israel', *Israel Journal of Zoology*, vol 32, pp21–35

Lo, C. T. and Lee, K. M. (1996) 'Pattern of emergence and the effects of temperature and light on the emergence and survival of heterophyid cercarieae *Centrocestus formosanus* and *Haplorchis pumilio*', *Journal of Parasitology*, vol 82, no 2, pp357–350

Loker, E., Moyo, H. G. and Gardner, S. L. (1981) 'Trematode-gastropod associations in nine non-lacustrine habitats in the Mwanza region of Tanzania', *Parasitology*, vol 83, pp381–399

Lydeard, C., Cowie, R. H., Ponder, W. F., Bogan, A. E., Bouchet, P., Clark, S. A., Cummings, K. S., Frest, T. J., Gargominy, O., Herbert, D. G., Hershler, R., Perez, K. E, Roth, B., Seddon, M., Strong, E. E. and Thompson, F. G. (2004) 'The global decline of nonmarine mollusks', *Bioscience*, vol 54, no 4, pp321–330

Madhavi, R. (1986) 'Distribution of metacercariae of *Centrocestus formosanus* (Trematode: Heterophyidae) on the gills of *Aplocheilus panchax*', *Journal of Fish Biology*, vol 29, pp685–690

Madsen, H. and Frandsen, F. (1989) 'The spread of freshwater snails including those of medical and veterinary importance', *Acta Tropica*, vol 46, pp139–146

Madsen, H., Daffalla, A. A., Karoum, K. O. and Frandsen, F. (1988) 'Distribution of freshwater snails in irrigation schemes in the Sudan', *Journal of Applied Ecology*, vol 25, pp853–866

Manfrin, A., Rubini, S., Caffara, M., Volpin, M., Alborali, L. and Fioravanti, M. L. (2002) 'Bacterial and parasitical diseases in ornamental fishes coming from extra EU countries: Preliminary results', *Bollettino Societa Italiana di Patologia Ittica*, vol 14, no 33, pp44–54

Martin, W. E. (1958) 'The life histories of some Hawaiian heterophyid trematodes', *Journal of Parasitology*, vol 44, no 3, pp305–318

Martin, W. E. (1959) 'Egyptian heterophyid trematodes', *Transactions of the American Microscopical Society*, vol 78, no 2, pp172–181

McCollough, F. S. (1964) 'Observations on bilharziasis and the potential snail hosts in the republic of the Congo (Brazzaville)', *Bulletin of the World Health Organization*, vol 30, pp375–388

McDermott, K. (2000) 'Distribution and infection relationships of an undescribed digenetic trematode, its exotic intermediate host, and endangered fishes in springs of west Texas', MSc Thesis, Southwest Texas State University, Texas

Mitchell, A. J. and Brandt, T. M. (2005) 'Temperature tolerance of red-rim melania (*Melanoides tuberculatus*), an exotic aquatic snail established in the United States', *Transactions of the American Fisheries Society*, vol 134, pp126–131

Mitchell, A. J. and Brandt, T. M. (2009) 'Use of ice-water and salt treatments to eliminate an exotic snail, the red-rim melania, from small immersible fisheries equipment', *North American Journal of Fisheries Management*, vol 29, pp828–828

Mitchell, A. J., Salmon, M. J., Huffman, D. G., Goodwin, A. E. and Brandt, T. M. (2000) 'Prevalence and pathogenicity of a heterophyid trematode infecting the gills of an endangered fish *Etheostoma fonticola* in two central Texas spring-fed rivers', *Journal of Aquatic Animal Health*, vol 12, pp283–289

Mitchell, A. J., Overstreet, R. M., Goodwin, A. E. and Brandt, T. M. (2005) 'Spread of an exotic fish-gill trematode: A far-reaching and complex problem', *Fisheries*, vol 30, no 8, pp11–16

Mitchell, A. J., Hobbs, M. S. and Brandt, T. M. (2007) 'The effect of chemical treatments of red-rim melania, *Melanoides tuberculata*, an exotic aquatic snail that serves as a vector of trematodes to fish and other species in the USA', *North American Journal of Fisheries Management*, vol 27, no 4, pp1287–1293

Mukaratirwa, S., Hove, T., Cindzi, Z. M., Manonga, D. B., Taruvinga, M. and Matenga, E. (2005) 'First report of a field outbreak of the oriental eye-fluke, *Philophthalmus gralli* (Mathis and Leger 1910), in commercially reared ostriches (*Struthio camelus*) in Zimbabwe', *Onderstepoort Journal of Veterinary Research*, vol 72, no 3, pp203–206

Muli, J. R. and Mavuti, K. M. (2001) 'The benthic macrofauna community of Kenyan waters of Lake Victoria', *Hydrobiologia*, vol 458, pp83–90

Murray, H. D. (1964) '*Tarebia granifera* and *Melanoides tuberculata* in Texas', *Annual Report to the American Malacological Union*, no 31, pp15

Murray, H. D. (1971) 'The Introduction and spread of thiarids in the United States', *The Biologist*, vol 53, no 3, pp133–135

Nazneen, S. and Begum, F. (1994) 'Distribution of molluscs in Layari River (Sindh), Pakistan', *Hydrobiologia*, vol 273, no 2, pp95–100

Ndifon, G. T. and Ukoli, F. M. A. (1989) 'Ecology of freshwater snails in south-western Nigeria. I: Distribution and habitat preferences', *Hydrobiologia*, vol 171, pp231–253

Nelson, J. S. (1976) *Fishes of the World*, Wiley-Interscience, New York

Ngonseu, E., Greer, G. and Mimpfoundi, R. (1992) 'Dynamique des populations et infection de *Bulinus truncates* et *Bulinus forskalii* par les larves de schistosomes en zone Soudano-Sahelienne au Cameroun', *Annales de la Societe Belge de Medecine Tropicale*, vol 72, pp311–320

Nishigori, M. (1924) 'On a new species of digenean, *Stamnosoma formosanum* and its life history', *Taiwan Igakkai Zasshi*, vol 234, pp181–238

Olivier, L. (1947) '*Cercaria koliensis*, a new fork-tailed cercaria from Guadalcanal', *Journal of Parasitology*, vol 33, pp234–240

Olson, R. E. and Pierce, J. R. (1997) 'A trematode metacercaria metacausing gill cartilage proliferation in steelhead trout from Oregon', *Journal of Wildlife Diseases*, vol 33, no 4, pp886–890

Ortega, C., Fajardo, R. and Enriquez, R. (2009) 'Trematode *Centrocestus formosanus* infection and distribution in ornamental fishes in Mexico', *Journal of Aquatic Animal Health*, vol 21, pp18–22

Perera, G., Yonh, M. and Sanchez, R. (1987) 'First record and ecological studies on *Melanoides tuberculata* in Cuba', *Walkerana, Trans. POETS Soc.*, vol 2, pp165–171

Pointier, J. P. and Augustin, D. (1999) 'Biological control and invading freshwater snails: A case study', *Life Sciences*, vol 322, pp1093–1098

Pointier, J. P. and Marquet, G. (1990) 'Taxonomy and distribution of freshwater molluscs of French Polynesia', *Japanese Journal of Malacology*, vol 49, pp215–231

Pointier, J. P., Thaler, L., Pernot, A. F. and Delay, B. (1993a) 'Invasion of the Martinique Island by the parthenogenetic snail *Melanoides tuberculata* and the succession of morphs', *Acta Ecologica*, vol 14, no 1, pp33–42

Pointier, J. P., Theron, A. and Borel, G. (1993b) 'Ecology of the introduced snail *Melanoides tuberculata* (Gastropoda: Thiaridae) in relation to *Biomphalaria glabrata* in the marshy forest zone of Guadeloupe, French West Indies', *Journal of Molluscan Studies*, vol 59, pp421–428

Pointier, J. P., Incani, R. N., Balzan, C., Chrosciechowski, P. and Prypchan, S. (1994) 'Invasion of the rivers of the littoral central region of Venezuela by *Thiara granifera* and *Melanoides tuberculata* (Mollusca: Prosobranchia: Thiaridae) and the absence of *Biomphalaria glabrata*, snail host of *Schistosoma mansoni*', *Nautilus*, vol 107, no 4, pp124–128

Poulin, R. (2006) 'Global warming and temperature-mediated increases in cercarial emergence in trematode parasites', *Parasitology*, vol 132, pp143–151

Prenter, J., MacNeil, C., Dick, J. T. A. and Dunn, A. M. (2004) 'Roles of parasites in animal invasions', *Trends in Ecology and Evolution*, vol 19, no 7, pp385–390

Rader, R. B., Belk, M. C. and Keleher, M. J. (2003) 'The introduction of an invasive snail (*Melanoides tuberculata*) to spring ecosystems of the Bonneville Basin, Utah', *Journal of Freshwater Ecology*, vol 18, pp647–657

Rekharani, R. and Madhavi, R. (1985) 'Digenetic trematodes from mullets of Visakhapatnam (India)', *Journal of Natural History*, vol 19, no 5, pp929–951

Ricciardi, A. (2007) 'Are modern biological invasions an unprecedented form of global change?', *Conservation Biology*, vol 21, no 2, pp329–336

Roessler, M. A., Beardsley, C. L. and Tabb, D. C. (1977) 'New records of the introduced snail *Melanoides tuberculata* (Mollusca: Thiaridae) in South Florida', *Florida Scientist*, vol 40, no 1, pp87–94

Salgado-Maldonado, G. (2006) 'Checklist of trematode parasites of Mexico', *Zootaxa*, vol 1324, pp1–357

Samadi, S., Mavarez, J., Pointier, J. P., Delay, B. and Jarne, P. (1999) 'Microsatellite and morphological analysis of population structure in the parthenogenetic freshwater snail *Melanoides tuberculata*: Insights into the creation of clonal variability', *Molecular Ecology*, vol 8, pp1141–1153

Sandford, K. S. (1936) 'Observations on the distribution of land and freshwater mollusca in the southern Libyan desert', *Quarterly Journal of the Geological Society*, vol 92, pp201–220

Sapp, K. K. and Loker, E. S. (2000a) 'Mechanisms underlying digenean-snail specificity: Role of miracidial attachment and host plasma factors', *Journal of Parasitology*, vol 86, no 5, pp1012–1019

Sapp, K. K. and Loker, E. S. (2000b) 'A comparative study of factors underlying digenean-snail specificity: In vitro interactions between hemocytes and digenean larvae', *Journal of Parasitology*, vol 86, no 5, pp1020–1029

Scholz, T. and Salgado-Maldonado, G. (2000) 'The introduction and dispersal of *Centrocestus formosanus* (Nishigori, 1924) (Digenea: Heterophyidae) in Mexico: A review', *American Midland Naturalist*, vol 143, pp185–200

Smith, B. D. (2003) 'Prosobranch gastropods of Guam', *Micronesica*, vol 35–36, pp244–270

Sohn, W., Keeseon, S., Duk-Young, M., Han-Jong, R., Eui-Hyug, H., Yichao, Y. and Xueming, L. (2009) 'Fishborne trematode metacercariae in freshwater fish from Guangxi Zhuang Autonomous Region, China', *Korean Journal of Parasitology*, vol 47, no 3, pp249–257

Soule, M. E. (1990) 'The onslaught of alien species, and other challenges in the coming decades', *Conservation Biology*, vol 4 no 3, pp233–239

Strayer, D. L. (2010) 'Alien species in fresh waters: ecological effects, interactions with other stressors, and prospects for the future', *Freshwater Biology*, vol 55, pp52–174

Strayer, D. L. and Dudgeon, D. (2010) 'Freshwater biodiversity conservation: Recent progress and future challenges', *Journal of the North American Benthological Society*, vol 29, no 1, pp344–358

Sukontason, K., Piangjai, S., Muangyimpong, Y., Sukontason, K., Methanitikorn, R. and Chaitong, U. (1999) 'Prevalence of trematode metacercariae in cyprinoid fish of Ban Pao District, Chiang Mai Province, Northern Thailand', *Southeast Asian Journal of Tropical Medicine and Public Health*, vol 30, no 2, pp365–370

Supian, Z. and Ikhwanuddin, A. M. (2002) *Population Dynamics of Freshwater Mollusks (Mollusca: Gastropoda: Melanoides) in Crocker Range Park, Sabah*, ASEAN Review of Biodiversity and Environmental Conservation, ASEAN, www.arbec.com.my/pdf/art13julysep02.pdf, accessed 23 July 2010

Taraschewski, H. (2006) 'Hosts and parasites as aliens', *Journal of Helminthology*, vol 80, pp99–128

Thien, P. C., Dalsgaard, A., Thanh, B. N., Olsen, A. and Murrell. K. D. (2007) 'Prevalence of fishborne zoonotic parasites in important cultured fish species in the Mekong Delta, Vietnam', *Parasitology Research*, vol 101, pp1277–1284

Thilakaratne, I. D. S. I. P., Rajapaksha, G., Hewakopara, A., Rajapakse, R. P. V. J. and Faizal, A. C. M. (2003) 'Parasitic infections in freshwater ornamental fish in Sri Lanka', *Diseases of Aquatic Organisms*, vol 54, pp157–162

Thomas, J. D. (1987) 'A holistic view of schistosomiasis and snail control', *Memórias do Instituto Oswaldo Cruz*, vol 82, suppl 4, pp183–192

Thornton, I. W. and New, T. (1988) 'Freshwater communities on the Krakatau Islands', *Philosophical Transactions of the Royal Society London B*, vol 322, pp487–492

Tolley-Jordan, L. R. and Owen, J. M. (2008) 'Habitat influences snail community structure and trematode infection levels in a spring-fed river, Texas, USA', *Hydrobiologia*, vol 600, pp29–40

Vélez-Hernández, E. M., Constantino-Casas, F., García-Màrquez, L. J. and Osorio-Sarabia, D. (1998) 'Gill lesions in common carp, *Cyprinus carpio* L., in Mexico due to the metacercariae of *Centrocestus formosanus*', *Journal of Fish Diseases*, vol 21, pp229–232

Vivar, R., Larrea, H. and Huamán, P. (1990) 'Un gasterópodo de la familia Thiaridae en el Perú: *Melanoides tuberculata* (Müller, 1774)', *Boletín de Lima*, vol 69, pp33–34

Vogelbein, W. and Overstreet, R. M. (1988) 'Life history and pathology of a heterophyid trematode infecting Florida-reared ornamental fishes', *International Association for Aquatic Animal Medicine Proceedings*, vol 19, p138

Wright, C. (1973) *Digeneans and Snails*, Macmillan, New York

Wu, S. K. (1989) 'Colorado freshwater mollusks', *Natural History Inventory of Colorado*, vol 11, no 1, pp1–17

Yanohara, Y., Hisatake, N. and Sato. A. (1987) 'Incidence of *Centrocestus formosanus* infection in snails', *Journal of Parasitology*, vol 73, no 2, pp424–436

Yaseen, A. (1995) 'The chromosomes of the Egyptian freshwater snail *Melanoides tuberculata* (Gastropoda: Prosobranchia)', *Journal of Molluscan Studies*, vol 61, pp499–500

Yildiz, H. Y. (2005) 'Infection with metacercariae of *Centrocestus formosanus* (Trematoda: Heterophyidae) in ornamental fish imported into Turkey', *Bulletin of the Association of European Fisheries Pathologists*, vol 25, no 5, pp244–246

Yildirim, M. Z. (1999) 'The Prosobranchia (Gastropoda: Mollusca) species of Turkey and their zoogeographic distribution 1. Fresh and brackish water', *Turkish Journal of Zoology*, vol 23, pp877–900

Youzhu, C. and Wang, W. (2000) 'Preliminary study of *Centrocestus formosanus* at Zhaoan Fujian (Trematoda: Heterophyidae)', *Wuji Science Journal*

Yu, S., Xu, L., Jiang, Z., Xu, S., Han, J., Zhu, Y., Chang, J., Lin, J. and Xu, F. (1994) 'Report on the first nationwide survey of the distribution of human parasites in China. 1. Regional distribution of parasite species', *Chinese Journal of Parasitology and Parasitic Diseases*, vol 12, no 4, pp241–247

Zeng, B. and Liao, X. (2000) 'Monthly changes of the metacercarial cyst infrapopulation of *Centrocestus formosanus* (Nishigori, 1924) on the gills of grass carps *Ctenopharyngodon idellus*', *Acta Hydrobiologica Sinica*, vol 24, pp137–144

34

Myxobolus cerebralis Höfer (whirling disease)

Jerri L. Bartholomew

History of the Species and its Dissemination

Reviews of aquatic nuisance species understandably concentrate on macroscopic organisms, but movements of aquatic species may also translocate their pathogens. A parasite that has been widely dispersed as a result of fish movements is *Myxobolus cerebralis*, one of the most pathogenic and intensely studied members of the Myxozoa. This parasite infects members of the Salmonidae (primarily salmon and trout), causing an affliction known as 'whirling disease'. The Myxozoa are a diverse group of metazoans, most of which are parasitic in fish, although amphibians, birds and even mammals can be hosts (Lom and Dykova, 2006). This broad, largely aquatic, host range provides numerous opportunities for dispersal of these parasites in aquatic environments.

Despite the importance of whirling disease in trout culture since the early 1900s, transmission of the parasite was not understood until the mid-1980s. The discovery that the aquatic oligochaete *Tubifex tubifex* was required for transmission (Wolf and Markiw, 1984) explains why this parasite is so difficult to control. In its two-host lifecycle (Figure 34.1) there are two waterborne spore stages. Myxospores, which develop in the fish host, are small and nearly oval (length 8.7µm, width 8.2µm and thickness 6.3µm) with two resilient valve shells that surround a sporoplasm and two polar capsules, each containing a single extrudible polar filament (Lom and Hoffman, 1971). Their small size probably facilitates sinking to the

substrate, where they are consumed by *T. tubifex*. The actinospore develops in the worm host, and once it exits the host and inflates, it is much larger and shaped like a grappling hook, with an axis (150µm), three caudal processes (each nearly 200 µm) and three polar capsules (El-Matbouli and Hoffman, 1998). This morphology allows the spore to float and attach to its salmon or trout host, completing the infection cycle.

The original report of *M. cerebralis* infection in 1903 (Höfer, 1903) described spores in the brain of the affected fish, hence the designation 'cerebralis'. However, subsequent studies of the infection revealed that the parasite localizes in the cartilage, primarily of the cranium and spine. Destruction of this tissue causes constriction of the spinal cord and brain stem, and severely infected fish are unable to maintain equilibrium. The resulting erratic whirling or tail chasing behaviour is the hallmark of the disease. Another common disease sign in young fish is darkening of the tail (blacktail), a condition that fades as the fish age. In older fish that survive infection, the disease is usually manifest as skeletal deformities such as a misshapen cranium, shortened opercula, misaligned jaws and spinal curvature (Hedrick and El-Matbouli, 2002).

Whirling disease was first reported in Germany in the late 1890s from farmed rainbow (*Oncorhynchus mykiss*) and brook trout (*Salvelinus fontinalis*) (Höfer, 1903). These species are not native to Europe, and eggs were imported from the US to supplement the culture of the native brown trout (*Salmo trutta*). Speculations on the origin of the outbreaks included feeding the

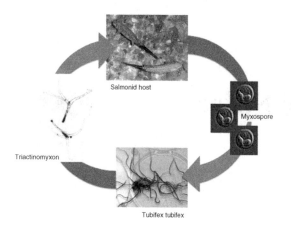

Salmonid host

Myxospore

Triactinomyxon

Tubifex tubifex

Figure 34.1 *Lifecycle of* Myxobolus cerebralis *showing the two alternating spores: the myxospore that is released from the salmonid host and the actinospore released from the invertebrate host,* Tubifex tubifex

trout with raw marine fish infected with the parasite (Phlen, 1924). However, given our current knowledge, the parasite observed in the marine fish was probably a closely related *Myxobolus* species (Hoffman, 1970, 1990). From its first recognition until the mid-1930s, the disease was reported only from Germany, Denmark (Bruhl, 1926) and Finland (Dogel, 1932, in Uspenskaya, 1957). The increasing popularity of rainbow trout as a food fish, coupled with the unrestricted transfers of infected trout throughout Europe following World War II, is believed to have resulted in dissemination of the parasite throughout the continent. An alternative explanation for this widespread occurrence is that uninfected, naïve rainbow trout were introduced into locations where the parasite was already present in the more resistant native brown trout. In his thoughtful review of *M. cerebralis* translocation, Hoffman (1970) considers both scenarios and suggests that the original range of *M. cerebralis* covered an area from central Europe to northeast Asia. In either case, whirling disease severely limited trout production where it became established. Effects on wild, predominantly brown, trout populations appeared to be negligible (Christensen, 1972) except in southern Finland and western Russia (Uspenskaya, 1957). In the latter region, the parasite was reported to be endemic in wild salmonids (*Onchorhynchus* and *Salvelinus*) of the Sakhalin Islands, Sea of Japan (Bogdanova, 1968).

The resistance of brown trout to clinical whirling disease is cited as evidence that *M. cerebralis* evolved as a parasite of this species in Eurasia (Hoffman, 1970; Halliday, 1976). This conclusion does not explain why other European salmonids such as Danube (*Hucho hucho*) and Atlantic (*Salmon salar*) salmon are susceptible to whirling disease, or why coho salmon (*Oncorhynchus kisutch*), which are native to North America, are resistant (Hedrick et al, 1998). However, the fact that the parasite was first observed in Europe and was not detected on other continents for over 50 years provides support for the Eurasian origin of *M. cerebralis*.

Establishment of *M. cerebralis* outside its original range has occurred primarily through transport of live fish or fish products (Hoffman, 1990). Introductions as a result of movements of live fish are most likely to result in establishment because fish are moved in large numbers (providing a large infective source), often multiple times, and into locations where other salmonid species may be present. In the US, trout showing signs similar to whirling disease were reported as early as the 1930s in New York, but the presence of *M. cerebralis* was not confirmed at this time (Bartholomew and Reno, 2002). The first confirmed detection of *M. cerebralis* in the US occurred at the Benner Spring Fish Research Station, Pennsylvania, in 1958 (Hoffman et al, 1962). Around the same time as the discovery in Pennsylvania, the parasite was also present in Nevada, although it went undetected until archived fish cranial tissue samples were examined over a decade later (Taylor et al, 1973), and in 1965, *M. cerebralis* was diagnosed from a private hatchery in California. There has been a great deal of speculation about the source of these initial introductions, with blame often placed on feeding infected frozen trout from Europe to hatchery fish (Hoffman, 1990). However, recent laboratory studies demonstrate that myxospores are not as impervious to environmental extremes as assumed (Hedrick et al, 2008), and their inability to survive freezing suggests that frozen fish are unlikely to have been the primary source of introduction. Early speculation also assumed a direct lifecycle, thus other sources of infection may not have been considered. Perhaps a more likely source of the parasite was live brown trout that were imported from Europe in the 1950s by private and government hatcheries for commercial and sport fishing use. It is also likely that multiple introductions occurred, perhaps by different routes, as new detections could not always be linked to transfers from known positive locations.

Importation of contaminated egg shipments has also been implicated in long distance movement of *M. cerebralis*. Although the parasite is not vertically transmitted (inside the eggs), the associated water and packing material for egg shipment may have introduced the parasite into New Zealand (Hewitt and Little, 1972). Infections of *M. cerebralis* were confirmed there in 1971, although whirling behaviour was reported as early as 1955 (Hewitt and Little, 1972). Because it may take years or even decades for *M. cerebralis* to propagate to levels high enough to cause overt disease, it is possible that the parasite was already established in New Zealand prior to importation bans on salmonid fish products enacted in 1952. The parasite was detected at a low prevalence in wild trout collected from rivers and lakes on the South Island, probably as a result of movements from affected hatcheries, although natural movements of fish (e.g. straying Chinook salmon) may have also played a role (Boustead, 1993).

Once established in the eastern and western US, subsequent spread of *M. cerebralis* is attributed to transfers of live fish (Hoffman, 1990). Most transfers occurred prior to knowledge that the parasite was present, as in Nevada. The localization of the parasite in

cartilage tissue, its prolonged developmental period in the fish and its ability to cause infection without overt disease make diagnosis extremely difficult. Thus despite the enactment of federal measures to prohibit further introductions and concerted efforts on the part of federal and state agencies to prevent further spread, the parasite continues to be detected at culture facilities across the country. *Myxobolus cerebralis* has been reported from 25 states, with most detections occurring since the late 1980s (Bartholomew and Reno, 2002; Arsan et al, 2007). The lack of variation in the genetic sequence (ITS-1 region) between samples collected from Germany and two geographically distant regions of the US (West Virginia and California) supports the view that the parasite was recently introduced into the US from Europe (Whipps et al, 2004). In many of these states the parasite became established, sometimes broadly, in state waters. However, reports of whirling disease in wild fish populations did not occur until the mid-1990s when large declines in free ranging trout were reported from Colorado and Montana (Nehring and Walker, 1996; Vincent, 1996). *Myxobolus cerebralis* has now been reported from five continents (Europe, Asia, Africa, North America and Oceania) (Figures 34.2 and 34.3).

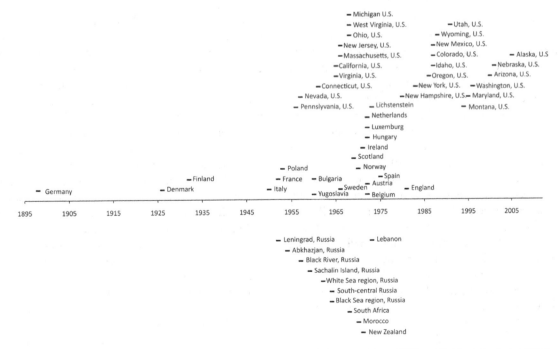

Source: Data represent a compilation of introductions listed in Hoffman (1990), Bartholomew and Reno (2002) and Steinbach Elwell et al (2009)

Figure 34.2 *Timeline for* Myxobolus cerebralis *detection both in Europe, from its native range, and into areas where it has been introduced*

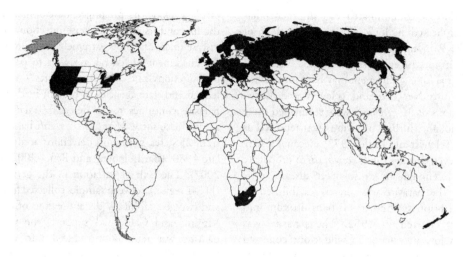

Note: Countries and states shaded black indicate confirmed detection of parasite spores at least once either in culture facilities or in the wild. This does not indicate the continued presence of the parasite or extent of distribution within that country or state. In Alaska (grey shading) parasite DNA was detected from trout at one hatchery without visual observance of parasite spores (Arsan et al, 2007).

Source: Data represent a compilation of introductions listed in Hoffman (1990), Bartholomew and Reno (2002) and Steinbach Elwell et al (2009)

Figure 34.3 *Distribution of* Myxobolus cerebralis

The lack of, and inconsistencies in, survey data, and differences in sensitivity of the parasite detection methods have resulted in confusion about where the parasite occurs. Many reports fail to distinguish between parasite introduction and establishment of the lifecycle. In addition, most surveys present the distribution of *M. cerebralis* cumulatively, so that locations that have tested positive at any time remain listed as such. For example, reports of the parasite in Morocco (Preudhomme, 1970) and South Africa (Hoffman, 1970) probably resulted from importations of infected fish, yet there is little evidence that the parasite became established in these regions. Reports of whirling behaviour alone are an unreliable indicator of *M. cerebralis* infection, as other pathogens may cause similar disease signs. As a result of descriptions of whirling disease from several South American countries, Margolis et al (1996) critically reviewed the data and concluded that the parasite is absent in South America. Similarly, reports of *M. cerebralis* from Mexico (Halliday, 1976) could not be confirmed.

Diagnostic methods used to detect and confirm *M. cerebralis* are variable in both specificity and

sensitivity. Accurate visual identification is confounded by the morphological similarity of *M. cerebralis* myxospores to those of many other species of the genus, including several that are neurotropic or infect tissues that easily contaminate cartilage preparations (Hogge et al, 2004). Because of the inability to confirm parasite infections, reports of the parasite from British Columbia, Canada (Bogdanova, in Halliday, 1976), and Japan (Halliday, 1976) are also suspect (Margolis et al, 1981); however, the proximity of these locations to areas of known establishment makes these detections plausible. Conversely, detection of parasite DNA alone, although indicative of parasite presence, should be followed up with further sampling to confirm establishment has occurred (Arsan et al, 2007; Zielinski et al, 2010). It is now generally accepted that identification of myxospores must be accompanied by a second confirmation test; either detection of parasite DNA or visualization of the parasite in the cartilage tissue. Reports of infections in non-salmonids or from areas where the parasite has not been confirmed by accepted diagnostic procedures (e.g. American Fisheries Society Fish Health Section 2007 Blue Book (AFS-FHS, 2007)) should be carefully considered.

Ecological Niche in Native and Introduced Ranges

As an obligate parasite, the ecological niche of *M. cerebralis* is defined by its hosts and thus it exploits different niches (fish host, water/sediment, oligochaete host) at different stages of its life history. If the hypothesis that the parasite is of Eurasian origin is accepted, then the parasite's original niche was probably as a freshwater parasite of the native brown trout. Within the fish the parasite has a unique tropism; although *M. cerebralis* traverses the epithelium and nervous system, its target is cartilage. When *M. cerebralis* is introduced into freshwater habitats that support salmon and trout, the parasite often finds a suitable fish host. Although most species of the family Salmonidae can serve as host for *M. cerebralis* (summarized by Hedrick et al, 1998; MacConnell and Vincent, 2002), susceptibility to disease varies. Highly susceptible species, such as rainbow and cutthroat trout (*O. clarki*), propagate the parasite in high numbers and may show clinical disease signs. Those with more resistance may not propagate as many parasites, but the host is also more likely to survive until myxospores mature. In natural habitats where trout population declines as a result of the disease have been documented, establishment of the parasite was facilitated by the co-occurrence of brown trout, which provide a reservoir host, and rainbow trout, a highly susceptible host (Baldwin et al, 1998, 2000).

Early speculations on *M. cerebralis* dissemination did not take into account the complex lifecycle of the parasite. We now know that *T. tubifex* is the parasite's definitive host (El-Matbouli and Hoffman, 1998). A cosmopolitan species, *T. tubifex* inhabits sediments of lakes and streams worldwide and is adapted to survive a broad range of conditions, from highly eutrophic environments to pristine alpine streams (reviewed by Steinbach Elwell et al, 2009). This widespread distribution and ability to occupy diverse habitats has permitted *M. cerebralis* to establish over a broad geographic range. In the oligochaete, the parasite is adapted to a broad temperature range, with parasite replication most productive between 10 and 15°C. At temperatures below 10°C and above 20°C the rate of parasite replication declines, and above 25°C it is inhibited (El-Matbouli et al, 1999; Kerans and Zale, 2002; Blazer et al, 2003). Although this may suggest

one reason why *M. cerebralis* is not problematic in warm waters, this temperature is also prohibitive to most species of salmon and trout.

Management Approaches

Management approaches for *M. cerebralis* vary worldwide according to the degree of parasite establishment and the perceived seriousness of the problem. Designing effective control strategies necessitates understanding how the parasite is dispersed. Certainly human movements of infected fish are a primary cause of dissemination, but other vectors, such as infected migratory fish (Engelking, 2002; Zielinski et al, 2010), piscivorous birds (El-Matbouli and Hoffmann, 1991a; Koel et al, 2010), anglers (Gates et al, 2008) and recreational activities (Arsan and Bartholomew, 2009) are also likely to play a role. Risk assessments that identify high probability introduction routes will be important for predicting and managing disease in the future (Bartholomew et al, 2005).

Diagnostic tools

Any strategy for management requires rapid detection and identification of the parasite. In the US, most states now require rigorous testing and adherence to recommended presumptive and confirmatory testing methods (AFS-FHS, 2007) prior to any fish movement. A variety of methods, including sensitive assays for detecting *M. cerebralis* DNA, are available for detecting the parasite in fish, worms and water (reviewed by Andree et al, 2002; Kelley et al, 2004; Steinbach Elwell et al, 2009).

Cultural control and sanitary measures

Trout culture in northern Europe was severely affected by whirling disease when it was first discovered and an intensive research programme resulted in fundamental changes in the trout industry. The transition from earthen to concrete ponds and raceways, disinfection of ponds with calcium cyanide, calcium cyanamide or sodium hypochlorite, and the practice of rearing young fish on parasite-free water sources were key to controlling whirling disease (Hoffman, 1990). It is interesting to note that the most effective control measures for whirling disease, those that targeted the invertebrate host by eliminating habitat, were enacted decades prior to knowledge of the parasite lifecycle. Treatment methods for incoming water by ozonation, chlorination, UV light or filtration are all effective in removing the fragile actinospore stages (Hoffman, 1974; Hedrick et al, 2000; Wagner et al, 2003; Arndt and Wagner, 2003). Hatcheries now also consider their effects on wild populations and include additional measures such as testing resident fish and surface waters for the parasite, and educating anglers.

A variety of chemical and physical methods have been tested for their ability to kill either myxospore or actinospore stages (reviewed by Wagner et al, 2003). Efficacy of many compounds depends on the life stage targeted and conditions under which it is applied. For example, chlorine used at 500mg L^{-1} inactivates purified myxospores, but in the presence of organics the effective concentration is tenfold higher (5000mg mL^{-1})(Hedrick et al, 2008). Actinospores are killed at a much lower chlorine concentration (130mg mL^{-1}) (Wagner et al, 2003).

There are currently no approved therapeutants for treating M. cerebralis infections in fish. Although a number of compounds have been tested (Hoffman et al, 1962; Taylor et al, 1973; Alderman, 1986; El-Matbouli and Hoffmann, 1991b), none have prevented disease or eliminated the parasite and some had toxic side effects. Because of the low priority for treating this disease in culture (due to the more effective sanitary measures), the impediments associated with drug registration, and the limited applicability to wild fish (Steinbach Elwell et al, 2009), it is unlikely that an effective commercial therapeutant will be widely available in the near future.

Control in the wild

Although parasite control and eradication are difficult in natural ecosystems, availability of more sensitive diagnostic methods and changes in fish stocking practices are reducing its further spread. Utilizing larger fish (Ryce et al, 2004), resistant species or resistant strains of fish (Hedrick et al, 2003; Schisler et al, 2006) may reduce both disease impacts and parasite abundance in rivers managed for sport fishing. Additionally, certain lineages of the worm host are unable to propagate the parasite (Beauchamp et al, 2005, reviewed in Steinbach Elwell et al, 2009), and these are being stocked experimentally (in Colorado, US) in attempts to reduce parasite levels (Winkelman et al, 2007; Thompson et al, 2008).

Modifications of streams to reduce T. tubifex habitat have been attempted but the benefits are difficult to assess and are often very localized (Thompson and Nehring, 2004). Although engineered stream modifications have not yielded the intended benefits, passive stream restoration modifications such as increasing riparian shading and reducing inputs of sediments and nutrients may reduce disease severity (Kerans and Zale, 2002; Gilbert and Granath, 2003). Areas with fine sediment are particularly associated with disease risk, probably a combined result of the suitability of this habitat for the oligochaete and the likelihood that microscopic myxospores would settle in the same area (Lemmon and Kerans, 2001; Gilbert and Granath, 2003; Kreuger et al, 2006). In rivers where flows are regulated, increased flows during certain times of the year may reduce fine sediment and thus decrease T. tubifex habitat (Modin, 1998; Hallett and Bartholomew, 2008).

Legislative measures

Worldwide, movement of animals is regulated by the World Organisation for Animal Health (formerly Office International des Epizooties) and M. cerebralis was removed from the list of injurious pathogens that require notification in 1993. However, some countries require imports to be certified free from M. cerebralis or limit imports of live fish to life stages at low risk for pathogen introduction (e.g. eggs). In the US, the concern that M. cerebralis and other pathogens were continuing to be introduced led to development of the first national fish disease control law, adopted in 1968 (Code of Federal Regulations, Title 50, Section 13.7). In addition to federal restrictions, most US states still require fish health monitoring and certification they are M. cerebralis free prior to transporting fish into the state or stocking fish in state waters. Illegal transfers of infected fish are now considered the highest risk human activity for spreading the parasite, largely because of increasing construction of private ponds and the public's lack of understanding of fish health regulations (Steinbach Elwell et al, 2009).

Education

Similar to programmes developed for other invasive species, engaging the public in resource management also plays a role in minimizing further spread of *M. cerebralis*. Recommended precautions include: not transporting live fish from one water body to another or using salmonid parts as bait; rinsing mud and debris from equipment and wading gear; draining water from boats before leaving a water body; allowing boats and gear to dry between trips, and disposing of fish entrails and skeletal parts away from streams or rivers. Individuals with fish ponds also have a responsibility to be aware of fish health regulations (Steinbach Elwell et al, 2009).

Predictive tools

Risk assessment and modelling approaches are useful tools for predicting future spread of *M. cerebralis* (Bruneau, 2001; Bartholomew et al, 2005). For example, high elevation streams in Colorado were predicted to have a low risk of parasite establishment because of limited *T. tubifex* habitat, large distances to locations where the parasite was established and changes in fish stocking protocols (Schisler and Bergersen, 2002). In Alaska, Arsan and Bartholomew (2008) predicted that locations at highest risk for *M. cerebralis* establishment were in the south-central region where *T. tubifex* was abundant; however, adverse effects on natural populations could be mitigated by the low water temperatures, non-susceptible strains of *T. tubifex* and predominance of less susceptible fish species.

Risk assessments can be structured to address specific management questions. For example, the potential risk for adult salmon from endemic areas to disseminate the parasite via straying during their return migration (Arsan and Bartholomew, 2009; Zielinski et al, 2010) or the likelihood that anglers are important vectors for parasite dissemination (Gates et al, 2008). Other modelling approaches evaluate the roles of environmental factors and host community diversity in influencing disease severity (Anlauf and Moffitt, 2008).

Controversies

Detection of *M. cerebralis* in the US Intermountain West during the 1990s (Nehring and Walker, 1996; Vincent, 1996) renewed controversies about potential impacts of parasite introduction that have been voiced each time it has been detected from a new location. The consequences of *M. cerebralis* introduction in the US have varied widely for reasons that are not fully understood. Initially,

there were criticisms that surveillance in many states was insufficient to detect declines in wild populations. This is true in part; however, accumulating evidence supported the absence of population declines in wild fish in areas in the eastern and coastal western US despite the persistence of the parasite in these regions (Modin, 1998; Kaeser et al, 2006). These different experiences with *M. cerebralis* introduction affect how the parasite is regulated in the US and also at the international scale. At present, the parasite is not regulated according to international standards, yet a number of countries (Australia, New Zealand, Canada) and most US states, have legislation to reduce risks of introducing *M. cerebralis*.

The use of a resistant strain of rainbow trout in stocking programmes in the US has also been controversial. Imported from Germany, this strain is highly resistance to whirling disease compared with other strains of rainbow trout (Hedrick et al, 2003). It is expected they would have higher survival rates in the wild and that stocking resistant fish would decrease the number of parasites in the ecosystem. Initial concerns that importing these fish might introduce new pathogens were resolved by quarantine and testing for infectious agents (Bartholomew et al, 2004). Concerns about interactions with other species and suitability in the sport fishery are being addressed by selective breeding with locally adapted rainbow trout strains to produce fish with increased resistance to whirling disease that retain genetic traits important for survival in the wild (Schisler et al, 2006). Perhaps the most valid argument against using these fish is that resistance should be allowed to develop naturally. Although this may take time in some rivers, there is evidence this may be occurring in the Madison River, Montana (Miller and Vincent, 2008). However, it has also been argued that this strain of rainbow trout will be stocked primarily in areas where rainbow trout are themselves an introduced species and already highly domesticated. Resistant trout are now being used experimentally in three states (Colorado, Utah and California). Initial results of stocking of crosses between the resistant trout and locally adapted rainbow trout are promising, with survival and reproduction occurring in one of the most heavily impacted rivers in Colorado (G. Schisler, Colorado Department of Fish and Game, pers. comm.).

Similarly, experimental introductions of resistant strains of *T. tubifex* have been conducted (Thompson et al, 2008; Winkelman et al, 2007). The complexity of natural ecosystems suggest that this could be a difficult method for altering the effects of whirling disease, although it may have some application for newly created habitats, such as private ponds. As with resistant fish, the benefits and risks of introducing resistant worms into the wild should be carefully considered.

Acknowledgements

I thank Stephen Atkinson for creating the distribution map, Adam Ray for assistance creating the timeline and Gerri Buckles for assistance with references. This chapter benefited from discussions with Sascha Hallett, who also provided photos of infected fish and oligochaetes, and with Ronald Hedrick, University of California-Davis. Barbara Nowak, University of Tasmania, and Simon Jones and Mark Higgins, Department of Fisheries and Oceans, Canada each provided current information on fish health requirements in their regions. I also acknowledge the idea for developing the timeline came from preview of Chapter 22 on *Gambusia* by Walton et al.

References

AFS-FHS (American Fisheries Society Fish Health Section) (2007) 'Suggested procedures for the detection and identification of certain finfish and shellfish pathogens', *Fish Health Section Blue Book,* 2007 edition, American Fisheries Society, Bethesda, Maryland

Alderman, D. J. (1986) 'Whirling disease chemotherapy', *Bulletin of the European Association of Fish Pathologists*, vol 6, pp38–40

Andree, K. B., Hedrick, R. P. and MacConnell, E. (2002) 'Review: A review of the approaches to detect *Myxobolus cerebralis*, the cause of salmonid whirling disease', in J. L. Bartholomew and J. C. Wilson (eds) *Whirling Disease: Reviews and Current Topics*, Symposium 29, American Fisheries Society, Bethesda, Maryland, pp197–212

Anlauf, K. and Moffitt, C. M. (2008) 'Models of stream habitat characteristics associated with tubificid populations in an intermountain watershed', *Hydrobiologia*, vol 603, pp147–158

Arndt, R. E. and Wagner, E. J. (2003) 'Filtering *Myxobolus cerebralis* triactinomyxons from contaminated water using rapid sand filtration', *Aquacultural Engineering*, vol 29, pp77–91

Arsan, E. L. and Bartholomew, J. L. (2008) 'Potential for dissemination of the nonnative salmonid parasite *Myxobolus cerebralis* in Alaska', *Journal of Aquatic Animal Health*, vol 20, pp136–149

Arsan, E. L. and Bartholomew, J. L. (2009) 'Potential dispersal of the non-native parasite *Myxobolus cerebralis* in the Willamette River Basin, Oregon: A qualitative analysis of risk', *Reviews in Fisheries Science*, vol 17, pp360–372

Arsan, E. L., Atkinson, S. D., Hallett, S. L., Meyers, T. and Bartholomew, J. L. (2007) 'Expanded geographical distribution of *Myxobolus cerebralis*: First detections from Alaska', *Journal of Fish Diseases*, vol 30, pp483–491

Baldwin, T. J., Peterson, J. E., McGhee, G. C., Staigmiller, K. D., Motteram, E. S., Downs, C. C. and Stanek, D. R. (1998) 'Distribution of *Myxobolus cerebralis* in salmonid fishes in Montana', *Journal of Aquatic Animal Health*, vol 10, pp361–371

Baldwin, T. J., Vincent, E. R., Silflow, R. M. and Stanek, D. (2000) '*Myxobolus cerebralis* infection in rainbow trout (*Oncorhynchus mykiss*) and brown trout (*Salmo trutta*) exposed under natural stream conditions', *Journal of Veterinary Diagnostic Investigation*, vol 12, pp312–321

Bartholomew, J. L. and Reno, P. W. (2002) 'The history and dissemination of whirling disease', in J. L. Bartholomew and J. C. Wilson (eds) *Whirling Disease: Reviews and Current Topics*, Symposium 29, American Fisheries Society, Bethesda, Maryland, pp3–24

Bartholomew, J. L., McDowell, T. S., Mattes, M., El-Matbouli, M. and Hedrick, R. P. (2004) 'Susceptibility of rainbow trout resistant to *Myxobolus cerebralis* to selected salmonid pathogens', in M. J. Nickum, P. M. Mazik, J. G. Nickum and D. D. MacKinlay (eds) *Propagated Fishes in Resource Management*, Symposium 44, American Fisheries Society, Bethesda, Maryland, pp549–557

Bartholomew, J. L., Kerans, B. L., Hedrick, R. P., MacDiarmid, S. C. and Winton, J. R. (2005) 'A risk assessment based approach for the management of whirling disease', *Reviews in Fisheries Science*, vol 13, pp205–230

Beauchamp, K. A., Kelley, G. O., Nehring, R. B. and Hedrick, R. P. (2005) 'The severity of whirling disease among wild trout corresponds to the differences in genetic composition of *Tubifex tubifex* populations in central Colorado', *Journal of Parasitology*, vol 91, pp53–60

Blazer, V. S., Waldrop, T. B., Schill, W. B., Densmore, C. L. and Smith, D. (2003) 'Effects of water temperature and substrate type on spore production and release in eastern *Tubifex tubifex* worms infected with *Myxobolus cerebralis*', *Journal of Parasitology*, vol 89, pp21–26

Bogdanova, E. A. (1968) 'Modern data on the distribution and biology of *Myxosoma cerebralis* (Protozoa, Cnidosporidia) as agent of whirling disease of salmonids', Troisiéme Symposium de la Commission de l'Office International des Épizooties pour l'Étude des Maladies des Poissons, Stockholm

Boustead, N. C. (1993) 'Detection and New-Zealand distribution of *Myxobolus cerebralis*, the cause of whirling disease of salmonids', *New Zealand Journal of Marine and Freshwater Research*, vol 27, pp431–436

Bruhl, L. (1926) 'Bericht über die Fischereitagung in Königsberg i. Pr. Von 2. Juli bis 1 August', *Fischereizeitung*, vol 29, pp813–815

Bruneau, N. A. (2001) 'A quantitative risk assessment for the introduction of *Myxobolus cerebralis* to Alberta, Canada, through the importation of live farmed salmonids', Office International des Epizooties, Paris

Christensen, N. O. (1972) *Fiskesygodomme. Forelaesninger holdt for veterinaerstuderende ved Den kgl*, Veterinaer-og Landbohøjskole, Copenhagen

El-Matbouli, M. and Hoffmann, R. W. (1991a) 'Effects of freezing, aging, and passage through the alimentary canal of predatory animals on the viability of *Myxobolus cerebralis* spores', *Journal of Aquatic Animal Health*, vol 3, pp260–262

El-Matbouli, M. and Hoffmann, R. W. (1991b) 'Prevention of experimentally induced whirling disease in rainbow trout *Oncorhynchus mykiss* by Fumagillin', *Diseases of Aquatic Organisms,* vol 10, pp109–113

El-Matbouli, M. and Hoffmann, R. W. (1998) 'Light and electron microscopic study on the chronological development of *Myxobolus cerebralis* to the actinosporean stage in *Tubifex tubifex*', *International Journal for Parasitology*, vol 28, pp195–217

El-Matbouli, M., McDowell, T. S., Antonio, D. B., Andree, K. B. and Hedrick, R. P. (1999) 'Effect of water temperature on the development, release and survival of the triactinomyxon stage of *Myxobolus cerebralis* in its oligochaete host', *International Journal for Parasitology*, vol 29, pp627–641

Engelking, H. M. (2002) 'Potential for introduction of *Myxobolus cerebralis* into the Deschutes River watershed in central Oregon from adult anadromous salmonids', in J. L. Bartholomew and J. C. Wilson (eds) *Whirling Disease: Reviews and Current Topics*, Symposium 29, American Fisheries Society, Bethesda, Maryland, pp25–31

Gates, K. K., Guy, C. S., Zale, A. V. and Horton, T. B. (2008) 'Adherence of *Myxobolus cerebralis* myxospores to waders: Implications for disease dissemination', *North American Journal of Fisheries Management*, vol 28, pp1453–1458

Gilbert, M. A. and Granath, W. O. Jr (2003) 'Whirling disease of salmonid fish: Life cycle, biology, and disease', *Journal of Parasitology*, vol 89, pp658–667

Hallett, S. L. and Bartholomew, J. L. (2008) 'Effects of water flow on the infection dynamics of *Myxobolus cerebralis*', *Parasitology*, vol 135, pp371–684

Halliday, M. M. (1976) 'The biology of *Myxosoma cerebralis*: The causative organism of whirling disease of salmonids', *Journal of Fish Biology*, vol 9, pp339–357

Hedrick, R. P. and El-Matbouli, M. (2002) 'Recent advances with taxonomy, life cycle, and development of *Myxobolus cerebralis* in the fish and oligochaete hosts', in J. L. Bartholomew and J. C. Wilson (eds) *Whirling Disease: Reviews and Current Topics*, Symposium 29, American Fisheries Society, Bethesda, Maryland, pp45–53

Hedrick, R. P., El-Matbouli, M., Adkison M. A. and MacConnell, E. (1998) 'Whirling disease: Re-emergence among wild trout', *Immunological Reviews*, vol 166, pp365–376

Hedrick, R. P., McDowell, T. S., Marty, G. D., Mukkatira, K., Antonio, D. B., Andree, K. B., Bukhari, Z. and Clancy, T. (2000) 'Ultraviolet irradiation inactivates the waterborne infective stages of *Myxobolus cerebralis*: A treatment for hatchery water supplies', *Diseases of Aquatic Organisms*, vol 42, pp53–59

Hedrick, R. P., McDowell, T. S., Marty, G. D., Fosgate, G. T., Mukkatira, K., Myklebust, K. and El-Matbouli, M. (2003) 'Susceptibility of two strains of rainbow trout (one with suspected resistance to whirling disease) to *Myxobolus cerebralis* infection', *Diseases of Aquatic Organisms*, vol 55, pp37–44

Hedrick, R. P., McDowell, T. S., Mukkatira, K., MacConnell, E. and Petri, B. (2008) 'The effects of freezing, drying, ultraviolet irradiation, chlorine and quaternary ammonium treatments in the infectivity of myxospores of *Myxobolus cerebralis* for *Tubifex tubifex*', *Journal of Aquatic Animal Health*, vol 20, pp116–125

Hewitt, G. C. and Little, R. W. (1972) 'Whirling disease in New Zealand trout caused by *Myxosoma cerebralis* (Hofer, 1903) (Protozoa: Myxosporidia)', *New Zealand Journal of Marine and Freshwater Research*, vol 6, pp1–10

Höfer, B. (1903) 'Ueber die Drehkrankheit der Regenbogenforelle', *Allgemeine Fischerei Zeitschrift*, vol 28, pp7–8

Hoffman, G. L. (1970) 'Intercontinental and transcontinental dissemination and transfaunation of fish parasites with emphasis on whirling disease', in S. F. Snieszko (ed) *Symposium on Diseases of Fishes and Shellfishes*, American Fisheries Society, Bethesda, MD, pp69–81

Hoffman, G. L. (1974) 'Disinfection of contaminated water by ultraviolet irradiation, with emphasis on whirling disease (*Myxosoma cerebralis*) and its effect on fish', *Transactions of the American Fisheries Society*, vol 103, pp541–550

Hoffman, G. L. (1990) '*Myxobolus cerebralis*, a worldwide cause of salmonid whirling disease', *Journal of Aquatic Animal Health*, vol 2, pp30–37

Hoffman, G. L., Dunbar, C.E. and Bradford, A. (1962) 'Whirling disease of trouts caused by *Myxosoma cerebralis* in the United States', United States Fish and Wildlife Service, Special Scientific Report, Fisheries No 427

Hogge, C., Campbell, M. and Johnson, K. (2004) 'Discriminating between a neurotropic *Myxobolus* sp. and *M. cerebralis*, the causative agent of salmonid whirling disease', *Journal of Aquatic Animal Health*, vol 16, pp137–144

Kaeser, A. J., Rasmussen, C. and Sharpe, W. E. (2006) 'An examination of environmental factors associated with *Myxobolus cerebralis* infection of wild trout in Pennsylvania', *Journal of Aquatic Animal Health*, vol 18, pp90–100

Kelley, G. O., Zagmutt-Vergara, F. J., Leutenegger, C. M., Myklebust, K. A., Adkison, M. A., McDowell, T. S., Marty, G. D., Kahler, A. L., Bush, A. L., Gardner, I. A. and Hedrick, R. P. (2004) 'Evaluation of five diagnostic methods of the detection and quantification of *Myxobolus cerebralis*', *Journal of Veterinary Diagnostic Investigation*, vol 16, pp202–211

Kerans, B. L. and Zale, A. V. (2002) 'Review: The ecology of *Myxobolus cerebralis*', in J. L. Bartholomew and J. C. Wilson (eds) *Whirling Disease: Reviews and Current Topics*, Symposium 29, American Fisheries Society, Bethesda, Maryland, pp145–166

Koel, T. M., Kerans, B. L., Barras, S. C., Hanson, K. C. and Wood, J. S. (2010) 'Avian piscivores as vectors for *Myxobolus cerebralis* in the greater Yellowstone ecosystem', *Transactions of the American Fisheries Society*, vol 139, pp976–988

Krueger, R. C., Kerans, B. L., Vincent, E. R. and Rasmussen, C. (2006) 'Risk of *Myxobolus cerebralis* infection to rainbow trout in the Madison River, Montana, USA', *Ecological Applications*, vol 16, pp770–783

Lemmon, J. C. and Kerans, B. L. (2001) 'Extraction of whirling disease myxospores from sediments using the plankton centrifuge and sodium hexametaphosphate', *Intermountain Journal of Sciences*, vol 7, pp57–62

Lom, J. and Dykova, I. (2006) 'Myxozoan genera: Definition and notes on taxonomy, life-cycle terminology and pathogenic species', *Folia Parasitologica*, vol 53, pp1–36

Lom, J. and Hoffman, G. L. (1971) 'Morphology of the spores of *Myxosoma cerebralis* (Höfer, 1903) and *M. cartilaginis* (Hoffman, Putz, and Dunbar, 1965)', *Journal of Parasitology*, vol 56, pp1302–1308

MacConnell, E. and Vincent, E. R. (2002) 'Review: The effects of *Myxobolus cerebralis* on the salmonid host', in J. L. Bartholomew and J. C. Wilson (eds) *Whirling Disease: Reviews and Current Topics*, Symposium 29, American Fisheries Society, Bethesda, Maryland, pp95–107

Margolis, L., McDonald, T. E. and Hoskins, G. E. (1981) 'Absence of the protozoan *Myxosoma cerebralis* (Myxocea: Myxosporea), the cause of whirling disease, in a survey of salmonids from British Columbia', *Canadian Journal of Fisheries and Aquatic Sciences*, vol 38, pp996–998

Margolis, M. L., Kent, M. L. and Bustos, P. (1996) 'Diseases of salmonids resembling myxosporean whirling disease, and the absence of *Myxosoma cerebralis*, in South America', *Diseases of Aquatic Organisms*, vol 25, pp33–37

Miller, M. P. and Vincent, E. R. (2008) 'Rapid natural selection for resistance to an introduced parasite of rainbow trout', *Evolutionary Applications*, vol 1, pp336–341

Modin, J. (1998) 'Whirling disease in California: A review of its history, distribution, and impacts, 1965–1997', *Journal of Aquatic Animal Health*, vol 10, pp132–142

Nehring, R. B. and Walker, P. G. (1996) 'Whirling disease in the wild: The new reality in the Intermountain West', *Fisheries*, vol 21, pp28–30

Phlen, M. (1924) *Praktikum der Fishkrankheiten*, Stuttgart, West Germany

Preudhomme, J. G. (1970) 'Whirling disease of trout in Morocco', *FAO Aquaculture Bulletin*, vol 2, p14

Ryce, E. K. N., Zale, A. V. and MacConnell, E. (2004) 'Effects of fish age and development of whirling parasite dose on the disease in rainbow trout', *Diseases of Aquatic Organisms*, vol 59, pp225–233

Schisler, G. J. and Bergersen, E. P. (2002) 'Evaluation of risk of high elevation Colorado waters to the establishment of *Myxobolus cerebralis*', in J. L. Bartholomew and J. C. Wilson (eds) *Whirling Disease: Reviews and Current Topics*, Symposium 29, American Fisheries Society, Bethesda, Maryland, pp95–107

Schisler, G. L., Myklebust, K. A. and Hedrick, R. P. (2006) 'Inheritance of *Myxobolus cerebralis* resistance among F_1-generation crosses of whirling disease resistant and susceptible rainbow trout strains', *Journal of Aquatic Animal Health*, vol 18, pp109–115

Steinbach Elwell, L. C., Stromberg, K. E., Ryce, E. K. N. and Bartholomew, J. L. (2009) 'Whirling disease in the United States: A summary of progress in research and management', Whirling Disease Initiative and Trout Unlimited, Bozeman, Montana

Taylor, R. E. L., Coli, S. J. and Junell, D. R. (1973) 'Attempts to control whirling disease by continuous drug feeding', *Journal of Wildlife Diseases*, vol 9, pp302–305

Thompson, K. G. and Nehring, R. B. (2004) 'Evaluating the efficacy of physical habitat modification to reduce the impacts of *Myxobolus cerebralis* infection in streams', in *Proceedings of the 10th Annual Whirling Disease Symposium, Salt Lake City, Utah*, Whirling Disease Foundation, Bozeman, Montana, pp46–47

Thompson, K. G., Winkelman, D. and Clapp, C. (2008) 'Investigating the potential for biological control of whirling disease in natural streams through introductions of resistant *Tubifex tubifex*', in *Proceedings of the 14th Annual Whirling Disease Symposium, Denver, Colorado*, Whirling Disease Foundation, Bozeman, Montana, pp1–2

Uspenskaya, A. V. (1957) 'The ecology and spreading of the pathogen of trout whirling disease-*Myxosoma cerebralis* (Höfer 1903, Plehn 1905) in the fish ponds of the Soviet Union', *All-Union Research Institute of Lake and River Fishery*, vol 42, pp47–55

Vincent, E. R. (1996) 'Whirling disease and wild trout: The Montana experience', *Fisheries*, vol 21, pp32–33

Wagner, E. J., Smith, M., Arndt, R. and Roberts, D. W. (2003) 'Physical and chemical effects on viability of the *Myxobolus cerebralis* triactinomyxon', *Diseases of Aquatic Organisms*, vol 53, pp133–142

Whipps, C. M., El-Matbouli, M., Hedrick, R. P., Blazer, V. and Kent, M. L. (2004) '*Myxobolus cerebralis* internal transcribed spacer 1 (ITS-1) sequences support recent spread of the parasite to North America and within Europe', *Diseases of Aquatic Organisms*, vol 60, pp105–108

Winkelman, D., Clapp, C. and Thompson, K. (2007) 'Investigating competition among lineages of *T. tubifex* and the potential for biological control of whirling disease in natural streams', in *Proceedings of the 13th Annual Whirling Disease Symposium, Denver, Colorado*, Whirling Disease Foundation, Bozeman, Montana, pp28–29

Wolf, K. and Markiw, M. E. (1984) 'Biology contravenes taxonomy in the Myxozoa: New discoveries show alternation of invertebrate and vertebrate hosts', *Science*, vol 225, pp1449–1452

Zielinski, C. M., Lorz, H. V. and Bartholomew, J. L. (2010) 'Detection of *Myxobolus cerebralis* in the lower Deschutes River basin, Oregon, USA', *North American Journal of Fishery Management*, vol 30, pp1032–1040

Conclusion

35

Management of freshwater invasive alien species

Robert A. Francis and Petr Pyšek

Introduction

Developing effective prevention and control techniques for the management of invasive alien species (IAS) is the ultimate aim of much research into invasion ecology (Pyšek and Richardson, 2010) and remains a substantial challenge because of the sheer volume of alien introductions (Chapter 1; Mack, 2003; Maynard and Nowell, 2009; Hulme et al, 2009a), the specificity of treatment required for most IAS (see e.g. Simberloff, 2008, and many of the case studies in this volume), and the both extensive and intensive effort that needs to go into controlling most established IAS populations (e.g. Byers et al, 2002; Simberloff, 2008; Boudjelas, 2009). As Simberloff (2008, p149) notes, 'there are at best rough guidelines rather than general rules about what approach to undertake'. This does not mean that prevention and control do not work of course, and successful examples of both can be found in the literature; though many remain unpublished (e.g. Simberloff, 2008; Parkes and Panetta, 2009). For IAS in general, several criteria for effective management have been established, which are essentially:

- early detection and action;
- allocation of sufficient resources for the entire management effort;
- coordination and cooperation of stakeholders, made easier by a governing body with overall responsibility for management;

- sufficient ecological knowledge of the species to guide techniques applied; and
- persistence, particularly following initial failures (see Simberloff, 2008).

In all cases, prevention is the most effective form of management (Maynard and Nowell, 2009; Hulme et al, 2009b).

Freshwater systems present an additional suite of complications for prevention and control efforts, partly for some of the reasons that freshwater ecosystems are vulnerable in the first place, as noted in Chapter 1 (see also Shine et al, 2000; Millennium Ecosystem Assessment, 2005). For example, some European freshwater invaders, such as muskrat (*Ondatra zibethicus*), or Chinese mitten crab (*Eriocheir sinensis*) are known to cause negative impacts in 20–50 regions, and overall freshwater invaders, along with terrestrial vertebrates, affect the widest range of ecosystem services of all taxonomic groups of invasive species in Europe (Vilà et al, 2010). Moreover, invasions in freshwater ecosystems often cause trophic cascades and introduced predators seem to have greater effects due to poor defence mechanisms and greater naïvety of native species towards novel predators (Cox and Lima, 2006). Prevention is difficult due to the scale of anthropogenic use of freshwater systems, which leads to high frequency, intensity and duration of introductions (e.g. Ricciardi and MacIsaac, 2000; Mack, 2003; Hulme et al, 2008; Pyšek et al, 2010). Once species are introduced, the inherent connectivity of freshwater systems makes management problematic, as this (1) allows rapid establishment and spread; (2) often allows rapid recolonization or repopulation following treatment if water bodies are not effectively isolated from propagule sources; and (3) enhances the risk of application of (for example) chemical or biological control methods that are not species-specific or which may easily spread through the system. This chapter briefly reviews the general forms of IAS management in relation to freshwater ecosystems, drawing where appropriate on some of the observations made in the case studies presented in this volume.

A Short Background to Freshwater IAS Prevention and Control

Freshwater IAS are some of the most high profile (see Chapter 1) and were some of the earliest IAS to be demonstrated to have an impact on their invaded ecosystem, whether this was ecological or in relation to human use (Elton, 1958; see discussion in Welcomme, 1988). Consequently, attempts to control freshwater IAS have taken place for many decades, from the local removal of invasive *Cyprinus carpio* (common carp) from isolated water bodies such as a Lake Erie marsh in Michigan in the 1950s (King and Hunt, 1967), to the complete eradication of the mosquito *Anopheles arabiensis* from South America in the 1940s (after introduction around 1929) due to a concerted effort by the Brazilian government and the Rockefeller Foundation (Elton, 1958). As with most attempts at IAS control, early efforts were often driven by self-motivated local groups rather than by government agencies; indeed, public participation remains important for invasive species management, and is increasingly being motivated along the lines of 'citizen science' (Delaney et al, 2008; Boudjelas, 2009; Chapters 25 and 26). Much early invasive species control took place particularly when key economic or recreation resources or human health were perceived to be at risk from the invaders (e.g. East, 1949; Elton, 1958). In the Lake Erie marsh example, control of common carp was conducted by the Erie Club, as the species was reducing macrophytes important for waterfowl (King and Hunt, 1967), while the spread of malaria via *A. arabiensis* provided the impetus for the three year eradication campaign in Brazil (Elton, 1958). Despite these early examples, many freshwater aliens were generally not recognized as potential threats until the 1980s or later (Minchin, 2006).

Governmental oversight of freshwater IAS control became more prevalent as understanding of impacts became apparent, such that in many countries governmental agencies are (in principle) responsible for the majority of IAS control (and general 'biosecurity'), either directly or via subsidiary organizations (e.g. Miller and Fabian, 2004; De Poorter, 2009). Governance in different regions can be spatially and temporally variable, and is a major issue for effective management and control of freshwater IAS; particularly where governmental intervention may conflict with important economic activities (see for example Chapter 20 for a discussion of the controversy surrounding the potential spread of bigheaded carps via the Chicago Sanitary and Ship Canal that connects the Mississippi River system and Lake Michigan, and the corresponding environmental and economic implications of closure, which remain unresolved). De Poorter (2009) also notes the growing importance of international and region-crossing instruments for effective prevention and control, as IAS frequently cannot be managed by national agencies in isolation. Practically, most actual control (via the broad methods described here) is performed at local scales by local (governmental or non-governmental) organizations or community groups (such as the 'toad busting' teams described in Chapter 25; see Boudjelas, 2009), though these methods may be informed by extensive research and resource investment at national and international levels.

Some of the most effective roles of governmental and international agencies have been in implementing legislation and protocols to prevent IAS introduction and spread, which remains the most effective means of management (e.g. Holcombe and Stohlgren, 2009; Maynard and Nowell, 2009). The earliest specific example of a diplomatic agreement on freshwater alien species was the Convention on Fishing in the Danube (Bucharest, 1958) (see Shine et al, 2000), which prohibited the introduction of alien species to the river. More recently, several national and international treaties and other instruments that require some form of prevention and control of freshwater IAS have been established, such as the Convention on Biological Diversity (CBD) (including COP 5 Decision V/8), the Convention on the Conservation of Migratory Species of Wild Animals (Bonn, 1979) and (specifically for wetlands and freshwater ecosystems) the Ramsar Conference of the Parties Resolution VII/14 and the Convention on the Law of Non-Navigational Uses of

International Watercourses (New York, 1997). Shine et al (2000) and De Poorter (2009) review these and other instruments, and discuss how they may feed into the establishment of legal frameworks for IAS. Work by international organizations and programmes such as the Global Invasive Species Programme (GISP), the IUCN (in particular the Invasive Species Specialist Group), the Global Invasive Species Database (GISD), the Delivering Alien Invasive Species Inventories for Europe (DAISIE) project and Global Invasive Species Information Network (GISIN) ensure that prevention and control information is publicly available and shared amongst relevant parties, informing the development of further legislation and empowering local organizations to make appropriate management decisions based on the best information available (e.g. De Poorter, 2009).

Although there is much acknowledgement that in many cases more needs to be done to regulate and prevent alien introductions both nationally and internationally (e.g. Shine et al, 2000; Naylor et al, 2001; Padilla and Williams, 2004), globally there is much awareness about IAS and most countries have legislation and protocols in place to prevent introductions, as well as to control introduced IAS. The complex issues of IAS legislation are not discussed in detail here; for more information on this area see Shine et al (2000), Miller and Fabian (2004), De Poorter (2009) and Kettunen et al (2009).

Prevention of Introduction and Spread

Most countries maintain some form of quarantine and screening systems to meet with national and international instruments on prevention of alien species introductions, whether intentional or unintentional (Maynard and Nowell, 2009). As well as pre-entry risk assessment, border (entry) checks are the key point of quarantine implementation (Maynard and Nowell, 2009; Pyšek and Richardson, 2010). Notable freshwater examples include the inspection or treatment of ballast water in ships (e.g. Minchin, 2006; Chapter 13), port or customs inspection of goods that may contain freshwater species (e.g. tyre inspection for *Aedes albopictus*; Chapter 12), and (though somewhat limited) screening of species introduced for industry, such as the aquarium and aquaculture industries (Padilla and Williams, 2004; Hewitt and Campbell, 2007; Whittington and Chong, 2007). These systems undoubtedly assist in preventing introductions; for

example Michin (2006) outlines several techniques for exterminating species contained in ballast water with varying effectiveness, though notes that the requirement for ships to change ballast in saline waters prior to arrival at a port is the only technique required as standard by the International Maritime Organization (though this can be effective; see Chapter 13). Quarantine efforts can be extensive, operate at many levels, and include substantial resources. Although quarantine systems, protocols and regulations can vary both nationally and internationally and can be somewhat irregular in their application (e.g. Chapter 13), this aspect of management is generally taken very seriously, and increasingly so in recent years (e.g. Cook et al, 2010).

Nevertheless, the amount of material to be screened or treated can be problematic, and represents a practical limit to the efficacy of quarantine and screening measures; this is observed for example in the small proportion of tyres that could actually be inspected for *A. albopictus* presence prior to entry into the US (Chapter 12). Once again, the aquatic vectors responsible for the introduction of many freshwater species means that quarantine protocols are difficult to follow; it is difficult for example to effectively screen ships for hull fouling by IAS, and extensive cleaning is not always possible, even to prevent secondary spread of IAS at the local scale (e.g. Johnson et al, 2001; Vander Zanden and Olden, 2008).

Once a species is effectively established within a region, prevention of further spread is a frequent management objective. This can be achieved in two main ways: effectively containing IAS populations by isolating water bodies or restricting transportation vectors, e.g. limiting access to water bodies to prevent spread by humans; or by controlling populations using some of the methods detailed below, with the aim of reducing movement of potential colonizers or eradicating nascent satellite populations (e.g. Wimbush et al, 2009; Vander Zanden et al, 2010). The rapid rate of spread of some freshwater organisms was noted in Chapter 1, and as a result both containment and control to prevent spread can be difficult and needs to be conducted swiftly (see e.g. Chapters 7 and 25). Isolation of populations to limit natural and anthropogenic spread (e.g. via the closing or installation of barriers in connecting waterways, or the fencing of standing water bodies) is difficult especially when human use of infested systems requires hydrological connectivity and/or direct access, and where freshwater resources are particularly important for regional economies or quality of life (as for example with the North American Great Lakes; see Chapter 20). As discussed in Chapter 1, the artificial interruption of connectivity in some freshwater systems is containing the spread of some IAS (Jackson and Pringle, 2010; Rood et al, 2010; Chapter 16), and it is suggested that the restoration of such connectivity (e.g. via dam removal) should in many cases only be conducted if there are very strong ecological justifications, due to the risk of further IAS spread.

Whether management objectives are eradication, containment or population limitation, there are three main forms of control that can be attempted: physical, chemical and biological. There is extensive literature on all aspects of these (see e.g. McFadyen, 1998; Culliney, 2005; Clout and Williams, 2009) and particular techniques vary according to both broad distinctions between general groups (e.g. aquatic plants vs. aquatic invertebrates) and individual species. Some general summaries of control applications are provided here.

Physical Control

Physical (or mechanical) control (i.e. the manual or automated removal or killing of individuals) of freshwater IAS populations is perhaps the simplest but most labour-intensive management option, and possibly the one least likely to succeed in substantial population reduction or eradication, at least when applied without other forms of supporting control or long term investment of effort and resources (e.g. Charudattan, 2001; Simberloff, 2008; though see e.g. Hein et al, 2007; Coetzee and Hill, 2009). A huge range of different physical control options exist, though some broad distinctions can be drawn between invasive freshwater plants and animals.

Physical control of invasive freshwater plants

Freshwater vegetation can be entirely free floating, emergent (rooted), entirely submerged or riparian (and usually tolerant of inundation). The sedentary nature of most aquatic and riparian vegetation makes physical control of populations somewhat easier than for animals, as individuals can be located and removed relatively easily; though Parkes and Panetta (2009) note that plant eradication (where possible) often takes

longer than for animals. Invasive freshwater plant populations are often controlled by hand-pulling or cutting (when rooted and where populations are accessible), removal via suction harvesting, or the use of heavy duty machinery to destroy or remove biomass, and which includes cutters, shredders, harvesters, dredgers and crushers; some of which are designed for specific invasive plants, for example *Eichhornia crassipes* (water hyacinth; see Chapter 4). Other methods include the covering of vegetation to prevent photosynthesis and ultimately lead to mortality, the draining of water bodies, burning of riparian areas and flooding with fresh or salt water (e.g. Chapters 3, 8 and 10). There are advantages to such forms of mechanical control, particularly for small populations, in that the effects are instant and it is relatively easy to target the IAS, so that collateral damage to native species is minimized (e.g. Greenfield et al, 2007).

However, limitations exist not just in the labour effort required for physical control, but also the associated costs if specialized machinery is used, the environmental disturbances that can result (such as disruption to soils and sediments, or increased water turbidity), and the relative slowness and inefficiency of physical removal. This is often made worse by the ability of many invasive freshwater plants to:

*   regrow very fast following disturbance, so that if even small populations remain rapid recolonization can take place (see for example Chapters 3, 5 and 7);
*   reproduce from vegetative fragments, so that many physical control techniques that break apart the plants can actually facilitate spread and establishment, if all fragments are not removed from the site and disposed of correctly (e.g. Chapters 7 and 9); and
*   remain in seed banks for long periods of time, so that regeneration is hard to prevent (e.g. Parkes and Panetta, 2009).

Often physical control of invasive plants must be timed very carefully to maximize biomass destruction and prevent dispersal of viable propagules (e.g. cutting towards the end of the growing season but before seed is set, so that there is limited energy for regrowth and seeds are not inadvertently moved to other locations; Pyšek et al, 2007).

There have more recently been attempts to encourage utilization of invasive plants as a form of physical control (harvesting), though these have generally not proven successful due to the large amount of plant material that often has to be removed for sufficient resource acquisition (e.g. because of high water content) and to make a notable difference to established populations (see Coetzee and Hill, 2009). Nevertheless, this potential may exist for some species, for example *Trapa natans* (water chestnut; Hummel and Kiviat, 2004).

Physical control of invasive freshwater animals

The large range of taxa, life histories and general vagility of invasive freshwater animals mean that physical control is more problematic than for plants (with correspondingly fewer published case studies), partly because substantial investment in finding individuals prior to control is required. The physical culling or removal of individuals is a common control method, and with sufficient investment of resources can be relatively successful; Hein et al (2007) for example report dramatic reductions in *Orconectes rusticus* (rusty crayfish) populations in a Wisconsin lake following intensive trapping combined with limiting fishing of species known to predate the crayfish, though eradication was not possible. Most physical control is conducted on larger organisms that are easier to observe/trap, and consists of trapping, fishing, electrofishing, killing by hand, shooting, use of explosives, draining, fencing water bodies to restrict access (e.g. for amphibians), the installation of barriers to prevent movement and access to (for example) spawning habitat for fish, and the installation of maceration devices in pipes that draw from water bodies (e.g. Hein et al, 2007; Simberloff, 2008; Ling, 2009; Chapters 22 and 27). The more dramatic physical interventions, such as draining, are more likely to result in successful control or eradication (e.g. O'Keeffe, 2005; Chapters 22 and 25), although these are of course equally harmful to other biota within the system and are only possible in relatively small water bodies (and are not a guarantee of success; see Chapter 17). Trapping and culling techniques are generally most effective in smaller, isolated water bodies and so may be combined with draining/dewatering to reduce the culling effort required and ensure high mortality (e.g. Chapter 27). In general, physical control achieves a temporary decrease in population, which can often be rapidly replenished by colonization from other locations or increased reproduction of the remaining individuals, in the absence of continued management. Physical control is often combined with other forms of control (as discussed below) in order to maximize effectiveness.

One of the main problems with physical control is the labour effort required (Simberloff, 2008) and the need to maintain this over long temporal scales (and often indefinitely). As noted above for plants, one way to solve this problem is to find some way to exploit the IAS for economic gain, for example harvesting of individuals for food; this approach has generally had more success with invasive freshwater animals than with plants. Ling (2009) notes that physical control of *Lates niloticus* (Nile perch) in Lake Victoria (East Africa) has occurred because a fishing industry has developed based around the species, with fillets being harvested for sale in western Europe. However, similar encouragement for utilization of other invasive freshwater animals (e.g. bigheaded carps and *Eriocheir sinensis* (Chinese mitten crab)) have been less successful (e.g. Chapters 16 and 20).

Chemical Control

Control of freshwater IAS by chemical means is very common, and is usually conducted in combination with physical control (e.g. Simberloff, 2008; Boyer and Burdick, 2010; see many of the case studies in this volume). Application of herbicides and pesticides can be an effective intervention, though this is usually most suitable for small and relatively isolated populations and over short timescales (Gopal, 1987; Britton and Brazier, 2006; Coetzee and Hill, 2009). Although a wide range of industrial herbicides and pesticides are available, their application can be somewhat sporadic as not all are authorized for use in all regions, and in some areas potential human health impacts, effects on native species or general public concern (e.g. chemophobia, see Chapter 7) can restrict use (e.g. Getsinger et al, 2008). Consequently, chemical application remains an often used, variably effective and sometimes controversial form of freshwater IAS control.

Chemical control of invasive freshwater plants

Herbicide application is usually conducted on large monospecific stands to help to minimize any non-target impacts on native species (e.g. Parsons et al, 2009), with 2,4-D, diquat, paraquat (N,N'-dimethyl-4,4'-bipyridinium dichloride), glyphosate, dichlobenil, fluridone, copper, endothall and triclopyr being among the most commonly used (Charudattan, 2001; Coetzee and Hill, 2009; Chapters 2–11). However, these have limitations in not just non-target impacts but also the environmental conditions in which they are effective (e.g. Coetzee and Hill, 2009; Chapters 3 and 7), and may not be registered in all invaded regions – for example, diquat and dichlobenil are both now unavailable in Europe, despite their efficacy (e.g. Chapter 3).

For species that can be controlled at low dosage, herbicide application may be relatively inexpensive, though plants that require higher concentrations or cover extensive areas may require more substantial investment of resources. This can be further complicated by the tendency of treated plants to develop resistance to herbicides (e.g. Dayan and Netherland, 2005; Getsinger et al, 2008). Even following successful population or biomass reduction, the same limitations for herbicide application exist for physical control, in that invasive plants can often regenerate from propagules remaining within the system, and may be able to do so faster than native species, which may also be decimated by the herbicide application. Occasionally a non-herbicidal option for plant treatment may be available (usually at fine spatial scales), for example the treatment of *Crassula helmsii* by liquid nitrogen applied directly to the plant (Chapter 3). Overall, however, herbicides are a key tool for invasive freshwater plant control and will remain so for some time.

Chemical control of invasive freshwater animals

As with invasive freshwater plants, freshwater animals are most effectively controlled by chemicals in or around relatively small and isolated water bodies and/or where populations are limited. The size and connectivity of a water body may also limit application potential due to the risk of collateral damage to non-target species. Most chemical control of animals involves the application of some form of pesticide – mainly piscicides, insecticides (also used on a range of invertebrates) and molluscicides – to reduce or eradicate populations: the most common piscicides include rotenone, antimycin, saponins and 3-trifluoromethyl-4-nitrophenol (Ling, 2009), while the most frequently used aquatic molluscicides are niclosamide/clonitralid, and to a lesser extent, sodium pentachlorophenate (e.g. Wang et al, 2006; Clearwater et al, 2008). A large number of insecticides are utilized, mainly based on organochlorine compounds, organophosphates, carbamates and pyrethroids. Although these can be effective, many of these are not species or taxa-specific and can consequently have substantial impacts on the wider ecosystem and native biotic communities. As with herbicides, seasonal, environmental or biological variation in species susceptibility is common, making successful application complex in many cases (e.g. Costa et al, 2008; Ling, 2009; Chapter 22). Further operational complications arise from attempts to calibrate the concentrations required for different sized water bodies and population densities, as well as a reported lack of consistency in the pesticide products applied (Ling, 2009). It is sometimes recommended that where possible infested water bodies should be isolated and the volume of water reduced to maximize application effectiveness and efficiency, where this is feasible (and where any collateral damage can be countered or is justified; e.g. Ling, 2009).

Although relatively uncommon, some taxon-specific chemical control techniques have been developed, with varying specificity. At broad levels, piscicides are usually developed to have limited impacts on non-fish species, though of course non-target fish are also placed at risk. Novel designs for other animals include 'BioBullets', which are aimed at filter feeders such as *Dreissena polymorpha* (zebra mussel) and work by 'encapsulating' a toxin (e.g. potassium chloride) with edible particles, so that the feeders accumulate the toxin without closing their valves, as they would for a toxin sensed in their immediate environment (Aldridge et al, 2006). This has the added advantage of reducing the amount of chemicals added to a water body to achieve the desired effectiveness (Aldridge et al, 2006). Semiochemicals (e.g. pheromones) have also been used as a form of taxon-specific freshwater IAS control, to either facilitate trapping (and killing) or for mating disruption (e.g.

Sorenson and Vrieze, 2003; Aquiloni and Gherardi, 2010; Chapter 17), though such techniques are not commonly employed and require a substantial research effort; as such, they mainly remain in the early stages of research, development and application.

Other forms of chemical control do not include pesticidal chemicals but rather utilize other chemicals to achieve environmental conditions that induce mortality; these may include deoxygenation of water bodies using dry ice, sodium sulphate or nitrogen, raising alkalinity to lethal levels via the addition of lime, blood chemistry disruption from ammonia addition, and chlorine poisoning (Clearwater et al, 2008). These methods are all not taxa-specific and can have ecosystem level impacts, so are generally discouraged except for particular situations, e.g. the treatment of ballast water (e.g. Tamburri et al, 2002; Aldridge et al, 2006; Cooper et al, 2007; Clearwater et al, 2008; Oplinger and Wagner, 2009).

Biological Control

Biological control is potentially effective but often controversial for two main reasons: (1) the potential problems of releasing a further alien species into a new environment, along with the extensive research required to reasonably assume that the species is host-specific and so unintended environmental impacts will not occur; and (2) the number of biological control agents that are released for a given IAS, which can be large (Pemberton, 2000; Culliney, 2005; Simberloff, 2008). As a result, biological control is often considered as something of a 'last resort' (Simberloff, 2008). Usually the aim of biological control is not eradication (as no control agent will completely destroy its food source), but rather population control as a form of impact limitation, or one of a combination of control techniques (i.e. alongside physical and/or chemical control). Although failures in biological control are often reported, there have been some successes (e.g. McFadyen, 1998; Coetzee and Hill, 2009; Simberloff, 2008; Chapter 2).

Biological control of invasive freshwater plants

McFadyen (1998; see also Coetzee and Hill, 2009) notes that biological control has in general been more successful against aquatic plants than terrestrial, though generally only a few case studies are cited and quantitative comparisons are lacking. Certainly dramatic successes have been demonstrated: the huge reductions in extent of *E. crassipes* in Lake Victoria by the weevils *Neochetina eichhorniae* and *N. bruchi*, and control of *Alternanthera philoxeroides* (alligator weed) by *Agasicles hygrophila* are frequently cited examples (Buckingham, 1996; Moorhouse et al, 2001; see Chapters 2 and 4). It should be noted, however, that these successes are not repeated throughout the introduced range of the IAS; variability in level of control is observed, and may be due to environmental variations and/or the general adaptability of invasive plants, which means they can, for example, exploit resources more effectively in some areas and thereby increase their resistance to control (e.g. Buckingham, 2002; Coetzee and Hill, 2009; Chapter 2).

Though they are generally not host-specific and so not technically regarded as biological control, the use of herbivorous fish or birds to reduce invasive plant populations should also be noted; the most commonly cited example being *Ctenopharyngodon idella* (grass carp), which is often used to control extensive patches of invasive weeds, with some success (Tanner et al, 1990; Clayton and Wells, 1999; Chapter 7). This kind of control can be particularly useful for submerged plants, which are generally harder to find control agents for (Chapter 7), though this presents a new range of problems in, for example, determining stocking densities and post-control removal of the species (Clayton and Wells, 1999). Consequently, although the use of biological agents for aquatic plant control can be highly effective, there are limitations, and of the plant species covered in this volume (Chapters 2–11), none have been successfully controlled throughout their invasive range.

Biological control of invasive freshwater animals

There are varying forms of biological control for freshwater animals, ranging from additions of predators or aquatic pathogens that may induce mortality or make individuals non-viable or reduce fitness, to the release of sterile males (sterile male release technique, or SMRT) to reduce breeding success (e.g. Aquiloni et al, 2009; Bellini et al, 2010; Davidson et al, 2010; Chapter 12). The release of predators is perhaps the most common form of biological control, with fish release being a common treatment for invasive freshwater invertebrates (e.g. Chapters 12 and 17), though this can be problematic: most introduced fish will also feed on native invertebrates, and in some cases the control species may in turn become invasive – this can be observed for example in the global spread of *Gambusia* spp. (mosquitofish), which were originally introduced to control mosquitoes acting as vectors for malaria (Chapter 22). Classical biological control requires species-specific predators or pathogens to be released, minimizing the risk of wider impacts on native species; for many invasive freshwater animals, such species-specific controls are more likely to be pathogens than predators, though use of pathogens is relatively novel and experimental, and effectiveness may be limited, or temporary (Ling, 2009; Chapters 19 and 25). For example, Ling (2009) notes success in reduction of *Cyprinus carpio* (common carp) biomass using carp herpes virus, but also that effectiveness varies with environmental conditions and that the development of some level of resistance is likely. In situations where a pathogen (e.g. parasite) is the invader, then the release of resistant hosts into the ecosystem may potentially represent a viable biological control method (Chapter 34).

The release of sterile males may have some success in controlling invasive freshwater animals, whether by release of laboratory-reared sterile males directly into populations (e.g. for invasive insects; Myers et al, 1998) or by sterilizing or otherwise reducing reproductive capacity of males obtained from invasive populations, for example by ionizing irradiation (see Aquiloni et al, 2009 for an example of the experimental testing of this technique for *Procambarus clarkia* (North American crayfish)). The SMRT may have notable potential based on successes with native pests (see e.g. Myers et al, 1998) and has been important for the management of invasive populations of *Petromyzon marinus* (sea lamprey) in the North American Great Lakes (e.g. Bergstedt and Twohey, 2007), though the technique is not always feasible (Chapter 17). More novel forms of biological control will be forthcoming with technological advancements, such as the use of recombinant genetics to engineer viruses or 'autocidal' genes (genes that reduce population size or impact by, for example, inducing mortality in response to an external trigger, or creating sterile or universal same-sex offspring) in invasive species (Thresher, 2008).

Conclusion

Prevention and control of freshwater IAS remain a major challenge, though international efforts are greater now than ever before. The same biological, ecological and environmental factors that make freshwater IAS successful invaders make population isolation, reduction and ultimately eradication difficult, but substantial progress is being made in both research and applications (e.g. Simberloff, 2008; Parkes and Panetta, 2009) as the proportion of papers addressing management-related issues has been steadily increasing over the last two decades (Pyšek and Richardson, 2010). Typically, physical and chemical control are attempted first, and then biological control; due to the potential risks involved, stringent regulations on release of alien species, and the level of research and risk analysis that must be performed prior to release.

Alongside these advances, there is now increasing recognition that environmental degradation facilitates the invasion and impacts of IAS, and that impacts often do not result solely from an IAS itself (Didham et al, 2005; MacDougall and Turkington, 2005). Consequently, the management of IAS in isolation from detrimental influences on freshwater ecosystems is likely to be more difficult or result in failure; this is further complicated by IAS impacts varying with different forms and levels of environmental degradation and at different stages of invasion (e.g. Bulleri et al, 2010; Chapter 3). The next decade will probably see a more explicit focus on the holistic management of IAS, as freshwater systems become increasingly recognized for their many ecological and societal services, and the continued focus of ecological improvement efforts (Dudgeon et al, 2006).

The majority of the case studies presented in this book demonstrate various forms of management, with variable effectiveness; almost all have limitations that preclude comprehensive treatment. This is partly because the chosen species are particularly successful IAS, but the same pattern is found for many of the freshwater IAS reported in the literature and discussed at conferences around the world. It is hoped that these case studies and the more general discussions presented in this volume will contribute to tackling this important global environmental issue, particularly given its importance for our continued use of essential freshwater ecosystems.

References

Aldridge, D. C., Elliott, P. and Moggridge, G. D. (2006) 'Microencapsulated BioBullets for the control of biofouling zebra mussels', *Environmental Science and Technology*, vol 40, pp975–979

Aquiloni, L. and Gherardi, F. (2010) 'The use of sex pheromones for the control of invasive populations of the crayfish *Procambarus clarkii*: a field study', *Hydrobiologia*, vol 649, pp249–254

Aquiloni, L., Becciolini, A., Berti, R., Porciani, S., Trunfio, C. and Gherardi, F. (2009) 'Managing invasive crayfish: Use of X-ray sterilisation of males', *Freshwater Biology*, vol 54, pp1510–1519

Bellini, R., Albieri, A., Balestrino, F., Carrieri, M., Porretta, D., Urbanelli, S., Calvitti, M., Moretti, R. and Maini, S. (2010) 'Dispersal and survival of *Aedes albopictus* (Diptera: Culicidae) males in Italian urban areas and significance for sterile insect technique application', *Journal of Medical Entomology*, vol 47, no 6, pp1082–1091

Bergstedt, R. A. and Twohey, M. B. (2007) 'Research to support sterile-male-release and genetic alteration techniques for sea lamprey control', *Journal of Great Lakes Research*, vol 33, pp48–69

Boudjelas, S. (2009) 'Public participation in invasive species management', in M. N. Clout and P. A. Williams (eds) *Invasive Species Management: A Handbook of Principles and Techniques*, Oxford University Press, Oxford, pp93–107

Boyer, K. E. and Burdick, A. P. (2010) 'Control of *Lepidium latifolium* (perennial pepperweed) and recovery of native plants in tidal marshes of the San Francisco Estuary', *Wetlands Ecology and Management*, vol 18, pp731–743

Britton, J. R. and Brazier, M. (2006) 'Eradicating the invasive topmouth gudgeon, *Pseudorasbora parva*, from a recreational fishery in northern England', *Fisheries Management and Ecology*, vol 13, pp329–335

Buckingham, G. R. (1996) 'Biological control of alligator weed, *Alternanthera philoxeroides*, the world's first aquatic weed success story', *Castanea*, vol 61, pp232–243

Buckingham, G. R. (2002) 'Alligator weed', in R. G. van Driesche, S. Lyon, B. Blossey, M. S. Hoddle and R. Reardon (eds) *Biological Control of Invasive Plants in the Eastern United States*, USDA Forest Service, Morgantown, pp5–15

Bulleri, F., Balata, D., Bertocci, I., Tamburello, L. and Benedetti-Cecchi, L. (2010) 'The seaweed *Caulerpa racemosa* on Mediterranean rocky reefs: From passenger to driver of ecological change', *Ecology*, vol 91, no 8, pp2205–2212

Byers, J. E., Reichard, S., Randall, J. M., Parker, I. M., Smith, C. S., Lonsdale, W. M., Atkinson, I. A. E, Seastedt, T. R., Williamson, M., Chornesky, E. and Hayes, D. (2002) 'Directing research to reduce the impacts of nonindigenous species', *Conservation Biology*, vol 16, no 3, pp630–640

Charudattan, R. (2001) 'Are we on top of aquatic weeds? Weed problems, control options, and challenges', *International Symposium on the World's Worst Weeds: British Crop Protection Council Symposium Proceedings*, vol 77, pp43–68

Clayton, J. S. and Wells, R. D. S. (1999) 'Some issues in risk assessment reports on grass carp and silver carp', Conservation Advisory Science Notes 257, Department of Conservation, Wellington

Clearwater, S. J., Hickey, C. W. and Martin, M. L. (2008) 'Overview of potential piscicides and molluscicides for controlling aquatic pest species in New Zealand', *Science for Conservation 283*, Department of Conservation, Wellington

Clout, M. N. and Williams, P. A. (2009) *Invasive Species Management: A Handbook of Principles and Techniques*, Oxford University Press, Oxford

Coetzee, J. A. and Hill, M. P. (2009) 'Management of invasive aquatic plants', in M. N. Clout and P. A. Williams (eds) *Invasive Species Management: A Handbook of Principles and Techniques*, Oxford University Press, Oxford, pp141–152

Cook, D. C., Liu, S. G., Murphy, B. and Lonsdale, W. M. (2010) 'Adaptive approaches to biosecurity governance', *Risk Analysis*, vol 30, no 9, pp1303–1314

Cooper, W. J., Jones, A. C., Whitehead, R. F. and Zika, R. G. (2007) 'Sunlight-induced photochemical decay of oxidants in natural waters: Implications in ballast water treatment', *Environmental Science and Technology*, vol 41, pp3728–3733

Costa, R., Aldridge, D. C. and Moggridge, G. D. (2008) 'Seasonal variation of zebra mussel susceptibility to molluscicidal agents', *Journal of Applied Ecology*, vol 45, no 6, pp1712–1721

Cox, J. G. and Lima, S. L. (2006) Naïvite and an aquatic-terrestrial dichotomy in the effects of introduced predators', *Trends in Ecology and Evolution*, vol 21, pp674–680

Culliney, T. W. (2005) 'Benefits of classical biological control for managing invasive plants', *Critical Reviews in Plant Sciences*, vol 24, pp131–150

Davidson, E. W., Snyder, J., Lightner, D., Ruthig, G., Lucas, J. and Gilley, J. (2010) 'Exploration of potential microbial control agents for the invasive crayfish, *Orconectes virilis*', *Biocontrol Science and Technology*, vol 20, no 3, pp297–310

Dayan, F. E. and Netherland, M. D. (2005) '*Hydrilla*, the perfect aquatic weed, becomes more noxious than ever', *Pesticide Outlook*, vol 16, pp277–282

Delaney, D. G., Sperling, C. D., Adams, C. S. and Leung, B. (2008) 'Marine invasive species: Validation of citizen science and implications for national monitoring networks', *Biological Invasions*, vol 10, pp117–128

De Poorter, M. (2009) 'International legal instruments and frameworks for invasive species', in M. N. Clout and P. A. Williams (eds) *Invasive Species Management: A Handbook of Principles and Techniques*, Oxford University Press, Oxford, pp108–125

Didham, R. K., Tylianakis, J. M., Hutchison, M. A., Ewers, R. M. and Gemmell, N. J. (2005) 'Are invasive species the drivers of ecological change?', *Trends in Ecology and Evolution*, vol 20, pp470–474

Dudgeon, D., Arthington, A. H., Gessner, M. O., Kawabata, Z. I., Knowler, D. J., Leveque, C., Naiman, R. J., Prieur-Richard, A. H., Soto, D., Stiassny, M. L. J. and Sullivan, C. A. (2006) 'Freshwater biodiversity: Importance, threats, status and conservation challenges', *Biological Reviews*, vol 81, pp163–182

East, B. (1949) 'Is the lake trout doomed?', *Natural History, New York*, vol 58, pp424–428

Elton, C. S. (1958) *The Ecology of Invasions by Animals and Plants*, Methuen and Co. Ltd, London

Getsinger, K. D., Netherland, M. D., Grue, C. E. and Koschnick, T. J. (2008) 'Improvements in the use of aquatic herbicides and establishment of future research directions', *Journal of Aquatic Plant Management*, vol 46, pp32–41

Gopal, B. (1987) *Water Hyacinth*, Elsevier, Amsterdam

Greenfield, B. K., Siemering, G. S., Andrews, J. C., Rajan, M., Andrews, S. P. and Spencer, D. F. (2007) 'Mechanical shredding of water hyacinth (*Eichhornia crassipes*): Effects on water quality in the Sacramento-San Joaquin River Delta, California', *Estuaries and Coasts*, vol 30, pp627–640

Hein, C. L., Vander Zanden, M. J. and Magnuson, J. J. (2007) 'Intensive trapping and increased fish predation cause massive population decline of an invasive crayfish', *Freshwater Biology*, vol 52, pp1134–1146

Hewitt, C. L. and Campbell, M. L. (2007) 'Mechanisms for the prevention of marine bioinvasions for better biosecurity', *Marine Pollution Bulletin*, vol 55, pp395–401

Holcombe, T. and Stohlgren, T. J. (2009) 'Detection and early warning of invasive species', in M. N. Clout and P. A. Williams (eds) *Invasive Species Management: A Handbook of Principles and Techniques*, Oxford University Press, Oxford, pp36–46

Hulme, P. E., Bacher, S., Kenis, M., Klotz, S., Kühn, I., Minchin, D., Nentwig, W., Olenin, S., Panov, V., Pergl, J., Pyšek, P., Roques, A., Sol, D., Solarz, W. and Vilà, M. (2008) 'Grasping at the routes of biological invasions: A framework for integrating pathways into policy', *Journal of Applied Ecology*, vol 45, pp403–414

Hulme, P. E., Pyšek, P., Nentwig, W. and Vilà, M. (2009a) 'Will threat of biological invasions unite the European Union?', *Science*, vol 324, pp40–41

Hulme, P. E., Nentwig, W., Pyšek, P. and Vilà, M. (2009b) 'Common market, shared problems: Time for a coordinated response to biological invasions in Europe?', *Neobiota*, vol 8, pp3–19

Hummel, M. and Kiviat, E. (2004) 'Review of world literature on water chestnut with implications for management in North America', *Journal of Aquatic Plant Management*, vol 42, pp17–28

Jackson, C. R. and Pringle, C. M. (2010) 'Ecological benefits of reduced hydrologic connectivity in intensively developed landscapes', *BioScience*, vol 60, pp37–46

Johnson, L. E., Ricciardi, A. and Carlton, J. T. (2001) 'Overland dispersal of aquatic invasive species: A risk assessment of transient recreational boating', *Ecological Applications*, vol 11, pp1789–1799

Kettunen, M., Genovesi, P., Gollasch, S., Pagad, S., Starfinger, U., ten Brink P. and Shine, C. (2009) *Technical Support to EU Strategy on Invasive Species (IAS): Assessment of the Impacts of IAS in Europe and the EU (Final Module Report for the European Commission)*, Institute for European Environmental Policy, Brussels

King, D. R. and Hunt, G. S. (1967) 'Effect of carp on vegetation in a Lake Erie Marsh', *Journal of Wildlife Management*, vol 31, pp181–188

Ling, N. (2009) 'Management of invasive fish', in M. N. Clout and P. A. Williams (eds) *Invasive Species Management: A Handbook of Principles and Techniques*, Oxford University Press, Oxford, pp185–204

MacDougall, A. S. and Turkington, R. (2005) 'Are invasive species the drivers or passengers of change in degraded ecosystems?', *Ecology*, vol 86, pp42–55

Mack, R. N. (2003) 'Global plant dispersal, naturalization, and invasion: Pathways, modes and circumstances', in G. M. Ruiz and J. T. Carlton (eds) *Invasive Species: Vectors and Management Strategies*, Island Press, Washington, DC, pp3–30

Maynard, G. and Nowell, D. (2009) 'Biosecurity and quarantine for preventing invasive species', in M. N. Clout and P. A. Williams (eds) *Invasive Species Management: A Handbook of Principles and Techniques*, Oxford University Press, Oxford, pp1–18

McFadyen, R. E. C. (1998) 'Biological control of weeds', *Annual Review of Entomology*, vol 43, pp369–393

Millennium Ecosystem Assessment (2005) *Ecosystems and Human Well-being: Synthesis*, Island Press, Washington, DC

Miller, M. L. and Fabian, R. N. (2004) *Harmful Invasive Species: Legal Responses*, Environmental Law Institute, Washington, DC

Minchin, D. (2006) 'The transport and the spread of living aquatic species', in J. Davenport and J. L. Davenport (eds) *The Ecology of Transportation: Managing Mobility for the Environment*, Springer, Amsterdam, pp77–97

Moorhouse, T. M., Agaba, P. and McNabb, T. J. (2001) 'Recent efforts in biological control of water hyacinth in the Kagera River headwaters of Rwanda', in M. H. Julien, M. P. Hill, T. D. Center and J.-Q. Ding (eds) *Biological and Integrated Control of Water Hyacinth*, Eichhornia crassipes, ACIAR Proceedings No 102, pp39–42

Myers, J. H., Savoie, A. and van Randen, E. (1998) 'Eradication and pest management', *Annual Review of Entomology*, vol 43, pp471–491

Naylor, R. L., Williams, S. L. and Strong, D. R. (2001) 'Aquaculture: A gateway for exotic species', *Science*, vol 294, pp1655–1656

O'Keeffe, S. (2005) 'Investing in conjecture: Eradicating the red-eared slider in Queensland', in *Proceedings of 13th Australasian Vertebrate Pest Conference*, Te Papa Wellington, New Zealand, 2–6 May 2005, Manaaki Whenua-Landcare Research, Lincoln, New Zealand

Oplinger, R. W. and Wagner, E. J. (2009) 'Toxicity of common aquaculture disinfectants to New Zealand mud snails and mud snail toxicants to rainbow trout eggs', *North American Journal of Aquaculture*, vol 71, pp229–237

Padilla, D. K. and Williams, S. L. (2004) 'Beyond ballast water: Aquarium and ornamental trades as sources of invasive species in aquatic ecosystems', *Frontiers in Ecology and the Environment*, vol 2, pp131–138

Parkes, J. P. and Panetta, F. D. (2009) 'Eradication of invasive species: Progress and emerging issues in the 21st century', in M. N. Clout and P. A. Williams (eds) *Invasive Species Management: A Handbook of Principles and Techniques*, Oxford University Press, Oxford, pp47–60

Parsons, J. K., Couto, A., Hamel, K. S. and Marx, G. E. (2009) 'Effect of fluridone on macrophytes and fish in a coastal Washington lake', *Journal of Aquatic Plant Management*, vol 47, pp31–40

Pemberton, R. W. (2000) 'Predictable risk to native plants in weed biological control', *Oecologia*, vol 125, pp489–494

Pyšek, P. and Richardson, D. M. (2010) 'Invasive species, environmental change and management, and health', *Annual Review of Environment and Resources*, vol 35, pp25–55

Pyšek, P., Perglová, I., Krinke, L., Jarošík, V., Pergl, J. and Moravcová, L. (2007) 'Regeneration ability of *Heracleum mantegazzianum* and implications for control', in P. Pyšek, M. J. W. Cock, W. Nentwig. and H. P. Ravn (eds) *Ecology and Management of Giant Hogweed* (Heracleum mantegazzianum), CAB International, Wallingford, pp112–125

Pyšek, P., Bacher, S., Chytrý, M., Jarošík, V., Wild, J., Celesti-Grapow, L., Gasso, N., Kenis, M., Lambdon, P. W., Nentwig, W., Pergl, J., Roques, A., Sádlo, J., Solarz, W., Vilà, M. and Hulme, P. E. (2010) 'Contrasting patterns in the invasions of European terrestrial and freshwater habitats by alien plants, insects and vertebrates', *Global Ecology and Biogeography*, vol 19, pp317–331

Ricciardi, A. and MacIsaac, H. J. (2000) 'Recent mass invasion of the North American Great Lakes by Ponto-Caspian species', *Trends in Ecology and Evolution*, vol 15, pp62–65

Rood, S. B., Braatne, J. H. and Goater, L. A. (2010) 'Favorable fragmentation: River reservoirs can impede downstream expansion of riparian weeds', *Ecological Applications*, vol 20, pp1664–1677

Shine, C., Williams, N. and Gündling, L. (2000) *A Guide to Designing Legal and Institutional Frameworks on Alien Invasive Species*, IUCN, Cambridge

Simberloff, D. (2008) 'We can eliminate invasions or live with them: Successful management projects', *Biological Invasions*, vol 11, no 1, pp149–157

Sorensen, P. W. and Vrieze, L. A. (2003) 'The chemical ecology and potential application of the sea lamprey migratory pheromone', *Journal of Great Lakes Research*, vol 29, pp66–84

Tamburri, M. N., Wasson, K. and Matsuda, M. (2002) 'Ballast water deoxygenation can prevent aquatic introductions while reducing ship corrosion', *Biological Conservation*, vol 103, pp331–341

Tanner, C. C., Wells, R. D. S. and Mitchell, C. P. (1990) 'Re-establishment of native macrophytes in Lake Parkinson following weed control by grass carp', *New Zealand Journal of Marine and Freshwater Research*, vol 24, pp181–186

Thresher, R. E. (2008) 'Autocidal technology for the control of invasive fish', *Fisheries*, vol 33, no 3, pp114–121

Vander Zanden, M. J. and Olden, J. D. (2008) 'A management framework for preventing the secondary spread of aquatic invasive species', *Canadian Journal of Fisheries and Aquatic Sciences*, vol 65, pp1512–1522

Vander Zanden, M. J., Hansen, G. J. A., Higgins, S. N. and Kornis, M. S. (2010) 'A pound of prevention, plus a pound of cure: Early detection and eradication of invasive species in the Laurentian Great Lakes', *Journal of Great Lakes Research*, vol 36, pp199–205

Vilà, M., Basnou, C., Pyšek, P., Josefsson, M., Genovesi, P., Gollasch, S., Nentwig, W., Olenin, S., Roques, A., Roy, D., Hulme, P. E. and DAISIE partners (2010) 'How well do we understand the impacts of alien species on ecosystem services? A pan-European, cross-taxa assessment', *Frontiers in Ecology and the Environment*, vol 8, pp135–144

Wang, H., Cai, W.-M., Wang, W.-X. and Yang, J.-M. (2006) 'Molluscicidal activity of *Nerium indicum* Mill, *Pterocarya stenoptera* DC, and *Rumex japonicum* Houtt on *Oncomelania hupensis*', *Biomedical and Environmental Sciences*, vol 19, pp245–248

Welcomme, R. L. (1988) *International Introductions of Inland Aquatic Species*, FAO Fisheries Technical Paper 294, FAO, Rome

Whittington, R. J. and Chong, R. (2007) 'Global trade in ornamental fish from an Australian perspective: The case for revised import risk analysis and management strategies', *Preventive Veterinary Medicine*, vol 81, pp92–116

Wimbush, J., Frischer, M. E., Zarzynski, J. W. and Nierzwicki-Bauer, S. A. (2009) 'Eradication of colonizing populations of zebra mussels (*Dreissena polymorpha*) by early detection and SCUBA removal: Lake George, NY', *Aquatic Conservation: Marine and Freshwater Ecosystems*, vol 16, pp703–713

Index

Page references to illustrations are in italics. Colour illustrations, which appear in the plate sections between pp 216 and 217 in Chapter 18 and pp 288 and 289 in Chapter 24, are referenced by plate number, e.g. *P18.1*.

2,4-D amine   51–52, 108, 440

Abramis brama   279
Abrothrix longipilis   346
Abrothrix xanthorhinus   346
Acartia spp.   165
Acheilognathus rhombeus   385
Actinemys marmorata   333
Adam, P.   118
Adamek, Z.   253
Adelaide River   8
Adhami, J.   139
Aedes aegypti   13, 137, 140–142, 262
Aedes albopictus   13, 137–144, *138*, 437;
    distribution and introduction   137, 139;
    ecological niche   137–138, 140–141;
    impacts   141–142; management
    methods   142–144
Aedes japonicus   13
Aedes triseriatus   140–142
Aequidens pulcher   409
Afghanistan   405
Africa   27, 49, 79, 91, 104, 125, 174, 187, 247,
    263, 285, 299, 358, 404, 439; *Aedes
    albopictus*   139, 142; apple snails   207–209,
    212, 215; *Bothriocephalus acheilognathi*   387,
    392; *Centrocestus formosanus*   406–407;
    Pseudorasbora parva   275, 279; *Trachemys
    scripta*   332, 334; *see also individual countries*
Afropomus spp.   207
Agasicles hygrophila   31, 441
Agonostomus monticola   412
Ahnelt, H.   276
Ahuriri River   83
Akhmerov, A. K.   391
Akodon hershkovitzi   346
Alabama   210–211, 236, 264, 387
Alaska   166, 427
Albania   139, 276
Aleutian disease   372
Alexandrov, B.   8
Algansea tincella   410
Algeria   91, 277
Allee effects   152
Allen, J. D.   346
alligator weed *see Alternanthera philoxeroides*
Alona weinecki   150
Alternanthera denticulata   29, 32

Alternanthera philoxeroides   25–32, *26–27*, 441;
    controversies   32; distribution and
    introduction   26–28; ecological niche   28–30;
    impacts   27–28; management methods   30–32
Alternanthera sessilis   29
Alternaria eichhorniae   52–53
Altica hemensis   72
Amazon basin   47, 49
American beaver   107, 343–353, *344*
American bullfrogs *see Rana [Lithobates] catesbeiana*
American mink *see Neovison vison*
aminote-T   119
Amori, G.   7
Amur sleeper   394
Amynothrips andersoni   31
Anabus testudineus   409
Anacharis crispa see *Lagarosiphon major*
Anderson, C. B.   346, 352
Anderson, M. G.   99
Andorra   332
Angiostrongylus cantonensis   215
Anguilla anguilla   8, 200, 279
Anguilla australis   256
Anguilla dieffenbachii   256
Anguilla japonica   8, 409
Anguillicola crassus   8, 279
Animal and Plant Health Inspection Service
    (APHIS)   130
Anopheles arabiensis   4, 13, 436
Antarctic   152
anthropogenic modification   10–11
antimycin   291
Aphanomyces astaci   196
Aphis fabae   71
Aplocheilus panchax   410
Aplochiton taeniatus   289
Aplochiton zebra   289
Apocyclops dengizicus   162
apple snails   207–216; distribution and
    introduction   208–211; ecological
    niche   212–214; management
    methods   212, 215–216
aquaculture   7, 13, 235–236, 239, 241, 247
aquarium trade   7, 79–80, 176, 195, 211, 226
Aquatic Nuisance Task Force   190
Aquiloni, L.   200
Arai, H. P.   391
Aral Sea   247, 387
Aranda, C.   139
arboviruses   141–142
Archocentrus nigrofasciatum   409
Arcola malloi   31
Arctodiaptomus dorsalis   162, 169
Argentina   127, 359; *Aedes albopictus*   139, 142;
    Alternanthera philoxeroides   26, 28; apple

snails   208–212; *Castor canadensis*   343–
353; *Centrocestus formosanus*   405; *Neovison
vison*   370, 372; *Salmo trutta*   285, 289,
292; *Tamarix* spp.   125
Aristichthys nobilis   410
Aristichthys spp.   235
Arizona   49, 210–211, 225
Arkansas   49, 169, 236–237, 387, 395
Arsan, E. L.   427
Arthington, A. H.   267–269
Arvicola terrestris   371–372
Asia   27, 49, 104, 185, 225, 299, 321, 358, 369,
    385, 404, 422; *Aedes albopictus*   137, 141–
    142; apple snails   207–209, 211–215;
    Centrocestus formosanus   401, 404–406, 408;
    copepods   162–163, 166; *Corbicula
    fluminea*   173–175, 180; *Cyprinus
    carpio*   247–248; *Gambusia affinis* and *G.
    holbrooki*   263–264; *Hypophthalmichthys*
    spp.   235–236, 238–239, 242; *Pseudorasbora
    parva*   275, 279; *Salmo trutta*   285–286;
    Trachemys scripta   331–332, 334; *see also
    individual countries*
Asian clam *see Corbicula fluminea*
Asian gill-trematode   401–413, *402*
Asian tapeworm *see Bothriocephalus acheilognathi*
Asian tiger mosquito *see Aedes albopictus*
Asolene spp.   207, 211, 213
Astacus astacus   196
Astyanax spp.   409
Atherina boyeri   387
Atherinella spp.   409
Atlantic salmon *see Salmo salar*
Atractoscion nobilis   279
Aujeszky's disease   372
Australia   8, 32, 37, 49, 80, 91, 116, 162, 174,
    180, 187, 280, 324, 412, 427; *Aedes
    albopictus*   140, 143; *Alternanthera
    philoxeroides*   25, 27–28, 31; apple
    snails   207–208, 210–211; *Bothriocephalus
    acheilognathi*   387, 394; *Centrocestus
    formosanus*   407–408; *Cyprinus carpio*   247–
    248, 250–253, 255; *Gambusia affinis* and *G.
    holbrooki*   263, 266–269; *Myriophyllum
    aquaticum*   104, 108; *Potamopyrgus
    antipodarum*   223–227; *Rhinella
    marina*   299–304, 306; *Salmo trutta*   285,
    289–292; *Trachemys scripta*   331–332, 334
Australian swamp stonecrop *see Crassula helmsii*
Austria   80, 197, 276, 332, 343, 405
Austropotamobius pallipes   12, 196, 198–199, 201

Bacillus thuringiensis israelensis (BTI)   144
Baer, J. C.   391
Bahamas   332

Bahrain 332
Bailey, S. A. 155
Bajer, P. G. 250
Baker, M. A. 227
Baker, S. J. 365
ballasting 7, 27, 93, 155, 161–163, 166, 169, 176, *186*, 187, 189–190, 239, 437 (*See also* shipping)
Balon, E. K. 7, 248
balsam bashing 75
Baltic Sea 150, 185, 187, 226
Bangladesh 209
Bañuelos, M. J. 71
Baratula barbatula 289
Barbados 139
Barbus spp. 249, 391
Barreto, R. 107
Barria, N. D. 346
Bartholomew, J. L. 427
Bartley, D. M. 208–209
Bastlová-Hanzélyová, D. 95
Bathybothrium rectangulum 392
Batrachochytrium dendrobatidis 315, 324–325
Bauer, O. N. 387
Bee-Bums *see Impatiens glandulifera*
Beerling, D. J. 71
Belarus 371, 375
Belews Lake 389–390
Belgium 80–81, 139, 149, 186, 197, 211, 276, 332, 369
Belk, M. C. 289
Bellamya chinensis 11
Benedict, M. Q. 139–140
Benin 49, 406
benthic zone 52
benzethonium chloride 227–228
Bermuda 142, 332
Berthold, T. 209
bighead carp *see Hypophthalmichthys nobilis*
bigmouth buffalo *see Ictiobus cyprinellus*
biocontrol *see* biological control
biodiversity 3–4, 11
biogeochemical cycles 50, 177–178, 290
biological control 7, 31, 42, 52–53, 62, 84–85, 97–98, 107, 120, 212, 215–216, 229, 276–278, 315, 436, 438, 441–442
Biosecurity Act (1993) 108, 291
Birks, J. D. S. 371
black bean aphid *see Aphis fabae*
black rat *see Rattus rattus*
Black Sea 95, 150, 247, 276
Blakeslee, S. 394
Blaustein, L. 139
bluetongue skink *see Tiliqua* spp.
bogs 10
Boiga irregularis 315
Boklund, R. J. 269
Bolivia 139, 286
Bond, R. 70
Bonesi, L. 373–374
Bonneville Dam 93
bonytail chub 393
Booth, B. D. 6
Borja, E. M. 139
Borneo 407
Bos tarus 346

Bosnia 139
Bossdorf, O. 6
Botanical Society of the British Isles (BSBI) 38
Bothriocephalus acheilognathi 385–395, *389*, *392*; distribution and introduction 385–388; ecological niche 388–390; identification 390–393; impacts 393; management methods 393–395
Bothriocephalus spp. 38–36, 390–392
Bouley, P. 168
Bouma, M. J. 13
Bowes, G. 106
Bowser, P. R. 268
Brachgalaxias bullocki 289
Brachydanio retrio 410
Bramocharax caballeroi 409
Branstrator, D. K. 153
Brazil 7, 13, 52, 104, 139, 141–142, 209–212, 332, 405, 436
bream 279
Bridges, L. 346
Bringsøe, H. 336
Brinsmead, J. 151
Broche, R. G. 139
brook trout 289–290, 421
Brown, D. S. 209
Brown, K. C. 5–6
Brown, M. L. 253–254
Brown, P. 251
brown tree snake 315
brown trout *see Salmo trutta*
Buchan, L. A. 155
Budy, P. 289
Bufo marinus see *Rhinella marina*
Bulgaria 176, 332
Bulleri, F. 43–44
Burma 209
Burton, R. M. 70
Burundi 406
Bush, G. W. 129
Bythotrephes longimanus 149–156, *150*, 161, 169; distribution and introduction 149–152; ecological niche 152–154; management methods 154–156

CABI UK 74
Cabrera, A. 346
Cadi, A. 333
caffeine 315
Cahn, A. R. 253
calcium concentration 49
California 49, 52, 93, 103–104, 108, 116, 125, 143, 153–154, 162–163, 168–169, 187, 189, 195–196, 210–211, 225, 263, 266–267, 291, 323–325, 333, 387, 390, 395, 422, 427
California red-legged tree frogs 323–324, 326
Cambaroides japonicus 196
Cambodia 207, 209–211
Cameroon 139, 406
Campos, H. 251
Canada 59, 67, 176, 187, 195, 225, 240, 255, 263, 280, 321, 332, 343, 358, 371, 395; *Bythotrephes longimanus* 149–151, 153, 155–156; copepods 162–163; *Lythrum salicaria* 91, 93; *Myxobolus cerebralis* 424, 427; *Salmo trutta* 286, 289

Canadian beaver *see Castor canadensis*
canals 93, 150, 166, 228, 240–241, 263, 364, 436
cane toad *see Rhinella marina*
Canis lupus familiaris 346
Cape Horn mouse 346
Capra hircus 346
Carassius auratus 247, 249–250, 255, 410
Carassius repasson 410
carbon dioxide 38, 41, 72, 83, 143–144
Carcinus aestuarii 187
Carcinus maenas 187, 189
Caribbean tree frog *see Eleutherodactylus coqui*
Carlton, J. T. 187
Caspian Sea 188, 247
Castor canadensis 107, 343–353, *344*
Castor fiber 343
Catastomus spp. 289
Caulerpa racemosa 44
Cayman Islands 139
Center, T. D. 52
Center for Biological Diversity 130
Centrocestus formosanus 401–413, *402*
Ceratophyllum demersum 82
Cercopagis pengoi 150, 152–153, 155
Cervus elaphus 346
Cervus nippon 41
Chadee, D. D. 139
Chaetophractus villosus 346
chairmaker's bullrush 362
Cháng Jiang *see* Yangtze River
Channa formosana 409
Chapman, D. C. 238
Chasiempis spp. 314
Chaunis marinus see *Rhinella marina*
chemical control 42–43, 51–52, 62, 74, 84, 97, 108, 120, 143, 180, 215, 227–228, 266, 280, 291, 305, 315–316, 412, 436, 438–441; *see also* herbicides; pesticides
Chen, P. F. 238
Chesapeake Bay 187
Chihuahuan Desert 166
chikungunya 137, 142
Chile 142, 209, 250–251, 286, 289, 292, 332, 343–353, 370, 376
Chiloé bat 346
China 57, 91, 142, 149, 247, 286, 370, 406, 409–412; *Alternanthera philoxeroides* 27, 29, 31; apple snails 208–209, 214; *Bothriocephalus acheilognathi* 385, 387; copepods 162–163, 169; *Corbicula fluminea* 173–175; *Eichhornia crassipes* 48–49; *Eriocheir sinensis* 185, 188, 190; *Hypophthalmichthys* spp. 235–238, 242; *Pseudorasbora parva* 275–279; *Spartina anglica* 113, 118, 120
Chinese mitten crab *see Eriocheir sinensis*
Chinese mystery snails 11
Chionespis spp. 128
Chirostoma humboldtianum 409
Chittka, L. 71
Chlamydia psittaci 362
chlorine 180, 227, 426
cholera 13, 167
Choudhury, A. 391–392
Chromalaena odorata 12

Chubb, J. C.   385, 387, 391
chytrid fungus   315, 324–325
Cichlasoma spp.   409–410
Cinncinnatia comalensis   408
Cirrhinus molitorella   410
citric acid   315–316
Clarias fuscus   410
Clark, P. F.   188, 191
Claudi, R.   176
climate change   53, 72, 81, 140
clopyralid   108
Coffinet, T.   139
Cohen, A. N.   187
Cohen, J.   163
coho salmon   422
Colautti, R. I.   5
Cole, R.   392
Collier, K. J.   210
Colombia   139, 208–210
Colorado   125–126, 130, 153–154, 225, 387,
    423, 426–427
Colorado River   129–130, 388, 394
Columbia River   93, 162–165, 168–169, 187
Comas, L.   96
common carp *see Cyprinus carpio*
common cord-grass *see Spartina anglica*
common floss flower   12
common watercress   42
Comoros Islands   406
Convention on Biological Diversity (CBD) 73–74,
    436
Conwy River   188
Cooke, S. L.   238
Cooper, A.   120
copepods   161–170, 237, 389; ecological
    effects   169–170; *Eurytemora affinis*
    162–163, 165–167, *166*, 169; *Limnoithona
    sinensis* and *L. tetraspina*   162, 167–168,
    170; *Pseudodiaptomus forbesi*   162–165,
    169
Copp, G. H.   280
copper sulphate   180, 227
Coqui Hawaiian Integration and Reeducation
    Project (CHIRP)   317
Corbicula fluminea   173–181, *174*;
    challenges   180–181; distribution and
    introduction   174–176; impacts   177–179;
    lifecycle   176–177; management
    methods   179–180
Corbula amurensis   165
Cordell, J. R.   162, 164, 169
Corethrella appendiculata   141
Cornel, A. J.   139
Costa Rica   142, 209
Côte d'Ivorie   49, 406
cottonwood   127–128
Cottus cognatus   289
Cottus gobio   289
couch grass   31
Cousinho, B.   140
Cowie, R. H.   8, 207–210
coypu *see Myocastor coypus*
Crandall, T. A.   268
Crassula helmsii   37–44, *38*, *40*, *44*, 440;
    biological attributes   38–40;
    controversies   43–44; distribution and

introduction   37–38; ecological
    niche   40–42; management methods   42–44
Crassula recurva see *Crassula helmsii*
crassulacean acid metabolism (CAM) 38, 41
Crawley, M. J.   5
crayfish plague   196–197
Creese, R. G.   280
Crivelli, A. J.   251
Croatia   139, 197, 251, 332, 405, 408, 410
Crocodylus johnsoni   299
Crocodylus niloticus   12
Cronk, Q. C. B.   72
Crooks, J. A.   5
Ctenomys magellanicus   346
Ctenopharyngodon idella   42, 52, 85, 107, 247,
    279, 385, 391, 393, 395, 411, 441
Cuba   139, 209, 262, 405
Cudmore, B.   238
Culex pipiens   137, 140–141
curly water weed *see Lagarosiphon major*
cutthroat trout   289, 291, 425
Cynodon dactylon   31
Cyprinella lutrensis   390, 410
Cyprinella venusta   410
Cyprinus carpio   6, 107, 242, 247–256, *248–249*,
    253, 279, 387–388, 390, 395, 410, 436, 442;
    controversies   256; distribution and
    introduction   247–249; ecological
    niche   249–253; impacts   253–254;
    management methods   254–256
Cyprinus rubrofuscus   248
Cyprus   263, 286
Czech Republic   59, 71–72, 176, 197, 200, 225,
    253, 276, 332, 405

Daehler, C. C.   5–6
Damasonium alisma   41
D'Amore, K.   322
dams   10, 83, 93, 164, 199, 238, 390
Daniel, A. J.   251–252
Danio rerio   411
Danube River   6, 176, 236, 247, 276
Danube salmon   422
Daphnia spp.   150, 152–153, 155
Darwin, C.   5
Darwish, A.   393
Dasyurus hallucatus   299
Data Archive for Seabed Species (DASSH) 188
David, L.   247
David, P.   209
Davydov, O. N.   389
Dawson, F. H.   38–42
Day of the Triffids, The   57
de Lafontaine, Y.   187
De Poorter, M.   436–437
Debussche, M.   8
Deegan, B. M.   41
Defra   73–75
Deilephila elpenor   71
dengue   137, 140–142
Denmark   71, 115–116, 197, 332, 372, 422
Dent, C. L.   13
Department for Environment, Food and Rural
    Affairs   73–75
Department of Agriculture (US) 130, 144, 315
dermatitis   215

Derraik, J. G. B.   140
Deters, J. E.   238
Dethier, M. N.   118
di Castri, F.   7
Diaz, A.   39
dichlobenil   43, 84, 108, 440
dichlorodiphenyltrichlotoethane (DDT) 263
Didymosphenia geminate   10
Dillion, R. T.   210
Dionda spp.   411
Diorhabda spp.   128, 130–131
diquat   42–43, 51, 84, 86, 108, 440
disease   4, 11, 13, 28, 62, 167, 176, 179, 201,
    215, 262–263, 268, 270, 324–325, 327, 362,
    371–372, 393; and *Aedes albopictus*   137,
    141–142, 144; crayfish plague   196–197;
    whirling disease *see Myxobolus cerebralis*; *see
    also* human health; *individual diseases*
Disonycha argentinensis   31
dissolved oxygen   41–2, 50–51, 53, 176
distemper   372
Dittel, A. I.   189
DNA analysis   207, 211, 236, 240–241, 372,
    395, 424–425
Dolores River   130
Dominican Republic   139, 209, 212, 215, 311, 405
Dormitator latifrons   411
Dorosoma cepedianum   237
Doubledee, R. A.   326
Douglas, M.   13
drainage   9, 27, 72, 103, 115, 164, 189, 388,
    391, 395, 439
Dray, F. A.   52
Dreissena polymorpha   13, 155, 225, 440
drinking water   3, 84, 179
Driver, P. D.   252
Drotz, M. K.   188
Dubinina, M. N.   391
Duchovnay, A.   162
Dukes, J. S.   72
Duncan, R. P.   5
Dunstone, N.   371
dykes   119, 357

eared bat   346
Eastern equine encephalitis virus   141
economic issues   3, 43, 50–51, 61, 73, 179, 292,
    314, 364, 372, 376
Edwards, K. R.   96
Egeria densa   79, 82, 84–85
Egypt   49, 91, 209, 212, 332, 391, 406
Eichhornia crassipes   13, 28, 44, 47–53, *48*, 438,
    441; controversies   53; distribution and
    introduction   49–50; ecological
    niche   47–49; impacts   50–51; management
    methods   51–53
El Salvador   139
Eldredge, L. G.   209
Eleagnus angustifolia   129
electrofishing   200, 255, 291
Eleocharis bonariensis   360
elephant hawkmoth   71
Eleutherodactylus coqui   311–317, *312*;
    distribution and introduction   311–313;
    ecological niche   313; impacts   314–315;
    management methods   315–316

Elimia comalensis 408
Elodea canadensis 79, 84
Elodea crispa see *Lagarosiphon major*
Elton, C. S. 4–5, 7, 13
Empidonax traillii extimus 129–130
Emys orbicularis 331
Endangered Species Act (US) 129–130
endothal 84, 108, 440
English Channel 186
English cord-grass see *Spartina anglica*
Environment Agency (UK) 42–43, 73–75, 280
environmental DNA (eDNA) 240–241
Environmental Protection Agency (EPA) 144, 153–154
Enydria aquatica see *Myriophyllum aquaticum*
Epifanio, C. E. 189
Epilobium hirsutum 41, 71
Equus caballus 346
Erie Canal 93
Eriocheir japonica 190
Eriocheir sinensis 4, 12, 185–191, *186, 188*, 435, 439; distribution and introduction 185–187; ecological niche 187–189; management methods 189–191
Eritja, R. 139–140
Esch, G. W. 389–390
Escobar, J. M. 346
Esox lucius 252
Estonia 196, 375
Etheostoma spp. 412
Ethiopia 285, 406
Eubosmina spp. 150
Eubothrium tulipai 390, 392
Eudiaptomus spp. 169
Euneomys chinchilloides 346
Euphorbia splendens 412
Europe 6, 27, 49, 104, 113, 149, 263, 321, 343, 404, 422, 439; *Aedes albopictus* 137, 139–141; *Bothriocephalus acheilognathi* 385, 387; *Centrocestus formosanus* 405, 408; copepods 162, 166; *Corbicula fluminea* 173–176, 181; *Cyprinus carpio* 247, 249–250; *Eriocheir sinensis* 185–187, 190; *Heracleum mantegazzianum* 57, 59–61; *Impatiens glandulifera* 67–68, 70–72, 74; *Lagarosiphon major* 79–81, 83–85; *Lythrum salicaria* 93, 95; *Myocastor coypus* 358, 361; *Neovison vison* 369, 371; *Pacifastacus leniusculus* 195–196, 201; *Potamopyrgus antipodarum* 223–226; *Pseudorasbora parva* 276–277, 279–280; *Salmo trutta* 285–288; *Trachemys scripta* 331–336
European eel 8, 200, 279
European mink 369, 372
European pond turtle 331
European rabbit 346
European Union 43, 61, 74, 179, 325, 331, 335
Eurytemora affinis 162–163, 165–167, *166*, 169
eutrophication 12–13, 49
evapotranspiration rates 27, 50
extinction 11, 196, 369, 408

Fabian, R. N. 437
Fain, A. 391
Falkland Islands 286

Fallopia japonica 9, 73–74
fathead minnow 279
Fausch, K. D. 10
Felipponea spp. 207, 213
Felis domesticus 346
fenuron 119
Fernández-Delgado, C. 251
Ferrari, F. D. 162
fertilizer 53
Ficetola, G. F. 332
Fiji 140, 407
Finger Lakes 150
Finland 149, 186, 188, 197, 225, 332, 343, 372, 374–375, 422
Fish and Wildlife Recovery Plan (US) 130
Fish Invasiveness Scoring Kit (FISK) 279
fisheries 179, 189–190, 227, 235–237, 255–256, 285, 292, 394–395, 412, 427
Fitter, A. 5
flame throwers 42
flooding 3, 9, 30, 96, 103, 119, 127–130, 211, 250, 287
Florida 37, 49, 51–52, 209–215, 264, 311, 321, 387
Florida apple snail 208, 213–214
fluridone 84, 108, 440
Fontenille, D. 139
Food and Agriculture Organization (FAO) 394
Forattini, O. P. 139
Formula-409 227–228
France 8, 27, 71, 80, 113, 115, 139, 173, 186, 197, 224, 251–252, 276, 325, 332, 369, 405
Freeman, T. E. 107
freshwater crocodiles 299
frost 71
Fuegian red fox 346
Fuller, J. 72
fur trade 343–344, 357–358, 369, 377

Gabon 139, 406
Galaxias auratus 255, 289
Galaxias maculatus 268, 289
Galaxias sp. 255, 268, 289
Galerucella spp. 97–98
Galium debile 41
Gambusia affinis 4, 261–270, 387, 389–390, 393, 395, 412, 442; controversies 270; distribution and introduction 261–264; ecological niche 264–265; management methods 265–270
Gambusia holbrooki 261–270, *262*, 387, 389–390, 393, 395, 442; controversies 270; distribution and introduction 261–264; ecological niche 264–265; management methods 265–270
Gambusia yucatana 412
Ganf, G. G. 41
García-Berthou, E. 8, 251
García-Prieto, L. 387
geographical information system (GIS) 140, 144
Georgia (US) 210
Gerberich, J. B. 262–3
Germany 8, 71, 80, 104, 115, 139, 149, 197, 225, 254, 276, 363, 369, 405; *Eriocheir sinensis* 186, 188–190; *Heracleum mantegazzianum* 59, 61; *Myxobolus*

cerebralis 421–423, 427; *Trachemys scripta* 331–332
Ghana 49, 406
Giant Alien project 61–62
giant hogweed see *Heracleum mantegazzianum*
giant ramshorn snail see *Marisa cornuarietis*
Gila cypha 388, 390, 394
Gila elegans 393
Gippoliti, S. 7
gizzard shad 237
Global Invasive Species Database 139–140, 437
globalization 149
Glossogobius giurnus 411
Glugea 268
glyphosate 30, 51–52, 97, 108, 119, 440
Gobio gobio 289
Gobiormorus spp. 411
golden apple snail see *Pomacea canaliculata*
golden galaxias 255, 289
golden shiners 392
goldfish see *Carassius auratus*
Goodea atripinnis 411
Gosling, L. M. 365
governments 32, 73–74, 208, 239, 292, 303, 314–315, 343, 345, 348–349, 436
Govindarajulu, P. 326
Gozlan, R. E. 276, 278–279, 281
Grabas, G. P. 99
Granath, W. O. 389–930
Grand Canyon 388, 390, 393–394
grass carp see *Ctenopharyngodon idella*
Gray, A. J. 115
Gray, C. J. 108
Great Depression 93
Great Lakes 11, 93, 149–150, 152–155, 162, 166, 187, 225, 238, 240–242, 395, 438, 442 (*See also individual lakes*)
Great Ouse River 188
Great Plains 166, 225
great pond snail 11
Greece 139, 197, 225, 332
green frogs 323
Green River 130
Grevstad, F. S. 97
Grey, I. 326
grey fox 346
Grime, J. P. 71–72
Guadeloupe 209, 212, 332, 405
Guam 137, 142, 208–209, 211, 332, 407
Guan, R.-Z. 199
guanaco 346
Guatemala 139
Guha, D. 250
Guichón, M. L. 361
Guinea 139
Gunnison River 130
guppies 263, 412
Gutiérrez-Aguirre, M. 162
Guyana 209, 332

Habit, E. 289
habitat loss 9
Hacker, S. D. 118
Hall, R. O. 226
Halliwell, E. C. 371
Hallows, H. B. 42

Halwart, M.   208–209
Hammond, M. E. R.   120
Harrell, J. P.   7
Hasler, A. D.   252
Hatler, D.   371
Hawaii   49, 137, 176, 207–216, 262, 299, 311–
    317, 385–387, 390, 404, 407–408, 412
Hawaiian duck   216
Hawaiian hoary bat   314
Hayes, K. A.   207, 209–211
health impacts *see* disease; human health
Hefti, D.   197
Hein, C. L.   439
Hejda, M.   72
Hemiamphus dussmieri   412
Henebry, M. S.   241
Heracleum mantegazzianum   9, 13, 57–63, *60*,
    70, 73–74; controversies   63;
    distribution   *58*, 59; ecological
    niche   59–61; management methods   61–63
Heracleum persicum   57
Heracleum sosnowskyi   57
Hérault River   8
herbicides   30–31, *48*, 50–53, 62–63, 74, 266,
    439–440; 2,4-D amine   51–52, 108, 440;
    aminote-T 119; clopyralid   108;
    dichlobenil   43, 84, 108, 440;
    diquat   42–43, 51, 84, 86, 108, 440;
    endothal   84, 108, 440; fenuron   119;
    fluridone   84, 108, 440; glyphosate   30,
    51–52, 91, 108, 119, 440; imazamox   108;
    imazapyr   31, 108, 130; Rodeo   97, 120;
    Roundup   51, 97; triclopyr   84, 108, 440;
    see also chemical control
Herborg, L.-M.   188, 395
Herichthys cyanoguttatus   410
Herpestes javanicus   314
Hesse, K.-J.   139
Heterandria bimaculata   412
Heterandria spp.   412
Hicks, B.   253
Higgins, S. I.   5
Higgs, S.   188
Hill, W. R.   238
Himalayan balsam *see Impatiens glandulifera*
Himalayas   68, 71–72, 74
Hiodon spp.   392
Histiotus montanus   346
HIV 13
Hoffman, G. L.   387
Hoffnagle, T. L.   394
Hofstra, D. E.   108
Holdich, D. M.   197
Holling, C. S.   12
Holzapfel, E. P.   7
Honduras   139
Hong Kong   185, 211–212, 332, 406
Horwath, J. L.   187
house mouse   346
hover-fly   39, *40*
Hribar, L. J.   162
Hucho hucho   422
Hudson, P. L.   162
Hudson River   93
Hughes, J. D.   7
Hulme, P. E.   7

human health   13, 141–142, 191, 215, 362; *see
    also* disease
human immunodeficiency virus   13
humpback chub   388, 390, 394
Hungary   57, 176, 197, 236, 252, 276, 332
Hunt, B. P.   209
Hunt, R. H.   139
hydrated lime   315–316
Hydrobia jenkinsi see *Potamorpyrgus antipodarum*
hydroelectric power   103
Hyla arborea   333
Hylobatis transversovitattus   97
Hyphessobrycon sp.   409
Hypophthalmichthys molitrix   235–242, *236*, *241*,
    247, 254, 279, 411
Hypophthalmichthys nobilis   235–242, *236–237*,
    254
Hypophthalmichthys spp.   235–242;
    controversies   240–241; distribution and
    introduction   235–237; ecological
    niche   237–238; management
    methods   238–240
Hypostomus plecostomus   412

Iceland   372, 374–375
Ictalurus punctatus   412
Ictiobus cyprinellus   237
Idaho   93, 130, 163–165, 169, 225
Illinois   59, 236–238, 240–242, 254–255, 263–
    264, 395
Illinois River   237–238, 240, 242
Ilyodon furcidens   411
Ilyodon whitei   411
imazamox   108
imazapyr   31, 108, 130
Imhoff, E.   196, 198, 201
Impatiens glandulifera   8–9, 67–75, *68*;
    adaptation   71–72; distribution and
    introduction   67–71; impacts   72–73;
    legislation   73–74; management
    methods   74–75
Impatientinum balsamines   71
Import of Live Fish Act (1980) 280
India   27, 67, 137, 141, 209, 264, 286, 332,
    405, 410, 412
Indian balsam *see Impatiens glandulifera*
Indian mongoose   314
Indian Ocean   137, 142
Indiana   169, 264
Indonesia   27, 208, 332, 407
Inland Fisheries Service   256
insecticides   263, 266, 440
integrated pest management (IPM) 267–268
Intergovernmental Panel on Climate Change
    (IPCC) 140
International Code of Zoological Nomenclature
    (ICZN) 404
International Union for Conservation of Nature
    (IUCN) 4, 321, 437
Invading Species Awareness Program   156
invertebrates   41–42, 52, 72, 87, 177, 199, 201,
    226–227, 253, 323; *see also individual species*
Iowa   130
Iran   57, 91, 188, 276, 286, 405
Iraq   57, 91, 188, 225, 405
Ireland   59, 79–81, 188

Iridomyrmex   304, *305*
irrigation   3, 27, 103, 179, 214, 228, 357, 364
island apple snail *see Pomacea insularum*
isopropyl alcohol   228
Israel   139, 332, 406, 410
Italy   27, 80, 139, 142, 149, 169, 188, 197, 277,
    332, 336, 364–365, 405

Japan   49, 67, 91, 104, 142, 185, 187, 358, 369,
    404, 406, 409–410, 412, 424; apple
    snails   208–209, 211–213, 215;
    Bothriocephalus acheilognathi   385, 387;
    copepods   163, 169; *Cyprinus carpio*   247–
    248; *Pacifastacus leniusculus*   195–196;
    Potamorpyrgus antipodarum   223, 225;
    Pseudorasbora parva   275–277; *Salmo
    trutta*   286, 289, 292; *Trachemys
    scripta*   331–332
Japanese crayfish   196
Japanese eel   8, 409
Japanese encephalitis virus   141
Japanese knotweed   9, 73–74
Java   104, 407
Johnsen, P. B.   252
Johnson, P. T. J.   11, 169
Joly, P.   333
Jones, M. J.   251–252
Jordan   406
Joyner, B. G.   107
Jug, T.   289
Jumping Jacks *see Impatiens glandulifera*

Kansas   125, 130
Karch, S.   139
Kay, B. H.   140
Keane, R. M.   5
Keawjam, R. S.   211
Kentucky   395
Kenya   49, 155, 253, 285, 332, 357–358, 406
Kercher, S.   10
Kerguelen Islands   286
Kettunen, M.   437
Kew Gardens   57, 59, 67, 71
Khan, T. A.   251
Kiesecker, J. M.   323–324
kikuyu   30
Kimmerer, W. J.   168
King, C.   373
Kinmen Islands   406
Kirpitchnikov, V. S.   247
Kitano, S.   289
Klavsen, S. K.   38
Klobucar, A.   139
Klobucar, A.   139
Kollmann, J.   71
Korea   168–169, 175, 208–210, 235, 275, 332,
    358, 404, 406
Korting, W.   385
Kottelat, M.   248
Kowarik, I.   93
Kraus, F.   299
Küpper, H.   42
Kwong, K.-L.   212

Labeo rohita   411
Labeobarbus kimberleyensis   390
LaCrosse encephalitis   140–142

Lagarosiphon major   79–87, *80*;
   controversies   86–87; distribution and
   introduction   79–81; ecological
   niche   81–83; management methods   83–86
Laio, H.   391
Laird, M.   262–263
Lake Champlain   150, 162
Lake Corrib   81, 86
Lake Crescent   255–256
Lake Dunstan   81–82
Lake Erie   93, 150, 187, 189, 225, 387, 436
Lake Michigan   9, 238, 240, 436
Lake Ontario   149, 225, 387
Lake Rotoma   81–82, 84
Lake Rotorua   79, 84
Lake Sorrell   252, 255–256
Lake Taupo   79, 81
Lake Victoria   11, 49, 439, 441
Lama guanicoe   346
Lampetra planeri   289
Langdon, S. J.   41–42
Languriophasma cyanea   72
Lanistes spp.   207, 215
Laos   209, 211, 406, 410–411
large river otter   346
larger hairy armadillo   346
Lasiurus cinereus semotus   314
Lates nilotica   11, 439
Latvia   57, 197, 332
Laurentian Great Lakes *see* Great Lakes
Laverty, T. M.   99
Lavoie, C.   99
Leach, J. H.   187
Lebanon   263
Lee, C. E.   162, 169
Lee, W.   116
Lee, Y.-C.   209
legislation   7, 73–74, 98, 108, 129, 189, 201,
   239, 241, 267, 269, 280, 291–292, 314, 325,
   335, 426, 436
Lemaitre, A.   139
Lepidomeda spp.   289
Lepomis spp.   392, 409
Leptodora kindtii   153
Leptospira spp.   362
Leslie, A. J.   12
Lesotho   285
lesser joy weed   29, 32
Letelier, S.   209
Leucaspius delineates   279
Lever, C.   299
Libya   91, 406
life history plasticity   286–287
Limnoithona sinensis   162, 167–168
Limnoithona tetraspina   162, 167–170, *167*
Limnoperna fortunei   179
Ling, N.   442
liquid nitrogen   42
Listronotus marginicollis   107
Lithuania   197, 276–277, 332, 374–375
Little Colorado River   388, 390, 393–394
littoral zones   95
Liu, H.   5
Liza macrolepis   412
Lizarralde, M. S.   346
Lloyd, L. N.   268–269

Lockwood, J. L.   5
Lolium multiflorum   360
long haired grass mouse   346
long-tailed mouse   346
Lonsdale, W. M.   7–8
Lontra felina   346
Lontra provocax   346
Loo, S. E.   226
López, M. A.   210
Louisiana   52, 210, 333, 362–365, 387, 395
Lounibos, L. P.   139
Lu, J. B.   49, 51
Lucas, J. R.   70
Ludwigia palustris   41
Lugo, E. D.   140
Lun, Z.   391
Luo, H. Y.   390–391
Luxembourg   197
Lymnaea columella   212
Lymnaea stagnalis   11
Lythrum salicaria   41, 71, 91–99, *92*, *94*;
   biological attributes   91–93;
   controversies   98–99; distribution and
   introduction   93–95; ecological niche   95;
   management methods   95–98

Maberly, S. C.   38
Mabuchi, K.   247–248
Maccullochella macquariensis   289
Macdonald, D. W.   371
Machino, Y.   197
MacIsaac, H. J.   4, 152, 154–155
Mack, R. N.   7
Mackie, G. L.   176
Macquaria australasica   289
Macrocystis pyrifera   73
macroinvertebrates   50
Madagascar   79, 285
Madison River   225, 427
Madsen, J. D.   108
Magellanic tuco tuco   346
Mal, T. K.   99
malaria   7, 262–263, 442
Malawi   49, 285, 407
Malaysia   209–211, 332, 387, 407
Mali   49
Malta   332, 405
management methods   30–32, 42–44, 51–53,
   61–63, 83–86, 95–98, 107–108, 119–120,
   129–130, 142–144, 154–156, 189–191, 212,
   215–216, 227–229, 265–270, 279–280, 303–
   305, 315–316, 364–365, 393–395, 425–428;
   see also biological control; chemical control;
   mechanical control
Mandrak, N. E.   238
Manuka Natural Area Reserve   316
marble trout   289–290
Marciochi, A.   289
Margolis, M. L.   424
Maricopa Audubon Society   130
Marisa cornuarietis P18.1, P18.3, 208–209, 211–
   212, 214–215
Marisa spp.   207, 213
Marks, T. C.   117
marsupial quolls   299
Marten, G. G.   162

Martinique   405
Maryland   93, 187, 363–364
Massemin, D.   209
Mauremys leprosa   331
Mauritius   385, 387
Mayama, H.   289
McDowall, R. M.   289
McFadyen, R. E. C.   441
McHugh, P.   289
McIntosh, A. R.   289
McMahon, R. F.   175
McMichael, A. J.   13
'meat ants'   304, *305*
mechanical control   30–31, 51, 85–86, 96–97,
   107–108, 119–120, 199–200, 216, 228, 267,
   291, 303–304, 315–356, 325–356, 335, 438–
   439
medicinal qualities   93, 120–121
Mediterranean Sea   44, 263
Megacyclops viridis   162
Megalobrama amblycephala   411
Melanoides tuberculata   402, 404, 408, 412
Melero, Y.   371
Mentha aquatica   41
Merrimac River   93
Mesocyclops spp.   162
metaldehyde   215
methyl bromide   143
Metialma suturella   72
metsulfuron   31
Mexico   125, 139, 162–163, 166, 225, 261,
   263–264, 286, 321, 331–332, 343, 369, 387,
   405, 409–412, 424
Michigan   150–151, 169, 436
Microphallus sp.   224
Micropterus spp.   409
Mignani, R.   139
Miller, M. L.   437
Mimosa pigra   8
Minchin, D.   8, 188
mink enteritis   372
Minnesota   150–151, 252
Misgurnus spp.   410
Mississippi River   187, 189, 225, 237–238, 240,
   263, 387
Missouri   130, 154, 169, 263, 395
Missouri River   225, 237, 387
Mitchell, A. J.   393
mitochondrial DNA (mtDNA)   211, 236
Mizzan, L.   188
Mohawk River   93
Molnar, K.   385, 391
Monaco   139
Montana   93, 125, 130, 195, 225, 423, 427
Montenegro   139
Mooney, H. A.   13, 72
Moquete, C.   209
Moravec, F.   391
Morocco   91, 407, 424
morphology   6
mosquitoes   7, 13, 28, 103, 262–264, 266, 270,
   436, 442; *see also* individual species
mosquitofish (eastern) *see Gambusia holbrooki*
mosquitofish (western) *see Gambusia affinis*
Moyle, P. B.   289
Mozambique   407

Mudry, D. R. 391
Mugil cephalus 412
Mukgerjee, D. 250
Murai, E. 391
Murkaeva, A. 247
Murphy, B. R. 50
Murray River 38, 248, 251–252
Mus musculus 346
Muskegon Lake 150
muskrat 346, 363, 435
Mustela furo 374
Mustela lutreola 369, 372
Myanmar 209
Mylopharyngodon piceus 411
Myocastor coypus 346, 357–365, *358*, *363*;
    distribution and introduction 357–358,
    362; ecological niche 358–361;
    impacts 362–364; management
    methods 364–365
Myosotis scorpioides 41
Myotis chiloensis 346
Myriophyllum aquaticum 103–108; biological
    attributes 103; distribution and
    introduction 104; ecological niche 104–
    106; management methods 107–108
Mytilopsis spp. 179–180
Mytilus spp. 179
Myxobolus cerebralis 421–428, *422*;
    controversies 428; distribution and
    introduction 421–424; ecological
    niche 425; management methods 425–428

Namibia 332, 407
Nannini, M. A. 289
Nasturtium officinale 42
National Environment Research Council (NERC)
    75
National Lakes Assessment (US) 153
National Road 93
Natural England 190
Nebraska 125, 130
Neck, R. W. 209
Nehring, S. 115
Nelson, D. 300
Neochanna burrowsius 289
Neochetina spp. 52, 441
Neogobius melanostomus 9
Neovison macrodon 369
Neovison vison 346, 369–377, *370*;
    controversies 376–378; distribution and
    introduction 369–370; impacts 370–372;
    management methods 372–376
Nepszy, S. J. 187
Netherlands 59, 115, 139, 143, 149, 186, 190,
    197, 276, 332, 369, 405
Network Rail 74–75
Nevada 125, 130, 225, 266, 422–423
New England 153–154
New Jersey 264
New Mexico 125, 128
New York 93, 150–151, 422
New Zealand 37, 67, 91, 210, 263, 280, 374,
    407; *Aedes albopictus* 140, 143; *Alternanthera
    philoxeroides* 27–28, 31; *Cyprinus
    carpio* 248–252, 256; *Lagarosiphon
    major* 79, 81–86; *Myriophyllum

aquaticum 104, 108; *Myxobolus
    cerebralis* 423, 427; *Potamorpyrgus
    antipodarum* 223–225; *Salmo trutta* 285–
    286, 292, 388–390; *Spartina anglica* 113,
    116; *Trachemys scripta* 331–332
New Zealand Biosecurity Act (NZBA) 108, 291
New Zealand mudsnail *see Potamorpyrgus
    antipodarum*
New Zealand pygmyweed *see Crassula helmsii*
Newfoundland 59
Nguma, J. F. M. 209
Nicaragua 140, 332
niclosamide 215
Niger 49
Nigeria 49, 139, 407
Nile crocodile 12
Nile perch 11, 439
Nile River 49, 391
Niphograpta albiguttalis 52
Nippotaenia mogurndae 394
Nishigori, M. 404
noble crayfish 196
normalized difference vegetation index (NDVI)
    128
North American beaver 107, 343–353, *344*
North American signal crayfish *see Pacifastacus
    leniusculus*
North Carolina 37, 263, 387
North Dakota 125, 130
North Sea 186–187
Norway 116, 149, 197
Norway rat 346
Notemigonus chrysoleucas 392
Nothofagus spp. 345
Notropis spp. 411
Nuphar lutea 362
nutria *see Myocastor coypus*
Nuttall, P. M. 254

Oberholzer, I. G. 107
O'Driscoll, P. 394
Ogata, K. 139
Ohio 150–151, 154, 167
Ohtsuka, S. 162, 168
Ojaveer, H. 188
Oklahoma 168
Oligoryzomys longicaudatus 346
Oliver, E. 75
Olrog, C. C. 346
Olsen, D. G. 289
Oman 406
Omora Ethnobotanical Park 347, 349–350
Oncorhynchus clarkii 289, 291, 425
Oncorhynchus kisutch 422
Oncorhynchus mykiss 10–11, 290, 412, 421–422,
    425, 427
Ondatra zibethica 346, 363, 435
Ophiocephalus straitus 412
Opsius stactgalus 128
Orconectes spp. 9, 11, 439
Oregon 125, 130, 225, 362
Oreochromis spp. 410
Ornamental Jewelweed *see Impatiens glandulifera*
Orsi, J. J. 162, 168
Oryctolagus cuniculus 346

Osborne, M. W. 251
Osorio Sarabia, D. 387
Osteochilus hasseltii 411
Ouse River 190
oxygen 83, 103, 117, 119, 212–213, 264

Pacifastacus fortis 196
Pacifastacus leniusculus 9, 11–12, 195–201, *196*,
    198; controversies 201; distribution and
    introduction 195–196; impacts 196–199;
    management methods 199–201
Pacifastacus nigrescens 11, 196
Padilla, D. K. 7, 155
Pakistan 286, 406
Palau 208–209
Panama 140, 209–210, 332
Panama Canal 263
Panetta, F. D. 438
Pangani River 49
Panzacchi, M. 365
Paperna, I. 387, 391, 394
Papes, M. 151
Papua New Guinea 209, 211, 285
Parachromis spp. 410
Paragonimus westermani 191, 408
Paraguay 140, 209, 212, 332
Paraná River 26
paraquat 440
Parasilurus asotus 412
parasites 11, 191, 215, 224, 229, 249, 279, 304,
    352, 390–395, 401, 404, 408, 412, 442; *see
    also individual species*
Parkes, J. P. 438
parrot feather *see Myriophyllum aquaticum*
Partridge, T. R. 116
Pascual, M. 289
Patagonia 371
Patagonian chinchilla mouse 346
Patterson, B. D. 346
Pearl, C. A. 324
Pearl River 239, 275
Peay, S. 199, 201
Pecos River 129
Peebles, C. R. 209
Pejchar, L. 13
Pekny, R. 197
Pelobates cultripes 333
Pelophylax perezi 333
Peña, C. J. 139
Peña, L. F. 346
Penczak, T. 289
Pennisetum clandestinum 30
Pennsylvania 59, 93, 98, 422
Perca flavescens 392
Percina caproides 412
Percottus glenii 394
Perera, G. 209
Perrins, J. M. 71
Perrone, A. 358
Persons, W. 388
Peru 104, 286, 405
pesticides 215, 315–316, 439–440
Petromyzon marinus 4, 442
Pflieger, W. L. 263
pH levels 47, 49, 71, 83, 103, 127, 153, 176,
    264, 280, 404

Phal-Wostl, C. 12
phenotypic plasticity 6, 29, 38, 166
Philadelphia 93
Philippines 208–209, 211, 214–215, 407, 409, 411–412
Philophthalmus gralli 408
Phleospora heraclei 62
phosphorus 29, 82, 106, 254
photosynthesis 41, 50, 83, 106, 117
Phoxinus spp. 289, 391
Phragmites australis 362
physical control *see* mechanical control
phytoplankton 82, 173, 177–178, 237
pike 252
pike-minnows 390–391
pikeperch 252
Pila conica 208–209, 212, 215
Pila lepoldvillensis 208–209, 211
Pila spp. 207–209, 211–212, 214–215
Pilica River 290
Pimental, D. 13
Pimephales promelas 279
Pineda-López, R. 387
Pink Peril *see Impatiens glandulifera*
Pinto-Coelho, R. M. 162
pipelines 150, 227, 267
piscicides 85, 266, 291, 440
planktonic copepods *see* copepods
Planty-Tabacchi, A. M. 11
Pöckl, M. 197
Poecilia sp. 263, 412
Poeciliopsis spp. 267
Pointer, J. P. 209
Poland 188, 197, 224–225, 277, 289–290, 332, 343, 369, 371
Policeman's Helmet *see Impatiens glandulifera*
Poljak, S. 346
pollination 39, 60, 71–73, 83, 91, 95, 99
pollution 3, 47, 49, 120, 199
Polyacanthus operculatus 409
Polyphemus pediculus 152
Pomacea bridgesii 208
Pomacea canaliculata 12, *P18.1–18.4*, 207–215
Pomacea cuprina 208
Pomacea diffusa P18.1, P18.3, 207–208, 210–213
Pomacea dolioides 214–215
Pomacea gigas 208
Pomacea glauca 212
Pomacea haustrum P18.1, P18.3, 208, 210, 212–213
Pomacea insularum P18.1, P18.3, 207–208, 210–215
Pomacea levior 208
Pomacea lineata 208, 211
Pomacea paludosa 208, 213–214
Pomacea scalaris P18.1, P18.3, 207–208, 210, 212
Pomacea urceus 214
Pomella spp. 207
Pool, D. W. 385, 391
Poor Man's Orchid *see Impatiens glandulifera*
Populus spp. 127–128
porcelain disease 201
Portugal 104, 197, 225, 332
Potamorpyrgus antipodarum 223–229, *224*; biological attributes 223–224; distribution and introduction 225–226; ecological

niche 224–225; impacts 226–227; management methods 227–229
Prach, K. 8
praziquantel 393
Predick, K. I. 10
Previtali, A. 371
Procambarus clarkia 200, 442
Prochelle, O. 251
Pseudalopex spp. 346
Pseudoceratodes spp. 207
Pseudodiaptomus forbesi 162–165, 169
Pseudodiaptomus inopinus 163
Pseudodiaptomus marinus 163
Pseudorasbora parva 275–281, *278*, 411; controversies 280–281; distribution and introduction 275–277; ecological niche 277–279; management methods 279–280
Pterophyllum scalare 410
Ptychocheilus spp. 390–391
Puccinia cf. komarovii 72
Puerto Rico 209, 212, 215, 311, 314–315, 317, 385
Puky, M. 197
Pullin, A. S. 119–120
Puntius brevis 411
purple loosestrife *see Lythrum salicaria*
Pyšek, P. 5–6, 8, 72

Qiu, J.-W. 216

Raab River 276
rabies 372
Rach, J. J. 238
rainbow trout *see Oncorhynchus mykiss*
Raleigh, R. F. 288
Ramey, V. 49
Ramsar 436
Rana [Lithobates] catesbeiana 321–327, *322*; controversies 327; distribution and introduction 321–322; ecological niche 321–325; management methods 325–327
Rana clamitans 323
Rana draytonii 323–324, 326
Ranamukhaarachchi, S. L. 209
Ranunculus sceleratus 41
Rattus norvegicus 346
Rattus rattus 314, 346
Rawinski, T. J. 97
Rawlings, T. A. 207–210
Ray, P. 52
Rayner, T. S. 280
red deer 346
red shiners 390, 410
red swamp crayfish 200, 442
Reed, E. B. 162
Reglone 42
Reid, J. W. 162, 168–169
Reise, D. 346
Reiter, P. 139
reproductive styles 6
Reynolds, J. C. 374
Rhabdias pseudosphaerocephala 304
Rhamdia guatamalensis 412
Rhine River 8

Rhinella marina 299–306, *300, 305*; controversies 306; distribution and introduction 299–301; ecological niche 301–303; management methods 303–305
Rhinogobius giurinus 411
Rhodeus spp. 409
Ricciardi, A. 4, 163
rice crops 27, 161, 185, 212, 214–216
rice grass *see Spartina anglica*
Richards, D. C. 227
Richardson, B. J. 254
Richardson, D. M. 5–6, 11
Richardsonius balteatus 289
Rift Valley virus 141
ringed crayfish 9
Río Bravo del Norte 387, 390
Río Grande 387, 390
Ritchie, S. A. 140
roach 279
Robbins, R. S. 188
Roberts, P. D. 115, 119–120
Robinson, D. G. 8
Rockefeller Foundation 263, 436
Rocky Mountains 321
Rödder, D. 332
Rodeo (herbicide) 97, 120
Roiz, D. 139
Romania 176, 276, 332
Rosario, J. 209
Rossi, G. C. 139
rotenone 85, 266, 280, 291
round goby 9
Roundup 51, 97
Royal Botanical Gardens 57, 59, 67, 71
Royle, J. F. 67
Rudnick, D. 187
Ruiz, G. M. 187
Rumex spp. 362
Russia 37, 57, 67, 149, 175, 186, 197, 225, 235, 276, 286, 332, 343, 369, 422
Russian olive 129
rusty crayfish 11, 439
Rutilus rutilus 279
Rwanda 407
Rysavy, B. 391

Sagittaria spp. 362–363
Salgado-Maldonado, G. 387
salinity 25, 42, 49, 104–105, 127–128, 130, 153, 163, 166, 168–169, 176, 199, 212, 264, 437
Salix spp. 127–128
Salmo marmoratus 289–290
Salmo salar 287, 289–290, 422
Salmo trutta 6, *P24.1–24.2*, 255, 285–292, 421–422; controversies 292; distribution and introduction 285–286; ecological niche 287–288; impacts 288–290; management methods 290–292
salt marsh grass *see Spartina anglica*
saltcedar *see Tamarix ramossisima*
Salvelinus fontinalis 289–290, 421
Salvelinus leucomaenis 289–290
Salvelinus malma 289
Salvinia molesta 28

Samayoa, A. L.   139
Sameodes albiguttalis see *Niphograpta albiguttalis*
Sampson, S. J.   237
San Joaquin River   168, 189
San Juan River   130
San Marino   139
Sargassum muticum   73
Sass, G.   242
Saudi Arabia   332, 406
Saulea spp.   207
Saunders, J. F.   162
Savage, H. M.   139
Savucci, M. E.   106
scanning electron microscopy (SEM) 391
Schaffner, F.   139
Schizocoytle fluviatilis   391
Scholte, E. J.   139
Schreiber, E. S. G.   226
Schürkens, S.   71
Scirpus americanus   362
Scirpus spp.   362
Scopus database   173
Scottish Executive   73–75
sea lamprey   4, 442
sea mink   369
sea otter   346
SeaKleen   155
Senegal   7
Septoria heracleicola   62
Serbia   176, 332
sessile joy weed   29
Seychelles   332
Shasta crayfish   196
Shemai, B.   289
Shine, C.   437
shipping   7, 27, 93, 113, 150, 155, 189–190; *see also* ballasting
Sicydium multipunctatum   411
Sidorovich, V.   371
Siefeld, W.   346
signal crayfish see *Pacifastacus leniusculus*
Sika deer   41
silver carp see *Hypophthalmichthys molitrix*
silver fox   346
Simberloff, D.   5
Singapore   209–211, 332, 407, 410, 412
Sinocalanus doerrii   162, 170
Sinodiaptomus sarsi   162
Sisler, S. P.   250
Skewes, O.   352
slider terrapin see *Trachemys scripta*
Slovakia   332, 405
Slovenia   139, 197, 289, 332
Smal, C. M.   371
Smith, B. D.   209
Snake River   163–165, 169, 225
sodium chloride   155
sodium hypochlorite   180
Somalia   407
sooty crayfish   11, 196
Sorensen, P. W.   250
Soto-Acuña, S.   209
Sousa, R.   174
South Africa   49, 79, 107, 139–140, 187, 209, 211, 263, 280, 285, 289, 291, 332, 387, 407

South America   113, 162, 187, 247, 263, 286, 321, 390, 405, 424, 436; *Aedes albopictus* 137, 139–142; apple snails   207–208; *Castor canadensis* 343–353, 347; *Corbicula fluminea* 173–176, 181; *Myocastor coypus* 357, 362; *Myriophyllum aquaticum* 103–104, 107; *Neovison vison* 369–370; *Rhinella marina* 299, 304; *Trachemys scripta* 332, 334
South Carolina   210
South Dakota   125, 130, 250
south-western willow flycatcher   129–130
Soviet Union see USSR
Spain   139, 141, 197, 210, 251, 263, 286, 332, 371–372, 375, 405
Spanish terrapin   331
Sparquat   228
Spartina alterniflora   113
Spartina anglica   113–121, *114*; controversies   120–121; distribution and introduction   113–116; ecological niche   116–119; management methods   119–120
Spartina maritima   113
Spartina townsendii   4
Sphaerothecum destruens   279–280
spike-topped apple snail see *Pomacea diffusa*
spinal arthritis   303
spiny water flea see *Bythotrephes longimanus*
spondylosis   303
Spotila, J. R.   12
Sprenger, D.   139
Sri Lanka   208, 210–211, 286, 332, 406, 410
St Lawrence River   93, 166, 187
Stachys palustris   71
Stamp, N. E.   70
Stapf, O.   118
Stebbing, P. D.   198
Stennomelania spp.   404
Stiling, P.   5
Stizostedion lucioperca   252
Stone, D. M.   393
stone crayfish   196
Stop Coqui Hawaii   314
Strongyloides myoptami   362
Stuart, I. G.   251–252
Stuart-Smith, R. D.   289
Stucki, P.   197
Suárez-Morales, E.   162
Sudan   49, 209, 407
Sumatra   407
sunbleak   279
Suriname   209, 332
Sus scrofa   346
Suzuki, T.   189
swamp stonecrop see *Crassula helmsii*
Swanson, C.   264
Swaziland   285
Sweden   71, 140, 149, 188, 196–197, 332, 372, 405
Switzerland   59, 71, 80, 139, 149, 197, 332
Syria   406
Syrphidae   39, *40*

Tabacchi, E.   11
Taiwan   49, 142, 208–11, 216, 262, 275, 277, 332, 387, 404, 406, 409–412
tamarisk see *Tamarix ramossisima*

Tamarix chinensis   125–131; controversies 130–131; ecological niche   126–129; management methods   129–130
Tamarix ramossisima   4, 13, 125–131, *126*; controversies   130–131; ecological niche   126–129; management methods   129–130
Tanner, R.   73
Tanzania   49, 209, 212, 285, 407
Tasmania   37, 252
Tempero G. W.   251
Tennessee   93, 153–154, 395
Texas   52, 125, 154, 168, 208–210, 262–264, 395, 408
Thailand   27, 31, 210–211, 332, 406
Thames River   74, 186–187, 190–191
Thébaud, C.   8
Thelohanellus nikolskii   249
Thelohania contejeani   201
Therapon plumbeus   412
Thermocyclops crassus   162
Thiara scabra   404
Thie, I.   72
Thomas, M. R.   236
Thomas, O.   346
Thompson, D. Q.   92
Thorichthys helleri   410
Three Gorges Dam   238
Thunder Bay   225
Tibet   275–276
Tiefenbach, O.   276
Tilapia zillii   410
Tiliqua spp.   299
Tillaea helmsii see *Crassula helmsii*
titan apple snail see *Pomacea haustrum*
topmouth gudgeon see *Pseudorasbora parva*
Torridge River   188
Tortanus dextrilobatus   169
Toto, J. C.   139
tourism   13, 179
Townsend's cordgrass   4
Toxoplasma gondii   362
Toxorhychites rutilus   141
Trachemys scripta   331–336, *332*; controversies 336; distribution and introduction   331–333; ecological niche   333–335; management methods   335
Tran, C. T.   207, 209, 211
Trapa natans   362, 439
trapping   143–144, 189–190, 199–201, 255, 291, 304, 315, 335, 373–374, *373–374*, 439
Treer, T.   251
Trichogaster trichopterus   409
triclopyr   84, 108, 440
Trinidad and Tobago   139, 332
Truscott, A. J.   117
Tubifex tubifex   421–422, 425–427
Tunisia   91, 407
Turkey   57, 225, 276, 332, 390, 406, 410, 412
Turner, M. G.   10
Tyne River   188
typhoons   211

Uganda   49, 407
UK 149, 225, 236, 332; *Crassula helmsii* 38–44; *Eriocheir sinensis* 186–191; *Heracleum*

mantegazzianum 57, 59; *Impatiens glandulifera* 67–74; *Lagarosiphon major* 80–81; *Myocastor coypus* 364–365; *Neovison vison* 371–372, 374–377; *Pacifastacus leniusculus* 196–197, 199–201; *Pseudorasbora parva* 276, 280; *Spartina anglica* 113, 115, 117–119
Ukraine 8, 57, 276, 405
ultraviolet (UV) radiation 61, 155, 327, 426
United Arab Emirates 406
United Nations (UN) 73, 394
United States Geological Survey (USGS) 130–131
Upatham, E. S. 211
Uredo eichhorniae 52
Urtica dioica 72, 74
Uruguay 209, 212
US 4, 8–11, 37, 59, 85, 195, 299, 321, 369, 409–412, 439; *Aedes albopictus* 137, 139, 141–143, 437; *Alternanthera philoxeroides* 27, 29; apple snails 207–214; *Bothriocephalus acheilognathi* 385–387, 390, 392–395; *Bythotrephes longimanus* 149–155; *Centrocestus formosanus* 405, 408; copepods 161–170; *Corbicula fluminea* 173–176, 181; *Cyprinus carpio* 247, 250, 252–255, 436; *Eichhornia crassipes* 49, 51–52; *Eleutherodactylus coqui* 311; *Eriocheir sinensis* 185, 187, 189–191; *Gambusia affinis* and *G. holbrooki* 261–267, 269–270; *Hypophthalmichthys* spp. 236–242; *Impatiens glandulifera* 67, 71, 73; *Lythrum salicaria* 91, 93, 95–99; *Myocastor coypus* 358, 362–365; *Myriophyllum aquaticum* 104, 108; *Myxobolus cerebralis* 421–423, 425–427; *Potamopyrgus antipodarum* 223, 225–226, 228; *Pseudorasbora parva* 279–280; *Salmo trutta* 285–286, 288–292; *Spartina anglica* 113, 116–118, 120; *Tamarix* spp. 125–131; *Trachemys scripta* 331–333
US Fish and Wildlife Service (USFWS) 130, 189
USSR 276, 279, 369, 387, 391; *see also* Russia
Utah 93, 130, 195, 225, 288, 291, 427

Vaal Dam 390
Valamugil cunnesius 412
Valentine, D. H. 71
Valéry, L. 163

Vatican City 139
Veija fenestra 410
Venegas, C. 346
Venezuela 142, 162, 208–210, 214, 285, 299, 405
Venturelli, C. 139
Veronica beccabunga 41
vertebrates 8
Vestiaria coccinea 314
Vibrio parahaemolyticus 191
Viet Nam 208, 210–211, 332, 406, 409–411
Vilazzi, L. 251
Villamanga, A. M. 50
Vinson, M. R. 227
Virgin Islands 49, 332
Virgin River 130
Virginia 93
Von Holle, B. 5
Vondracek, B. 289
Vörösmarty, C. J. 13
Vulpes vulpes 346

Wadden Sea 115–116
Wagley, S. 191
Waikato River 251–252
Walker, K. F. 251
Walker, P. 188
Walls, J. G. 209
Walter, T. C. 162
Ward, D. L. 388, 392–393
Warman, E. A. 39–42
Washington 49, 93, 116–18, 120, 125, 130, 169, 225
wastewater 49, 267, 364
water chestnut 362, 439
Water Framework Directive (WFD) 74, 179
water hyacinth *see Eichhornia crassipes*
water hyacinth moth 52
water levels 96, 106, 115, 238, 241, 255
water vole 371–372
watercrassula *see Crassula helmsii*
Waters, T. F. 289
Weber, M. J. 253–254
Welcomme, R. L. 7
Welsh Assembly 73
Wersal, R. M. 108
Weser River 186, 189
West Nile virus 13, 137, 140–142
western pond terrapin 333

Westman, K. 198
whirling disease *see Myxobolus cerebralis*
White River 130, 237
white-clawed crayfish 12, 196, 198–199, 201
Wikramasinghe, S. 209
Wildlife and Countryside Act (1981) 73
Wiles, P. R. 199
Williams, P. A. 5
Williams, S. L. 7
Williamson, M. 5–6, 68
willow 127–128
Wisconsin 9, 11, 91, 150–151, 253, 439
Wisconsin River 10
World Organisation for Animal Health 426
World War I 263
World War II 142, 263, 275–276, 422
Wu, W.-L. 208–210, 214
Wuithiranyagool, T. 139
Wymann, M. N. 139
Wyndham, J. 57
Wyoming 93, 125, 130, 225

Xenotaca variata 411
Xie, Y. 208–209, 214
Xiphophorus spp. 412

Yamaguti, S. 385
Yampa River 387
Yangtze River 49, 162–163, 167–168, 236, 238, 275, 277–278, 387, 391
Yangzi River *see* Yangtze River
yellow fever 262–263
yellow perch 392
Yellow River 275, 277–278
yellowfish 390
yellow-nosed mouse 346
Yellowstone National Park 223
Yemen 406
Yodo River 385
Yuriria alta 411
Yusa, Y. 215–216

Zabala, J. 376–377
Zacco platypus 411
Zambia 79, 407
Zander vitreus 392
zebra mussels 13, 155, 225, 440
Zedler, J. B. 10
Zimbabwe 49, 79, 285, 407
Zimmerman, J. K. H. 289